Geophysical Monograph Series

Including
IUGG Volumes
Maurice Ewing Volumes
Mineral Physics Volumes

Geophysical Monograph 199

Dynamics of the Earth's Radiation Belts and Inner Magnetosphere

Danny Summers
Ian R. Mann
Daniel N. Baker
Michael Schulz

Editors

Ⓢ American Geophysical Union
Washington, DC

Library of Congress Cataloging-in-Publication Data

Dynamics of the Earth's radiation belts and inner magnetosphere / Danny Summers, Ian R. Mann, Daniel N. Baker, and Michael Schulz, editors.
    pages cm.—(Geophysical monograph, ISSN 0065-8448 ; 199)
  Includes bibliographical references and index.
  ISBN 978-0-87590-489-4 (alk. paper)
  1. Magnetosphere. 2. Van Allen radiation belts. I. Summers, Danny, editor of compilation.
  QC809.M35D96 2012
  538'.766–dc23

                    2012041727

    ISBN: 978-0-87590-489-4
    ISSN: 0065-8448

Cover Image: Schematic illustration of Earth's radiation belts and orbiting satellites. Image credit: Andy Kale, University of Alberta, Edmonton, Alberta, Canada.

Copyright 2012 by the American Geophysical Union
2000 Florida Avenue, N.W.
Washington, DC 20009

CONTENTS

PREFACE

These are exciting times for radiation belt research. After 11 years of planning and preparation, NASA's Radiation Belt Storm Probes (RBSP) blasted off from Cape Canaveral Air Force Station, Florida, at 4:05 A.M. EDT on 30 August 2012. The identical twin spacecraft (A and B) were launched aboard a United Launch Alliance Atlas V rocket from Space Launch Complex 41 following a smooth countdown. The probes, identically equipped with state-of-the-art instrumentation and heavily shielded against the effects of space radiation, were released from the Centaur upper stage of the rocket one at a time into different orbits, thus beginning a 2 year prime mission to study Earth's radiation belts. During the RBSP mission the twin spacecraft will traverse both the inner and outer Van Allen radiation belts that encircle the Earth twice per 9 hour orbit. Slightly different apogees and orbital periods will cause one spacecraft to lap the other 4 to 5 times per year. The wealth of data expected to be collected by RBSP on particles, plasma waves, and electric and magnetic fields should provide unprecedented insight into how the radiation belts evolve in time and space. The overall goal of RBSP is to understand, ideally to the level of predictability, how Earth's radiation belts vary in response to dynamical inputs originating from the Sun. This will require detailed understanding of particle acceleration mechanisms, particle loss processes, and particle transport in the inner magnetosphere. It will also require understanding how radiation belt behavior links to or couples with the other important plasma components in the inner magnetosphere, namely, the ring current, plasmasphere, and ionosphere.

An important facet of the RBSP mission is to explore the extremes of space weather, namely, the extreme conditions in the space environment surrounding Earth that can disrupt human technologies and possibly endanger astronauts. For instance, magnetic storms induced by solar events such as coronal mass ejections or high-speed solar wind streams can generate highly energetic ("killer") electrons that can damage or even shut down Earth-orbiting satellites. Magnetic storms can also give rise to geomagnetically induced electric currents in the Earth that can interfere with technologies on the ground such as electric power grids. Energetic protons produced by solar storms can pose a serious hazard to both satellite electronics and astronauts. RBSP will produce a 24 hour space weather broadcast using selected data from its suite of instruments that will provide researchers a check on current conditions near Earth. RBSP data will be used by engineers to design radiation-hardened spacecraft and will enable forecasters to predict space weather events in order to alert astronauts and operators of spaceborne and ground-based technologies to potential hazards. The development of space weather science has intensified over the last 10–15 years, fueled by our increasing reliance on space technologies. In parallel with the coming of age of space weather science, there has been a resurgence in radiation belt research in the last decade that indeed has served as a prelude to the launch of RBSP.

In anticipation of the RBSP mission the AGU Chapman Conference Dynamics of the Earth's Radiation Belts and Inner Magnetosphere was held during 17–22 July 2011 in St. John's, Newfoundland and Labrador, Canada. This volume is based largely on the material presented at this Chapman Conference. The conference was held with the aim of drawing together radiation belt knowledge and refining science questions for RBSP and other upcoming missions; summaries of the conference are given by D. Summers, I. R. Mann, and D. N. Baker (*Eos*, *92*(49), 6 December 2011) and D. N. Baker, D. Summers, and I. R. Mann (*Space Weather*, *9*, S10008, doi:10.1029/2011SW000725, 2011). Prevailing themes of the conference and this volume include radiation belt particle acceleration and loss processes, particle transport in the inner magnetosphere, radiation belt responses to different solar wind drivers, and the control of radiation belt dynamics by wave-particle interactions. A key conclusion of the conference is that, despite more than 50 years of radiation belt investigations, our knowledge of the radiation belts, both from observational and theoretical points of view, is far from complete. The RBSP era promises to significantly improve our understanding of the dramatic and puzzling aspects of radiation belt behavior. We hope that the present volume will serve as a useful benchmark at this exciting and pivotal period in radiation belt research in advance of the new discoveries that the RBSP mission will surely bring.

Dynamics of the Earth's Radiation Belts and Inner Magnetosphere
Geophysical Monograph Series 199
© 2012. American Geophysical Union. All Rights Reserved.
10.1029/2012GM001435

We would like to thank Brenda Weaver and Cynthia Wilcox of the AGU Meetings Department for their great help in ensuring the success of the Chapman Conference. We also thank Maxine Aldred, Colleen Matan, Maria Lindgren, and Telicia Collick of the AGU Books Department for their work in the production and timely completion of this book. Finally, we are most grateful to the more than 60 referees who reviewed the articles submitted to this volume.

Note added in proof: NASA has recently renamed the Radiation Belt Storm Probes (RBSP) mission. At a special ceremony held at the Johns Hopkins University Applied Physics Laboratory, Laurel, Maryland, on 9 November 2012, NASA renamed the mission as the Van Allen Probes in honor of James Van Allen, the discoverer of Earth's radiation belts. The ceremony also highlighted the successful commissioning of the spacecraft.

Danny Summers
Memorial University of Newfoundland
Kyung Hee University

Ian R. Mann
University of Alberta

Daniel N. Baker
University of Colorado

Introduction

Danny Summers

Department of Mathematics and Statistics, Memorial University of Newfoundland, St. John's, Newfoundland, Canada
School of Space Research, Kyung Hee University, Yongin, South Korea

Ian R. Mann

Department of Physics, University of Alberta, Edmonton, Alberta, Canada

Daniel N. Baker

Laboratory for Atmospheric and Space Physics, University of Colorado, Boulder, Colorado, USA

Earth's radiation belts have attracted much experimental and theoretical investigation since their discovery by James Van Allen in 1958. In this introductory article, we briefly scan developments in radiation belt science since 1958, both with respect to satellite observations and theory and modeling. We then provide an overview of the articles in this book, which mainly derive from the 2011 Chapman Conference on Dynamics of the Earth's Radiation Belts and Inner Magnetosphere. In the past decade, there has been a resurgence in radiation belt studies in parallel with the rapid development of space weather science. NASA's Radiation Belt Storm Probes (RBSP) mission, which has just been launched at the time of writing, promises to provide unprecedented measurements of the particles, electric and magnetic fields, and plasma waves in the Earth's radiation belts. This volume provides a timely state-of-the-art account of radiation belt science prior to the start of the RBSP era.

The discovery by James Van Allen in 1958 of the Earth's radiation belts (now "Van Allen belts"), using Explorer 1 data [*Van Allen et al.*, 1958], was a momentous event in space physics. The intense radiation environment around the Earth has since attracted much scientific interest. Here we recount some important developments in radiation belt research, a selection of which we include in Table 1. The late 1950s and 1960s heralded the birth of the space age. Sputnik

Dynamics of the Earth's Radiation Belts and Inner Magnetosphere
Geophysical Monograph Series 199
© 2012. American Geophysical Union. All Rights Reserved.
10.1029/2012GM001434

1, launched on 4 October 1957 by the Soviet Union, was the first successful Earth-orbiting satellite. The first commercial telecommunications satellite Telstar-1, launched on 10 July 1962, carried a set of solid-state detectors to characterize the radiation environment that the vehicle would encounter. A day before the launch of Telstar, the Starfish high-altitude nuclear explosion greatly enhanced the trapped electron fluxes, thereby creating an artificial radiation belt in the vicinity of Telstar's orbit. The radiation environment was further enhanced by a Soviet nuclear test in October 1962. The resulting intense radiation caused the premature demise of Telstar-1 in February 1963. The detectors onboard Telstar were able to monitor the artificial radiation belt and record its degradation via particle precipitation losses attributed to

Table 1. Developments in Radiation Belt Science

Satellites and Observations		Theory and Modeling	
Timeline	Event	Timeline	Description
1958	Discovery of the Earth's radiation belts by James Van Allen using Explorer 1	1907, 1933, 1955	Stormer: motion of a charged particle in a dipole magnetic field
1958	Creation of artificial radiation belts by Argus high-altitude nuclear bombs	1960s	Development of fundamental radiation belt theory (particle sources, losses, transport, diffusion)
1958	Sputnik 3 confirms the existence of the Earth's radiation belts	1960s	Development of AE, AP models of radiation belt electron, proton environment
1960s	Polar Operational Environmental Satellites (POES) program of polar orbiting satellites begins	1961	McIlwain L shell parameter
1962	Starfish high-altitude nuclear explosion: artificial radiation belt produced	1963	Northrop: adiabatic invariants
1962–1963	Telstar-1	1964	Andronov and Trakhtengerts: kinetic instability of the outer radiation belt
1966	Launch of (geosynchronous) Applications Technology Satellite-1	1965	Falthammar: radial diffusion produced by time-varying electric field
1970s–1980s–1990s	Radiation belts observed at Jupiter, Saturn, Uranus, Neptune (Pioneer 11, Voyagers 1, 2)	1966	Kennel-Petschek theory
1975	Launch of GOES geostationary satellites begin	1970	Roederer $L*$ parameter
1976	Los Alamos National Laboratory geosynchronous satellite energetic particle measurements begin	1974	Schulz and Lanzerotti monograph on radiation belt particle diffusion
1990–1991	CRRES	1981	Gendrin: wave-particle interactions
1992	Launch of Solar, Anomalous, and Magnetospheric Particle Explorer (SAMPEX)	1991	Vette AE-8 model of trapped electron environment
1994	Term "killer electrons" coined	1996	AGU Geophysical Monograph-*Radiation Belts: Models and Standards*
1996–2008	Polar	2000s	Resurgence of radiation belt studies
2000s	Coming of age of space weather science	2011	AGU Chapman Conference: Dynamics of the Earth's Radiation Belts and Inner Magnetosphere, St. John's, Canada
2012	Launch of Radiation Belt Storm Probes (RBSP)		

natural wave-particle interactions. Much theoretical work has since been carried out to evaluate wave-particle interaction processes in the Earth's (natural) radiation belts. It is interesting to note that prior to the discovery of the natural radiation belts by Van Allen, the U.S. Air Force was preparing to carry out an experiment, code-named Argus, to study the trapping of energetic particles by the Earth's magnetic field. It was actually suggested during the planning sessions for Argus that a natural radiation belt might exist around the Earth. Then, immediately following Van Allen's discovery of the Earth's radiation belts, the U.S. Air Force exploded the Argus high-altitude nuclear bombs in order to create artificial radiation belts. These artificial belts were studied by the satellite Explorer 4, built for this purpose by Van Allen and his group.

The Soviet spacecraft Sputniks 2 and 3 also contributed to the early measurements of the Earth's radiation belts. An experiment by S. N. Vernov et al. on board Sputnik 2, launched on 3 November 1957 before Explorer 1, might have discovered the radiation belts, but the orbit was in far northern latitudes. It was thus beneath most of the outer radiation belt when it was monitored in the USSR. Moreover, published data from Sputnik 2, which showed increased detector count rates above the USSR, were not reported as unusual, nor interpreted as geomagnetically trapped particles [*Vernov et al.*, 1959]. Subsequently, Sputnik 3 confirmed the existence of the Earth's inner and outer radiation belts, which had already been found and documented by Van Allen.

With the launches of Syncom 3 in 1964 and Intelsat 1 in 1966, geosynchronous orbit (GEO) soon became the preferred orbit for commercial communications and television broadcasts. NASA launched the Applications Technology Satellite (ATS-1) in December 1966. In addition to its communications experiments, ATS-1 carried three separate charged particle instrument suites designed to characterize the GEO radiation environment. These instruments provided new and fundamental information on the dynamics of the radiation belts that would impact commercial and government space systems at GEO, including the discovery of the dramatic changes that can occur during geomagnetic storms (both the intense particle enhancements and depletions) and the rapid access of solar energetic particles to GEO.

Early in the space program, the charged particle data being gathered by numerous satellites circling the Earth (such as Interplanetary Monitoring Platforms (IMPs), Explorers (especially Explorer 26), Orbiting Geophysical Observatories (OGOs), International Sun-Earth Explorer (ISEE) 1, 2) were incorporated into the so-called AE, AP models of the radiation belt electron and proton environment. The NOAA Polar Operational Environmental Satellites (POES) program began in the 1960s, while the Geostationary Operational Environmental Satellites (GOES) program began in 1975. Los Alamos National Laboratory (LANL) geosynchronous satellite energetic particle measurements began in 1976. The POES, GOES, and LANL satellites continue to monitor the radiation environment today.

A particularly valuable source of radiation belt data is the Combined Release and Radiation Effects Satellite (CRRES), which was launched on 25 July 1990 and functioned until 12 October 1991 [*Johnson and Kierein*, 1992]. The spacecraft had a geosynchronous transfer orbit, namely, an elliptical orbit with a perigee of 1.05 R_E and apogee of 6.26 R_E, with respect to the Earth's center, with an inclination of 18.15°. The outermost L shell reached by CRRES was about $L = 8$. The orbital period was about 9 h 55 min, and the apogee precessed from 10:00 magnetic local time (MLT) to 14:00 MLT through midnight before the mission ended. The satellite was, thus, able to provide excellent coverage of the radiation belts for nearly 15 months since it traversed the inner magnetosphere on average about 5 times per day. The CRRES particle, field, and wave data remain a valuable (and still not fully mined) radiation belt resource today.

In 1992, NASA launched the Solar, Anomalous, and Magnetospheric Particle Explorer (SAMPEX) satellite with a nearly circular, low-Earth (600 km), high-inclination (82°) orbit with a period of about 100 min [*Baker et al.*, 1993]. While the primary mission of SAMPEX was to measure cosmic rays and solar energetic particle enhancements, the spacecraft also carried highly sensitive electron detectors. The latter instruments served to reveal new insights into radiation belt dynamics and to provide monitoring of the radiation belts on a continuous basis. SAMPEX has, in fact, proved to be one of the best NASA radiation belt missions. The SAMPEX spacecraft is scheduled to end its mission and reenter the Earth's atmosphere by December 2012. The contribution of SAMPEX to our knowledge of the radiation belts

over its 20 year mission is the subject of an article by *Baker and Blake* [this volume].

The NASA spacecraft Polar, carrying charged particle, field and electromagnetic wave instruments, was launched in 1996 and primarily designed to study the polar magnetosphere and aurora [*Acuña et al.*, 1995]. Polar was placed in a highly elliptical orbit with perigee 1.8 R_E and apogee 9.0 R_E, with an inclination of 86°, and a period of about 17.5 h. Notwithstanding its primary mission, as its orbit precessed over time, Polar made valuable observations of the Earth's radiation belts and equatorial inner magnetosphere. The Polar mission was terminated in April 2008.

The NASA twin-spacecraft Radiation Belt Storm Probes (RBSP) mission, successfully launched in August 2012, should provide the most complete measurements hitherto made of the charged particles, fields, and waves in the Earth's radiation belts. The overall objective of RBSP is to provide understanding, ideally to the point of predictability, of the dynamics of radiation belt particles in response to variable energy inputs from the Sun. Equivalently, this will require a complete understanding of the energization, transport, and loss processes of radiation belt particles in time and space. The RBSP spacecraft will have 600 km perigee, 5.8 R_E geocentric apogee, and will traverse both inner and outer radiation belts twice per 9 h orbit. The low-inclination (10°) orbit will permit RBSP to access essentially all magnetically trapped particles. The sunward oriented spin axis will permit full particle pitch angle distribution sampling twice per 12 s spin and ensure accurate measurements of the dominant dawn-to-dusk electric field component. The identically instrumented spacecraft will carry five suites of instruments: Energetic Particle Composition and Thermal Plasma Suite (ECT), Electric and Magnetic Field Instrument Suite and Integrated Science (EMFISIS), Electric Field and Waves (EFW) Instrument, Radiation Belt Storm Probes Ion Composition Experiment (RBSPICE), and Proton Spectrometer Belt Research (PSBR). Collaborations are planned for RBSP with the Time History of Events and Macroscale Interactions during Substorms (THEMIS) and Balloon Array for Radiation belt Relativistic Electron Losses (BARREL) missions. Details of the mission design, instruments, and science questions related to RBSP are given in an article by *Kessel* [this volume].

In the right-hand side of Table 1, selected developments in radiation belt theory and modeling are listed. We comment here, briefly, on some of these developments.

C. Stormer was the first to calculate the different types of trajectory of a charged particle moving in a dipole magnetic field [*Stormer*, 1907, 1933, 1955]. His landmark work, which was carried out before the discovery of the Earth's radiation belts, is still relevant today. Stormer determined "allowed regions" and "forbidden regions" in the vicinity of the Earth that can or cannot be reached by a charged particle traveling toward the Earth. Calculation of such regions is relevant to determining the regions near the Earth that can be accessed by solar energetic particles (SEP) and galactic cosmic rays (GCR) and can be applied to the problem of creating a "magnetic shield" around a spacecraft to protect it from highly energetic particles. It is clearly of interest to determine the extent to which Stormer's theory is applicable to a time-dependent geomagnetic field. For instance, *Kress et al.* [2005] show that a generalization of Stormer's theory, to a time dependently perturbed dipole magnetic field, is relevant to SEP geomagnetic trapping

The fundamental physics of the radiation belts was developed in the 1960s in parallel with the early radiation belt particle observations. The main topics addressed then were precisely those with which we are concerned today, namely, radiation belt particle sources and losses, transport, and diffusion. *Andronov and Trakhtengerts* [1964] and *Kennel and Petschek* [1966] showed that radiation belt particle distributions can be unstable to the generation of plasma waves. Thus, there was early recognition of the importance of wave-particle interactions and, in particular, of particle precipitation resulting from resonant pitch angle scattering by electromagnetic waves. Seminal approaches to radiation belt modeling were established by *Northrop* [1963], who showed that radiation belt particle dynamics could be described in terms of three adiabatic invariants, and *Fälthammar* [1965], who developed a radial diffusion equation. Subsequently, *Schulz and Lanzerotti* [1974] published a comprehensive account of what might be called the classical theory of particle diffusion in the radiation belts. The framework of radiation belt dynamics was simplified by the introduction of the magnetic shell parameter L by *McIlwain* [1961] and, later, the (adiabatically invariant) generalized L shell parameter L^* by *Roederer* [1970]. A synthesis of the early literature on radiation belt studies, both theoretical and experimental, is provided by *Hess* [1968]. *Gendrin* [1981] devised a method for treating wave-particle interactions by evaluating wave growth and particle diffusion in terms of the relative configuration of specified curves in the particle velocity space. *Van Allen* [1983] provides a retrospective view of the discovery of Earth's radiation belts and the early developments in magnetospheric physics.

The 1970s to the 1980s was largely a period of stagnation in studies of the Earth's radiation belts. The prevailing sentiments were that the radiation belts were static, the necessary theory had been developed, and that the radiation belts were well understood. A dramatic change in thinking in the radiation belt science community occurred in the 1990s as a result of observations by the CRRES and SAMPEX satellites. The new radiation belt of highly relativistic electrons created in minutes by the March 1991 shock injection, observed by CRRES, was without precedent. The complexity of radiation belt dynamics revealed by the CRRES particle and wave instruments was unexpected. Moreover, SAMPEX provided a totally new insight into how the radiation belts evolve in time with respect to solar events, geomagnetic storms, and over the solar cycle. The 1990s, and especially the 2000s, saw an intensification of radiation belt studies, with attempts to create realistic models of the dynamical radiation belts that were now known to change over many time scales. The AGU volume edited by *Lemaire et al.* [1996] provides a useful account of radiation belt modeling at that time.

The resurgence in radiation belt studies, continuing today, is partly the result of the emergence of space weather science. Increasing human reliance on spacecraft technology has caused an increased need to protect spacecraft against such space weather hazards as "killer electrons," the highly energetic (~MeV) electrons that can be generated in the outer radiation belt during geomagnetic storms (the evocative term "killer electrons" was coined in an article by *Graham* [1994]). Understanding exactly how, and under what circumstances, killer electrons are generated remains a great challenge in radiation belt physics. As well, notwithstanding considerable efforts over the last decade, radiation belt modeling attempts face significant difficulties, e.g., multidimensional, time-dependent models are required but lack global spatiotemporal information on the various plasma wave modes that control radiation belt particle dynamics, reliable magnetic field models that are valid under disturbed conditions are unavailable, classical radial diffusion theory may not be as widely applicable as previously thought, and quasilinear theory of wave-particle interactions may need to be augmented or replaced by nonlinear theories. Further, radiation belt dynamics couples with the ring current, plasmasphere, and ionosphere, so realistic radiation belt models should incorporate such coupling. The papers in this volume speak to these challenges in radiation belt modeling, as well as providing a current state-of-the-art account of radiation belt science.

The contents of this monograph are as follows.

As a historical prelude, *Lanzerotti* [this volume] describes in a narrative style the development of technologies on the Earth and in space and how the space environment around the Earth has affected their design and operation. Lanzerotti's article is written partly with reference to Newfoundland, Canada, the location of the Chapman Conference from which this book derives. Newfoundland played key roles in world communication "firsts," including the landing of the first trans-Atlantic telegraph and telecommunications cables, and the first reception of trans-Atlantic wireless signals at Signal Hill in St. John's.

The second section addresses our current state of knowledge of the Earth's radiation belts. At the approach to sunspot minimum, recurrent solar wind streams typically cause geomagnetic storms and enhanced relativistic electrons in the outer radiation belt. On the other hand, strong magnetic storms associated with aperiodic coronal mass ejections occur most frequently around sunspot maximum. Such disturbances can also cause significant radiation belt enhancements. *Baker and Blake* [this volume] illustrate these phenomena by using SAMPEX particle data to characterize the differences in radiation belt behavior over the course of the 11 year solar activity cycle. The remaining papers in the second section relate to particular types of plasma wave in the inner magnetosphere that are considered to be instrumental in controlling radiation belt particle dynamics. *Cattell et al.* [this volume] review measurements from the STEREO and Wind Waves waveform capture instruments of large-amplitude whistler mode waves. These waves, which are different from whistler mode chorus, are shown by simulations to result in rapid electron energization by many MeV as a result of nonlinear processes including phase trapping. A key implication of the work of Cattell et al. is that quasilinear treatments of electron energization and scattering may not be adequate for understanding radiation belt dynamics. *Fraser et al.* [this volume] use GOES data to classify various types of electromagnetic ion cyclotron (EMIC) waves at geosynchronous orbit ($L = 6.6$). EMIC wave scattering is thought to contribute significantly to radiation belt MeV electron losses and localized ring current ion decay. *Mann et al.* [this volume] present an overview of the role of ULF waves in radiation belt dynamics. These authors characterize ULF waves around electron drift orbits on a mesoscale and global scale by using the ground-based magnetometer arrays CARISMA and THEMIS GMAG. ULF waves are highlighted for their role in influencing a wide range of radiation belt acceleration, transport, and loss processes, including cross-energy coupling of the radiation belts with the ring current and plasmasphere.

The third section addresses satellite missions that will probe the Earth's radiation belts and inner magnetosphere, specifically NASA's RBSP mission, the Japanese Energization and Radiation in Geospace (ERG) project, and the Russian RESONANCE mission. We have already provided information on the RBSP mission in the Preface and earlier in the Introduction. *Kessel* tracks the progress of RBSP from formulation and development of the science objectives through instrument selection to mission design, integration, and testing, with an emphasis on how the chosen measurements address the science objectives. Following the successful launch of RBSP on 30 August 2012 (see Figure 1), commissioning of the instruments will take place during September and October in 2012, with the start of normal operations scheduled for 1 November 2012. While the initial period of operation of RBSP will be 2 years, both RBSP spacecraft carry sufficient propellant for up to 5.5 years of normal operations. *Miyoshi et al.* [this volume] describe the ERG mission, which is a science satellite program of the Institute of Space and Astronautical Science (ISAS)/Japan Aerospace Exploration Agency (JAXA). The ERG satellite will explore how relativistic electrons in the radiation belts are generated during space storms. ERG will measure the plasma distribution function, electric and magnetic fields, and plasma waves. A new and innovative technique will be used for wave-particle interactions that directly measures the energy exchange process between particles and waves. ERG has been approved for implementation with a nominal launch date of December 2015. *Mogilevsky et al.* [this volume] provide an overview of the current state of the Russian RESONANCE project. RESONANCE is a four-satellite mission designed to investigate wave processes in the inner magnetosphere related to resonant wave-particle interactions. A characteristic feature of this mission is that the four satellites will make long-term simultaneous measurements of electromagnetic fields and particle fluxes at different locations in the same magnetic flux tube. The RESONANCE spacecraft will be launched in two steps. The first pair of satellites is due to be launched at the end of 2014; the launch of the second pair is planned for March–April 2015.

In the fourth section, the modeling of radiation belt electron dynamics using global MHD simulations is addressed. *Elkington et al.* [this volume] analyze the azimuthal mode structure of ULF waves during the 24–26 September 1998 geomagnetic storm and find that the bulk power in the fluctuating electric and magnetic field components can be described by low ($m < 3$) mode numbers during the storm recovery phase, but that there was significant power in the higher mode numbers ($m > 3$) during the main phase. These results have implications for the commonly made assumption $m = 1$ used in studies of the interaction of radiation belt particles with global ULF waves. *Ozeke et al.* [this volume] employ ULF wave-driven radial diffusion simulations of outer radiation belt electrons and demonstrate that the radial diffusion coefficients hitherto typically used in such models may not actually be correct. Ozeke et al. rederive the radial diffusion coefficient as the sum of a term due to electric field fluctuations and a term due to magnetic field fluctuations. They then show that corresponding electron flux enhancements at lower L shells can be several orders lower than the flux enhancements obtained using the commonly employed diffusion coefficients. In the final paper of the fourth section,

Figure 1. Launch of the United Launch Alliance Atlas V 401 rocket carrying the NASA Radiation Belt Storm Probes at Cape Canaveral Air Force Station on 30 August 2012. Photo credit: Pat Corkery, United Launch Alliance.

Kress et al. [this volume] call into question the modeling of the radial transport of radiation belt electrons by the very process of electron diffusion. Kress et al. compare radial diffusion with transport modeled by computing electron-guiding center trajectories in MHD magnetospheric model fields and claim that the radial transport of MeV outer radiation belt electrons due to moderate solar wind fluctuations is not well modeled by a diffusion equation.

The fifth section mainly addresses the topics of localized radiation belt particle injection and radiation belt flux "dropouts." Aside from the artificial injection of energetic electrons into the radiation belts by such means as high-altitude

nuclear explosions and relativistic electron beam injections from a satellite, there are two main forms of radiation belt particle injection. These are particle injection from the magnetotail and rapid particle energization by a shock front passing through the inner magnetosphere. *Liemohn et al.* [this volume] analyze time scales for localized particle injections to spread into a thin shell and find, for instance, that at $L = 2.8$ (in the slot region) during quiet driving conditions, it takes 4–6 h for a narrow MLT initial distribution of 3 MeV electrons to transform into a uniformly distributed ring. This result implies that the common assumption of instant symmetrization with respect to MLT of a localized injection of

~MeV electrons is incorrect and clearly has ramifications for radiation belt modeling. Using particle observations from the spacecraft Akebono, *Nagai* [this volume] reports a rapid storm time rebuilding of the central part of the outer radiation belt over a timescale of a few hours. Since this rebuilding coincides with a large-scale dipolarization of the magnetic field caused by storm time substorms, it is likely that electrons are transported from the magnetotail by an intense substorm-associated electric field. *Blake* [this volume] presents hitherto unreported CRRES data on the well-known shock injection of 24 March 1991, that produced a new radiation belt in the slot region. It is hoped that analysis of these unpublished CRRES observational details of the first minutes of the rapid injection may prove useful in the analysis of data from the RBSP mission. *Turner et al.* [this volume] provide a current understanding of the sudden depletion of the outer radiation belt electron fluxes known as a flux dropout. Dropouts are characterized by the depletion of electron fluxes by up to several orders of magnitude over a broad range of L shells, energies, and equatorial pitch angles in just a few hours. *Hendry et al.* [this volume] emphasize the importance of energetic electron precipitation in flux dropouts occurring during high-speed solar wind stream-driven storms. In the final paper of this section, *Thomson* [this volume] analyzes the background variability of the equatorial magnetosphere as inferred from a long time series of GOES data and suggests that the dominant component of the variability in the solar wind, the driver of the magnetosphere, comes from the discrete normal modes of the Sun.

Whistler mode chorus waves are known to play an important role in controlling radiation belt dynamics. In the sixth section, on wave-particle interactions, *Omura et al.* [this volume] discuss the generation processes of chorus emissions and summarize the current status of nonlinear wave growth theory. The generation mechanisms of chorus involve the nonlinear dynamics of resonant electrons and the formation of electromagnetic "holes" or "hills" that result in resonant currents generating rising-tone emissions or falling-tone emissions, respectively. *Summers et al.* [this volume] analyze the generation of a whistler mode rising-tone chorus element and are able to construct complete time-profiles for the wave amplitude that smoothly match at the interface of the linear and nonlinear growth phases. *Albert et al.* [this volume] analyze nonlinear, test-particle behavior under the action of a coherent quasimonochromatic wave using a Hamiltonian approach, with a view toward practical long-term modeling of nonlinear wave-particle interactions in the radiation belts. *Ripoll and Mourenas* [this volume] adopt quasilinear diffusion theory and, by using effective analytical approximations, determine precipitation lifetimes for radiation belt electrons due to resonant scattering by whistler

mode hiss waves. *Ni and Thorne* [this volume] discuss how wave-particle interactions, specifically resonant electron interactions with whistler mode chorus and electrostatic electron cyclotron harmonic waves, play a dominant role in the scattering of injected plasma sheet electrons leading to diffuse auroral precipitation.

The seventh section relates to how cross-energy coupling of the particle populations of the radiation belts, ring current, plasmasphere, and ionosphere influences the dynamics of the inner magnetosphere. Using numerical modeling, *Jordanova* [this volume] finds that storm time development of the ring current affects radiation belt dynamics in three ways: it depresses the background magnetic field on the nightside, provides a low-energy seed population for the radiation belts, and generates electromagnetic wave modes that scatter radiation belt particles. *Siscoe and Fok* [this volume] analyze the recently formulated Love-Gannon relation connecting the storm time ring current to the dawn-dusk asymmetry in the geomagnetic field observations on the ground at low latitudes. The (possibly controversial) relation, which states that the dawn-dusk asymmetry is proportional to the *Dst* index, may cause a revision of some classical ideas of magnetospheric dynamics and magnetosphere-ionosphere coupling.

Moldwin and Zou [this volume] argue that the plasmapause, which acts as a separator between different wave and particle environments in the inner magnetosphere, should be considered the plasmasphere boundary layer (PBL). The PBL concept captures the complexity and local dynamics of the plasmapause with respect to both L shell and azimuth. The PBL modulates ULF and plasma waves, which in turn modulate the higher-energy particle populations in the inner magnetosphere.

It is accepted that ion outflow from the polar ionosphere is a significant supplier of plasma to the terrestrial plasma sheet and ring current. *Yau et al.* [this volume] model the transit of polar wind oxygen O^+ ions to the storm time inner magnetosphere and find that such outflow could explain the prompt presence of energetic O^+ ions in the plasma sheet and ring current at the storm onset. *Haaland et al.* [this volume] use measurements from the Cluster mission to quantify the amount of cold plasma supplied to the magnetosphere from the polar ionosphere for various geomagnetic disturbance levels and solar wind conditions.

Reversal of the geomagnetic field polarity has dramatic effects on the radiation belts and ring current, as well as on the access to the magnetosphere of GCR and SEP. *Lemaire and Singer* [this volume] use an adaptation of Stormer's theory to determine the depletion, rebuilding, and characteristic properties of the radiation belts over the course of a geomagnetic field reversal. These results may provide insight, for instance, on the role of the northward/southward turning of the interplanetary magnetic field on the dynamics of today's inner magnetosphere as well as in paleomagnetospheres.

In the penultimate section, three papers discuss particular issues related to space weather and the radiation belts. *O'Brien et al.* [this volume] provide details of the various types of space weather hazards from energetic electron and ion populations that affect spacecraft and describe the type of information that the satellite design community needs from the radiation belt science community. In general, the satellite design community needs worst case and mean radiation environment specifications. O'Brien et al. make recommendations on how the science community can improve the quality and quantity of knowledge transfer to the satellite designers. *Fennel et al.* [this volume] examine energetic electron responses to storms that occurred during 1998–2008 in the inner magnetosphere, $2 \leq L \leq 4$, using HEO3 data. They conclude that a definitive explanation of the electron flux response to storms, namely, "flux increase," "no response," or "flux decrease" requires a better combination (than is currently available) of electron observations and supporting information on plasma waves, plasmapause position, magnetopause position, and ring current penetration. *Li et al.* [this volume] describe the space weather mission CSSWE, which was launched on 13 September 2012. The science objectives of CSSWE are twofold: to determine the precipitation loss and evolution of the energy spectrum of radiation belt electrons, and to investigate the relationship of solar flare properties to the timing, duration, and energy spectrum of SEPs reaching the Earth. This NSF-funded Cubesat mission will not only provide valuable space weather data but also provide training for the next generation of engineers and scientists.

Finally, in the last section, we examine radiation belts beyond the Earth. All the strongly magnetized planets of the solar system have robust radiation belts extending to relativistic energies. It is natural to ask what we have learned about the Earth's radiation belts that can immediately carry over to the other planetary radiation belts, and what lessons can be learned about the Earth's radiation belts from a comparative study of solar system radiation belts. *Mauk* [this volume] uses the Kennel-Petschek differential flux limit to compare the radiation belts at Earth, Jupiter, Saturn, Uranus, and Neptune and further provides a cautionary tale about attempts to apply radiation belt physics to hyperenergetic radiation regions outside the solar system.

Plasma wave emissions have been detected at all of the planets that have been visited by spacecraft equipped with plasma wave instruments. Wave-particle interactions involving whistler mode chorus, hiss, equatorial noise, and electron cyclotron harmonic waves are implicated in the acceleration and loss of radiation belt particles and are expected to play a major role in radiation belt dynamics throughout the solar system. *Hospodarsky et al.* [this volume] summarize the properties of these wave modes and discuss the similarities and differences of the plasma waves detected at the Earth, Jupiter, and Saturn.

To conclude, we thank all the authors for their stimulating contributions and hope that this volume will prove to be a valuable resource for experienced researchers and beginning graduate students alike.

Note added in proof: NASA has recently renamed the Radiation Belt Storm Probes (RBSP) mission. At a special ceremony held at the Johns Hopkins University Applied Physics Laboratory, Laurel, Maryland, on 9 November 2012, NASA renamed the mission as the Van Allen Probes in honor of James Van Allen, the discoverer of Earth's radiation belts. The ceremony also highlighted the successful commissioning of the spacecraft.

REFERENCES

Acuña, M. H., K. W. Ogilvie, D. N. Baker, S. A. Curtis, D. H. Fairfield, and W. H. Mish (1995), The Global Geospace Science Program and its investigations, *Space. Sci. Rev., 71*, 5–21.

Albert, J. M., X. Tao, and J. Bortnik (2012), Aspects of nonlinear wave-particle interactions, in *Dynamics of the Earth's Radiation Belts and Inner Magnetosphere, Geophys. Monogr. Ser.*, doi:10.1029/2012GM001324, this volume.

Andronov, A. A., and V. Y. Trakhtengerts (1964), Kinetic instability of the Earth's outer radiation belt, *Geomag. Aeron., 4*, 181–188.

Baker, D. N., and J. B. Blake (2012), SAMPEX: A long-serving radiation belt sentinel, in *Dynamics of the Earth's Radiation Belts and Inner Magnetosphere, Geophys. Monogr. Ser.*, doi:10.1029/2012GM001368, this volume.

Baker, D. N., G. M. Mason, O. Figueroa, G. Colon, J. G. Watzin, and R. M. Aleman (1993), An overview of the Solar, Anomalous, and Magnetospheric Particle Explorer (SAMPEX) mission, *IEEE Trans. Geosci. Remote Sensing, 31*(3), 531–541.

Blake, J. B. (2012), The shock injection of 24 March 1991: Another look, in *Dynamics of the Earth's Radiation Belts and Inner Magnetosphere, Geophys. Monogr. Ser.*, doi:10.1029/2012GM001311, this volume.

Cattell, C. A., A. Breneman, K. Goetz, P. J. Kellogg, K. Kersten, J. R. Wygant, L. B. Wilson III, M. D. Looper, J. B. Blake, and I. Roth (2012), Large-amplitude whistler waves and electron acceleration in the Earth's radiation belts: A review of STEREO and Wind observations, in *Dynamics of the Earth's Radiation Belts and Inner Magnetosphere, Geophys. Monogr. Ser.*, doi:10.1029/2012GM001322, this volume.

Elkington, S. R., A. A. Chan, and M. Wiltberger (2012), Global structure of ULF waves during the 24–26 September 1998 geomagnetic storm, in *Dynamics of the Earth's Radiation Belts and Inner Magnetosphere, Geophys. Monogr. Ser.*, doi:10.1029/2012GM001348, this volume.

Fälthammar, C.-G. (1965), Effects of time-dependent electric fields on geomagnetically trapped radiation, *J. Geophys. Res.*, *70*(11), 2503–2516.

Fennell, J. F., S. Kanekal, and J. L. Roeder (2012), Storm responses of radiation belts during solar cycle 23: HEO satellite observations, in *Dynamics of the Earth's Radiation Belts and Inner Magnetosphere, Geophys. Monogr. Ser.*, doi:10.1029/2012GM 001356, this volume.

Fraser, B. J., S. K. Morley, R. S. Grew, and H. J. Singer (2012), Classification of Pc1-2 electromagnetic ion cyclotron waves at geosynchronous orbit, in *Dynamics of the Earth's Radiation Belts and Inner Magnetosphere, Geophys. Monogr. Ser.*, doi:10.1029/ 2012GM001353, this volume.

Gendrin, R. (1981), General relationships between wave amplification and particle diffusion in a magnetoplasma, *Rev. Geophys.*, *19*(1), 171–184.

Graham, R. (1994), Killer electrons on rise, *Albuquerque J.*, *B8*, 15 Aug.

Haaland, S., et al. (2012), Cold ion outflow as a source of plasma for the magnetosphere, in *Dynamics of the Earth's Radiation Belts and Inner Magnetosphere, Geophys. Monogr. Ser.*, doi:10.1029/ 2012GM001317, this volume.

Hendry, A. T., C. J. Rodger, M. A. Clilverd, N. R. Thomson, S. K. Morley, and T. Raita (2012), Rapid radiation belt losses occurring during high-speed solar wind stream–driven storms: Importance of energetic electron precipitation, in *Dynamics of the Earth's Radiation Belts and Inner Magnetosphere, Geophys. Monogr. Ser.*, doi:10.1029/2012GM001299, this volume.

Hess, W. N. (1968), *The Radiation Belt and Magnetosphere*, Blaisdell, Waltham, Mass.

Hospodarsky, G. B., K. Sigsbee, J. S. Leisner, J. D. Menietti, W. S. Kurth, D. A. Gurnett, C. A. Kletzing, and O. Santolík (2012), Plasma wave observations at Earth, Jupiter, and Saturn, in *Dynamics of the Earth's Radiation Belts and Inner Magnetosphere, Geophys. Monogr. Ser.*, doi:10.1029/2012GM001342, this volume.

Johnson, M. H. and J. Kierein (1992), Combined Release and Radiation Effects Satellite (CRRES): Spacecraft and mission, *J. Spacecr. Rockets*, *29*(4), 556–563.

Jordanova, V. K. (2012), The role of the Earth's ring current in radiation belt dynamics, in *Dynamics of the Earth's Radiation Belts and Inner Magnetosphere, Geophys. Monogr. Ser.*, doi:10. 1029/2012GM001330, this volume.

Kennel, C. F., and H. E. Petschek (1966), Limit on stably trapped particle fluxes, *J. Geophys. Res.*, *71*(1), 1–28.

Kessel, R. L. (2012), NASA's Radiation Belt Storm Probes mission: From concept to reality, in *Dynamics of the Earth's Radiation Belts and Inner Magnetosphere, Geophys. Monogr. Ser.*, doi:10. 1029/2012GM001312, this volume.

Kress, B. T., M. K. Hudson, and P. L. Slocum (2005), Impulsive solar energetic ion trapping in the magnetosphere during geomagnetic storms, *Geophys. Res. Lett.*, *32*, L06108, doi:10.1029/ 2005GL022373.

Kress, B. T., M. K. Hudson, A. Y. Ukhorskiy, and H.-R. Mueller (2012), Nonlinear radial transport in the Earth's radiation belts, in

Dynamics of the Earth's Radiation Belts and Inner Magnetosphere, Geophys. Monogr. Ser., doi:10.1029/2012GM001333, this volume.

Lanzerotti, L. J. (2012), Space weather: Affecting technologies on Earth and in space, in *Dynamics of the Earth's Radiation Belts and Inner Magnetosphere, Geophys. Monogr. Ser.*, doi:10.1029/ 2012GM001372, this volume.

Lemaire, J. F., and S. F. Singer (2012), What happens when the geomagnetic field reverses?, in *Dynamics of the Earth's Radiation Belts and Inner Magnetosphere, Geophys. Monogr. Ser.*, doi:10.1029/2012GM001307, this volume.

Lemaire, J. F., D. Heynderickx, and D. N. Baker (Eds.) (1996), *Radiation Belts: Models and Standards, Geophys. Monogr. Ser.*, vol. 97, 322 pp., AGU, Washington, D. C., doi:10.1029/GM097.

Li, X., S. Palo, R. Kohnert, D. Gerhardt, L. Blum, Q. Schiller, D. Turner, W. Tu, N. Sheiko, and C. Shearer Cooper (2012), Colorado Student Space Weather Experiment: Differential flux measurements of energetic particles in a highly inclined low Earth orbit, in *Dynamics of the Earth's Radiation Belts and Inner Magnetosphere, Geophys. Monogr. Ser.*, doi:10.1029/2012GM 001313, this volume.

Liemohn, M. W., S. Xu, S. Yan, M.-C. Fok, and Q. Zheng (2012), Time scales for localized radiation belt injections to become a thin shell, in *Dynamics of the Earth's Radiation Belts and Inner Magnetosphere, Geophys. Monogr. Ser.*, doi:10.1029/2012GM 001335, this volume.

Mann, I. R., K. R. Murphy, L. G. Ozeke, I. J. Rae, D. K. Milling, A. Kale, and F. Honary (2012), The role of ultralow frequency waves in radiation belt dynamics, in *Dynamics of the Earth's Radiation Belts and Inner Magnetosphere, Geophys. Monogr. Ser.*, doi:10.1029/2012GM001349, this volume.

Mauk, B. H. (2012), Radiation belts of the solar system and universe, in *Dynamics of the Earth's Radiation Belts and Inner Magnetosphere, Geophys. Monogr. Ser.*, doi:10.1029/2012GM 001305, this volume.

McIlwain, C. E. (1961), Coordinates for mapping the distribution of magnetically trapped particles, *J. Geophys. Res.*, *66*(11), 3681–3691.

Miyoshi, Y., et al. (2012), The Energization and Radiation in Geospace (ERG) project, in *Dynamics of the Earth's Radiation Belts and Inner Magnetosphere, Geophys. Monogr. Ser.*, doi:10.1029/ 2012GM001304, this volume.

Mogilevsky, M. M., L. M. Zelenyi, A. G. Demekhov, A. A. Petrukovich, D. R. Shklyar, and RESONANCE Team (2012), RESONANCE project for studies of wave-particle interactions in the inner magnetosphere, in *Dynamics of the Earth's Radiation Belts and Inner Magnetosphere, Geophys. Monogr. Ser.*, doi:10.1029/ 2012GM001334, this volume.

Moldwin, M. B., and S. Zou (2012), The importance of the plasmasphere boundary layer for understanding inner magnetosphere dynamics, in *Dynamics of the Earth's Radiation Belts and Inner Magnetosphere, Geophys. Monogr. Ser.*, doi:10.1029/2012GM 001323, this volume.

Nagai, T. (2012), Rebuilding process of the outer electron radiation belt: The spacecraft Akebono observations, in *Dynamics of the*

Earth's Radiation Belts and Inner Magnetosphere, Geophys. Monogr. Ser., doi:10.1029/2012GM001281, this volume.

Ni, B., and R. M. Thorne (2012), Recent advances in understanding the diffuse auroral precipitation: The role of resonant wave-particle interactions, in *Dynamics of the Earth's Radiation Belts and Inner Magnetosphere, Geophys. Monogr. Ser.*, doi:10.1029/2012GM001337, this volume.

Northrop, T. G. (1963), *The Adiabatic Motion of Charged Particles*, Interscience, New York.

O'Brien, T. P., J. E. Mazur, and T. B. Guild (2012), What the satellite design community needs from the radiation belt science community, in *Dynamics of the Earth's Radiation Belts and Inner Magnetosphere, Geophys. Monogr. Ser.*, doi:10.1029/2012GM001316, this volume.

Omura, Y., D. Nunn, and D. Summers (2012), Generation processes of whistler mode chorus emissions: Current status of nonlinear wave growth theory, in *Dynamics of the Earth's Radiation Belts and Inner Magnetosphere, Geophys. Monogr. Ser.*, doi:10.1029/2012GM001347, this volume.

Ozeke, L. G., I. R. Mann, K. R. Murphy, I. J. Rae, and A. A. Chan (2012), ULF wave–driven radial diffusion simulations of the outer radiation belt, in *Dynamics of the Earth's Radiation Belts and Inner Magnetosphere, Geophys. Monogr. Ser.*, doi:10.1029/2012GM001332, this volume.

Ripoll, J.-F., and D. Mourenas (2012), High-energy electron diffusion by resonant interactions with whistler mode hiss, in *Dynamics of the Earth's Radiation Belts and Inner Magnetosphere, Geophys. Monogr. Ser.*, doi:10.1029/2012GM001309, this volume.

Roederer, J. G. (1970), *Dynamics of Geomagnetically Trapped Radiation*, Springer, New York.

Schulz, M., and L. J. Lanzerotti (1974), *Particle Diffusion in the Radiation Belts*, Springer, New York.

Siscoe, G. L., and M.-C. Fok (2012), Ring current asymmetry and the Love-Gannon relation, in *Dynamics of the Earth's Radiation Belts and Inner Magnetosphere, Geophys. Monogr. Ser.*, doi:10.1029/2012GM001350, this volume.

Stormer, C. (1907), Sur les trajectoires des corpuscules électrisés dans l'espace sous l'action du magnétisme terrestre avec application aux aurores boréales, *Arch. Sci. Phys. Nat.*, *24*, 317–364.

Stormer, C. (1933), On the trajectories of electrical particles in the field of a magnetic dipole with applications to the theory of cosmic radiation, *Avh. Norske Videnskap. Akad. Oslo, Mat. Nat.*, *K1*, 11.

Stormer, C. (1955), *The Polar Aurora*, Oxford Univ. Press, New York.

Summers, D., R. Tang, and Y. Omura (2012), Linear and nonlinear growth of magnetospheric whistler mode waves, in *Dynamics of the Earth's Radiation Belts and Inner Magnetosphere, Geophys. Monogr. Ser.*, doi:10.1029/2012GM001298, this volume.

Thomson, D. J. (2012), Background magnetospheric variability as inferred from long time series of GOES data, in *Dynamics of the Earth's Radiation Belts and Inner Magnetosphere, Geophys. Monogr. Ser.*, doi:10.1029/2012GM001318, this volume.

Turner, D. L., S. K. Morley, Y. Miyoshi, B. Ni, and C.-L. Huang (2012), Outer radiation belt flux dropouts: Current understanding and unresolved questions, in *Dynamics of the Earth's Radiation Belts and Inner Magnetosphere, Geophys. Monogr. Ser.*, doi:10.1029/2012GM001310, this volume.

Van Allen, J. A. (1983), *Origins of Magnetospheric Physics*, Smithsonian Inst. Press, Washington, D. C.

Van Allen, J. A., G. H. Ludwig, E. C. Ray, and C. E. McIlwain (1958), Observation of high intensity radiation by satellites 1958 Alpha and Gamma, *Jet Propul.*, *28*, 588–592.

Vernov, S. N., A. E. Chudakov, P. V. Vakulov, and Y. I. Logachev (1959), Study of terrestrial corpuscular radiation and cosmic rays during flight of a cosmic rocket, *Dokl. Akad. Nauk SSSR*, *125*, 304–307.

Yau, A. W., A. Howarth, W. K. Peterson, and T. Abe (2012), The role of quiet time ionospheric plasma in the storm time inner magnetosphere, in *Dynamics of the Earth's Radiation Belts and Inner Magnetosphere, Geophys. Monogr. Ser.*, doi:10.1029/2012GM001325, this volume.

D. N. Baker, Laboratory for Atmospheric and Space Physics, University of Colorado, Boulder, CO 80309, USA. (daniel.baker@lasp.colorado.edu)

I. R. Mann, Department of Physics, University of Alberta, Edmonton, AB T6G 2J1, Canada. (ian.mann@ualberta.ca)

D. Summers, Department of Mathematics and Statistics, Memorial University of Newfoundland, St. John's, NL A1C 5S7, Canada. (dsummers@mun.ca)

Space Weather: Affecting Technologies on Earth and in Space

Louis J. Lanzerotti

Center for Solar Terrestrial Research, New Jersey Institute of Technology, Newark, New Jersey, USA

Beginning with the era of development of electrical telegraph systems in the early nineteenth century, the space environment around Earth has influenced the design and operations of ever-increasing and sophisticated technical systems, both on the ground and now in space. Newfoundland had key roles in important first events in communications, including the landing of the first working telegraph cable in Heart's Content (1866), the first reception of trans-Atlantic wireless signals at Signal Hill in St. John's (1901), and the North American location of the first trans-Atlantic telecommunications cable in Clarenville (1956). All of the systems represented by these "firsts" suffered from effects of space weather. This paper reviews some of the historical effects of space weather on technologies from the telegraph to the present, describing several events that impacted communications and electrical power systems in Canada. History shows that as electrical technologies changed in nature and complexity over the decades, including their interconnectedness and interoperability, many important ones continue to be susceptible to space weather effects. The effects of space weather on contemporary technical systems are described.

1. INTRODUCTION

The aurora has been observed and marveled at for as long as humans have existed. The aurora has been viewed with awe not only in polar regions but also has been visible at various times at low geomagnetic latitudes such as Hawaii, Cuba, Rome, and Bombay. An understanding of the origins of the aurora only began to become scientifically rigorous in the late nineteenth century. Slowly, the aurora became to be understood as somehow related to the Sun, and therefore, solar activity might influence the Earth. That is, the Sun could influence the space environment around the Earth [*Chapman and Bartels*, 1940; *Soon and Yaskell*, 2003]. The Sun, in fact, does influence the "weather" in the space environment around the Earth, although the terminology of "space weather" did not become commonly employed until near the end of the twentieth century.

Any noticeable effects of the Sun and the Earth's space environment on human technologies had to wait until the first large-scale electrical technologies began to be deployed and used. This first large-scale technology was the electrical telegraph, first put into use some 160 or so years ago, several generations ago, and yet short in the course of human existence. W. H. Barlow, the company engineer for the Midland Railroad in England reported "spontaneous deflections" of the needles of the telegraph lines running aside the railroad tracks [*Barlow*, 1849]. Barlow's data for the Derby to Birmingham line is shown in Figure 1 for about 2 weeks of measurements in May 1847. The hourly variations in the "deflections" are clearly evident, as is an approximately daily variation throughout the interval. *Barlow* [1849, p. 66] further noted that "... in every case that [came under his] observation, the telegraph needles [were] deflected whenever aurora [were] visible."

Less than two decades after Barlow's observations, the white light solar flare observed by *Carrington* [1863] on 1 September 1959, ejected what we now know to be a coronal

Dynamics of the Earth's Radiation Belts and Inner Magnetosphere
Geophysical Monograph Series 199
© 2012. American Geophysical Union. All Rights Reserved.
10.1029/2012GM001372

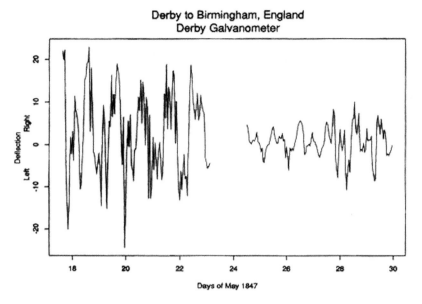

Figure 1. Galvanometer deflections on the telegraph cable along the Midland Railroad line from Derby to Birmingham for a 2 week interval in May 1847.

mass ejection (CME) into the interplanetary medium. Less than 20 h later, the Earth's space environment was struck. Huge changes in the geomagnetic field were observed wherever measurements were being made: "During the great auroral display ... disturbances of the magnetic needle [at] Toronto, in Canada, the declination of the needle changed nearly four degrees in half an hour" [*Loomis*, 1869, p. 12]. Auroras were observed on Earth from the north to as low latitude as Hawaii.

In the years since Barlow, the telegraph made large strides in its development and deployment across many locales on Earth, greatly enabling faster communications across large distances. Carrington's solar event produced greater disturbances in telegraph systems than Barlow had experienced. As reported by *Prescott* [1875, p. 322], on the telegraph line from Boston to Portland (Maine) on Friday, 2 September 1859, "... the line was worked [without batteries] more than two hours when, the aurora having subsided, the batteries were resumed." Between South Braintree and Fall River (Massachusetts; a distance of about 40 miles) "such was the state of the line ... when for more than an hour [the operators] held communication over the wire with the aid of the celestial batteries alone."

Prescott [1875], in his treatise on the electric telegraph, records the observations and experiences of many eastern U.S. telegraph operators during the Carrington event (as well as observations of effects on telegraphs from other auroral events in Europe and the United States). Prescott subtitles one section of his chapter on "Terrestrial Magnetism" as

"Working Telegraph Lines with Auroral Magnetism." *Shea and Smart* [2006] published a compendium of eight contemporary, published U.S. articles attributed to, or written by, Elias Loomis (Professor at Yale) related to the aurora and magnetic observations during the Carrington event.

The decades following the Carrington event found much work by electrical and telegraph engineers in attempts to understand the origins of the "spontaneous" electrical currents in their systems and to mitigate against them. In parallel, scientists worked to attempt to understand how an event on the Sun, such as the one Carrington reported, could affect the Earth so far away.

Newfoundland in the late nineteenth and early twentieth century was the site of many firsts in electrical communications technologies. The first trans-Atlantic telegraph cable from Valencia, Ireland, to Bull Arm (Trinity Bay) failed after 1 month. Cyrus Field, using the huge *Great Eastern* ship, was successful in establishing the first operating cable from Valencia, landing it in Heart's Content (Trinity Bay) in 1868 (Figure 2). This cable was operational for a century, until 1965. Each year, a celebration is held in the small village of Heart's Content on the anniversary day of the landing, 27 July. The 145th anniversary party occurred the week following the 2011 Chapman Conference on Dynamics of the Earth's Radiation Belts and Inner Magnetosphere in St. John's (Figure 3). As recounted by *Rowe* [2009, p. 45], "... the beginning of cable service was far from reliable ... Earth's magnetic currents, lightning, the aurora borealis ... sent the [galvanometer] into wild and rapid gyrations." The "Earth's

Figure 2. Still existent trans-Atlantic telegraph cables landing from Trinity Bay, Newfoundland, with Heart's Content cable station (original station in red) in the background. The station is now a Canadian Provincial Historical Site.

magnetic currents" and the aurora borealis were evidence of "space weather," although not known to be such at the time.

The year 1956 saw the inaugural call on the first trans-Atlantic voice telephone cable (TAT-1) across the Atlantic. The cable, from Oban, Scotland, was landed in Clarenville, Newfoundland (Trinity Bay), and saw service until 1978. TAT-2, from Penmarch, France, to Clarenville, was placed in service in 1959 and was retired in 1982. The large magnetic storm of February 1958 produced havoc on TAT-1. As John Brooks wrote in the *New Yorker* magazine "At almost the exact moment when the magnetograph traces leaped and the aurora flared up, huge currents in the earth . . . manifested themselves not only in power lines in Canada but in cables under the north Atlantic" [*Brooks*, 1959, p. 56].

Axe [1968] reported that an induced voltage swing larger than ~1.5 kV was measured at Oban, Scotland, during the most intense portion of the geomagnetic storm that affected TAT-1. Voltage excursions larger than 1 kV were measured

on the telephone cable from Clarenville to Sydney Mines, Nova Scotia, at the peak of the storm [*Winckler et al.*, 1959].

2. ADVANCES IN ELECTRICAL TECHNOLOGIES

Over the following century and a half, to today, as humans continued to develop electrical technologies for communications, electric power, and other uses, the effects of the Sun and the Earth's space environment continued to be felt and had to be dealt with. Engineers and scientists did not understand solar-terrestrial phenomena, or believe, until the early twentieth century, that the Sun could actually disturb the Earth in the ways that were implied by the coincidences observed between solar activity and geomagnetic storms, aurora, and disturbances on electrical systems.

In 1885, Guglielmo Marconi began experimental studies of wireless transmissions on his father's estate near Bologna, Italy. This work evolved into extensive experiments on land, shore-to-sea, and between-ship transmissions, much carried out in England. In 1901, Marconi established transmitting stations at Poldhu and The Lizard in Cornwall, and receiving stations at Wellfleet on Cape Cod, Massachusetts, and on Signal Hill, St. John's in Newfoundland (there are many books and articles relating to Marconi's life and work, e.g., the work of *Bussey* [2001]). He encountered many hurdles, including wind damage and destruction of transmission and receiving towers, in his attempts to cross the Atlantic with a radio wave.

On 12 December 1901, Marconi had success, receiving the Morris Code letter "S" at Signal Hill (Figure 4) as transmitted from Poldhu. For this, he was awarded, with Karl Braun of Germany, the Nobel Prize in Physics for 1909. The beginnings of communications through the "air" had begun. Such communications had the potential to provide more bandwidth, did not require the laying of long cables across a deep ocean, and could evade the pesky "spontaneous" currents in the telegraph cables.

The Marconi Company established a wireless station at Cape Race, Newfoundland, in 1904 (Figure 5). It was at this station that the distress signals from the Titanic were received on 14 April 1912.

While the anomalous (and often large) electrical currents experienced in the telegraph wires were avoided by Marconi's wireless innovation, the solar-terrestrial environment had surprises for the new electrical technology. As *Marconi* [1928] himself wrote

". . . times of bad fading [of the wireless signals] always coincide with the appearance of large sun-spots and intense aurora-boreali usually accompanied by magnetic storms. . . .." These are ". . . the same periods when cables and land lines experience difficulties or are thrown out of action."

Figure 3. Celebration cake for the 145th year of landing of the first successful trans-Atlantic telegraph cable, 27 July 2011.

Figure 4. Cabot Tower, Signal Hill, St. John's, Newfoundland, August 2011, a Canadian National Historic Site.

Figure 5. Cape Race, Newfoundland, on the Avalon Peninsula, the site of the Marconi Company wireless station established in 1904.

In the early days of wireless, and as the understanding of the relationship of solar activity to successful operations of the technology increased, the need for better understanding of the causes of wireless "fading" and other anomalies became more important. In fact, the data shown in Figure 6 (reproduced from the work of *Lanzerotti*

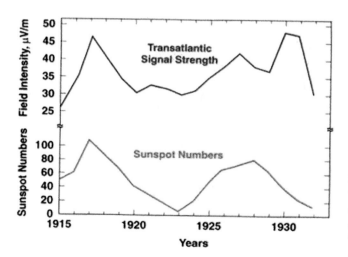

Figure 6. Trans-Atlantic daylight wireless transmission signal levels (15–23 kHz) measured by the AT&T Company during sunspot cycles 15 and 16.

[2004]) of the relationship between daylight trans-Atlantic signal strength (15–23 kHz) and sunspot numbers shortly after the AT&T company began trans-Atlantic transmissions represent one of the earlier efforts at what might today be called "space weather predictions." It is clear that the field strength of the signal appears to be "correlated" with sunspot number, with higher strength when there was more solar activity during these two solar cycles (numbers 15 and 16).

The experience of "unexpected" effects of the Earth's space environment on new electrical technologies (such as cable and then wireless) is a theme that persists to the current day. As new technologies are introduced, their successful operations can often be impeded by surprises from the solar-terrestrial environment. The characteristics of the environment must often be used in the making of design decisions. Since some of the most intense space environmental changes occur infrequently, past events of operational failure can be forgotten; design and operation decisions might then be made on the basis of more benign assumptions as to possible space environmental impacts. The fact that the most intense space environmental effects occur so infrequently also means that design decisions have to be made on imperfect knowledge of the solar-terrestrial environment [e.g., *Riley*, 2012]. All of these types of considerations have been operative throughout the twentieth century as one after the other of electrical technologies have been introduced and employed, for civilian and for national defense purposes.

While not recognized until their discovery by James Van Allen in 1958, the trapped radiation belt particles are centrally involved in producing many of the foregoing historical impacts on electrical technologies. In particular, the depletion of the trapped radiation under geomagnetic storm conditions greatly increases the conductivity of the ionosphere at the foot points of the magnetic field flux tubes that contained the formerly trapped particles. These changes in ionosphere conductivity, and the spatial differences in the conductivity that are produced because of different intensities of trapped particle loss, are responsible for large changes in magnetic fields at the Earth's surface. These Earth surface magnetic field changes are those that give rise to the telluric currents that flow in and disrupt long conductor systems. The enhanced ionosphere conductivities also produce the anomalous propagation conditions for wireless signal transmissions.

3. CONTEMPORARY SPACE WEATHER EFFECTS

Table 1 (adapted from the work of *Lanzerotti* [2004]) lists a majority of the solar terrestrial processes that are understood today that can affect contemporary technical systems. These processes, and some of the impacted technologies, are illustrated in Figure 7. Many of these physical processes are coupled. For example, the magnetic field variations as measured on the Earth's surface that can affect systems consisting of long conductors (such as the first telegraph cables and contemporary electric power grids and transocean cables) are produced by variations in the electrical currents flowing in the ionosphere. This, in fact, is the physical basis behind the observation made by Marconi in the quote above from one of his publications. Different types of ionosphere disturbances (such as the "plasma bubbles" in Figure 7), and confined principally to some regions on the Earth such as near the equator and in the auroral zones, are the source of disturbances (signal fading and signal scintillations) on other "wireless" signals such as modern-day navigation and satellite-to-ground signals.

3.1. Magnetic Field Variations

The magnetic field variations in Table 1 that produce electrical currents in the Earth that can affect electricity grids, long communications cables, and pipelines primarily result from geomagnetic storms. The largest of these storms, those most likely to produce the largest currents in the Earth,

Table 1. Solar-Terrestrial Processes and Their Consequences[a]

Solar-Terrestrial Process	Impact	Technologies Affected
Magnetic field variations	induction of electrical currents in the Earth	power distribution systems long communications cables pipelines
	directional variations	spacecraft attitude control compasses
Ionosphere variations	reflection, propagation attenuation	wireless communication systems
	interference, scintillation	communication satellites geophysical prospecting
Solar radio bursts	excess radio noise	wireless systems radar systems GPS transmissions
Particle radiation	solar cell damage	spacecraft power
	semiconductor damage/failure	spacecraft control
	faulty operation of semiconductor devices	spacecraft attitude control
	charging of surface and interior materials	spacecraft electronics
	human radiation exposure	astronauts Airline passengers
Micrometeoroids and artificial space debris	physical damage	Solar cells Orbiting mirrors, surfaces, materials Entire vehicles
Atmosphere	increased drag	Low altitude satellites
	attenuation/scatter of wireless signals	Wireless communication systems

[a]Adapted from *Lanzerotti* [2004].

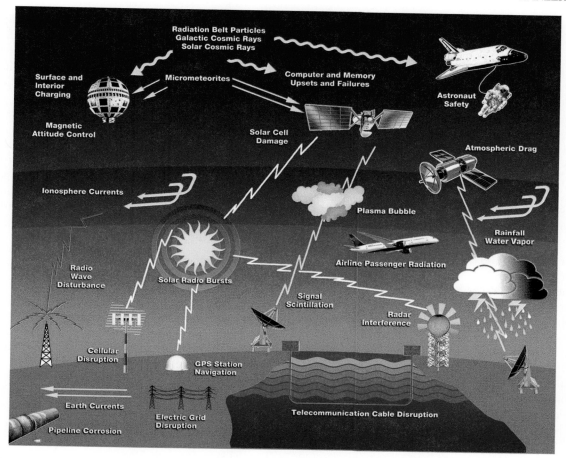

Figure 7. Illustration of many space weather effects and some of the modern-day electrical technologies and systems that can be adversely affected by them.

usually result from CME events striking the Earth's magnetosphere. These geomagnetic storms produce large changes in the electrical currents in the ionosphere. These fluctuating ionosphere currents produce fluctuating magnetic fields at the Earth's surface, which in turn cause electric currents to flow in the Earth's crust. These "telluric currents" seek the highest conductivity paths in which to flow, and long grounded conductors such as power grids, communications cables, and pipelines provide such paths [e.g., *Lanzerotti and Gregori*, 1986].

The magnitude and locations of the flowing telluric currents depend very much on the locations of the currents in the ionosphere and the conductivity of the underlying Earth. The ionosphere currents vary in location from geomagnetic storm to geomagnetic storm; current models of geomagnetic activity cannot predict these locations very precisely. The conductivity profile of the Earth is not known well in many locales. A given ionosphere current variation (if there was a "standard" variation) could pro-

duce very different telluric currents in regions that had different Earth conductivities.

The telluric currents that flow in the long conductors can produce damage, and even failure, of power system transformers, overwhelm constant current powering systems on long communication cables, and render inoperative pipeline corrosion protection circuits. Mitigation procedures depend upon the particular technical system and are implemented in some systems. There are costs for mitigation, and business decisions are made as to the cost/benefit results that might be achieved by implementations.

Some spacecraft use the Earth's magnetic field for orientation and guidance. A very intense solar wind shock wave, as from a strong CME event, can push the Earth's magnetopause inside geosynchronous orbit. The magnetic field just outside the magnetopause is of opposite polarity to that inside; a magnetically oriented spacecraft that crosses the magnetopause will thus suddenly be misoriented, and operators will likely have to intervene.

3.2. Ionosphere Fluctuations

The fluctuating ionosphere currents can cause havoc in the propagation of radio signals over a wide bandwidth. At HF and VHF frequencies, frequencies used, for example, by civil emergency agencies and by commercial airlines flying over the north polar region, signals transmitted from a location on the Earth can be absorbed or reflected anomalously from the ionosphere and, thus, not reach their destinations. At higher frequencies, such as those used for satellite-to-ground (and ground-to-satellite) transmissions and for GPS navigation signals, the radio waves can be severely distorted in phase and amplitude, affecting severely communications and navigation.

3.3. Solar Radio Noise

A scientific area that has become of more importance in recent years is that of the effects of solar radio noise on navigation technologies in the form of GPS. Solar radio bursts produced by solar flares were discovered in 1942 when British radars were rendered inoperable by jamming signals. These radars were being used to warn of enemy aircraft launched from continental Europe. The initial thinking was that the enemy was purposefully jamming the radars. J. S. Hay, a British scientist, identified the source as actually coming from the Sun, and not from across the English Channel [e.g., *Hey*, 1975].

While radars are still susceptible to solar radio noise, the vast proliferation of technologies that operate in the near-GHz and GHz frequency ranges, such as GPS and cell phones, means that this aspect of solar phenomena becomes of more significance in terms of its influence on technical systems [e.g., *Cerruti et al.*, 2006, 2008]. Solar radio noise, as well as bursts of solar X-rays, arrives at the Earth at the speed of light. Thus, there is no warning of their occurrence as there is for a possible encounter of a CME with the Earth to produce a magnetic storm. Accurate predictions of solar flares are still rather rudimentary, and predictions of how intense a radio burst or (an X-ray burst) might be are nonexistent. Radio noise and bursts, and X-rays are produced by electrons trapped and propagating in the Sun's magnetic fields. Until the intensities of the trapped electrons and of the magnetic fields can be readily measured and/or predicted, the forecast of the occurrence and intensity of solar radio and X-ray events remains a major unsolved problem.

3.4. Particle Radiation

When Sir Arthur Clark and John Pierce proposed Earth-orbiting communications satellites (at geosynchronous and low Earth-orbiting altitudes, respectively), they did not anticipate that the space environment around the Earth was not benign. However, the charged particle environment in space determines the design of space systems in many important ways. These charged particles are those trapped in the Earth's magnetosphere (the radiation belts), solar energetic particles outside the magnetosphere and those that penetrate into the magnetosphere, and galactic cosmic rays.

Even though the spatial extent and intensities of the trapped radiation were not delineated for a number of years following Van Allen's discovery in 1958, it was, nevertheless, recognized that the radiation presented a formidable environmental constraint to system designs and to human occupation of space. The design and build of the first active telecommunications satellite Telstar 1, conceived and promoted by John Pierce of Bell Laboratories, could not provide substantial shielding for many reasons, including the launch vehicle available (a Delta), and the size and weight of the small spacecraft (about 77 kg; 87.6 cm in diameter).

Telstar 1 was built at Bell Laboratories of largely discrete components including transistors (no integrated circuits or microprocessors available in those days), paid for by AT&T, including the launch costs reimbursed to NASA. The spacecraft carried several transistor solid-state detectors with different front aperture thicknesses to measure the radiation environment encountered by the satellite. Telstar was launched on 10 July 1962 into a low Earth orbit (perigee 952 km; apogee 5933 km). On the prior day, the United States had conducted the Starfish Prime high-latitude nuclear test over the Pacific. In addition to large disturbances in the electric grids in Hawaii and New Zealand, the test injected into the low-altitude magnetosphere fluxes of electrons more than 100 times the natural background radiation. Less than 8 months after launch, Telstar succumbed to the radiation environment. The electron data obtained from the radiation detectors on Telstar are still referenced today for the information obtained on electron lifetimes at those altitudes.

The radiation environment at geosynchronous altitude was unknown at the time of launch (July 1963) of the first operational satellite at that location, Syncom 2. The NASA Applications Technology Satellite ATS-1, launched to geosynchronous orbit in December 1966, carried particle detector instruments from three U.S. institutions: Aerospace Corporation, University of Minnesota, and Bell Laboratories. Discovery data from these instruments showed that the radiation environment at this altitude, where almost all communications satellites reside today, is highly variable in time and location along the orbit. These data also provided the first evidence that in some solar-produced events, the magnetopause can be pushed inside geosynchronous, exposing space assets to the interplanetary environment.

Table 1 lists many of the ways that charged particles can affect space systems. The lowest-energy particles, in the hundreds of eV to KeV energy range, are the sources of charging on spacecraft surfaces and solar arrays. If spacecraft surfaces are not adequately grounded to one another, discharges (similar to lightning) can occur between these areas, discoloring surface materials and producing electromagnetic noise interference in electronic systems. Very low energy neutral atoms of oxygen and nitrogen, as encountered in low Earth orbits, can sputter and discolor spacecraft surfaces.

At hundreds of keV and MeV and higher energies, charged particles, dominantly the trapped population, can damage solar cells in the arrays (thus, decreasing over time the electrical output of the arrays), damage and cause upsets in semiconductor components, and produce charging in dielectric materials deep within a spacecraft. Dielectrics that experience sufficient charging will also discharge, with the resultant damage to materials and the emission of electromagnetic noise interference.

Charged particles will also produce noise and, at times, complete obscuration of optical instruments such as star trackers and scientific telescopes. The loss of lock from a temporarily unavailable star tracker signal will affect spacecraft control.

As is clear from the first days of the space age, radiation trapped in the Earth's magnetosphere is of central importance to the design and operations of space systems, from communication satellites at geosynchronous orbit, to elliptical and low circular orbits of national security systems, to low orbit communication and navigation systems. Deeper understanding of the dynamics of these trapped populations is required as space systems become more complex and as component parts decrease in size and increase in density. Better and more comprehensive measurements will result in deeper understanding that will then result in better models for making wise design decisions and for potential forecasting of the occurrence of deleterious space weather conditions.

In addition to the trapped populations, low-energy magnetosphere particles, from ambient plasma conditions to the ring current population, are also critical for describing and modeling the radiation environment that space systems encounter. Much better understanding of the temporal and spatial distributions of these populations are necessary in their own right for design and modeling purposes and, importantly, as seed populations for the trapped particles. The acceleration mechanism(s) of these seed populations to the high energies that can penetrate deeply into space systems is(are) yet to be understood. These acceleration mechanisms are, thus, important from both the aspect of fundamental cosmical plasma physics and for their importance in applications to practical engineering problems.

3.5. Micrometeoroids and Artificial Space Debris

Solid materials in the space environment, natural or artificial, will produce physical damage to space vehicles and will cause disturbances to spacecraft orientation. The micrometeoroid environment at geosynchronous orbit, where communications satellites operate, is poorly known, and better understanding could aid in the operations of such spacecraft.

3.6. Atmosphere

The neutral atmosphere in low Earth orbit varies appreciably over a solar cycle. The atmosphere density increases with increasing altitude during solar maximum conditions when the enhanced solar EUV and UV emissions heat the atmosphere, causing it to rise. Therefore, the drag from the atmosphere on low Earth-orbiting satellites will increase during solar maximum years, with the resultant that orbits for the International Space Station and for any other satellites that have active orbit control will need to be raised more frequently than they have to be at other times in the solar cycle. Other, not actively controlled, spacecraft in low Earth orbit will see their orbit altitude decrease with time.

The increased atmosphere density also causes increased drag on artificial space debris, causing pieces to descend and burn up in the ionosphere. This "cleansing" effect helps to lower the danger from some debris objects (but not those at altitudes too high to be affected by the atmosphere).

It is well understood that water vapor and rainfall can seriously affect the propagation and transmission of microwaves through the atmosphere. It is now acknowledged that there is coupling between clouds and the ionosphere under at least some (not well understood) meteorological and ionosphere conditions. While it is only speculation at the moment as to whether space weather conditions that change the ionosphere might somehow be related to cloud nucleation and electrification processes, and thus to atmospheric weather that can affect microwave signals, the idea warrants continued examination.

4. CONCLUSION

Over the last 160 years, there have been striking advances in electrical technologies. Many of these advances are related in one way or the other to communications, since the time of the recording of space weather disturbances on the first telegraph lines, initially on short distances, and then on the first trans-Atlantic cable that landed in Hearts Content, Newfoundland. The historical record demonstrates that space weather processes often provide surprises in the implementation and operation of new electrical technologies, such as the wireless receiver that Marconi employed at Signal Hill, to early radar receivers in Great Britain, to spacecraft. The historical record also demonstrates that as the complexity of systems increase, including their interconnectedness and interoperability, they can become more susceptible to space weather effects. This is especially true for electrical grids in developed nations, upon which depends the energy for most aspects of modern society.

Central to understanding many physical processes in the solar-terrestrial environment that can affect technical systems are the trapped and lower-energy plasma populations of the Earth's magnetosphere. Deeper understanding, at fundamental physical levels, are required of these populations if more reliable models are to be achieved and if more accurate forecasting of potentially damaging conditions are liable to occur following a solar event.

REFERENCES

Axe, G. A. (1968), The effects of Earth's magnetism on submarine cables, *Electr. Eng. J.*, *61*, 37–43.

Barlow, W. H. (1849), On the spontaneous electrical currents observed in the wires of the electric telegraph, *Philos. Trans. R. Soc. London*, *139*, 61–72.

Brooks, J. (1959), A reporter at large: The subtle storm, *New Yorker*, 19 Feb.

Bussey, G. (2001), *Marconi's Atlantic Leap*, Radio Soc. of G. B., London, U. K.

Carrington, R. C. (1863), *Observations of the Spots on the Sun From November 9, 1853, to March 24, 1861, Made at Red Hill*, Williams and Norgate, London, U. K.

Cerruti, A. P., P. M. Kintner, D. E. Gary, L. J. Lanzerotti, E. R. de Paula, and H. B. Vo (2006), Observed solar radio burst effects on GPS/Wide Area Augmentation System carrier-to-noise ratio, *Space Weather*, *4*, S10006, doi:10.1029/2006SW000254.

Cerruti, A. P., P. M. Kintner Jr., D. E. Gary, A. J. Mannucci, R. F. Meyer, P. Doherty, and A. J. Coster (2008), Effect of intense December 2006 solar radio bursts on GPS receivers, *Space Weather*, *6*, S10D07, doi:10.1029/2007SW000375.

Chapman, S., and J. Bartels (1940), *Geomagnetism*, Clarendon Press, Oxford, U. K.

Hey, J. S. (1975), *The Evolution of Radio Astronomy*, Watson Intl., London, U. K.

Lanzerotti, L. J. (2004), Solar and solar radio effects on technologies, in *Solar and Space Weather Radiophysics*, edited by D. E. Gary and C. U. Keller, pp. 1–16, Springer, Heidelberg, Germany.

Lanzerotti, L. J., and G. P. Gregori (1986), Telluric currents: The natural environment and interactions with man-made systems, in *The Earth's Electrical Environment*, pp. 232–258, Natl. Acad. Press, Washington, D. C.

Loomis, E. (1869), The aurora borealis or polar light, *Harper's New Mon. Mag.*, *39*, 1–21.

Marconi, G. (1928), Radio communications, *Proc. IRE*, *16*, 40–49.

Prescott, G. B. (1875), *History, Theory and Practice of the Electric Telegraph*, Osgood, Boston, Mass.

Riley, P. (2012), On the probability of occurrence of extreme space weather events, *Space Weather*, *10*, S02012, doi:10.1029/2011SW000734.

Rowe, T. (2009), *Connecting the Continents: Heart's Content and the Atlantic Cable*, Creative Book, St. John's, Newfoundland, Canada.

Shea, M. A., and D. Smart (2006), Compendium of the eight articles on the "Carrington Event" attributed to or written by Elias Loomis in the *American Journal of Science*, 1859–1861, *Adv. Space Res.*, *38*, 313–385.

Soon, W. W.-H., and S. H. Yaskell (2003), *The Maunder Minimum and the Variable Sun-Earth Connection*, World Sci., Singapore.

Winckler, J. R., L. Peterson, R. Hoffman, and R. Arnoldy (1959), Auroral x-rays, cosmic rays, and related phenomena during the storm of February 10–11, 1958, *J. Geophys. Res.*, *64*(6), 597–610.

L. J. Lanzerotti, Center for Solar Terrestrial Research, New Jersey Institute of Technology, Newark, NJ 07102, USA. (ljl@adm.njit.edu)

SAMPEX: A Long-Serving Radiation Belt Sentinel

Daniel N. Baker

Laboratory for Atmospheric and Space Physics, University of Colorado Boulder, Boulder, Colorado, USA

J. Bernard Blake

SSL, The Aerospace Corporation, Los Angeles, California, USA

The near-Earth region of the magnetosphere responds powerfully to changes of driving forces from the Sun and the solar wind. The Earth's radiation belts and inner magnetosphere show substantial differences in their characteristics as the Sun's magnetic field and solar wind plasma properties change over the approximately 11 year solar activity cycle. Solar coronal holes produce regular, recurrent fast solar wind streams in geospace, often enhancing highly relativistic electrons and causing recurrent geomagnetic storms. These phenomena are characteristic of the approach to sunspot minimum. On the other hand, major geomagnetic disturbances associated with aperiodic coronal mass ejections occur most frequently around sunspot maximum. Such disturbances also can often produce significant radiation belt enhancements. We describe the observational results that characterize the differences throughout the inner part of geospace during the course of the solar activity cycle. We place particular emphasis on long-term, homogeneous data sets from the Solar, Anomalous, and Magnetospheric Particle Explorer (SAMPEX) mission. The NASA SAMPEX spacecraft launched in 1992 is expected to end its mission by December 2012. This space platform has revolutionized our views of the dynamic radiation belt environment. We conclude that SAMPEX has been a most successful and impactful mission for radiation belt studies.

1. INTRODUCTION

In the summer of 1992, a small spacecraft called the Solar, Anomalous, and Magnetospheric Particle Explorer (SAMPEX) was ready to be launched from NASA's Western Test Range into a low-Earth, high-inclination orbit on board a Scout launch vehicle. SAMPEX was the first mission in a new line of Small Explorer (SMEX) projects that were initiated by NASA in the late 1980s [*Baker et al.*, 1991]. The primary expressed goals of the SAMPEX program were to study (anomalous) galactic cosmic rays and solar energetic particle (SEP) enhancements [*Baker et al.*, 1993]. The SAMPEX payload was tailored for high and medium energy ion measurements. Little was it imagined that the tertiary objectives of measuring magnetospheric electron populations with the SAMPEX instrument complement would establish the spacecraft in a position of key prominence as one of the finest NASA radiation belt missions.

Despite the modest role initially envisioned for SAMPEX in the magnetospheric particle detection part of the program, it quickly became apparent after launch on 3 July 1992 that SAMPEX had remarkable potential to study radiation belt processes. The 600 km, nearly circular 82° inclination orbit was ideal for sampling essentially all magnetic field lines

Dynamics of the Earth's Radiation Belts and Inner Magnetosphere
Geophysical Monograph Series 199
10.1029/2012GM001368

threading the inner and outer Van Allen belts [*Baker et al.*, 1993]. The instrument payload included sensors that measured high-energy ($E > 1$ MeV) electrons with two separate systems, namely, the Proton-Electron Telescope (PET) [*Cook et al.*, 1993] and Heavy-Ion Large Telescope (HILT) [*Klecker et al.*, 1993]. There was also an important ability to measure medium-energy electron populations with a portion of the HILT investigation, as well as with the low-energy ion composition analyzer (LICA) [*Mason et al.*, 1993].

From the outset of the SAMPEX mission, the large-area, highly sensitive electron (and proton) detectors on board the spacecraft gave new insights into the dynamics of the Earth's radiation belts [e.g., *Baker et al.*, 1994a]. Issues of acceleration, transport, and loss of relativistic electrons could be assessed with high time-resolution ability [e.g., *Blake et al.*, 1996], and the continuous monitoring of the radiation belts gave a valuable new space weather analysis tool [e.g., *Baker et al.*, 1994b].

For the purposes of this paper (and in the context of this radiation belt monograph), we have chosen to review some of the considerable contributions that SAMPEX has made to magnetospheric particle physics. The SAMPEX spacecraft, which was launched in 1992, is now close to termination and is expected to reenter the Earth's atmosphere in December 2012 at the latest. Available space does not allow an exhaustive assessment or documentation of all that the mission has done to advance the discipline of radiation belt science. When the range and depth of magnetospheric studies are further considered in the context of all SAMPEX has contributed in solar and galactic cosmic ray studies, it becomes clear that this small and relatively unprepossessing spacecraft has been one of the most successful programs (especially when considered dollar for dollar and pound for pound) that NASA has ever flown.

2. RADIATION BELT STRUCTURE AND DYNAMICS

Figure 1 shows a schematic diagram of the Earth, its strong dipolar magnetic field region, and the regions of trapped energetic particles known as the inner (blue) and outer (purple) Van Allen radiation zones. Also shown in Figure 1 is a schematic diagram of the SAMPEX orbit. As is evident from Figure 1, the SAMPEX spacecraft samples essentially all of the relevant radiation belt regions from the vantage point of low-Earth orbit by cutting through various magnetic field lines. In a single ~100 min orbit, SAMPEX would be expected to cut twice through all field lines threading the northern and southern extensions of the inner and outer radiation belts.

Figure 2 shows an approximately 12 year record of electron ($E = 2$–6 MeV) measurements made by the SAMPEX/PET sensors [adapted from *Baker et al.*, 2005a]. The format of the data display, commonly used in radiation belt studies, shows intensity of electron fluxes color coded according to the logarithmic scale to the right. The vertical axis of Figure 2 is L value (i.e., the geocentric distance scaled in Earth radii, R_E, at which magnetic field lines would cross the magnetic equatorial plane). An L value of 1.0 would be near the Earth's surface, while an L value near 6.6 would be the region of space where geostationary Earth orbit spacecraft operate. The horizontal axis in Figure 1 is time measured in years: SAMPEX launched in July 1992 and the data are shown through early 2004.

Figure 2 clearly illustrates the double-belt structure of the Earth's electron radiation zones. The inner belt ($1 < L < 2$) is relatively weak in electron flux strength and varies (generally) over long time scales (an exception is the period in late 2003 known as the Halloween Storm period; see below). The

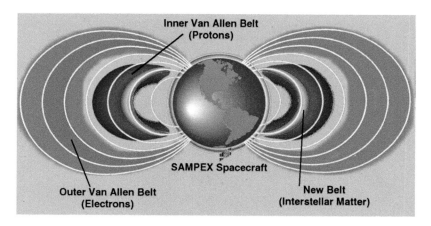

Figure 1. A schematic cross-sectional diagram of the Earth's inner and outer Van Allen radiation belts superimposed on dipolar magnetic field lines shown as white curves emanating from the Earth. Also shown is the SAMPEX low-altitude orbit and (in yellow) the trapped belt of anomalous cosmic rays.

Common Data Presentation

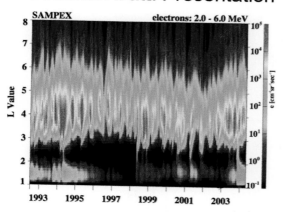

Figure 2. Color-coded flux levels of $E = 2–6$ MeV electrons measured by SAMPEX spacecraft instruments from July 1992 through March 2004. The color bar is shown to the right and is logarithmic in electron intensity. The vertical axis in the figure is magnetic L shell parameter (as described in the text), and the horizontal axis is time (in years). Adapted from *Baker et al.* [2005a].

electron "slot region" ($2 < L < 3$) is usually devoid of relativistic electrons [see, e.g., *Lyons and Thorne*, 1973; *Schulz and Lanzerotti*, 1974]. The outer electron belt ($L \sim 3$ to $L > 6$) is highly variable and often shows electron intensities ($E > 2$ MeV) that are 5 orders of magnitude higher than the inner zone population.

Figure 2 also demonstrates clearly that outer zone electrons are much higher in certain intervals of time that are about a decade apart (~1994 and ~2004 in the data shown). This will be discussed further below.

3. INNER ZONE PROPERTIES

Figure 2 makes clear that the inner zone electron fluxes were relatively elevated in 1993 through 1995 and then were at quite weak levels for much of the period from 1996 into the year 2000. An obvious enhancement of inner zone electrons occurred again in late 2003 in association with the Halloween Storm period [see *Baker et al.*, 2004a]. It was the established position of the SAMPEX team [*Blake et al.*, 2001b; *Baker et al.*, 2004a, 2005a, 2005b] that the higher 1992–1995 fluxes around $L \sim 1.5$ were residual effects of the famous March 1991 geomagnetic storm that was observed by the CRRES [*Blake et al.*, 1992].

A key fact to recall is that the weak electron inner zone population is spatially commingled with a high-intensity proton (ion) population to form the inner Van Allen belt. SAMPEX was designed innovatively to use the Earth's magnetic field as a charge-state analyzer to study high-energy

heavy ions and to confirm the source of the so-called anomalous cosmic ray (ACR) component [e.g., *Fisk et al.*, 1974; *Blake and Freisen*, 1977]. As shown in Figure 3, SAMPEX was able to measure with high precision both the interplanetary flux of ACR ions (such as oxygen nuclei and nitrogen nuclei) on open field lines over the Earth's polar caps as well as the trapped ACR ions confined on closed magnetic field lines in the Earth's inner Van Allen zone (see Figure 1). The work completed by the SAMPEX team [e.g., *Klecker et al.*, 1998; *Mewaldt et al.*, 1996] confirmed the interplanetary acceleration mechanism for the ACRs [*Fisk et al.*, 1974] as well as the magnetospheric trapping and flux concentration mechanism proposed by *Blake and Freisen* [1977].

4. OUTER ZONE ELECTRON DEPENDENCE ON SOLAR WIND FORCING

Figure 4 shows the record of slot and outer zone electron ($E = 2–6$ MeV) behavior from the time of SAMPEX launch (July 1992) through the middle of 2009 (when the PET sensor on SAMPEX ceased returning data). Figure 4 [from *Li et al.*, 2011] is a record of continuous, homogeneous data for nearly two solar sunspot cycles. As was hinted in Figure 2 above, and as discussed in detail by *Li et al.* [2011], the intensity and spatial extent of outer zone electrons is controlled in clear ways by solar and solar wind forcing. Figure 4 (top) shows smoothed sunspot number (black curve) and plots solar wind speed (red spiky curve with axis to the

Anomalous Cosmic Rays

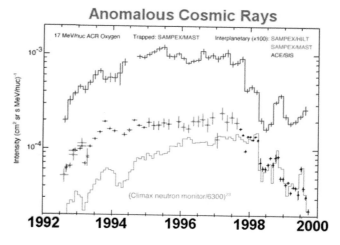

Figure 3. Directional differential fluxes of 17 MeV (nucleon oxygen cosmic rays)$^{-1}$ measured from 1992 to 2000. The lower trace is from the Climax neutron monitor, while the middle set of data points is the interplanetary flux levels (as shown by the colors indicated) of the anomalous cosmic ray (ACR) component. The highest trace is the trapped ACR flux measured by SAMPEX. From *Selesnick et al.* [2000].

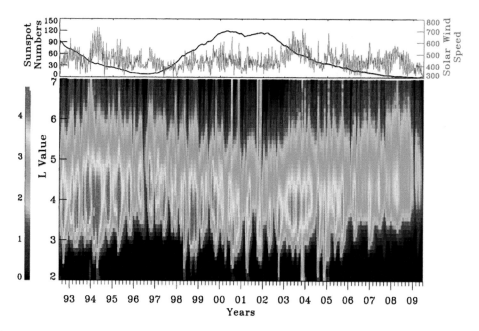

Figure 4. (bottom) A 17 year record of high-energy electrons measured by SAMPEX similar in format to Figure 2. (top) A trace of smoothed sunspot number (black line) and solar wind speed (red line) for the same period from July 1992 through July 2009. The black trace in the bottom plot shows the estimated plasmapause location at any given time. From *Li et al.* [2011].

right) for this extended period. Obviously, times of higher solar wind speed (e.g., 1994–1995 and 2003–2004) tended to be times of elevated electron flux throughout the outer zone [*Paulikas and Blake*, 1979; *Baker et al.*, 1994a, 1994b, 1997a]. But the interplanetary magnetic field also plays a key, indispensable role [*Blake et al.*, 1997], and this point has been clarified in recent papers [e.g., *Li et al.*, 2011].

Note in Figure 4 the excellent correlation of the inward extent of the outer zone electrons and the concurrent location of the plasmapause boundary shown by the superimposed black trace (scale to the right of bottom plot). As described by *Li et al.* [2006, 2011] and as seen in extreme events such as the Halloween Storm period of 2003 [*Baker et al.*, 2004a], the coldest plasmas in the Earth's magnetosphere and plasmasphere really do exert a significant influence on the radiation belts.

5. TRANSIENT SOLAR DISTURBANCES AND OUTER BELT RESPONSES

Figure 5 shows results from an early event analysis in which SAMPEX data played a key role [*Baker et al.*, 1998a]. Figure 5a is the familiar *L*-versus-time color spectrogram plot for SAMPEX 2–6 MeV electrons during the

first 200 days of 1997. On day of year (DOY) 135 (15 May 1997), a powerful coronal mass ejection (CME) struck the Earth and produced a quite abrupt acceleration of the Earth's entire outer zone electron population. *Baker et al.* [1998a] were able to study this event in detail and place it into the context of solar, interplanetary, and other magnetospheric measurements. In particular, as shown by Figures 5b and 5c, the orbit-by-orbit data of SAMPEX and the corresponding detailed data from the Polar spacecraft High-Sensitivity Telescope [*Blake et al.*, 1995] were able to show the coherent and nearly simultaneous global acceleration of electrons throughout much of the outer radiation belt on time scales of minutes to hours.

This kind of work using correlated SAMPEX and Polar measurements was subsequently pursued in other related studies [*Baker et al.*, 1998b, 2000, 2001; *Blake et al.*, 2001b]. The idea of remarkable "global coherence" in radiation belt acceleration was especially well established by the studies of *Kanekal et al.* [2001]. Figure 6 (top) [from *Kanekal et al.*, 2001] shows data from SAMPEX and closely analogous data from Polar (Figure 6, bottom) for all of 1998. Virtually every feature seen by Polar at high altitudes (relatively close to the magnetic equator) was also seen, quite comparably, by SAMPEX at low altitudes near the foot of corresponding magnetic field lines.

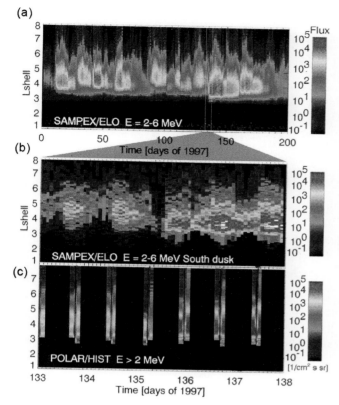

Figure 5. (a) L sorted electron fluxes measured at low altitudes by SAMPEX instruments for $2 < E < 6$ MeV. Data are coded according to the color bar to the right. L values from 1 to 8 are shown for the first 200 days of 1997, and several electron enhancement events are seen, notably one commencing on Day 135 (May 15). (b) Similar to Figure 5a but showing orbit-by-orbit data from SAMPEX for Day 133 (May 13) through Day 137 (May 17) of 1997. Data are for Southern Hemisphere, duskside portions of the SAMPEX orbit. (c) Similar to Figures 5a and 5b but for $E > 2$ MeV electrons measured by the Polar spacecraft. From *Baker et al.* [1998a].

6. ACCELERATION MECHANISMS

SAMPEX data have proven to be important to study the systematics of electron acceleration throughout the outer radiation belt. Figure 7 is an example of such data. Figure 7 (top) is the usual format of electron data for 1 year from mid-2000 through to mid-2001. As can be seen, many specific enhancements of outer-zone electrons can be identified throughout late 2000 and again in the spring season of 2001. Figure 7 (bottom) shows a higher PET energy channel of SAMPEX ($E = 3.5–16$ MeV). It is seen from this channel that only a few of the events that were so clear in the $E \sim 2$ MeV channel of the top plot were really prominent in the multi-MeV energy range of the bottom plot. This means that

some magnetospheric electron acceleration events produce very "hard" energy spectra, but most do not.

Elkington et al. [2004, 2005] have used sophisticated methods to study high-energy electron transport and acceleration in the Earth's magnetosphere. This work starts with a global simulation of the solar wind-magnetosphere system using the Lyon-Fedder-Mobarry (LFM) numerical simulation code. *Elkington et al.* then "push" energetic electrons in the self-consistent electric and magnetic fields of the LFM code. The result is an ability to simulate specific solar-driven geomagnetic storms and to see by direct simulations how the magnetosphere methodically transports and accelerates relativistic electrons within the inner magnetosphere.

An example of these results is shown in Figure 8 [see *Elkington et al.*, 2005 and *Baker et al.*, 2005b]. The powerful geomagnetic storm of 31 March 2001 [see *Baker et al.*, 2002] was simulated by the LFM model. (This storm and the resulting relativistic electron enhancement are, it might be noted, seen to be quite prominent in Figure 7, top). Figure 8 shows "snapshots" of the inward transport (Figure 8a), the radial diffusive acceleration (Figure 8b), and ultimately the strong trapping (Figure 8c) of $E > 1$ MeV electrons during the course of the March 2001 storm [see *Baker et al.*, 2005b].

The test-particle MHD simulation codes of *Elkington et al.* [2004, 2005] and other authors produce electron acceleration by means of earthward radial transport, which essentially equates to the betatron mechanism. Energy diffusion due to gyroresonant interaction with VLF chorus waves can also be effective in generating relativistic (>1 MeV) radiation belt electrons [*Summers et al.*, 1998, 2002, 2007; *Roth et al.*, 1999; *Varotsou et al.*, 2005; *Horne et al.*, 2005]. While radial diffusion and transport is particularly effective for energizing electrons outside geosynchonous orbit, there is considerable evidence that an additional local acceleration mechanism (e.g., VLF chorus diffusion) is required to explain observed relativistic electron flux increases inside geosynchronous orbit [e.g., *Miyoshi et al.*, 2004; *Iles et al.*, 2006; *Shprits et al.*, 2006]. Since geomagnetic storms can result in both net acceleration and net loss of energetic radiation belt electrons [*Reeves et al.*, 2003], it is important to incorporate electron loss mechanisms in radiation belt electron dynamical models. An important radiation belt loss process is due to pitch angle scattering into the atmospheric loss cone due to electron cyclotron resonance with VLF chorus, ELF hiss, and electromagnetic ion cyclotron waves [*Summers et al.*, 1998; *Summers and Thorne*, 2003; *Summers et al.*, 2007]. Construction of realistic 3-D time-dependent radiation belt models requires comprehensive observational data on the spatiotemporal properties of these waves. While such data sets are currently limited, it is expected that the imminent Radiation Belt Storm Probes mission [*Kessel*, this volume]

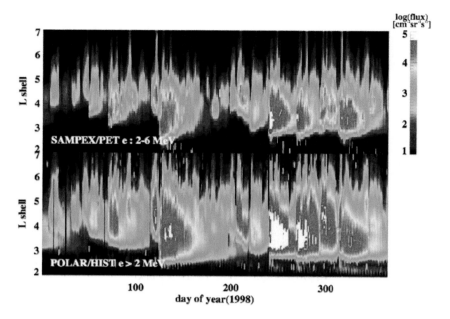

Figure 6. (top) Data similar in format to Figure 2 from SAMPEX for the year 1998. (bottom) Similar data from the Polar/ High-Sensitivity Telescope instrument showing remarkable global coherence of electron throughout the outer radiation belt. From *Kanekal et al.* [2001].

will provide valuable new data on these and other plasma waves that control radiation belt electron behavior.

7. HIGH-SPEED SOLAR WIND STREAM ACCELERATION

Immediately prior sections of this review have emphasized the key and obvious role of aperiodic, transient solar disturbances in accelerating high-energy magnetospheric electrons. This role of CME-driven events is undeniable in many cases. However, the strongest and most methodical acceleration of relativistic electrons in the Earth's outer radiation belts really is associated with recurrent high-speed solar wind streams [e.g., *Paulikas and Blake*, 1979; *Baker et al.*, 1990]. SAMPEX observations [*Baker et al.*, 1997a, 1997b] have proven to play a very key role in understanding this aspect (see Figures 2 and 4 above).

Figure 9 shows SAMPEX data for the entire year of 1994 in the *L* versus time spectrogram format. The data show regular, episodic enhancements of 2–6 MeV electrons for essentially the entire year. The white vertical arrows in Figure 9 delineate 27 day recurrent periods. As is evident, there is a strong tendency for the electron flux enhancements to occur with an obvious 27 day period.

Baker et al. [1997a, 1997b] showed evidence that prototypical relativistic electron acceleration in the Earth's radiation belts often occurs with several repeatable steps: (1) High-

speed solar wind streams drive strong magnetospheric substorm activity. (2) Substorms produce a large population of "seed" electrons extending up to several hundreds of keV in energy. (3) The enhanced radial diffusion of the seed electrons associated with high solar wind speeds, ULF waves drivers, and VLF wave heating produced by such streams inside the magnetosphere leads regularly to high-intensity electron radiation belt populations.

Many details of this complex process are not fully understood, however, and construction of realistic 3-D models of radiation belt electron dynamics remains a challenge, as we have implied above.

8. RADIATION BELT CONTENT AND STATISTICAL STUDIES

The continuous, homogeneous radiation belt information provided by SAMPEX has proven to be essential for long-term and statistical studies. This data set has given substantial insight into how the connected solar wind-magnetosphere-atmosphere system works.

A realization [*Baker et al.*, 1999] was that one could use the SAMPEX measurements such as those shown in Figure 2 or Figure 4 and integrate across all outer zone *L* values (say $2.5 < L < 6.5$) to estimate the entire average flux of electrons within the outer zone. One can do this on a daily, monthly, seasonal, or annual basis. Figure 10, for example, shows

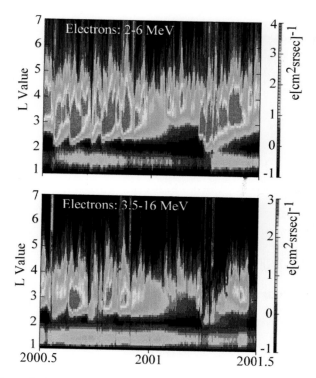

Figure 7. (top) Data similar in format to Figure 2 showing SAMPEX electron fluxes from July 2000 through June 2001. (bottom) SAMPEX data for the same time period as top plot but showing electron fluxes in the energy range 3.5–16 MeV.

results from SAMPEX for a study of seasonal flux variations [*Baker et al.*, 1999]. These results, sorted according to equinox and solstice periods, show well that outer radiation belt fluxes around the Spring and Fall equinox periods (roughly March and September) are clearly three or more times higher than are those around the solstice periods (roughly June and December). Close inspection of Figure 2 or Figure 4 above will confirm this result. More recent studies [e.g., *McPherron et al.*, 2009] have confirmed these earlier results and have discussed the phenomenon in terms of the semiannual Russell-McPherron effect.

Baker et al. [2004b] extended the 1999 seasonal-dependence study and introduced the radiation belt content index (often shortened to radiation belt content (RBC)). Figure 11a and 11b show examples of the RBC as described by *Baker et al.* [2004b]. Figure 11a shows the total estimated number of electrons in the outer radiation belt (2.5 < L < 6.5) as a function of time from 1992 through 2002. The computation of RBC takes into account the entire flux tube content and also integrates the particle spectrum in energy above $E = 2$ MeV. It is seen in Figure 11a that the peak RBC content was ~5 × 10^{23} electrons (mid-1994), and the minimum content of

the outer belt (occurring in 1996 and again in 1999) was ~3 × 10^{20} electrons.

Another use of the RBC, shown by Figure 11b, is to assess probabilities of encountering certain flux levels or certain total content levels [see *Baker et al.*, 2004b]. It is seen in Figure 11b that the content index follows a log-normal Gaussian probability distribution function (over several orders of magnitude). The 50% probability value of the RBC

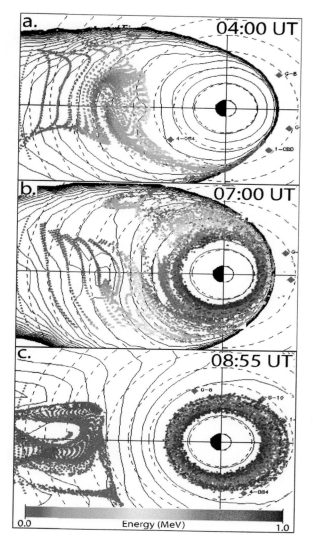

Figure 8. Snapshots of numerical simulation results [*Elkington et al.*, 2004, 2005] shown for a geomagnetic storm that occurred on 31 March 2001. Energetic electrons are "pushed" in the self-consistent electric and magnetic fields of the Lyon-Fedder-Mobarry (LFM) MHD simulation code (as described in the text). Three different times on 31 March are shown: (a) 04:00 UT; (b) 07:00 UT; and (c) 08:55 UT. From *Baker et al.* [2005b].

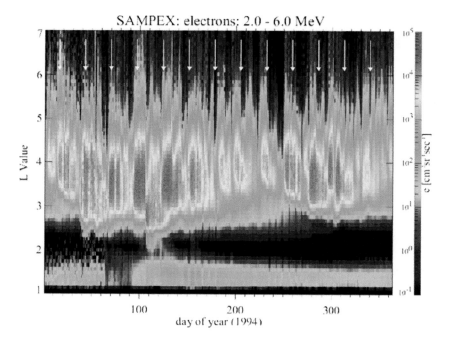

Figure 9. Data similar in format to Figure 2 but showing electron measurements from the SAMPEX spacecraft for the year 1994. The white vertical arrows are placed at recurrent 27 day intervals. The data show that electron flux enhancements recur with a clear 27 day period.

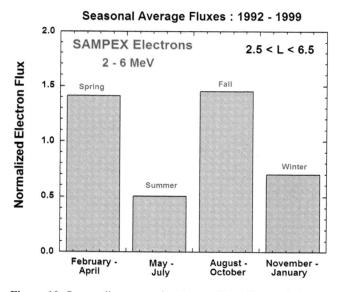

Figure 10. Seasonally averaged and normalized fluxes of electron fluxes measured by SAMPEX over the period 1992–1999. The data have been integrated over the entire outer radiation belt (2.5 < L < 6.5) and show a much higher average flux level for the spring and fall equinox periods compared to the summer and winter solstice periods. From *Baker et al.* [1999].

for 1992–2002 was about 10^{22} electrons contained within the outer belt.

Finally, the statistical studies using the RBC index formalism led to some other insights about the radiation belt electron behavior. For example, Figure 12 from *Baker et al.* [2004b] shows the scatter of daily RBI values versus concurrent solar wind speed values (V_{SW}).

Rather than following a linear or log-linear relationship as might have been expected from earlier work [e.g., *Paulikas and Blake*, 1979; *Baker et al.*, 1979], the scatter of points in Figure 12 is quite "triangular." *Baker et al.* [2004b] argued that enhanced solar wind speeds clearly increase the probability of higher radiation belt fluxes, but fluxes can remain quite elevated even after the solar wind "forcing" speed has diminished again. Thus, in many cases, the RBC index remained high for days or even weeks after the solar wind speed had decreased. The substance of this kind of correlation work has been revisited recently using geostationary orbit data [see *Li et al.*, 2011, and references therein; *Kellerman and Shprits*, 2012]

9. RADIATION BELT ENHANCEMENT AND DECAY RATES

As has been noted in several ways above, the sudden appearance of high-energy radiation belt electrons is an

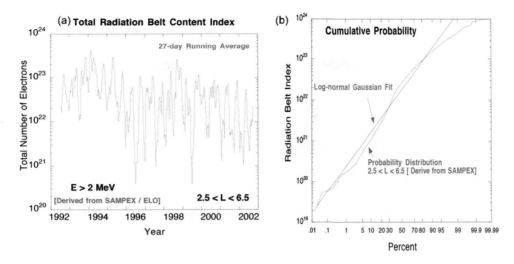

Figure 11. (a) Daily values for the radiation belt content (RBC) index for the period 1992 through 2002. Adapted from *Baker et al.* [2004b]. (b) Cumulative probability curve for the radiation belt content index as discussed in the text. Modified from *Baker et al.* [2004b].

important and intriguing scientific problem. It is also quite clear that such electron enhancements can cause significant spacecraft operational problems [*Baker*, 1987, 2002] through the mechanism of deep dielectric charging. Obviously, the flux of electrons at a given point in the magnetosphere is a delicate balance of source strength and local loss rates. Many

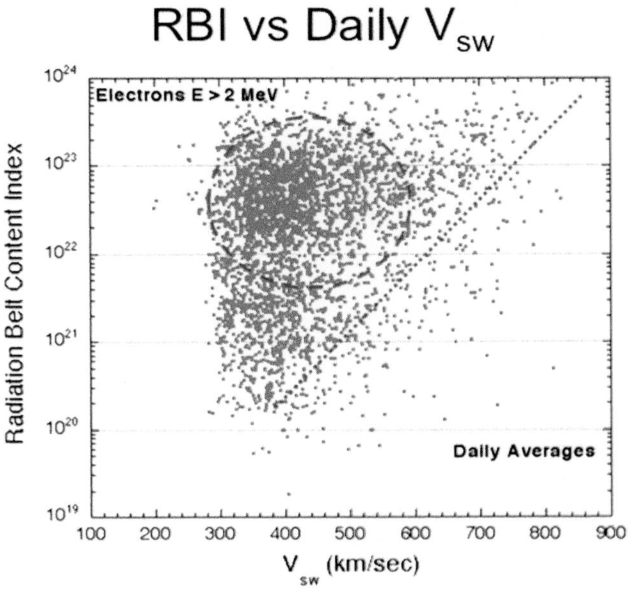

important "space weather" incidents related to radiation belt enhancements have been studied using the SAMPEX data sets [e.g., *Baker et al.*, 1994b, 1998c].

One of the largest and most studied radiation belt enhancement events occurred in late October and early November 2003. This Halloween Storm event (or set of events) was touched on several times previously in this paper. Figure 13 shows an expanded record of SAMPEX data (and solar wind speed data in the top plot) from a paper by *Baker et al.* [2007]. The period of time covered is 2003 to the end of 2005. Figure 13 (bottom) shows quite obviously the sudden appearance (or at least, powerful reemergence) of a "new" belt of electrons associated with the Halloween Storm [*Baker et al.*, 2004a]. Over at least the next 2 years, the inner zone belt decayed away. But superimposed on the gradual decay were punctuated episodes of further enhancement of the inner zone associated with the outer zone events (such as around DOY 600 and again around DOY 700).

The original powerful acceleration event in late October 2003 was analyzed in detail by *Baker et al.* [2004a]. As shown in the interpretive sketches of Figure 14 (from that paper), the Earth's outer Van Allen zone was virtually annihilated by the Halloween Storm. The complete disappearance of multi-MeV electrons at $L \sim 4$ was quickly followed by the new generation of MeV electrons in the heart of the slot region (which usually is devoid of such electrons). *Baker et al.* [2004a] showed that the plasmasphere was scoured away in this event, and this allowed a complete (if temporary) reconfiguration of the radiation belts.

Figure 12. A scatter diagram of the RBC index versus solar wind speed (V_{SW}) based on daily average values. From *Baker et al.* [2004b].

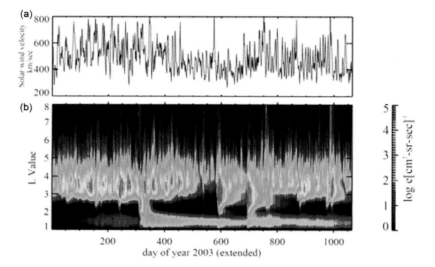

Figure 13. Daily averaged data for the years 2003–2005 inclusive. (a) Solar wind speeds measured by instruments onboard the ACE spacecraft upstream of the Earth's magnetosphere. (b) Relativistic (2–6 MeV) electron fluxes for the range $1 < L < 8$ in a logarithmic color-coded format as shown by the color bar to the right. Data in Figure 13b were obtained from instruments onboard the SAMPEX spacecraft ($E = 2$–6 MeV electrons). From *Baker et al.* [2007].

Following the formation of the new inner zone belt of electrons, we were able to study episodic losses of electrons from low L shells for the next several years. Figure 15 is from *Baker et al.* [2007]. It shows "cuts" through the data portrayed in Figure 13 at several fixed L values (indicated by the different colors). The result is traces of electron flux levels versus time. *Baker et al.* [2007] were able to use the exponential decay rates of each flux spike in Figure 15 to deduce the inner zone electron lifetimes from $L \sim 1.5$ to $L \sim 2.2$. Such estimates of lifetime were made possible by the continuous SAMPEX monitoring.

10. ELECTRON LOSSES: ATMOSPHERIC COUPLING

Figure 16 is another form of display of electron flux measurements from SAMPEX. These global polar maps were developed largely by *Callis et al.* [1996a, 1996b, 1998] to show the regions of the Earth's atmosphere that would be affected by electrons being precipitated and lost during radiation belt events. What is fascinating about Figure 16 is how dramatically the precipitating electron flux can alter from just 1 day to the next. As shown by *Callis et al.*, the energy into the Earth's middle atmosphere could easily change by 5 or more orders of magnitude in a day. SAMPEX has been an adept tool for studying long-term loss of radiation belt particles and assessing the effects of these particles in the atmosphere [*Callis et al.*, 1996a, 1996b, 1998].

Figure 17 shows the altitude profiles where different precipitating particles would be effective. At altitudes from ~80 to >150 km, the low-energy electrons measured by the LICA

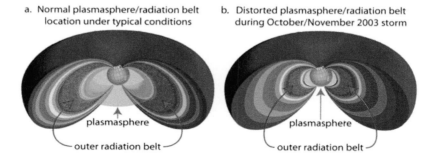

Figure 14. Schematic diagrams of radiation belt changes occurring during the Halloween 2003 solar storms. From *Baker et al.* [2004a].

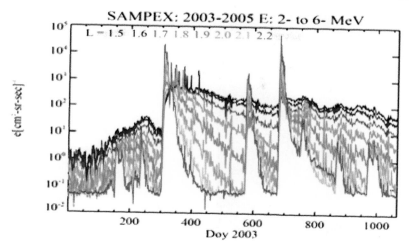

Figure 15. SAMPEX 2–6 MeV electron fluxes. Cuts are taken for several different fixed L values (delineated at the top) for the period 2003–2005 inclusive. The vertical axis is directional particle intensity, and the horizontal axis is time reckoned in days (day of year) from the beginning of 2003. Several flux enhancement events are evident as discussed in the text. From *Baker et al.* [2007].

sensors [*Mason et al.*, 1993] have been compared to nitric oxide measurements from the Student Nitric Oxide Explorer (SNOE) [*Barth et al.*, 2001, 2002, 2004; *Baker et al.*, 2001]. Higher-energy electrons measured by PET and HILT (see Figure 17) can have dramatic ionization effects in the range of altitudes from ~50 to ~100 km. Of considerable interest from an atmospheric chemistry standpoint, SEPs (also measured by SAMPEX) can greatly affect middle and lower atmospheric ionization and chemistry [e.g., *Callis et al.*, 1996a].

Given the importance of minor atmospheric constituents such as NO, we have used SAMPEX capabilities to study energetic particle precipitation and coupling. In Figure 18, for example, we show a comparison between SAMPEX precipitating electrons ($E > 25$ keV) and zonally resolved altitude profiles of NO from the SNOE spacecraft [*Baker et al.*, 2001]. We found strong evidence that precipitating auroral electrons were exercising quite powerful control on nitric oxide production at auroral latitudes.

11. MICROBURSTS AND OTHER TRANSIENT ELECTRON PRECIPITATION

At launch, SAMPEX had five integral electron channels that were sampled 10 times per second from the HILT instrument [*Klecker et al.*, 1993]. The lowest energy channel, $E > 150$ keV, used the rear proportional counter in the HILT instrument. The gas counter in HILT was of the flow type, and SAMPEX carried a tank of gas sufficient for 3 years of operation. Subsequently, this channel was no longer opera-

tional. The rest of the channels were from a pixelated array in the HILT sensor and measured $E > 1$ MeV. Sixteen detectors were grouped into rows of four. The field-of-view of each row had a different pitch angle view and was used to tell when the local fluxes were isotropic.

The combination of a very large geometric factor and high sampling rate gave SAMPEX an unprecedented view of bursty, energetic electron precipitation. A small burst was seen at $E > 150$ keV just before 02:06 in Figure 19. The $E > 1$ MeV channel saw only galactic cosmic ray background. At that time, SAMPEX was at $L \sim 1.8$. In the slot region, the $E > 150$ keV channel count rate quickly rose until the counter was saturated, whereas the $E > 1$ MeV channel saw few electrons until SAMPEX reached $L \sim 3$, where the count rate greatly increased. Note that the lower-energy electrons decreased, while the higher energy electrons continued to increase. The different count rates in the four $E > 1$ MeV channels indicated a trapped pitch angle distribution. The count rate peaked at $L \sim 4$ and began to decrease until at $L = 5$; there was an abrupt increase to the highest count rates seen during this pass (Figure 20). Within this precipitation band, the four $E > 1$ MeV channels showed the same count rate, indicating that the electron fluxes were isotropic within the band. This isotropy has been found to be characteristic of precipitation bands.

Clear microburst precipitation, temporal structure, short compared to a second, also was seen from the beginning of the SAMPEX mission [*Nakamura et al.*, 1995; *Blake et al.*, 1996] (Figure 21). Both types of precipitation were seen around the same time.

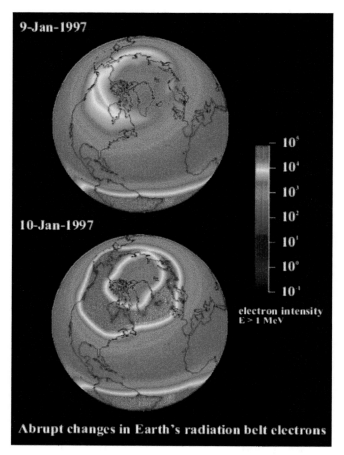

Figure 16. Global electron plots of SAMPEX Northern Hemisphere electron fluxes (2–6 MeV) comparing 9 January 1997 with the following day, 10 January 1997. Dramatic changes of precipitating electron fluxes are demonstrated by such data.

Note also the overall behavior of the count rates as SAMPEX moves to lower L and into the slot region. The MeV electrons go away, the half MeV electrons drop by an order of magnitude but show a small, smooth bump around $L = 3$, yet the rates in the proportional counter soar to the highest level seen. All seem to go away about the time that SAMPEX enters the inner zone.

SAMPEX observations of microbursts continued over many years including correlation with VLF chorus and quantifying losses during geomagnetic storms [*Blanchard et al.*, 1998; *Nakamura et al.*, 2000; *Lorentzen et al.*, 2001; *O'Brien et al.*, 2004]. Surveys were carried out to determine the microburst dependence upon geomagnetic activity, local time, and electron energy. Results of an early such survey [*Lyons*, 1997] revealed that microburst precipitation peaked in occurrence from midnight to past dawn and had a strong dependence upon geomagnetic activity.

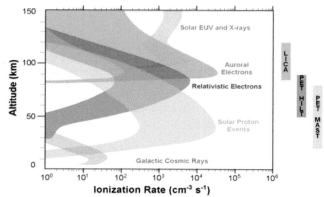

Figure 17. Ionization rates due to different types of precipitating energetic particles as a function of atmospheric altitude. The colored bands show where various particle types are most effective in stopping and ionizing atmospheric constituents. The similarly colored vertical bars to the right show which SAMPEX detectors best measure the relevant particle types.

SAMPEX frequently observed bursty precipitation associated with lightning; a dramatic example is shown in Figure 22. SAMPEX passed several hundred kilometers east of an area of substantial thunderstorm activity over the central Pacific. The lightning discharges caused electron precipitation into the drift

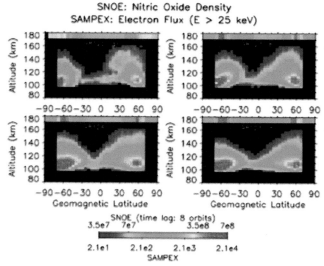

Figure 18. A comparison of Student Nitric Oxide Explorer (SNOE) nitric oxide measurements and the Solar, Anomalous, and Magnetospheric Particle Explorer (SAMPEX) electron ($E >25$ keV) measurements for days 122 and 125 of 1998 (2–5 May). SNOE data are plotted on a logarithmic scale according to the color scales at the bottom of the figure. The SAMPEX electron counting rates range over several orders of magnitude, as also shown by the color bar. Further details are provided in the text. From *Baker et al.* [2001].

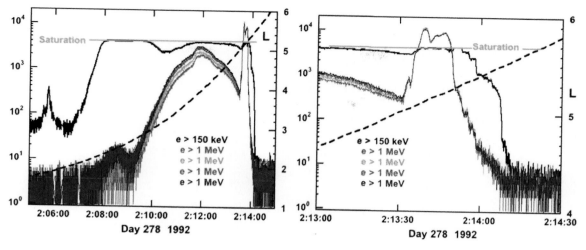

Figure 19. SAMPEX poleward pass from the inner zone until SAMPEX exited the trapping boundary and moved over the polar cap. The right-hand plot is an expanded view of the time period when a large precipitation band was just below the trapping boundary. Throughout this time period, SAMPEX was in the drift loss cone. From *Blake et al.* [1996].

loss cone over a broad region but localized in L as can be seen by the detailed congruence of the electron intensity peaks in the Northern and Southern Hemispheres. Drifting electrons populated narrow L shells, creating several curtains of electrons traversed by SAMPEX.

It is well known that MeV radiation belt electrons are seldom found inside $L \sim 3$ and then only during or immediately after large geomagnetic storms. This situation is understood as being a result of strong loss processes in the slot region for relativistic electrons. Therefore, the following ob-

servations of electrons $E > 1$ MeV at 50 samples s^{-1} came as a surprise. Remarkably, the three MeV microbursts occurred just outside $L = 2$, at the lower edge of the slot region (Figure 23a). The first two bursts occurred when SAMPEX was in the bounce loss cone; these two bursts rose out of the galactic cosmic ray background. The third and final microburst occurred when the mirror altitude was less than 120 km. The three microbursts show differences in their temporal structure.

The first microburst consisted of a single peak, broader than the subsequent two, the second showed one return

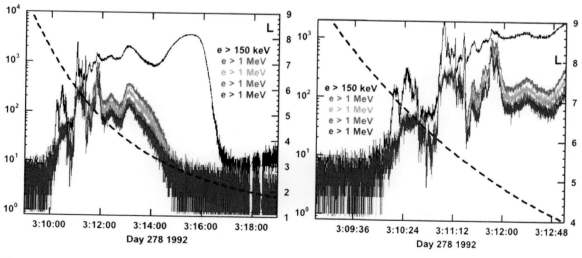

Figure 20. Complex burst structure is seen at higher L followed by much smoother variability at lower L. A loss cone pitch angle distribution also is indicated at lower L by differing count rates in the four silicon solid-state detectors (SSDs). The right-hand plot shows the high L portion of the plot above on an expanded scale to show how the variability differs with electron energy. There are correlations to be sure but sometimes negative as well as positive.

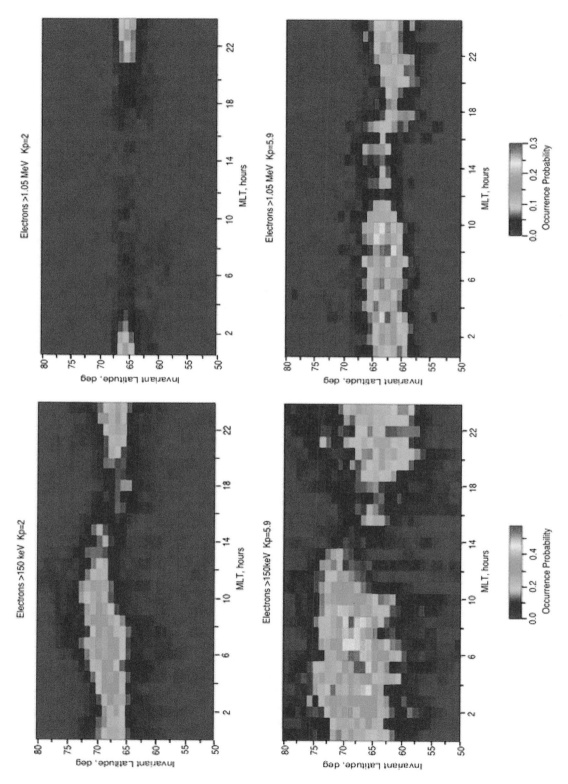

Figure 21. The occurrence probability of microburst precipitation for $E > 150$ keV and $E > 1.05$ MeV is shown as a function of latitude and LT for (top) quiet ($Kp = 2$) and (bottom) disturbed ($Kp = 5.9$) conditions.

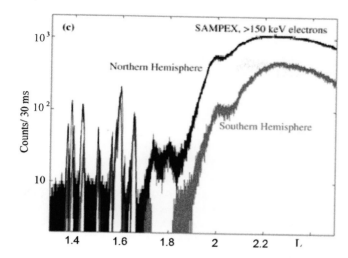

Figure 22. The time history of the $E > 150$ keV electrons as measured along the SAMPEX orbital track as the spacecraft moved from the Southern Hemisphere to the Northern Hemisphere [*Blake et al.*, 2001a].

"bounce," and the third microburst showed multiple returns (Figure 23b). This can be understood as due to the rising mirror altitude for the particles. For the first peak, the upper atmospheric density heavily smeared the burst, the second peak was smeared somewhat, and for the third peak, multiple returns were possible. The bouncing microburst packet was seen with excellent statistics (Figure 23c), and the bounce frequency derived from the observation is in excellent agreement with that calculated using the formulae given in the work of *Schulz and Lanzerotti* [1974]. In summary, SAMPEX observed three isolated, MeV-energy microbursts over a time period of about 1 min inside and at the edge of the bounce loss cone and just above $L = 2$.

The SAMPEX payload also included a sensor that measured electrons with energy $E > 500$ keV. The data sample rate with this sensor (LICA) was 10 s^{-1} rather than 50 s^{-1} for the HILT sensor. At the time of the first microburst, the local $E > 500$ keV electron flux (Figure 23d) had already begun to rise unlike the situation for electrons $E > 1$ MeV. Since SAMPEX was in the bounce loss cone at the time, this observation would suggest that precipitation was ongoing just to the west of SAMPEX, and residual electrons that had bounced a few times were being observed before loss into the atmosphere was complete.

Figure 23. (right) (a) A trio of microbursts, $E > 1$ MeV, at $L = 2$ as SAMPEX exited the bounce loss cone moving to higher latitude. (b) The microburst trio on an expanded time scale. (c) The bouncing packet of electrons over a 1 s time period. (d) The trio of microbursts as seen in the $E > 500$ keV channel.

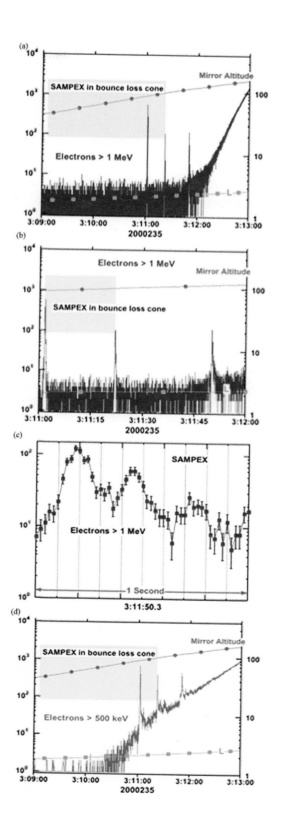

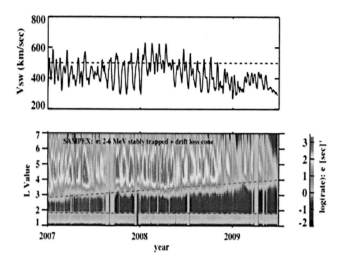

Figure 24. Similar to Figure 13 but for the profound solar activity minimum period of 2007 through 2009.

The lower sample rate of the $E > 500$ keV channel makes a frequency analysis as was done in the higher-energy channel uncertain. Nonetheless, the width of these bursts is consistent with that of the bursts seen at higher energy. In addition to the three primary microbursts, other precipitation enhancements can be seen whose widths and separation do not fit the bouncing packet model. The first two microbursts were followed by enhancements, whereas the third microburst had a precursor as well as being followed by enhancements.

This trio of microbursts is highly unusual in the SAMPEX database of almost 175,000 h of observations [see *Blake et al.*, 1996; *Nakamura et al.*, 1995; *O'Brien et al.*, 2004]. Although several search techniques for bursty precipitation have been employed, it is hard to be sure how complete the searches have been. Nonetheless, some infrequent strong scattering and, perhaps, even acceleration process took place at this time.

12. THE PROFOUND 2007–2010 SOLAR MINIMUM

During the period from roughly 2007 well into the year 2010, the Sun went through its deepest minimum of activity in nearly two centuries [e.g., *Turner*, 2011]. As a consequence, there were essentially no significant CME events during that interval and the solar wind forcing of the Earth's radiation belts diminished by an extraordinary degree. Especially in the late 2008 and 2009, the solar wind speed remained at historically low average levels as well (see Figure 4). Thus, radiation belt driving was virtually nonexistent.

Figure 24 shows a detail of radiation belt fluxes (bottom plot) measured by SAMPEX and solar wind speed (top plot) for 2007, 2008, and part of 2009. As can be seen, by the end of 2008, the slot region has broadened to a dramatic extent. In fact, the slot was seen by SAMPEX to extend from $L \sim 2$ to $L > 4$. For all intents and purposes, the outer radiation belt disappeared entirely from November 2008 and all through 2009. This resulted because (top plot) the solar wind speed almost never became greater than $V_{SW} \sim 400$ km s^{-1}.

13. THE LAST CHAPTER OF THE SAMPEX SAGA

The resurgence of sunspot activity on the Sun in 2011 (and up to the present time) after the extraordinary 4 year sunspot minimum just described means that the uppermost layers of the Earth's atmosphere are again being strongly heated and are expanding outward [see *Baker et al.*, 2012]. This is causing increased atmospheric drag on low-Earth orbit satellites. The SAMPEX spacecraft is now soon expected to succumb to this important space weather phenomenon. After the two decades of achievements documented in this review, SAMPEX is falling victim to the very solar activity and enhanced space weather disturbances that it has helped for so long to assess and understand. The prediction of drag experts is that SAMPEX will reenter the atmosphere most probably in September 2012. The earliest likely entry is August 2012, and the latest likely atmospheric entry is December 2012. Figure 25 [from *Baker et al.*, 2012] shows the expected orbit time profile.

When SAMPEX was launched in July 1992, the Sun was just coming over the peak of its ~11 year sunspot activity cycle (see Figure 25) and was moving toward much more quiet solar minimum conditions. Early after its launch, it was regarded as likely that SAMPEX would reenter due to atmospheric drag forces by 1998 or so and would very probably not make it through the solar activity maximum around the year 2000.

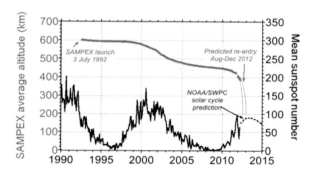

Figure 25. Daily averaged SAMPEX altitude (red trace) in the context of the past 22 years of solar activity shown by the monthly averaged sunspot number (black trace). From *Baker et al.* [2012].

However, contrary to the direst predictions, the spacecraft made it through the 2000–2001 solar maximum. SAMPEX was, therefore, able to go on to become one of the longest-serving and most productive platforms in NASA's inventory. The long, continuous record of monitoring the highest-energy particles in the Earth's neighborhood revealed more than we might ever have hoped or expected about how particles are accelerated, transported, and ultimately lost from the Earth's magnetosphere.

The situation for SAMPEX shows both the inevitability of the effects of the space environment on operating systems as well as the inherent uncertainties that relate to our proximity to a highly variable star. Early in its lifetime, SAMPEX suffered operational anomalies and upsets due to the energetic particles that it was put into space to study. In particular, there were single-event effects in the newly flown optical fiber data bus and electrostatic discharges from surface charging that affected one of the science instruments [*Baker et al.*, 2012]. Fortunately, none of these effects had serious impacts on the mission. However, the Sun's relentless heating of the Earth's atmosphere and the small, but irresistible, drag of that atmosphere will end SAMPEX's run. But, as noted, until now, no one could be sure exactly when the inevitable end would be reached.

14. SUMMARY

The official SAMPEX NASA science mission ended in June 2004. As part of an engineering and education program, NASA continued to track the satellite and operate it through the Bowie State University Operations and Control Center. One result of the extended unofficial operation of the mission was the continued collection of science data and continued processing at The Aerospace Corporation for archiving and future use by the space science community. A soon-to-be completed update of the SAMPEX data center with the postscience mission data means that new science and understanding from SAMPEX will continue long after its entry.

Our understanding of the magnetosphere, of space weather, and our technical capabilities to monitor the environment have come a long way in the two decades since SAMPEX first went on duty in space. We have developed highly capable new tools for observing the Sun and its turbulent corona. We have achieved ever-improving understanding of solar flares and CMEs. New space weather forecasting tools now in place are giving operational warnings of impending solar energetic particle events and geomagnetic storms. The unparalleled data from SAMPEX have given us profound new insights about how the radiation environment around the Earth waxes and wanes. The science community will greatly miss the monitoring prowess of SAMPEX as a sentinel warning of magneto-

spheric and solar events. But the impact of this modest mission will long be felt even after space weather and atmospheric drag claims another low-altitude space asset.

Acknowledgments. The SAMPEX mission has been a long and remarkably successful one owing to the efforts of countless scientists, engineers, and managers. The authors thank Glenn Mason and the entire science team for development of outstanding instruments and for a stimulating scientific environment over the more than two decades of the program. The authors also thank colleagues at NASA Headquarters, especially Chuck Holmes, for helping to keep the program going through many difficult circumstances. Finally, sincere thanks to Todd Watson and colleagues, past and present, at Bowie State University and Goddard Space Flight Center for their heroic efforts to maintain operations over the entire project lifetime.

REFERENCES

Baker, D. N. (1987), Effects of the solar terrestrial environment on satellite operations, *Artif. Satell.*, *22*, 103.

Baker, D. N. (2002), How to cope with space weather, *Science*, *297*(5586), 1486–1487, doi:10.1126/science.1074956.

Baker, D. N., P. R. Higbie, R. D. Belian, and E. W. Hones Jr. (1979), Do Jovian electrons influence the terrestrial outer radiation zone?, *Geophys. Res. Lett.*, *6*(6), 531–534.

Baker, D. N., R. L. McPherron, T. E. Cayton, and R. W. Klebesadel (1990), Linear prediction filter analysis of relativistic electron properties at 6.6 R_E, *J. Geophys. Res.*, *95*(A9), 15,133–15,140.

Baker, D. N., G. Chin, and R. F. Pfaff Jr. (1991), NASA's Small Explorer Program, *Phys. Today*, *44*, 44.

Baker, D. N., G. M. Mason, O. Figueroa, G. Colon, J. Watzin, and R. Aleman (1993), An overview of the SAMPEX mission, *IEEE Trans. Geosci. Remote Sens.*, *31*, 531–541.

Baker, D. N., J. B. Blake, L. B. Callis, J. R. Cummings, D. Hovestadt, S. Kanekal, B. Klecker, R. A. Mewaldt, and R. D. Zwickl (1994a), Relativistic electron acceleration and decay time scales in the inner and outer radiation belts: SAMPEX, *Geophys. Res. Lett.*, *21*(6), 409–412.

Baker, D. N., S. Kanekal, J. B. Blake, B. Klecker, and G. Rostoker (1994b), Satellite anomalies linked to electron increase in the magnetosphere, *Eos Trans. AGU*, *75*(35), 401.

Baker, D. N., et al. (1997a), Recurrent geomagnetic storms and relativistic electron enhancements in the outer magnetosphere: ISTP coordinated measurements, *J. Geophys. Res.*, *102*(A7), 14,141–14,148.

Baker, D. N., H. E. Spence, and J. B. Blake (1997b), ISTP: Relativistic particle acceleration and global energy transport, *Adv. Space Res.*, *20*, 1075–1080.

Baker, D. N., et al. (1998a), A strong CME-related magnetic cloud interaction with the Earth's Magnetosphere: ISTP observations of rapid relativistic electron acceleration on May 15, 1997, *Geophys. Res. Lett.*, *25*(15), 2975–2978.

Baker, D. N., T. I. Pulkkinen, X. Li, S. G. Kanekal, J. B. Blake, R. S. Selesnick, M. G. Henderson, G. D. Reeves, H. E. Spence, and G. Rostoker (1998b), Coronal mass ejections, magnetic clouds, and relativistic magnetospheric electron events: ISTP, *J. Geophys. Res.*, *103*(A8), 17,279–17,291.

Baker, D. N., J. H. Allen, S. G. Kanekal, and G. D. Reeves (1998c), Pager satellite failure may have been related to disturbed space environment, *Earth Space*, *11*(4), 6–11.

Baker, D. N., S. G. Kanekal, T. I. Pulkkinen, and J. B. Blake (1999), Equinoctial and solstitial averages of magnetospheric relativistic electrons: A strong semiannual modulation, *Geophys. Res. Lett.*, *26*(20), 3193–3196.

Baker, D. N., S. G. Kanekal, J. B. Blake, B. Klecker, G. M. Mason, and R. A. Mewaldt (2000), Magnetospheric relativistic electron response to magnetic cloud events of 1997, *Adv. Space Res.*, *25*(7–8), 1387–1392.

Baker, D. N., C. A. Barth, K. E. Mankoff, S. G. Kanekal, S. M. Bailey, G. M. Mason, and J. E. Mazur (2001), Relationships between precipitating auroral zone electrons and lower thermospheric nitric oxide densities: 1998 – 2000, *J. Geophys. Res.*, *106*(A11), 24,465–24,480.

Baker, D. N., R. E. Ergun, J. L. Burch, J.-M. Jahn, P. W. Daly, R. Friedel, G. D. Reeves, T. A. Fritz, and D. G. Mitchell (2002), A telescopic and microscopic view of a magnetospheric substorm on 31 March 2001, *Geophys. Res. Lett.*, *29*(18), 1862, doi:10.1029/2001GL014491.

Baker, D. N., S. G. Kanekal, X. Li, S. P. Monk, J. Goldstein, and J. L. Burch (2004a), An extreme distortion of the Van Allen belt arising from the 'Hallowe'en' solar storm in 2003, *Nature*, *432*(7019), 878–881, doi:10.1038/nature03116.

Baker, D. N., S. G. Kanekal, and J. B. Blake (2004b), Characterizing the Earth's outer Van Allen zone using a radiation belt content index, *Space Weather*, *2*, S02003, doi:10.1029/2003SW000026.

Baker, D., S. Kanekal, J. Blake, and J. Allen (2005a), Radiation belt responses to the solar events of October—November 2003, in *Inner Magnetosphere Interactions: New Perspectives From Imaging, Geophys. Monogr. Ser.*, vol. 159, edited by J. Burch, M. Schulz and H. Spence, pp. 251–259, AGU, Washington, D. C., doi:10.1029/159GM19.

Baker, D., S. Elkington, X. Li, and M. Wiltberger (2005b), Particle acceleration in the inner magnetosphere, in *The Inner Magnetosphere: Physics and Modeling, Geophys. Monogr. Ser.*, vol. 155, edited by T. I. Pulkkinen, N. A. Tsyganenko and R. H. Friedel, pp. 73–85, AGU, Washington, D. C., doi:10.1029/155GM09.

Baker, D. N., S. G. Kanekal, R. B. Horne, N. P. Meredith, and S. A. Glauert (2007), Low-altitude measurements of 2–6 MeV electron trapping lifetimes at $1.5 \leq L \leq 2.5$, *Geophys. Res. Lett.*, *34*, L20110, doi:10.1029/2007GL031007.

Baker, D. N., J. E. Mazur, and G. M. Mason (2012), SAMPEX to reenter atmosphere: Twenty-year mission will end, *Space Weather*, *10*, S05006, doi:10.1029/2012SW000804.

Barth, C. A., D. N. Baker, K. D. Mankoff, and S. M. Bailey (2001), The northern auroral region as observed in nitric oxide, *Geophys. Res. Lett.*, *28*(8), 1463–1466.

Barth, C. A., D. N. Baker, K. D. Mankoff, and S. M. Bailey (2002), Magnetospheric control of the energy input into the thermosphere, *Geophys. Res. Lett.*, *29*(13), 1629, doi:10.1029/2001GL014362.

Barth, C. A., D. N. Baker, and S. M. Bailey (2004), Seasonal variation of auroral electron precipitation, *Geophys. Res. Lett.*, *31*, L04809, doi:10.1029/2003GL018892.

Blake, J. B., and L. M. Freisen (1977), A technique to determine the charge state of the anomalous low-energy cosmic rays, *Proc. Int. Conf. Cosmic Rays 15th*, *2*, 341–346.

Blake, J. B., W. A. Kolasinski, R. W. Fillius, and E. G. Mullen (1992), Injection of electrons and protons with energies of tens of MeV into L < 3 on 24 March 1991, *Geophys. Res. Lett.*, *19*(8), 821–824, doi:10.1029/92GL00624.

Blake, J. B., et al. (1995), Comprehensive energetic particle and pitch angle distribution experiment on POLAR, *Space Sci. Rev.*, *71*(1–4), 531–562.

Blake, J. B., M. D. Looper, D. N. Baker, R. Nakamura, B. Klecker, and D. Hovestadt (1996), New high temporal and spatial resolution measurements by SAMPEX of the precipitation of relativistic electrons, *Adv. Space Res.*, *18*(8), 171–186.

Blake, J. B., D. N. Baker, N. Turner, K. W. Ogilvie, and R. P. Lepping (1997), Correlation of changes in the outer-zone relativistic-electron population with upstream solar wind and magnetic field measurements, *Geophys. Res. Lett.*, *24*(8), 927–929, doi:10.1029/97GL00859.

Blake, J. B., U. S. Inan, M. Walt, T. F. Bell, J. Bortnik, D. L. Chenette, and H. J. Christian (2001a), Lightning-induced energetic electron flux enhancements in the drift loss cone, *J. Geophys. Res.*, *106*(A12), 29,733–29,744.

Blake, J. B., R. S. Selesnick, D. N. Baker, and S. Kanekal (2001b), Studies of relativistic electron injection events in 1997 and 1998, *J. Geophys. Res.*, *106*(A9), 19,157–19,168.

Blanchard, G. T., L. R. Lyons, J. B. Blake, and F. J. Rich (1998), SAMPEX observations of energetic electron precipitation in the dayside low-latitude boundary layer, *J. Geophys. Res.*, *103*(A1), 191–198.

Callis, L. B., R. E. Boughner, D. N. Baker, R. A. Mewaldt, J. B. Blake, R. S. Selesnick, J. R. Cummings, M. Natarajan, G. M. Mason, and J. E. Mazur (1996a), Precipitating electrons: Evidence for effects on mesospheric odd nitrogen, *Geophys. Res. Lett.*, *23*(15), 1901–1904.

Callis, L. B., D. N. Baker, M. Natarajan, J. B. Bernard, R. A. Mewaldt, R. S. Selesnick, and J. R. Cummings (1996b), A 2-D model simulation of downward transport of NO_y into the stratosphere: Effects on the 1994 austral spring O_3 and NO_y, *Geophys. Res. Lett.*, *23*(15), 1905–1908.

Callis, L. B., M. Natarajan, J. D. Lambeth, and D. N. Baker (1998), Solar atmospheric coupling by electrons (SOLACE): 2. Calculated stratospheric effects of precipitating electrons, 1979–1988, *J. Geophys. Res.*, *103*(D21), 28,421–28,438.

Cook, W. R., et al. (1993), PET: A proton/electron telescope for studies of magnetospheric, solar, and galactic particles, *IEEE Trans. Geosci. Remote Sens.*, *31*(3), 565–571.

Elkington, S. R., M. J. Wiltberger, A. A. Chan, and D. N. Baker (2004), Physical models of the geospace radiation environment, *J. Atmos. Sol. Terr. Phys.*, *66*(15–16), 1371–1387.

Elkington, S. R., D. N. Baker, and M. Wiltberger (2005), Injection of energetic ions during the 31 March 0630 substorm, in *The Inner Magnetosphere: Physics and Modeling, Geophys. Monogr. Ser.*, vol. 155, edited by T. I. Pulkkinen, N. A. Tsyganenko and R. H. Friedel, pp. 147–154, AGU, Washington, D. C., doi:10.1029/155GM17.

Fisk, L. A., B. Kozlovsky, and R. Ramaty (1974), An interpretation of the observed oxygen and nitrogen enhancements in low-energy cosmic rays, *Astrophys. J.*, *190*, L35–L37.

Horne, R. B., et al. (2005), Wave acceleration of electrons in the Van Allen radiation belts, *Nature*, *437*, 227–230, doi:10.1038/nature03939.

Iles, R. H. A., N. P. Meredith, A. N. Fazakerley, and R. B. Horne (2006), Phase space density analysis of the outer radiation belt energetic electron dynamics, *J. Geophys. Res.*, *111*, A03204, doi:10.1029/2005JA011206.

Kanekal, S. G., D. N. Baker, and J. B. Blake (2001), Multisatellite measurements of relativistic electrons: Global coherence, *J. Geophys. Res.*, *106*(A12), 29,721–29,732.

Kellerman, A. C., and Y. Y. Shprits (2012), On the influence of solar wind conditions on the outer-electron radiation belt, *J. Geophys. Res.*, *117*, A05217, doi:10.1029/2011JA017253.

Kessel, R. L. (2012), NASA's Radiation Belt Storm Probes mission: From concept to reality, in *Dynamics of the Earth's Radiation Belts and Inner Magnetosphere, Geophys. Monogr. Ser.*, doi:10.1029/2012GM001312, this volume.

Klecker, B., et al. (1993), HILT: A heavy ion large area proportional counter telescope for solar and anomalous cosmic rays, *IEEE Trans. Geosci. Remote Sens.*, *31*(3), 542–548.

Klecker, B., R. A. Mewaldt, M. Oetliker, R. S. Selesnick, and J. R. Jokipii (1998), The ionic charge composition of anomalous cosmic rays, *Space Sci. Rev.*, *83*, 294–299.

Li, X., D. N. Baker, T. P. O'Brien, L. Xie, and Q. G. Zong (2006), Correlation between the inner edge of outer radiation belt electrons and the innermost plasmapause location, *Geophys. Res. Lett.*, *33*, L14107, doi:10.1029/2006GL026294.

Li, X., M. Temerin, D. N. Baker, and G. D. Reeves (2011), Behavior of MeV electrons at geosynchronous orbit during last two solar cycles, *J. Geophys. Res.*, *116*, A11207, doi:10.1029/2011JA016934.

Lorentzen, K. R., J. B. Blake, U. S. Inan, and J. Bortnik (2001), Observations of relativistic electron microbursts in association with VLF chorus, *J. Geophys. Res.*, *106*(A4), 6017–6027.

Lyons, L. R. (1997), Magnetospheric processes leading to precipitation, *Space Sci. Rev.*, *80*, 109–132.

Lyons, L. R., and R. M. Thorne (1973), Equilibrium structure of radiation belt electrons, *J. Geophys. Res.*, *78*(13), 2142–2149.

Mason, G. M., D. C. Hamilton, P. H. Walpole, K. F. Heuerman, T. L. James, M. H. Lennard, and J. E. Mazur (1993), LEICA: A low energy ion composition analyzer for the study of solar and magnetospheric heavy ions, *IEEE Trans. Geosci. Remote Sens.*, *31*(3), 549–556.

McPherron, R. L., D. N. Baker, and N. U. Crooker (2009), Role of the Russell-McPherron effect in the acceleration of relativistic electrons, *J. Atmos. Sol. Terr. Phys.*, *71*(10–11), 1032–1044.

Mewaldt, R. A., R. S. Selesnick, J. R. Cummings, E. C. Stone, and T. T. von Rosenvinge (1996), Evidence for multiply charged anomalous cosmic rays, *Astrophys. J.*, *466*, L43–L46.

Miyoshi, Y. S., V. K. Jordanova, A. Morioka, and D. S. Evans (2004), Solar cycle variations of the electron radiation belts: Observations and radial diffusion simulation, *Space Weather*, *2*, S10S02, doi:10.1029/2004SW000070.

Nakamura, R., D. N. Baker, J. B. Blake, S. Kanekal, B. Klecker, and D. Hovestadt (1995), Relativistic electron precipitation enhancements near the outer edge of the radiation belt, *Geophys. Res. Lett.*, *22*(9), 1129–1132.

Nakamura, R., M. Isowa, Y. Kamide, D. N. Baker, J. B. Blake, and M. Looper (2000), SAMPEX observations of precipitation bursts in the outer radiation belt, *J. Geophys. Res.*, *105*(A7), 15,875–15,885.

O'Brien, T. P., M. D. Looper, and J. B. Blake (2004), Quantification of relativistic electron microburst losses during the GEM storms, *Geophys. Res. Lett.*, *31*, L04802, doi:10.1029/2003GL018621.

Paulikas, G. A., and J. B. Blake (1979), Effects of the solar wind on magnetospheric dynamics: Energetic electrons at the synchronous orbit, in *Quantitative Modeling of Magnetospheric Processes, Geophys. Monogr. Ser.*, vol. 21, edited by W. P. Olson, pp. 180–202, AGU, Washington, D. C., doi:10.1029/GM021p0180.

Reeves, G. D., et al. (2003), Acceleration and loss of relativistic electrons during geomagnetic storms, *Geophys. Res. Lett.*, *30*(10), 1529, doi:10.1029/2002GL016513.

Roth, I., M. Temerin, and M. K. Hudson (1999), Resonant enhancement of relativistic electron fluxes during geomagnetically active periods, *Ann. Geophys.*, *17*(5), 631–638.

Schulz, M., and L. J. Lanzerotti (1974), *Particle Diffusion in the Radiation Belts*, Springer, New York.

Selesnick, R. S., A. C. Cummings, R. A. Leske, R. A. Mewaldt, E. C. Stone, and J. R. Cummings (2000), Solar cycle dependence of the geomagnetically trapped anomalous cosmic rays, *Geophys. Res. Lett.*, *27*(15), 2349–2352, doi:10.1029/2000GL000049.

Shprits, Y. Y., R. M. Thorne, R. B. Horne, S. A. Glauert, M. Cartwright, C. T. Russell, D. N. Baker, and S. G. Kanekal (2006), Acceleration mechanism responsible for the formation of the new radiation belt during the 2003 Halloween solar storm, *Geophys. Res. Lett.*, *33*, L05104, doi:10.1029/2005GL024256.

Summers, D., and R. M. Thorne (2003), Relativistic electron pitch-angle scattering by electromagnetic ion cyclotron waves during geomagnetic storms, *J. Geophys. Res.*, *108*(A4), 1143, doi:10.1029/2002JA009489.

Summers, D., R. M. Thorne, and F. Xiao (1998), Relativistic theory of wave-particle resonant diffusion with application to electron acceleration in the magnetosphere, *J. Geophys. Res.*, *103*(A9), 20,487–20,500.

Summers, D., C. Ma, N. P. Meredith, R. B. Horne, R. M. Thorne, D. Heynderickx, and R. R. Anderson (2002), Model of the energization of outer-zone electrons by whistler-mode chorus during the October 9, 1990 geomagnetic storm, *Geophys. Res. Lett.*, *29*(24), 2174, doi:10.1029/2002GL016039.

Summers, D., B. Ni, and N. P. Meredith (2007), Timescales for radiation belt electron acceleration and loss due to resonant wave-particle interactions: 2. Evaluation for VLF chorus, ELF hiss, and electromagnetic ion cyclotron waves, *J. Geophys. Res.*, *112*, A04207, doi:10.1029/2006JA011993.

Turner, R. (2011), Solar cycle slow to get going: What does it mean for space weather?, *Space Weather*, *9*, S04004, doi:10.1029/2011SW000671.

Varotsou, A., D. Boscher, S. Bourdarie, R. B. Horne, S. A. Glauert, and N. P. Meredith (2005), Simulation of the outer radiation belt electrons near geosynchronous orbit including both radial diffusion and resonant interaction with Whistler-mode chorus waves, *Geophys. Res. Lett.*, *32*, L19106, doi:10.1029/2005GL023282.

D. N. Baker, Laboratory for Atmospheric and Space Physics, University of Colorado, 3365 Discovery Drive, Boulder, CO 80303-7820, USA. (Daniel.Baker@LASP.colorado.edu)

J. B. Blake, SSL, The Aerospace Corporation, MS 259, PO Box 92957, Los Angeles, CA 90245-4691, USA. (jbernard.blake@aero.org)

Large-Amplitude Whistler Waves and Electron Acceleration in the Earth's Radiation Belts: A Review of STEREO and Wind Observations

C. A. Cattell,[1] A. Breneman,[1] K. Goetz,[1] P. J. Kellogg,[1] K. Kersten,[1] J. R. Wygant,[1] L. B. Wilson III,[2] M. D. Looper,[3] J. B. Blake,[3] and I. Roth[4]

One of the critical problems for understanding the dynamics of Earth's radiation belts is determining the physical processes that energize and scatter relativistic electrons. We review measurements from the Wind/Waves and STEREO S/Waves waveform capture instruments of large-amplitude whistler mode waves. These observations have provided strong evidence that large amplitude (100s of mV m^{-1}) whistler mode waves are common during magnetically active periods. The large-amplitude whistler mode waves are usually nondispersive and obliquely propagating, with a large longitudinal electric field and significant parallel electric field. These characteristics are different than those of typical chorus, though it remains to be seen whether the large-amplitude whistler mode waves are a subpopulation of chorus or something else entirely. We will also review comparisons of STEREO and Wind wave observations with SAMPEX observations of electron microbursts. Simulations show that the waves can result in electron energization by many MeV and/or scattering by large pitch angles during a single wave packet encounter due to coherent, nonlinear processes including trapping. The experimental observations combined with simulations suggest that quasilinear theoretical models of electron energization and scattering via small-amplitude waves, with timescales of hours to days, may be inadequate for understanding radiation belt dynamics.

1. INTRODUCTION

The importance of whistler mode waves for understanding the Earth's radiation belts has been a subject of continuing interest since the initial work by *Kennel and Petscheck* [1966] that examined limits on stably trapped fluxes due to

the interaction of resonant electrons with whistler waves. Most of the early work focused on whistler mode waves as a loss mechanism, due to scattering of electrons into the loss cone, and dealt primarily with waves propagating parallel to the geomagnetic field. Later work recognized the potentially important role for whistler mode waves in energization of electrons [*Summers et al.*, 1998; *Roth et al.*, 1999; *Meredith et al.*, 2001, and references therein] and the effect of oblique propagation on the resonance conditions [*Kennel*, 1966; *Roth et al.*, 1999]. Because observational studies of whistler waves using spectral data, often averaged over long periods, suggested that wave amplitudes were small (the order of ~0.1 mV m^{-1} and ~0.01 nT), quasilinear approaches are most often used. There were several studies [*Omura and Matsumoto*, 1982; *Bell*, 1984; *Roth et al.*, 1999; *Albert*, 2002] that examined coherent acceleration via small amplitude whistler mode waves and/or the effect of waves with wave vectors oblique to the geomagnetic field. Motivated by observations

[1]School of Physics and Astronomy, University of Minnesota, Minneapolis, Minnesota, USA.

[2]NASA Goddard Space Flight Center, Greenbelt, Maryland, USA.

[3]Aerospace Corporation, El Segundo, California, USA.

[4]Space Sciences Lab, University of California, Berkeley, California, USA.

Dynamics of the Earth's Radiation Belts and Inner Magnetosphere
Geophysical Monograph Series 199
10.1029/2012GM001322

of an association between storm-time enhancement of radiation belt relativistic electron flux and chorus wave activity [e.g., *Meredith et al.*, 2001], there has been renewed interest in the radiation belts resulting in numerous theoretical studies of scattering and energization of radiation belt particles via whistler waves, primarily focusing on quasilinear approaches (see reviews by *Friedel et al.* [2002], *Millan and Thorne* [2007], *Bortnik and Thorne* [2007], and *Thorne* [2010]).

A study by *Santolík et al.* [2003] utilizing Cluster waveform data provided the first indication that wave amplitudes could at times be much larger (30 mV m^{-1}) than usually assumed. The four-satellite Cluster mission also provided unique opportunities to investigate propagation effects, source locations, and frequency structure of wave packets [*Parrot et al.*, 2003; *Chum et al.*, 2007; *Santolík*, 2008; *Breneman et al.*, 2009, and references therein]. New studies motivated by the Cluster 30 mV m^{-1} observation found nonlinear acceleration/trapping for parallel propagating waves [e.g., *Omura et al.*, 2007]. With the addition of large-amplitude wave observations from STEREO [*Cattell et al.*, 2008] and Time History of Events and Macroscale Interactions during Substorms (THEMIS) [*Cully et al.*, 2008], recent years have seen an explosion of studies focusing on nonlinear processes that address many aspects of whistler mode waves in the radiation belts, including wave-generation mechanisms, propagation, electron energization, and wave packet structure [*Bortnik et al.*, 2008; *Yoon*, 2011; *Omura and Nunn*, 2011; *Tao et al.*, 2012].

In this paper, we review observations of large-amplitude whistler mode waves made by waveform capture instruments on the Wind and STEREO spacecraft in the Earth's radiation belts. Comparisons with simultaneous measurements of relativistic electrons from the SAMPEX HILT instrument are presented to address the origin of microbursts. Particle tracing simulations of the large-amplitude waves provide evidence that both energization and large-angle scattering can occur in a single-wave encounter. Other results on wave properties are briefly summarized. We show that large-amplitude whistler waves are common [*Wilson et al.*, 2011; *Kellogg et al.*, 2011; L. B. Wilson et al., A statistical study of the properties of large amplitude whistler waves and their association with few eV to 30 keV electron distributions observed in the magnetosphere by Wind, 2011, http://arxiv.org/abs/1101.3303, hereinafter referred to as Wilson et al., online report, 2011] and are correlated with microbursts [*Cattell et al.*, 2008; *Kersten et al.*, 2011]. The electric field waveforms show evidence of electron trapping in the electric potential [*Kellogg et al.*, 2010]. Note that large-amplitude whistler mode waves have also been observed in the inner belt in association with lightning and high-powered VLF transmitters [*Breneman et al.*, 2011, 2012].

2. STEREO AND WIND EVENTS

Although neither Wind nor STEREO was designed to study the Earth's radiation belts, these spacecraft did have passages through the belts; we have examined 13 passes from Wind and 4 on each of the two STEREO satellites. Both Wind/Waves [*Bougeret et al.*, 1995] and S/Waves [*Bougeret et al.*, 2008] had waveform capture instruments (called "TDS" or time-domain sampler) designed to store and transmit the largest amplitude waves observed in a given time period, within data rate constraints. S/Waves also stores the largest amplitude electric field observed in each minute (TDSMax) to provide additional diagnostics on wave amplitudes. The data sets obtained by the two missions are complementary; Wind had both search coil and electric field data, while STEREO had longer and more frequent TDS samples and the TDSMax measurement.

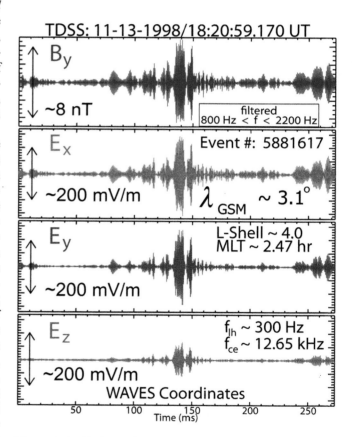

Figure 1. A whistler-mode waveform obtained by Wind in the near-equatorial outer radiation belt. (top to bottom) One component of the magnetic field and three components of the electric field in spacecraft coordinates. Adapted from *Wilson et al.* [2011].

3. LARGE-AMPLITUDE WHISTLER MODE WAVES ARE COMMON

An example of a near-equatorial (magnetic latitude ~3°) large-amplitude whistler mode waveform obtained by Wind at L ~4 and MLT ~2:30 is presented in Figure 1, which plots the one measured component of the magnetic field and the three components of the electric field in spacecraft coordinates. The waveform was bursty with a maximum magnetic field amplitude of >8 nT peak-to-peak (to our knowledge, the largest ever reported in the radiation belt) and electric field amplitude of >200 mV m^{-1}. The frequency was ~0.15 f_{ce}, the electron cyclotron frequency. These amplitudes are 2 to 3 orders of magnitude larger than those that have typically been reported in the radiation belt, based on frequency domain (filter bank) data. The estimated Poynting flux for this event was >300 μW m^{-2} [*Wilson et al.*, 2011], roughly 4 orders of magnitude above estimates from previous

satellite measurements [*Santolík et al.*, 2010]. Note that because Wind obtained only one component of the search coil measurement in this mode, and because both the search coil and electric field data were saturated, this is a lower bound on both the magnetic perturbation and the energy flux.

That such large-amplitude waves are common is demonstrated by the fact that every satellite sampling, the radiation belts with an appropriate "waveform capture" instrument observes large-amplitude whistler mode waves during intervals with substorms or other active periods. It is often the case that TDS events are found in bursts, filling the instrument memory buffer, rather than in isolation. Illustrative examples are shown in Figures 2 and 3. Figure 2 [*Wilson et al.*, 2011] presents an overview of a Wind traversal of the magnetosphere on 13 November 1998; the top panel plots the magnitude and the three GSM components of the magnetic field [*Lepping et al.*, 1995], and the bottom two panels plot the energetic electron (labeled "SST Foil") and ion (labeled

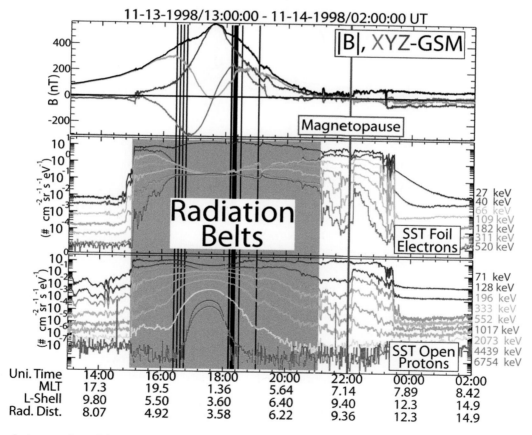

Figure 2. An overview of the Wind pass through the magnetosphere on 13 November 1998 (top to bottom) plotting the three components of the magnetic field in GSM and the magnitude of the field, the energetic electrons and energetic ions. The black vertical lines indicate the times of whistler mode waveforms in the time-domain sampler (TDS). Adapted from *Wilson et al.* [2011].

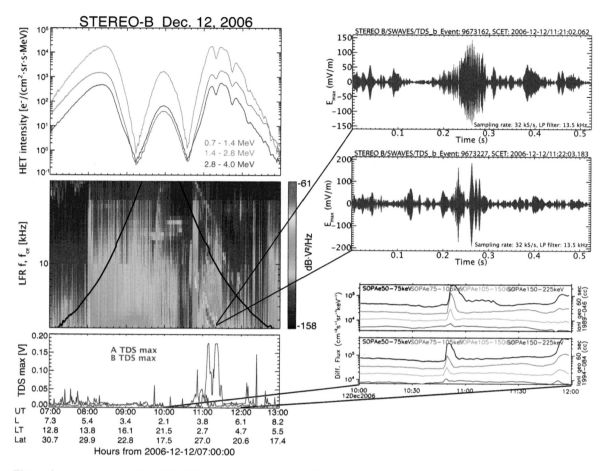

Figure 3. An overview of the STEREO passage through the radiation belts on 12 December 2006. (left top) Plot of fluxes of electrons seen on STEREO-B in three energy bands from the IMPACT High-Energy Telescope [*Luhmann et al.*, 2008], (left middle) plots of the field spectrum from 2.5 to 60 kHz on STEREO-B, with f_{ce} overplotted in black, and (left bottom) plots of the peak electric field in one dipole channel seen in each minute by the TDS peak detector (sampled at 125 kHz) from STEREO-B in red and STEREO-A (time-lagged) in green. (right top) Plot of one electric field component from two different whistler mode waveforms seen on STEREO-B, (right bottom) plot of the fluxes of 50–225 keV electrons from two Los Alamos Geosynchronous satellites. Adapted from *Cattell et al.* [2008].

"SST Open") fluxes [*Lin et al.*, 1995]. During 6 min from 18:15:08 to 18:21:17 UT in the postmidnight outer radiation belt, Wind obtained 14 waveform captures containing whistler mode waves (11 with only electric field components and 3 with either one or three wave magnetic field components), indicated by the vertical black lines. All 14 had wave electric fields >100 mV m^{-1}, and all 3 with wave magnetic field components had amplitudes >4 nT. Note that there were also four large-amplitude whistler waveforms seen on the evening side outer belt between ~16:24:42 and 16:44:40 UT).

A similar interval of large-amplitude whistler mode waves can be seen in Figure 3. The left-hand panels plot 6 h of data as STEREO encountered the radiation belts; the top left panel shows the energetic electron fluxes at five energies

from ~700 keV to 4 MeV observed on STEREO-B (note that the modulations on the morningside are due to spacecraft rolls). The middle left panel shows the power in the electric field from 2.5 kHz to 60 kHz (the black line indicates the value of f_{ce}) on STEREO-B, and the bottom panel shows the peak electric field in one dipole channel seen in each minute by the TDS peak detector (sampled at 125 kHz) on STEREO-A (green) and STEREO-B (red). Note that the STEREO-A data have been time-lagged by the 1.4 h difference between the two satellite orbits. It can be seen that the large-amplitude waves observed on STEREO-B occurred just after a substorm injection (lower panel, right-hand side). In contrast, no large-amplitude waves were observed for the STEREO-A passage through the same L-shells ~1.4 h earlier, before the substorm onset and electron injection. This is consistent with the usual picture that the waves are excited by newly injected electrons that drift around to the morning-side. During the traversal of the morningside outer belt on STEREO-B, large-amplitude waves at $\sim 0.2 f_{ce}$ were observed for ~ 20 min, covering $L \sim 3.5$ to 4.8, as indicated by the electric field spectrum and the peak detector. Note that the peak detector saturated during several minutes in the region of interest on STEREO-B. During ~ 4 min (11:20:21–11:24:26 UT), there were 24 half-second duration waveform captures; 87% had electric field amplitudes >100 mV m^{-1} with maximum amplitudes >240 mV m^{-1} [*Cattell et al.*, 2008]. Two examples are shown in the upper right-hand panels.

Consistent with previous research on the association of chorus with magnetic activity [*Tsurutani and Smith*, 1974], the Wind and STEREO studies [*Cattell et al.*, 2008; *Wilson et al.*, 2011; *Kellogg et al.*, 2011; Wilson et al., online report, 2011] indicate that the large-amplitude waves occur during intervals of enhanced geomagnetic activity. *Wilson et al.* [2011] found that almost all the whistler events were associated with geosynchronous injections. They also showed that $\sim 70\%$ of the observed large-amplitude whistler mode waves occurred when the auroral electrojet index AE > 200 nT and that the amplitudes were weakly correlated with AE. An analysis of THEMIS chorus events also shows that the wave amplitude increases with AE [*Li et al.*, 2009].

Wilson et al. [online report, 2011] examined the electron distributions from the Wind 3DP EESA detectors [*Lin et al.*, 1995] observed in association with large-amplitude whistler wave events. Note that the particle distributions were obtained in ~ 3 s, whereas the waveform captures typically had durations of ~ 0.25 s with wave packet durations on the order of ~ 0.1 s. Because the Wind particle detectors were not designed for radiation belt conditions, the higher-energy SST detectors often saturated and had significant penetrating radiation, at least at some energies (see bottom two panels of Figure 2). Essentially, all the whistler wave packets that occurred within the radiation belts were associated with anisotropic electron distributions ($T_\perp/T_\parallel > 1$ for energies from ~ 400 eV to ≥ 30 keV). Using the distributions observed during several events that occurred close to the magnetic equator as input to the warm plasma dispersion solver OSCARS, *Kellogg et al.* [2011] found that the plasma was unstable to oblique whistler mode waves. However, calculated wave characteristics from the electron distributions did not match all of those observed. Possible explanations are that the waves were generated elsewhere, the characteristics of the distributions changed over the measurement interval, or that wave characteristics are significantly modified by nonlinear effects.

4. CHARACTERISTICS OF WAVES DIFFERENT FROM TYPICAL CHORUS

Whistler mode waves observed in the radiation belts have usually been identified as chorus, lightning-associated whistler mode waves, VLF transmitters, and hiss. Chorus occurs in two bands, above and below $0.5 f_{ce}$. Lower-band chorus is generally observed to have a dispersive signature, usually with frequency increasing with time. Studies have shown that a majority of this frequency separation is a property of the source region, while a smaller amount occurs due to propagation from the source to the spacecraft [*Burtis and Helliwell*, 1976; *Breneman et al.*, 2009]. In addition, propagation angles are most often close to parallel to the geomagnetic field ($\leq 20°$), but larger propagation angles, up to the resonance cone, are also observed [*Goldstein and Tsurutani*, 1984; *Hayakawa et al.*, 1984; *Santolík et al.*, 2009; *Li et al.*, 2011a, 2011b].

In contrast to the typical chorus burst, most of the large-amplitude whistler mode waves observed by STEREO and Wind propagate obliquely and are usually not dispersive over the timescale of the waveform capture. This suggests that either the observed large-amplitude waves may be a different phenomenon than chorus or that they are emitted by chorus sources that do not emit a changing frequency with time. The large wave normal angles may be from effects due to trapped electrons as suggested by *Kellogg et al.* [2010]. The detailed data set that will be obtained from the Radiation Belt Storm Probes mission may solve this question.

Statistical studies of large-amplitude whistler mode waves observed by Wind inside 15 Re have been described by *Wilson et al.* [2011, online report, 2011] and *Kellogg et al.* [2011] found that $\sim 90\%$ of the large-amplitude whistler mode waves occurred in the radiation belts and $\sim 70\%$ during active times. Almost all are in the lower band, i.e., have frequencies at peak power below $0.5 \, f_{ce}$. Only a few ($\sim 20\%$) of the observed wave packets show the frequency changes with time that are typical of chorus [*Wilson et al.*, 2011]. *Kellogg et al.* [2011] shows one example of a waveform that does have a rise in frequency with time (their Figure 2). No dispersion was seen in any of the 24 12 December 2006 STEREO-B waveforms. An example is shown in Figure 4; the waveforms are on the left-hand side, and corresponding wavelet transforms are on the right. It can be seen that the frequency is constant throughout the packet. The second harmonic is a signature of electron trapping as described by *Kellogg et al.* [2010].

Figure 5 shows a Wind waveform in minimum variance coordinates for one event when vector wave magnetic field data were transmitted; the panels on the left plot the time series of the three components, and the panels on the right

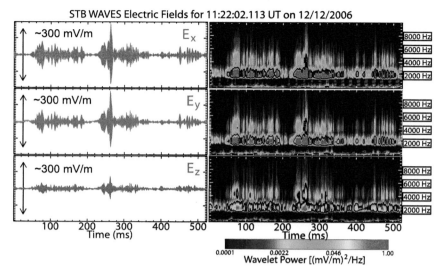

Figure 4. Three components of (left) the wave electric field for one whistler mode waveform from STEREO-B and (right) the wavelet power spectra. No dispersion in frequency occurred.

present the corresponding hodograms. The expected right-hand polarization is clear in the upper hodogram, and the wave vector, $\vec{k}$, was 44° to the geomagnetic field. Of the 46 events for which three components of the wave magnetic field were obtained, 35% had a propagation angle greater than 20°. Note that most of the waveform samples on both Wind and STEREO were close to the equator and that *Wilson*

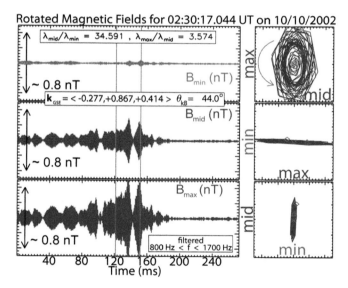

Figure 5. (left) Plots of the three components of the wave magnetic field for a waveform obtained by Wind; (right) plots of the hodograms for the interval between the magenta lines. The wave is right-hand polarized. Minimum variance parameters and wave vector direction are given.

et al. [2011] found no correlation between the propagation angle and magnetic latitude. It is, therefore, unlikely that the oblique propagation is due only to propagation effects.

Kellogg et al. [2010] described an alternative approach to determining the direction of wave vector for waveforms that show distortion that is consistent with electron trapping in the electrostatic potential (discussed in more detail in section 6). The wave normal angle is found by noting that the electric field of the trapped layer of electrons must be parallel to $\vec{k}$. For the 24 STEREO-B TDS in the outer belt on 12 December 2006, the average angle between $\vec{k}$ and $\vec{B}$ varied from ~46° to 51°, consistent with angles determined from the electric field polarization and the cold plasma dispersion relation, as well as from the those determined from wave magnetic field measurements and a minimum variance analysis. This provides further confirmation of the very oblique propagation.

5. WAVES ARE ASSOCIATED WITH MICROBURSTS

For more than 20 years, observations of "microbursts," short bursts of relativistic electrons, have been seen by balloon detectors or low-altitude satellites imaging the loss cone [*Blake et al.*, 1996; *Millan and Thorne*, 2007; *Millan*, 2011, for review]. Studies by *Lorentzen et al.* [2001], *O'Brien et al.* [2003], and others provided evidence that suggested that oblique whistler mode waves could provide the needed scattering.

There was a fortuitous conjunction between SAMPEX and STEREO during the intervals when both STEREO-A and STEREO-B encountered the radiation belts on 12 December 2006 [*Cattell et al.*, 2008]. Figure 6 [*Kersten et al.*, 2011] summarizes the wave and microburst observations; the top

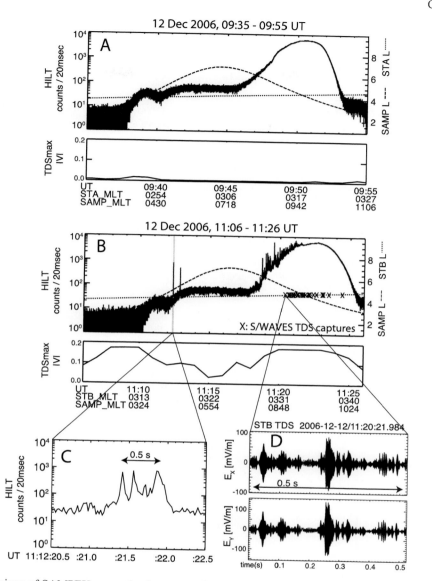

Figure 6. Comparison of SAMPEX energetic electrons and STEREO whistler mode wave seen on 12 December 2006 [*Kersten et al.*, 2011]. (a) Plots of the SAMPEX HILT counts at the time of (top) the STEREO-A conjunction and (bottom) the STEREO-A TDS-Max, showing that no microbursts are seen when there are no strong waves. (b) Plots of the HILT counts at the time of the (top) STEREO-B conjunction and the STEREO-B TDS-Max (the times of the TDS waveforms are indicated by x's). (c) A high time resolution snapshot of the HILT data is shown and (d) a whistler-waveform. Strong microbursts were seen in association with the intense waves.

plot in both Figures 6a and 6b shows the counts in the SAMPEX HILT detector [*Blake et al.*, 1996] with the L-shells of both satellites overplotted and the bottom plot (TDSMax) shows the value of the maximum wave electric field in each minute, Figure 6c shows an expanded interval of the relativistic electron counts in HILT, and Figure 6d shows one of the STEREO-B waveform captures. As described above, no large-amplitude waves occurred during

the STEREO-A passage, which occurred when the magnetic activity was low. Figure 6a shows that there was no microburst activity observed by SAMPEX during the STEREO-A traversal of the radiation belt. During the STEREO-B encounter, intense microbursts occurred (top plot in Figure 6b) in conjunction with large amplitude waves. The time variations in the microbursts (Figure 6c) during the indicated 0.5 s timespan are similar to those seen in the waves (Figure 6d).

Near conjunctions between SAMPEX and Wind have been studied [*Kersten et al.*, 2011] to further explore the relationship between relativistic electron loss and waves. Figure 7 presents an example that occurred on 13 November 1998. Intense microbursts are visible in the 4 min of relativistic electron count rates plotted in the upper panel. During this time period, Wind Waves observed large-amplitude whistler mode waves; one example waveform is plotted in the bottom panel.

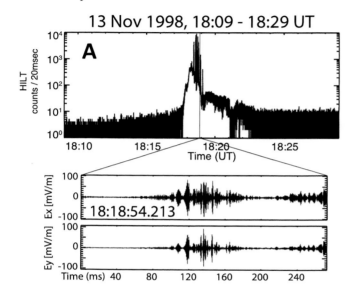

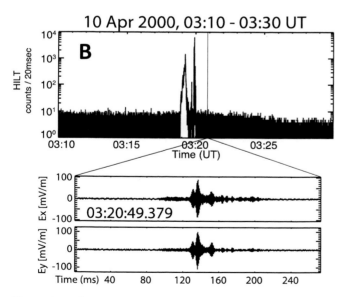

Figure 7. A close conjunction between SAMPEX and Wind. (top) HILT counts are plotted, (middle) the L-shell and MLT of Wind and SAMPEX are plotted, and (bottom) the Wind whistler waveform is plotted.

The observed correlation between large-amplitude whistler mode waves and microbursts is consistent with results of a particle-tracing code that traces electrons in a straightened dipole geometry interacting with a whistler wavefield consistent with the observed characteristics. Results show that resonant electrons can be scattered by very large angles (~40°) into the loss cone in a single wave encounter [*Cattell et al.*, 2008; *Kersten et al.*, 2011].

6. DISCUSSION AND CONCLUSIONS

We have briefly reviewed a set of studies using Wind and STEREO data to characterize the occurrence and properties of whistler mode waves in the outer radiation belt and the association with relativistic electron microbursts as seen on SAMPEX. The results of particle tracing codes modeling the interaction of electrons with waves similar to those observed show that these large-amplitude, oblique waves can energize resonant electrons by the order of MeV during a single wave packet encounter and/or scatter electrons by large angles (40°). Energization and scattering are not due to slow quasi-linear, stochastic processes. The process is nonlinear, associated with either or both cyclotron or electrostatic trapping, resulting in rapid changes in pitch angle and energy. This suggests that usual theoretical models of electron energization and scattering via small-amplitude waves, with timescales of hours to days, cannot provide a complete understanding of radiation belt dynamics. Instead, further development of nonlinear models such as those developed by *Albert* [2002] and *Omura et al.* [2008, 2009] is needed.

Both STEREO and Wind have measured whistler mode waves in the Earth's radiation belt with amplitudes of greater than 240 mV m^{-1} in the electric field and 8 nT in the magnetic field, 2 to 3 orders of magnitude larger than previously reported. Many wave properties are distinctly different from usual lower band chorus, including large amplitudes, oblique propagation, lack of frequency dispersion, large longitudinal electric fields, and significant electric field components parallel to the geomagnetic field. It remains to be seen whether these large-amplitude waves are a subpopulation of chorus that grows to large amplitude (with a source that emits a constant frequency with time) or are an entirely new wave phenomenon altogether. Similar large-amplitude whistler mode waves have been reported using THEMIS data [*Cully et al.*, 2008; *Li et al.*, 2011a, 2011b]. The Wind and STEREO data indicate that these coherent large-amplitude oblique whistler mode waves are common in the outer radiation belt during magnetically active periods. During a STEREO B encounter with the outer belt after a substorm, large-amplitude waves occurred for ~20 min, the time taken for the passage through the belt.

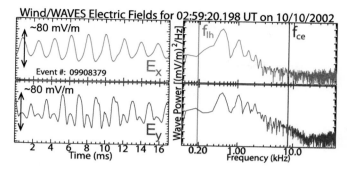

Wind/WAVES Electric Fields for 02:59:20.198 UT on 10/10/2002

Figure 8. An example waveform capture observed by Wind showing the distorted electric fields due to the electron trapping described by *Kellogg et al.* [2010].

STEREO observations provide evidence that the large-amplitude whistler mode waves are associated with rapid enhancement (timescale of ~10 minutes) in fluxes of relativistic electrons in the outer radiation belt [*Cattell et al.*, 2008]. These observations were made after a small substorm injection, but were not associated with a magnetic storm. The waves may be larger during major storms or after stronger injections. Waves can scatter electrons by large angles, consistent with the fact that microbursts were seen by SAMPEX when large-amplitude wave events occurred in close conjunction [*Kersten et al.*, 2011].

The interaction of the waves with electrons can be highly nonlinear; both electrostatic [*Kumagai et al.*, 1980] and cyclotron trapping of electrons can occur. *Kellog et al.* [2010, 2011] described the modifications in the whistler electric field associated with trapping of electrons in the electric potential of an oblique whistler (see Figure 8, for example). This mechanism can only occur in obliquely propagating waves, consistent with the fact that waveform distortion is not seen in large-amplitude parallel or near-parallel propagating waves. The density of trapped electrons can be large enough to significantly modify the whistler wave electric field and large enough to transport kilovolts of potential to new regions in the magnetosphere. Note that this trapping is distinctly different from cyclotron resonance trapping, which can also occur and is associated with the rapid energization and scattering seen in both the observations and particle tracing results.

The STEREO and Wind waveform capture data, in conjunction with SAMPEX particle data and simulations, have provided unique measurements that indicate the potentially critical role of large-amplitude whistler mode waves in radiation belt physics. A complete description of radiation belt dynamics will involve the use of nonlinear mechanisms. Because neither mission was designed for radiation belt studies, their coverage of the region is limited. A detailed and more complete picture of the waves and their roles in radiation belt dynamics will be provided by the Radiation Storm Belt Probes [*Ukhorskiy et al.*, 2011] and the BARREL mission [*Millan*, 2011].

Acknowledgments. The authors thank Benoit Lavraud and Michelle Thomsen for providing the LNL data, Robert Lin for Wind particle data, R. Lepping for Wind magnetic field data, and J. Luhmann for STEREO IMPACT data. This research was supported by NASA grant NNX07AI05G, NNX07AF23G, NNX08AT81G and a contract from APL for the development of RBSP/EFW. L. B. Wilson III was supported by NESSF grant NNX07AU72H and a Leonard Burlaga/Arctowski Medal Fellowship, and K. Kersten was partially supported by a Leonard Burlaga/Arctowski Medal Fellowship. Test particle simulations were performed at the University of Minnesota Supercomputing Institute for Digital Simulation and Advanced Computation.

REFERENCES

Albert, J. M. (2002), Nonlinear interaction of outer zone electrons with VLF waves, *Geophys. Res. Lett.*, *29*(8), 1275, doi:10.1029/2001GL013941.

Bell, T. F. (1984), The nonlinear gyroresonance interaction between energetic electrons and coherent VLF waves propagating at an arbitrary angle with respect to the earth's magnetic field, *J. Geophys. Res.*, *89*(A2), 905–918, doi:10.1029/JA089iA02p00905.

Blake, J. B., M. D. Looper, D. N. Baker, R. Nakamura, B. Klecker, and D. Hovestadt (1996), New high temporal and spatial resolution measurements by SAMPEX of the precipitation of relativistic electrons, *Adv. Space Res.*, *18*, 171–186.

Bortnik, J., and R. M. Thorne (2007), The dual role of ELF/VLF chorus waves in the acceleration and precipitation of radiation belt electrons, *J. Atmos. Sol. Terr. Phys.*, *69*, 378–386, doi:10.1016/j.jastp.2006.05.030.

Bortnik, J., R. M. Thorne, and U. S. Inan (2008), Nonlinear interaction of energetic electrons with large amplitude chorus, *Geophys. Res. Lett.*, *35*, L21102, doi:10.1029/2008GL035500.

Bougeret, J.-L., et al. (1995), Waves: The radio and plasma wave investigation on the WIND spacecraft, *Space Sci. Rev.*, *71*, 231–263, doi:10.1007/BF00751331.

Bougeret, J. L., et al. (2008), S/WAVES: The radio and plasma wave investigation on the STEREO Mission, *Space Sci. Rev.*, *136*, 487–528, doi:10.1007/s11214-007-9298-8.

Breneman, A. W., C. A. Kletzing, J. Pickett, J. Chum, and O. Santolik (2009), Statistics of multispacecraft observations of chorus dispersion and source location, *J. Geophys. Res.*, *114*, A06202, doi:10.1029/2008JA013549.

Breneman, A., C. Cattell, J. Wygant, K. Kersten, L. B. Wilson III, S. Schreiner, P. J. Kellogg, and K. Goetz (2011), Large-amplitude transmitter-associated and lightning-associated whistler waves in the Earth's inner plasmasphere at L < 2, *J. Geophys. Res.*, *116*, A06310, doi:10.1029/2010JA016288.

Breneman, A., C. Cattell, J. Wygant, K. Kersten, L. B. Wilson III, L. Dai, C. Colpitts, P. J. Kellogg, K. Goetz, and A. Paradise (2012), Explaining polarization reversals in STEREO wave data, *J. Geophys. Res.*, *117*, A04317, doi:10.1029/2011JA017425.

Burtis, W. J., and R. A. Helliwell (1976), Magnetospheric chorus: Occurrence patterns and normalized frequency, *Planet. Space. Sci.*, *24*, 1007–1024, doi:10.1016/0032-0633(76)90119-7.

Cattell, C., et al. (2008), Discovery of very large amplitude whistler-mode waves in Earth's radiation belts, *Geophys. Res. Lett.*, *35*, L01105, doi:10.1029/2007GL032009.

Chum, J., O. Santolík, A. W. Breneman, C. A. Kletzing, D. A. Gurnett, and J. S. Pickett (2007), Chorus source properties that produce time shifts and frequency range differences observed on different Cluster spacecraft, *J. Geophys. Res.*, *112*, A06206, doi:10.1029/2006JA012061.

Cully, C. M., J. W. Bonnell, and R. E. Ergun (2008), THEMIS observations of long-lived regions of large-amplitude whistler waves in the inner magnetosphere, *Geophys. Res. Lett.*, *35*, L17S16, doi:10.1029/2008GL033643.

Friedel, R. H. W., G. D. Reeves, and T. Obara (2002), Relativistic electron dynamics in the inner magnetosphere—A review, *J. Atmos. Sol. Terr. Phys.*, *64*, 265–282, doi:10.1016/S1364-6826(01)00088-8.

Goldstein, B. E., and B. T. Tsurutani (1984), Wave normal directions of chorus near the equatorial source region, *J. Geophys. Res.*, *89*(A5), 2789–2810, doi:10.1029/JA089iA05p02789.

Hayakawa, M., Y. Yamanaka, M. Parrot, and F. Lefeuvre (1984), The wave normals of magnetospheric chorus emissions observed on board GEOS 2, *J. Geophys. Res.*, *89*(A5), 2811–2821, doi:10.1029/JA089iA05p02811.

Kellogg, P. J., C. A. Cattell, K. Goetz, S. J. Monson, and L. B. Wilson III (2010), Electron trapping and charge transport by large amplitude whistlers, *Geophys. Res. Lett.*, *37*, L20106, doi:10.1029/2010GL044845.

Kellogg, P. J., C. A. Cattell, K. Goetz, S. J. Monson, and L. B. Wilson III (2011), Large amplitude whistlers in the magnetosphere observed with Wind-Waves, *J. Geophys. Res.*, *116*, A09224, doi:10.1029/2010JA015919.

Kennel, C. (1966), Low-frequency whistler mode, *Phys. Fluids*, *9*, 2190–2202, doi:10.1063/1.1761588.

Kennel, C. F., and H. E. Petscheck (1966), Limit on stably trapped particle fluxes, *J. Geophys. Res.*, *71*(1), 1–28.

Kersten, K., C. A. Cattell, A. Breneman, K. Goetz, P. J. Kellogg, J. R. Wygant, L. B. Wilson III, J. B. Blake, M. D. Looper, and I. Roth (2011), Observation of relativistic electron microbursts in conjunction with intense radiation belt whistler-mode waves, *Geophys. Res. Lett.*, *38*, L08107, doi:10.1029/2011GL046810.

Kumagai, H., K. Hashimoto, I. Kimura, and H. Matsumoto (1980), Computer simulation of a Cerenkov interaction between obliquely propagating whistler mode waves and an electron beam, *Phys. Fluids*, *23*, 184–193, doi:10.1063/1.862837.

Lepping, R., et al. (1995), The WIND magnetic field investigation, *Space Sci. Rev.*, *71*, 207–229, doi:10.1007/BF00751330.

Li, W., R. M. Thorne, V. Angelopoulos, J. Bortnik, C. M. Cully, B. Ni, O. LeContel, A. Roux, U. Auster, and W. Magnes (2009), Global distribution of whistler-mode chorus waves observed on the THEMIS spacecraft, *Geophys. Res. Lett.*, *36*, L09104, doi:10.1029/2009GL037595.

Li, W., J. Bortnik, R. M. Thorne, and V. Angelopoulos (2011a), Global distribution of wave amplitudes and wave normal angles of chorus waves using THEMIS wave observations, *J. Geophys. Res.*, *116*, A12205, doi:10.1029/2011JA017035.

Li, W., R. M. Thorne, J. Bortnik, Y. Y. Shprits, Y. Nishimura, V. Angelopoulos, C. Chaston, O. Le Contel, and J. W. Bonnell (2011b), Typical properties of rising and falling tone chorus waves, *Geophys. Res. Lett.*, *38*, L14103, doi:10.1029/2011GL047925.

Lin, R. P., et al. (1995), A three-dimensional plasma and energetic particle investigation for the WIND spacecraft, *Space Sci. Rev.*, *71*, 125–153, doi:10.1007/BF00751328.

Lorentzen, K. R., J. B. Blake, U. S. Inan, and J. Bortnik (2001), Observations of relativistic electron microbursts in association with VLF chorus, *J. Geophys. Res.*, *106*(A4), 6017–6027, doi:10.1029/2000JA003018.

Luhmann, J. G., et al. (2008), STEREO IMPACT investigation goals, measurements, and data products overview, *Space Sci. Rev.*, *136*, 117–184, doi:10.1007/s11214-007-9170-x.

Meredith, N. P., R. B. Horne, and R. R. Anderson (2001), Substorm dependence of chorus amplitudes: Implications for the acceleration of electrons to relativistic energies, *J. Geophys. Res.*, *106*(A7), 13,165–13,178, doi:10.1029/2000JA900156.

Millan, R. M. (2011), Understanding relativistic electron losses with BARREL, *J. Atmos. Sol. Terr. Phys.*, *73*, 1425–1434, doi:10.1016/j.jastp.2011.01.006.

Millan, R. M., and R. M. Thorne (2007), Review of radiation belt relativistic electron losses, *J. Atmos. Sol. Terr. Phys.*, *69*, 362–377, doi:10.1016/j.jastp.2006.06.019.

O'Brien, T. P., K. R. Lorentzen, I. R. Mann, N. P. Meredith, J. B. Blake, J. F. Fennell, M. D. Looper, D. K. Milling, and R. R. Anderson (2003), Energization of relativistic electrons in the presence of ULF power and MeV microbursts: Evidence for dual ULF and VLF acceleration, *J. Geophys. Res.*, *108*(A8), 1329, doi:10.1029/2002JA009784.

Omura, Y., and H. Matsumoto (1982), Computer simulations of basic processes of coherent whistler wave-particle interactions in the magnetosphere, *J. Geophys. Res.*, *87*(A6), 4435–4444, doi:10.1029/JA087iA06p04435.

Omura, Y., and D. Nunn (2011), Triggering process of whistler mode chorus emissions in the magnetosphere, *J. Geophys. Res.*, *116*, A05205, doi:10.1029/2010JA016280.

Omura, Y., N. Furuya, and D. Summers (2007), Relativistic turning acceleration of resonant electrons by coherent whistler mode waves in a dipole magnetic field, *J. Geophys. Res.*, *112*, A06236, doi:10.1029/2006JA012243.

Omura, Y., Y. Katoh, and D. Summers (2008), Theory and simulation of the generation of whistler-mode chorus, *J. Geophys. Res.*, *113*, A04223, doi:10.1029/2007JA012622.

Omura, Y., M. Hikishima, Y. Katoh, D. Summers, and S. Yagitani (2009), Nonlinear mechanisms of lower-band and upper-band VLF chorus emissions in the magnetosphere, *J. Geophys. Res.*, *114*, A07217, doi:10.1029/2009JA014206.

Parrot, M., O. Santolýk, N. Cornilleau-Wehrlin, M. Maksimovic, and C. C. Harvey (2003), Source location of chorus emissions observed by Cluster, *Ann. Geophys.*, *21*, 473–480.

Roth, I., M. Temerin, and M. K. Hudson (1999), Resonant enhancement of relativistic electron fluxes during geomagnetically active periods, *Ann. Geophys.*, *17*, 631–638, doi:10.1007/s005850050791.

Santolík, O. (2008), New results of investigations of whistler-mode chorus emissions, *Nonlin. Processes Geophys.*, *15*, 621–630.

Santolík, O., D. A. Gurnett, J. S. Pickett, M. Parrot, and N. Cornilleau-Wehrlin (2003), Spatio-temporal structure of storm-time chorus, *J. Geophys. Res.*, *108*(A7), 1278, doi:10.1029/2002JA009791.

Santolík, O., D. A. Gurnett, J. S. Pickett, J. Chum, and N. Cornilleau-Wehrlin (2009), Oblique propagation of whistler mode waves in the chorus source region, *J. Geophys. Res.*, *114*, A00F03, doi:10.1029/2009JA014586.

Santolík, O., J. S. Pickett, D. A. Gurnett, J. D. Menietti, B. T. Tsurutani, and O. Verkhoglyadova (2010), Survey of Poynting flux of whistler mode chorus in the outer zone, *J. Geophys. Res.*, *115*, A00F13, doi:10.1029/2009JA014925.

Summers, D., R. M. Thorne, and F. Xiao (1998), Relativistic theory of wave-particle resonant diffusion with application to electron acceleration in the magnetosphere, *J. Geophys. Res.*, *103*(A9), 20,487–20,500, doi:10.1029/98JA01740.

Tao, X., J. Bortnik, R. M. Thorne, J. M. Albert, and W. Li (2012), Effects of amplitude modulation on nonlinear interactions between electrons and chorus waves, *Geophys. Res. Lett.*, *39*, L06102, doi:10.1029/2012GL051202.

Thorne, R. M. (2010), Radiation belt dynamics: The importance of wave-particle interactions, *Geophys. Res. Lett.*, *37*, L22107, doi:10.1029/2010GL044990.

Tsurutani, B. T., and E. J. Smith (1974), Postmidnight chorus: A substorm phenomenon, *J. Geophys. Res.*, *79*(1), 118–127, doi:10.1029/JA079i001p00118.

Ukhorskiy, A. Y., B. H. Mauk, N. J. Fox, D. G. Sibeck, and J. M. Grebowsky (2011), Radiation belt storm probes: Resolving fundamental physics with practical consequences, *J. Atmos. Sol. Terr. Phys.*, *73*, 1417–1424, doi:10.1016/j.jastp.2010.12.005.

Wilson, L. B., III, C. A. Cattell, P. J. Kellogg, J. R. Wygant, K. Goetz, A. Breneman, and K. Kersten (2011), The properties of large amplitude whistler mode waves in the magnetosphere: Propagation and relationship with geomagnetic activity, *Geophys. Res. Lett.*, *38*, L17107, doi:10.1029/2011GL048671.

Yoon, P. H. (2011), Large-amplitude whistler waves and electron acceleration, *Geophys. Res. Lett.*, *38*, L12105, doi:10.1029/2011GL047893.

J. B. Blake and M. D. Looper, Aerospace Corporation, El Segundo, CA, USA.

A. Breneman, C. A. Cattell, K. Goetz, P. J. Kellogg, K. Kersten, and J. R. Wygant, School of Physics and Astronomy, University of Minnesota, Minneapolis, MN, USA. (awbrenem@hotmail.com)

I. Roth, Space Sciences Lab, University of California, Berkeley, CA, USA.

L. B. Wilson III, NASA Goddard Space Flight Center, Greenbelt, MD, USA.

Classification of Pc1-2 Electromagnetic Ion Cyclotron Waves at Geosynchronous Orbit

B. J. Fraser,[1] S. K. Morley,[2] R. S. Grew,[1] and H. J. Singer[3]

Electromagnetic ion cyclotron (EMIC) waves are known to play an important role in contributing to localized ion ring current decay and radiation belt MeV electron losses during geomagnetic storms. It is therefore important to understand the distribution of EMIC Pc1-2 wave types and occurrence in the plasmasphere and magnetosphere. Results from a statistical study of EMIC waves seen by GOES 10, 11, and 12 at geosynchronous orbit ($L = 6.6$) over 2007 in the 0.1–1 Hz frequency band are considered. Over this interval, GOES 10 was located at 60°W (geographic longitude), GOES 11 at 135°W, and GOES 12 at 75°W. The various types of waves seen are classified based on their dynamic spectral structure into six groups ranging from periodic emissions, hydromagnetic (Hm) chorus, narrowband Pc1-2 emissions, substorm-associated emissions (intervals of pulsations with diminishing period), less organized Hm emission bursts, and irregular emissions. Each category of wave has a unique local time occurrence distribution and frequency range at geosynchronous orbit. Statistical results also consider the similarities and differences in the waves seen by GOES 11 and 12 located at different geomagnetic latitudes.

1. INTRODUCTION

Electromagnetic ion cyclotron waves, propagating in the Earth's magnetosphere at frequencies below the proton cyclotron frequency are considered to be excited in the equatorial region by the cyclotron instability involving ~10–100 keV anisotropic ($T_\perp > T_\parallel$) protons [*Cornwall*, 1965; *Kozyra et al.*, 1984; *Jordanova et al.*, 2001]. These waves have been observed on the ground at all latitudes [e.g., *Saito*, 1969; *Campbell and Stiltner*, 1965] and directly in the magneto-

sphere and plasmasphere by satellites [*Anderson et al.*, 1992a, 1992b; *Fraser and Nguyen*, 2001; *Pickett et al.*, 2010]. Recent studies have shown that EMIC waves play an important role in rapid ion heating, pitch angle scattering, and precipitation leading to ring current decay [*Jordanova et al.*, 2001; *Fok et al.*, 2005]. They are also considered to contribute to the precipitation of MeV electrons from the radiation belts during geomagnetic storms [*Summers and Thorne*, 2003; *Loto'aniu et al.*, 2009]. Satellite observations show these waves propagate predominantly in the left-hand polarized mode [*Mauk and McPherron*, 1980]. The dispersion relation for EMIC waves propagating parallel to the geomagnetic field shows that the parallel energy of the interacting protons has a minimum at the geomagnetic equator, suggesting the equatorial region is favored for EMIC wave generation [*Mauk*, 1982; *Hu and Fraser*, 1994]. This is also where wave amplification maximizes, and the Alfven velocity is lowest along the field-aligned path [*Fraser et al.*, 1992].

Observations of Pc1-2 (0.1–5 Hz) ULF waves, the ground signature of EMIC waves at middle and low latitudes, show a

[1]Centre for Space Physics, University of Newcastle, Callaghan, New South Wales, Australia.

[2]Los Alamos National Laboratory, Los Alamos, New Mexico, USA.

[3]Space Weather Prediction Center, NOAA, Boulder, Colorado, USA.

Dynamics of the Earth's Radiation Belts and Inner Magnetosphere
Geophysical Monograph Series 199
10.1029/2012GM001353

regular fine structure embedded in dynamic frequency-time spectra of the waves, indicating wave packet propagation [*Tepley*, 1964; *Obayashi*, 1965]. In contrast, at high latitude, with exception of the intervals of pulsations with diminishing period (IPDP), the waves are generally unstructured and do not show an ordered fine structure [*Fukunishi et al.*, 1981; *Menk et al.*, 1992, 1993]. In the magnetosphere, much the same is seen with a recognizable fine structure seen in emissions in the plasmasphere or plasmapause, particularly in Cluster satellite observations [*Pickett et al.*, 2010; *Usanova et al.*, 2010; Y. H. Liu et al., EMIC waves observed by Cluster near the plasmapause, manuscript in preparation, 2012, hereinafter referred to as Liu et al., manuscript in preparation, 2012]. At high latitudes in the plasma trough and at geosynchronous orbit, EMIC wave emissions are mostly unstructured [*Fraser and Nguyen*, 2001; *Fraser et al.*, 2006; *Clausen et al.*, 2011]. The most comprehensive study of EMIC wave properties in the magnetosphere was undertaken by *Anderson et al.* [1992a, 1992b] using AMPTE-CCE observations, and recently, *Min et al.* [2012], using Time History of Events and Macroscale Interactions during Substorms (THEMIS) data, over $L = 3.5–9$ and all local times. *Anderson et al.* showed EMIC waves were found with a 10%–70% probability over 11–18 magnetic local time (MLT), while occurrence at $L = 5$, near the nominal location of the plasmapause is far less frequent at $\leq 1\%$ and spread relatively evenly over the day. Similar events were found at THEMIS, while at CRRES, most waves were seen in the 14–18 MLT sector [*Fraser and Nguyen*, 2001].

Ion cyclotron instability growth rates integrated along field-aligned propagation paths may maximize over a wide range of L values both inside and outside the plasmapause [*Horne and Thorne*, 1993; *Hu and Fraser*, 1994; *Chen et al.*, 2009]. It is suggested that more amplified waves are seen in the plasma trough region than the plasmasphere, particularly in the He^+ and H^+ bands. With this in mind, it is important to characterize the EMIC wave population seen outside the plasmapause, and geosynchronous orbit data provide this opportunity.

In addition to single satellite studies of EMIC waves, there have been satellite missions where multiple satellites have observed EMIC waves in the magnetosphere. These include the ISEE pair of elliptically orbiting satellites [*Fraser et al.*, 1989], the Cluster mission with four satellites flying in tetrahedron formation [*Pickett et al.*, 2010; *Liu*, 2011], the THEMIS multiple satellite mission [*Usanova et al.*, 2008], and two or three GOES satellites at geosynchronous orbit [*Fraser et al.*, 2005, 2006; *Clausen et al.*, 2011]. However, so far, researchers have not made significant intersatellite comparisons of EMIC wave properties within these missions. There are significant differences between EMIC wave

events seen by different satellites within a constellation. For example, an event seen by the Cluster mission was similar at three reasonably close-spaced satellites, but the following satellite showed a much longer emission event when traveling through the same region of the outer plasmasphere (Liu et al., manuscript in preparation, 2012).

Studies of EMIC waves at geosynchronous orbit using the GOES satellite constellation data have been reported by *Fraser et al.* [2005, 2006, 2010] in relationship to plasmaspheric plumes and geomagnetic storm occurrence. More recently, *Clausen et al.* [2011] looked at GOES 10, 11, and 12 geosynchronous EMIC wave data and found through studying the simultaneous occurrence at pairs of satellites that wave occurrence was essentially temporal with waves occurring at a particular time rather than a satellite moving through an active region. In contrast to this, *Fraser et al.* [2005] found that two satellites, following each other 5 h apart, saw EMIC wave events associated with a stationary radially extended plasma plume, suggesting a spatial relationship.

It is important to understand the properties of EMIC waves that interact with convected ring current and plasma sheet ions associated with subauroral proton arcs [*Jordanova et al.*, 2007; *Spasojevic et al.*, 2011] and similarly the precipitation of MeV electrons from the radiation belt [*Summers and Thorne*, 2003], which may have implications for the atmosphere, affecting electron density and ozone concentrations [*Verronen et al.*, 2005]. Elliptically orbiting satellites [e.g., *Anderson et al.*, 1992a, 1992b; *Fraser and Nguyen*, 2001] only provide a single point observation of the waves in space. These and other similar studies have shown a large peak in EMIC wave activity in the local afternoon sector, which is attributed to the coexistence of increased cold/cool plasma from an inflated plasmasphere or plasmaspheric plumes enhancing the cold/cool plasma density, thereby lowering the threshold of EMIC wave instability. However, EMIC waves do occur at other times of the day in space, and it is important to identify all Pc1-2 band wave activity in space in order to improve our understanding of the physical processes involved. GOES geosynchronous observations offer a partial solution to the continuous monitoring problem in that they provide an opportunity to use multiple satellites to document EMIC wave emissions at all local times albeit at one radial location.

In this study, we use GOES 10, 11, and 12 high-resolution fluxgate magnetometer data over January–December 2007 to define the various categories of EMIC waves seen at geosynchronous orbit and study the solar wind and magnetospheric conditions under which the waves occur and note the differences between the satellites due to their location. Section 2 outlines the observations made by the three GOES satellites over 2007 and associated analysis procedures. In section 3, the various Pc1-2 wave types seen at geosynchronous orbit and their spectral and occurrence characteristics are described. Upstream solar wind and interplanetary magnetic field (IMF) conditions under which the waves were seen over the quiet year 2007 by GOES 11 and 12 are considered statistically in section 4. This is followed by a discussion of the importance of the wave types identified and a conclusion.

2. GOES OBSERVATIONS

The GOES 10, 11, and 12 geosynchronous satellite magnetometers [*Singer et al.*, 1996] provide a rich source of data to study ULF waves. The satellites are three-axis stabilized, and fluxgate magnetometer data are typically analyzed in the spacecraft coordinate frame. In this coordinate system (Hp, He, Hn), Hp is parallel to the Earth's spin axis for zero inclination orbit or approximately parallel to the field. The He component is perpendicular to Hp and directed earthward. The Hn component completes the orthogonal system and is directed eastward. The resultant total wavefield is Ht. For this study in 2007, GOES 12 was located at 75°W geographic longitude, GOES 10 at 90°W, and GOES 11 at 135°W. At these locations, GOES 10, GOES 12, and GOES 11 are 4, 5, and 9 h in local time (LT), respectively, behind UT. The high-resolution data, sampled at 0.512 s are able to provide information on the low end of the Pc1-2 EMIC wave spectrum from 0.1 Hz up to the 1 Hz Nyquist frequency. However, as noted earlier by *Fraser et al.* [2005], the upper end of this response is reduced by a five-pole Butterworth low-pass antialiasing filter. Although the filter 3 dB point is at 0.5 Hz, the slow dropoff in filter response allows EMIC waves with amplitudes ≥1 nT to be seen up to 0.8 Hz. Below 0.5 Hz, the noise level is ~0.1 nT. This frequency response allows the observation of the lower end of the EMIC wave spectrum, which is generally below ~1 Hz in the local afternoon and evening sector [*Fraser*, 1968; *Anderson et al.*, 1992a; *Fraser and Nguyen*, 2001]. Although the greatest power in EMIC waves in space is seen in the azimuthal magnetic component [*Fraser*, 1985], this study uses the total wave power in the event observations and statistical studies.

EMIC wave events described in section 3 were identified from dynamical spectral analysis of GOES 0.512 s Ht data undertaken over approximately hourly steps of 7000 data points or 58.33 min. A fast Fourier transform (FFT) was performed on a window of 200 points (1.6 min) with an overlap of 50 points (0.4 min), providing 140 FFTs over the hour. The FFT frequency resolution was 10 mHz. Input parameters for the statistical study described in section 4 were obtained from 4 h spectral plots using 120 point FFTs

with a 120 point step and no overlap, resulting in a frequency resolution of 16.6 mHz. For all analyses, the initial main field was high pass filtered at 0.01 Hz using a 200 point smoothing function. The FFT intervals were linearly detrended, tapered with a Hanning window in the time domain, compensated for the Hanning window in the frequency domain and then normalized by dividing the spectrum by the frequency resolution.

Although the three satellites are located at geosynchronous orbit on the equator, they are at different geographic longitudes and local times and will be at different geomagnetic latitudes due to dipole tilt. Figure 1, adapted from *Onsager et al.* [2004], illustrates the geometry of the situation. Here we see that GOES 12 is located at geomagnetic latitude 11°, GOES 11 at 4° and GOES 10 at 5°, all north of the geographic equator. Because the satellites are located at different geomagnetic latitudes, they will be observing different geomagnetic fields and EMIC waves on different *L* shells, particularly GOES 11 compared to GOES 10 and 12. This may explain, for example, why we see different wave powers at the same time at different satellites.

3. WAVE SPECTRAL PROPERTIES

Although it is commonly agreed that Pc1-2 waves in the 0.2–5 Hz band can be broadly divided into two groups, structured and unstructured on the basis of identification in frequency-time spectra, there are a number of subcategories observed in ground data, which show differing spectral structure [*Fukunishi et al.*, 1981; *Kangas et al.*, 1998], but are considered to be associated with the ion cyclotron instability in the magnetosphere. Since it is not known what role, if any, some of these play in ring current decay, radiation belt loss mechanisms or other dynamic processes, it is important to note their characteristics and occurrence through in situ satellite observations. In this section, we outline the various spectral structures seen at geosynchronous orbit by GOES 10, 11, and 12 over 2007. Some of the spectral structures identified are similar to those presented by *Fukunishi et al.* [1981] and *Menk et al.* [1992, 1993], while others are based on descriptions originally defined for VLF emissions in the magnetosphere [*Helliwell*, 1965].

3.1. Hydromagnetic Whistlers and Periodic Emissions

Here we consider the dispersive hydromagnetic whistlers defined by *Obayashi* [1965], which consist of a train of dispersed fine structure elements, similar to VLF whistlers but which exhibit a decreasing positive slope with time. Although they relate to classical dispersion along a geomagnetic field line in the left-hand polarized mode, they are

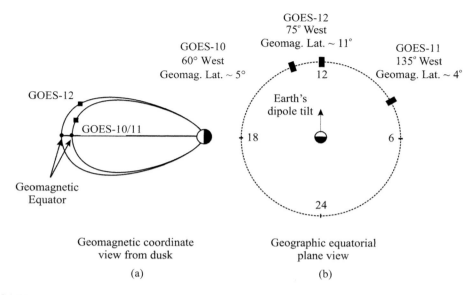

Figure 1. GOES 10, GOES 11, and GOES 12, located at 5°, 4°, and 11° geomagnetic latitudes, respectively, are shown (a) on a coplanar display of their field lines and (b) at their locations in the geographic equatorial plane. Adapted from *Onsager et al.* [2004].

rarely seen in Pc1-2 emission events as it is considered that the wave generation instability also contributes to the slope of the frequency-time structure [*Gendrin et al.*, 1971]. Dispersion, a propagation effect is most likely to be observed toward the end of a wave train when the generation mechanism process has ceased and only dispersion takes place. Because of this, we have combined hydromagnetic whistlers with the most frequently seen group of waves defined as periodic emissions and first studied by *Tepley* [1964]. These emissions may show regular or irregular fine structure.

Periodic emissions can comprise a single wave train lasting for a few to many minutes or a series of wave trains, which may last for hours and overlap in frequency and time at frequencies up to about 4 Hz in space [*Anderson et al.*, 1992a; *Fraser and Nguyen*, 2001]. They exhibit fine structure elements, which mostly do not show dispersion and exhibit a constant positive slope or broken bands of organized dots [*Gendrin et al.*, 1971]. An example is illustrated in Figure 2a where a periodic emission on 13 September 2007 seen over 06:15–09:00 UT at GOES 12 in the frequency band 0.4–0.6 Hz. The emission is seen in the H^+ band above the local helium cyclotron frequency (f_{He+}). This is typical of a long duration emission, which is not too often seen at geosynchronous orbit [*Clausen et al.*, 2011].

The 10 min diurnal occurrence rates for these events seen over 2007 by the three GOES satellites are shown in Figure 2b. The periodic emissions show a broad peak over 12–16 LT, and there is also a secondary peak in the early morning

hours 06–07 LT and suggestion of a prenoon minimum around 09–10 LT, with low occurrence after 18 LT.

3.2. Hydromagnetic Emission Bursts

These are broader band and more impulsive than periodic emissions and are generally seen at frequencies above f_{He+}. An example is illustrated in Figure 3a seen by GOES 11 on 29 January over 12:00–14:00 UT, at 0.2–0.8 Hz mainly in the H^+ band. There may be trains of pulses lasting up to 2 h, or on some occasions, only one to three band-limited pulses. They show relatively wide band fine structure elements, which may or may not show dispersion and exhibit variable fine structure element spacing, which is not typical of a bouncing wave packet or regular wave train propagation, but more typical of an impulsive process. The diurnal distribution of occurrence in Figure 3b shows both a morning peak at 06–07 LT and an afternoon peak at 14–16 LT.

3.3. Hydromagnetic Chorus

This is, by far, the most common emission seen at geosynchronous orbit, and an example is shown in Figure 4a. This emission was named because of its similarity to VLF chorus emissions and comprises a mix of band-limited unstructured emissions and discrete elements appearing at irregular intervals. There are often sequences of periodic emissions embedded within chorus, and on the ground, these are a

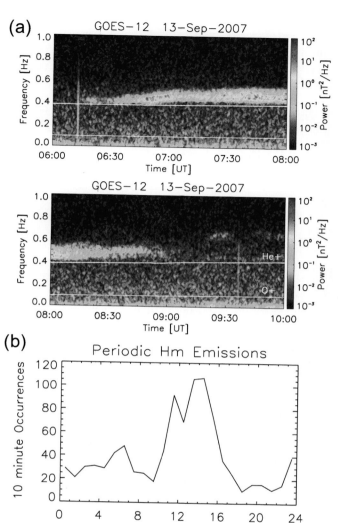

Figure 2. (a) Typical example of a periodic emission with both dispersive and nondispersive fine structure seen by GOES 12 about 11° off the geomagnetic equator. (b) Combined diurnal distribution of 10 min occurrences in local time over January–December 2007, seen by GOES 10, 11, and 12.

common feature in the plasmatrough region between the plasmapause and the cusp [*Menk et al.*, 1993]. The diurnal occurrence rate of hydromagnetic chorus plotted in Figure 4b shows a broad peak in the afternoon over 14–18 LT indicating that it is primarily an afternoon early evening phenomenon.

3.4. Pc1-2 Band

These waves generally are seen below 0.3 Hz [*Fukunishi et al.*, 1981] and include the narrowband 0.1–0.2 Hz Pc2 emissions identified by *Menk et al.* [1992] and probably the

4 s band identified by *Heacock* [1966]. An example on 26 June 2007, over 16–18 UT is illustrated in Figure 5a. The diurnal distribution of occurrence is shown in Figure 5b where a broad peak is seen around the middle of the day 12–15 LT. These are the longest duration events observed both on the ground and at geosynchronous orbit and are the narrowest band emissions seen, always in the He⁺ band. From DMSP LEO satellite and ground-based observations, *Menk et al.* [1992] located their source under and a few degrees of latitude equatorward of the plasma sheet boundary layer.

3.5. Intervals of Pulsations With Diminishing Period

These are the classical irregular-structured storm and substorm-associated emissions seen in the Pc1-2 band. They typically comprise a superposition of band-limited noise and irregularly spaced rising tone elements similar to hydromagnetic chorus or dispersive irregular hydromagnetic emissions [*Fraser and Wawrzyniak*, 1978; *Arnoldy et al.*, 1979]. Figure 6a shows a typical IPDP emission seen at GOES 10 on 18 April 2007 over 20:40–21:20 UT in the H⁺ band. A second emission is seen commencing at 21:30 UT in the He⁺

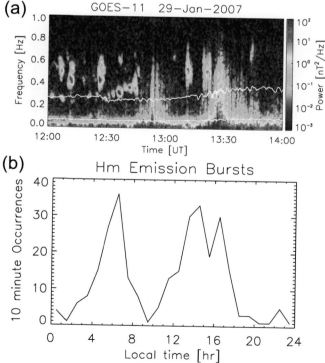

Figure 3. (a) Example of an hydromagnetic (Hm) emission burst observed by GOES 11. (b) The double-peaked diurnal distribution of 10 min Hm emission bursts seen by GOES 10, 11, and 12 over January–December 2007.

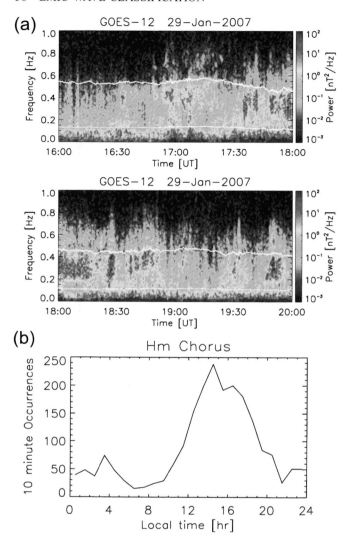

(b)

Figure 4. (a) Example of an Hm chorus emission observed by GOES 12 on 26 June 2007. (b) The afternoon peak in the diurnal distribution of 10 min chorus emissions seen by GOES 10, 11, and 12 over January–December 2007.

band. The diurnal occurrence is plotted in Figure 6b and shows an evening peak occurrence at 18 LT, which is typical of more geomagnetically active times [*Kangas et al.*, 1998]. In addition, IPDP events are also seen in the early morning hours on the ground. Morning IPDP comprises a series of isolated rising elements and occurs during intervals of weak or moderately disturbed geomagnetic activity corresponding to a decrease in the nightside auroral zone H component [*Fukunishi and Toya*, 1981]. The diurnal distributions of morning IPDP are included in Figure 6b. Although it is unusual to see a greater occurrence of IPDP in the morning than in the afternoon [*Fukunishi et al.*, 1981], this may be

associated with the quiet year of geomagnetic activity in 2007 or the limited location of the GOES satellites at $L = 6.6$.

3.6. Irregular Hydromagnetic Emissions

There is a group of discrete band-limited emission events, which occur at irregular intervals. They may show an irregular frequency-time structure with dispersive or nondispersive elements or appear as a bunch of dots. Dynamic spectral examples are shown in Figure 7a. Similar to the periodic emissions in section 3.1, these emissions show a double peaked diurnal distribution with an early morning peak over 03–05 LT and a maximum in the afternoon at 14–15 LT as seen in Figure 7b.

The results of the wave occurrence study described above are summarized in Table 1. Hm chorus contributes the greatest occurrence probability at 37% with a total event duration time of 28.4% compared to the lower-frequency Pc1-2 emission band, which shows a longer total event duration time of 39.5%, due to the very long individual event duration centered near noon. In contrast, over this year of the unusually quiet Sun, there were very few occurrences of evening IPDP events, typically associated with substorm activity.

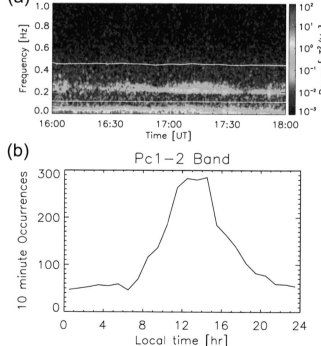

Figure 5. (a) Example of narrowband Pc1-2 emissions observed by GOES 12 on 26 June. (b) The postnoon peak in the diurnal distribution of 10 min Pc1-2 band emissions seen by GOES 10, 11, and 12 over January–December 2007.

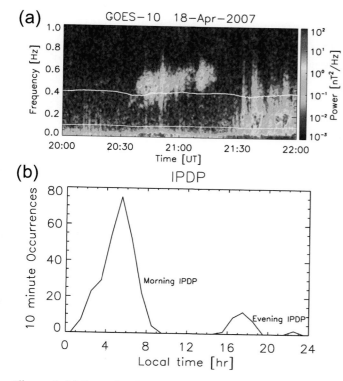

Figure 6. (a) Example of an afternoon interval of pulsations with diminishing period (IPDP) observed by GOES 10 on 18 April 2007. (b) The double-peaked IPDP diurnal distribution of 10 min Hm emission bursts seen by GOES 10, 11, and 12 showing separated morning and afternoon emission groupings.

On looking at the different emission types defined above, the categorization of a specific emission may seem confusing. However, the broad groups identified above become self-evident with experience in analyzing dynamic spectrograms with appropriate resolution. The bandwidth limitation of 0–0.8 Hz in the GOES data means some emissions at higher frequencies in the H^+ band may be missed. However, *Fraser* [1968] and *McPherron and Ward* [1965], using ground and geosynchronous data, respectively, showed wave occurrences peak at 10–18 LT for frequencies below 1 Hz. It will be shown later that any missed data most likely occurs over 00–08 LT.

4. STATISTICAL RESULTS

For the purposes of undertaking a statistical analysis, we will consider data from GOES 11 and GOES 12, separated by 5 h LT as shown in Figure 1. GOES 10 located 1 h in local time away from GOES 11 will be the subject of later three-satellite research. In addition to the GOES data, it is important to relate the wave activity observed to coincident upstream solar wind

and interplanetary field properties. Although this has been the subject of a recent paper by *Clausen et al.* [2011], it is important to identify the different EMIC wave response observed by satellites at different latitudes. Also, we use ACE plasma and IMF observations over 2007 to represent upstream conditions (Coordinated Data Analysis Workshops; http://www.srl.caltech.edu/ACE/ASC/level2/lvl2DATA_SWEPAM.html).

The relationship between EMIC wave power seen at GOES 11 and 12 and solar wind conditions are shown in Figure 8. Time has been adjusted for the propagation from ACE using the average of the solar wind velocity following the work of *Fraser et al.* [2010]. For solar wind velocities <500 km s^{-1}, there is a gradual increase in wave power with velocity at both GOES 11 and 12. Above this velocity, the continued increase is somewhat variable due to the small number of high velocities observed, but the increase is seen to continue, on average, at GOES 12 where wave power reaches 80 nT^2 Hz^{-1}. The wave power increases similarly with solar wind pressure up to 3–4 nPa. GOES 11 at higher pressures shows some wave power above 100 nT^2 Hz^{-1}, while GOES 12, further off the equator at 11° geomagnetic

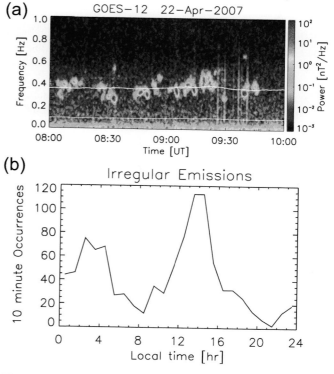

Figure 7. (a) Example of irregular Hm emissions including dispersive and dotted irregular emissions observed by GOES 12 on 22 April 2007. (b) The double-peaked diurnal distribution of 10 min irregular Hm emissions seen by GOES 10, 11, and 12 over January–December 2007.

Table 1. Occurrence Probabilities of the Various Types of EMIC Pc1-2 Wave Emission Events Seen by GOES 10, 11, and 12 Over 2007

Emission Type	Number of Events	Event Occurrence Probability (%)	Total Duration Time (min)	Event Duration Occurrence Probability (%)
Periodic hydromagnetic (Hm) emission/whistlers	182	18.6	9570	13.0
Hm emission bursts	81	8.3	2930	4.0
Hm chorus	362	37.0	20,920	28.4
Pc1-2 band	153	15.6	29,140	39.5
Intervals of pulsations with diminishing period, morning/evening	60	6.1	2210	3.0
Irregular Hm emissions	142	14.5	8920	12.1
Total	980	100	73,690	100

latitude shows lower power and does not see any waves at pressures above 8 nPa. With respect to solar wind pressure effects, *McCollough et al.* [2010] reported an increase in temperature anisotropy during magnetospheric compression events due to drift shell splitting and/or Shabansky orbits of ions. Since temperature anisotropy drives EMIC wave growth, this effect may explain the increase in EMIC wave amplitude seen with increasing solar wind pressure.

In investigating the IMF relationship, we confined the study to B_z component variations. Figure 9a shows that the variation in EMIC wave power is similar at both satellites being rather flat over 50–100 nT^2 Hz^{-1} corresponding to ±6 nT in B_z with a falling trend with increasing B_z at GOES 11. The higher power seen when B_z goes more south below < -6 nT relates to less reliable data. Figure 9b indicates the distribution of wave power with the *Dst* index, which shows a falling trend with increasing *Dst*. Again, data outside ±30 nT is considered unreliable. In this year of the quiet Sun, there were no EMIC waves seen with storms with *Dst* < -40 nT. Like B_z, there was a large occurrence at *Dst* > 0, indicating EMIC waves observed under quiet, storm onset and recovery phase conditions.

Of great interest is the diurnal variation in the wave power, which supplements the occurrence results for the various wave categories defined in section 3. Figure 10a shows the annual distribution of wave power with local time with an afternoon peak at 16 and 17 LT, respectively, for GOES 11 and 12. Higher power, at 120 nT^2 Hz^{-1}, is observed at GOES 12 located 11° off the geomagnetic equator compared with GOES 11 at 100 nT^2 Hz^{-1}. In this quiet year, the afternoon power peaks well above the nighttime-morning power at 20–40 nT^2 Hz^{-1} may be an indication of an expanded high-density plasmasphere with the plasmapause beyond geosynchronous orbit or radially extending plasmaspheric plumes [*Fraser et al.*, 2005]. It may also relate to the development of ring current ion anisotropy during westward convection, primarily due to pitch angle charge exchange losses [*Chen et al.*, 2010].

A more detailed breakdown of the diurnal statistics can be seen in the scatterplots in Figure 10b, where the local time variation in event peak wave power is plotted. Here the most intense wave power is seen over 12–20 LT, up to a peak power of 10^3 nT^2 Hz^{-1}, while power above 100 nT^2 Hz^{-1} was seen at all LT. These results are similar to CRRES observations over a radial range $L = 3.5$–8, which show a peak occurrence over 12–20 MLT centered on geosynchronous orbit [*Halford et al.*, 2010].

It is well known that the presence of heavy ions (He$^+$, O$^+$) in the cold/cool ambient magnetospheric plasma will result in additional plasma characteristic frequencies including cutoff frequencies, resonances, and crossover frequencies [*Young et al.*, 1981; *Fraser*, 1985]. The characteristic frequencies have profound effects on the propagation of EMIC waves along geomagnetic field lines. For example, in a two-ion (He$^+$, H$^+$) plasma, there is a nonpropagation stop band between the He$^+$ cyclotron frequency (f_{He+}), and the He$^+$ cutoff frequency (f_{co}). By observing this stop band in experimental wave spectra, it is then possible to invert the problem and obtain an indication of the He$^+$ plasma density [*Gurnett and Shawhan*, 1966; *Fraser et al.*, 2006]. In order to identify the stop band, the wave frequency must be normalized to the H$^+$ cyclotron frequency in order to provide an organized spectral slot. The spectral slot created by the nonpropagation stop band at GOES 11 and 12 combined over 6 h segments of LT and plotted against peak wave power is shown in the four panels in Figure 11. In Figure 11, the slot is best defined in the 12–18 LT interval over a bandwidth $f/f_{He+} = 0.21$–0.28, but is also seen over the 06–12 and 18–24 LT sectors. There is only a suggestion of a slot in the nightside 00–06 LT sector indicating that for the few EMIC waves seen here, the He$^+$ ions did not create a slot to the same extent as seen in the other sectors. Also, most waves here are seen above f_{He+}. The less obvious slot might indicate a low He$^+$ concentration or a warmer He$^+$ temperature [*Chen et al.*, 2011]. Measurements from the spectral

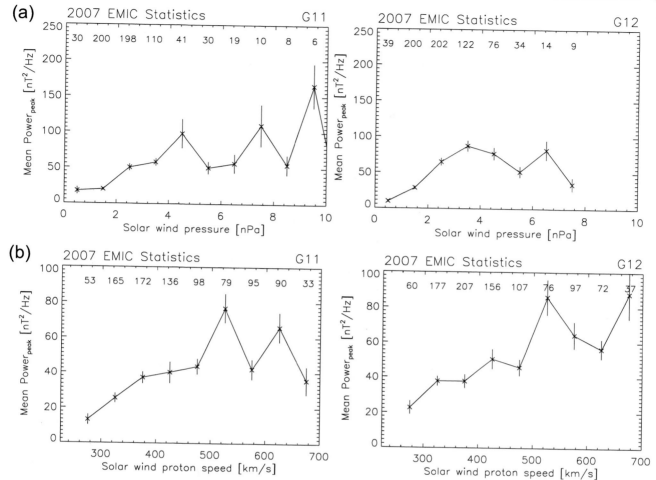

Figure 8. (a) Relationship between electromagnetic ion cyclotron (EMIC) wave power spectral density and associated upstream solar wind parameters observed by ACE. (a) Variation in wave power seen by GOES 11 and GOES 12 with solar wind speed. (b) Variation in wave power seen by GOES 11 and GOES 12 with solar wind dynamic pressure. The numbers at the top indicate the number of data points used in the averages.

slot in the 12–18 LT sector shows the cutoff frequency at $f_{co}/f_{H+} = 0.28$ and $f_{He+}/f_{H+} = 0.21$. Using the relationship $f_{co} = (1 + 3\beta)f_{He+}$ where β is the fractional He^+ concentration [*Young et al.*, 1981] provides $\beta = 0.11$ or 11%. Here we have used local conditions where $f/f_{He+} = 0.21$ rather than the expected 0.25. This relates to the deduction that the waves are generated at the equator where the characteristic frequencies are established, and propagate in a field aligned direction away from the generation region to off equator regions of higher magnetic fields [*Mauk and McPherron*, 1980]. This result agrees with those obtained at geosynchronous orbit by *Mauk et al.* [1981] and *Fraser* [1985] for a two-ion plasma and suggests an average He^+ concentration in the plasmapause bulge region over a quiet Sun year.

The diurnal variation in local normalized wave frequency at GOES 11 and 12 plotted in Figure 12 shows that waves are seen, on average, in the H^+ band in the midnight-dawn sector, while over the remainder of the day, they are observed in the He^+ band. Also, the lower normalized frequencies are seen at GOES 12 further off the equator as expected. Here the 1 Hz Nyquist frequency limits f/f_{H+} to <0.6 and indicates the upper frequencies seen in the morning may be limited by the spectral bandwidth of the data.

Finally, Figure 13 illustrates the variation of wave power with *Dst* where data from both satellites are combined and divided into two groups, *Dst* < −5 nT and >−5 nT. Waves in the He^+ band dominate at powers >50 nT^2 Hz^{-1}, particularly under more active geomagnetic conditions. Under quieter

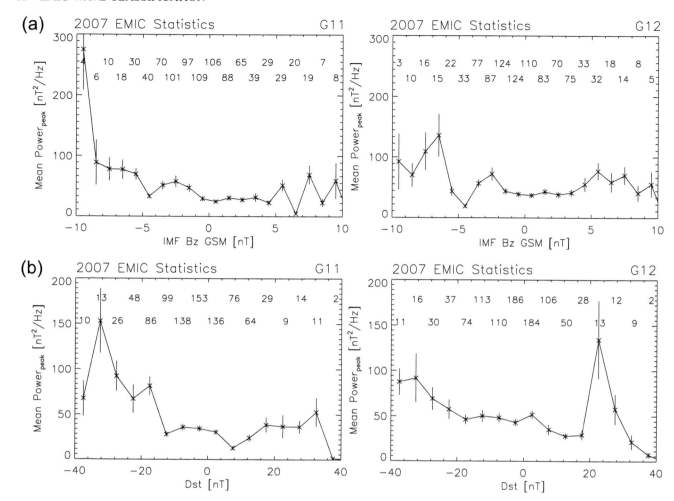

Figure 9. Relationship between EMIC wave activity and geomagnetic activity seen by GOES 11 and GOES 12. (a) Variation in mean wave power spectral density and the interplanetary magnetic field B_z component at ACE. (b) Variation in mean wave power spectral density with the *Dst* index. The numbers at the top indicate the number of data points used in the averages.

conditions, the He$^+$ band wave power is more evenly distributed, and more power is seen in the lower end of the H$^+$ band.

5. DISCUSSION

5.1. Spectral Properties

The identification of the various spectral forms of EMIC wave emissions seen on the ground in the Pc1-2 (0.1–5 Hz) band commenced in the 1960s and has been well-described by *Fukunishi et al.* [1981]. With satellite observations, it is not easy to undertake a comprehensive latitude, local time survey on a continuous basis, and results have been obtained by single-point measurements or close-spaced constellations of satellites (ATS-6, GEOS, AMPTE, ISEE, Polar, CRRES,

Cluster, GOES, THEMIS). An exception to this is a very recent comprehensive global study of EMIC wave properties using THEMIS data [*Min et al.*, 2012]. Within the spectral data from satellites, there have been very few observations of the classical fine structured "pearl" type dispersive emissions, which are seen frequently on the ground. For example, *Perraut* [1982] reported seeing only one over the GEOS mission. Instead, most emissions seen in space are the periodic emissions, Hm chorus, and Pc1-2 band emissions defined earlier. Only a few of the easily identified evening IPDP emission events recognized from their geomagnetic substorm association were seen.

With the discovery of the association of EMIC waves in the middle magnetosphere with enhanced density plasmaspheric plasma drainage plumes in the afternoon sector and

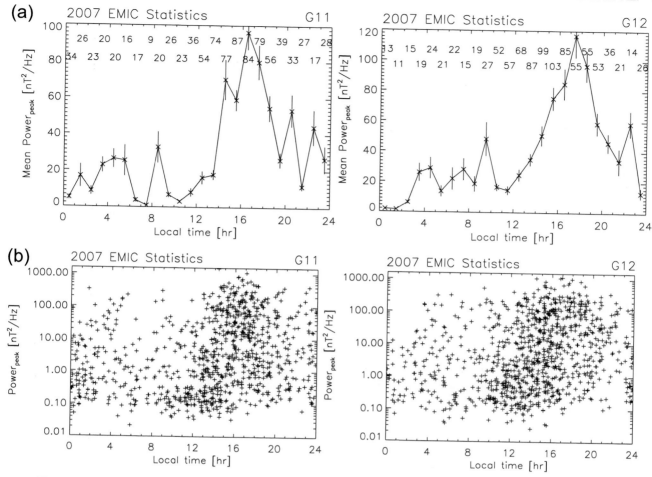

Figure 10. Diurnal variation in EMIC wave power seen by GOES 11 and GOES 12. (a) Mean wave power illustrating an afternoon maximum. The numbers at the top indicate the number of data points used in the averages. (b) Scatterplots of peak wave power illustrating the upper power limits observed during the day.

sometimes subauroral proton arcs [*Fraser et al.*, 2005; *Immel et al.*, 2005; *Spasojevic et al.*, 2005, 2011], it is of interest to identify the types of EMIC wave emissions involved in these interactions. In comparing the plume-associated emissions from the literature, it appears that plumes are typically associated with afternoon Hm emission bursts and patches of Hm chorus. In contrast, the narrow-band long duration Pc1-2 emissions have been shown in ground data to be associated with the noon passage of the cusp and seen a few degrees of latitude equatorward of the cusp [*Menk et al.*, 1992, 1993] or the plasma sheet boundary layer [*Engebretson et al.*, 2002]. However, it is not understood how the GOES satellites (at $L = 6.6$ (~67°) geomagnetic latitude) may see cusp/PSBL-associated waves, especially under the generally quiet geomagnetic conditions experienced in 2007.

5.2. GOES Location

Although the GOES fleet of satellites orbit around the geographic equator, they are located at different geomagnetic latitudes depending on longitude, due to the geomagnetic dipole tilt offset. This maximizes near the GOES 12 longitude, which is 11° geomagnetic north. Likewise, GOES 10 and GOES 11 are at 5° and 4° geomagnetic, respectively. Satellite latitude has a noticeable effect on EMIC wave observations. For example, with the EMIC wave generation source centered on the geomagnetic equator and extending to about ±11° in latitude [*Fraser et al.*, 1996; *Loto'aniu et al.*, 2005], GOES 10 and 11 will be well within the source region, while GOES 12 will be just inside or outside the source region. The effect of this is seen in Figure 12 where the local normalized wave frequency (f/f_{He^+}) is lower at

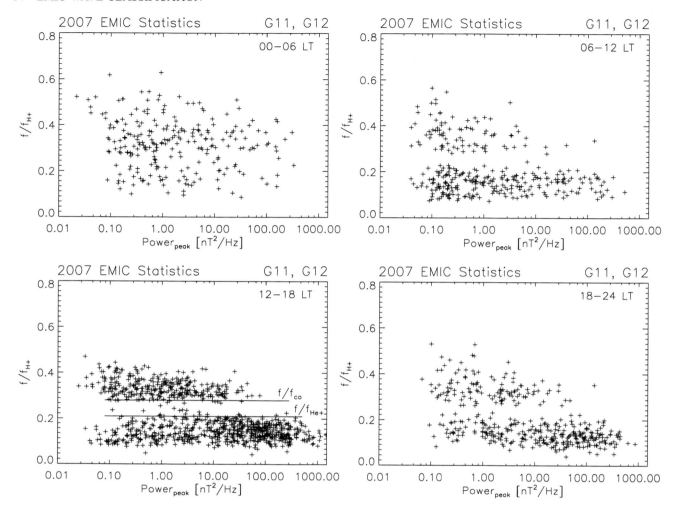

Figure 11. Scatterplots of EMIC peak wave power against the wave frequency normalized to the proton cyclotron frequency over four 6 h local time intervals during the day for combined GOES 11 and GOES 12 wave power. The He$^+$ nonpropagation stop band between the local He$^+$ cyclotron frequency and the He$^+$ cutoff frequency is well defined in all 6 h plots with the exception of 00–06 LT. (left bottom) The local He$^+$ cutoff frequency (f_{co}) and cyclotron frequency (f_{He+}) are represented by the black lines.

GOES 12 than 11 due to its location in a higher geomagnetic field assuming a near-equator source. Also, the higher peak mean wave power at 120 nT^{-2} Hz^{-1} at GOES 12 in Figure 10 compared to GOES 11 (100 nT^{-2} Hz), supports the integrated amplification of EMIC waves when propagating away from the source region [*Horne and Thorne*, 1993; *Hu and Fraser*, 1994].

6. SUMMARY

The GOES geosynchronous satellite high-resolution data set with a Nyquist frequency of 1 Hz, and a useable bandwidth for ULF waves up to ~0.8 Hz provides a unique

opportunity to study the properties of EMIC waves in the outer ring current and radiation belt. With continuous coverage at specific local times with multiple satellites, it is possible to study both spatial and temporal properties of the waves. In this study, we have used data from GOES 10 located at 90° west geographic longitude and ~5° north of the geomagnetic equator, GOES 11 at 135° west and ~4° north of the geomagnetic equator, and GOES 12 at 75° west and ~11° off the equator, to identify the different spectral characteristics of EMIC wave emission types seen at geosynchronous orbit. Statistical studies on the similarities and differences in the waves seen by GOES 11 and 12 over 2007 were also undertaken. Major results may be summarized as follows.

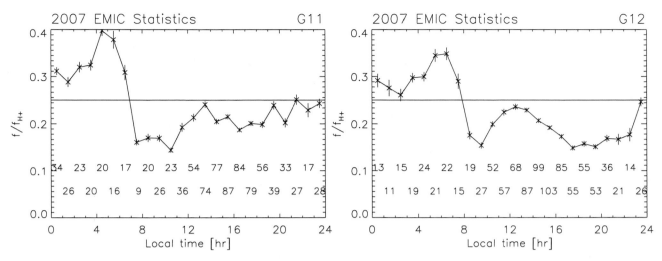

Figure 12. Diurnal variation in local normalized wave frequency over 2007 for GOES 11 and GOES 12. The horizontal lines represent the He$^+$ cyclotron frequency (f_{He+}). (bottom) The numbers indicate the number of data points used in the averages.

1. Six different categories of EMIC wave spectral structures were identified over 0.1–0.8 Hz. Some show the characteristic fine structure often associated with interhemisphere bouncing wave packets, while the other shows a messy disorganized or irregular fine structure or no structure at all. However, they are all band-limited discrete emissions presumed to be generated by ion cyclotron instability within the magnetosphere. The wave types include, in order of highest to lowest occurrence probability over the quiet year 2007: Pc1-2 band emissions, Hm chorus, irregular Hm emissions, Hm whistlers, and periodic emissions IPDP and Hm emission bursts.

2. With respect to upstream solar wind and IMF conditions, a gradual increase is seen in EMIC wave power with both increasing solar wind velocity up to about 500 km s^{-1}, and dynamic pressure up to 4 nPa. Beyond these limits, wave power continues to increase, but the data are less reliable.

3. With regard to internal magnetospheric conditions, maximum wave power decreases with Dst over −20 nT < Dst < 20 nT. Increased wave power is also seen under more active and quieter conditions, but the data are less reliable.

4. Both maximum wave occurrence and power are seen over 12–20 LT, corresponding to the times of enhanced cold/

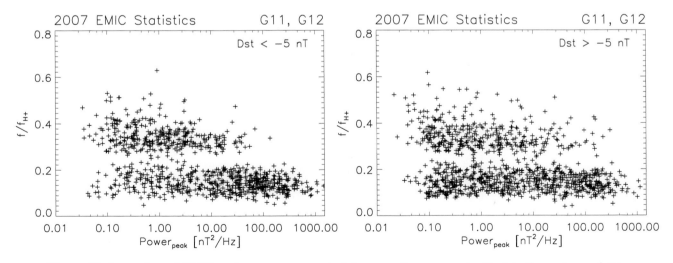

Figure 13. Scatterplots of EMIC peak wave power against the wave frequency normalized to the proton cyclotron frequency for GOES 11 and GOES 12 combined data under differing geomagnetic conditions represented by *Dst*. In this solar minimum year, quiet conditions are defined by *Dst* > −5 nT and more active conditions by *Dst* < −5 nT.

cool plasma density at geosynchronous orbit associated with plasmaspheric plumes or an inflated plasmasphere and the injection of plasma sheet keV ions.

5. Most H^+ band waves are seen over 00–06 LT, and the He^+ nonpropagation stop band is not well defined here, suggesting the absence of cold He^+ ions. He^+ band waves dominated over 06–24 LT.

6. Observations of the He^+ spectral slot suggests the average He^+ cold plasma abundance over 06–24 LT in 2007 was ~11%.

7. GOES 12 located some 11° off the geomagnetic equator at the edge of the wave-generation region sees waves with higher power and lower normalized wave frequencies than GOES 11 located nearer the equator and source region.

This study has provided a window on the spectral characteristics of the various types of EMIC waves seen in the Pc1-2 band at geosynchronous orbit. It also noted some of the differences in wave characteristics attributed to the different geomagnetic locations of the GOES satellites.

Acknowledgments. This research was supported by Australian Research Council Discovery Project grants DP0772504 and DP0663643 and Linkage International grant LX0882515. Infrastructure support has been provided by the University of Newcastle and the Space Weather Prediction Center, NOAA, Boulder.

REFERENCES

Anderson, B. J., R. E. Erlandson, and L. J. Zanetti (1992a), A statistical study of Pc 1–2 magnetic pulsations in the equatorial magnetosphere: 1. Equatorial occurrence distributions, *J. Geophys. Res.*, *97*(A3), 3075–3088, doi:10.1029/91JA02706.

Anderson, B. J., R. E. Erlandson, and L. J. Zanetti (1992b), A statistical study of Pc 1–2 magnetic pulsations in the equatorial magnetosphere: 2. Wave properties, *J. Geophys. Res.*, *97*(A3), 3089–3101, doi:10.1029/91JA02697.

Arnoldy, R. L., P. B. Lewis Jr., and L. J. Cahill Jr. (1979), Polarization of Pc 1 and IPDP pulsations correlated with particle precipitation, *J. Geophys. Res.*, *84*(A12), 7091–7098, doi:10.1029/JA084iA12p07091.

Campbell, W. H., and E. C. Stiltner (1965), Some characteristics of geomagnetic pulsations at frequencies near 1 c/s, *J. Res. Natl. Bur. Stand. U.S., Sect. D*, *69*, 1117–1132.

Chen, L., R. M. Thorne, and R. B. Horne (2009), Simulation of EMIC wave excitation in a model magnetosphere including structured high-density plumes, *J. Geophys. Res.*, *114*, A07221, doi:10.1029/2009JA014204.

Chen, L., R. M. Thorne, V. K. Jordanova, C.-P. Wang, M. Gkioulidou, L. Lyons, and R. B. Horne (2010), Global simulation of EMIC wave excitation during the 21 April 2001 storm from coupled RCM-RAM-HOTRAY modeling, *J. Geophys. Res.*, *115*, A07209, doi:10.1029/2009JA015075.

Chen, L., R. M. Thorne, and J. Bortnik (2011), The controlling effect of ion temperature on EMIC wave excitation and scattering, *Geophys. Res. Lett.*, *38*, L16109, doi:10.1029/2011GL048653.

Clausen, L. B. N., J. B. H. Baker, J. M. Ruohoniemi, and H. J. Singer (2011), EMIC waves observed at geosynchronous orbit during solar minimum: Statistics and excitation, *J. Geophys. Res.*, *116*, A10205, doi:10.1029/2011JA016823.

Cornwall, J. M. (1965), Cyclotron instabilities and electromagnetic emission in the ultra low frequency and very low frequency ranges, *J. Geophys. Res.*, *70*(1), 61–69, doi:10.1029/JZ070i001p00061.

Engebretson, M. J., W. K. Peterson, J. L. Posch, M. R. Klatt, B. J. Anderson, C. T. Russell, H. J. Singer, R. L. Arnoldy, and H. Fukunishi (2002), Observations of two types of Pc 1–2 pulsations in the outer dayside magnetosphere, *J. Geophys. Res.*, *107*(A12), 1451, doi:10.1029/2001JA000198.

Fok, M.-C., Y. Ebihara, T. E. Moore, D. M. Ober, and K. A. Keller (2005), Geospace storm processes coupling the ring current, radiation belt and plasmasphere, in *Inner Magnetosphere Interactions: New Perspectives From Imaging, Geophys. Monogr. Ser.*, vol. 159, edited by J. Burch, M. Schulz and H. Spence, pp. 207–220, AGU, Washington, D. C., doi:10.1029/159GM16.

Fraser, B. J. (1968), Temporal variations in Pc1 geomagnetic micropulsations, *Planet. Space Sci.*, *16*, 111–124, doi:10.1016/0032-0633(68)90048-2.

Fraser, B. J. (1985), Observations of ion cyclotron waves near synchronous orbit and on the ground, *Space Sci. Rev.*, *42*, 357–374.

Fraser, B. J., and T. S. Nguyen (2001), Is the plasmapause a preferred source region of electromagnetic ion cyclotron waves in the magnetosphere?, *J. Atmos. Sol. Terr. Phys.*, *63*(11), 1225–1247, doi:10.1016/S1364-6826(00)00225-X.

Fraser, B. J., and S. Wawrzyniak (1978), Source movement associated with IPDP pulsations, *J. Atmos. Terr. Phys.*, *40*, 1281–1288.

Fraser, B. J., W. J. Kemp, and D. J. Webster (1989), Ground-satellite study of a Pc 1 ion cyclotron wave event, *J. Geophys. Res.*, *94*(A9), 11,855–11,863, doi:10.1029/JA094iA09p11855.

Fraser, B. J., J. C. Samson, Y. D. Hu, R. L. McPherron, and C. T. Russell (1992), Electromagnetic ion cyclotron waves observed near the oxygen cyclotron frequency by ISEE 1 and 2, *J. Geophys. Res.*, *97*(A3), 3063–3074, doi:10.1029/91JA02447.

Fraser, B. J., H. J. Singer, W. J. Hughes, J. R. Wygant, R. R. Anderson, and Y. D. Hu (1996), CRRES Poynting vector observations of electromagnetic ion cyclotron waves near the plasmapause, *J. Geophys. Res.*, *101*(A7), 15,331–15,343, doi:10.1029/95JA03480.

Fraser, B. J., H. J. Singer, M. L. Adrian, D. L. Gallagher, and M. F. Thomsen (2005), The relationship between plasma density structure and EMIC waves at geosynchronous orbit, in *Inner Magnetosphere Interactions: New Perspectives From Imaging, Geophys. Monogr. Ser.*, vol. 159, edited by J. Burch, M. Schulz and H. Spence, pp. 55–70, AGU, Washington, D. C., doi:10.1029/159GM04.

Fraser, B. J., T. M. Loto'aniu, and H. J. Singer (2006), Electromagnetic ion cyclotron waves in the magnetosphere, in *Magnetospheric ULF Waves: Synthesis and New Directions, Geophys. Monogr. Ser.*, vol. 169, edited by K. Takahashi et al., pp. 195–212, AGU, Washington, D. C., doi:10.1029/169GM13.

Fraser, B. J., R. S. Grew, S. K. Morley, J. C. Green, H. J. Singer, T. M. Loto'aniu, and M. F. Thomsen (2010), Storm time observations of electromagnetic ion cyclotron waves at geosynchronous orbit: GOES results, *J. Geophys. Res.*, *115*, A05208, doi:10.1029/2009JA014516.

Fukunishi, H., and T. Toya (1981), Morning IPDP events observed at high latitudes, *J. Geophys. Res.*, *86*(A7), 5701–5709, doi:10.1029/JA086iA07p05701.

Fukunishi, H., T. Toya, K. Koike, M. Kuwashima, and M. Kawamura (1981), Classification of hydromagnetic emissions based on frequency-time spectra, *J. Geophys. Res.*, *86*(A11), 9029–9039, doi:10.1029/JA086iA11p09029.

Gendrin, R., S. Lacourly, A. Roux, J. Solomon, F. Z. Feïgin, M. V. Gokhberg, V. A. Troitskaya, and V. L. Yakimenko (1971), Wave packet propagation in an amplifying medium and its application to the dispersion characteristics and to the generation mechanisms of Pc 1 events, *Planet. Space Sci.*, *19*, 165–194.

Gurnett, D. A., and S. D. Shawhan (1966), Determination of hydrogen ion concentration, electron density, and proton gyrofrequency from the dispersion of proton whistlers, *J. Geophys. Res.*, *71*(3), 741–754, doi:10.1029/JZ071i003p00741.

Halford, A. J., B. J. Fraser, and S. K. Morley (2010), EMIC wave activity during geomagnetic storm and nonstorm periods: CRRES results, *J. Geophys. Res.*, *115*, A12248, doi:10.1029/2010JA015716.

Heacock, R. R. (1966), The 4-second summertime micropulsation band at College, *J. Geophys. Res.*, *71*(11), 2763–2775, doi:10.1029/JZ071i011p02763.

Helliwell, R. A. (1965), *Whistlers and Related Ionospheric Phenomena*, 349 pp, Stanford Univ. Press, Palo Alto, Calif.

Horne, R. B., and R. M. Thorne (1993), On the preferred source location for the convective amplification of ion cyclotron waves, *J. Geophys. Res.*, *98*(A6), 9233–9247, doi:10.1029/92JA02972.

Hu, Y. D., and B. J. Fraser (1994), Electromagnetic ion cyclotron wave amplification and source regions in the magnetosphere, *J. Geophys. Res.*, *99*(A1), 263–272, doi:10.1029/93JA01897.

Immel, T. J., S. B. Mende, H. U. Frey, J. Patel, J. W. Bonnell, M. J. Engebretson, and S. A. Fuselier (2005), ULF waves associated with enhanced subauroral proton precipitation, in *Inner Magnetosphere Interactions: New Perspectives From Imaging*, *Geophys. Monogr. Ser.*, vol. 159, edited by J. Burch, M. Schulz and H. Spence, pp. 71–84, AGU, Washington, D. C., doi:10.1029/159GM05.

Jordanova, V. K., C. J. Farrugia, R. M. Thorne, G. V. Khazanov, G. D. Reeves, and M. F. Thomsen (2001), Modeling ring current proton precipitation by electromagnetic ion cyclotron waves during the May 14–16, 1997, storm, *J. Geophys. Res.*, *106*(A1), 7–22, doi:10.1029/2000JA002008.

Jordanova, V. K., M. Spasojevic, and M. F. Thomsen (2007), Modeling the electromagnetic ion cyclotron wave-induced formation of detached subauroral proton arcs, *J. Geophys. Res.*, *112*, A08209, doi:10.1029/2006JA012215.

Kangas, J., A. Guglielmi, and O. Pokhotelov (1998), Morphology and physics of short-period magnetic pulsations, *Space Sci. Rev.*, *83*, 435–512, doi:10.1023/A:1005063911643.

Kozyra, J. U., T. E. Cravens, A. F. Nagy, E. G. Fontheim, and R. S. B. Ong (1984), Effects of energetic heavy ions on electromagnetic ion cyclotron wave generation in the plasmapause region, *J. Geophys. Res.*, *89*(A4), 2217–2233, doi:10.1029/JA089iA04p02217.

Loto'aniu, T. M., B. J. Fraser, and C. L. Waters (2005), Propagation of electromagnetic ion cyclotron wave energy in the magnetosphere, *J. Geophys. Res.*, *110*, A07214, doi:10.1029/2004JA010816.

Loto'aniu, T. M., B. J. Fraser, and C. L. Waters (2009), The modulation of electromagnetic ion cyclotron waves by Pc 5 ULF waves, *Ann. Geophys.*, *27*, 121–130.

Liu, Y. H. (2011), Multi-satellite observations of ULF waves in the Earth's magnetosphere, Ph.D. thesis, Univ. of Newcastle, Callaghan, New South Wales, Australia.

Mauk, B. H. (1982), Helium resonance and dispersion effects on geostationary Alfven/ion cyclotron waves, *J. Geophys. Res.*, *87*(A11), 9107–9119, doi:10.1029/JA087iA11p09107.

Mauk, B. H., and R. L. McPherron (1980), An experimental test of the electromagnetic ion cyclotron instability within the Earth's magnetosphere, *Phys. Fluids*, *23*(10), 2111–2127, doi:10.1063/1.862873.

Mauk, B. H., C. E. McIlwain, and R. L. McPherron (1981), Helium cyclotron resonance within the Earth's magnetosphere, *Geophys. Res. Lett.*, *8*(1), 103–106, doi:10.1029/GL008i001p00103.

McCollough, J. P., S. R. Elkington, M. E. Usanova, I. R. Mann, D. N. Baker, and Z. C. Kale (2010), Physical mechanisms of compressional EMIC wave growth, *J. Geophys. Res.*, *115*, A10214, doi:10.1029/2010JA015393.

McPherron, R. L., and S. H. Ward (1965), Auroral-zone pearl pulsations, *J. Geophys. Res.*, *70*(23), 5867–5882, doi:10.1029/JZ070i023p05867.

Menk, F. W., B. J. Fraser, H. J. Hansen, P. T. Newell, C.-I. Meng, and R. J. Morris (1992), Identification of the magnetospheric cusp and cleft using Pc1-2 pulsations, *J. Atmos. Terr. Phys.*, *54*, 1021–1042.

Menk, F. W., B. J. Fraser, H. J. Hansen, P. T. Newell, C.-I. Meng, and R. J. Morris (1993), Multi-station observations of Pc1-2 ULF pulsations between the plasmapause and polar cap, *J. Geomagn. Geoelectr.*, *45*, 1159–1173.

Min, K., J. Lee, K. Keika, and W. Li (2012), Global distribution of EMIC waves derived from THEMIS observations, *J. Geophys. Res.*, *117*, A05219, doi:10.1029/2012JA017515.

Obayashi, T. (1965), Hydromagnetic whistlers, *J. Geophys. Res.*, *70*(5), 1069–1078, doi:10.1029/JZ070i005p01069.

Onsager, T. G., A. A. Chan, Y. Fei, S. R. Elkington, J. C. Green, and H. J. Singer (2004), The radial gradient of relativistic electrons at geosynchronous orbit, *J. Geophys. Res.*, *109*, A05221, doi:10.1029/2003JA010368.

Perraut, S. (1982), Wave-particle interactions in the ULF range – GEOS-1 and -2 results, *Planet. Space Sci.*, *30*, 1219–1227.

Pickett, J. S., et al. (2010), Cluster observations of EMIC triggered emissions in association with Pc1 waves near Earth's plasmapause, *Geophys. Res. Lett.*, *37*, L09104, doi:10.1029/2010GL042648.

Saito, T. (1969), Geomagnetic pulsations, *Space Sci. Rev.*, *10*, 319–412.

Singer, H., L. Matheson, R. Grubb, A. Newman, and D. Bouwer (1996), Monitoring space weather with the GOES magnetometers, in *Society of Photo-Optical Instrumentation Engineers (SPIE) Conference Series*, vol. 2812, edited by E. R. Washwell, pp. 299–308, Soc. of Photo-Opt. Instrum. Eng., Bellingham, Wash.

Spasojevic, M., M. R. Thomsen, P. J. Chi, and B. R. Sandel (2005), Afternoon subauroral proton precipitation resulting from ring current—Plasmasphere interaction, in *Inner Magnetosphere Interactions: New Perspectives From Imaging, Geophys. Monogr. Ser.*, vol. 159, edited by J. Burch, M. Schulz and H. Spence, pp. 85–99, AGU, Washington, D. C., doi:10.1029/159GM06.

Spasojevic, M., L. W. Blum, E. A. MacDonald, S. A. Fuselier, and D. I. Golden (2011), Correspondence between a plasma-based EMIC wave proxy and subauroral proton precipitation, *Geophys. Res. Lett.*, *38*, L23102, doi:10.1029/2011GL049735.

Summers, D., and R. M. Thorne (2003), Relativistic electron pitch-angle scattering by electromagnetic ion cyclotron waves during geomagnetic storms, *J. Geophys. Res.*, *108*(A4), 1143, doi:10.1029/2002JA009489.

Tepley, L. (1964), Low-latitude observations of fine-structured hydromagnetic emissions, *J. Geophys. Res.*, *69*(11), 2273–2290, doi:10.1029/JZ069i011p02273.

Usanova, M. E., I. R. Mann, I. J. Rae, Z. C. Kale, V. Angelopoulos, J. W. Bonnell, K.-H. Glassmeier, H. U. Auster, and H. J. Singer (2008), Multipoint observations of magnetospheric compression-related EMIC Pc1 waves by THEMIS and CARISMA, *Geophys. Res. Lett.*, *35*, L17S25, doi:10.1029/2008GL034458.

Usanova, M. E., et al. (2010), Conjugate ground and multisatellite observations of compression-related EMIC Pc1 waves and associated proton precipitation, *J. Geophys. Res.*, *115*, A07208, doi:10.1029/2009JA014935.

Verronen, P. T., A. Seppälä, M. A. Clilverd, C. J. Rodger, E. Kyrölä, C.-F. Enell, T. Ulich, and E. Turunen (2005), Diurnal variation of ozone depletion during the October–November 2003 solar proton events, *J. Geophys. Res.*, *110*, A09S32, doi:10.1029/2004JA010932.

Young, D. T., S. Perraut, A. Roux, C. de Villedary, R. Gendrin, A. Korth, G. Kremser, and D. Jones (1981), Wave-particle interactions near Ω_{He+} observed on GEOS 1 and 2 1. Propagation of ion cyclotron waves in He^+-rich plasma, *J. Geophys. Res.*, *86*(A8), 6755–6772, doi:10.1029/JA086iA08p06755.

B. J. Fraser and R. S. Grew, Centre for Space Physics, University of Newcastle, Callaghan, NSW 2308, Australia. (brian.fraser@newcastle.edu.au)

S. K. Morley, Los Alamos National Laboratory, Los Alamos, NM 87545, USA.

H. J. Singer, Space Weather Prediction Center, NOAA, Boulder, CO 80305, USA.

The Role of Ultralow Frequency Waves in Radiation Belt Dynamics

Ian R. Mann, Kyle R. Murphy, Louis G. Ozeke, I. Jonathan Rae, David K. Milling, and Andy Kale

Department of Physics, University of Alberta, Edmonton, Alberta, Canada

Farideh Honary

Department of Physics, Lancaster University, Lancaster, UK

This paper reviews the role of long-period ultralow frequency (ULF) waves in contributing to the acceleration, transport, and loss of electrons in the outer zone Van Allen radiation belt. We place particular emphasis on the mesoscale and global-scale characterization of ULF waves around electron drift orbits available from ground-based magnetometer arrays, which complement conjugate single-point in situ satellite measurements. We examine the time domain relationship between 1–10 mHz ULF wave power and driving solar wind speed, demonstrating a close connection, which could contribute to explaining the Paulikas and Blake (1979) correlation between radiation belt electron flux and solar wind speed. We review both coherent and stochastic transport processes, such as radial diffusion, which can arise from resonant ULF wave-particle interactions. We examine the potential role of plasmaspheric cold plasma density on radiation belt morphology as effected through the intermediary of ULF waves. Significantly, we show that there is a close connection between Dst and the storm time penetration of enhanced ULF wave power to low L and thereby also to the position of the plasmapause. Finally, we introduce the concept of enhanced ULF wave magnetopause shadowing, whereby ULF wave advection can bring the magnetopause into contact with closed drift shells at lower altitudes than estimated from equilibrium magnetopause models. Overall, we highlight the important role for ULF waves in influencing a large range of radiation belt acceleration, transport, and loss processes, including the cross-energy coupling of the populations in the plasmasphere, ring current, and radiation belts.

1. INTRODUCTION

Despite the fact that it is more than 50 years since the accidental discovery by James Van Allen of the radiation belts which surround the Earth and which now bear his name [e.g., *Van Allen et al.*, 1958; *Van Allen and Frank*, 1959], the

processes controlling radiation belt dynamics and morphology remain relatively poorly understood. In this paper, we examine the roles which ULF waves can play in influencing the dynamic behavior of the outer zone electron radiation belt. The outer zone electron fluxes can change by many orders of magnitudes on time scales from minutes, to hours, days, months and years. As discussed by, for example, *Reeves et al.* [2003] (see also the review by *Friedel et al.* [2002]), the overall response of radiation belt electron flux is determined from a "delicate and complicated balance" between competing processes, which act to produce overall acceleration or loss. Apparently similar geomagnetic storms,

Dynamics of the Earth's Radiation Belts and Inner Magnetosphere
Geophysical Monograph Series 199
10.1029/2012GM001349

defined, for example, in terms of the ring current response in the form of *Dst*, can display a markedly different overall radiation belt flux response [cf. *Reeves et al.*, 2003].

This is further complicated by the fact that multiple wave-particle interaction processes can affect radiation belt dynamics at the same time, the overall response arising as the result of the cumulative effects of all of these processes, which can act to affect the electrons anywhere around their drift orbit. Even with data from the eagerly awaited new inner magnetosphere missions, such as the NASA Radiation Belt Storm Probes (RBSP) [see, e.g., *Kessel*, this volume], additional complementary coverage elsewhere around the electron drift orbits will be important. Data from ground-based networks can make a key contribution in helping to disentangle the effects and impacts of the multiple different, and likely coexisting, wave-particle interactions. In this paper, we make extensive use of the data available from ground-based magnetometer networks and highlight the significant contribution that such arrays can make to studies of radiation belt dynamics, especially in regard to the effects of ULF waves.

There are numerous wave-particle interactions, which can produce either acceleration or loss, including interactions with whistler mode chorus, plasmaspheric hiss, electromagnetic ion cyclotron (EMIC) waves, magnetosonic modes, as well as the interactions with ULF waves upon which we focus here [cf., for example, *Friedel et al.*, 2002; *Summers et al.*, 1998, 2007; *Mann et al.*, 2006; *Millan and Thorne*, 2007]. These processes typically operate in different local time sectors and have different geoeffectiveness under different solar wind driving and during various magnetospheric conditions such as during storm versus nonstorm times. Also of significance is the observation that the cold plasma populations in the plasmasphere (~eV energy), and medium energy ion and electron ring current populations (~1–400 keV energy), influence Alfven and other plasma wave propagation characteristics as well as particle-driven plasma wave mode growth rates. Since these modes themselves influence the radiation belt (~0.5–10+ MeV energy) particles, it is apparent that there is a cross-energy coupling, which spans many (>6) orders of magnitude in energy, which is at the core of determining and controlling the dynamical morphology of the outer electron radiation belt (see, e.g., the discussion of *Mann et al.* [2006]). We suggest that such cross-energy coupling between the plasmasphere, ring current, and radiation belts, manifest through the intermediary of numerous wave-particle interactions with plasma waves, should be adopted as a new paradigm for understanding inner magnetosphere dynamics. Only with a comprehensive suite of particle instruments, which span this range of eV to MeV energies, combined with wave instruments that monitor all of the relevant modes, can good progress be made in understanding the response of the radiation belts to solar wind forcing.

The upcoming launch of the two-satellite NASA RBSP mission [e.g., *Kessel*, this volume], scheduled for 2012, will enable the geoeffectiveness of many of these wave-particle processes to be reexamined in unprecedented detail. Many important discoveries about radiation belt wave-particle interaction processes were revealed by the Combined Release and Radiation Effects Satellite (CRRES) [e.g., *Johnson and Kierein*, 1992] using the 15 months of data from the equatorial inner magnetosphere in 1990–1991, but many challenges remain. Perhaps most significant is that in terms of establishing the geoeffectiveness of specific solar wind drivers, needed for future radiation belt models to be advanced to the point of predictability, the RBSP mission will not suffer from the challenges of the limited solar wind coverage which spanned the CRRES era. The increased temporal cadence available from the inbound and outbound passes of the two RBSP spacecraft, available at varying along-track interspacecraft separations, will also provide a new window on the significance of the role of ULF and other wave-particle interaction processes in shaping the morphology of the outer-zone Van Allen radiation belts. In this paper, we examine the role of ULF waves in changing the morphology of the radiation belts through a variety of resonant and nonresonant acceleration, transport, and loss processes.

2. ULF WAVE EXCITATION MECHANISMS

2.1. External Solar Wind Drivers

Since the early observations of *Paulikas and Blake* [1979], it has been clear that the enhancement of outer zone electron flux is correlated with intervals of enhanced solar wind speed. A similar feature of enhanced ULF wave power is also seen in the magnetosphere during fast solar wind streams [e.g., *Mathie and Mann*, 2000; *Engebretson et al.*, 1998], and a number of authors have suggested a causal link between enhanced populations of ULF waves and MeV electron acceleration [e.g., *Rostoker et al.*, 1998; *Baker et al.*, 1998; *Mathie and Mann*, 2000]. It is an observational fact that the outer zone electron radiation belt is at its most intense during the declining phase of the solar cycle, associated with corotating interaction regions (CIRs) generated in interplanetary space by repetitive fast solar wind streams emanating from transequatorial solar coronal holes. Conversely, although enhanced compared to quiet times, the belts are less intense during solar maximum than during the declining phase; solar maximum being characterized by the aperiodic impact of interplanetary coronal mass ejections (ICMEs) [e.g., *Kataoka and Myoshi*, 2006]. Storms associated with both ICMEs and CIRs are associated with the generation of enhanced levels of ULF wave power, with the ULF wave power generated by CIRs typically being longer lived than that arising from ICMEs [e.g., *Borovsky and Denton*, 2006]. During the declining phase of the solar cycle, the CIRs generated at fast-slow solar wind stream interfaces often repetitively impact the Earth at the approximately 27 day rotation period of the Sun in the Earth's frame [e.g., *Baker et al.*, 1979]. Since such fast solar wind speed streams are thought to be efficient drivers of ULF waves [e.g., *Engebretson et al.*, 1998; *Mathie and Mann*, 2000, 2001], it remains important to establish whether ULF waves excited in the magnetosphere represent the intermediary required to explain the *Paulikas and Blake* [1979] correlation between outer zone electron flux and fast solar wind speeds.

A series of papers have highlighted the relationship between fast solar wind speeds, and the generation of intense ULF wave power [e.g., *Mathie and Mann*, 2000; *Kessel et al.*, 2004] and an energetic electron response [e.g., *Rostoker et al.*, 1998; *Mathie and Mann*, 2000; *Mann et al.*, 2004; *O'Brien et al.*, 2003]. In particular, *Mathie and Mann* [2000] showed that during the declining phase of the solar cycle ULF wave power is strongly correlated with recurrent fast solar wind streams and that, in each case, a period of high ULF wave power was also followed by a MeV energy electron response at geosynchronous orbit. Similarly, *Li et al.* [2001] have shown that the flux of electrons at geosynchronous orbit can be predicted using models that drive the flux using solar wind inputs, the response being dominantly controlled by solar wind speed. Note there is also a connection to southward interplanetary magnetic field (IMF) [*Li et al.*, 2006, 2011], as well as the impact of the *Russell-McPherron* effect [*Russell and McPherron*, 1973] (see, e.g., the analysis in the work of *Miyoshi and Kataoka* [2008]). Since ULF waves are expected to play a role in the transport of electrons into the outer zone, for example, through ULF wave radial diffusion [e.g., *Fälthammar*, 1965; *Schultz and Lanzerotti*, 1974; *Elkington et al.*, 2003; *Fei et al.*, 2006], ULF waves hence provide the basis for a physical mechanism through which the observed strong dependence of outer-zone radiation belt electron flux on solar wind speed might be explained.

Since magnetospheric ULF waves can also be indirectly monitored using networks of ground-based magnetometers, predominantly due to the currents that they drive in the ionosphere, such networks provide an excellent basis for compiling the characteristics of ULF waves over epochs much longer than typical scientific satellite missions and with much denser multipoint latitudinal and longitudinal coverage. The virtual combination of data from multiple arrays into a global supernetwork, e.g., through initiatives such as SuperMAG [*Gjerloev*, 2009] or the Ultra Large

International Terrestrial Magnetometer Array (ULTIMA) initiatives (see, e.g., http://www.serc.kyushu-u.ac.jp/ultima/ultima.html) can provide truly global coverage. This can span all local times around a radiation belt electron drift orbit and across many *L*-shells and which can supplement the in situ single-point coverage provided by individual satellite observations. Figure 1 shows a map of the stations in the Canadian Array for Real-time Investigations of Magnetic Activity (CARISMA) array [*Mann et al.*, 2008] (triangles) as well as selected stations from the Time History of Events and Macroscale Interactions during Substorms Ground-based Observatory (GBO) array [*Russell et al.*, 2008]. Overplotted are illustrative orbit tracks for the two RBSP satellites magnetically traced to 100 km altitude using the *Tsyganenko* [1989, T89] magnetic field model with $Kp = 3$; tick marks along the orbits are 10 minutes apart. As can be clearly seen, the ground arrays provide excellent coverage magnetically to conjugate to the RBSP satellites. In the case of CARISMA [*Mann et al.*, 2008], this now corresponds to a network of 28 fluxgate magnetometers with a standard 1 sample per second data product (raw sampling at 8 samples per second, which is also stored) as well as eight stations comprising additional pairs of induction coils deployed in the horizontal plane (100 samples per second raw sampling). Note that due to the RBSP orbital period, the ground trace of the apogee will drift in local time with respect to the ground stations from orbit to orbit.

Figure 2 shows the variations of the H and D component summed 1–10 mHz ULF wave power derived from the local noon (6 h from 09–15 MLT) sector from the GILL station (see also Table 1) from the CARISMA array and their relation to solar *F*10.7 radio flux and solar wind speed during solar minimum. In this paper, the ascending, maximum, descending, and minimum phases of the solar cycle have been defined based on the quartiles of observed *F*10.7 solar radio flux in the manner described by *Murphy et al.* [2011]. In the case of Figure 2, the early part of this solar minimum period contains a series of 27 day recurrent fast solar wind streams, which characterize the period when both sunspot number and *F*10.7 flux are still decreasing in 1994 and 1995. Figures 2a–2f show the solar wind speed, solar *F*10.7 radio flux in solar flux units, the summed noon sector H component 1–10 mHz ULF wave power (log scale), an overplot of the 3 day average of both the H component ULF power (black, from Figure 2c) and solar wind speed (red; both log scale), the summed 1–10 mHz D component ULF wave power (log scale), and finally an overplot of the 3 day average of both the D component ULF power (black, from Figure 2e) and solar wind speed (red; both log scale). In the plots of the Figure 2, the blue dots show the daily data and the continuous lines show a running 3 day average. The solar wind data used are hourly data, propagated to the magnetosphere, from the

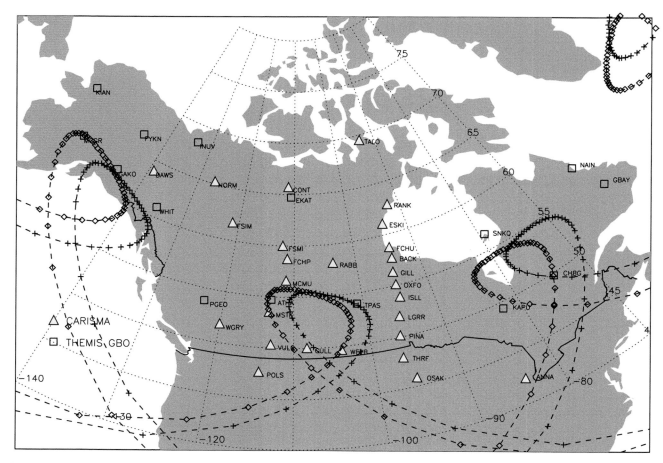

Figure 1. Stations from the Canadian Array for Real-time Investigations of Magnetic Activity (CARISMA) (triangles; www.carisma.ca) and the Time History of Events and Macroscale Interactions during Substorms GBO (squares) arrays. Overplotted are example trajectories for the NASA Radiation Belt Storm Probes satellites mapped to the ground using the *Tsyganenko* [1989] (T89) magnetic field model.

OMNI-2 database [*King and Papitashvili*, 2005]. The same data is used to characterize the solar wind speed throughout the rest of this paper. Very significantly, Figure 2 shows extremely clearly that the solar wind speed and the noon sector ULF wave power in both the H and D components are very strongly correlated and that their logarithms follow each other in the time domain for year after year. This is an extension of the result shown, for example, by *Mathie and Mann* [2000] and *Mann et al.* [2004].

Figures 3 and 4 show a summary of the relationship between the 3 day averaged solar wind speed and H and D component noon sector 1–10 mHz ULF wave power, respectively, from the GILL station for four phases of the solar cycle (for each plot, the solar cycle phase boundaries were defined in terms of the quartiles of the $F10.7$ flux, as in the work of *Murphy et al.* [2011]). In each plot of Figures 3 and 4, i.e., for each solar cycle phase, the traces are shown independently

normalized to span the maximum and minimum range of log solar wind and log ULF wave power. These figures clearly show that the very close relationship between both H and D component ULF wave power at GILL is maintained throughout the whole solar cycle. Table 2 shows the rank-order correlation coefficients (ROCCs) between the 3 day averages of the log of the H and D component 1–10 mHz noon ULF power from GILL and log solar wind speed for these four solar cycle phases and averaged across four different ranges of dayside local times. As is clearly shown, the ROCC are slightly larger in the H than the D component in all local time ranges in Table 2, reaching 0.8 in the H component for some solar cycle phases. The ROCCs are generally the largest at noon and in the morning sector and are lowest at dusk, but even there remain relatively high around 0.6.

Table 3 shows the L dependence of the ROCC between log ULF wave power and log solar wind speed for selected

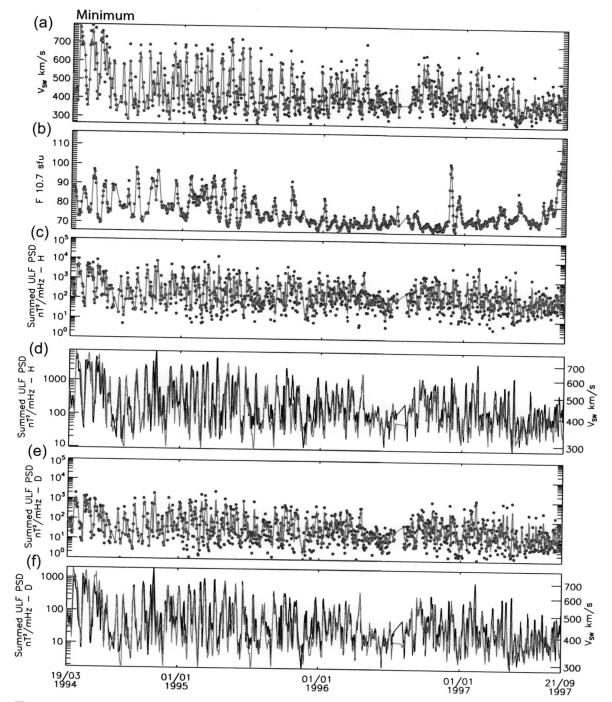

Figure 2. The solar minimum: (a) solar wind speed, (b) solar $F10.7$ radio flux in solar flux units, (c) summed noon sector GILL H component 1–10 mHz ULF wave power (log scale), (d) an overplot of the 3 day average GILL H component ULF power (black, from Figure 2c) and solar wind speed (red; both log scale), (e) 1–10 mHz summed GILL D component ULF wave power (log scale), and (f) an overplot of the 3 day average of both the GILL D component ULF power (black, from Figure 2e) and solar wind speed (red; both log scale). In all plots, the blue dots show the daily data and the continuous lines show a running 3 day average.

Table 1. Magnetometer Stations[a,b]

Stations		Geographic		Corrected Geomagnetic			
Station	Code	Latitude (deg)	Longitude (deg)	Latitude (deg)	Longitude (deg)	L-shell and Ranges	Data Interval Used and Central Year
Fort Churchill	FCHU	58.76	265.91	69.04 (69.36–68.63)	331.81 (330.76–332.81)	7.94 (8.18–7.65)	1997 (1990–2004)
Gillam	GILL	56.38	265.36	66.73 (67.04–66.33)	331.41 (330.41–332.36)	6.51 (6.68–6.30)	1997 (1990–2004)
Island Lake	ISLL	53.86	265.34	64.31 (64.60–63.92)	331.83 (330.89–332.73)	5.40 (5.52–5.26)	1997 (1990–2004)
Pinawa	PINA	50.20	263.96	60.60 (60.86–60.24)	330.27 (329.39–331.11)	4.21 (4.28–4.12)	1997 (1990–2004)
Glenmore Lodge	GML	57.16	3.68	54.31 (54.37–54.26)	84.13 (84.41–83.56)	2.98 (2.99–2.98)	1994 (1987–2002)
York	YOR	53.95	1.05	50.83 (50.93–50.73)	80.73 (80.98–80.20)	2.55 (2.56–2.54)	1994 (1987–2002)

[a]The six stations used in this study, together with their station code, geographic and corrected geomagnetic latitudes and longitudes.

[b]Values are shown corresponding to the middle year of the study, and the range of values is shown in parentheses [see also *Rae et al.*, 2012].

stations (Table 1) from the CARISMA (FCHU, GILL, ISLL and PINA) [*Mann et al.*, 2008] and the UK Sub-Auroral Magnetometer Network (SAMNET) (GML and YOR) [e.g., *Yeoman et al.*, 1990]. The ROCC are calculated for the noon (09–15 MLT) sector ULF power and show that these correlations are the highest in both H and D components at the higher L, but remain high even down to $L = 2.5$ where the ROCC can still reach 0.7 during the descending and minimum phases of the solar cycle (e.g., D component at GML and YOR). Interestingly, at the lowest L-shells during the descending and minimum solar cycle phases, the correlations with the D component become larger than those with the H component. These results are significant since they show that the relationship between ULF wave power and solar wind speed is not only maintained throughout the solar cycle, but also exists deep into the radiation belts at least as far as the nominal inner edge of the outer zone around $L \sim 2.5$. Of course, the fact that these parameters are well correlated on day [cf. *Mann et al.*, 2004] and 3 day averaged time scales (as shown here) demonstrates a very close connection and correlation between both the H and D component ULF waves and solar wind speed and therefore also in solar wind–correlated ULF wave energy input to the magnetosphere. Calculating the ROCC for averages of ULF wave power and solar wind speed determined from longer periods of data (e.g., 27 days (not shown)) retains the high levels of correlation. This can be compared, for example, to the comparisons between the radiation belt response and the plasmapause location on such time scales presented by *Li et al.* [2006, 2009], and the potential implications of this are discussed further in section 4 below.

We note in passing here that the correlation between solar wind speed and geosynchronous relativistic electron flux originally reported by *Paulikas and Blake* [1979] was recently revisited in work by *Reeves et al.* [2011]. Reeves et al. suggest, based on a long and recent epoch of data from

1989 to 2010, that the correlation with solar wind speed is not as clear as Paulikas and Blake originally suggested. Specifically, *Reeves et al.* [2011, p. 1] report that scatterplots of the Vsw-flux relationships tend to show evidence for a triangular distribution, such that the "fluxes have a distinct velocity-dependent lower limit but a velocity-independent upper limit" and that the "highest electron fluxes can occur for any value of Vsw with no indication of a Vsw threshold". On other hand, recent analysis of the geosynchronous flux variations by *Kellerman and Shprits* [2012] using 20 years of data have shown that when the distributions are normalized by solar wind speed occurrence into probability distribution functions, with the well-known and appropriate 2 day flux lag included, then a clear correlation between relativistic electron flux enhancements and solar wind speed is again reproduced. The *Kellerman and Shprits* [2012] analysis hence verifies that the *Paulikas and Blake* [1979] conclusion remains robust.

A similar analysis by *Li et al.* [2011] of the relationship between MeV electrons and solar wind parameters also found a close connection between MeV electron acceleration events and fast solar wind speeds, especially during the declining phase of the solar cycle. However, these authors also highlighted a strong connection to southward IMF. In the *Li et al.* [2001] model for MeV electron flux at geosynchronous orbit, "removing the southward component [of the IMF] from the model and relying only on the solar wind speed (with a readjustment of the parameters) only slightly degrades the performance of the model". However, in discussing these results *Li et al.* [2011, p. 9] suggest that the observed correlation between MeV electron enhancements and solar wind speed might be explained by a correlation between times of (fluctuating and hence) southward IMF and fast solar wind speed. Of course a full explanation for these correlations may only be possible once the dominant acceleration processes, and their detailed relationships to solar wind drivers, are understood.

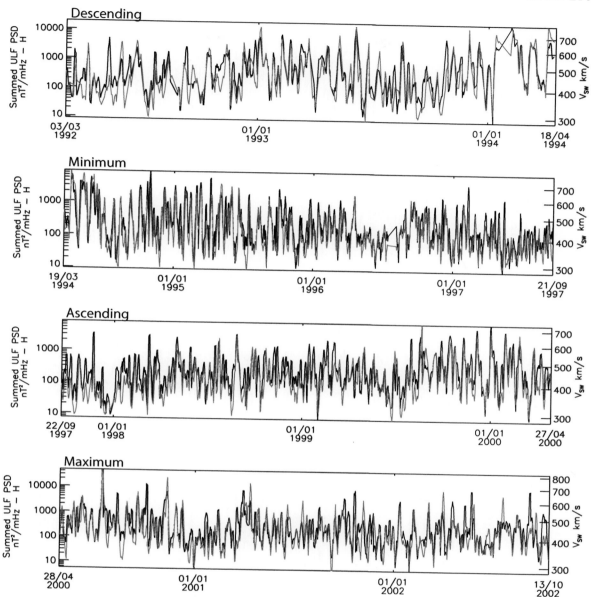

Figure 3. Three day averaged solar wind speed (red, right axis) and H component noon sector 1–10 mHz ULF wave power (black, left axis) from GILL for four phases of the solar cycle (for each plot, the solar cycle phase boundaries were defined in terms of the quartiles of the $F10.7$ flux, as given by *Murphy et al.* [2011]). The traces are independently normalized to span the maximum and minimum range of log solar wind and log ULF wave power in each plot.

2.2. Internal Energetic Particle Drivers

ULF waves can also be excited by energetic ring current ions via free energy available in distribution function-positive energy gradients [cf. *Hughes et al.*, 1978], or spatial gradients [e.g., *Southwood et al.*, 1969], through the processes of drift or drift-bounce resonance [e.g., *Southwood et al.*, 1969; *Southwood*, 1976; *Southwood and Kivelson*, 1982]. Under certain conditions, these energization processes can also be coupled to free energy sources from plasma pressure spatial gradients through ballooning and drift-ballooning modes [e.g., *Cheng and Qian*, 1994]. Such processes are thought to excite storm time high azimuthal wave number (m) modes, which are predominantly poloidally polarized, and which have maximum occurrence in the afternoon local time sector along ring current ion injection paths especially during magnetic storms

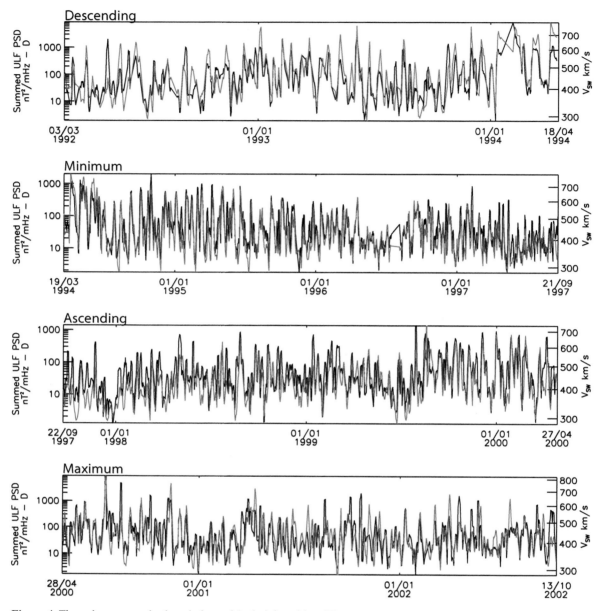

Figure 4. Three day averaged solar wind speed (red, right axis) and D component noon sector 1–10 mHz ULF wave power (black, left axis) from GILL for four phases of the solar cycle (same format as Figure 3).

(e.g., the "storm time Pc5 waves" described in analysis of AMPTE/CCE satellite data by *Anderson et al.* [1990]). Such poloidal wave modes, especially if they are excited with only moderate *m* values, can potentially be efficient energizers of radiation belt electrons since their electric fields are polarized with a significant component parallel to the direction of radiation belt electron drift velocity (see, e.g., the polarization discussion in the work of *Mann et al.* [1997]). This is in contrast to toroidal mode ULF waves, which have lower *m*

values and which are polarized such that their electric fields are largely polarized perpendicular to the electron drift velocity [e.g., *Southwood*, 1974].

Poloidally polarized modes can interact with MeV electrons through the dominant symmetric drift resonance; in contrast, toroidal modes can only interact through the asymmetric ($m \pm 1$) resonances, which occur once compression of the background magnetic field in the magnetosphere by the incident solar wind is taken into account [*Elkington et al.*,

Table 2. ULF Wave Power–Solar Wind Speed Correlations at GILL[a,b]

		1 Jan 1990 to 1 Jan 2006	Descending 3 Mar 1992 to 18 Mar 1994	Minimum 19 Mar 1994 to 21 Sep 1997	Ascending 22 Sep 1997 to 27 Apr 2000	Maximum 28 Apr 2000 to 13 Oct 2002
Morning	H	0.69	0.69	0.71	0.65	0.56
3–9 MLT	D	0.60	0.58	0.64	0.58	0.47
Noon	H	0.78	0.80	0.80	0.74	0.72
9–15 MLT	D	0.75	0.76	0.78	0.70	0.68
Dusk	H	0.65	0.58	0.69	0.60	0.56
15–21 MLT	D	0.59	0.55	0.62	0.56	0.49
Dayside	H	0.76	0.79	0.79	0.73	0.68
6–18 MLT	D	0.70	0.71	0.76	0.67	0.58

[a]Rank order correlation coefficients between log 1–10 mHz H and D component ULF wave power from GILL and log solar wind speed, for the period 1 January 1990 to 1 January 2006, as well as four phases of the solar cycle, in four different magnetic local time (MLT) sectors.

[b]See Figures 3 and 4. For each local time sector, the mean of the ULF power from each local time range was computed.

2003]. Consequently, the electric fields from poloidal modes will likely be more efficient in effecting radiation belt electron energy change than toroidal modes, especially under conditions of comparable wave amplitude.

As shown in Figures 2, 3, and 4, there is a close connection between solar wind speed and ULF wave power generated in both the H and D components. Under the action of a 90° perpendicular polarization rotation, which occurs upon transmission of an Alfven wave through a uniform ionosphere [e.g., *Hughes and Southwood*, 1976a, 1976b], the H component would correspond to a predominantly toroidal mode, while the D component might correspond to a predominantly poloidal mode in space (note that satellite data typically shows that even coherent waves typically have a somewhat

Table 3. ULF Wave Power–Solar Wind Speed Correlations With Solar Cycle Phase[a,b]

		1 Jan 1990 to 1 Jan 2006	Descending 3 Mar 1992 to 18 Mar 1994	Minimum 19 Mar 1994 to 21 Sep 1997	Ascending 22 Sep 1997 to 27 Apr 2000	Maximum 28 Apr 2000 to 13 Oct 2002
FCHU	H	0.77	0.82	0.79	0.75	0.70
$L = 7.94$	D	0.74	0.78	0.77	0.71	0.66
GILL	H	0.78	0.80	0.80	0.74	0.72
$L = 6.51$	D	0.75	0.76	0.78	0.70	0.68
ISLL	H	0.74	0.75	0.77	0.71	0.67
$L = 5.4$	D	0.71	0.73	0.73	0.68	0.63
PINA	H	0.72	0.74	0.71	0.67	0.68
$L = 4.21$	D	0.69	0.74	0.70	0.66	0.61
GML	H	0.66	0.64	0.66	0.68	0.62
$L = 2.98$	D	0.66	0.70	0.71	0.70	0.54
YOR	H	0.66	0.62	0.65	0.69	0.64
$L = 2.55$	D	0.63	0.71	0.72	0.68	0.55

[a]Rank order correlation coefficients between log 1 and 10 mHz H and D component ULF wave power and log solar wind speed, for the period 1 January 1990 to 1 January 2006, as well as for four phases of the solar cycle for selected stations from the CARISMA and SAMNET arrays for the noon (09-15 MLT) sector.

[b]See Figures 3 and 4.

mixed perpendicular in situ polarization in the magnetosphere) [e.g., *Clemmons et al.*, 2000; *Rae et al.*, 2005]. Hence, the observations in Figures 2, 3, and 4 suggest that solar wind correlated power in both polarizations likely exist in the magnetosphere.

Note that the mapping of ground magnetometer ULF wave power in the equatorial magnetosphere must be done carefully [see, e.g., *Ozeke et al.*, 2005, 2009, and references therein]. This mapping from the ground, through the ionosphere, to the equatorial magnetosphere can also be sensitive to the structure of ionospheric conductivity, especially gradients [e.g., *Glassmeier*, 1984], and this will affect this transmission and mapping process. Nonetheless, ULF wave excitation processes, which are correlated with solar wind speed, which might include both external (direct solar wind driving) and internal (resonance with energetic ion) excitation mechanisms, provide a plausible physical mechanism for explaining at least some of the observed correlation between solar wind speed and MeV electron response in the outer radiation belt.

3. ULF WAVE-DRIVEN RADIAL TRANSPORT AND RADIAL DIFFUSION

As described above, ULF waves can provide the fluctuation fields which can drive the radial transport of MeV energy electrons in the outer zone radiation belt. The standard picture of radial diffusion postulates changes in the energy of the particles as a result of their interaction with wavefields. The energy exchange between the particles and the wave can be considered to arise from the work done by the fields. Equation (1) below, from *Northrop* [1963], shows the energy change experienced by a charged particle of (relativistic) first adiabatic invariant M, charge q, and relativistic correction factor γ in the presence of electric and magnetic fields

$$\frac{dW}{dt} = q\vec{E} \cdot \vec{v_d} + \frac{M}{\gamma}\frac{\partial B}{\partial t} \qquad (1)$$

[see also, e.g., *Ozeke and Mann*, 2008]. The first term represents the work done by the electric field, and the second term the effects of inductive magnetic field changes. As the particles interact with the wave, their first adiabatic invariant will be conserved if the time scale of the interaction (for example, the wave period) is long compared to the particles' gyroperiod. In general, acceleration (or deceleration) due to the action of the wavefields doing positive (or negative) work on the particle leads to their inward (or outward) radial transport toward regions of higher (lower) magnetic field. As a result of the interaction with a wave packet of narrow frequency and narrow azimuthal wave-

number spectrum, individual particles can be accelerated or decelerated depending upon the phase of the wave which they interact with. A nice illustration of this, and the resulting inward and outward radial transport, accompanied by azimuthal drift phase bunching, is presented by *Elkington et al.* [2003]. Note that azimuthal localization of ULF wave power in a specific local time sector can also be considered to represent an effective broadening of the m value spectrum through the form of the envelope, even if the phase advance of the wavefield with azimuth is well-described by a single m value [e.g., *Degeling et al.*, 2007].

In order for an electron to exhibit a significant and continuous change in its energy, it is required to be in resonance with the wave. For 90° pitch angle particles, this requires the particles to be in drift resonance such that

$$\omega = m\omega_d,$$

where ω is the wave frequency, ω_d is the particle drift frequency, and m is the (dimensionless) azimuthal wave number. If the particle exhibits bounce motion, with an equatorial pitch angle less than 90°, the drift-bounce resonance condition

$$\omega - m\overline{\omega}_d = N\omega_b$$

must be satisfied, where ω_b is the bounce frequency, and $\overline{\omega}_d$ is the bounce-averaged drift frequency [e.g., *Southwood et al.*, 1969; see also *Southwood and Kivelson*, 1982; *Mann and Chisham*, 2000]. Such resonances represent the most efficient basis for the interaction between ULF waves and energetic radiation belt particles, interactions allowing energy changes with single wave packets, which increase proportional to time.

Note that finite length ULF wave trains can produce interesting and nonmonotonic radial profiles of phase space density (PSD), since particles can be transported along Liouville trajectories in L (or alternatively energy) and $(m\varphi - \omega t)$ space for a finite time. For example, such coherent interactions can generate local peaks in PSD as preexisting radial gradients in PSD are transported and mixed around the resonance islands, the individual particle trajectories from single-particle simulations tracing paths of constant PSD according to Liouville's theorem [e.g., *Degeling et al.*, 2008]. In such a case, the particles will circulate around resonance islands, but the phase mixing, which develops during these periodic trajectories around the resonance island, is terminated after perhaps only a few periods due to the finite length of the perturbing ULF wave train. Such single wave resonance interactions can also be understood and examined using Poincaré maps [e.g., *Elkington et al.*, 2003].

In the magnetosphere, the signatures of flux modulation due to these relatively coherent interactions with a coherent single ULF wave packet may be observable with satellites (cf. I. R. Mann et al., Coherent ULF wave acceleration in the Earth's radiation belts, manuscript in preparation, 2012, hereinafter referred to as Mann et al., manuscript in preparation, 2012). As discussed by E. A. Lee et al. (Energetic electron and Pc5 ULF wave interactions during great geomagnetic storms, manuscript in preparation, 2012, hereinafter referred to as Lee et al., manuscript in preparation, 2012), similar modulations of energetic particle flux can also be generated in the frame of the satellite due to the advection of (typically) radial spatial gradients of particle flux across the satellite. This could also be periodic, for example, in the case of the advection of flux tubes across the satellite due to a large-amplitude long-period coherent ULF wave. The active resonant acceleration of the electrons by an individual ULF wave packet can be distinguished from a simple ULF wave advection by the observation that, in general, the advection process should produce flux modulations which are in phase across different energy channels since particles of all energies are all largely being advected by the motion of the flux tube in the same way. Conversely, in the case of an active drift resonance, there will be a 180° phase shift in particle flux across the resonant particle energy (e.g., *Mann et al.*, manuscript in preparation, 2012; Lee et al., manuscript in preparation, 2012).

Of course, as a particle is accelerated or decelerated, it will gradually drift out of resonance such that, in general, a spectrum of waves is needed. Although note that two sufficiently close resonance islands can overlap for waves of sufficiently large amplitude leading to a transition to a more chaotic scattering of particle trajectories [e.g., *Elkington et al.*, 2003; *Degeling et al.*, 2007]. Such behavior can be considered to be a step toward a more stochastic transport process such as radial diffusion. Since the fastest energy transfer occurs through resonant wave-particle interactions, in our opinion, it remains instructive to think about such stochastic processes in terms of a superposition of a series of multiple, but phase decorrelated, resonant interactions, which act on the particle distributions. How does this process relate to radial diffusion? The answer is, of course, that the individual wave packet interactions with an azimuthally distributed ensemble of particles will result in both inward and outward transport (and acceleration and deceleration, respectively) due to the action and work done by the wavefields on the particles. In a flat initial radial distribution, there would be no net ensemble energization since equal numbers of particles move inward and outward. However, in a preexisting radial gradient, with higher PSD at higher L, the net result is that more particles are accelerated than decelerated, resulting in an ensemble energization of the distribution. This is in essence the basis of radial diffusion.

As described by *Fälthammar* [1965] and *Schulz and Lanzerotti* [1974], radiation belt particles interacting with long-period electric and magnetic perturbations in the a dipole magnetosphere can be described through a 1-D radial diffusion equation, which can be expressed in terms of L-shell by the equation

$$\frac{\partial f}{\partial t} = L^2 \frac{\partial}{\partial L}\left[\frac{D_{LL}}{L^2}\frac{\partial f}{\partial L}\right] - \frac{f}{\tau}.$$

Here f represents the PSD of the electrons, and it is assumed that the first and second adiabatic invariants, M and J, are conserved [see *Schulz and Lanzerotti*, 1974], D_{LL} is the diffusion coefficient, and τ is a loss time scale. Typically, the radial diffusive transport is specified through the form of the diffusion coefficients, which are typically separated into an electrostatic and an electromagnetic term (the latter of which includes the effects of inductive electric fields). For example, *Brautigam and Albert* [2000] specified these as a function of Kp based on an assimilation of various space- and ground-based fluctuation power spectra and assumptions about the inherent L dependence of the radial diffusion coefficients in background dipole field following the approach of *Schulz and Lanzerotti* [1974]. Similarly, *Brautigam et al.* [2005] used in situ CRRES electric field observations to specify the form of electric field diffusion as a function of Kp.

More recently, *Brizard and Chan* [2001] [see also *Fei et al.*, 2006] derived a different approach to the specification of magnetic and electric diffusion coefficients, D_{LL}^E and D_{LL}^B, which captures the functional form of these coefficients multiplied by either electric field or compressional magnetic field ULF wave power spectral densities, respectively. In recent work, *Rae et al.* [2012] used more than a solar cycle of ground-based magnetometer data to specify the spectral shape, and the dependence on either solar wind speed or Kp, of the magnetic ULF wave power as seen on the ground. Mapping this ground-based magnetic ULF wave power to the equatorial plane using an Alfven eigensolution [*Ozeke et al.*, 2009] allows an estimate of the equatorial electric field, which compares very favorably to the in situ measurements from CRRES from *Brautigam et al.* [2005]. *Ozeke et al.* [2012] then took these results, together with a statistical analysis of in situ compressional magnetic field power input to the *Brizard and Chan* formalism, to specify new ULF wave power–based formulae for D_{LL}^E and D_{LL}^B as a function of solar wind speed or Kp. Note that implicit in this formalism is the assumption that the electric field and compressional magnetic field ULF wave powers are not correlated, which would be appropriate for decoupled Alfven and fast magnetoacoustic modes, respectively. In general, however, this is unlikely to be true since the electric and magnetic fields may be coupled through Faraday's law.

It is appealing to consider the assumption that the ULF fluctuations, which perturb the trajectories of the radiation belts electron, have the form of fundamental mode field-aligned harmonics. Under such an assumption decoupled Alfvenic modes have an equatorial node in magnetic field and an antinode in electric field, justifying the use of mapped ground-based magnetic fields to specify purely electric field perturbations in the equatorial magnetosphere as an input to parameterized D_{LL}^E [cf. *Ozeke et al.*, 2009]. Similarly, decoupled fast mode compressional perturbations would have an antinode in magnetic field in the equatorial plane enabling their power to be incorporated into D_{LL}^B. If these two modes are uncorrelated, then the overall diffusive response can be calculated using a diffusion coefficient, which is the sum of both D_{LL}^E and D_{LL}^B [see *Ozeke et al.*, this volume]. In reality, in terms of observed electric and magnetic field power in the magnetosphere, it is probable that the modes are not only coupled but also that a fraction of the electric field wave power is inductive. Future work is required to establish the size of this effect and its impact in the possible overestimation of the combined diffusion coefficients. In a companion paper, *Ozeke et al.* [this volume] examine these further and present examples of the diffusion coefficients based on the *Ozeke et al.* [2012] approach. Significantly, *Ozeke et al.* [2012] conclude that the electric diffusion D_{LL}^E term likely dominates over the magnetic diffusion term. This can be contrasted to the conclusion reached by *Brautigam and Albert* [2000] in their formalism that the electromagnetic diffusion term dominated over the electrostatic one; the electric field diffusion, which dominates the *Ozeke et al.* [2012] diffusion model, is also around an order of magnitude slower than the (faster electromagnetic) diffusion estimated by *Brautigam and Albert* [2000]. See the work of *Ozeke et al.* [this volume] for further discussion.

Overall, the high correlation between solar wind speed and ULF wave power presented in section 2 above validates the link between these two parameters. More importantly, it further highlights the potential value of using either a statistical representation of ULF wave power, especially as a function of solar wind speed, or a time series of measured ULF powers during a magnetic storm interval, as an input to the specification of model radial diffusion coefficients. Moreover, since the *Ozeke et al.* [2012] [see also *Ozeke et al.*, this volume] approach suggests that D_{LL}^E is dominant, then radial diffusion simulations could perhaps be driven using only estimates of equatorial electric field power. Such

wave power estimates can be derived across *L*-shells using the continuous coverage of ULF waves, which is available from multiple stations in ground-based magnetometer arrays, but which is not, in general, available from in situ satellite data. Note that since the fast mode dispersion relation is, in general, controlled by the wave numbers in all three directions, rather than only in the field aligned direction as in the case of Alfven waves, it is much more difficult to derive an appropriate method for estimating the equatorial compressional magnetic ULF wave power using only ground-based magnetometer data inputs.

It is also interesting to compare to the results presented by *Mann et al.* [2004] who examined the relationship between 1–10 mHz ULF wave power, not only to solar wind speed but also to ~MeV energy electron flux both at geosynchronous orbit and at $L = 5.5$ and $L = 4.5$ from the HEO satellites. Figure 5, reproduced from *Mann et al.* [2004], shows the details of this relationship from 1991 to mid-2000. Figure 5a shows the yearly running average daily geosynchronous 1.8–3.5 MeV differential flux from the Los Alamos satellites, as well as the integrated >1.5 MeV flux from the HEO satellites from $L = 4.5$ and $L = 5.5$. Figure 5b shows the yearly running average of the solar wind speed (bold line) and the dawn sector 1–10 mHz ULF wave power from ground magnetometer stations at Sørøya (SOR, $L = 6.66$, from the IMAGE array [*Viljanen and Hakkinen*, 1997]; X component, top line), Faroes (FAR, $L = 4.27$, SAMNET array; H component multiplied by 4, middle line), and GML (GML, $L = 3.08$, SAMNET array; H component multiplied by 8, bottom line). Note ULF power in the 1–10 mHz ULF range is computed from a 2 h FFT tapered with a Hanning window. Figure 5c shows the monthly and yearly average sunspot number demonstrating that the data presented cover essentially an entire solar cycle. Figure 5d shows the yearly running ROCC of the daily GEO differential (with 2 day lag) and HEO integral ($L = 4.5$ and $L = 5.5$, both with a 3 day lag) fluxes from Figure 5a with solar wind speed. Finally, Figure 5e shows the yearly running ROCC of the daily dawnside 1–10 mHz ULF wave power from the SOR station ($L = 6.66$) with the fluxes at GEO, and the fluxes at $L = 4.5$ and $L = 5.5$ (see *Mann et al.* [2004] for more details about the data sources and analysis), as well as with solar wind speed. Note that the correlations with daily GEO flux peak with a 2 day lag, while those using the HEO integral flux data from lower *L*-shells at $L = 4.5$ and $L = 5.5$ both peak with a 3 day lag. This indicates that ULF wave (and solar wind speed)-correlated flux enhancements are associated with an inward transport of information, and most likely also an inward transport of energy, which would be expected under the action of ULF wave-driven inward radial diffusion. Figure 5 also shows that there is a good correlation between the ULF wave power and the lagged

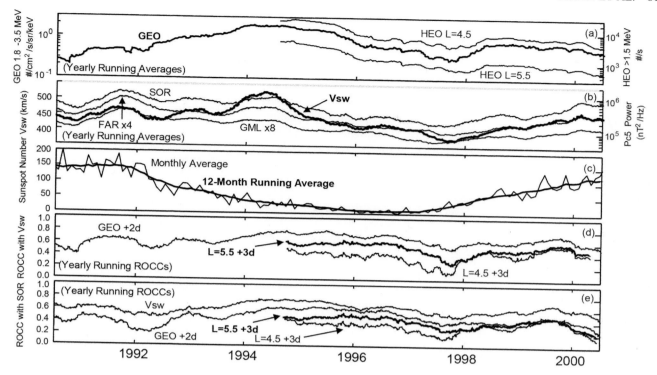

Figure 5. (a) Yearly running average daily geosynchronous 1.8–3.5 MeV differential flux from the Los Alamos satellites, as well as the integrated >1.5 MeV flux from the HEO satellites from $L = 4.5$ and $L = 5.5$. (b) Yearly running average of the solar wind speed (bold line) and the H component dawn sector 1–10 mHz ULF wave power from ground magnetometer stations at SOR ($L = 6.66$; X component, top line), FAR ($L = 4.27$; H component multiplied by 4, middle line), and GML ($L = 3.28$; H component multiplied by 8, bottom line). (c) The monthly and yearly average sunspot number. (d) The yearly running rank-order correlation coefficient of the daily GEO (with 2 day lag) and HEO ($L = 4.5$ and $L = 5.5$, both with 3 day lag) differential and integral fluxes with solar wind speed. (e) The yearly running rank-order correlation coefficient of the daily dawnside 1–10 mHz ULF wave power from SOR ($L = 6.66$) with the fluxes at GEO and $L = 4.5$ and $L = 5.5$, and the yearly daily average solar wind speed Vsw. Reproduced from *Mann et al.* [2004] with kind permission from Elsevier.

radiation belt fluxes, the latter being most intense during the declining phase of the solar cycle. These observations are consistent with a potentially causal role for inward transport and acceleration of radiation belt electrons through ULF wave-particle interactions.

4. ROLE OF THE PLASMAPAUSE

Recent work by *Li et al.* [2006] using data from the Solar, Anomalous, and Magnetospheric Particle Explorer (SAMPEX) satellite extended the initial ideas of *Tverskaya* [1986] [see also *Tversyka et al.*, 2003; *Tverskaya*, 1996] to show clearly that the inner edge of the radiation belts follows the storm time disturbance index *Dst* [*Sugiura*, 1964] (see also the discussion in the work of *O'Brien et al.* [2003]). If the location of the plasmapause is also correlated with *Dst*, then the plasmapause and inner edge of the outer radiation belts might track each other. For example, *Li et al.* [2006]

show that the 10 day average minimum L of the plasmapause (Lpp, based on an empirical model of *O'Brien and Moldwin* [2003]) follows the inner edge of the (monthly window averaged) 2–6 MeV electron radiation belt as observed by SAMPEX. This suggests that the plasmasphere is exerting a strong influence on the electron radiation belt: specifically its penetration to low L. This is rather remarkable since the radiation belt electrons have energies 6 orders of magnitude larger than the particles in the plasmasphere! This could be explained by enhanced losses, which are expected inside the plasmasphere, for example, as a result of scattering loss through interactions with plasmaspheric hiss [e.g., *Lyons et al.*, 1972; *Lyons and Thorne*, 1973]. Interestingly, however, the dynamical erosion and refilling of the plasmasphere and the consequent motion of the plasmapause and/or the presence of a heavy ion torus outside the plasmapause [e.g., *Fraser et al.*, 2005] will also change the radial profiles of the Alfven speed in the magnetosphere. Such radial profiles of

Alfven speed will also influence the extent to which ULF waves, which might perturb the radiation belts, can penetrate to such low L-shells [cf. *Loto'aniu et al.*, 2006; see also *Kale et al.*, 2009] offering a potential causal role for ULF waves in contributing to the correlation between the plasmapause position and the inner edge of the outer zone electron belt. Of course, it could also be possible that both the plasmapause location and the inner edge of the outer zone radiation belts are themselves each correlated with *Dst*, but that the processes responsible for each are distinct such that they are not directly causally linked to each other.

Specifically, in relation to the role of ULF waves in the transport of MeV electrons to low L-shells, *Loto'aniu et al.* [2006] studied the penetration of MeV electrons into the slot during the beginning of the 2003 Halloween storm period on 29 October 2003. *Loto'aniu et al.* [2006] showed enhanced penetration of ULF wave power deep into the magnetosphere down to $L < 2$, with sufficient power to be able to explain the initial electron transport into the slot on the 29–30 October 2003 [see also *Baker et al.*, 2004]. Modeling of the event by *Li et al.* [2009] further suggests, using a radial diffusion model, that ULF waves could have been responsible for the slot filling during the Halloween storms. Significantly, during this period, the L-shell profile of the Alfven eigenfrequency was significantly reduced compared to quiet times [*Loto'aniu et al.*, 2006; *Kale et al.*, 2009] over a time period of only several hours following storm onset. Given the low L extent of the Alfven frequency reduction, the obvious explanation offered by Loto'aniu et al. was that heavy O+ ions had been injected into the equatorial magnetosphere from the ionosphere. Overall, these works suggest that low-energy (eV) and very high energy (MeV) particle populations can be connected via the intermediary of ULF waves.

An interesting question is whether the ULF wave power penetration to low L also follows the location of Lpp, i.e., is it strongly related to *Dst*, in a manner similar to the *Li et al.* [2006] connection between Lpp and the inner edge of the outer zone electron radiation belt. If it does, it suggests that ULF waves may perhaps have an ongoing and active role to play in controlling the low L penetration of the radiation belts. Figure 6 shows the relationship between the noon sector (9–15 MLT) 1–10 mHz H and D component ULF wave power (in the form of a daily value of a 3 day average) from six selected ground-based magnetometer stations, spanning L-shells from $L = 2.5$ to $L = 7.9$. The top plot shows the logarithm of this daily noon sector H component ULF wave power from the six stations from Table 1. The second plot shows an interpolated color contour of the logarithm of the H component ULF wave power from the top plot, derived from these six stations, as a function of L

and time. Overplotted in the second plot are daily lowest *Dst* (black) and daily lowest Lpp (purple) (both dots) and their 3 day average (solid lines). The third and fourth plots show the results for the D component, in the same form as the top and second plots, respectively.

Immediately obvious is both the enhancement and low L penetration of ULF wave power during the storms, especially during the periods characterized by reductions in *Dst* (which typically represent the main phases of the storm events). In Figure 6, when *Dst* gets below -30 nT, this seems to be a good approximate boundary delineating periods of enhanced ULF wave power and low L ULF wave power penetration. Overall, throughout the calendar year of 1999, it is clear that the ULF power in both the H and D components penetrates to lower L during the periods of low (and especially decreasing, i.e., storm main phase) *Dst*. This is true both for individual storms, as well as through the year on longer monthly time scales; for example, the power is at higher L-shells during the middle of the year compared to both the beginning and end of 1999. Note that there is one obvious exception, on 30 July 1999 (marked with a red triangle in Figure 6). At this time, the power in the Canadian stations (PINA to FCHU; $L = 4.2$ to $L = 7.9$) is enhanced, but a similar enhancement is not seen at the lower-L European stations (GML and YOR). Since the data from the Canadian stations is taken from a different MLT sector, their noon sector power does not come from the same UT. During this activation, there appears to be significant power at $L > 4$ in the Canadian sector, but similar power enhancements are not seen in the lower-L European stations from around 7 h earlier when they were at local noon. This suggests that this day was characterized by some relatively short-lived ULF wave power activation, which had a time scale significantly less than 1 day in duration, such that it was seen in the Canadian noon sector but not the European noon sector on this day. Nonetheless, the relationship between the ULF wave power penetration and decreases in *Dst*, and hence lower Lpp, seems to be very clear and robust from storm to storm in this data set. Significantly, this suggests that ULF wave power penetration to low L may have a role to play in explaining the link between the inner edge of the radiation belt and the plasmapause. Further work examining this relationship is required.

Interestingly, *Ozeke and Mann* [2008] recently examined the possibility that energy from ring current ions could be transferred to radiation belt electrons via the intermediary of ULF waves and found an important role for the plasmasphere in this potential energy exchange as well. They examined the interesting possibility that free energy from ring current ions could internally excite eastward propagating Pc4-5 ULF waves, which themselves could then be resonant with radiation belt electrons. The overall response, in effect, represents an energization of radiation belt electrons from the much larger reservoir of ring current ion energy during magnetic storms. *Ozeke and Mann* [2008] concluded that this was possible for intermediate m value waves and that the preferred conditions for this to operate in the heart of the outer zone radiation belt were a depleted plasmasphere and a low plasmatrough density outside the plasmapause. The energetic ring current ion drift-bounce resonances, which could generate the ULF waves required to be additionally drift resonant

with ~1–10 MeV radiation belt electrons, were ~keV H+ or several hundreds of keV O+ ions. The latter of these two ion populations is known to be prevalent in the form of an energetic oxygen torus, which exists outside a depleted plasmasphere during magnetic storms [e.g., *Krimigis et al.*, 1985; *Fraser et al.*, 2005]. Consequently, it is interesting to note that there is the further possibility to tap ring current ion energy to feed radiation belt flux enhancements via the intermediary of high-m ULF waves. Significantly, these

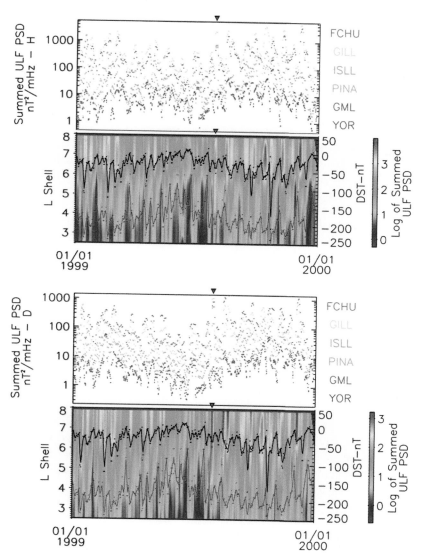

Figure 6. Daily noon sector (9–15 magnetic local time) 1–10 mHz H (top two plots) and D component (bottom two plots) ULF wave power from six selected ground-based magnetometer stations spanning L-shells from $L = 2.5$ to $L = 7.9$ for calendar year 1999. The first and third plots show the daily H and D component power, respectively. The second and fourth plots show interpolated color contours of the logarithm of the H and D component ULF wave power, respectively, as a function of L and time, overplotted with daily minimum *Dst* (black) and daily lowest L of the plasmapause (purple) (both dots) and their 3 day average (solid lines).

interactions appear to be more favorable during conditions characterized by a depleted plasmasphere in the presence of several hundreds of keV O+ ions in a torus outside the depleted plasmasphere, exactly the conditions that are known to occur during magnetic storms.

5. ENHANCED ULF WAVE MAGNETOPAUSE SHADOWING

Radiation belt loss is also of critical importance for controlling radiation belt dynamics and morphology since its effects compete with those of acceleration processes [see, e.g., *Millan and Thorne*, 2007]. Numerous radiation belt radiation loss processes have been proposed, including scattering into the loss cone by a variety of plasma wave modes such as electromagnetic ion cyclotron (EMIC) waves [e.g., *Summers and Thorne*, 2003], chorus waves [e.g., *Horne and Thorne*, 2003], and plasmaspheric hiss [e.g., *Lyons et al.*, 1972], which may also be related to penetration of chorus elements into the plasmasphere [*Bortnik et al.*, 2008]. Compression of the magnetopause to low-L values can also cause losses where drift paths intersect with the magnetopause [e.g., *Li et al.*, 1997], and this loss can be enhanced through fully adiabatic outward transport due to the *Dst* effect [e.g., *Kim and Chan*, 1997]. These magnetopause shadowing losses can also be enhanced on radial diffusion time scales by the action of outward ULF wave-driven radial diffusion [e.g., *Loto'aniu et al.*, 2010], which can be especially fast on the sharp PSD gradients and which may be created by recent and proximal magnetopause shadowing [e.g., *Turner et al.*, 2012].

One aspect of the influence of ULF waves upon loss, which has, up to now, been essentially ignored, is the possibility for coherent, and nondiffusive, loss of radiation belt electrons due to ULF waves. Flux tubes oscillating under the action of poloidal ULF waves will advect in and out with the radial wave displacement amplitude local to the wave; since MeV energy electrons will make a complete drift orbit on a similar time scale under the action of gradient-curvature drift, and are not advected by the wave, they can encounter the magnetopause and, hence, be lost from magnetopause shadowing at a radius significantly inside the equilibrium magnetopause position. This also means that at the time when large-amplitude ULF waves are present, magnetopause location models based on equilibrium magnetosphere configurations [e.g., *Kim et al.*, 2008] will not represent good field models in which to trace and estimate the full extent of shadowing losses to the magnetopause.

Fundamental field-aligned ULF wave harmonics have an antinode in displacement in the equatorial plane. This suggests that during dynamic times, especially, for example,

following sudden increases (or decreases) in dynamic pressure and in the regions where the radiation belt electron drift paths are closest to the magnetopause, this ULF wave advection can act to enhance magnetopause shadowing losses. These periodic encounters with the magnetopause will erode parts of the closed electron drift paths even for drift paths that do not intersect with the predicted equilibrium magnetopause location. In addition to these enhanced ULF wave magnetopause shadowing losses, which occur on wave period time scales, further losses on the L-shells even deeper inside the equilibrium magnetopause can also occur on longer time scales through the action of outward radial diffusion.

Figure 7 shows plots of the (left) median power spectral density of the azimuthal electric field in the equatorial plane of the magnetosphere as a function of Kp, derived from the D component of around 15 years of ground-based magnetometer data from the GILL and FCHU stations in the CARISMA array [*Rae et al.*, 2012] (see Table 1). The D component is used as it is assumed to be rotated by 90° through the ionosphere into a poloidal mode in the magnetosphere and is mapped to the equatorial plane using the Alfvenic approximation of *Ozeke et al.* [2009]. The middle plot shows the median electric field amplitude inferred from the median magnetic power in the left plot under the same standing Alfven eigenmode approximation. Finally, the right plot shows the equatorial radial displacement spectrum derived from the electric field in the middle plot under the assumption that the amplitude at each frequency corresponds to a fundamental mode standing Alfven wave oscillating with that discrete frequency. Each plot also includes a fit to a simple functional form of the Kp dependence of the power spectral density in the left plot as presented by *Ozeke et al.* [2012]. As can be clearly seen, the median displacement for $Kp = 9$ from this analysis is expected to produce displacements $\sim\pm1\,R_E$ and $\sim\pm0.5$–$0.6\,R_E$ at frequencies around 1 mHz on field lines that map to the ground magnetometer stations at FCHU and GILL, respectively (note, of course, that the mapping of the field lines from these stations to the magnetosphere will vary with geomagnetic activity and solar wind parameters, dynamic pressure, perhaps, being especially important in this context). Such frequencies around 1 mHz are expected to be close to the natural eigenfrequencies of field lines close to the magnetopause [e.g., *Mathie et al.*, 1999], although Figure 7 shows that higher-frequency modes around a few mHz would be expected to, in general, have lower statistical displacements.

The upper and lower quartiles have power spectral densities which are increased/decreased from the median by a factor ~ 3.5 (not shown). Consequently, the expected radial displacements would be increased by a factor $<\sim 2$ by

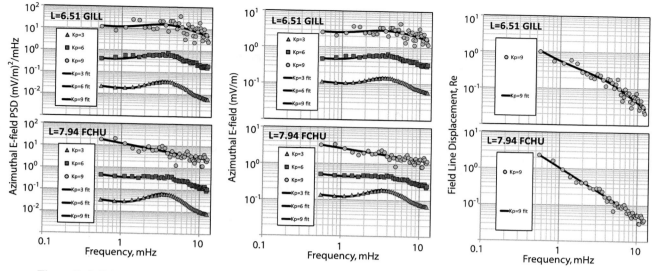

Figure 7. (left) Median power spectral density of the azimuthal electric field in the equatorial plane of the magnetosphere as a function of Kp, derived from the D component of around 15 years of ground-based magnetometer data from the GILL and FCHU stations in the CARISMA array [*Rae et al.*, 2012] (see Table 1). (middle) Median electric field amplitude inferred from the median magnetic power in the left plot. (right) Equatorial radial displacement spectrum derived from the electric field in the middle plot under the assumption that the amplitude at each frequency corresponds to a fundamental mode standing Alfven wave oscillating with that discrete frequency.

using the upper quartile rather that the median powers which are shown in Figure 7. Time domain case studies of events with large-amplitude ULF waves, perhaps following sudden impulses or stream interfaces [cf. *Morley et al.*, 2010], would be needed in order to validate the hypothesis that ULF wave advection can produce enhanced magnetopause shadowing losses inside the equilibrium magnetopause location, and that will be the subject of future work. Nonetheless, it appears that ULF wave flux tube advection on high L-shells close to the magnetopause could be as large as $\sim\pm1$–2 R_E based on the statistical characteristics of ULF wave power in terms of Kp. During large dynamic pressure changes and sudden impulses, the effects of enhanced ULF wave magnetopause shadowing would be expected to be especially strong. Interestingly, the study by *Kellerman and Shprits* [2012] found that dynamic pressure was, in general, correlated with radiation belt electron flux decreases.

The "enhanced ULF wave magnetopause shadowing" processes discussed here would operate on the time scale of ULF wave periods, i.e., also on MeV energy electron drift time scales. Future studies should examine the potential impact of this essentially coherent ULF wave process on radiation belt loss, specifically the role it may play in depleting the radiation belts at locations (perhaps at times significantly) inside the nominal equilibrium magnetopause location.

6. SUMMARY AND CONCLUSIONS

In this paper, we have presented an overview of some of the mechanisms by which ULF waves can significantly influence the acceleration, transport, and loss of relativistic electrons in the outer zone radiation belt. Since these electrons are influenced by wave-particle interactions all around their drift orbits, it is important to be able to characterize the global nature of these interactions. In terms of long-period ULF waves, wave monitoring from multiple global magnetometer arrays provides an important complement to the single-point measurements available from satellites such as the upcoming two-spacecraft NASA RBSP mission. Many of these magnetometer arrays, such as the CARISMA array, have the capabilities for real-time monitoring of waves from Pc1 to Pc5 frequencies across many hours of longitude spanning from the polar cap to deep inside the plasmasphere.

In recent years, there has been a definite tendency to suggest that processes that violate the first adiabatic invariant might be dominant in terms of driving radiation belt acceleration, a strong candidate process being resonance with lower-band whistler mode chorus [e.g., *Meredith et al.*, 2002]. Such inferences have been supported by observations of PSD peaks in the heart of the radiation belt around $L \sim 5$ [cf. *Green and Kivelson*, 2004; *Chen et al.*, 2007]. In data from repeated satellite passes through the inner magnetosphere, these PSD peaks appear to rise monotonically from the background. This can be contrasted with the response expected from an inward transport process such as radial diffusion where a PSD increase propagates inward and is followed by rapid high L decay, which could also create a localized PSD peak at a specific instant of time. Consequently, the behavior reported by *Green and Kivelson* [2004] and *Chen et al.* [2007] cannot be easily explained by inward transport processes such as radial diffusion [e.g., *Shprits et al.*, 2007]. Nonetheless, it will be interesting to reexamine the flux profiles arising from radial diffusion with new formulations of diffusion coefficients such as those developed by *Ozeke et al.* [2012] [see also *Ozeke et al.*, this volume]. Overall, we believe that ULF waves have an important role to play in controlling radiation belt dynamics and morphology, and this paper focuses on elements of such dynamics. For example, even in the scenario whereby local acceleration, such as through resonance with lower-band chorus, provides a source of relativistic electrons in the heart of the outer radiation belt, ULF wave radial diffusion would certainly play an important role in redistributing this flux through both inward and outward radial diffusion.

In this paper, we have shown that there is a very close correspondence between 1–10 mHz ULF wave power and solar wind speed across the entire radiation belt. This suggests that ULF waves may offer an explanation or at least contribution to the long-standing observation that radiation belt fluxes are correlated with solar wind speed [*Paulikas and Blake*, 1979]. Interestingly, at the low-L-shells in the heart and lower-altitude regions of the outer radiation belts during the declining and minimum phases of the solar cycle, the D component is slightly better correlated with solar wind speed than the H component. Under an assumption of a 90° Alfvenic polarization rotation through the ionosphere, the D component would correspond to an azimuthal electric field in the magnetosphere, which can strongly interact with the radiation belt electrons via drift resonance.

Also of interest is the observation presented here that ULF wave power enhancements, and their penetration to low-L-shells, seem to be strongly influenced by *Dst* especially during times when *Dst* is decreasing during storm main phase. This relation suggests that ULF wave power penetration into the inner magnetosphere can be important and, indeed, that perhaps this penetration has a role to play in explaining the observation that the inner edge of the radiation belts follows the plasmasphere. There is no doubt that the background radial density profiles can affect the penetration of ULF wavefields to low L [cf. *Loto'aniu et al.*, 2006; *Kale et al.*, 2009], and there is evidence that this can explain even some extreme dynamics such as the slot filling [*Baker et al.*, 2004] during the Halloween 2003

storms [*Loto'aniu et al.*, 2006; *Li et al.*, 2009]. Changes in background plasma density can also affect the drift-resonance condition itself, and this may also be important [e.g., *Ozeke and Mann*, 2008]. An interesting question is also whether there are solar wind conditions when radial diffusion provides the dominant acceleration and more generally to determine the extent to which ULF wave transport provides the energy source for the outer zone electron radiation belt through the work done by the ULF wavefields.

Despite the observations of *Green and Kivelson* [2004] and *Chen et al.* [2007], it is also important to consider the latitude of electron phase space density observations. It is possible that a radial diffusion process, which acts preferentially on large pitch angle particles at the equator, could appear to an off-equatorial observer to be a local rise in phase space density if a secondary equatorial pitch angle scattering process is responsible for decreasing the pitch angles, hence enabling the enhancement to be seen by the off-equatorial platform. For example, *Fennel and Roeder* [2008] used SCATHA data to analyze the phase space density structure as a function of the first and second adiabatic invariants for the 2 May 1986 storm ($Dst = -95$ nT). *Fennel and Roeder* [2008, p. 1] obtained

"results (which) imply that radial diffusion is a reasonable explanation of the near-equatorial post-storm PSD enhancements for L* > 5.2 for this storm but that either significant electron pitch angle transport, losses, and/or acceleration of off-equatorial mirroring electrons by waves play an important role in the evolution off-equator PSD profiles during the storm recovery."

Observations by the RBSP spacecraft will help to further determine the relative importance of local acceleration and ULF wave radial diffusion in shaping the morphology of the outer zone radiation belt.

Finally, we have examined the possible role of ULF waves in enhancing magnetopause shadowing losses in the magnetosphere through the ULF wave advection of the magnetopause from its equilibrium position to lower L values. We estimated the possible amplitude of this motion, finding on the basis of statistical ULF wave power measured by ground-based magnetometers that it might reach $\sim\pm1$–2 R_E. Such "enhanced ULF wave magnetopause shadowing losses" may be important in enhancing the losses seen during storm main phase, especially under conditions such as at solar wind stream interfaces where dynamics pressure increases or sudden impulses can be expected to excite large-amplitude ULF waves. Certainly, the results from the superposed epoch analysis of solar wind stream interface radiation belt losses observed by the GPS satellites presented by *Morley et al.* [2010] appear to show a close correspondence between model magnetopause location and the L-shell locus of the observed GPS energetic particle flux loss. Despite their close correspondence in shape, Morley et al. argued that magnetopause shadowing was not likely important since the model magnetopause location and the GPS loss profiles were separated by several Earth radii. In our opinion, ULF wave advection of the magnetopause may provide a basis on which to bridge this gap. The enhanced ULF wave magnetopause shadowing losses would occur coherently on ULF wave (and hence also electron drift) time scales; of course, outward radial diffusion may further enhance such magnetopause losses on the longer radial diffusion time scales especially if ULF wave power remains high following the stream interface.

Very recently, *Kellerman and Sphrits* [2012] revisited the relationship between solar wind parameters and geosynchronous radiation belt electron flux first studied by *Paulikas and Blake* [1979]. *Kellerman and Shprits*, using a probability distribution function approach, showed that the 2 day lagged electron flux was well correlated with solar wind speed, just as had *Paulikas and Blake* many years before. Our results show that there is a very close time domain correspondence between long-period ULF wave power and solar wind speed, suggesting that ULF waves may represent an important intermediary, which could explain some of the Paulikas and Blake correlation through ULF wave-particle interactions. Certainly, the ULF wave power increases by several orders of magnitude during fast solar wind speed intervals (Figures 2, 3, and 4) [see also, e.g., *Mathie and Mann*, 2000, 2001]. Such increases might be explained by the increased trapping of ULF waves inside the magnetopause [e.g., *Mann and Wright*, 1999] or by the action of the Kelvin-Helmholtz instability at the magnetopause or through overreflection of ULF waves there [e.g., *Mann et al.*, 1999], which may be prevalent for solar wind speeds $>\sim500$ km s^{-1}. Perhaps equally interesting is the observation by *Kellerman and Shprits* [2012, p. 1] that geosynchronous fluxes decrease with increases in solar wind dynamic pressure and that "ULF waves driven by dynamic pressure variations may act as a significant cause of loss for electrons in the 100's of keV to MeV range". Of course, such losses may be explained by magnetopause shadowing, but in our opinion, this is perhaps at times significantly enhanced by the action of ULF wave advection in the form of ULF wave-enhanced magnetopause shadowing as described above.

Overall, in this paper, we have highlighted an important role for ULF waves in influencing a large range of radiation belt acceleration, transport, and loss processes. We believe that a full understanding of inner magnetosphere dynamics can only be achieved through a full appreciation of the paradigm of cross-energy coupling. Such cross-energy coupling, spanning 6 orders of magnitude or more in energy via the populations in the plasmasphere, ring current, and radiation belts, occurs via the intermediary of plasma waves such as the long-period ULF waves studied here.

Acknowledgments. I.R.M. is supported by a Canadian NSERC Discovery Grant. K.R.M. is funded by Alberta Innovates and an NSERC Canadian Graduate Scholarship. I.J.R., L.G.O., D.K.M., and A.K. are funded by the Canadian Space Agency. CARISMA is operated by the University of Alberta and funded by the Canadian Space Agency. This work was supported, in part, by participation in the MAARBLE (Monitoring, Analyzing and Assessing Radiation Belt Loss and Energization) consortium. MAARBLE has received funding from the European Community's Seventh Framework Programme (FP7-SPACE-2010-1, SP1 Cooperation, Collaborative project) under grant agreement 284520. This paper reflects only the authors' views, and the Union is not liable for any use that may be made of the information contained herein. The Sub-Auroral Magnetometer Network (SAMNET) is operated by the Department of Communications Systems at Lancaster University (UK) and is funded by the Science and Technology Facilities Council (STFC). The authors are indebted to D. Wallis for the management and operation of the CANOPUS magnetometer network for the majority of the period studied within this manuscript and D. M. Pahud for the initial compilation and analysis of CARISMA data within the ULF database. The authors thank the WDC for Geomagnetism, Kyoto for the *Dst* data. The IMAGE data are collected as a Finnish-German-Norwegian-Polish-Russian-Swedish project.

REFERENCES

Anderson, B. J., M. J. Engebretson, S. P. Rounds, L. J. Zanetti, and T. A. Potemra (1990), A statistical study of Pc 3–5 pulsations observed by the AMPTE/CCE Magnetic Fields Experiment, 1. Occurrence distributions, *J. Geophys. Res.*, 95(A7), 10,495–10,523, doi:10.1029/JA095iA07p10495.

Baker, D. N., P. R. Higbie, R. D. Belian, and E. W. Hones Jr. (1979), Do Jovian electrons influence the terrestrial outer radiation zone?, *Geophys. Res. Lett.*, 6(6), 531–534.

Baker, D. N., et al. (1998), A strong CME-related magnetic cloud interaction with the Earth's Magnetosphere: ISTP observations of rapid relativistic electron acceleration on May 15, 1997, *Geophys. Res. Lett.*, 25(15), 2975–2978.

Baker, D. N., S. G. Kanekal, X. Li, S. P. Monk, J. Goldstein, and J. L. Burch (2004). An extreme distortion of the Van Allen belt arising from the 'Hallowe'en' solar storm in 2003, *Nature*, 432(7019), 878–881, doi:10.1038/nature03116.

Borovsky, J. E., and M. H. Denton (2006), Differences between CME-driven storms and CIR-driven storms, *J. Geophys. Res.*, 111, A07S08, doi:10.1029/2005JA011447.

Bortnik, J., R. M. Thorne, and N. P. Meredith (2008), The unexpected origin of plasmaspheric hiss from discrete chorus emissions, *Nature*, 452, 62–66, doi:10.1038/nature06741.

Brautigam, D. H., and J. M. Albert (2000), Radial diffusion analysis of outer radiation belt electrons during the October 9, 1990, magnetic storm, *J. Geophys. Res.*, 105(A1), 291–309.

Brautigam, D. H., G. P. Ginet, J. M. Albert, J. R. Wygant, D. E. Rowland, A. Ling, and J. Bass (2005), CRRES electric field power spectra and radial diffusion coefficients, *J. Geophys. Res.*, *110*, A02214, doi:10.1029/2004JA010612.

Brizard, A. J., and A. A. Chan (2001), Relativistic bounce-averaged quasilinear diffusion equation for low-frequency electromagnetic fluctuations, *Phys. Plasmas*, *8*(11), 4762–4771.

Chen, Y., G. D. Reeves, and R. H. W. Friedel (2007), The energization of relativistic electrons in the outer Van Allen radiation belt, *Nat. Phys.*, *3*, 614–617, doi:10.1038/nphys655.

Cheng, C. Z., and Q. Qian (1994), Theory of ballooning-mirror instabilities for anisotropic pressure plasmas in the magnetosphere, *J. Geophys. Res.*, *99*(A6), 11,193–11,209.

Clemmons, J. H., et al. (2000), Observations of traveling Pc5 waves and their relation to the magnetic cloud event of January 1997, *J. Geophys. Res.*, *105*(A3), 5441–5452, doi:10.1029/1999JA900418.

Degeling, A. W., R. Rankin, K. Kabin, R. Marchand, and I. R. Mann (2007). The effect of ULF compressional modes and field line resonances on relativistic electron dynamics, *Planet. Space Sci.*, *55*(6), 731–742, doi:10.1016/j.pss.2006.04.039.

Degeling, A. W., L. G. Ozeke, R. Rankin, I. R. Mann, and K. Kabin (2008), Drift resonant generation of peaked relativistic electron distributions by Pc 5 ULF waves, *J. Geophys. Res.*, *113*, A02208, doi:10.1029/2007JA012411.

Elkington, S. R., M. K. Hudson, and A. A. Chan (2003), Resonant acceleration and diffusion of outer zone electrons in an asymmetric geomagnetic field, *J. Geophys. Res.*, *108*(A3), 1116, doi:10.1029/2001JA009202.

Engebretson, M., K.-H. Glassmeier, M. Stellmacher, W. J. Hughes, and H. Lühr (1998), The dependence of high-latitude PcS wave power on solar wind velocity and on the phase of high-speed solar wind streams, *J. Geophys. Res.*, *103*(A11), 26,271–26,283.

Fälthammar, C.-G. (1965), Effects of time-dependent electric fields on geomagnetically trapped radiation, *J. Geophys. Res.*, *70*(11), 2503–2516, doi:10.1029/JZ070i011p02503.

Fei, Y., A. A. Chan, S. R. Elkington, and M. J. Wiltberger (2006), Radial diffusion and MHD particle simulations of relativistic electron transport by ULF waves in the September 1998 storm, *J. Geophys. Res.*, *111*, A12209, doi:10.1029/2005JA011211.

Fennell, J. F., and J. L. Roeder (2008), Storm-time phase space density radial profiles of energetic electrons for small and large *K* values – SCATHA results, *J. Atmos. Sol. Terr. Phys.*, *70*(14), 1760–1773, doi:10.1016/j.jastp.2008.03.014.

Fraser, B. J., J. L. Horwitz, J. A. Slavin, Z. C. Dent, and I. R. Mann (2005), Heavy ion mass loading of the geomagnetic field near the plasmapause and ULF wave implications, *Geophys. Res. Lett.*, *32*, L04102, doi:10.1029/2004GL021315.

Friedel, R. H., G. D. Reeves, and T. Obara (2002), Relativistic electron dynamics in the inner magnetosphere – A review, *J. Atmos. Sol. Terr. Phys.*, *64*(2), 265–282, doi:10.1016/S1364-6826(01)00088-8.

Gjerloev, J. W. (2009), A global ground-based magnetometer initiative, *Eos Trans. AGU*, *90*(27), 230, doi:10.1029/2009EO270002.

Glassmeier, K.-H. (1984), On the influence of ionospheres with non-uniform conductivity distribution on hydromagnetic waves, *J. Geophys.*, *54*(2), 125–137.

Green, J. C., and M. G. Kivelson (2004), Relativistic electrons in the outer radiation belt: Differentiating between acceleration mechanisms, *J. Geophys. Res.*, *109*, A03213, doi:10.1029/2003JA010153.

Horne, R. B., and R. M. Thorne (2003), Relativistic electron acceleration and precipitation during resonant interactions with whistler-mode chorus, *Geophys. Res. Lett.*, *30*(10), 1527, doi:10.1029/2003GL016973.

Hughes, W. J., and D. J. Southwood (1976a), The screening of micropulsation signals by the atmosphere and ionosphere, *J. Geophys. Res.*, *81*(19), 3234–3240, doi:10.1029/JA081i019p03234.

Hughes, W. J., and D. J. Southwood (1976b), An illustration of modification of geomagnetic pulsation structure by the ionosphere, *J. Geophys. Res.*, *81*(19), 3241–3247.

Hughes, W. J., D. J. Southwood, B. Mauk, R. L. McPherron, and J. N. Barfield (1978) Alfven waves generated by an inverted plasma energy distribution, *Nature*, *275*, 43–45.

Johnson, M. H., and J. Kierein (1992), Combined release and radiation effects satellite (CRRES) – Spacecraft and mission, *J. Spacecr. Rockets*, *29*(4), 556–563.

Kale, Z. C., I. R. Mann, C. L. Waters, M. Vellante, T. L. Zhang, and F. Honary (2009), Plasmaspheric dynamics resulting from the Hallowe'en 2003 geomagnetic storms, *J. Geophys. Res.*, *114*, A08204, doi:10.1029/2009JA014194.

Kataoka, R., and Y. Miyoshi (2006), Flux enhancement of radiation belt electrons during geomagnetic storms driven by coronal mass ejections and corotating interaction regions, *Space Weather*, *4*, S09004, doi:10.1029/2005SW000211.

Kellerman, A. C., and Y. Y. Shprits (2012), On the influence of solar wind conditions on the outer-electron radiation belt, *J. Geophys. Res.*, *117*, A05217, doi:10.1029/2011JA017253.

Kessel, R. L. (2012), NASA's Radiation Belt Storm Probes mission: From concept to reality, in *Dynamics of the Earth's Radiation Belts and Inner Magnetosphere*, *Geophys. Monogr. Ser.*, doi:10.1029/2012GM001312, this volume.

Kessel, R. L., I. R. Mann, S. F. Fung, D. K. Milling, and N. O'Connell (2004), Correlation of Pc5 wave power inside and outside the magnetosphere during high speed streams, *Ann. Geophys.*, *22*(2), 629–641, doi:10.5194/angeo-22-629-2004.

Kim, H.-J., and A. A. Chan (1997), Fully adiabatic changes in storm time relativistic electron fluxes, *J. Geophys. Res.*, *102*(A10), 22,107–22,116.

Kim, K. C., D.-Y. Lee, H.-J. Kim, L. R. Lyons, E. S. Lee, M. K. Úztürk, and C. R. Choi (2008), Numerical calculations of relativistic electron drift loss effect, *J. Geophys. Res.*, *113*, A09212, doi:10.1029/2007JA013011.

King, J. H., and N. E. Papitashvili (2005), Solar wind spatial scales in and comparisons of hourly Wind and ACE plasma and magnetic field data, *J. Geophys. Res.*, *110*, A02104, doi:10.1029/2004JA010649.

Krimigis, S. M., G. Gloeckler, R. W. McEntire, T. A. Potemra, F. L. Scarf, and E. G. Shelley (1985), Magnetic storm of September 4, 1984: A synthesis of ring current spectra and energy densities measured with AMPTE/CCE, *Geophys. Res. Lett.*, *12*(5), 329–332.

Li, X., D. N. Baker, M. Temerin, T. E. Cayton, E. G. D. Reeves, R. A. Christensen, J. B. Blake, M. D. Looper, R. Nakamura, and S. G. Kanekal (1997), Multisatellite observations of the outer zone electron variation during the November 3–4, 1993, magnetic storm, *J. Geophys. Res.*, *102*(A7), 14,123–14,140.

Li, X., M. Temerin, D. N. Baker, G. D. Reeves, and D. Larson (2001), Quantitative prediction of radiation belt electrons at geostationary orbit based on solar wind measurements, *Geophys. Res. Lett.*, *28*(9), 1887–1890.

Li, X., D. N. Baker, T. P. O'Brien, L. Xie, and Q. G. Zong (2006), Correlation between the inner edge of outer radiation belt electrons and the innermost plasmapause location, *Geophys. Res. Lett.*, *33*, L14107, doi:10.1029/2006GL026294.

Li, X., A. B. Barker, D. N. Baker, W. C. Tu, T. E. Sarris, R. S. Selesnick, R. Friedel, and C. Shen (2009), Modeling the deep penetration of outer belt electrons during the "Halloween" magnetic storm in 2003, *Space Weather*, *7*, S02004, doi:10.1029/2008SW000418.

Li, X., M. Temerin, D. N. Baker, and G. D. Reeves (2011), Behavior of MeV electrons at geosynchronous orbit during last two solar cycles, *J. Geophys. Res.*, *116*, A11207, doi:10.1029/2011JA016934.

Loto'aniu, T. M., I. R. Mann, L. G. Ozeke, A. A. Chan, Z. C. Dent, and D. K. Milling (2006), Radial diffusion of relativistic electrons into the radiation belt slot region during the 2003 Halloween geomagnetic storms, *J. Geophys. Res.*, *111*, A04218, doi:10.1029/2005JA011355.

Loto'aniu, T. M., H. J. Singer, C. L. Waters, V. Angelopoulos, I. R. Mann, S. R. Elkington, and J. W. Bonnell (2010), Relativistic electron loss due to ultralow frequency waves and enhanced outward radial diffusion, *J. Geophys. Res.*, *115*, A12245, doi:10.1029/2010JA015755.

Lyons, L. R., and R. M. Thorne (1973), Equilibrium structure of radiation belt electrons, *J. Geophys. Res.*, *78*(13), 2142–2149, doi:10.1029/JA078i013p02142.

Lyons, L. R., R. M. Thorne, and C. F. Kennel (1972), Pitch-angle diffusion of radiation belt electrons within the plasmasphere, *J. Geophys. Res.*, *77*(19), 3455–3474.

Mann, I. R., and G. Chisham (2000), Comment on "Concerning the generation of geomagnetic giant pulsations by drift-bounce resonance ring current instabilities" by K.-H. Glassmeier et al., Ann. Geophysicae, 17, 338–350, (1999), *Ann. Geophys.*, *18*, 161–166.

Mann, I. R., and A. N. Wright (1999), Diagnosing the excitation mechanisms of Pc5 magnetospheric flank waveguide modes and FLRs, *Geophys. Res. Lett.*, *26*(16), 2609–2612, doi:10.1029/1999GL900573.

Mann, I. R., A. N. Wright, and A. W. Hood (1997), Multiple-time-scales analysis of ideal poloidal Alfvén waves, *J. Geophys. Res.*, *102*(A2), 2381–2390, doi:10.1029/96JA03034.

Mann, I. R., A. N. Wright, K. J. Mills, and V. M. Nakariakov (1999), Excitation of magnetospheric waveguide modes by magnetosheath flows, *J. Geophys. Res.*, *104*(A1), 333–353, doi:10.1029/1998JA900026.

Mann, I. R., T. P. O'Brien, and D. K. Milling (2004), Correlations between ULF wave power, solar wind speed, and relativistic electron flux in the magnetosphere: Solar cycle dependence, *J. Atmos. Sol. Terr. Phys.*, *66*(2), 187–198, doi:10.1016/j.jastp.2003.10.002.

Mann, I. R., et al. (2006), The outer radiation belt injection, transport, acceleration and loss satellite (ORBITALS): A Canadian small satellite mission for ILWS, *Adv. Space Res.*, *38*(8), 1838–1860, doi:10.1016/j.asr.2005.11.009.

Mann, I. R., et al. (2008), The upgraded CARISMA magnetometer array in the THEMIS era, *Space Sci. Rev.*, *141*(1–4), 413–451, doi:10.1007/s11214-008-9457-6.

Mathie, R. A., and I. R. Mann (2000), A correlation between extended intervals of ULF wave power and storm-time geosynchronous relativistic electron flux enhancements, *Geophys. Res. Lett.*, *27*(20), 3261–3264, doi:10.1029/2000GL003822.

Mathie, R. A., and I. R. Mann (2001), On the solar wind control of Pc5 ULF pulsation power at mid-latitudes: Implications for MeV electron acceleration in the outer radiation belt, *J. Geophys. Res.*, *106*(A12), 29,783–29,796, doi:10.1029/2001JA000002.

Mathie, R. A., F. W. Menk, I. R. Mann, and D. Orr (1999), Discrete field line resonances and the Alfvén continuum in the outer magnetosphere, *Geophys. Res. Lett.*, *26*(6), 659–662, doi:10.1029/1999GL900104.

Meredith, N. P., R. B. Horne, R. H. A. Iles, R. M. Thorne, D. Heynderickx, and R. R. Anderson (2002), Outer zone relativistic electron acceleration associated with substorm-enhanced whistler mode chorus, *J. Geophys. Res.*, *107*(A7), 1144, doi:10.1029/2001JA900146.

Millan, R. M., and R. M. Thorne (2007), Review of radiation belt relativistic electron losses, *J. Atmos. Sol. Terr. Phys.*, *69*(3), 362–377, doi:10.1016/j.jastp.2006.06.019.

Miyoshi, Y., and R. Kataoka (2008), Flux enhancement of the outer radiation belt electrons after the arrival of stream interaction regions, *J. Geophys. Res.*, *113*, A03S09, doi:10.1029/2007JA012506.

Morley, S. K., R. H. W. Friedel, E. L. Spanswick, G. D. Reeves, J. T. Steinberg, J. Koller, T. Cayton, and E. Noveroske (2010), Dropouts of the outer electron radiation belt in response to solar wind stream interfaces: Global positioning system observations, *Proc. R. Soc. A*, *466*, 3329–3350, doi:10.1098/rspa.2010.0078.

Murphy, K. R., I. R. Mann, I. J. Rae, and D. K. Milling (2011), Dependence of ground-based Pc5 ULF wave power on F10.7 solar radio flux and solar cycle phase, *J. Atmos. Sol. Terr. Phys.*, *73*(11–12), 1500–1510, doi:10.1016/j.jastp.2011.02.018.

Northrop, T. G. (1963), *The Adiabatic Motion of Charged Particles*, Wiley-Interscience, New York.

O'Brien, T. P., and M. B. Moldwin (2003), Empirical plasmapause models from magnetic indices, *Geophys. Res. Lett.*, *30*(4), 1152, doi:10.1029/2002GL016007.

O'Brien, T. P., K. R. Lorentzen, I. R. Mann, N. P. Meredith, J. B. Blake, J. F. Fennell, M. D. Looper, D. K. Milling, and R. R. Anderson (2003), Energization of relativistic electrons in the presence of ULF power and MeV microbursts: Evidence for dual ULF and VLF acceleration, *J. Geophys. Res.*, *108*(A8), 1329, doi:10.1029/2002JA009784.

Ozeke, L. G., and I. R. Mann (2008), Energization of radiation belt electrons by ring current ion driven ULF waves, *J. Geophys. Res.*, *113*, A02201, doi:10.1029/2007JA012468.

Ozeke, L. G., I. R. Mann, and J. T. Mathews (2005), The influence of asymmetric ionospheric Pedersen conductances on the field-aligned phase variation of guided toroidal and guided poloidal Alfvén waves, *J. Geophys. Res.*, *110*, A08210, doi:10.1029/2005JA011167.

Ozeke, L. G., I. R. Mann, and I. J. Rae (2009), Mapping guided Alfvén wave magnetic field amplitudes observed on the ground to equatorial electric field amplitudes in space, *J. Geophys. Res.*, *114*, A01214, doi:10.1029/2008JA013041.

Ozeke, L. G., I. R. Mann, K. R. Murphy, I. J. Rae, D. K. Milling, S. R. Elkington, A. A. Chan, and H. J. Singer (2012), ULF wave derived radiation belt radial diffusion coefficients, *J. Geophys. Res.*, *117*, A04222, doi:10.1029/2011JA017463.

Ozeke, L. G., I. R. Mann, K. R. Murphy, I. J. Rae, and A. A. Chan (2012), ULF wave–driven radial diffusion simulations of the outer radiation belt, in *Dynamics of the Earth's Radiation Belts and Inner Magnetosphere*, Geophys. Monogr. Ser., doi:10.1029/2012GM001332, this volume.

Paulikas, G. A., and J. B. Blake (1979), Effects of the solar wind on magnetospheric dynamics: Energetic electrons at the synchronous orbit, in *Quantitative Modeling of Magnetospheric Processes*, Geophys. Monogr. Ser., vol. 21, edited by W. P. Olson, pp. 180–202, AGU, Washington, D. C., doi:10.1029/GM021p0180.

Rae, I. J., et al. (2005), Evolution and characteristics of global Pc5 ULF waves during a high solar wind speed interval, *J. Geophys. Res.*, *110*, A12211, doi:10.1029/2005JA011007.

Rae, I. J., I. R. Mann, K. R. Murphy, L. G. Ozeke, D. K. Milling, A. A. Chan, S. R. Elkington, and F. Honary (2012), Ground-based magnetometer determination of in situ Pc4–5 ULF electric field wave spectra as a function of solar wind speed, *J. Geophys. Res.*, *117*, A04221, doi:10.1029/2011JA017335.

Reeves, G. D., K. L. McAdams, R. H. W. Friedel, and T. P. O'Brien (2003), Acceleration and loss of relativistic electrons during geomagnetic storms, *Geophys. Res. Lett.*, *30*(10), 1529, doi:10.1029/2002GL016513.

Reeves, G. D., S. K. Morley, R. H. W. Friedel, M. G. Henderson, T. E. Cayton, G. Cunningham, J. B. Blake, R. A. Christensen, and D. Thomsen (2011), On the relationship between relativistic electron flux and solar wind velocity: Paulikas and Blake revisited, *J. Geophys. Res.*, *116*, A02213, doi:10.1029/2010JA015735.

Rostoker, G., S. Skone, and D. N. Baker (1998), On the origin of relativistic electrons in the magnetosphere associated with some geomagnetic storms, *Geophys. Res. Lett.*, *25*(19), 3701–3704.

Russell, C. T., and R. L. McPherron (1973), Semiannual variation of geomagnetic activity, *J. Geophys. Res.*, *78*(1), 92–108.

Russell, C. T., P. J. Chi, D. J. Dearborn, Y. S. Ge, B. Kuo-Tiong, J. D. Means, D. R. Pierce, K. M. Rowe, and R. C. Snare (2008), THEMIS ground-based magnetometers, *Space Sci. Rev.*, *141*, 389–412, doi:10.1007/s11214-008-9337-0.

Schulz, M., and L. J. Lanzerotti (1974), *Particle Diffusion in the Radiation Belts*, Phys. Chem. Space, vol. 7, 215 pp., Springer, New York.

Shprits, Y., D. Kondrashov, Y. Chen, R. Thorne, M. Ghil, R. Friedel, and G. Reeves (2007), Reanalysis of relativistic radiation belt electron fluxes using CRRES satellite data, a radial diffusion model, and a Kalman filter, *J. Geophys. Res.*, *112*, A12216, doi:10.1029/2007JA012579.

Southwood, D. J. (1976), A general approach to low-frequency instability in the ring current plasma, *J. Geophys. Res.*, *81*, 3340–3348.

Southwood, D. J. (1974), Some features of field line resonances in the magnetosphere, *Planet. Space Sci.*, *22*, 482–491.

Southwood, D. J., and M. G. Kivelson (1982), Charged particle behavior in low-frequency geomagnetic pulsations, 2. Graphical approach, *J. Geophys. Res.*, *87*(A3), 1707–1710.

Southwood, D. J., J. W. Dungey, and R. J. Etherington (1969), Bounce resonant interaction between pulsations and trapped particles, *Planet. Space Sci.*, *17*, 349–361.

Sugiura, M. (1964), Hourly values of equatorial Dst for the IGY, *Ann. Int. Geophys. Year*, *35*, 9–45.

Summers, D., and R. M. Thorne (2003), Relativistic electron pitch-angle scattering by electromagnetic ion cyclotron waves during geomagnetic storms, *J. Geophys. Res.*, *108*(A4), 1143, doi:10.1029/2002JA009489.

Summers, D., R. M. Thorne, and F. Xiao (1998), Relativistic theory of wave-particle resonant diffusion with application to electron acceleration in the magnetosphere, *J. Geophys. Res.*, *103*(A9), 20,487–20,500, doi:10.1029/98JA01740.

Summers, D., B. Ni, and N. P. Meredith (2007), Timescales for radiation belt electron acceleration and loss due to resonant wave-particle interactions: 2. Evaluation for VLF chorus, ELF hiss, and electromagnetic ion cyclotron waves, *J. Geophys. Res.*, *112*, A04207, doi:10.1029/2006JA011993.

Tsyganenko, N. A. (1989), A magnetospheric magnetic field model with a warped tail current sheet, *Planet. Space Sci.*, *37*, 5–20.

Turner, D. L., Y. Shprits, M. Hartinger, and V. Angelopoulos (2012), Explaining sudden losses of outer radiation belt electrons during geomagnetic storms, *Nat. Phys.*, *8*, 208–212, doi:10.1038/nphys2185.

Tverskaya, L. V. (1986), On the boundary of electron injection into the magnetosphere, *Geomagn. Aeron.*, *26*(5), 864–865.

Tverskaya, L. V. (1996), The latitude position dependence of the relativistic electron maximum as a function of Dst, *Adv. Space Res.*, *18*, 135–138.

Tverskaya, L. V., N. N. Pavlov, J. B. Blake, R. S. Selesnick, and J. F. Fennell (2003), Predicting the L-position of the storm-injected relativistic electron belt, *Adv. Space Res.*, *31*(4), 1039–1044, doi:10.1016/S0273-1177(02)00785-8.

Van Allen, J. A., and L. A. Frank (1959). Radiation measurements to 658,300 km with Pioneer-IV, *Nature, 184,* 219–224.

Van Allen, J. A., G. H. Ludwig, E. C. Ray, and C. E. McIlwain (1958), Observation of high intensity radiation by satellites 1958 Alpha and Gamma, *Jet Propul., 28,* 588–592.

Viljanen, A., and L. Hakkinen (1997), IMAGE magnetometer network, in *Satellite-Ground Based Coordination Sourcebook*, edited by M. Lockwood, M. N. Wild and H. Opgenoorth, *ESA Spec. Publ., ESA SP-1198,* 111–117.

Yeoman, T. K., D. K. Milling, and D. Orr (1990), Pi2 pulsation polarization patterns on the U.K. sub-auroral magnetometer network (SAMNET), *Planet. Space Sci., 38*(5), 589–602, doi:10.1016/0032-0633(90)90065-X.

F. Honary, Department of Physics, Lancaster University, Lancaster LA1, UK.

A. Kale, I. R. Mann, D. K. Milling, K. R. Murphy, L.G. Ozeke, and I. J. Rae, Department of Physics, University of Alberta, Edmonton, AB T6G 2J1, Canada. (imann@ualberta.ca)

NASA's Radiation Belt Storm Probes Mission: From Concept to Reality

R. L. Kessel

NASA Headquarters, Washington, District of Columbia, USA

NASA's Radiation Belt Storm Probes (RBSP) mission is a two-satellite mission designed to measure charged particle populations, fields, and waves in the Van Allen radiation belts in order to provide understanding of how these change in response to variable inputs of energy from the Sun. This paper tracks the progress of the mission from formulation and the development of science objectives through instrument selection to mission design, integration, and testing, with an emphasis on how the chosen measurements address the science objectives. At the start of normal operations (~60 days after launch), the spacecraft will be positioned to address a number of science questions such as the nature of whistler mode interactions and their roles in electron energization and loss, the large-scale dynamics and structure of the magnetosphere during geomagnetic storms, and the source, structure, and dynamics of the inner ($L < 2$) ion and electron belts. Collaborations with other missions such as Time History of Events and Macroscale Interactions during Substorms and Balloon Array for Radiation-belt Relativistic Electron Losses will enable studies of dawn-dusk differences in magnetospheric particle populations and will quantify particle losses through the magnetopause for comparison with losses by precipitation into the ionosphere. A real-time space weather broadcast will enable nowcasting of radiation belt conditions. The RBSP Key Messages summarize the high-level impacts expected from the mission.

1. INTRODUCTION

The Living With a Star program was established in 2000 by the United States Congress with the goal "to better study solar variability and understand its effect on humanity." One objective of the program was an extension of the existing Sun-Earth connections theme: to understand basic natural processes. The other objective stressed "investigations into how solar variability affects humans and technology." A Geospace Mission Definition Team (GMDT) was formed and charged with developing Geospace unique science objectives, prioritizing these and then developing an implementation plan. One emphasis was on energetic particle populations trapped in the Van Allen radiation belts because of their impact on our technology-based society. The GMDT issued a report [*Kintner et al.*, 2002] that identified three questions, the answers to which define the broad science objectives of the Radiation Belt Storm Probes (RBSP) mission:

1. Which physical processes produce radiation belt enhancement events?

2. What are the dominant mechanisms for relativistic electron loss?

3. How do ring current and geomagnetic processes affect radiation belt behavior?

The NASA RBSP Payload Announcement of Opportunity was issued in 2005. A Mission Concept Review for RBSP was successfully completed in January 2007, followed by a Mission Design Review in October 2007, a Preliminary Design review a year later, and a Confirmation review in January 2009. The Johns Hopkins University Applied Physics Laboratory (APL) has been responsible for the overall implementation and instrument management for the mission.

Dynamics of the Earth's Radiation Belts and Inner Magnetosphere
Geophysical Monograph Series 199
Published in 2012 by the American Geophysical Union.
10.1029/2012GM001312

RBSP has passed all of its milestones including the Mission Readiness Review in July 2012 and was launched successfully on 30 August 2012.

The overall, guiding RBSP Mission Objective as defined in the level-one (L1) mission requirements is to provide understanding, ideally to the point of predictability, of how populations of relativistic electrons and penetrating ions in space form or change in response to variable inputs of energy from the Sun. The impacts of the RBSP mission are broad; the fundamental processes that energize, transport, and cause the loss of the time-varying charged-particle populations in the Earth's inner magnetosphere also operate throughout the plasma universe at locations as diverse as magnetized and unmagnetized planets, the solar wind, our Sun, and other stars. In addition, the charged particle populations pose hazards to human and robotic explorers, as well as to spacecraft operating in or passing through the region.

Details of the mission design and the selected instruments, including the measurements needed to address the objectives, are discussed in sections 2 and 3. Section 4 traces the progress of the two spacecraft through integration and testing. Section 5 discusses some science endeavors that will take advantage of the early mission configurations and satellite locations, including collaborative studies. Section 6 briefly discusses the Space Weather Broadcast.

2. MISSION DESIGN

In order to address the three primary science objectives, the two-satellite RBSP mission was designed to measure the appropriate particle populations, fields and waves, in the inner magnetospheric regions of interest. The orbit was designed to cover all local times and interaction regions with varying separations between the two satellites. Specific measurements needed to address each of the three science objectives are discussed below, followed by orbital parameters and impacts. Detailed information about the RBSP science, mission, and spacecraft design are described in the RBSP Mission Book, a special issue of *Space Science Reviews* [*Mauk et al.*, 2012; J. Stratton et al., RBSP mission design and operations, submitted to *Space Science Reviews*, 2012; K. H. Kirby et al., RBSP spacecraft and environments, submitted to *Space Science Reviews*, 2012].

2.1. Measurements for Objective 1

To determine which physical processes produce radiation belt enhancement events, measurements of both the particle source region and the accelerated particles are needed with good energy and pitch angle coverage. In addition, electric and magnetic background and fluctuations over a broad range of frequencies are needed. Among the many measurements that RBSP must obtain to address this objective, it is particularly important that RBSP (1) simultaneously measure the radiation belt particle fluxes at a variety of radial separations over the course of the mission, (2) measure near-equatorial pitch angle distributions and magnetic fields to determine two-point phase space densities and radial phase space density profiles, (3) measure the local convection electric field as well as electrostatic and electromagnetic waves, which produce particle acceleration and transport, and (4) simultaneously measure the magnetic field and its variation at two points in the magnetosphere.

2.2. Measurements for Objective 2

The dynamics of the radiation belts are governed by both gains and losses in energetic particle populations, triggered by storm processes and/or internal wave particle interactions. In order to quantify the losses, RBSP must, at a minimum, be able to (1) measure the energetic electron pitch angle distributions and variability during solar storms and (2) measure the power spectral intensity of plasma waves responsible for pitch angle scattering of energetic electrons. RBSP will measure losses in the radiation belts; once particles leave the radiation belts, other missions can trace them by observing losses through the dayside magnetopause or through energetic electron precipitation (see section 5).

2.3. Measurements for Objective 3

The ring current exists in approximately the same spatial region as the radiation belts, though the ring current particles are at lower energies. Measurements of the ring current populations will aid in understanding the radiation belts. In order to address this objective, it is particularly important that RBSP be able to (1) measure the in situ composition of the ring current ions, both to understand their sources and to determine their energy density and pressure gradients, and (2) determine the global distribution of ring current ion composition, energy density, and pressure gradients.

2.4. Orbit Parameters and Impacts

Orbital precession will carry the RBSP spacecraft through all local times and interaction regions during the course of the planned 2 year prime mission. The 600 km perigee and 5.8 R_E geocentric apogee enables the spacecraft to traverse both inner and outer radiation belts twice per 9 h orbit. The low-inclination (10°) orbit permits the spacecraft to access almost all magnetically trapped particles. The sunward oriented spin axis permits full particle pitch angle distribution sampling twice per 12 s spin and ensures accurate measurements of the dominant dawn-dusk electric field component. Two identically instrumented spacecraft are needed to resolve space/time separation: for example, one spacecraft can measure the source region, and one spacecraft can measure the region to which accelerated particles are transported. Variable separations of the satellites throughout the mission permit the spacecraft to observe phenomena covering a wide range of spatial scales. Slightly different orbits cause one spacecraft to lap the other 4–5 times per year. Figure 1 shows the separation between the two satellites for one lapping period in 2012. Because each satellite moves more quickly at perigee, most orbits exhibit variable spacing between the two satellites. The top three panels show 9 h (orbit plots) on three different days during this period with satellite A in blue and satellite B in red and a line drawn between them showing separation distance (R_E). In Figure 1a, day 321, the satellites are near closet approach to each other. In Figure 1b, day 337, the satellites have variable spacing between them. Figure 1c on day 360 shows that the satellite separation is nearly fixed

for the orbit. In Figure 1d, separation distances between spacecraft A and B are shown in R_E. Vertical dotted lines indicate times corresponding to the orbit plots shown in Figures 1a, 1b, and 1c.

3. RBSP INSTRUMENTS

The instruments shown in Table 1 were selected to provide comprehensive measurements of the phenomena relevant to the Earth's ring current and radiation belts. Figure 2 illustrates the instrument placement on each RBSP spacecraft. Articles in the RBSP Mission Book describe each instrument in detail [*Mazur et al.*, 2012; H. Funsten et al., ECT-HOPE instrument, submitted to *Space Science Reviews*, 2012, hereinafter referred to as Funsten et al., submitted manuscript, 2012; J. B. Blake et al., ECT-MagEIS instrument, submitted to *Space Science Reviews*, 2012, hereinafter referred to as Blake et al., submitted manuscript, 2012; D. N. Baker et al. ECT-REPT instrument, submitted to *Space Science Reviews*, 2012, hereinafter referred to as Baker et al., submitted manuscript, 2012;

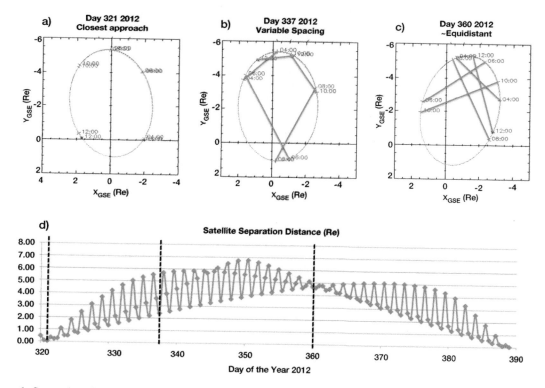

Figure 1. Separations between the two satellites for one representative lapping period in 2012 are displayed (predicted orbits). (top) The 9 h (orbit plots) on three different days during this period are shown: (a) the satellites are near closet approach to each other, (b) satellites have variable spacing between them, and (c) satellite separation is nearly fixed for the orbit. (d) The entire lapping period in 2012. Each satellite moves more quickly at perigee resulting in variable spacing between the two satellites (R_E) for most orbits. Vertical dotted lines indicate times corresponding to the orbit plots shown in Figures 1a–1c.

Table 1. RBSP Instruments and Principal Investigators (PI)

Investigation	Instruments	PI
Energetic Particle Composition and Thermal Plasma Suite (ECT)	HOPE, MagEIS, and REPT	H. Spence University of New Hampshire
Electric and Magnetic Field Instrument Suite and Integrated Science (EMFISIS)	MAG and Waves	C. Kletzing University of Iowa
Electric Field and Waves Instrument (EFW)	EFW	J. Wygant University of Minnesota
Radiation Belt Storm Probes Ion Composition Experiment (RBSPICE)	RBSPICE	L. Lanzerotti New Jersey Institute of Technology
Proton Spectrometer Belt Research (PSBR)	RPS	D. Byers National Reconnaissance Office

D. G. Mitchell et al., Radiation Belt Storm Probes Ion Composition Experiment (RBSPICE), submitted to *Space Science Reviews*, 2012, hereinafter referred to as Mitchell et al., submitted manuscript, 2012; C. Kletzing et al., EMFISIS investigation and instruments, submitted to *Space Science Reviews*, 2012, hereinafter referred to as Kletzing et al., submitted manuscript, 2012; J. Wygant et al., RBSP mission design and operations, submitted to *Space Science Reviews*, 2012, hereinafter referred to as Wygant et al., submitted manuscript, 2012].

Figure 3 shows the combined energy ranges for all the particle instruments to illustrate the comprehensive coverage of electrons, protons, and ion composition. The capability of the instruments exceeds the requirements established in the Mission Requirements Document as well as the L1 requirements. The Helium Oxygen Proton Electron (HOPE) instrument measures low-energy electrons and ions (required $\sim$20 eV–45 keV q^{-1}) (Funsten et al., submitted manuscript, 2012). These measurements are needed to fully characterize the source population for radiation belt electrons and ring current ions. The Radiation Belt Storm Probes Ion Composition Experiment (RBSPICE) measures the ions that carry the bulk of the ring current and produce the largest perturbations to

the magnetic field during solar storms (required 20–1000 keV for H, 30–1000 keV for He, 50–1000 keV for O) (Mitchell et al., submitted manuscript, 2012). RBSPICE will also help understand how and why the ring current and associated phenomena vary during storms. The Magnetic Electron Ion Spectrometer (MagEIS) will measure the heart of the radiation belt population (required 30 keV to 4 MeV for electrons and 20 keV to 1 MeV for ions) (Blake et al., submitted manuscript, 2012). In order to distinguish between different types of acceleration processes, MagEIS is designed to measure (1) locally mirroring, 90°, particles under all conditions, (2) the shape of the pitch angle distribution and how it varies under the influence of acceleration and loss, and (3) measurement of near-field-aligned portion of the distribution to quantify losses. The Relativistic Electron Proton Telescope (REPT) will capture the most intense events and the high-energy tail (required 4–10 MeV for electrons and 20–75 MeV for protons) (Baker et al., submitted manuscript, 2012). The Relativistic Proton Spectrometer (RPS) will measure ions at the very highest energies (>50 MeV) in the inner radiation belt and show how the inner belt varies in response to solar storms [*Mazur et al.*, 2012].

Figure 4 shows the combined frequency ranges for the magnetic and electric field instruments, again with ranges corresponding to capability of the instruments, Mission Requirements, and L1 requirements. The Electric and Magnetic Field Instrument Suite and Integrated Science Instrument (EMFISIS) will provide vector magnetic field measurements (DC magnetic) needed to provide the background magnetic

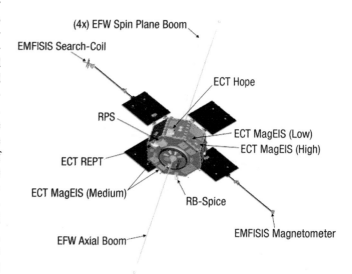

Figure 2. Model drawing of Radiation Belt Storm Probes (RBSP) spacecraft and instruments.

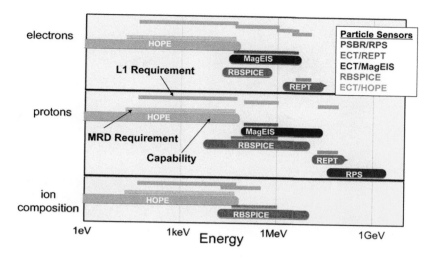

Figure 3. The combined energy ranges for all of the RBSP particle instruments showing the comprehensive coverage of electrons, protons, and ion composition, from a few eV to more than 50 MeV. Capability of each instrument exceeds the requirements of the Mission Requirements Document (MRD) as well as of the level-one (L1) Requirements.

field, determine ULF waves, and calculate particle pitch angles and phase space densities (Kletzing et al., submitted manuscript, 2012). EMFISIS also provides AC magnetic and electric fields that cover the range of waves observed in the inner magnetosphere. The EMFISIS investigation will focus on the important role played by magnetic fields and plasma waves in the processes of radiation belt particle acceleration and loss. The Electric Field and Waves Suite (EFW) will measure electric fields associated with a variety of mechanisms causing particle energization, scattering, and transport in the inner magnetosphere, both DC and AC electric fields

(Wygant et al., submitted manuscript, 2012). EFW will study the electric fields in near-Earth space that energize radiation particles and modify the structure of the inner magnetosphere.

4. INTEGRATION AND TESTING

Figure 5a–d shows snapshots of the integration and testing of instruments and subsystems on the two RBSP satellites at different stages. Both spacecraft A and B boxes were delivered to APL in November 2010; spacecraft A is shown as delivered in Figure 5a. Immediately thereafter, integration began on both

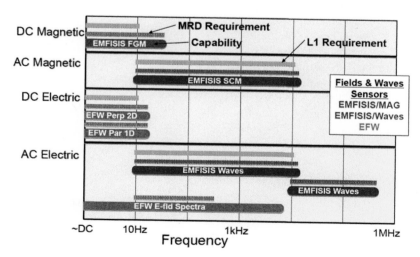

Figure 4. The combined frequency ranges for the RBSP magnetic and electric field instruments from DC up to tens of kHz and nearly a MHz for single-component AC electric. Capability of each instrument exceeds the requirements of the MRD as well as of the L1 requirements.

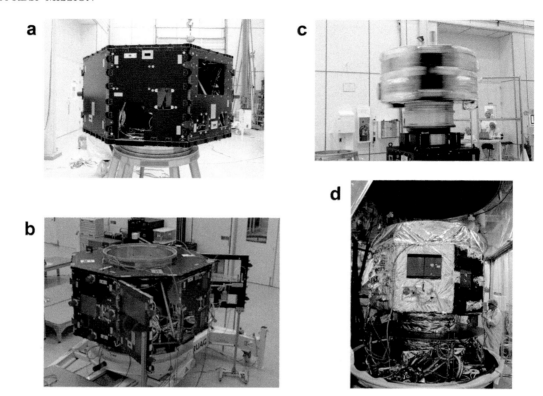

Figure 5. (a) RBSP spacecraft A on 17 November 2010 at the beginning of integration. (b) Spacecraft B in April 2011 with the Relativistic Proton Spectrometer instrument installed, along with Sun sensors, propulsion system, integrated electronics module, and power distribution unit. (c) Spacecraft A in September 2011 undergoing spin balance testing. (d) Spacecraft A in preparation for thermal vacuum testing in February 2012.

spacecraft, starting with the harness, avionics, and power distribution unit (PDU). Extensive testing of all subsystems, mission simulations, and reviews were held throughout the integration and testing phase. Figure 5b shows spacecraft B months later in April 2011 with the RPS instrument installed, along with Sun sensors, propulsion system, the integrated electronics module (IEM), and the PDU. Later in April, RBSPICE was installed and integrated on both spacecraft and in May 2011, EMFISIS was installed on both. REPT and EFW were integrated in June 2011. MagEIS has four separate units; the low and two medium sensors were integrated on spacecraft A in early August 2011. Later that month, the HOPE instrument was integrated on both spacecraft. Figure 5c shows spacecraft A in September 2011 undergoing spin balance testing. By January 2012, all four MagEIS sensors were integrated on both spacecraft. Figure 5d shows spacecraft A in preparation for thermal vacuum testing in February 2012, the last major hurdle before shipment to the launch site in May 2012. The RBSP mission launched on 30 August 2012 and, after 2 months of commissioning, will begin science operations.

5. EARLY SCIENCE ENDEAVORS

By the start of normal operations (nominally 60 days after launch), the orbit of the RBSP spacecraft will have an apogee in the dawn sector at about 6 LT, and the satellites will have a variable separation similar to Figure 1b. At a Science Working Group meeting in March 2012, the instrument teams addressed the following science questions that are likely to yield results early in the mission. Each of these questions relate directly to the broad science objectives listed in section 1.

1. What issues can be resolved about whistler mode interactions and their roles in electron energization and loss in the first 3 months? (objectives 1, 2)

2. What issues can be resolved about the large-scale dynamics and structure with just the first few major geomagnetic storms? (objective 3)

3. What issues can be resolved about the source, structure, and dynamics of the inner ($L < 2$) ion and electron belts in the first 3 months? (objectives 1, 2, 3)

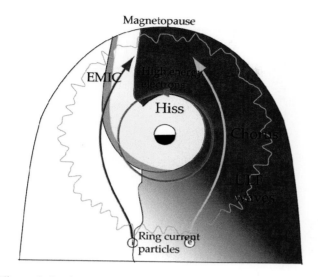

Figure 6. A schematic of Earth's inner magnetosphere showing the locations where various waves are typically found in the magnetosphere. The dawn sector is dominated by chorus and hiss, while EMIC waves are found primarily on the duskside. ULF waves occur throughout the magnetosphere. From *Shprits et al.* [2006].

5.1. Science Question 1

Figure 6 shows the locations where various waves are typically found in the magnetosphere. The dawn sector is dominated by chorus and hiss, while EMIC waves are found primarily on the duskside. ULF waves occur throughout the magnetosphere. Electron energization and loss is largely driven by interactions with these waves (review by *Thorne* [2010]). The early emphasis of the RBSP mission will be on chorus and hiss, possible relationships between the two, and the resulting electron energization and loss.

Plasmaspheric hiss is broadband noise between 100 Hz and a few kHz. Hiss is confined primarily to high-density regions: the plasmasphere and dayside drainage plumes. The generation mechanism is not yet understood. At the high-frequency end (>1 kHz) and at low L, the source could be lightning. At more typical frequencies (100–300 Hz), the source probably lies in the magnetosphere. We care about plasmaspheric hiss because hiss can deplete electron populations in the slot region between the inner and outer radiation belts through pitch angle scattering [*Millan and Thorne*, 2007, and references therein].

Whistler mode chorus waves have frequencies between 100 Hz and 5 kHz, a similar range to hiss. Chorus is typically found outside the plasmasphere and primarily on the dawnside near the equator. Chorus is generated by the electron cyclotron instability near the equator operating on particle anisotropies associated with injected plasma sheet electrons. The intensity of chorus waves increases during substorms

and the recovery phase. Scattering by chorus is likely responsible for relativistic electron microburst precipitation [*Lorentzen et al.*, 2001]. We care about chorus waves because they are a major potential mechanism for electron acceleration. Losses associated with microburst precipitation are capable of emptying the outer radiation belt in a day or less [*O'Brien et al.*, 2004].

Early questions to be addressed by the EMFISIS team include: What wave modes happen at both spacecraft as a function of separation and location? What is the spatial coherence of chorus for small separation? Is chorus the parent wave for hiss? Early questions to be addressed by the ECT and EMFISIS teams, working together, include: What waves are dominant when there is evidence of particle enhancement or loss? Do plasma density changes modulate the intensity of chorus wave activity?

5.2. Science Question 2

Following the occurrence of a few geomagnetic storms during the first few months of RBSP operation, the EFW team will be able to explore the connection between large-amplitude whistler mode chorus waves and microburst precipitation during the storms. *Santolík et al.* [2003] identified large-amplitude chorus waves in Cluster observations; lower band chorus (<0.5 f_{ce}) wave electric fields approached 30 mV m^{-1}. The chorus wave activity may cause electrons to precipitate into the ionosphere, perhaps in the form of microbursts, brief (<1 s) increases in the flux of precipitating MeV electrons that were first noted by *Imhof et al.* [1992] in S 81-1 spacecraft observations. Figure 7 shows observations of microbursts from the Heavy Ion Large Telescope on the SAMPEX satellite. Microbursts are usually observed from $L = 4$ to 6 and near dawn, though they may extend from near midnight past dawn. Statistically, the connection between large-amplitude whistler mode chorus and microburst precipitation is strong. Questions remain about the physical

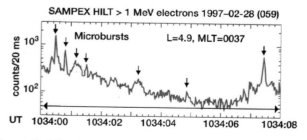

Figure 7. Microbursts observed on the SAMPEX Heavy Ion Large Telescope instrument in 1997. The microbursts indicated by black arrows are observed at $L = 4.9$ on the dawnside of the magnetosphere.

mechanism connecting chorus and microbursts and their variability during geomagnetic storms when the occurrence of both are at their highest value.

After a few geomagnetic storms during the first few months of RBSP operation, the RBSPICE team will explore the composition of the ring current during storms. *Hamilton et al.* [1988] showed that the energy density of oxygen ions dominated that of protons during an intense geomagnetic storm near solar minimum, using data from the Charge Composition Explorer instrument on the AMPTE satellite. The RBSPICE team will able to address the following question: How do the contributions to current density from protons, helium ions, and oxygen ions differ during weak and strong geomagnetic storms?

5.3. Science Question 3

Because few spacecraft have spent significant time near the magnetic equator and at the peak intensities of the inner belt, the RPS experiment on RBSP is positioned to discover the energy spectrum of the inner belt protons. The dominant source for protons above 50 MeV in the inner belt is the decay of albedo neutrons from galactic cosmic ray protons that collide with nuclei in the atmosphere and ionosphere (cosmic ray albedo neutron decay). The ion energy spectrum is known to extend beyond 1 GeV, but the spectral details are not well established: the shape, maximum energy, and time dependence are poorly quantified. Figure 8 shows two different models (AP8 MIN [*Sawyer and Vette*, 1976] and AD2005 [*Selesnick et al.*, 2007] that show a wide variation). RPS will supply the upper-energy range of the spectrum; MagEIS and REPT will fill in the lower-energy range.

5.4. Collaborations With Other Missions

5.4.1. Time History of Events and Macroscale Interactions during Substorms (THEMIS).
During early operations of RBSP, with the two RBSP spacecraft on the dawnside, the three THEMIS spacecraft will be on the duskside. An early science campaign will explore dawn-dusk differences and could address the following questions: What are the cause(s) of dawn-dusk differences in ion fluxes during geomagnetic storms? What are the relative roles of EMIC waves in the dusk magnetosphere, chorus waves in the dawn magnetosphere, and hiss deep within the magnetosphere?

5.4.2. Balloon Array for Radiation-belt Relativistic Electron Losses (BARREL).
In January 2013, balloons from the BARREL mission will be positioned in the southern hemisphere on magnetic field lines where electron precipitation is expected. The apogees of the three THEMIS spacecraft will be

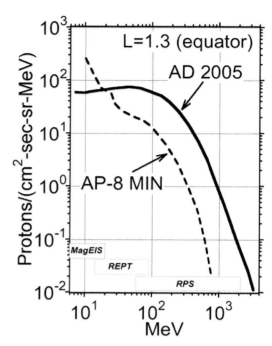

Figure 8. Example of the wide variation in modeled inner belt spectra. AP8 MIN [*Sawyer and Vette*, 1976]; AD2005 [*Selesnick et al.*, 2007].

on the dayside near the magnetopause and can measure losses. The two RBSP spacecraft, with apogee on the nightside, will be passing through all interior L shells on the dayside during the orbit, measuring changes in the in situ trapped electron intensity. This configuration can quantify the losses due to precipitation versus the losses through the magnetopause. This study is motivated by the recent results of *Turner et al.* [2012] showing that a sudden electron depletion was primarily a result of outward transport through the magnetopause, rather than precipitation loss to the atmosphere as in the earlier results from, e.g., *Selesnick* [2006] and *O'Brien et al.* [2004].

6. SPACE WEATHER BROADCAST

Space weather data on RBSP will be broadcast nearly 24/7; whenever one of the RBSP spacecraft is not downlinking science data or engaged in maneuvers, it will be broadcasting space weather data in real time that will be received by a number of ground stations spread around the world. Extensive ground station coverage is planned, although there are intervals along the orbit when spacecraft antennas are not in view, which will result in seasonally dependent data coverage. In addition, the real time coverage will be reduced by an average of 2.5 h for each spacecraft per day due to the primary science downlink. The 2.5 h of missing data will be backfilled later from the science data. The space weather data will be available within minutes of reception from the ground stations. More details on the space weather broadcast can be found in the work of R. L. Kessel (RBSP and space weather, submitted to *Space Science Reviews*, 2012).

The space weather data products were chosen to provide a quick look at the state of the radiation belts. Energies within the space weather data set covered by the particle detectors span the ranges expected for the two belts, 25 eV to >400 MeV for protons, 25 eV to 10 MeV for electrons. HOPE provides the lower range of both the inner-belt ions and outer-belt electrons, and the ion composition for the lower-energy ions. MagEIS provides the mid and upper range of outer belt electrons. REPT provides the upper range of inner belt protons and high range of outer belt electrons, while RPS covers the extreme upper range of inner belt protons. RBSPICE provides the mid energy protons intensities and composition.

The EFW instrument will provide the large-scale electric field, and the magnetometer on EMFISIS will provide the large-scale magnetic field. The Waves instrument portion of EMFISIS will provide three values of the high-frequency electric and magnetic field observations every 12 s for the RBSP space weather data set: (1) f_{ce} to 0.5 f_{ce} (lower-band chorus), (2) 0.5 f_{ce} to 0.7 f_{ce} (upper-band chorus), and (3) 10 Hz to $f_{ce}/\sqrt{(1836)}$ (magnetosonic waves). Here f_{ce} is the so-called electron cyclotron frequency, and 1836 is the proton to electron mass ratio.

7. SUMMARY: KEY MESSAGES

The RBSP key messages summarize the high-level impacts expected from the mission:

1. RBSP will advance our understanding of dramatic and puzzling aspects of Earth's radiation belts. The RBSP mission will provide unprecedented insight into how Earth's radiation belts change in both space and time.

2. RBSP will enable the prediction of extreme and dynamic space conditions. RBSP will enable the development of accurate models that can predict how the belts will react to constantly varying solar and solar wind drivers.

3. RBSP will provide understanding needed to design satellites to survive in space. RBSP will explore space weather that can disable satellites, cause power grid failures, and disrupt GPS communications, with instruments that are "hardened" to continue working even in the harshest conditions.

Acknowledgments. Special thanks to the RBSP and BARREL team members who contributed to this article: Barry Mauk, David Sibeck, Nicola Fox, Shri Kanekal, Kris Kersten, Craig Kletzing, Lou Lanzerotti, Tony Lui, Joe Mazur, Robyn Millan, Geoff Reeves, John Wygant, Jenny Rumburg, Rick Fitzgerald, Jim Stratton, and Michelle Weiss.

REFERENCES

Hamilton, D. C., G. Gloeckler, F. M. Ipavich, W. Stüdemann, B. Wilken, and G. Kremser (1988), Ring current development during the great geomagnetic storm of February 1986, *J. Geophys. Res.*, *93*(A12), 14,343–14,355.

Imhof, W. L., H. D. Voss, J. Mobilia, D. W. Datlowe, E. E. Gaines, J. P. McGlennon, and U. S. Inan (1992), Relativistic electron microbursts, *J. Geophys. Res.*, *97*(A9), 13,829–13,837.

Kintner, P. M., and Geospace Mission Definition Team (2002), The LWS Geospace storm investigations: Exploring the extremes of space weather, *NASA/TM-2002-211613*, NASA STI Program Off., Hanover, Md.

Lorentzen, K. R., J. B. Blake, U. S. Inan, and J. Bortnik (2001), Observations of relativistic electron microbursts in association with VLF chorus, *J. Geophys. Res.*, *106*(A4), 6017–6027, doi:10.1029/2000JA003018.

Mauk, B. H., N. J. Fox, S. G. Kanekal, R. L. Kessel, D. G. Sibeck, and A. Ukhorskiy (2012), Science objectives and rationale for the Radiation Belt Storm Probes mission, *Space Sci. Rev.*, doi:10.1007/s11214-012-9908-y, in press.

Mazur, J., et al. (2012), The Relativistic Proton Spectrometer (RPS) for the Radiation Belt Storm Probes mission, *Space Sci. Rev.*, doi:10.1007/s11214-012-9926-9, in press.

Millan, R. M., and R. M. Thorne (2007), Review of radiation belt relativistic electron losses, *J. Atmos. Sol. Terr. Phys.*, *69*, 362–377.

O'Brien, T. P., M. D. Looper, and J. B. Blake (2004), Quantification of relativistic electron microburst losses during the GEM storms, *Geophys. Res. Lett.*, *31*, L04802, doi:10.1029/2003GL018621.

Santolík, O., D. A. Gurnett, J. S. Pickett, M. Parrot, and N. Cornilleau-Wehrlin (2003), Spatio-temporal structure of storm-time chorus, *J. Geophys. Res.*, *108*(A7), 1278, doi:10.1029/2002JA009791.

Sawyer, D. M., and J. I. Vette (1976), AP-8 trapped proton environment for solar maximum and solar minimum, *NASA-TM-X-72605*, Goddard Space Flight Cent., Greenbelt, Md.

Selesnick, R. S. (2006), Source and loss rates of radiation belt relativistic electrons during magnetic storms, *J. Geophys. Res.*, *111*, A04210, doi:10.1029/2005JA011473.

Selesnick, R. S., M. D. Looper, and R. A. Mewaldt (2007), A theoretical model of the inner proton radiation belt, *Space Weather*, *5*, S04003, doi:10.1029/2006SW000275.

Shprits, Y. Y., W. Li, and R. M. Thorne (2006), Controlling effect of the pitch angle scattering rates near the edge of the loss cone on electron lifetimes, *J. Geophys. Res.*, *111*, A12206, doi:10.1029/2006JA011758.

Thorne, R. M. (2010), Radiation belt dynamics: The importance of wave-particle interactions, *Geophys. Res. Lett.*, *37*, L22107, doi:10.1029/2010GL044990.

Turner, D. L., Y. Shprits, M. Hartinger, and V. Angelopoulos (2012), Explaining sudden losses of outer radiation belt electrons during geomagnetic storms, *Nat. Phys.*, *8*, 208–212, doi:10.1038/nphys2185.

R. L. Kessel, NASA Headquarters, 300 E Street, SW, Washington, DC 20546, USA. (mona.kessel@nasa.gov)

The Energization and Radiation in Geospace (ERG) Project

Y. Miyoshi,[1] T. Ono,[2] T. Takashima,[3] K. Asamura,[3] M. Hirahara,[1] Y. Kasaba,[2] A. Matsuoka,[3] H. Kojima,[4] K. Shiokawa,[1] K. Seki,[1] M. Fujimoto,[3] T. Nagatsuma,[5] C. Z. Cheng,[6] Y. Kazama,[6] S. Kasahara,[3] T. Mitani,[3] H. Matsumoto,[7] N. Higashio,[7] A. Kumamoto,[2] S. Yagitani,[8] Y. Kasahara,[8] K. Ishisaka,[9] L. Blomberg,[10] A. Fujimoto,[3] Y. Katoh,[2] Y. Ebihara,[4] Y. Omura,[4] M. Nosé,[11] T. Hori,[1] Y. Miyashita,[1] Y.-M. Tanaka,[12] T. Segawa,[1] and ERG Working Group

The Energization and Radiation in Geospace (ERG) project for solar cycle 24 will explore how relativistic electrons in the radiation belts are generated during space storms. This geospace exploration project consists of three research teams: the ERG satellite observation team, the ground-based network observation team, and the integrated data analysis/simulation team. Satellite observation will provide in situ measurements of features such as the plasma distribution function, electric and magnetic fields, and plasma waves, whereas remote sensing by ground-based observations using, for example, HF radars, magnetometers, optical instruments, and radio wave receivers will provide the global state of the geospace. Various kinds of data will be integrated and compared with numerical simulations for quantitative understanding. Such a synergetic approach is essential for comprehensive understanding of relativistic electron generation/loss processes through cross-energy and cross-regional coupling in which different plasma populations and regions are dynamically coupled with each other. In addition, the ERG satellite will utilize a new and innovative measurement technique for wave-particle interactions that can directly measure the energy exchange process between particles and plasma waves. In this paper, we briefly review some of the profound problems regarding relativistic electron accelerations and losses that will be solved by the ERG project, and we provide an overview of the project.

[1]Solar-Terrestrial Environment Laboratory, Nagoya University, Nagoya, Japan.

[2]Department of Geophysics, Tohoku University, Sendai, Japan.

[3]Institute of Space and Astronautical Science, Japan Aerospace Exploration Agency, Sagamihara, Japan.

[4]Research Institute for Sustainable Humanosphere, Kyoto University, Uji, Japan.

[5]National Institute of Information and Communications Technology, Koganei, Japan.

[6]Plasma and Space Science Center, National Cheng Kung University, Tainan, Taiwan.

[7]Aerospace Research and Development Directorate, Japan Aerospace Exploration Agency, Tsukuba, Japan.

[8]Graduate School of Natural Science and Technology, Kanazawa University, Kanazawa, Japan.

[9]Department of Information Systems Engineering, Toyama Prefectural University, Toyama, Japan.

[10]Royal Institute of Technology, Stockholm, Sweden.

[11]Data Analysis Center for Geomagnetism and Space Magnetism, Graduate School of Science, Kyoto University, Kyoto, Japan.

[12]National Institute of Polar Research, Tachikawa, Japan.

Dynamics of the Earth's Radiation Belts and Inner Magnetosphere
Geophysical Monograph Series 199
10.1029/2012GM001304

1. INTRODUCTION

High-energy particles (ions and electrons) are trapped in the Earth's magnetic field and form the radiation belts. MeV electrons in the radiation belts are the highest energy particles in geospace. Recent satellite (CRRES, Akebono, Time History of Events and Macroscale Interactions during Substorms (THEMIS), etc.), ground-based observations, and modeling studies have revealed the detailed structure and variations of the inner magnetosphere. As shown later, the cross-energy coupling and cross-regional coupling between plasmasphere, plasma sheet, ring current, and radiation belts, and magnetosphere-ionosphere coupling become important concept for understanding on the radiation belts and inner magnetosphere [*Ebihara and Miyoshi*, 2011].

In space storms, the outer belt electrons decrease significantly during the main phase, then recover to, and often increase over, prestorm levels during the recovery phase [e.g., *Baker et al.*, 1986; *Nagai*, 1988; *Reeves et al.*, 1998, 2003; *Miyoshi and Kataoka*, 2005]. During huge magnetic storms, the radiation belts are largely deformed, and large flux enhancement are observed in the slot region and the inner belt [*Baker et al.*, 2004]. Two possible mechanisms have been proposed for the acceleration of relativistic electrons (see reviews, e.g., of *Friedel et al.* [2002], *Shprits et al.* [2008a, 2008b], *Hudson et al.* [2008], and *Ebihara and Miyoshi* [2011]). One is the external source process via quasi-adiabatic acceleration [*Schulz and Lanzerotti*, 1974]. In this process, the energy of electrons increases due to the conservation of their first and second adiabatic invariants, when electrons are transported from the plasma sheet to the inner magnetosphere. This process has been modeled as the stochastic radial diffusion process, which is a fundamental transportation mode of energetic electrons. The ULF Pc5 pulsations with periods of a few minutes have been considered a main driver for the radial transportation via drift-resonance with electrons [e.g., *Rostoker et al.*, 1998; *Hudson et al.*, 2001; *Elkington et al.*, 1999; *Elkington*, 2006; *Mathie and Mann*, 2000].

Another candidate is termed the internal acceleration process. It has been suggested that resonant interactions by whistler mode waves cause relativistic electron acceleration inside the radiation belts [e.g., *Summers et al.*, 1998; *Miyoshi et al.*, 2003; *Horne et al.*, 2005]. The free energy for generating whistler mode waves is the temperature anisotropy of electrons of tens of keV [e.g., *Kennel and Petscheck*, 1966; *Jordanova et al.*, 2010], and subsequent wave-particle interactions including nonlinear process will generate chorus waves [e.g., *Katoh and Omura*, 2007; *Omura et al.*, 2008] that accelerate relativistic electrons of the outer belt [e.g., *Summers and Ma*, 2000; *Summers et al.*, 2007].

Wave generation and resonant conditions are affected by the cold plasma distribution in the inner magnetosphere. Satellite observations have shown that MeV electron flux enhancement of the outer belt occurs outside the plasmapause where the intense whistler mode chorus waves are generated (Figure 1), suggesting the importance of whistler mode waves for electron accelerations [e.g., *Meredith et al.*, 2003; *Miyoshi et al.*, 2003, 2007; *Horne et al.*, 2005; *Y. Kasahara et al.*, 2009]. Thus, the whistler mode waves work as a mediating agent that can convert the energy from a low-energy electron population to a higher-energy one, and dynamical cooperation of plasma/particles of wide energy range from eV to MeV via wave-particle interactions is important. Magnetosonic mode waves are also plausible candidates for internal accelerations [*Horne et al.*, 2007, *Meredith et al.*, 2008].

The electron-flux enhancement process is the result of a delicate balance between acceleration and loss of relativistic electrons [*Reeves et al.*, 2003], so that loss processes are as important as acceleration processes. Several possibilities have been proposed as loss processes (see reviews of *Millan and Thorne* [2007] and *Turner et al.* [this volume]). Although adiabatic deceleration always operates during ring current evolutions [*Kim and Chan*, 1997], nonadiabatic loss processes also work during space storms. Magnetopause shadowing and subsequent outward radial diffusion may cause rapid loss of outer belt electrons [e.g., *Brautigam and Albert*, 2000; *Miyoshi et al.*, 2003; *Shprits et al.*, 2006; *Matsumura et al.*, 2011; *Turner et al.*, 2012]. Pitch angle scatterings by electromagnetic ion cyclotron waves (EMIC) and whistler mode waves are important in that they cause precipitation of relativistic electrons in the atmosphere [e.g., *Thorne and Kennel*, 1971; *Lyons et al.*, 1972; *Abel and Thorne*, 1998; *Li et al.*, 2007; *Miyoshi et al.*, 2008; *Jordanova et al.*, 2008]. The loss processes associated with pitch angle scattering are expected to be localized, and therefore, multipoint observations by satellites and ground-based network stations should be important.

Figure 2 summarizes the transport/acceleration mechanisms in the L energy diagram of the inner magnetosphere. In radial diffusion (blue arrows), the electrons move earthward with increasing energy due to the conservation of the first two adiabatic invariants. ULF Pc5 waves will be the main driver for the radial diffusion. On the other hand, in the in situ acceleration by waves (red arrows), subrelativistic electrons are accelerated to MeV energies by whistler mode/magnetosonic waves that are generated by the plasma instability of the ring current electrons and ions. Plasma waves such as whistler waves and magnetosonic waves transport energies from the population of ring current electrons/ions to the population of subrelativistic electrons via wave-particle

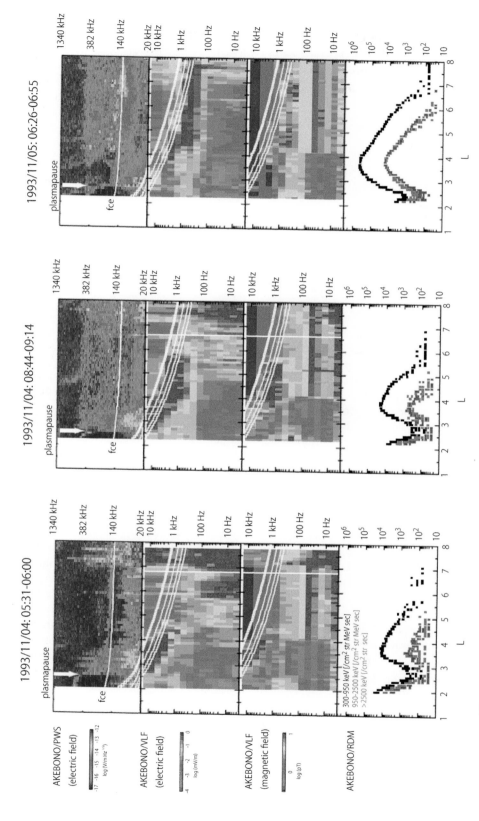

Figure 1. Akebeono measurement on plasma waves and relativistic electrons as a function of L shell during the 3 November 1993 magnetic storm. (top) White arrows indicate the plasmapause, which are derived from the upper-hybrid resonant waves. (middle two) VLF wave observations for electric/magnetic fields, which show intense chorus emissions outside the plasmapause. (bottom) The energetic electrons of three channels (300–950 keV, 950–2500 keV, >2500 keV) and the evolution of the outer belt were observed. After *Miyoshi et al.* [2003].

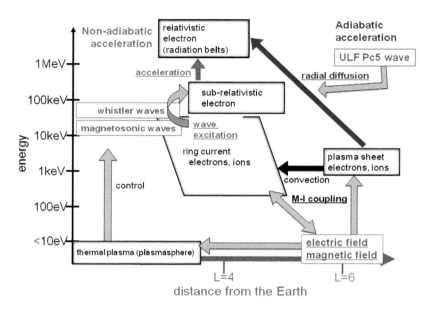

Figure 2. Diagram of cross-energy coupling of the inner magnetosphere.

interactions, and relativistic electron flux increases inside the outer belt. In this process, thermal plasma density also plays an important role as the ambient medium. Because the transport of ring current electrons and thermal plasma is predominantly controlled by convective electric fields, the process of convection may affect the relativistic electron dynamics in the internal acceleration process. The cross-regional coupling between magnetosphere and ionosphere drives the dynamical evolution of convective electric fields in the inner magnetosphere [e.g., *Ebihara et al.*, 2004]. Moreover, the cross-energy coupling among widely differing energy plasma/particle populations more than 6 orders (eV–MeV) in the inner magnetosphere plays to generate MeV electrons of the outer belt via wave-particle interactions. Therefore, the formation of the radiation belt is one of the manifestations of cross-energy/cross-regional couplings in geospace [*Mann et al.*, 2006; *Mann*, 2008, *Ebihara and Miyoshi*, 2011], and these couplings are the key concept of the Energization and Radiation in Geospace (ERG) project in understanding how relativistic electrons are generated in geospace during space storms.

A study of relativistic electrons in the radiation belts is also important for understanding space weather [*Baker*, 2002]. Space infrastructure such as GPS, meteorological satellites, and telecommunications satellites that operate in the radiation belts is indispensable in our modern society. The high-energy particles can cause operational anomalies with satellites. In fact, a close relationship has been suggested between satellite anomalies and the large flux enhancement of relativistic electrons of the outer belt [*Pilipenko et al.*,

2006]. Deep dielectric charging in the satellite by the enhancement of relativistic electrons is one of the major causes of satellite anomalies. Further, the International Space Station (ISS) also operates at the bottom of the inner and outer radiation belt. An astronaut during extravehicular activity (EVA) at ISS is exposed to radiation not only by high-energy protons but also by relativistic electrons because MeV electrons can penetrate into a spacesuit [*National Research Council*, 2000]. To control the exposed dose of an astronaut during EVA at ISS, information about current and future condition of relativistic electron environment is important.

To examine what mechanisms mainly contribute to the evolution of the outer belt electrons, direct measurements of plasma, fields, and waves are important around the magnetic equator. CRRES and Akebono satellites suggested the importance of observations at the equatorial plane [*Seki et al.*, 2005]. Moreover, the phase space density profile is a key to discriminate between the external supply process and the internal acceleration process [*Green and Kivelson*, 2004; *Chen et al.*, 2007]. Comprehensive observations, however, have never been realized in the inner magnetosphere because the strong radiation environment causes serious contamination of particle measurements. Thus, a clear understanding of the acceleration mechanism has not yet been obtained.

2. ERG PROJECT

As noted in the previous section, the acceleration mechanisms that cause the large flux enhancement of the outer belt as well as the loss mechanisms have not been proven. The comprehensive observations on plasma/particles for wide-energy range and electric/magnetic fields for wide-frequency range are essential for the understanding of cross-energy coupling. The complement observations among satellites and ground networks are necessary for the understanding of cross-regional coupling. To understand particle acceleration/loss mechanisms and dynamical evolution of space storms in the context of cross-energy and cross-regional coupling, the ERG project has been designed.

The following teams are involved in the ERG project: the satellite observation team, the ground-based network observations team, and the integrated data analysis/simulation team. The project science team and the project science center also work with project management. The preliminary concept study for the ERG project has been reported [*Shiokawa et al.*, 2006].

3. THE ERG SATELLITE

3.1. Overview of the ERG Satellite

The ERG satellite is being developed as a science satellite by the Institute of Space and Astronautical Science (ISAS)/Japan Aerospace Exploration Agency (JAXA). Figure 3 shows the appearance of the ERG satellite, which will be launched around the declining phase of solar cycle 24

Figure 3. Appearance of the Energization and Radiation in Geospace (ERG) satellite.

(~2015–2016). The nominal mission life is planned to be longer than 1 year. The satellite is designed to be Sun-oriented and spin-stabilized with a rotation rate of 7.5 rpm (8 s). The designed apogee altitude is 4.5 R_E ($L \sim 5.5$ at equatorial plane), located at the average peak flux L shell of the outer belt during the declining phase [*Miyoshi and Kataoka*, 2011], and the perigee altitude is ~300 km. The planned orbital inclination is 31°, while the satellite does not always observe at the magnetic equator. However, the satellite will have many chances for the observations near the equator and will measure the phase space density, generation of plasma waves, and accelerations of relativistic electrons, etc., near the equator. Moreover, the off-equator observations are important to discuss the propagation of plasma waves from the equator, so that comparative studies with the observations by other satellites near the equator are also important.

Comprehensive observations of plasma/particles, fields, and waves are important for understanding the cross-energy coupling for relativistic electron accelerations and dynamics of space storms. As described below, instruments for plasma and particle experiment (PPE) and for electric fields and plasma wave experiment (PWE)/magnetic field experiment (MGF) are installed in the satellite. A cutting-edge detection technique for wave-particle interactions (software-wave particle interaction analyzer (S-WPIA)) will be realized for the first time by the ERG satellite, which will be able to directly measure the energy exchange processes between particles and plasma waves.

3.2. Plasma and Particle Experiment

The PPE consists of four electron sensors (LEP-e, MEP-e, HEP-e, and XEP-e) and two ion sensors (LEP-i and MEP-i). The electron sensors can measure electrons from 12 eV to 20 MeV, while the ion sensors can measure ions from 10 eV q^{-1} to 180 keV q^{-1} with mass discrimination. Figure 4a shows the energy range of each plasma/particle instrument together with plasma/particle populations in the inner magnetosphere. The energy ranges of detectors are designed to overlap each other to provide a seamless energy spectrum for a wide energy range. Table 1 shows the specifications of PPE sensors.

Regarding electron observations, both HEP-e and XEP-e instruments mainly observe relativistic electrons of the radiation belts, and these instruments are used to derive the radial profile of the phase space density. Since the anisotropy of the distribution function of hot electrons is a free energy of plasma waves, observations of the distribution function by LEP-e/MEP-e are important to clarify how plasma waves are generated in the inner magnetosphere. Measurements of particles with energies lower than tens of keV have been difficult inside the radiation belts because of serious

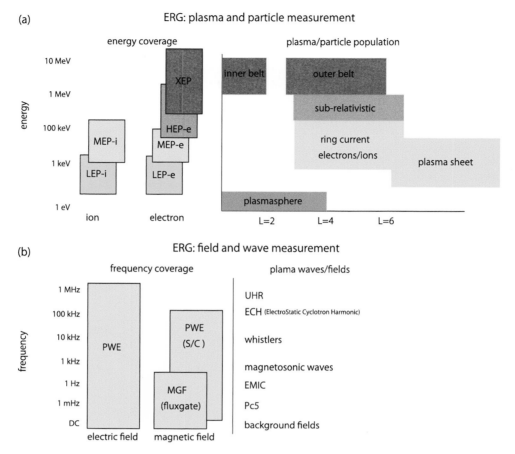

Figure 4. Science instruments onboard the ERG satellite. (a) Energy range of the plasma and particle experiment instruments. (b) Frequency range of the plasma wave experiment and magnetic field experiment instruments.

contamination of the energetic particles [*Liu et al.*, 2005]. Newly developed technologies to reduce the background contamination can be applied in ERG/PPE to allow for detailed observations of tens of keV electrons [*S. Kasahara et al.*, 2009a].

The LEP-i and MEP-i instruments are ion energy-mass spectrometers. Several ion species can be discriminated with a time-of-flight (TOF) method which also provides effective reduction of the contamination noise due to high-energy particles. Especially at tens of keV energies, a new technology enables full solid angle observations of ions up to 180 keV q^{-1} [*Kasahara et al.*, 2006]. These ion observation data will be used to study the evolution of ring current ions. Recent computer simulations showed that the ion hole contributes the nonlinear evolution of EMIC waves [*Shoji and Omura*, 2011], and the ring distribution of ions causes magnetosonic mode waves [*Chen et al.*, 2010]. The measurement by the ERG satellite will reveal what plasma instabilities generate these waves. Moreover, ion observations with mass

discrimination are necessary for studying the composition of ring current particles needed to understand how plasma pressure distributions change the field configuration of the inner magnetosphere during space storms.

3.3. Electric Fields and Plasma Wave Experiment

The PWE observes electric fields at the frequency range from DC to 10 MHz and magnetic fields at the frequency range from a few Hz to 100 kHz. It is based on the design slightly modified from the plasma wave investigations aboard the BepiColombo Mercury Magnetospheric Orbiter (MMO) [*Kasaba et al.*, 2010]. The electric field component is measured by two pairs of wire dipole antennas with ~32 m tip-to-tip length. The magnetic field component is measured by the three-axial search coils. They are served two receivers, EWO (electric field detector (EFD), waveform capture (WFC), and onboard frequency analyzer (OFA)) and high-frequency analyzer (HFA). The EWO always produces the

Table 1. Specification of Plasma and Particle Experiment Sensors Onboard ERG Satellite

	Sensor	Energy Range	$\Delta E/E$	Angular Resolution (Sensor Elevation (deg) × Sensor Azimuth (deg))	Field of View (Sensor Elevation (deg) × Sensor Azimuth (deg))	Measurement Techniques[a]
Electron	XEP-e	200 keV to 20 MeV	0.1–0.2 (200 keV to 2 MeV) 0.25 (2 MeV to 20 MeV)	20 × 20	20 × 20 (full width of cone)	Si solid-state detector (SSD) + gadolinium silicon oxide scintillator telescope Anticoincidence detection for noise rejection
	HEP-e	70 keV to 2 MeV	50 keV (<200 keV) 0.1 (>200 keV)	10 × 10	10 × 180	Multistacked Si SSD telescope Coincidence detection for noise rejection (Refer to HEP on Mercury Magnetospheric Orbiter (1) for the SSD type and read out)
	MEP-e	10–80 keV	0.1	5 × 5	5 × 360	Cusp-type electrostatic energy analyzer (2) with Si avalanche photo diode (3) Energy coincidence detection for noise rejection (4)
	LEP-e	12 eV to 20 keV	0.08	2.9 × 21.7	2.9 × 270	Tophat-type electrostatic energy analyzer Radiation shields for noise suppression
Ions	MEP-i	10–180 keV q^{-1}	0.1	5 × 22.5	5 × 360	Cusp-type electrostatic energy (2) +Time-of-flight (TOF) type mass analyzer + Si SSD (5) TOF coincidence detection for noise rejection
	LEP-i	10 eV q^{-1} to 25 keV^{-1}	0.13	5 × 22.5	5 × 290	Tophat-type electrostatic energy analyzer + TOF type mass analyzer TOF coincidence detection for noise rejection

[a]References are as follows: 1, *Y. Saito et al.* [2010]; 2, *Kasahara et al.* [2006]; 3, *Kasahara et al.* [2012]; 4, *S. Kasahara et al.* [2009a]; 5, *S. Kasahara et al.* [2009b].

electric and magnetic field data as low-frequency waveforms (512 Hz sampling), wave spectrum (10 Hz to 20 kHz), and their spectral matrices, which can be used to determine wave normal angles and Poynting fluxes. The spacecraft potential (256 Hz sampling) will also be observed for the detection of fast electron density variations. The EWO also stores raw waveforms of electric field and magnetic fields (60 kHz sampling), which are partially downloaded to the ground within the telemetry limit after the selection by researchers checking low-resolution data. The HFA produces wave spectrum of electric field (10 kHz to 10 MHz) and magnetic field (10–100 kHz). By the onboard data identification, the electron density is identified from the upper-hybrid resonance frequency. Electric and magnetic waveforms from EWO and electron density from HFA will be provided to the S-WPIA

for the identification of wave modes and the characteristics of wave particle interactions (see section 3.5).

Figure 4b shows the frequency range of the PWE and the MGF (see section 3.4) with expected key plasma waves observed on the ERG orbit. Whistler mode chorus waves [e.g., *Santolík et al.*, 2003; *Miyoshi et al.*, 2007; *Omura et al.*, 2008; *Li et al.*, 2011] and magnetosonic mode waves [e.g., *Kokubun et al.*, 1991; *Kasahara et al.*, 1994; *Meredith et al.*, 2008] will be important for the study of nonadiabatic acceleration of generate relativistic electrons. EMIC waves [e.g., *Sawada et al.*, 1991; *Kasahara et al.*, 1992; *Jordanova et al.*, 2008] generated from ring current ions will cause the rapid pitch angle scattering of these relativistic electrons [e.g., *Summers and Thorne*, 2003; *Albert*, 2003; *Miyoshi et al.*, 2008]. Inside the plasmapause, whistler mode hiss waves work for the pitch angle scattering of high-energy electrons [e.g., *Lyons et al.*, 1972; *Meredith et al.*, 2009].

The PWE observations will also reveal how the electric fields in the inner magnetosphere evolve during space storms. Convective electric field causes global-scale transportation of plasma sheet and ring current components. It should contain large electric fields observed in the inner magnetosphere by CRRES and Akebono [e.g., *Rowland and Wygant*, 1998; *Nishimura et al.*, 2007], the localized electric fields of subauroral polarization streams affecting the evolution of the plasmasphere [e.g., *Goldstein et al.*, 2003], and the evolution of the electric fields in the inner magnetosphere largely controlled by the magnetosphere-ionosphere coupling via the region 2 current system [e.g., *Ebihara et al.*, 2004]. With the MGF, the PWE will also detect the ULF Pc5 pulsation with ~5 min periods, which is a driver for quasi-adiabatic acceleration by radial diffusion.

3.4. Magnetic Field Experiment

A fluxgate magnetometer for the MGF is used to observe the ambient magnetic field, ULF Pc5 pulsations, and Pc1/EMIC waves in the inner magnetosphere. Since the ring current produces the distortion of the ambient magnetic field, and the distortion affects the particle distribution and trajectories in the inner magnetosphere, accurate measurement of the magnetic field deviation from the intrinsic magnetic field is important to evaluate the effect and evolution of the ring current. The MGF instrument is designed and will be calibrated to measure the field deviation with the accuracy of better than 3° during space storms. MGF has two dynamic ranges, ±60000 and ±8000 nT. The measurement resolution is 18 digits, and therefore, the accuracy is 460 and 61 pT for two dynamic ranges, respectively. The observation frequency range of MGF is from DC to several tens Hz (the sampling frequency is 256 Hz and the cut-off frequency will be tuned to be 120 Hz), including the local ion cyclotron frequency, which is key to understand the interaction between the MHD waves and the ambient plasma. The waves at frequencies of EMIC waves may be observed by MGF as well as the PWE instrument. The design of the MGF instrument for ERG is based on that of MGF-I for BepiColombo MMO [*Baumjohann et al.*, 2010; *Matsuoka et al.*, 2012]. The sensor of MGF is mounted at the tip of the 5 m boom deployed from the spacecraft to reduce the noise in the measurement data.

3.5. Software-Wave Particle Interaction Analyzer

To understand the energy exchange process between plasma/particles and waves through wave-particle interactions, the newly developed S-WPIA system is installed in the ERG satellite. The time evolution of the kinetic energy of electrons via wave-particle interactions is determined as follows:

$$\frac{d(m_0 c^2(\gamma - 1))}{dt} = q\mathbf{E}(t)\cdot\mathbf{v}(t) = q|E||v|cos\,\theta$$

where m_0 is the electron rest mass, c is the speed of light, γ is the relativistic factor, t is time, q is the charge of electrons, $\mathbf{E}$ is the instantaneous electric field vector of plasma waves, $\mathbf{v}$ is the electron velocity, and θ is the relative phase between $\mathbf{E}$ and $\mathbf{v}$. Positive $q\mathbf{E}(t)\cdot\mathbf{v}(t)$ indicates the acceleration of particles by waves, whereas negative $q\mathbf{E}(t)\cdot\mathbf{v}(t)$ means the growth of waves. The relative phase θ determines the direction of energy flow.

Recently, an innovative method for direct measurement of $\mathbf{E}\cdot\mathbf{v}$ has been proposed [*Fukuhara et al.*, 2009]. Based on this idea, S-WPIA directly calculates the relative phase θ at each event of particle detections by the onboard particle instruments. S-WPIA has the capability to identify the fraction in the $\mathbf{v} - \theta$ phase space. The fraction corresponds to the so-called electron hole that is theoretically expected [*Omura et al.*, 2008]. Thus, S-WPIA will discriminate quantitatively which electrons contribute to chorus generation and which electrons are actually accelerated by chorus waves. This will be the first observation to unambiguously identify how energy conversion takes place via wave-particle interactions in space.

Figure 5 schematically shows the data processing of the S-WPIA onboard the ERG. All velocity vectors (**v**) for every particle caught by the particle detector as well as the waveforms (**E**) observed by the plasma wave receiver are stored to the onboard memory in the real-time basis. The onboard CPU reads out the stored data from the onboard memory and calculates the phase between a velocity vector and corresponding instantaneous amplitude. Once we sort the data according to θ, the distribution of the number of counts can be obtained as shown in (a). On the other hand, the time variation of $\sum\mathbf{E}\cdot\mathbf{v}$ provides important information of the energy exchange due to wave-particle interactions. We can compare the energy exchange depending on different pitch angles.

4. THE ERG GROUND NETWORK OBSERVATIONS AND INTEGRATED STUDIES/SIMULATION

Multipoint ground-based network observations will act as complement of the in situ satellite observations. Networks of SuperDARN HF-radars, fluxgate and induction magnetometers, airglow/aurora imagers, riometers, VLF/ELF receivers, and standard LF wave receivers will be part of the ERG ground-network observations. These ground-network observations can

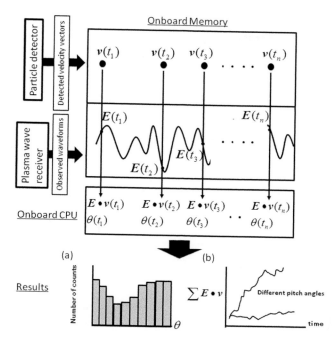

(a)

(b)

Results

Figure 5. Schematic diagram of the measurement of software-wave particle interaction analyzer system onboard the ERG.

subauroral latitudes. Since single spacecraft observations cannot determine the spatial distributions of the electric fields, the HF radar observations help to resolve the global evolutions of the electric fields of both the convection and MHD waves.

Integrated analysis using many types of data from both satellite and ground-based observations is inevitable for a quantitative understanding of the physical processes affecting particle accelerations and complex dynamics of space storms. Macroprocess and microprocess simulations, which can be compared with the observations, are also important for quantitative understanding of the causal relationship. As shown in Figure 7, several physical models of the inner magnetosphere have been newly developed in Japan and will be compared with the observation data from the ERG project. Figure 7a is the Geospace Environment Modeling System for Integrated Studies (GEMSIS)-ring current model [*Amano et al.*, 2011] that solves 5-D distribution function of ions with the Maxwell equations. The excitation of Pc5 waves due to the drift-bounce resonance of ring current ions and the ULF-relativistic electron interactions are subject for this simulation. Figure 7b shows the GEMSIS-radiation belt model [*S. Saito et al.*, 2010] that can solve the trajectory of relativistic electrons with any magnetic field. The model will be used to estimate radial transportation of the energetic electrons and the magnetopause shadowing in the realistic magnetic field. Figure 7c shows dynamic spectra of chorus waves from the self-consistent PIC simulation [*Hikishima et al.*, 2009]. The simulation reveals chorus-generation process as well as the accelerations of relativistic electrons via

provide the global state of the potential electric field, magnetic field, and current system, particle precipitation, and ULF/ELF/VLF wave activities during space storms. Figure 6 illustrates the conjugate observations by the ERG satellite and the Super-DARN Hokkaido Radar that can measure ionospheric flow in

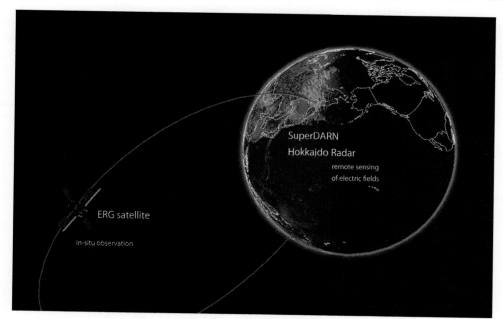

Figure 6. Schematic picture for conjunction observations between the ERG satellite and the SuperDARN Hokkaido Radar.

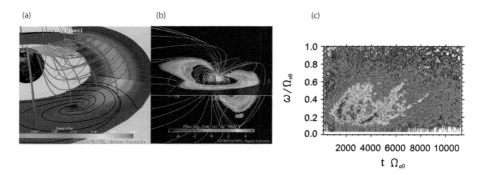

Figure 7. Examples of simulations: (a) GEMSIS-ring current [after *Amano et al.*, 2011], (b) GEMSIS-radiation belt model [after *S. Saito et al.*, 2010], and (c) PIC simulation for chorus waves [after *Hikishima et al.*, 2009].

nonlinear wave-particle interactions, and the results will be compared with S-WPIA measurement.

5. PROJECT SCIENCE COORDINATION TEAM AND PROJECT SCIENCE CENTER

The ERG project science coordination team and project science center has been established. The project science center is developing the integrated data analysis tool for analyzing all kinds of project data in cooperation with the THEMIS project [*Angelopoulos*, 2008] and Inter-university Upper Atmosphere Global Observation NETwork (IUGONET) project in Japan. Several types of software to analyze the ERG ground-based network observation data have already been implemented in the THEMIS Data Analysis Suite (TDAS) as a plug-in package. All of the ERG project observation data will be archived in the Common Data Format (CDF), which can be read easily by TDAS and other modern software.

6. INTERNATIONAL COLLABORATIONS

Solar cycle 24 offers a valuable opportunity to comprehensively study radiation belts and geospace, and satellite missions have been planned in several countries. For example, the Radiation Belt Storm Probes (United States) [*Reeves*, 2007; *Ukhorskiy et al.*, 2011] and Resonance (Russia) projects are planned for geospace exploration around the solar maximum. The THEMIS mission will continuously observe the plasma sheet, which is a source population region of the inner magnetosphere plasma. The monitoring satellites POES, GOES, LANL, and GPS will also provide continuous information on the global particle environment of geospace. Simultaneous observations at different radial distances and different local times are possible by the international fleet of satellites, which provides quantitative understanding of the radiation belts and geospace.

Besides satellite observations, international ground-based network observations are very powerful tools to understand the state of the geospace system. As mentioned above, the SuperDARN HF-radar network, ground-magnetometer networks, and optical imager network observations will be important to resolve the global plasma environment of the geospace. The balloon-based X-ray observations by the Balloon Array for RBSP Relativistic Electron Losses (BARREL) [*Millan and the BARREL Team*, 2011], and the worldwide consortium of the standard VLF/LF radio wave observations to monitor ionization at the D layer [*Clilverd et al.*, 2009], will identify when and where precipitations of relativistic electrons occur. The synergetic observations among satellites and several ground network observations will be important for revealing the complex dynamics of the inner magnetosphere as a nonlinear, compound system.

The data analysis environment is a key to effectively obtain scientific outputs during this excellent opportunity. Different kinds of data sets including those of satellite and ground-based observations and simulations should be analyzed seamlessly. The ERG project will use TDAS as the project data analysis software. TDAS has already been widely used in the space physics community. Moreover, a quick-look tool to find conjunctions among satellites and ground-based observations [e.g., *Miyashita et al.*, 2011] is quite useful to establish cooperation for various projects. International collaborations, which include sharing information on data format and data analysis tools, as well as development of common data analysis tools, will provide significant advantages for understanding the whole picture of the geospace dynamics.

7. FUTURE PERSPECTIVES

The Earth's radiation belts are a unique natural laboratory for developing our understanding of the generation processes of relativistic particles, which operate across the universe.

All magnetized planets in our solar system except for Mercury have radiation belts [*Mauk and Fox*, 2010]. The latest knowledge on the Earth's radiation belts will contribute to the understanding on the planetary radiation belts. For example, in Jovian radiation belts where ultrarelativistic electrons are generated [*Bolton et al.*, 2002; *Ezoe et al.*, 2010], radial diffusion would be important for understanding the electron acceleration process in the Jovian magnetosphere [e.g., *Goertz et al.*, 1979], while the nonadiabatic acceleration process via wave-particle interactions has been proposed based on the recent studies in the Earth's radiation belts [*Horne et al.*, 2008]. Detailed understanding of the relativistic electron acceleration process in the Earth's radiation belts will shed light on how high-energy electrons are generated in planetary magnetospheres and the universe by the ERG project.

Acknowledgments. The ERG is a science satellite program of ISAS/JAXA. The ERG-Science Center is operated by the joint research program of the Solar-Terrestrial Environment Laboratory, Nagoya University. The ERG-Science Center acknowledges close partnerships with the THEMIS mission, the SuperDARN project, and the IUGONET project in Japan for the development of the data analysis software. The development of the S-WPIA system and simulation studies are supported by Grants-in-Aid for Scientific Research (23340146, 23224011) of the Japan Society for the Promotion of Science (JSPS). The ground network observations are supported by Grants-in-Aid for Scientific Research (20244080, 23403009) of JSPS.

REFERENCES

Abel, B., and R. M. Thorne (1998), Electron scattering loss in Earth's inner magnetosphere 1. Dominant physical processes, *J. Geophys. Res.*, *103*(A2), 2385–2396.

Albert, J. M. (2003), Evaluation of quasi-linear diffusion coefficients for EMIC waves in a multispecies plasma, *J. Geophys. Res.*, *108*(A6), 1249, doi:10.1029/2002JA009792.

Amano, T., K. Seki, Y. Miyoshi, T. Umeda, Y. Matsumoto, Y. Ebihara, and S. Saito (2011), Self-consistent kinetic numerical simulation model for ring current particles in the Earth's inner magnetosphere, *J. Geophys. Res.*, *116*, A02216, doi:10.1029/2010JA015682.

Angelopoulos, V. (2008), The THEMIS mission, *Space Sci. Rev.*, *141*, 5–34, doi:10.1007/s11214-008-9336-1.

Baker, D. N. (2002), How to cope with space weather, *Science*, *297*, 1486–1487, doi:10.1126/science.1074956.

Baker, D. N., J. B. Blake, R. W. Klebesadel, and P. R. Higbie (1986), Highly relativistic electrons in the Earth's outer magnetosphere, 1. Lifetimes and temporal history 1979–1984, *J. Geophys. Res.*, *91*(A4), 4265–4276, doi:10.1029/JA091iA04p04265.

Baker, D. N., S. G. Kanekal, X. Li, S. P. Monk, J. Goldstein, and J. L. Burch (2004), An extreme distortion of the Van Allen belt arising from the 'Halloween' solar storm in 2003, *Nature*, *432*, 878–881.

Baumjohann, W., et al. (2010), Magnetic field investigation of mercury's magnetosphere and the inner heliosphere by MMO/MGF, *Planet. Space Sci.*, *58*, 279–286.

Bolton, S., et al. (2002), Ultra-relativistic electrons in Jupiter's radiation belt, *Nature*, *415*, 987–991, doi:10.1038/415987a.

Brautigam, D. H., and J. M. Albert (2000), Radial diffusion analysis of outer radiation belt electrons during the October 9, 1990, magnetic storm, *J. Geophys. Res.*, *105*(A1), 291–309, doi:10.1029/1999JA900344.

Chen, Y., G. D. Reeves, and R. H. W. Friedel (2007), The energization of relativistic electrons in the outer Van Allen radiation belt, *Nat. Phys.*, *3*, 614–617.

Chen, L., R. M. Thorne, V. K. Jordanova, and R. B. Horne (2010), Global simulation of magnetosonic wave instability in the storm time magnetosphere, *J. Geophys. Res.*, *115*, A11222, doi:10.1029/2010JA015707.

Clilverd, M. A., et al. (2009), Remote sensing space weather events: Antarctic-Arctic Radiation-belt (Dynamic) Deposition-VLF Atmospheric Research Konsortium network, *Space Weather*, *7*, S04001, doi:10.1029/2008SW000412.

Ebihara, Y., and Y. Miyoshi (2011), Dynamic inner magnetosphere: A tutorial and recent advances, in *The Dynamic Magnetosphere*, *IAGA Spec. Sopron Book Ser.*, vol. 3, edited by W. Liu and M. Fujimoto, pp. 145–187, Springer, Dordrecht, The Netherlands, doi:10.1007/978-94-007-0501-2_9.

Ebihara, Y., M.-C. Fok, R. A. Wolf, T. J. Immel, and T. E. Moore (2004), Influence of ionosphere conductivity on the ring current, *J. Geophys. Res.*, *109*, A08205, doi:10.1029/2003JA010351.

Elkington, S. R. (2006), A review of ULF interactions with radiation belt electrons, in *Magnetospheric ULF Waves: Synthesis and New Directions*, *Geophys. Monogr. Ser.*, vol. 169, edited by K. Takahashi et al., pp. 177–193, AGU, Washington, D. C., doi:10.1029/169GM12.

Elkington, S. R., M. K. Hudson, and A. A. Chan (1999), Acceleration of relativistic electrons via drift-resonant interaction with toroidal-mode Pc-5 ULF oscillations, *Geophys. Res. Lett.*, *26*(21), 3273–3276.

Ezoe, Y., K. Ishikawa, T. Ohashi, Y. Miyoshi, N. Terada, Y. Uchiyama, and H. Negoro (2010), Discovery of diffuse hard X-ray emission around Jupiter with Suzaku, *Astrophys. J. Lett.*, *709*, L178–L182, doi:10.1088/2041-8205/709/2/L178.

Friedel, R. H. W., G. D. Reeves, and T. Obara (2002), Relativistic electron dynamics in the inner magnetosphere—A review, *J. Atmos. Sol. Terr. Phys.*, *64*, 265–282.

Fukuhara, H., H. Kojima, Y. Ueda, Y. Omura, Y. Katoh, and H. Yamakawa (2009), A new instrument for the study of wave-particle interactions in space: One-chip Wave-Particle Interaction Analyzer, *Earth Planets Space*, *61*(6), 765–778.

Goertz, C. K., J. A. Van Allen, and M. F. Thomsen (1979), Further observational support for the Lossy radial diffusion model of the inner Jovian magnetosphere, *J. Geophys. Res.*, *84*(A1), 87–92, doi:10.1029/JA084iA01p00087.

Goldstein, J., B. R. Sandel, M. R. Hairston, and P. H. Reiff (2003), Control of plasmaspheric dynamics by both convection and sub-auroral polarization stream, *Geophys. Res. Lett.*, *30*(24), 2243, doi:10.1029/2003GL018390.

Green, J. C., and M. G. Kivelson (2004), Relativistic electrons in the outer radiation belt: Differentiating between acceleration mechanisms, *J. Geophys. Res.*, *109*, A03213, doi:10.1029/2003JA010153.

Hikishima, M., S. Yagitani, Y. Omura, and I. Nagano (2009), Full particle simulation of whistler-mode rising chorus emissions in the magnetosphere, *J. Geophys. Res.*, *114*, A01203, doi:10.1029/2008JA013625.

Horne, R. B., R. M. Thorne, S. A. Glauert, J. M. Albert, N. P. Meredith, and R. R. Anderson (2005), Timescale for radiation belt electron acceleration by whistler mode chorus waves, *J. Geophys. Res.*, *110*, A03225, doi:10.1029/2004JA010811.

Horne, R. B., R. M. Thorne, S. A. Glauert, N. P. Meredith, D. Pokhotelov, and O. Santolík (2007), Electron acceleration in the Van Allen radiation belts by fast magnetosonic waves, *Geophys. Res. Lett.*, *34*, L17107, doi:10.1029/2007GL030267.

Horne, R. B., R. M. Thorne, S. A. Glauert, J. D. Menietti, Y. Y. Shprits, and D. A. Gurnett (2008) Gyro-resonant acceleration at Jupiter, *Nature*, *4*, 301–304, doi:10.1038/nphys897.

Hudson, M. K., S. R. Elkington, J. G. Lyon, M. Wiltberger, and M. Lessard (2001), Radiation belt electron acceleration by ULF wave drift resonance: Simulation of 1997 and 1998 storms, in *Space Weather, Geophys. Monogr. Ser.*, vol. 125, edited by P. Song, H. J. Singer and G. L. Siscoe, pp. 289–296, AGU, Washington, D. C., doi:10.1029/GM125p0289.

Hudson, M. K., B. T. Kress, H.-R. Mueller, J. A. Zastrow, and J. B. Blake (2008), Relationship of the Van Allen radiation belts to solar wind drivers, *J. Atmos. Sol. Terr. Phys.*, *70*, 708–729.

Jordanova, V. K., J. Albert, and Y. Miyoshi (2008), Relativistic electron precipitation by EMIC waves from self-consistent global simulations, *J. Geophys. Res.*, *113*, A00A10, doi:10.1029/2008JA013239. [Printed 114(A3), 2009].

Jordanova, V. K., R. M. Thorne, W. Li, and Y. Miyoshi (2010), Excitation of whistler mode chorus from global ring current simulations, *J. Geophys. Res.*, *115*, A00F10, doi:10.1029/2009JA014810.

Kasaba, Y., et al. (2010), The Plasma Wave Investigation (PWI) onboard the BepiColombo/MMO: First measurement of electric fields, electromagnetic waves, and radio waves around Mercury, *Planet. Space Sci.*, *58*, 238–278.

Kasahara, S., K. Asamura, Y. Saito, T. Takashima, M. Hirahara, and T. Mukai (2006), Cusp type electrostatic analyzer for measurements of medium energy charged particles, *Rev. Sci. Instrum.*, *77*, 123303, doi:10.1063/1.2405358.

Kasahara, S., K. Asamura, K. Ogasawara, Y. Kazama, T. Takashima, M. Hirahara, and Y. Saito (2009a), A noise attenuation method for the medium-energy electron measurements in the radiation belt, *Adv. Space Res.*, *43*, 792–801.

Kasahara, S., T. Mitani, K. Ogasawara, T. Takashima, M. Hirahara, and K. Asamura (2009b), Application of single-sided silicon strip detector to energy and charge state measurements of medium energy ions in space, *Nucl. Instrum. Methods Phys. Res., Sect. A*, *603*, 355–360, doi:10.1016/j.nima.2009.02.004.

Kasahara, S., T. Takashima, and M. Hirahara (2012), Variability of the minimum detectable energy of an APD as an electron detector, *Nucl. Inst. Methods Phys. Res., Sect. A*, *664*(1), 282–288.

Kasahara, Y., A. Sawada, M. Yamamoto, I. Kimura, S. Kokubun, and K. Hayashi (1992), Ion cyclotron emissions observed by the satellite Akebono in the vicinity of the magnetic equator, *Radio Sci.*, *27*(2), 347–362.

Kasahara, Y., H. Kenmochi, and I. Kimura (1994), Propagation characteristics of the ELF emissions observed by the satellite Akebono in the magnetic equatorial region, *Radio Sci.*, *29*(4), 751–767.

Kasahara, Y., Y. Miyoshi, Y. Omura, O. P. Verkhoglyadova, I. Nagano, I. Kimura, and B. T. Tsurutani (2009), Simultaneous satellite observations of VLF chorus, hot and relativistic electrons in a magnetic storm "recovery" phase, *Geophys. Res. Lett.*, *36*, L01106, doi:10.1029/2008GL036454.

Katoh, Y., and Y. Omura (2007), Computer simulation of chorus wave generation in the Earth's inner magnetosphere, *Geophys. Res. Lett.*, *34*, L03102, doi:10.1029/2006GL028594.

Kennel, C. F., and H. E. Petschek (1966), Limit on stably trapped particle fluxes, *J. Geophys. Res.*, *71*(1), 1–28.

Kim, H.-J., and A. A. Chan (1997), Fully adiabatic changes in storm time relativistic electron fluxes, *J. Geophys. Res.*, *102*(A10), 22,107–22,116.

Kokubun, S., M. Takami, K. Hayashi, H. Fukunishi, I. Kimura, A. Sawada, and Y. Kasahara (1991), Triaxial search coil measurements of ELF waves in the plasmasphere: Initial results from EXOS-D, *Geophys. Res. Lett.*, *18*(2), 301–304.

Li, W., Y. Y. Shprits, and R. M. Thorne (2007), Dynamic evolution of energetic outer zone electrons due to wave-particle interactions during storms, *J. Geophys. Res.*, *112*, A10220, doi:10.1029/2007JA012368.

Li, W., R. M. Thorne, J. Bortnik, Y. Y. Shprits, Y. Nishimura, V. Angelopoulos, C. Chaston, O. Le Contel, and J. W. Bonnell (2011), Typical properties of rising and falling tone chorus waves, *Geophys. Res. Lett.*, *38*, L14103, doi:10.1029/2011GL047925.

Liu, S., M. W. Chen, J. L. Roeder, L. R. Lyons, and M. Schulz (2005), Relative contribution of electrons to the stormtime total ring current energy content, *Geophys. Res. Lett.*, *32*, L03110, doi:10.1029/2004GL021672.

Lyons, L. R., R. M. Thorne, and C. F. Kennel (1972), Pitch-angle diffusion of radiation belt electrons within the plasmasphere, *J. Geophys. Res.*, *77*(19), 3455–3474, doi:10.1029/JA077i019p03455.

Mann, I. R. (2008), The outer radiation belt injection, transport, acceleration and loss satellite (ORBITALS): A Canadian mission to the Van Allen belts, *Eos Trans. AGU*, *89*(53), Fall Meet. Suppl., Abstract U21B-05.

Mann, I. R., et al. (2006), The outer radiation belt injection, transport, acceleration and loss satellite (ORBITALS): A Canadian small satellite mission for ILWS, *Adv. Space Res.*, *38*, 1838–1860, doi:10.1016/j.asr.2005.11.009.

Mathie, R. A., and I. R. Mann (2000), A correlation between extended intervals of Ulf wave power and storm-time geosynchronous relativistic electron flux enhancements, *Geophys. Res. Lett.*, 27(20), 3261–3264, doi:10.1029/2000GL003822.

Matsumura, C., Y. Miyoshi, K. Seki, S. Saito, V. Angelopoulos, and J. Koller (2011), Outer radiation belt boundary location relative to the magnetopause: Implications for magnetopause shadowing, *J. Geophys. Res.*, 116, A06212, doi:10.1029/2011JA016575.

Matsuoka, A., M. Shinohara, Y. Tanaka, A. Fujimoto, and K. Iguchi (2012), Development of fluxgate magnetometers and applications to the space science missions, in *Science Instruments for Sounding Rocket and Satellite*, edited by K.-I. Oyama and C. Z. Cheng, Terra Sci., Tokyo, Japan, in press.

Mauk, B. H., and N. J. Fox (2010), Electron radiation belts of the solar system, *J. Geophys. Res.*, 115, A12220, doi:10.1029/2010JA015660.

Meredith, N. P., M. Cain, R. B. Horne, R. M. Thorne, D. Summers, and R. R. Anderson (2003), Evidence for chorus-driven electron acceleration to relativistic energies from a survey of geomagnetically disturbed periods, *J. Geophys. Res.*, 108(A6), 1248, doi:10.1029/2002JA009764.

Meredith, N. P., R. B. Horne, and R. R. Anderson (2008), Survey of magnetosonic waves and proton ring distributions in the Earth's inner magnetosphere, *J. Geophys. Res.*, 113, A06213, doi:10.1029/2007JA012975.

Meredith, N. P., R. B. Horne, S. A. Glauert, D. N. Baker, S. G. Kanekal, and J. M. Albert (2009), Relativistic electron loss timescales in the slot region, *J. Geophys. Res.*, 114, A03222, doi:10.1029/2008JA013889.

Millan, R. M., and the BARREL Team (2011), Understanding relativistic electron losses with BARREL, *J. Atmos. Sol. Terr. Phys.*, 73, 1425–1434.

Millan, R. M., and R. M. Thorne (2007), Review of radiation belt relativistic electron losses, *J. Atmos. Sol. Terr. Phys.*, 69, 362–377.

Miyashita, Y., I. Shinohara, M. Fujimoto, H. Hasegawa, K. Hosokawa, T. Takada, and T. Hori (2011), A powerful tool for browsing quick-look data in solar-terrestrial physics: "Conjunction Event Finder", *Earth Planet Space*, 63, e1–e4, doi:10.5047/eps.2011.01.003.

Miyoshi, Y., and R. Kataoka (2005), Ring current ions and radiation belt electrons during geomagnetic storms driven by coronal mass ejections and corotating interaction regions, *Geophys. Res. Lett.*, 32, L21105, doi:10.1029/2005GL024590.

Miyoshi, Y., and R. Kataoka (2011), Solar cycle variations of outer radiation belt and its relationship to solar wind structure dependences, *J. Atmos. Sol. Terr. Phys.*, 73, 77–87.

Miyoshi, Y., A. Morioka, H. Misawa, T. Obara, T. Nagai, and Y. Kasahara (2003), Rebuilding process of the outer radiation belt during the 3 November 1993 magnetic storm: NOAA and Exos-D observations, *J. Geophys. Res.*, 108(A1), 1004, doi:10.1029/2001JA007542.

Miyoshi, Y., A. Morioka, R. Kataoka, Y. Kasahara, and T. Mukai (2007), Evolution of the outer radiation belt during the November 1993 storms driven by corotating interaction regions, *J. Geophys. Res.*, 112, A05210, doi:10.1029/2006JA012148.

Miyoshi, Y., K. Sakaguchi, K. Shiokawa, D. Evans, J. Albert, M. Connors, and V. Jordanova (2008), Precipitation of radiation belt electrons by EMIC waves, observed from ground and space, *Geophys. Res. Lett.*, 35, L23101, doi:10.1029/2008GL035727.

Nagai, T. (1988), "Space weather forecast": Prediction of relativistic electron intensity at synchronous orbit, *Geophys. Res. Lett.*, 15(5), 425–428, doi:10.1029/GL015i005p00425.

National Research Council (2000), Radiation and the International Space Station: Recommendations to reduce risk, 92 pp., Natl. Acad. Press, Washington, D. C.

Nishimura, Y., A. Shinbori, T. Ono, M. Iizima, and A. Kumamoto (2007), Evolution of ring current and radiation belt particles under the influence of storm-time electric fields, *J. Geophys. Res.*, 112, A06241, doi:10.1029/2006JA012177.

Omura, Y., Y. Katoh, and D. Summers (2008), Theory and simulation of the generation of whistler-mode chorus, *J. Geophys. Res.*, 113, A04223, doi:10.1029/2007JA012622.

Pilipenko, V., N. Yagova, N. Romanova, and J. Allen (2006), Statistical relationships between satellite anomalies at geostationary orbit and high-energy particles, *Adv. Space Res.*, 37, 1192–1205.

Reeves, G. D. (2007), Radiation belt storm probes: A new mission for space weather forecasting, *Space Weather*, 5, S11002, doi:10.1029/2007SW000341.

Reeves, G. D., R. H. W. Friedel, R. D. Belian, M. M. Meier, M. G. Henderson, T. Onsager, H. J. Singer, D. N. Baker, X. Li, and J. B. Blake (1998), The relativistic electron response at geosynchronous orbit during the January 1997 magnetic storm, *J. Geophys. Res.*, 103(A8), 17,559–17,570, doi:10.1029/97JA03236.

Reeves, G. D., K. L. McAdams, R. H. W. Friedel, and T. P. O'Brien (2003), Acceleration and loss of relativistic electrons during geomagnetic storms, *Geophys. Res. Lett.*, 30(10), 1529, doi:10.1029/2002GL016513.

Rostoker, G., S. Skone, and D. N. Baker (1998), On the origin of relativistic electrons in the magnetosphere associated with some geomagnetic storms, *Geophys. Res. Lett.*, 25(19), 3701–3704, doi:10.1029/98GL02801.

Rowland, D. E., and J. R. Wygant (1998), Dependence of the large-scale, inner magnetospheric electric field on geomagnetic activity, *J. Geophys. Res.*, 103(A7), 14,959–14,964.

Saito, S., Y. Miyoshi, and K. Seki (2010), A split in the outer radiation belt by magnetopause shadowing: Test particle simulations, *J. Geophys. Res.*, 115, A08210, doi:10.1029/2009JA014738.

Saito, Y., J. A. Sauvaud, M. Hirahara, S. Barabash, D. Delcourt, T. Takashima, K. Asamura, and BepiColombo MMO/MPPE Team (2010), Scientific objectives and instrumentation of Mercury Plasma Particle Experiment (MPPE) onboard MMO, *Planet. Space Sci.*, 58, 182–200.

Santolík, O., D. A. Gurnett, J. S. Pickett, M. Parrot, and N. Cornilleau-Wehrlin (2003), Spatio-temporal structure of storm-time chorus, *J. Geophys. Res.*, 108(A7), 1278, doi:10.1029/2002JA009791.

Sawada, A., Y. Kasahara, M. Yamamoto, I. Kimura, S. Kokubun, and K. Hayashi (1991), ELF emissions observed by the EXOS-D satellite around the geomagnetic equatorial region, *Geophys. Res. Lett.*, *18*(2), 317–320.

Schulz, M., and L. Lanzerotti (1974), *Particle Diffusion in the Radiation Belts*, Springer, Berlin.

Seki, K., Y. Miyoshi, D. Summers, and N. P. Meredith (2005), Comparative study of outer-zone relativistic electrons observed by Akebono and CRRES, *J. Geophys. Res.*, *110*, A02203, doi:10.1029/2004JA010655.

Shiokawa, K., et al. (2006), ERG – A small-satellite mission to investigate dynamics of the inner magnetosphere, *Adv. Space Res.*, *38*, 1861–1869.

Shoji, M., and Y. Omura (2011), Simulation of electromagnetic ion cyclotron triggered emissions in the Earth's inner magnetosphere, *J. Geophys. Res.*, *116*, A05212, doi:10.1029/2010JA016351.

Shprits, Y. Y., R. M. Thorne, R. Friedel, G. D. Reeves, J. Fennell, D. N. Baker, and S. G. Kanekal (2006), Outward radial diffusion driven by losses at magnetopause, *J. Geophys. Res.*, *111*, A11214, doi:10.1029/2006JA011657.

Shprits, Y. Y., S. R. Elkington, N. P. Meredith, and D. A. Subbotin (2008a), Review of modeling of losses and sources of relativistic electrons in the outer radiation belt I: Radial transport, *J. Atmos. Sol. Terr. Phys.*, *70*, 1679–1693, doi:10.1016/j.jastp.2008.06.008.

Shprits, Y. Y., S. R. Elkington, N. P. Meredith, and D. A. Subbotin (2008b), Review of modeling of losses and sources of relativistic electrons in the outer radiation belt II: Local acceleration and loss, *J. Atmos. Sol. Terr. Phys.*, *70*, 1694–1713, doi:10.1016/j.jastp.2008.06.014.

Summers, D., and C. Ma (2000), A model for generating relativistic electrons in the Earth's inner magnetosphere based on gyroresonant wave-particle interactions, *J. Geophys. Res.*, *105*(A2), 2625–2639, doi:10.1029/1999JA900444.

Summers, D., and R. M. Thorne (2003), Relativistic electron pitch-angle scattering by electromagnetic ion cyclotron waves during geomagnetic storms, *J. Geophys. Res.*, *108*(A4), 1143, doi:10.1029/2002JA009489.

Summers, D., R. M. Thorne, and F. Xiao (1998), Relativistic theory of wave-particle resonant diffusion with application to electron acceleration in the magnetosphere, *J. Geophys. Res.*, *103*(A9), 20,487–20,500.

Summers, D., B. Ni, and N. P. Meredith (2007), Timescales for radiation belt electron acceleration and loss due to resonant wave-particle interactions: 1. Theory, *J. Geophys. Res.*, *112*, A04206, doi:10.1029/2006JA011801.

Thorne, R. M., and C. F. Kennel (1971), Relativistic electron precipitation during magnetic storm main phase, *J. Geophys. Res.*, *76*(19), 4446–4453.

Turner, D. L., Y. Shprits, M. Hartinger, and V. Angelopoulos (2012), Explaining sudden losses of outer radiation belt electrons during geomagnetic storms, *Nat. Phys.*, *8*, 208–212.

Turner, D. L., S. K. Morley, Y. Miyoshi, B. Ni, and C.-L. Huang (2012), Outer radiation belt flux dropouts: Current understanding and unresolved questions, in *Dynamics of the Earth's Radiation Belts and Inner Magnetosphere, Geophys. Monogr. Ser.*, doi:10.1029/2012GM001310, this volume.

Ukhorskiy, A. Y., B. H. Mauk, N. J. Fox, D. G. Sibeck, and J. M. Grebowsky (2011), Radiation belt storm probes: Resolving fundamental physics with practical consequences, *J. Atmos. Sol. Terr. Phys.*, *73*, 1417–1424.

K. Asamura, A. Fujimoto, M. Fujimoto, S. Kasahara, A. Matsuoka, T. Mitani and T. Takashima, Institute of Space and Astronautical Science, Japan Aerospace Exploration Agency, Sagamihara 252-5210, Japan.

L. Blomberg, Royal Institute of Technology, SE-100 44 Stockholm, Sweden.

C. Z. Cheng and Y. Kazama, Plasma and Space Science Center, National Cheng Kung University, 70101 Tainan, Taiwan.

Y. Ebihara, H. Kojima and Y. Omura, Research Institute for Sustainable Humanosphere, Kyoto University, Uji 611-0011, Japan.

N. Higashio and H. Matsumoto, Aerospace Research and Development Directorate, Japan Aerospace Exploration Agency, Tsukuba 305-8505, Japan.

T. Hori, M. Hirahara, Y. Miyashita, Y. Miyoshi, T. Segawa, K. Seki and K. Shiokawa, Solar-Terrestrial Environment Laboratory, Nagoya University, Nagoya 464-8601, Japan. (miyoshi@stelab.nagoya-u.ac.jp)

K. Ishisaka, Department of Information Systems Engineering, Toyama Prefectural University, Toyama 939-0311, Japan.

Y. Kasaba, Y. Katoh, A. Kumamoto and T. Ono, Department of Geophysics, Tohoku University, Sendai 980-8578, Japan.

Y. Kasahara and S. Yagitani, Graduate School of Natural Science and Technology, Kanazawa University, Kanazawa 920-1192, Japan.

T. Nagatsuma, National Institute of Information and Communications Technology, Koganei 184-8795, Japan.

M. Nosé, Data Analysis Center for Geomagnetism and Space Magnetism, Graduate School of Science, Kyoto University, Kyoto 606-8502, Japan.

Y.-M. Tanaka, National Institute of Polar Research, Tachikawa 190-8518, Japan.

RESONANCE Project for Studies of Wave-Particle Interactions in the Inner Magnetosphere

M. M. Mogilevsky,[1] L. M. Zelenyi,[1] A. G. Demekhov,[2] A. A. Petrukovich,[1] D. R. Shklyar,[1] and RESONANCE Team

We present an overview of the current state of the RESONANCE project. It is planned as a space mission consisting of two pairs of satellites each staying in the same magnetic flux tube over a substantial part of orbit, referred to as magneto-synchronous orbit. Such an orbit and four-satellite constellation permit long simultaneous measurements of electromagnetic fields and particle fluxes at different locations in the same flux tube to study self-consistent wave-particle dynamics in the inner magnetosphere. This dynamics, in which resonance wave-particle interactions play a key role, includes wave generation and propagation, particle energization, pitch angle scattering, and precipitation. In particular, the RESONANCE mission should provide new insights into such critical issues of magnetospheric plasma physics as chorus generation, particle acceleration in auroral region and wave modes in action, mechanism(s) of relativistic electron precipitation bursts, the nature and detailed propagation properties of auroral kilometric radiation, the role of ion cyclotron waves in precipitation of ring current (RC) ions and radiation belt electrons, the connection between RC and the radiation belts, and others.

1. INTRODUCTION

Dynamics of energetic electrons in the Earth's radiation belts remains a hot topic of magnetospheric plasma physics since the discovery of the radiation belts in 1958 [*Vernov et al.*, 1958; *Van Allen et al.*, 1958]. A profound scientific talk on the physics of the radiation belts, including an exciting story of their discovery, was presented by Michael Schulz in his invited talk "Foundations of Radiation-Belt Physics" at AGU Chapman Conference on Dynamics of the Earth's Radiation Belts and Inner Magnetosphere (St. John's, Newfoundland and Labrador, Canada, 17–22 July 2011). This

dynamics is closely connected with geomagnetic activity and solar wind flux, involving processes of various spatial and time scales [*Tverskoy*, 1969; *Tverskaya et al.*, 2005; *Summers et al.*, 2007a, 2007b]. These include, although are not limited to, generation of various plasma modes [*Andronov and Trakhtengerts*, 1964; *Cornwall et al.*, 1970; *Horne and Thorne*, 1998; *Santolík et al.*, 2002; *Trakhtengerts and Rycroft*, 2008, and references therein], particle energization due to resonant wave-particle interactions [e.g., *Albert*, 2000, 2002; *Trakhtengerts et al.*, 2003; *Bortnik et al.*, 2008; *Shklyar*, 2011a], energy transfer between different populations of energetic particles possibly mediated by rising unstable waves [*Horne et al.*, 2005; *Shklyar*, 2011b], particle energy and pitch angle diffusion caused by interaction with a wide wave spectrum [*Andronov and Trakhtengerts*, 1964; *Kennel and Petschek*, 1966; *Albert and Shprits*, 2009], particle precipitation due to resonant interaction with quasi-monochromatic waves [*Karpman and Shklyar*, 1977; *Shklyar*, 1986; *Lauben et al.*, 2001], and others. (The number of works that should have been cited to these points is indeed very large. Those mentioned above, however, contain further

[1]Space Research Institute of Russian, Academy of Sciences, Moscow, Russia.

[2]Institute of Applied Physics, Nizhny, Novgorod, Russia.

Dynamics of the Earth's Radiation Belts and Inner Magnetosphere
Geophysical Monograph Series 199
10.1029/2012GM001334

references to important studies in this field.) These processes, which are usually intensified during geomagnetically active periods, lead both to enhanced energization and losses of radiation belt particles, so that the energetic particle fluxes may either increase (more often) or decrease after geomagnetic storm [*Reeves et al.*, 2003]. As was mentioned above, the characteristic times of energetic particle flux variations may be very different: from milliseconds (microbursts of relativistic electron precipitations [*Blake et al.*, 1996]), to minutes (formation of a new radiation belt [*Li et al.*, 1993]), hours (decrease of electron flux during the main phase of geomagnetic storm [*Desorgher et al.*, 2000]), and days (increase of electron flux during the recovery phase of geomagnetic storm [*Bakhareva et al.*, 2011]).

It has been understood long ago that the processes of wave generation and variations of energetic particle distribution (including particle losses) constitute strongly interdependent sides of self-consistent dynamics of the radiation belts [*Bespalov and Trakhtengerts*, 1986]. This comprehension has lead to the concept of magnetospheric cyclotron maser, which is formed in magnetic flux tubes filled by cold plasma, with energetic charged particles as an active substance [*Trakhtengerts*, 1963; *Bespalov*, 1981]. This concept, which has proven to be fruitful in studying the wave-particle interactions for different wave modes and for both noise-like and discrete emissions [*Trakhtengerts and Rycroft*, 2008, and references therein] has been the basis of a space project RESONANCE whose main goals and current state are briefly described in this paper.

2. SCIENTIFIC GOALS OF RESONANCE

V. Yu. Trakhtengerts was a scientific leader of RESONANCE at the conceptional phase of the project, when its main features, including specific orbits, and scientific objectives had been suggested [*Demekhov et al.*, 2003]. In most general terms, RESONANCE is destined to investigate wave processes in the inner magnetosphere related to resonant wave-particle interactions. We will formulate scientific problems of RESONANCE in terms of unresolved problems of wave processes in the inner magnetosphere.

1. Generation of whistler mode chorus [e.g., *Burtis and Helliwell*, 1976; *Tsurutani and Smith*, 1977; *Santolík et al.*, 2003] remains the matter in question. Numerical simulations [*Nunn et al.*, 1997; *Katoh and Omura*, 2007, 2011] are now capable of reproducing chorus elements, but they require unrealistically high energetic electron fluxes and anisotropies, which also follows from estimates based on the linear theory. The model of chorus generation suggested by *Trakhtengerts* [1995, 1999] allowed to explain the large growth rate of chorus by suggesting that a sharp gradient can be

formed in the election velocity space due to the interactions with noise-like emissions. In this case, the cyclotron generation occurs in the regime of backward wave oscillator (BWO), and such a regime allows to explain not only the growth rate but also small spatial scale (on the order of a few hundred kilometers) of chorus source near the geomagnetic equator revealed by Polar and Cluster measurements [*Le-Docq et al.*, 1998; *Santolík et al.*, 2003]. Estimates for the nonlinear stage of the BWO regime are also in agreement with data [*Titova et al.*, 2003; *Trakhtengerts et al.*, 2004], and chorus elements were successfully reproduced in numerical simulations based on this model [*Demekhov and Trakhtengerts*, 2008]. However, experimental detection of a sharp velocity space gradient, which is necessary for the BWO regime, remains a very difficult task, so the model needs further experimental check. Moreover, mechanisms of formation of rising and falling tones in chorus, as well as the origin of two chorus bands separated by a gap at a half gyrofrequency remain undetermined in both theory and experiment, although the above-mentioned papers, as well as the works by *Omura et al.* [2009], *Omura and Nunn* [2011], *Bell et al.* [2009], *Demekhov* [2011], and *Cully et al.* [2011] constitute essential contributions into investigation of these problems. Note also the important and still unresolved problems of the reasons dictating chorus generation in quasi-parallel or oblique propagation modes and relative roles of parallel and oblique chorus emissions in self-consistent wave-particle dynamics.

2. Investigation of the mechanisms for particle energization remains a topical problem in physics of the Earth's radiation belts [e.g., *Summers et al.*, 2007a, 2007b; *Trakhtengerts and Rycroft*, 2008, and references therein]. A good deal of suggested mechanisms is related to resonant wave-particle interactions. Analytical and numerical studies of these interactions are usually performed in two limiting cases of a wide or quasimonochromatic wave spectrum. The investigations of the first case that can be traced back to classical works by *Andronov and Trakhtengerts* [1964] and *Kennel and Petschek* [1966] commonly proceed from the quasilinear theory as applied to magnetospheric conditions. In the second case, the consideration is usually based on the analysis of particle motion in the field of a single wave. The number of works in both directions is immense. We will cite the reviews by *Millan and Thorne* [2007], *Shprits et al.* [2008], and *Shklyar and Matsumoto* [2009] where further references to earlier and more recent papers may be found. In general, the considerations of both cases assume energy conservation in the interacting system consisting of wave(s) and resonant particles. However, calculations of diffusion coefficients in the quasilinear case, or particle energization in the monochromatic case, are often performed assuming a

given wave spectrum, or a given wavefield, respectively. Then, the resulting energy increase of a certain group of particles may be larger than the wave energy. This apparent contradiction does not necessarily mean that the calculations are wrong. It can well be that, in those cases, the dominant process is not the energy exchange between wave(s) and resonant particles, but the energy transfer between different groups of particles mediated by the wave(s). Such a possibility was pointed out by *Horne et al.* [2005] and recently has been investigated in detail by two examples of resonant wave-electron interactions in the case of monochromatic whistler mode waves [*Shklyar*, 2011a, 2011b]. Those considerations are readily generalized to proton interaction with ion-cyclotron waves.

An important question related to electron resonant energization in the radiation belts is that of the most efficient wave mode(s) involved. Basically, all plasma modes with sufficiently large index of refraction, namely, magnetosonic, ion-cyclotron, whistler, and Z mode waves have been considered as possible candidates [see, e.g., *Ni and Summers*, 2010]. Particularly, quasi-electrostatic domains of these branches may be responsible for energization and losses of particles with relatively low energies [*Shklyar*, 1986; *Shklyar and Kliem*, 2006].

3. Generation mechanism, fine structure, and propagation features of auroral kilometric radiation (AKR) will also be investigated in the frame of RESONANCE mission. AKR was first observed as ~1 MHz extraterrestrial emission from the Earth's auroral zone by *Benedictov et al.* [1965]. D.A. Gurnett and colleagues contributed essentially to the study of this fascinating phenomenon [*Gurnett*, 1974]. According to present-day notion, AKR is generated by the cyclotron maser instability, driven by energetic electron beams in the Earth's auroral regions [e.g., *Treumann*, 2006; *Burinskaya and Rauch*, 2007, and references therein]. The free energy source for the instability is the electron distribution, called horseshoe distribution, which is formed while the electron beam moving to the Earth is accelerated by the parallel electric field and, at the same time, electron parallel kinetic energy is adiabatically transformed to transversal kinetic energy in the increasing geomagnetic field [*Strangeway et al.*, 2001]. AKR generation takes place in auroral cavities where the electron plasma frequency is much less than the cyclotron frequency [*Calvert*, 1981; *Delory et al.*, 1998]. As was argued by many authors, energetic electrons (and even relativistic effects) play an important role not only in defining the emission growth rates, but also its dispersion characteristics [*Strangeway et al.*, 2001]. In spite of much progress in understanding the nature of AKR, there remain some open questions. They concern the exit of AKR from plasma cavity and conversion to free space mode. Fine structure of the emission

in frequency-time domain with frequency scales ≲1 kHz and time scales ~100–150 ms [*Ergun et al.*, 1998] also remains an open issue. These questions will be addressed by RESONANCE mission as the spacecraft orbit (see below) crosses the region of AKR generation.

4. Ring current (RC) ion dynamics is known to be largely influenced by ULF waves. In particular, electromagnetic ion cyclotron (EMIC) waves are generated by RC ions via the cyclotron instability and cause the ion losses due to pitch angle diffusion [*Cornwall et al.*, 1970]. The efficiency of cyclotron resonant interactions of RC ions with EMIC waves depends strongly on the plasma density, since the Doppler-shifted cyclotron resonance condition has to be satisfied. Therefore, such interactions are localized in the regions with sufficiently dense cold plasma where the wave refractive index is large enough, i.e., in the plasmasphere or detached cold-plasma regions [*Cornwall et al.*, 1970]. Furthermore, conditions of guiding the EMIC waves along the geomagnetic field facilitate their generation at the plasma-density gradients such as the plasmapause [*Mazur and Potapov*, 1983]. This conception has recently been confirmed by comparing energetic-ion data from low-orbiting NOAA spacecraft with Pc1 wave data at ground-based observatories, IMAGE data on proton aurora, and on He$^+$ EUV emission from the plasmasphere, and on LANL geosynchronous spacecraft data on energetic-ion injections and on cold-plasma density [e.g., *Spasojevic et al.*, 2005; *Yahnin and Yahnina*, 2007; *Yahnin et al.*, 2009; *Usanova et al.*, 2010, and references therein]. The localized precipitation of RC ions can have a strong effect on the global current system in the magnetosphere and ionosphere since a significant part of the RC (about 1 MA) is driven into the ionosphere by this process. In particular, the current system related to such precipitation can generate ionospheric electric fields of up to 40 mV m^{-1} [*Trakhtengerts and Demekhov*, 2005], which is consistent with enhanced westward drift of ionospheric ions known as subauroral polarization streams [*Foster and Vo*, 2002]. These findings do not solve the problems of RC formation and losses [e.g., *Ukhorskiy et al.*, 2006] and its role in the magnetospheric electric and magnetic field disturbances. RESONANCE project shall contribute to the study of these problems.

We should mention that widely known and extensively discussed space mission RBSP (http://www.rbsp.jhuapl.edu/) will begin before RESONANCE. These two missions have most scientific objectives in common and target basically the same processes. While RBSP will give the equatorial section of the processes, RESONANCE should give the meridional one, and in this sense, the missions will complement each other. At the moment, a coordinated program of measurements is being worked out.

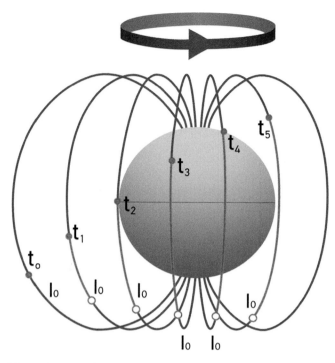

Figure 1. Formation of magnetosynchronous orbit: view from an inertial reference frame. The satellite stays in the same flux tube during a substantial time. The point on the field line where the satellite was at the moment $t = t_0$ is marked by I_0.

3. SPACECRAFT ORBITS

One of the main features of RESONANCE project is the choice of specific (magnetosynchronous) orbits that ensure the presence of four satellites in different places of the same magnetic flux tube during a long period of time.

The original design with two spacecraft on slightly separated orbits [*Demekhov et al.*, 2003] has been modified, and the new planned orbital set will consist of two pairs of spacecraft, R1(A,B) and R2(A,B), each pair on its own orbit. The parameters of these orbits are as follows: orbital period is 8 h 15 min, apogee ~27,341 km, perigee ~500 km, inclination is 63.4°, and the corotation time in a magnetic flux tube at $L \sim 4$–5 is up to 3 h if the transverse scale of the flux tube is about 100 km at the ionosphere. The spacecraft within a pair will be separated by several hundred kilometers on average. As mentioned above, according to theory and experiment, a few hundred kilometers is the characteristic scale on which chorus generation is developed. This is the reason why the time during which two satellites remain within this distance is important. On the other hand, larger scales and effects of wave propagation far from the generation region are also of interest. That is why another pair of spacecraft

located sufficiently far from the first one should be useful for the investigation of multiple scale phenomena in wave-particle interactions.

The apogees and perigees of these two orbits are located in different hemispheres. Two satellites in each pair move along the same orbit, although the distance between satellites varies and is controlled by telemetric system. It is worth mentioning that a part of each orbit lies in the subauroral region of the magnetosphere where important processes that control geomagnetic storms take place. Moreover, relative position of two spacecraft pairs is such that one is in the auroral zone when the other is in the equatorial region. A view of magnetosynchronous orbit from an inertial reference frame is shown in Figure 1. As a satellite moves along its orbit, the magnetic flux tube rotates together with the Earth. With an appropriate choice of the orbit parameters, the satellite remains in the same flux tube for an extended period of time. The satellite position at the moment when it enters the flux tube is marked by t_0, while t_1 through t_5 mark the positions at the corresponding moments of time. Bold red lines show the distance covered during that time. Spacecraft trajectories in a meridian plane rotating with the Earth are shown in Figure 2.

4. SCIENTIFIC INSTRUMENTATION

RESONANCE project is carried out under the aegis of Space Agency of the Russian Federation in the frame of federal space program. Satellite preparation and testing is performed by S. A. Lavochkin Scientific Production Association. The lead institution responsible for the set of scientific equipment is Space Research Institute of Russian Academy of Sciences. Apart from

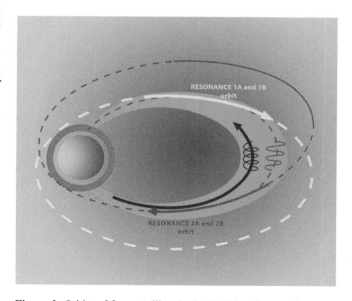

Figure 2. Orbits of four satellites in the Earth's reference frame.

researches from Russia, scientists and specialists from a few countries also participate in the project. Those countries are Austria, Bulgaria, Czech Republic, Finland, France, Germany, Greece, Poland, Slovakia, Ukraine, and the United States.

RESONANCE spacecraft will be launched in two steps. The first pair of satellites is due to be launched at the end of 2014; the launch of the second pair is planned for March–April 2015. Each pair of satellites will be equipped identically, although devices installed on satellites A and B in each pair will be somewhat different. Specific instruments, their placement on satellites, and measured parameters are given in Table 1. The drawing of the spacecraft with its instruments is shown in Figure 3.

The fluxgate magnetometer (fluxgate magnetometer for RESONANCE (FM-7R)) is designed according to the classical scheme, repeatedly used in many projects. It is intended to measure three components of quasiconstant magnetic field, field-aligned currents in the auroral region, and also for orientation of the satellite. Unlike the previous experiments, this device has an extended frequency range for measurements of geomagnetic pulsations, from DC up to 10 Hz. This instrument has been developed at IZMIRAN (acronym for Russian: Institut Zemnogo Magnetizma, Ionosfery i Rasprostraneniya Radiovoln Akademii Nauk, i.e., Institute of Terrestrial Magnetism, Ionosphere, and Radio Wave Propagation of the Academy of Sciences), Moscow region, Russia.

Three instruments, Wide Band Analyzer of Multicomponent Electric Field (AMEF-WB), ElectroMagnetic Waves Analyzer (ELMAVAN), and High-Frequency Analyzer (HFA)

are intended for 3-D measurements of electromagnetic fields in the frequency range from fractions of Hz to 1 MHz. AMEF-WB is a three-component ELF electric field receiver. Its antennas represent spherical sensors placed at the booms, which are 15 m long in the rotation plane of the satellite and 4 m long in the direction of the rotation axis. Electronic compensation of capacity is implemented, which allows to perform measurements in the wide range of frequencies and to use the same sensors for all three receivers. The developer of the device is the Institute of Space Research in Sofia, Bulgaria.

ELMAVAN is a multicomponent low-frequency receiver in the frequency range from 20 Hz to 20 kHz. The main objective of this experiment is the investigation of chorus emissions in the equatorial region and of various signals in the auroral zone. This receiver contains an analyzer, which permits onboard calculations of wave characteristics. The electronics of this device has been developed at the Institute of Atmospheric Physics in Prague, Czech Republic. For measurements of three electric components, the AMEF-WB sensors are used, and magnetic components are measured by their own three-component search-coil sensor developed in the Lvov branch of the Institute of Space Research, Ukraine.

HFA is a high-frequency analyzer (10 kHz to 1 MHz) whose main objective is studying of the AKR. Besides, the device includes the receiver-analyzer for calculation of the phase characteristics of radiation from RIK (acronym for Russian: Radio Izmerenie Koncentracii, i.e., radio measurements of concentration, see below). The electronic unit is developed at the Center of Space Research in Warsaw,

Table 1. RESONANCE Scientific Instrumentation

Instrument	Measured Parameter or Device Function	Location on Spacecraft
Fluxgate Magnetometer for RESONANCE (magnetometer)	Quasi-DC magnetic field in the frequency band 0–10 Hz	A + B
Wide Band Analyzer of Multicomponent Electric Field (ULF receiver)	Quasi-DC electric field in the frequency band 0–10 Hz	A + B
ElectroMagnetic Waves Analyzer (VLF receiver)	Electric and magnetic field components in the frequency band 10 Hz to 20 kHz	A + B
High-Frequency Analyzer (HF receiver)	Electric and magnetic field components in the frequency band 10 kHz to 1 MHz	A + B
RESONANCE Plasma Instrument	Cold plasma density	A + B
CAMERA-E	Density and distribution of plasma electrons with energy up to 20 keV	A + B
CAMERA-I	Density, distribution, and composition of plasma ions with energy up to 20 keV	A + B
BELA	Fast electron analyzer in the energy range 5–50 keV	A + B
DOC-M	Spectrometer of energetic electrons (30–400 keV) and ions (30–1000 keV)	A
Relativistic Electrons in the Magnetosphere	Spectrometer of relativistic electrons (0.3–5 MeV)	A
RIC	Radio interferometer	B
SUSPI	System for instrument control, data collection, and transmission	A + B

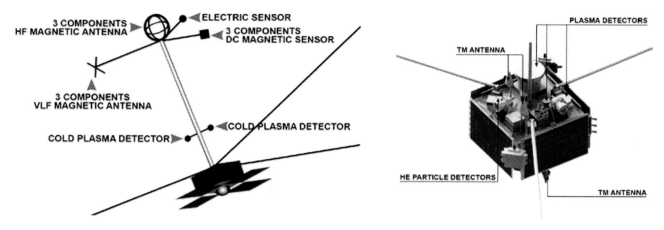

Figure 3. Composition of RESONANCE satellites. One axis of the satellite is directed to the Sun. Three electric sensors (not shown here) are mounted at the tips of 15 m long booms, which are placed in the rotation plane. The fourth 4 m long boom is used for magnetic antennas, cold plasma detectors, and for one electric sensor mounting. Particle and plasma instruments are installed at the core of the satellite. Complete angular spectrum of high-energy particles will be measured during one rotation period of about 2 min.

Poland. Measurements of electric field components use AMEF-WB sensors, while magnetic field components are registered by its own induction sensors (which are being developed at the Research Institute for Radio Physics, Nizhny Novgorod, Russia), and for receiving the RIK radiation, special antennas are used.

The RESONANCE Plasma Instrument (REPIN) device is intended for measurement of parameters of cold plasma. The technique is based on measurement of the collector current of charged-particle traps (Faraday's cylinders). A prototype of the sensor of REPIN experiment is the Faraday's cylinder of plasma experiment ROMAP of the European ROSETTA-LANDER project. For a full coverage of 4π sphere angle, six identical sensors are installed, integrated into two groups, oriented in solar and antisolar directions. The instrument is manufactured in the Institute of Atmospheric Physics in Prague, Czech Republic.

CAMERA-E instrument aims to study the fine structure of electron distribution. The design of the device allows to obtain 2-D distribution function with high-time resolution, up to 0.1 s, in the energy range from 50 eV to 20 keV.

Analysis of ionic composition and fine structure of ion distribution is carried out by CAMERA-I. The technique of measurement is similar to that used in CAMERA-E and allows measurements of ionic composition and 2-D ion distribution function during the time of the order of 1 s, in the energy range from 50 eV to 20 keV. The development of CAMERA-E and CAMERA-I is lead by IKI (acronym for Russian: Institut Kosmicheskih Issledovanii, i.e., Space Research Institute), Moscow, Russia.

Energetic electrons (30–500 keV) and ions (30–1000 keV) will be measured by the DOK-M (acronym for Russian: Detektor Okhlazhdaemyi Kremnievyi, i.e., Silicon-Cooled Detector) instrument provided by the Institute of Experimental Physics, Košice, Slovakia. This is an improvement of the successful experiment DOK-2 first implemented in the Interball project. It is designed for measurements of protons and electrons of the RC and substorm injections. The device represents a set of semiconductor detectors and has an energy resolution below 10% (except the lowest energy range).

Relativistic Electrons in the Magnetosphere (REM) is a spectrometer of relativistic electrons (300–5000 keV) based on a coincidence scheme of two semiconductor detectors and a scintillator. It is provided by IKI.

The BELA (acronym for Russian: Bystryi Electronnyi Analizator, i.e., Fast Analyzer of Electrons) spectrometer, also developed at IKI, which has high energy and time resolution, is meant for the analysis of electron distribution function in near-equatorial region where wave-particle interactions leading to chorus generation take place. This sets special requirements to the instrument, namely, fast (during ~1 s) measurements of electrons in the range from 5 to 50 keV with a high-energy resolution. Such characteristics of the device are achieved due to the design of the sensor and special logic of value sampling of the detector elements. More specifically, BELA sensor provides electron measurements with $3° \times 10°$ angular resolution. Electrons with energies within the indicated range are detected by BELA sensor, which covers the full energy range by 24 sampling steps. Each sampling step lasts for 1 ms. Maximum time resolution of one energy sweep is 24 ms, while energy resolution is ~1%–2%.

We should mention that, in the context of particle measurements, 2-D means a single slice of the sky covered by a slit camera, with the second dimension acquired using collimation and spacecraft spin in such a way that the 2π or full 4π angular distribution of particles would be measured during each spacecraft turn. On three-axis stabilized spacecraft, a scanning platform can be used to provide rotation of the slit camera. The second dimension can also be acquired by means of a deflection system, whose purpose is to deflect particles coming from angles lying between β_1 and β_2 (where β_1, β_2 are different polar angles) with respect to the symmetry axis, into the sensor.

The RIK device represents two-frequency transmitter radiating coherent signals at frequencies of 5 and 10 MHz. The device operates together with the HFA receiver and is intended for the measurement of plasma density (total electron content) between satellites of one pair by means of phase comparison for two frequencies of 5 and 10 MHz. The device is developed at the Institute of Applied Physics in Nizhny Novgorod, Russia.

5. RESONANCE-HAARP JOINT EXPERIMENTS

A long-duration stay of the RESONANCE spacecraft in a certain geomagnetic flux tube corotating with the Earth allows us to plan coordinated space-ground observations and active experiments. In particular, the High Frequency Active Auroral Research Program facility for HF heating of the ionosphere (HAARP, http://www.haarp.alaska.edu/) can be a suitable tool for such coordinated experiments. Apart from the above-given website, some information about HAARP may be found in the references given below. HAARP location conjugate with the RC and outer radiation belt ($L = 4.9$) is particularly suitable for joint experiments with RESO-NANCE. Owing to this feature and the high effective radiated power of the HAARP heater ($ERP_{max} \sim 1$ GW), the produced ionospheric disturbances can influence significantly the magnetospheric processes. In particular, if the ionosphere heating is modulated at a frequency matching one of the plasma eigen modes, then the heated region can serve as an antenna radiating this plasma mode into space [see, e.g., *Getmantsev et al.*, 1977; *Papadopoulos and Chang*, 1985; *Inan et al.*, 2004]. Experiments on such wave injection are performed in a wide range of frequencies from ULF to VLF frequencies. The heated ionospheric plasma moves to higher altitudes and thus forms density ducts above the heated region [e.g., *Milikh et al.*, 2008, 2010]. Such ducts can affect propagation of both natural and artificial signals. Another mechanism of the ionosphere-magnetosphere coupling, which can be modified by the ionosphere heating, is the reflection of magnetospheric waves from the ionosphere. Modulated HF heating has also been considered as a tool for activating substorm-like phenomena [e.g., *Trakhtengerts et al.*, 2000; *Streltsov et al.*, 2010]. The goals of RESONANCE-HAARP joint experiments may be summarized as follows: (1) artificial excitation and/or stimulation of various electromagnetic wave modes, in particular, ELF/VLF and ULF waves excited by the modulated heating, (2) modification of precipitation particle fluxes caused by nonlinear interactions between excited waves and energetic charged particles, and (3) variation of magnetospheric maser Q-parameter (quality factor) by modification of the reflection coefficient from the ionosphere in selected magnetic flux tube and, possibly, formation of artificial ducts.

The advantage of RESONANCE mission in attaining these objects is that the spacecraft will spend a long time in the flux tube conjugate with HAARP heater. This affords an opportunity to perform an active experiment with a long observation time of the above-mentioned phenomena in the magnetosphere induced by ionospheric heating.

6. CONCLUSIONS

In conclusion, we point out that the main scientific objectives of the RESONANCE project remain topical, and the planned spacecraft constellation should help in obtaining valuable experimental data on the natural and artificial processes in the inner magnetosphere. RESONANCE should be one of several forthcoming projects devoted to inner magnetospheric studies, and a close cooperation between these projects, both spaceborne and ground-based, is anticipated.

Acknowledgments. This work was supported by the Russian Academy of Sciences (Program P-22) and the Russian Foundation for Basic Research (grant 10-02-93115-CNRS and 11-02-00654).

REFERENCES

Albert, J. M. (2000), Gyroresonant interactions of radiation belt particles with a monochromatic electromagnetic wave, *J. Geophys. Res.*, *105*(A9), 21,191–21,209.

Albert, J. M. (2002), Nonlinear interaction of outer zone electrons with VLF waves, *Geophys. Res. Lett.*, *29*(8), 1275, doi:10.1029/2001GL013941.

Albert, J. M., and Y. Y. Shprits (2009), Estimates of lifetimes against pitch angle diffusion, *J. Atmos. Sol. Terr. Phys.*, *71*, 1647–1652, doi:10.1016/j.jastp.2008.07.004.

Andronov, A. A., and V. Y. Trakhtengerts (1964), Kinetic instability of the earth's outer radiation belt, *Geomagn. Aeron.*, English Transl., *4*, 181–188.

Bakhareva, M. F., L. V. Tverskaya, and K. G. Orlova (2011), Peculiarities of relativistic electron flux dynamics in recovery phase of magnetic storm of 6 April 2000, *Geomagn. Aeron.*, *51*(3), 303–308.

Bell, T. F., U. S. Inan, N. Haque, and J. S. Pickett (2009), Source regions of banded chorus, *Geophys. Res. Lett.*, *36*, L11101, doi:10.1029/2009GL037629.

Beneditkov, E. A., G. G. Getmansev, Y. A. Sazonov, and A. F. Tarasov (1965), Preliminary results of measurement of the intensity of disturbed extraterrestrial radio-frequency emission at 725 and 1525 kHz frequencies by the satellite Electron-2, *Kosm. Issled.*, *3*, 614–617.

Bespalov, P. A. (1981), Self-modulation of radiation of a plasma cyclotron maser, *JETP Lett.*, *33*(4), 182–185.

Bespalov, P. A., and V. Y. Trakhtengerts (1986), The cyclotron instability in the Earth radiation belts, in *Reviews of Plasma Physics*, vol. 10, edited by M. A. Leontovich, pp. 155–192, Plenum, New York.

Blake, J. B., M. D. Looper, D. N. Baker, R. Nakamura, B. Kleker, and D. Hovestadt (1996), New high temporal and spatial resolution measurements by SAMPEX of the precipitation of relativistic electrons, *Adv. Space Res.*, *18*(8), 171–186.

Bortnik, J., R. M. Thorne, and U. S. Inan (2008), Nonlinear interaction of energetic electrons with large amplitude chorus, *Geophys. Res. Lett.*, *35*, L21102, doi:10.1029/2008GL035500.

Burinskaya, T. M., and J. L. Rauch (2007), Waveguide regime of cyclotron maser instability in plasma regions of depressed density, *Plasma Phys. Rep.*, *33*(1), 28–37, doi:10.1134/S1063780X-07010047.

Burtis, W. J., and R. A. Helliwell (1976), Magnetospheric chorus: Occurrence patterns and normalised frequency, *Planet. Space. Sci.*, *24*, 1007–1024.

Calvert, W. (1981), The auroral plasma cavity, *Geophys. Res. Lett.*, *8*(8), 919–921.

Cornwall, J. M., F. V. Coroniti, and R. M. Thorne (1970), Turbulent loss of ring current protons, *J. Geophys. Res.*, *75*(25), 4699–4709.

Cully, C. M., V. Angelopoulos, U. Auster, J. Bonnell, and O. Le Contel (2011), Observational evidence of the generation mechanism for rising-tone chorus, *Geophys. Res. Lett.*, *38*, L01106, doi:10.1029/2010GL045793.

Delory, G. T., R. E. Ergun, C. W. Carlson, L. Muschietti, C. C. Chaston, W. Peria, J. P. McFadden, and R. Strangeway (1998), FAST observations of electron distributions within AKR source regions, *Geophys. Res. Lett.*, *25*(12), 2069–2072.

Demekhov, A. G. (2011), Generation of VLF emissions with the increasing and decreasing frequency in the magnetospheric cyclotron maser in the backward wave oscillator regime, *Radiophys. Quantum Electron.*, *53*(11), 609–622.

Demekhov, A. G., and V. Y. Trakhtengerts (2008), Dynamics of the magnetospheric cyclotron ELF/VLF maser in the backward-wave-oscillator regime. II. Influence of the magnetic-field inhomogeneity, *Radiophys. Quantum Electron.*, *51*(11), 880–889.

Demekhov, A. G., V. Y. Trakhtengerts, M. M. Mogilevsky, and L. M. Zelenyi (2003), Current problems in studies of magneto-spheric cyclotron maser and new space project RESONANCE, *Adv. Space Res.*, *32*(3), 355–374.

Desorgher, L., P. Bühler, A. Zehnder, and E. O. Flückiger (2000), Simulation of the outer radiation belt electron flux decrease during the March 26, 1995, magnetic storm, *J. Geophys. Res.*, *105*(A9), 21,211–21,223, doi:10.1029/2000JA900060.

Ergun, R. E., et al. (1998), FAST satellite wave observations in the AKR source region, *Geophys. Res. Lett.*, *25*(12), 2061–2064.

Foster, J. C., and H. B. Vo (2002), Average characteristics and activity dependence of the subauroral polarization stream, *J. Geophys. Res.*, *107*(A12), 1475, doi:10.1029/2002JA009409.

Getmantsev, G. G., A. V. Gul'el'mi, B. I. Klain, D. S. Kotik, S. M. Krylov, N. A. Mityakov, V. O. Rapoport, V. Y. Trakhtengerts, and V. A. Troitskaya (1977), Excitation of magnetic pulsations under the action on the ionosphere of radiation of a powerful short-wave transmitter, *Radiophys. Quantum Electron.*, *20*, 703–705, doi:10.1007/BF01040635.

Gurnett, D. A. (1974), The Earth as a radio source: Terrestrial kilometric radiation, *J. Geophys. Res.*, *79*(28), 4227–4238.

Horne, R. B., and R. M. Thorne (1998), Potential waves for relativistic electron scattering and stochastic acceleration during magnetic storms, *Geophys. Res. Lett.*, *25*(15), 3011–3014, doi:10.1029/98GL01002.

Horne, R. B., R. M. Thorne, S. A. Glauert, J. M. Albert, N. P. Meredith, and R. R. Anderson (2005), Timescale for radiation belt electron acceleration by whistler mode chorus waves, *J. Geophys. Res.*, *110*, A03225, doi:10.1029/2004JA010811.

Inan, U. S., M. Golkowski, D. L. Carpenter, N. Reddell, R. C. Moore, T. F. Bell, E. Paschal, P. Kossey, E. Kennedy, and S. Z. Meth (2004), Multi-hop whistler-mode ELF/VLF signals and triggered emissions excited by the HAARP HF heater, *Geophys. Res. Lett.*, *31*, L24805, doi:10.1029/2004GL021647.

Karpman, V. I., and D. R. Shklyar (1977), Particle precipitation caused by a single whistler-mode wave injected into the magnetosphere, *Planet. Space Sci.*, *25*, 395–403.

Katoh, Y., and Y. Omura (2007), Computer simulation of chorus wave generation in the Earth's inner magnetosphere, *Geophys. Res. Lett.*, *34*, L03102, doi:10.1029/2006GL028594.

Katoh, Y., and Y. Omura (2011), Amplitude dependence of frequency sweep rates of whistler mode chorus emissions, *J. Geophys. Res.*, *116*, A07201, doi:10.1029/2011JA016496.

Kennel, C. F., and H. E. Petschek (1966), Limit on stably trapped particle fluxes, *J. Geophys. Res.*, *71*(1), 1–28, doi:10.1029/JZ071i001p00001.

Lauben, D. S., U. S. Inan, and T. F. Bell (2001), Precipitation of radiation belt electrons induced by obliquely propagating lightning-generated whistlers, *J. Geophys. Res.*, *106*(A12), 29,745–29,770.

Li, X., I. Roth, M. Temerin, J. R. Wygant, M. K. Hudson, and J. B. Blake (1993), Simulation of the prompt energization and transport of radiation belt particles during the March 24, 1991 SSC, *Geophys. Res. Lett.*, *20*(22), 2423–2426.

Mazur, V. A., and A. S. Potapov (1983), The evolution of pearls in the Earth magnetosphere, *Planet. Space Sci.*, *31*, 859–863.

Milikh, G. M., K. Papadopoulos, H. Shroff, C. L. Chang, T. Wallace, E. V. Mishin, M. Parrot, and J. J. Berthelier (2008), Formation of artificial ionospheric ducts, *Geophys. Res. Lett.*, *35*, L17104, doi:10.1029/2008GL034630.

Milikh, G. M., A. G. Demekhov, K. Papadopoulos, A. Vartanyan, J. D. Huba, and G. Joyce (2010), Model for artificial ionospheric duct formation due to HF heating, *Geophys. Res. Lett.*, *37*, L07803, doi:10.1029/2010GL042684.

Millan, R. M., and R. M. Thorne (2007), Review of radiation belt relativistic electron losses, *J. Atmos. Sol. Terr. Phys.*, *69*(3), 362–377.

Ni, B., and D. Summers (2010), Resonance zones for electron interaction with plasma waves in the Earth's dipole magnetosphere. II. Evaluation for oblique chorus, hiss, electromagnetic ion cyclotron waves, and magnetosonic waves, *Phys. Plasmas*, *17*, 042903, doi:10.1063/1.3310835.

Nunn, D., Y. Omura, H. Matsumoto, I. Nagano, and S. Yagitani (1997), The numerical simulation of VLF chorus and discrete emissions observed on the Geotail satellite using a Vlasov code, *J. Geophys. Res.*, *102*(A12), 27,083–27,097, doi:10.1029/97JA02518.

Omura, Y., and D. Nunn (2011), Triggering process of whistler mode chorus emissions in the magnetosphere, *J. Geophys. Res.*, *116*, A05205, doi:10.1029/2010JA016280.

Omura, Y., M. Hikishima, Y. Katoh, D. Summers, and S. Yagitani (2009), Nonlinear mechanisms of lower-band and upper-band VLF chorus emissions in the magnetosphere, *J. Geophys. Res.*, *114*, A07217, doi:10.1029/2009JA014206.

Papadopoulos, K., and C. L. Chang (1985), Generation of ELF/ULF waves in the ionosphere by dynamo processes, *Geophys. Res. Lett.*, *12*(5), 279–282.

Reeves, G. D., K. L. McAdams, R. H. W. Friedel, and T. P. O'Brien (2003), Acceleration and loss of relativistic electrons during geomagnetic storms, *Geophys. Res. Lett.*, *30*(10), 1529, doi:10.1029/2002GL016513.

Santolík, O., J. S. Pickett, D. A. Gurnett, M. Maksimovic, and N. Cornilleau-Wehrlin (2002), Spatiotemporal variability and propagation of equatorial noise observed by Cluster, *J. Geophys. Res.*, *107*(A12), 1495, doi:10.1029/2001JA009159.

Santolík, O., D. A. Gurnett, J. S. Pickett, M. Parrot, and N. Cornilleau-Wehrlin (2003), Spatio-temporal structure of storm-time chorus, *J. Geophys. Res.*, *108*(A7), 1278, doi:10.1029/2002JA009791.

Shklyar, D. R. (1986), Particle interaction with an electrostatic VLF wave in the magnetosphere with an application to proton precipitation, *Planet. Space Sci.*, *34*, 1091–1099.

Shklyar, D. R. (2011a), On the nature of particle energization via resonant wave-particle interaction in the inhomogeneous magnetospheric plasma, *Ann. Geophys.*, *29*, 1179–1188.

Shklyar, D. R. (2011b), Wave-particle interactions in marginally unstable plasma as a means of energy transfer between energetic particle populations, *Phys. Lett. A*, *375*, 1583–1587.

Shklyar, D. R., and B. Kliem (2006), Relativistic electron scattering by electrostatic upper hybrid waves in the radiation belt, *J. Geophys. Res.*, *111*, A06204, doi:10.1029/2005JA011345.

Shklyar, D. R., and H. Matsumoto (2009), Oblique whistler-mode waves in the inhomogeneous magnetospheric plasma: Resonant interactions with energetic charged particles, *Surv. Geophys.*, *30*, 55–104.

Shprits, Y. Y., D. A. Subbotin, N. P. Meredith, and S. R. Elkington (2008), Review of modeling of losses and sources of relativistic electrons in the outer radiation belts: II. Local acceleration and loss, *J. Atmos. Sol. Terr. Phys.*, *70*, 1694–1713.

Spasojevic, M., M. R. Thomsen, P. J. Chi, and B. R. Sandel (2005), Afternoon subauroral proton precipitation resulting from ring current–plasmasphere interaction, in *Inner Magnetosphere Interactions: New Perspectives From Imaging, Geophys. Monogr. Ser.*, vol. 159, edited by J. Burch, M. Schulz and H. Spence, pp. 85–99, AGU, Washington, D. C., doi:10.1029/159GM06.

Strangeway, R. J., R. E. Ergun, C. W. Carlson, J. P. McFadden, G. T. Delory, and P. L. Pritchett (2001), Accelerated electrons as the source of Auroral Kilometric Radiation, *Phys. Chem. Earth, Part C*, *26*, 145–149.

Streltsov, A. V., T. R. Pedersen, E. V. Mishin, and A. L. Snyder (2010), Ionospheric feedback instability and substorm development, *J. Geophys. Res.*, *115*, A07205, doi:10.1029/2009JA014961.

Summers, D., B. Ni, and N. P. Meredith (2007a), Timescales for radiation belt electron acceleration and loss due to resonant wave-particle interactions: 1. Theory, *J. Geophys. Res.*, *112*, A04206, doi:10.1029/2006JA011801.

Summers, D., B. Ni, and N. P. Meredith (2007b), Timescales for radiation belt electron acceleration and loss due to resonant wave-particle interactions: 2. Evaluation for VLF chorus, ELF hiss, and electromagnetic ion cyclotron waves, *J. Geophys. Res.*, *112*, A04207, doi:10.1029/2006JA011993.

Titova, E. E., B. V. Kozelov, F. Jiricek, J. Smilauer, A. G. Demekhov, and V. Y. Trakhtengerts (2003), Verification of the backward wave oscillator model of VLF chorus generation using data from MAGION 5 satellite, *Ann. Geophys.*, *21*(5), 1073–1081.

Trakhtengerts, V. Y. (1963), On generation mechanism of VLF emissions in external Earth's radiation belts, *Geomagn. Aeron.*, *3*, 442–451.

Trakhtengerts, V. Y. (1995), Magnetosphere cyclotron maser: Backward wave oscillator generation regime, *J. Geophys. Res.*, *100*(A9), 17,205–17,210.

Trakhtengerts, V. Y. (1999), A generation mechanism for chorus emission, *Ann. Geophys.*, *17*, 95–100.

Trakhtengerts, V. Y., and A. G. Demekhov (2005), Discussion paper: Partial ring current and polarization jet, *Int. J. Geomagn. Aeron.*, *5*, GI3007, doi:10.1029/2004GI000091.

Trakhtengerts, V. Y., and M. J. Rycroft (2008), *Whistler and Alfvén Mode Cyclotron Masers in Space*, Cambridge Univ. Press, Cambridge, U. K.

Trakhtengerts, V. Y., P. P. Belyaev, S. V. Polyakov, A. G. Demekhov, and T. Bosinger (2000), Excitation of Alfven waves and vortices in the ionospheric Alfven resonator by modulated powerful radio waves, *J. Atmos. Sol. Terr. Phys.*, *62*, 267–276.

Trakhtengerts, V. Y., M. J. Rycroft, D. Nunn, and A. G. Demekhov (2003), Cyclotron acceleration of radiation belt electrons by whistlers, *J. Geophys. Res.*, *108*(A3), 1138, doi:10.1029/2002JA 009559.

Trakhtengerts, V. Y., A. G. Demekhov, E. E. Titova, B. V. Kozelov, O. Santolik, D. Gurnett, and M. Parrot (2004), Interpretation of Cluster data on chorus emissions using the backward wave oscillator model, *Phys. Plasmas*, *11*(4), 1345–1353.

Treumann, R. A. (2006), The electron–cyclotron maser for astrophysical application, *Astron. Astrophys. Rev.*, *13*, 229–315.

Tsurutani, B. T., and E. J. Smith (1977), Two types of magnetospheric ELF chorus and their substorm dependences, *J. Geophys. Res.*, *82*(32), 5112–5128.

Tverskaya, L. V., T. A. Ivanova, N. N. Pavlov, S. Y. Reizman, I. A. Rubinstein, E. N. Sosnovets, and N. N. Veden'kin (2005), Storm-time formation of a relativistic electron belt and some relevant phenomena in other magnetosphere plasma domains, *Adv. Space Res.*, *36*(12), 2392–2400.

Tverskoy, B. A. (1969), Main mechanisms in the formation of the Earth's radiation belts, *Rev. Geophys.*, *7*(1–2), 219–231.

Ukhorskiy, A. Y., B. J. Anderson, P. C. Brandt, and N. A. Tsyganenko (2006), Storm time evolution of the outer radiation belt: Transport and losses, *J. Geophys. Res.*, *111*, A11S03, doi:10. 1029/2006JA011690.

Usanova, M. E., et al. (2010), Conjugate ground and multisatellite observations of compression-related EMIC Pc1 waves and asso-ciated proton precipitation, *J. Geophys. Res.*, *115*, A07208, doi:10.1029/2009JA014935.

Van Allen, J. A., G. H. Ludwig, E. C. Ray, and C. E. McIlwain (1958), Observation of high intensity radiation by satellites 1958 Alpha and Gamma, *Jet Propul.*, *28*, 588–592.

Vernov, S. N., N. L. Grigorov, Y. I. Logachev, and A. E. Chudakov (1958), Measurement of cosmic radiation on Earth Orbiter [in Russian], *Rep. Acad. Sci. USSR*, *120*(6), 1231–1233.

Yahnin, A. G., and T. A. Yahnina (2007), Pc1 geomagnetic pulsations and related energetic particle precipitation, *J. Atmos. Sol. Terr. Phys.*, *69*(14), 1690–1706.

Yahnin, A. G., T. A. Yahnina, H. U. Frey, T. Bösinger, and J. Manninen (2009), Proton aurora related to intervals of pulsations of diminishing periods, *J. Geophys. Res.*, *114*, A12215, doi:10. 1029/2009JA014670.

A. G. Demekhov, Institute of Applied Physics, Russian Academy of Sciences, Ulyanov str. 46, 603950, Nizhny Novgorod, Russia. (andrei@appl.sci-nnov.ru)

M. M. Mogilevsky, A. A. Petrukovich, D. R. Shklyar, and L. M. Zelenyi, Space Research Institute, Russian Academy of Sciences, Profsoyuznaya str. 84/32, 117997, Moscow, Russia. (mogilevs-mogilevsky@romance.iki.rssi.ru; apetruko@iki.rssi.ru; david@iki. rssi.ru; zelenyi@iki.rssi.ru)

Global Structure of ULF Waves During the 24–26 September 1998 Geomagnetic Storm

Scot R. Elkington

Laboratory for Atmospheric and Space Physics, University of Colorado, Boulder, Colorado, USA

Anthony A. Chan

Department of Physics and Astronomy, Rice University, Houston, Texas, USA

Michael Wiltberger

High Altitude Observatory, NCAR, Boulder, Colorado, USA

The geomagnetic storm event occurring during the 24–26 September 1998 interval was characterized by significant ULF wave activity in the inner magnetosphere. Here we undertake an analysis of the ULF wave activity occurring in global MHD simulations of the event, paying particular attention to the azimuthal mode structure characterizing the waves. We find that most of the power in the fluctuation components of the electric and magnetic fields can be described by low ($m < 3$) mode numbers during the recovery phase of the storm but that there is appreciable activity at finer azimuthal spatial scales in the electric fields during the main phase of the storm. The results of this work have particular implications for the common $m = 1$ assumption frequently used in studies of the interactions of radiation belt particles with global ULF waves.

1. INTRODUCTION

Radial diffusion, a result of drift-resonant interactions between energetic particles and large-scale pulsations in the global geomagnetic field, provides an effective mechanism for transporting and energizing relativistic electrons in the outer zone radiation belts [*Fälthammar*, 1968; *Schulz and Lanzerotti*, 1974; *Elkington et al.*, 2003; *Ukhorskiy et al.*, 2005]. A review of the physical mechanisms by which electrons may be affected by global waves is given in the work of *Elkington* [2006]. It is generally understood that large-scale magnetohydrodynamic variations in the electric and magnetic fields surrounding the Earth, occurring at frequencies commensurate with the drift frequency of the trapped energetic electrons, break the drift invariant while maintaining the gyro and bounce invariants associated with trapped particle motion [*Roederer*, 1970]. The result is a radial drift across field lines into regions of either stronger or weaker magnetic field, with those particles drifting radially inward to stronger field regions gaining energy through conservation of the gyro invariant. Radial diffusion describes the stochastic motion of an ensemble of particles undergoing this drift resonant interaction. Depending on the global particle distribution with respect to the radial coordinate, radial diffusion may act to either increase flux levels in the inner magnetosphere as particles diffuse inward from regions of higher phase space density near the trapping boundary; or, if there is a paucity of electrons at the trapping boundary, act as a loss

Dynamics of the Earth's Radiation Belts and Inner Magnetosphere
Geophysical Monograph Series 199
10.1029/2012GM001348

process as the aggregate particles drift outward and are lost to the magnetopause [*Green et al.*, 2004; *Shprits et al.*, 2006].

As the drift timescales of interest in radiation belt physics are typically in the range of a few to tens of minutes, the pulsations that most effectively drive radial diffusion occur at mHz frequencies, in the so-called Pc-5 category of magnetospheric ULF waves [*Jacobs et al.*, 1964]. Alfvénic field line resonances represent a particular class of Pc-5 ULF variations, characterized by quasisinusoidal signals that exist for several wave periods. The broader class of ULF variations may be either broadband or quasisinusoidal in nature and may originate in a variety of internal (magnetospheric) or external (solar wind-driven) processes. Internally generated ULF waves may result from mirror [*Hasegawa*, 1969] or drift-bounce [*Southwood et al.*, 1969; *Chen and Hasegawa*, 1991] instabilities, or as a result of anisotropies in the perpendicular ring current [*Takahashi et al.*, 1985]. Externally driven ULF waves may result from shear flow instabilities along the flanks of the magnetopause [*Kivelson and Pu*, 1984; *Cahill and Winckler*, 1992; *Mann et al.*, 1999; *Claudepierre et al.*, 2008], as a result of variations in solar wind pressure [*Kivelson and Southwood*, 1988; *Kepko and Spence*, 2003; *Claudepierre et al.*, 2009], or as a result of variations in the large-scale convective motion of the magnetosphere driven by changes in the global magnetospheric reconnection rate and the flow of solar wind around the magnetopause [*Ridley et al.*, 1998; *Ruohoniemi and Greenwald*, 1998]. The origin and character of these waves has particular implications for the nature and effectiveness of the drift resonant interaction leading to radial diffusion [*Elkington*, 2006].

The rate of energy change, dW/dt, of a particle acted on by an electric field E and a time-varying magnetic field B is given by

$$\frac{\mathrm{d}W}{\mathrm{d}t} = q\mathbf{E}\cdot\mathbf{v_d} + M\frac{\partial B}{\partial t} \qquad (1)$$

[*Northrop*, 1963]. Here q is the particle charge, $\mathbf{v_d}$ is the guiding center drift velocity, and M is the particle gyro invariant. If the ULF variations driving radial diffusion come strictly from changes in the global convection electric field, only the first term in equation (1) is important [*Fälthammar*, 1966]; in the electromagnetic Alfvénic variations characterizing field line resonances, both terms in equation (1) must be considered in the calculation of the rate of change of the energy of a particle [*Perry et al.*, 2005].

There is a clear association between the bulk speed of the plasma in the solar wind and the occurrence of ULF variations in the magnetosphere [*Mann et al.*, 1999; *Mathie and Mann*, 2000]. Extreme variations in the flux levels in the outer radiation belts are well correlated with geomagnetic

activity [*Reeves and Thorne*, 2002]. The geomagnetic storm of 24–26 September 1998 was characterized by high solar wind speeds and an order-of-magnitude increase in outer zone electron fluxes occurring over a period of less than 18 h. *Fei et al.* [2006] examined this event from the point of view of the energetic particle response to ULF-driven radial diffusion. In this effort, they conducted test particle simulations of the trapped radiation belt electron population consisting of particles with a first adiabatic invariant of 1870 MeV G^{-1}, corresponding approximately to a 1 MeV geosynchronous electron in a dipole magnetic field. Radial diffusion calculations were concurrently carried out, with radial diffusion coefficients based on the ULF wave power observed in the MHD results that were used to drive the test particle simulations. They considered the basic drift-resonant interaction,

$$\omega = m\omega_d, \qquad (2)$$

where ω_d is the bounce-averaged drift frequency of the electrons, ω is the frequency of the magnetic and electric variations driving the drift, and m denotes the azimuthal mode number of the global ULF wave. They also considered asymmetric drift-resonant interactions,

$$\omega = (m \pm 1)\omega_d, \qquad (3)$$

which are the additional resonances present in a compressed dipole magnetic field and are the result of the $m = 1$ asymmetry in the global magnetic field induced by the dynamic pressure of the solar wind [*Hudson et al.*, 1999; *Elkington et al.*, 1999].

Spacecraft observations can provide considerable insight into the nature and distribution of ULF wave power in the radiation belt region [e.g., *Brautigam and Albert*, 2000; *Brautigam et al.*, 2005; *Liu et al.*, 2009, 2010; *Huang et al.*, 2010a; *Hartinger et al.*, 2010; *Ozeke et al.*, 2012; *Rae et al.*, 2012]. For example, *Brautigam et al.* [2005] used electric field measurements from CRRES to calculate radial diffusion coefficients based on a functional fit of the electric field power spectral density as a function of radial position, L, and the magnetospheric activity parameter K_p. They found that the diffusion coefficient, D_{LL}, could vary by 1–2 orders of magnitude between low activity ($K_p = 1$) and high activity ($K_p = 6$), depending on L and the first (gyro) invariant. However, the nature of the single-point electric field measurements forced them to consider only the diffusive effects of the curl-free part of the electric field variations and to assume only the $m = 1$ azimuthal resonance.

In contrast to the single-point measurements provided by spacecraft such as CRRES, MHD simulations of the magnetosphere provide a global picture of the changing electric and magnetic fields driving the radial diffusion [*Elkington et al.*,

2002, 2004]. For example, *Huang et al.* [2010a, 2010b] used global MHD simulations to examine simulated wave activity and radiation belt dynamics as a function of solar wind and geomagnetic activity, and found that simulated ULF wave activity produced in the Lyon-Fedder-Mobarry global MHD code [*Lyon et al.*, 2004] agreed well with statistical observations compiled at geosynchronous orbit via the GOES spacecraft. Because global MHD simulations are resolved in azimuth to a high degree compared to the limited data points available from in situ measurements, they provide a valuable means of inferring the azimuthal mode structure of the ULF waves induced by the actions of the solar wind without the $m = 1$ assumption that was necessary in the work of *Brautigam et al.* [2005]. *Fei et al.* [2006] found that the best agreement between the MHD-particle simulations and the radial diffusion simulations resulted when they considered not only the effect of the combined symmetric and asymmetric resonances but also the higher-order mode structure of the ULF waves available from the global MHD simulations.

This paper describes in detail the ULF wave environment seen in the MHD simulations of the 24 September 1998 geomagnetic storm used in the particle and diffusion simulations by *Fei et al.* [2006]. In the sections that follow, we discuss the sequence and effect of the September 1998 event, give an overview of the MHD simulations of this storm, and describe the power spectrum of the waves observed in the simulations. We pay particular attention to the global mode structure of the waves and to the implications for the commonly used $m = 1$ assumption necessitated by single-point in situ measurements.

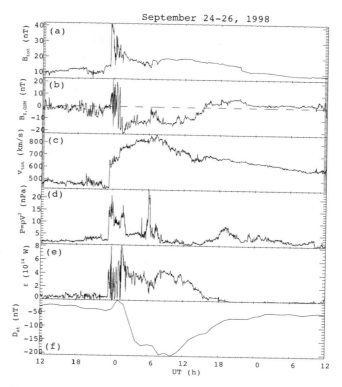

Figure 1. (a)–(e) Solar wind conditions observed at the spacecraft WIND in the period between 12 UT, 24 September 1998, and 12 UT, 26 September 1998. (f) The geomagnetic response as characterized by the hourly D_{st} index is plotted.

2. THE 24–26 SEPTEMBER 1998 GEOMAGNETIC EVENT

Solar wind conditions observed during this storm appear characteristic of a class of coronal mass ejections associated with arbitrary closed magnetic field topologies and high solar wind speeds [*McComas et al.*, 1989; *Gosling et al.*, 1991; *Tsurutani et al.*, 1992]. In these events, a coronal mass ejection traveling at speeds exceeding those of the ambient solar wind induces the formation of a shock in the interplanetary medium ahead of the coronal mass ejection (CME). Behind the shock, compression and draping of interplanetary magnetic field (IMF) lines around the body of the CME leads to strong magnetic fields at the leading edge, where solar wind velocities likewise jump to relatively high values before gradually tapering off.

Solar wind conditions observed by the spacecraft WIND for the period of 24–26 September, 1998 are indicated in Figure 1. At approximately 23:20 UT on 24 September 1998, the spacecraft observed a strong interplanetary shock ~185 R_E

upstream of the Earth. Associated with this shock was a large increase in the total strength of the IMF, which turned sharply northward for approximately 30 min before entering a period of north-south fluctuation. At about 1 UT, the IMF turned steadily southward, reaching negative values as low as −20 nT, and maintaining a southward orientation for ~14 h before relaxing northward. Solar wind velocities behind the shock jumped from 400 to 700 km s^{-1} at the shock, gradually increasing to a maximum value of over 870 km s^{-1} at 7 UT on 25 September, after which they steadily declined. A large increase in the kinetic pressure of the solar wind, from 2 to 15 nPa, and solar wind densities of 20–30 cm^{-3}, were observed at the leading edge of the shock. There were two further pressure increases of significance, one occurring around 1:30 UT on 25 September, and a larger pressure increase, associated with a jump in solar wind densities from ~5 cm^{-3} to over 20 cm^{-3}, occurring at around 6 UT on the 25th. Transit time of the initial shock from WIND to the dayside magnetopause was a little over 20 min, the resulting compression inducing a storm sudden commencement at approximately 23:45 UT [*Russell et al.*, 1999].

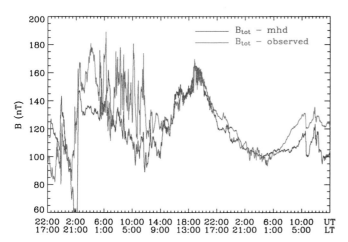

Figure 2. Simulated and observed magnetic fields at the location of GOES-8 during the 24–26 September 1998 geomagnetic storm.

The solar wind coupling parameter, $\varepsilon = v_{sw}B_{IMF}^2L^2\sin^4(\theta/2)$, is plotted in Figure 1e and reached peak values well over 10^{14} W, with time-integrated values over 10^{17} J through the course of the storm. ε is determined by the solar wind velocity v_{sw}, magnetic field B_{IMF} and its clock angle relative to the Earth's dipole, θ, and the scale length L (typically $L = 7\ R_E$). This indication of the energy input is more than sufficient to cause strong geomagnetic disturbance [*Perreault and Akasofu*, 1978]. In keeping with this, the D_{st} index reached maximum negative values of -207 nT, making this a significant geomagnetic event.

3. MHD SIMULATIONS

MHD simulations of the period between ~22 UT on 24 September 1998, and ~16 UT on 26 September 1998 were undertaken and the results used to drive particle simulations as described in the work of *Fei et al.* [2006]. The MHD

simulation code used in this event, the Lyon-Fedder-Mobarry global MHD code, is described in detail in the work of *Lyon et al.* [2004]. The degree to which the MHD simulations were able to reproduce the geomagnetic effects of this storm are indicated in Figure 2, where we have plotted magnetic fields measured by GOES-8, a geosynchronous spacecraft orbiting in the geographic equatorial plane, against the simulated fields seen by a spacecraft at the same azimuthal location and radial position as GOES-8 but in the geomagnetic equatorial plane. There was generally good agreement between measured and simulated fields seen at this spacecraft over the ~36 h period of the simulations. The initial shock, just prior to 00 UT on the 25th, is captured by both the real and virtual spacecraft, as well as a second large increase in magnetic field strength at 1:30–2:00 UT. This second discontinuity in the magnetic field corresponds in time to a local maximum in pressure occurring at a time when solar wind velocities exceeded 750 km s^{-1} and, likewise, at a time of maximum geomagnetic coupling with the solar wind as indicated by the parameter ε.

A third pressure impulse in the solar wind, occurring at ~6 UT, does not show up as clearly in magnetic field seen at GOES-8, in spite of its more prominent nature when compared to the 1:30 UT impulse. Contour plots of the magnetic field strength at this time, shown in Figure 3, indicate the magnetosphere to be in a somewhat less compressed state than during the previous impulse. This, coupled with the position of the satellite near local midnight, may diminish the magnetic signature of this pressure pulse at the spacecraft.

Large-scale magnetic fluctuations seen in Figure 2 in the period between 6 and 12 UT, where the spacecraft enters the morning sector, suggests significant wave power in a range of frequencies at geosynchronous during this period. These fluctuations subside significantly with the northward turning of the IMF at ~15 UT. The predominant "humps" in the magnetic field subsequent to the northward turning may be

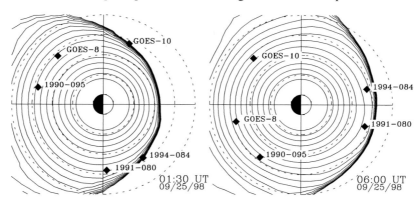

Figure 3. Global equatorial magnetic fields during two periods of high solar wind pressure, based on MHD simulations of the September 1998 event. The positions of several geosynchronous spacecraft are indicated.

largely attributed to diurnal variations in the magnetic field, the spacecraft seeing larger fields when it is in the compressed local noon sector of the magnetosphere.

4. ULF ANALYSIS

ULF waves in the magnetosphere might be expected to result from variations in the relevant solar wind parameters. Several important magnetospheric coupling parameters that might be considered include vB_z, ε, and ρV_{sw}^2 [*Gonzales et al.*, 1994; *Baker et al.*, 1999; *Newell et al.*, 2007]. Figure 4 shows the power spectral density of two of these parameters, ε and ρV_{sw}^2, for a 48 h period centered around the time of storm sudden commencement, ~23:45 on 24 September. These parameters were selected based on the speculation that ULF wave power may be effectively transmitted into the magnetosphere either through variations in the Poynting flux through the magnetopause (ε) or via the buffeting effect of

compressional pulses due to variations in the kinetic pressure in the solar wind.

In Figure 4a (top), the power spectrum of ε is plotted at frequencies of up to 8 mHz and shows significant power in frequency ranges corresponding to Pc-5 oscillations, compared to that seen in the recovery phase or prior to storm onset. Because the ε parameter is functionally dependent on the solar wind velocity and the southward component of the IMF, we use ε to infer the relative strength time-varying convection electric field in the inner magnetosphere. Below this is the integrated power between 1 and 8 mHz, indicating that most of the power in the Pc-5 frequency range came during the initial phases of the storm, before the IMF settled into a steady southward state. Figure 4b shows the power spectral density and integrated power for the solar wind kinetic pressure, in the same format as (a). There is significant power in the Pc-5 range of frequencies throughout the main phase of the storm.

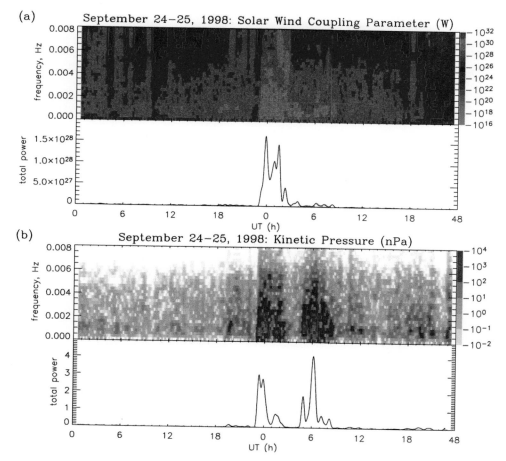

Figure 4. Power spectral density and frequency-integrated power as a function of time for (a) ε, the solar wind coupling parameter, and (b) solar wind kinetic pressure observed at the spacecraft WIND. The spectral density plots were created using a sliding window 64 min in width, for frequency resolution $df = 0.26$ mHz.

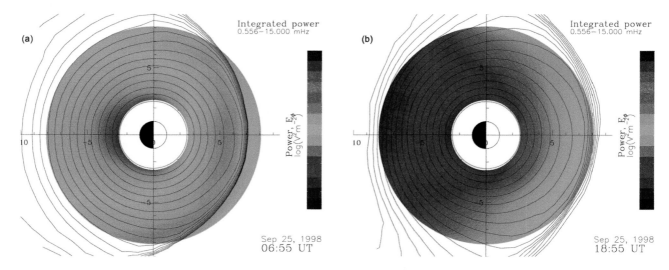

Figure 5. Equatorial power spectral density in the electric field during the (a) main phase and (b) recovery phase of the geomagnetic storm.

Spectral analyses of the time-varying electric and magnetic fields for the MHD simulations were undertaken in the inner magnetosphere. Figure 5 shows the power in the azimuthal electric field for two periods of time during the simulation, one during the peak main phase of the geomagnetic storm and a second period during the recovery phase. In each case, the power spectral density was calculated on a grid of individual points within the magnetosphere for a 2 h period around the time indicated. Clearly, there is significantly more wave power during the main phase of the storm. In both cases, the power spectral density in the electric field is seen to peak around local noon, increasing with increasing radial distance from the center of the Earth.

5. ULF MODE SPECTRUM

Rates of inward or outward radial diffusion of energetic particles drifting with drift frequency ω_d are directly dependent on the power of the waves acting at frequency ω, as suggested by equations (2) and (3). If we can determine the amplitude and the mode structure, m, of the waves at a particular frequency, we can identify the particle populations that will be most effectively transported by those ULF waves. Conversely, if we identify a particle population of interest, say, geosynchronous electrons of a particular energy [*Fei et al.*, 2006], we can inquire about the frequency and structure of the waves acting on those particles and, thus, predict their rates of energization. Here we use the simulations of the September 1998 event to characterize the power as a function of mode number during the geomagnetic storm,

to determine the extent to which one is justified in making low-mode number assumptions such as required in the efforts of *Brautigam et al.* [2005].

We assume that the global ULF activity in the MHD simulations can be described in the fashion used by

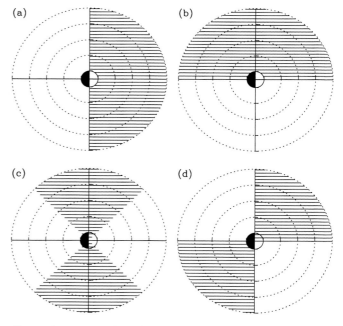

Figure 6. Interpretation of the nodal structure of the cosine (P_A) and sine (P_B) powers for the first two azimuthal mode numbers. (a) $P_A(m = 1)$, (b) $P_B(m = 1)$, (c) $P_A(m = 2)$ (d) $P_B(m = 2)$.

Fälthammar [1965] in his derivation of a functional form for D_{LL}, namely,

$$E_\phi(L, \phi, t) = \sum_{m=1}^{\infty} E_m(L, t)\cos(m\phi + \xi_m). \qquad (4)$$

Since we are unlikely to have phase $\xi_m = 0$ for all m throughout the simulation, we determine the mode structure of the ULF waves in the MHD simulations using the methods

outlined in the work of *Holzworth and Mozer* [1979] and rewrite equation (4) as the equivalent series expansion

$$E_\phi(L, \phi, t) = \sum_{m=1}^{\infty} E_{\phi_m}^{A}(L, t)\cos m\phi$$
$$+ \sum_{m=1}^{\infty} E_{\phi_m}^{B}(L, t)\sin m\phi, \qquad (5)$$

where $\phi = 0$ corresponds to noon magnetic local time. Because the time-dependent parameters of the transformation

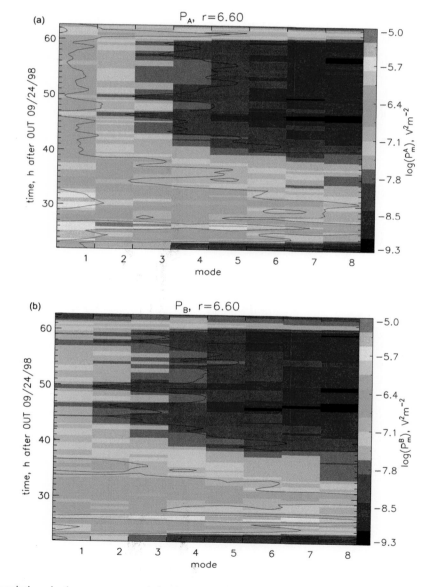

Figure 7. Time variations in the power spectral density by global azimuthal mode number in the equatorial electric field during the September 1998 geomagnetic storm for the azimuthal electric field E_ϕ, corresponding to power in the (a) cosine and (b) sine azimuthal mode components. Fields in this calculation were taken at a constant radial distance of 6.6 R_E.

are independent of each other, we can calculate the relevant power as the sum $P_m = P_m^A + P_m^B$, where P_m^A and P_m^B re the power spectral densities of the fluctuating electric fields $E_{\phi_m}^A$ and $E_{\phi_m}^B$. Physically, one might expect more power in the dayside-symmetric P^A terms as a result of (for example) direct driving by solar wind velocity and magnetic perturbations, or shear interactions at the flanks, in contrast to the asymmetric P^B terms, as illustrated in Figure 6.

The estimation of the power spectral densities P_m^B and P_m^B requires two transforms. At each time step in the simulation, we determine the fluctuation electric fields $E_{\phi_m}^A$ and $E_{\phi_m}^B$ in

equation (5) for successive radial distances based on real and imaginary parts of the transform, respectively. The coefficients of this transform indicate the amplitude of each azimuthal mode at the given snapshot in time. Subsequently, a transform in the time domain at each individual mode number is taken. The end result is the global wave power spectral density as a function of mode number and frequency. The MHD simulations used in this analysis were interpolated to a polar grid of 48 uniform azimuthal locations at 0.2 R_E radial distances from 2.5 to 8 R_E, a resolution commensurate with the underlying (nonuniform) LFM computational grid.

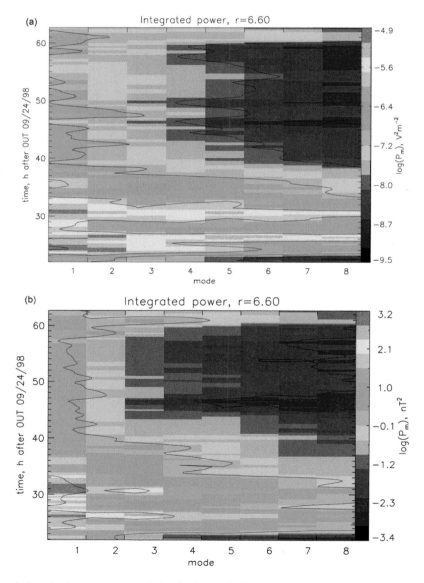

Figure 8. Time variations in the power spectral density by mode for the September 1998 geomagnetic storm for the (a) azimuthal electric field and (b) compressional magnetic field.

Interpolating to this uniform polar grid allows calculation of the power spectral density throughout the inner magnetosphere up to an azimuthal Nyquist mode number of $m = 24$.

Figure 7 indicates the geosynchronous power spectral density calculated as a function of mode number and time for the azimuthal electric field in the equatorial plane, integrated over all frequencies to the Nyquist frequency corresponding to the 60 s MHD dump interval, 8.3 mHz. As expected from Figure 6, there is generally more power in the cosine than sine terms. In Figure 8, we look at the summed

azimuthal power $P_A + P_B$ at geosynchronous orbit, for both electric and magnetic field fluctuations. The power in both components is generally larger at all mode numbers during the main phase of the storm.

In Figure 9, we plot the fraction of the total power spectral density in each mode, for the equatorial azimuthal electric field and magnetic field. The integrated power represents the power spectral density summed over all mode numbers representable on the azimuthal MHD grid, $\sum_{m=0}^{24} P_m$, integrated in frequency up to the temporal Nyquist frequency. The

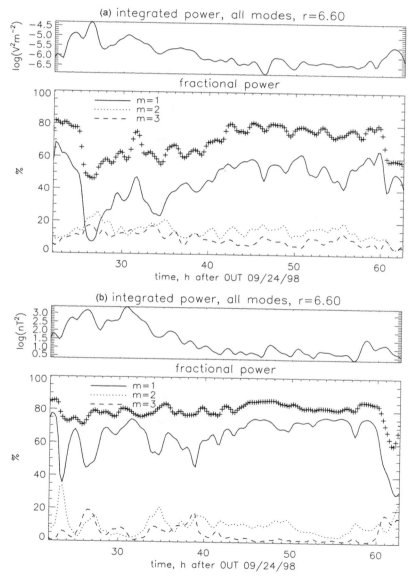

Figure 9. Fraction of the power in the lowest three mode numbers for the (a) electric and (b) magnetic ULF wave fields. (top) The total power spectral density over all azimuthal modes as a function of time is indicated. (bottom) The symbols indicate the fraction of the power in the sum of the first three modes.

fractional power in each mode for the $m = 1$, 2, and 3 modes are indicated in the lower of each pair of plots by the solid, dotted, and dashed lines, respectively, and the fractional power of the sum of these three modes plotted using symbols.

We see in Figure 9 that the $m = 1$ mode is generally dominant in both the electric and magnetic fields during the recovery phase, with much smaller contributions from the $m = 2$ and 3 modes. However, during the main phase, the electric field power exhibits relatively large contributions from the $m = 2$ and 3 modes; for a period of time the power in these modes is actually larger than the power in the $m = 1$ mode. In this event, we see much more structure in the electric field than the magnetic field during the main phase, with the power in the first three mode numbers accounting for only ~60% of the total power, in contrast to closer to ~80% in the case of the magnetic power. During the recovery phase, the power is concentrated in the lower mode numbers for both the electric and magnetic fields, with the lowest three mode structures accounting for ~80% of the power in both cases. These results are consistent in the simulations both at geosynchronous altitudes and at equatorial radial distances of 4.2 R_E, corresponding to altitudes of the GPS constellation of spacecraft.

6. ANALYSIS AND CONCLUSIONS

In this paper, we used the geomagnetic storm of 24–26 September 1998 as a case study in examining the characteristics of ULF waves throughout a geomagnetic event. Owing to the importance of the drift-resonant nature of the interaction of an electron with ULF waves indicated in equations (2) and (3), we undertook this study paying particular attention to the power as a function of the global azimuthal mode number of the waves, m.

Good agreement was seen between the observed and simulated fields, as evidenced by Figure 2 and consistent with the comparisons described in *Huang et al.* [2010a], suggesting that the MHD simulations were capable of resolving the global magnetospheric configuration to a good degree of accuracy. Spectral analysis of the event showed significant power at Pc-5 frequency ranges. This was seen both in observations of the solar wind, where power in relevant coupling parameters such as ε and ρV_{sw}^2 might be expected to directly couple into wave power in the magnetosphere [*Huang et al.*, 2010a, 2010b] and in the spectral analysis of the MHD fields during the event.

The power spectral density analysis of the global mode structure indicates that the bulk of the ULF power during the recovery phase of this storm could be described by examining the lowest-order mode numbers. However, there was

significant power in the mode numbers $m > 3$ during the main phase, particularly in the equatorial electric fields. This may be related to the fine structure present in the convection electric field during the main phase of a geomagnetic storm [e.g., *Lopez et al.*, 1999; *Farr et al.*, 2008]. This has particular implications in analyses of satellite field observations as they may be related to particle dynamics in the inner magnetosphere. For example, the analysis of CRRES electric field data by *Brautigam et al.* [2005] explicitly assumed that all observed power could be ascribed to an $m = 1$ mode. The MHD simulations used in this study, however, indicate that some of this power was likely in a finer azimuthal mode structure.

The implications of the distribution of power in azimuthal mode number for particle dynamics are not explicitly addressed by this paper. We note that large azimuthal mode numbers may resonantly interact with low-energy particles; conversely, power in higher azimuthal mode numbers require higher wave frequencies to effectively resonate with a given population of particles. However, since ULF wave power generally decreases with increasing frequency [e.g., *Bloom and Singer*, 1995; *Perry et al.*, 2005; *Rae et al.*, 2012; *Ozeke et al.*, 2012], we expect the importance of power at higher mode numbers to be mitigated somewhat in the case of electric field variations. By contrast, formulations explicitly considering the mode structure in the derivation of the magnetic diffusion coefficient indicate an m^2 dependence in the rate of radial transport [*Fei et al.*, 2006]. While this study indicates rather less power at mode numbers $m > 1$ for the magnetic field, the explicit dependence on m could make these higher-order azimuthal modes relatively more important. The detailed effects of wave mode structure versus frequency on particle dynamics will be the subject of ongoing work.

Finally, we wish to note the importance of global models in interpreting the effect of ULF wave activity on energetic particles. Observations from a single satellite cannot determine the mode structure of a given disturbance; observations from N ideally spaced satellites can unambiguously determine the mode structure only up to $m = N/2$. We, therefore, expect analyses like these, using state-of-the-art global magnetospheric simulation models, to play an important role in understanding the effect of ULF waves on energetic particles in the foreseeable future.

Acknowledgments. Work has been supported by NASA grants NNX09AI05G, NNX10AQ50G, and NNX10AL02G at LASP, and NASA grants NNX10AL02G, NNX11AJ38G, and NNX08AM36G at Rice University. The National Center for Atmospheric Research is sponsored by the National Science Foundation. Computational work was completed with the aid of NCAR computers.

REFERENCES

Baker, D. N., S. G. Kanekal, T. I. Pulkkinen, and J. B. Blake (1999), Equinoctial and solstitial averages of magnetospheric relativistic electrons: A strong semiannual modulation, *Geophys. Res. Lett.*, 26(20), 3193–3196.

Bloom, R. M., and H. J. Singer (1995), Diurnal trends in geomagnetic noise power in the Pc 2 through Pc 5 bands at low geomagnetic latitudes, *J. Geophys. Res.*, 100(A8), 14,943–14,953.

Brautigam, D. H., and J. M. Albert (2000), Radial diffusion analysis of outer radiation belt electrons during the October 9, 1990, magnetic storm, *J. Geophys. Res.*, 105(A1), 291–309.

Brautigam, D. H., G. P. Ginet, J. M. Albert, J. R. Wygant, D. E. Rowland, A. Ling, and J. Bass (2005), CRRES electric field power spectra and radial diffusion coefficients, *J. Geophys. Res.*, 110, A02214, doi:10.1029/2004JA010612.

Cahill, L. J., Jr., and J. R. Winckler (1992), Periodic magnetopause oscillations observed with the GOES satellites on March 24, 1991, *J. Geophys. Res.*, 97(A6), 8239–8243.

Chen, L., and A. Hasegawa (1991), Kinetic theory of geomagnetic pulsations, 1. Internal excitations by energetic particles, *J. Geophys. Res.*, 96(A2), 1503–1512.

Claudepierre, S. G., S. R. Elkington, and M. Wiltberger (2008), Solar wind driving of magnetospheric ULF waves: Pulsations driven by velocity shear at the magnetopause, *J. Geophys. Res.*, 113, A05218, doi:10.1029/2007JA012890.

Claudepierre, S. G., M. Wiltberger, S. R. Elkington, W. Lotko, and M. K. Hudson (2009), Magnetospheric cavity modes driven by solar wind dynamic pressure fluctuations, *Geophys. Res. Lett.*, 36, L13101, doi:10.1029/2009GL039045.

Elkington, S. R. (2006), A review of ULF interactions with radiation belt electrons, in *Magnetospheric ULF Waves: Synthesis and New Directions*, Geophys. Monogr. Ser., vol. 169, edited by K. Takahashi et al., pp. 177–193, AGU, Washington, D. C., doi:10.1029/169GM12.

Elkington, S. R., M. K. Hudson, and A. A. Chan (1999), Acceleration of relativistic electrons via drift-resonant interaction with toroidal-mode Pc-5 ULF oscillations, *Geophys. Res. Lett.*, 26(21), 3273–3276.

Elkington, S. R., M. K. Hudson, M. J. Wiltberger, and J. G. Lyon (2002), MHD/particle simulations of radiation belt dynamics, *J. Atmos. Sol. Terr. Phys.*, 64, 607–615.

Elkington, S. R., M. K. Hudson, and A. A. Chan (2003), Resonant acceleration and diffusion of outer zone electrons in an asymmetric geomagnetic field, *J. Geophys. Res.*, 108(A3), 1116, doi:10.1029/2001JA009202.

Elkington, S. R., M. Wiltberger, A. A. Chan, and D. N. Baker (2004), Physical models of the geospace radiation environment, *J. Atmos. Sol. Terr. Phys.*, 66, 1371–1387, doi:10.1016/j.jastp.2004.03.023.

Fälthammar, C.-G. (1965), Effects of time-dependent electric fields on geomagnetically trapped radiation, *J. Geophys. Res.*, 70(11), 2503–2516.

Fälthammar, C.-G. (1966), Coefficients of diffusion in the outer radiation belts, in *Radiation Trapped in the Earth's Magnetic Field, Astrophys. Space Sci. Lab.*, vol. 5, edited by B. M. McCormac, p. 398, Springer, New York.

Fälthammar, C.-G. (1968), Radial diffusion by violation of the third adiabatic invariant, in *Earth's Particles and Fields*, edited by B. M. McCormac, p. 157, Reinhold, New York.

Farr, N. L., D. N. Baker, and M. Wiltberger (2008), Complexities of a 3-D plasmoid flux rope as shown by an MHD simulation, *J. Geophys. Res.*, 113, A12202, doi:10.1029/2008JA013328.

Fei, Y., A. A. Chan, S. R. Elkington, and M. J. Wiltberger (2006), Radial diffusion and MHD particle simulations of relativistic electron transport by ULF waves in the September 1998 storm, *J. Geophys. Res.*, 111, A12209, doi:10.1029/2005JA011211.

Gonzalez, W. D., J. A. Joselyn, Y. Kamide, H. W. Kroehl, G. Rostoker, B. T. Tsurutani, and V. M. Vasyliunas (1994), What is a geomagnetic storm?, *J. Geophys. Res.*, 99(A4), 5771–5792.

Gosling, J. T., D. J. McComas, J. L. Phillips, and S. J. Bame (1991), Geomagnetic activity associated with Earth passage of interplanetary shock disturbances and coronal mass ejections, *J. Geophys. Res.*, 96(A5), 7831–7839.

Green, J. C., T. G. Onsager, T. P. O'Brien, and D. N. Baker (2004), Testing loss mechanisms capable of rapidly depleting relativistic electron flux in the Earth's outer radiation belt, *J. Geophys. Res.*, 109, A12211, doi:10.1029/2004JA010579.

Hartinger, M., M. B. Moldwin, V. Angelopoulos, K. Takahashi, H. J. Singer, R. R. Anderson, Y. Nishimura, and J. R. Wygant (2010), Pc5 wave power in the quiet-time plasmasphere and trough: CRRES observations, *Geophys. Res. Lett.*, 37, L07107, doi:10.1029/2010GL042475.

Hasegawa, A. (1969), Drift mirror instability in the magnetosphere, *Phys. Fluids*, 12(12), 2642–2650.

Holzworth, R. H., and F. S. Mozer (1979), Direct evaluation of the radial diffusion coefficient near L = 6 due to electric field fluctuations, *J. Geophys. Res.*, 84(A6), 2559–2566.

Huang, C.-L., H. E. Spence, H. J. Singer, and W. J. Hughes (2010a), Modeling radiation belt radial diffusion in ULF wave fields: 1. Quantifying ULF wave power at geosynchronous orbit in observations and in global MHD model, *J. Geophys. Res.*, 115, A06215, doi:10.1029/2009JA014917.

Huang, C.-L., H. E. Spence, M. K. Hudson, and S. R. Elkington (2010b), Modeling radiation belt radial diffusion in ULF wave fields: 2. Estimating rates of radial diffusion using combined MHD and particle codes, *J. Geophys. Res.*, 115, A06216, doi:10.1029/2009JA014918.

Hudson, M. K., S. R. Elkington, J. G. Lyon, C. C. Goodrich, and T. J. Rosenberg (1999), Simulation of radiation belt dynamics driven by solar wind variations, in *Sun-Earth Plasma Connections*, Geophys. Monogr. Ser., vol. 109, edited by L. Burch, L. Carovillano and K. Antiochos, pp. 171–182, AGU, Washington, D. C., doi:10.1029/GM109p0171.

Jacobs, J. A., Y. Kato, S. Matsushita, and V. A. Troitskaya (1964), Classification of geomagnetic micropulsations, *J. Geophys. Res.*, 69(1), 180–181.

Kepko, L., and H. E. Spence (2003), Observations of discrete, global magnetospheric oscillations directly driven by solar wind density variations, *J. Geophys. Res.*, *108*(A6), 1257, doi:10.1029/2002JA009676.

Kivelson, M. G., and Z.-Y. Pu (1984), The Kelvin-Helmholtz instability on the magnetopause, *Planet. Space Sci.*, *32*(11), 1335–1341.

Kivelson, M. G., and D. J. Southwood (1988), Hydromagnetic waves in the ionosphere, *Geophys. Res. Lett.*, *15*(11), 1271–1274.

Liu, W., T. E. Sarris, X. Li, S. R. Elkington, R. Ergun, V. Angelopoulos, J. Bonnell, and K. H. Glassmeier (2009), Electric and magnetic field observations of Pc4 and Pc5 pulsations in the inner magnetosphere: A statistical study, *J. Geophys. Res.*, *114*, A12206, doi:10.1029/2009JA014243.

Liu, W., T. E. Sarris, X. Li, R. Ergun, V. Angelopoulos, J. Bonnell, and K. H. Glassmeier (2010), Solar wind influence on Pc4 and Pc5 ULF wave activity in the inner magnetosphere, *J. Geophys. Res.*, *115*, A12201, doi:10.1029/2010JA015299.

Lopez, R. E., M. Wiltberger, J. G. Lyon, C. C. Goodrich, and K. Papadopoulos (1999), MHD simulations of the response of high-latitude potential patterns and polar cap boundaries to sudden southward turnings of the interplanetary magnetic field, *Geophys. Res. Lett.*, *26*(7), 967–970.

Lyon, J. G., J. A. Fedder, and C. M. Mobarry (2004), The Lyon–Fedder–Mobarry (LFM) global MHD magnetospheric simulation code, *J. Atmos. Sol. Terr. Phys.*, *66*(15–16), 1333–1350, doi:10.1016/j.jastp.2004.03.020.

Mann, I. R., A. N. Wright, K. J. Mills, and V. M. Nakariakov (1999), Excitation of magnetospheric waveguide modes by magnetosheath flows, *J. Geophys. Res.*, *104*(A1), 333–353.

Mathie, R. A., and I. R. Mann (2000), Observations of Pc5 field line resonance azimuthal phase speeds: A diagnostic of their excitation mechanism, *J. Geophys. Res.*, *105*(A5), 10,713–10,728.

McComas, D. J., J. T. Gosling, S. J. Bame, E. J. Smith, and H. V. Cane (1989), A test of magnetic field draping induced B_z perturbations ahead of fast coronal mass ejecta, *J. Geophys. Res.*, *94*(A2), 1465–1471.

Newell, P. T., T. Sotirelis, K. Liou, C.-I. Meng, and F. J. Rich (2007), A nearly universal solar wind-magnetosphere coupling function inferred from 10 magnetospheric state variables, *J. Geophys. Res.*, *112*, A01206, doi:10.1029/2006JA012015.

Northrop, T. G. (1963), *The Adiabatic Motion of Charged Particles*, 109 pp., Interscience, New York.

Ozeke, L. G., I. R. Mann, K. R. Murphy, I. J. Rae, D. K. Milling, S. R. Elkington, A. A. Chan, and H. J. Singer (2012), ULF wave derived radiation belt radial diffusion coefficients, *J. Geophys. Res.*, *117*, A04222, doi:10.1029/2011JA017463.

Perreault, P., and S.-I. Akasofu (1978), A study of geomagnetic storms, *Geophys. J. R. Astron. Soc.*, *54*, 547–573.

Perry, K. L., M. K. Hudson, and S. R. Elkington (2005), Incorporating spectral characteristics of Pc5 waves into three-dimensional radiation belt modeling and the diffusion of relativistic electrons, *J. Geophys. Res.*, *110*, A03215, doi:10.1029/2004JA010760.

Rae, I. J., I. R. Mann, K. R. Murphy, L. G. Ozeke, D. K. Milling, A. A. Chan, S. R. Elkington, and F. Honary (2012), Ground-based magnetometer determination of in situ Pc4–5 ULF electric field wave spectra as a function of solar wind speed, *J. Geophys. Res.*, *117*, A04221, doi:10.1029/2011JA017335.

Reeves, G., and R. Thorne (2002), Working Group reports: Inner magnetosphere and storms campaign Working Group 2—Radiation belts, *The GEMstone*, *12*(1), 16–19.

Ridley, A. J., G. Lu, C. R. Clauer, and V. O. Papitashvili (1998), A statistical study of the ionospheric convection response to changing interplanetary magnetic field conditions using the assimilative mapping of ionospheric electrodynamics technique, *J. Geophys. Res.*, *103*(A3), 4023–4039.

Roederer, J. G. (1970), *Dynamics of Geomagnetically Trapped Radiation*, 166 pp., Springer, New York.

Ruohoniemi, J. M., and R. A. Greenwald (1998), The response of high-latitude convection to a sudden southward IMF turning, *Geophys. Res. Lett.*, *25*(15), 2913–2916.

Russell, C. T., X. W. Zhou, P. J. Chi, H. Kawano, T. E. Moore, W. K. Peterson, J. B. Cladis, and H. J. Singer (1999), Sudden compression of the outer magnetosphere associated with an ionospheric mass ejection, *Geophys. Res. Lett.*, *26*(15), 2343–2346.

Schulz, M., and L. J. Lanzerotti (1974), *Particle Diffusion in the Radiation Belts*, *Phys. Chem. Space*, vol. 7, 215 pp., Springer, New York.

Shprits, Y. Y., R. M. Thorne, R. Friedel, G. D. Reeves, J. Fennell, D. N. Baker, and S. G. Kanekal (2006), Outward radial diffusion driven by losses at magnetopause, *J. Geophys. Res.*, *111*, A11214, doi:10.1029/2006JA011657.

Southwood, D. J., J. W. Dungey, and R. J. Etherington (1969), Bounce resonant interaction between pulsations and trapped particles, *Planet. Space Sci.*, *17*, 349–361.

Takahashi, K., C. T. Russell, and R. R. Anderson (1985), ISEE 1 and 2 observation of the spatial structure of a compressional Pc5 wave, *Geophys. Res. Lett.*, *12*(9), 613–616.

Tsurutani, B. T., W. D. Gonzalez, F. Tang, and Y. T. Lee (1992), Great magnetic storms, *Geophys. Res. Lett.*, *19*(1), 73–76.

Ukhorskiy, A. Y., K. Takahashi, B. J. Anderson, and H. Korth (2005), Impact of toroidal ULF waves on the outer radiation belt electrons, *J. Geophys. Res.*, *110*, A10202, doi:10.1029/2005JA011017.

A. A. Chan, Department of Physics and Astronomy, Rice University, Houston, TX 77005, USA.

S. R. Elkington, Laboratory for Atmospheric and Space Physics, University of Colorado, Boulder, CO 80303, USA (scot.elkington@lasp.colorado.edu).

M. Wiltberger, HAO/NCAR, Boulder, CO 80301, USA.

ULF Wave–Driven Radial Diffusion Simulations of the Outer Radiation Belt

Louis G. Ozeke, Ian R. Mann, Kyle R. Murphy, and I. Jonathan Rae

Department of Physics, University of Alberta, Edmonton, Alberta, Canada

Anthony A. Chan

Department of Physics and Astronomy, Rice University, Houston, Texas, USA

MeV energy electrons are thought to be transported inward from the plasma sheet into the radiation belts under the action of radial diffusion driven by waves in the ULF band. Current radiation belt models use radial diffusion coefficients, which are separated into a term due to electromagnetic fluctuations and a term due to electrostatic fluctuations, such as those presented by Brautigam and Albert (2000). In contrast, here we model the dynamics of the outer radiation belt electrons during the CRRES era using radial diffusion coefficient derived from 15 years of ground-based magnetometer and ~10 years of in situ spacecraft measurements, which are separated into a term due to electric field fluctuations and a term due to magnetic field fluctuations. These simulations are compared with electron flux simulations over the same time interval with the diffusion coefficients presented by Brautigam and Albert (2000). Interestingly, our results indicate that on low L shells, the electron flux enhancements resulting from the electric and magnetic field diffusion coefficients are several orders of magnitude weaker than the electron flux enhancements produced using the electrostatic and electromagnetic diffusion coefficients presented by Brautigam and Albert (2000).

1. INTRODUCTION

After over 50 years of in situ measurements, there is still no universally accepted mechanism responsible for the energization and dynamics of relativistic electrons trapped in the Earth's geomagnetic field. However, inward radial diffusion driven by fluctuations of the geomagnetic field is one of the earliest processes identified as playing a critical role in the dynamics and energization of outer radiation belt electrons [*Fälthammar*, 1968; *Schulz and Lanzerotti*, 1974; *Fei et al.*, 2006]. In this model, electrons gain energy as they are radially transported inward from the plasma sheet to the radiation belts under the action of fluctuating electric and magnetic fields conserving their first adiabatic invariant [*Fälthammar*, 1968; *Schulz and Lanzerotti*, 1974; *Fei et al.*, 2006].

Early treatments of radial diffusion separated the diffusion coefficient into an electromagnetic term D_{LL}^{EM} and an electrostatic term D_{LL}^{static}. Here D_{LL}^{EM} is expressed in terms of the power spectral density (PSD) of the azimuthally symmetric component of the compressional magnetic fluctuations produced by a sudden magnetic impulse [*Schulz and Lanzerotti*, 1974], while D_{LL}^{static} is expressed in terms of the PSD of the azimuthal electric field produced by fluctuations in the convection electric field [*Schulz and Lanzerotti*, 1974]. For both the electromagnetic and electrostatic terms, the PSD of the ULF wave fluctuations must be known in space in the magnetic equatorial plane. Several different approaches have been used to estimate these ULF wave electric and magnetic field PSDs in space.

Dynamics of the Earth's Radiation Belts and Inner Magnetosphere
Geophysical Monograph Series 199
10.1029/2012GM001332

Holzworth and Mozer [1979] characterized the electrostatic diffusion term by using balloon measurements of ULF wave electric field PSDs at $L = 6$ in the ionosphere. By assuming that the field lines were electric equipotentials, these authors mapped the ionospheric electric fields to the magnetic equatorial plane to obtain an equatorial electric field PSD in space. Using these electric field PSD estimates, the electrostatic diffusion coefficient was derived at $L = 6$ [*Holzworth and Mozer*, 1979]. *Cornwall* [1968] derived an analytic expression for the electric field diffusion coefficient modeling the electric field spectrum as a substorm convection electric field characterized by a rapid rise time and an exponential decay. A similar approach was used by *Brautigam and Albert* [2000] to estimate the electrostatic diffusion coefficients.

Lanzerotti and Morgan [1973] used 18 days of ground measurements taken at two conjugate magnetometer stations close to $L = 4$ to determine the electromagnetic diffusion term. Here the magnetic PSD on the ground was determined as a function of the 3 h averaged magnetic index K_{FR} from the Fredricksburg magnetometer ($L \sim 2.7$). These ground PSD values were mapped to the compressional PSD in the equatorial plane in space using a mapping model based on symmetric oscillations of a dipole magnetic field with no wave propagation or ionospheric effects included [*Lanzerotti and Morgan*, 1973]. The electromagnetic diffusion term was then expressed as a function of L shell by assuming an L^{10} variation in D_{LL}^{EM}. *Lanzerotti et al.* [1978] determined the electromagnetic diffusion term using 1 month of in situ ATS-6 geosynchronous compressional PSD measurements expressed as a function of the 12 h summed Kp. To obtain the required azimuthally symmetric component of the compressional magnetic PSD, the PSD values measured over the nightside were multiplied by a factor of 5.71, and the dayside PSD values were multiplied by a factor of 0.314 [*Lanzerotti et al.*, 1978]. Again, the electromagnetic diffusion term was then expressed as a function of L shell by assuming an L^{10} variation in D_{LL}^{EM}.

More recently, *Brautigam and Albert* [2000] derived an analytic expression for the electromagnetic diffusion coefficient as a function of L shell by fitting the diffusion coefficient at $L = 6.6$ taken from the work of *Lanzerotti et al.* [1978] and at $L = 4$ taken from the work of *Lanzerotti and Morgan* [1973] to a line proportional to L^{10} for each value of Kp from 1 to 6.

There are two difficulties with obtaining the electrostatic diffusion coefficient, D_{LL}^{static} and the electromagnetic diffusion coefficient D_{LL}^{EM} from measurements of the electric and magnetic field PSD. In order to determine D_{LL}^{static}, the component of the measured azimuthal electric field PSD resulting from only electrostatic fluctuations needs to be obtained, with all electromagnetic electric field fluctuations being ignored. Similarly, in order to determine D_{LL}^{EM}, the component of the measured compressional magnetic field PSD resulting from only azimuthally symmetric fluctuations also needs to be known. These two required components of the electric and magnetic field PSDs may be difficult or even impossible to separate out from the in situ measured electric and magnetic field PSDs. An alternative expression for the diffusion coefficient has been derived by *Brizard and Chan* [2001] and *Fei et al.* [2006]. These authors express the diffusion coefficients explicitly in terms of the azimuthal electric field and the compressional magnetic field PSD, without having to separate the PSD into electromagnetic and electrostatic terms. In the work of *Ozeke et al.* [2012], the expressions from *Brizard and Chan* [2001] and *Fei et al.* [2006] are used to determine the diffusion coefficients due to the azimuthal electric field D_{LL}^{E} and the compressional magnetic field D_{LL}^{E}. Here the compressional magnetic field diffusion term was determined as a function of Kp and solar wind speed using GOES East and West data from 1996 to 2005 as well as Active Magnetospheric Particle Tracer Explorers (AMPTE) data [*Ozeke et al.* 2012]. The D_{LL}^{E} term was derived from 15 years of Canadian Array for Realtime Investigations of Magnetic Activity (CARISMA) and Sub-Auroral Magnetometer Network (SAMNET) ground magnetometer data mapped to the magnetic equatorial plane using a guided Alfven wave solution [*Ozeke et al.*, 2012; *Rae et al.*, 2012].

Several models have been developed to understand the physical processes responsible for the energization and dynamics of relativistic electrons trapped in the Earth's geomagnetic field, such as VERB [*Subbotin and Shprits*, 2009], SALAMBO [*Beutier and Boscher*, 1995], and DREAM [*Koller et al.*, 2007]. All of these models currently use the electromagnetic and electrostatic diffusion coefficients derived by *Brautigam and Albert* [2000], which is based on only 18 days of ground-based magnetometer data and 1 month of in situ ATS6 data for D_{LL}^{EM} and the empirical convection model for D_{LL}^{static}. In this paper, we determine what impact using the diffusion coefficients determined in the work of *Ozeke et al.* [2012] (determined using more than a solar cycle of data both from ground-based magnetometers and in situ from satellites) has on the outer radiation belt electron flux variation compared with the electromagnetic and electrostatic diffusion coefficients presented by *Brautigam and Albert* [2000], which are typically used in radiation belt models.

2. MODELING APPROACH

Our approach for modeling the influence of ULF wave radial diffusion on the dynamics and energization of outer radiation belt electrons involves solving the radial diffusion equation [*Fälthammar*, 1968] in a dipole field with the addition of a loss term, which gives equation (1)

$$\frac{\partial f}{\partial t} = L^2 \frac{\partial}{\partial L}\left[\frac{D_{LL}}{L^2}\frac{\partial f}{\partial L}\right] - \frac{f}{\tau}. \tag{1}$$

In equation (1), f represents the phase space density of the electrons, and it is assumed that the first and second adiabatic invariants, M and J, are conserved [see *Schulz and Lanzerotti*, 1974].

The lifetime of the electrons arising as a result of loss processes is given by the parameter τ [*Shprits et al.*, 2005]. In our approach, following *Fei et al.* [2006], the diffusion coefficient D_{LL} is the sum of the diffusion coefficients due to the azimuthal electric and compressional magnetic field perturbations D_{LL}^E and D_{LL}^B. This approach assumes random phases between the electric and magnetic perturbations [see *Fei et al.*, 2006].

In a dipole magnetic field, the diffusion coefficients due to the electric and magnetic field perturbations D_{LL}^E and D_{LL}^B can be expressed as

$$D_{LL}^E = \frac{1}{8B_E^2 R_E^2} L^6 \sum_m P_m^E(m\omega_d), \tag{2}$$

$$D_{LL}^B = \frac{M^2}{8q^2\gamma^2 B_E^2 R_E^4} L^4 \sum_m m^2 P_m^B(m\omega_d), \tag{3}$$

$$M = \frac{p_\perp^2 L^3}{2m_e B_E} \tag{4}$$

[see *Brizard and Chan*, 2001; *Fei et al.*, 2006]. Here the constants B_E, R_E, and q represent the equatorial magnetic field strength at the surface of the Earth, the Earth's radius, and the electron charge, respectively. The relativistic correction factor, γ, given by

$$\gamma = (1 - v^2/c^2)^{-1/2}, \tag{5}$$

where v is the total speed of the electron, and c is the speed of light, does not remain constant but increases or decreases as the electrons diffuse radially inward or outward, respectively. M represents the first adiabatic invariant, which depends on the electron mass, m_e, and the perpendicular momentum $p_\perp$ (see equation (4)).

In equations (2) and (3), the terms $P_m^E(m\omega_d)$ and $P_m^B(m\omega_d)$ represent the PSD of the electric and magnetic field perturbations with azimuthal wave number m at wave frequencies ω, which satisfy the drift resonance condition given by

$$\omega - m\omega_d = 0. \tag{6}$$

Here ω_d represents the bounce-averaged angular drift frequency of the electron [see *Southwood and Kivelson*, 1982; *Brizard and Chan*, 2001]. Since ω_d is a function of the electron's energy and L shell, this introduces both an energy and L shell dependence into the PSD terms $P_m^E(m\omega_d)$ and $P_m^B(m\omega_d)$ in equations (2) and (3). The values of the PSD terms $P_m^E(m\omega_d)$ and $P_m^B(m\omega_d)$ will, hence, vary with time as the ULF wave power increases and decreases.

In order for space weather models to forecast or nowcast the influence of ULF wave-driven radial diffusion on the dynamics and energization of the outer radiation belts, the models require the time variation of the diffusion coefficients to be known. Our approach to quantifying this time variation involves calculating hourly compressional magnetic PSD values from in situ equatorial spacecraft measurements over a range of L shells and local time sectors spanning the outer radiation belt region. Similarly, the hourly azimuthal electric field PSD should also be calculated either from in situ measurements or from mapped ground magnetometer measurements, again spanning the outer radiation belt region. Ideally, these measurements should be made over a long time scale, ideally a solar cycle or longer. For each hour, the PSD values should be binned with a geomagnetic or solar index, I. This could be Kp, solar wind speed, the dynamic solar wind pressure, P_{dyn}, or its temporal variation on some time scale ΔP_{dyn}. The resulting statistical database of electric and magnetic PSD values binned with the index, I, can then be used with equations (2) and (3) to determine an average value of D_{LL}^E and D_{LL}^B for any given value of I. These average (median) values of D_{LL}^E and D_{LL}^B for each index I can then be used to estimate the temporal variation of the diffusion coefficient and, hence, the outer radiation belt electron flux using the known time variation of the index, I. This general approach to forecasting or nowcasting the impact of the ULF wave-driven radial diffusion on the outer radiation belt electron flux is illustrated in Figure 1.

In this paper, we use the P_m^E and P_m^B values as a function of Kp as determined by the fits presented in the supplementary material of *Ozeke et al.* [2012]. In general, the values of P_m^E and P_m^B will depend on the m value of the ULF wave. To obtain the ULF waves, PSD at each m value from ground magnetometer observations requires coherent measurements of the same wave by multiple longitudinally spaced instruments at the same time and latitude on the ground [see, e.g., *Chisham and Mann*, 1999]. Obtaining the waves' m value spectrum from in situ measurements is even more difficult requiring multiple longitudinal spaced spacecraft observations at the same radius and latitude. An alternative approach to estimate the likely m value spectrum of ULF waves in the radiation belt region is to analyze the spectrum of waves produced by global MHD simulations of the Earth's

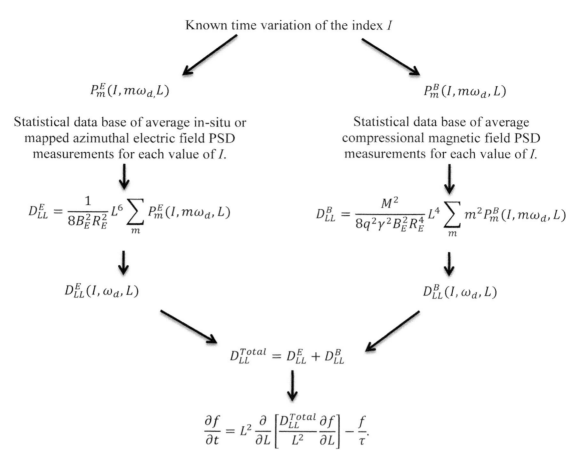

Known time variation of the index I

$$P_m^E(I, m\omega_d, L)$$

$$P_m^B(I, m\omega_d, L)$$

Statistical data base of average in-situ or mapped azimuthal electric field PSD measurements for each value of I.

Statistical data base of average compressional magnetic field PSD measurements for each value of I.

$$D_{LL}^E = \frac{1}{8B_E^2 R_E^2} L^6 \sum_m P_m^E(I, m\omega_d, L)$$

$$D_{LL}^B = \frac{M^2}{8q^2\gamma^2 B_E^2 R_E^4} L^4 \sum_m m^2 P_m^B(I, m\omega_d, L)$$

$$D_{LL}^E(I, \omega_d, L)$$

$$D_{LL}^B(I, \omega_d, L)$$

$$D_{LL}^{Total} = D_{LL}^E + D_{LL}^B$$

$$\frac{\partial f}{\partial t} = L^2 \frac{\partial}{\partial L} \left[\frac{D_{LL}^{Total}}{L^2} \frac{\partial f}{\partial L} \right] - \frac{f}{\tau}.$$

Figure 1. Modeling approach to determining the impact of ULF wave-driven radial diffusion on the dynamics and energization of outer radiation belt electrons.

magnetosphere. These global MHD simulations indicate that, generally, most of the ULF wave PSD is contained in waves with $|m| = 1$, although during intervals with high AE indices, the contributions from the $|m| > 1$ waves become more significant [see *Tu et al.*, 2012].

Here we assume for simplicity that all of the ULF wave power is contained in the $P_{m=1}^E$ and $P_{m=1}^B$ terms. This approach is also used in the works of *Brautigam et al.* [2005] and *Ozeke et al.* [2012] to determine the diffusion coefficients, which simplifies equations (2) and (3) to

$$D_{LL}^E = \frac{1}{8B_E^2 R_E^2} L^6 P_{m=1}^E(\omega_d) \tag{7}$$

$$D_{LL}^B = \frac{M^2}{8q^2\gamma^2 B_E^2 R_E^4} L^4 P_{m=1}^B(\omega_d). \tag{8}$$

3. DIFFUSION COEFFICIENT COMPARISON

In the work of *Ozeke et al.* [2012], the $P_m^E(m\omega_d)$ is determined from CARISMA and SAMNET ground magnetometer measurements over 15 years mapped to the azimuthal electric field in space assuming an Aflvenic wave solution along the field line [*Ozeke et al.*, 2012; *Rae et al.*, 2012]. However, the value of $P_m^E(m\omega_d)$ as a function of Kp has also been determined from 9 months of in situ CRRES measurements [*Brautigam et al.*, 2005]. Figure 2 illustrates a comparison between the azimuthal electric field PSD derived from CRRES in situ measurements and that derived from the ground magnetometer measurements at $L = 4.2$ and $L = 6.5$, respectively. The CRRES and ground magnetometer inferred electric field PSD values show a similar behavior with frequency, each having a peak in the PSD at approximately 4 mHz [see Figure 2]. However, the amplitudes of the

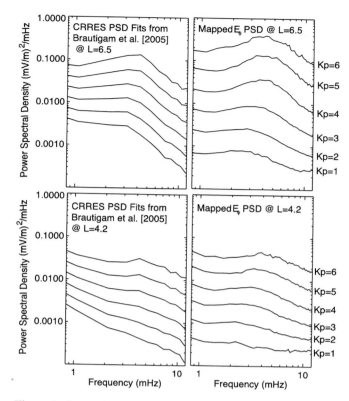

Figure 2. Comparison of the mapped equatorial azimuthal electric field power spectral density (PSD) with the electric field PSD derived from the CRRES analytic fits presented by *Brautigam et al.* [2005].

mapped PSD values from the work of *Ozeke et al.* [2012] are slightly higher at $L = 6.5$ for high Kp values than those derived from the in situ CRRES PSD measurements. One likely cause for this difference is that the ground-based magnetometer measurements taken at these high L shells (Gillam (GILL) and Fort Churchill (FCHU)) show enhanced power in the prenoon local time sector (0600–1200 magnetic local time (MLT)) particularly during disturbed times [see *Pahud et al.*, 2009; *Rae et al.*, 2012], while in the work of *Brautigam et al.* [2005], no electric field data was used as the apogee of CRRES passed through the prenoon local time, reducing the median PSD derived from the CRRES statistical database compared to that derived from the ground magnetometer statistical database with full local time coverage.

In this paper, the azimuthal electric field PSD fits presented by *Ozeke et al.* [2012] are used to determine the electric diffusion coefficients. The compressional magnetic field PSD derived from the in situ AMPTE and GOES PSD measurements taken between 1996 and 2005 are used by *Ozeke et al.* [2012] to determine the magnetic diffusion coefficients as a function of Kp or solar wind speed. We

adopt the same approach in this paper, again choosing to examine the radial diffusion as a function of Kp.

Figure 3 illustrates a comparison between the electric field diffusion coefficient derived from the mapped ground PSDs (dashed line) and from the in situ CRRES-measured electric field PSD presented by *Brautigam et al.* [2005] (dotted line). Also shown in Figure 3 is the magnetic diffusion coefficient derived from the compressional magnetic field PSD fits also presented by *Ozeke et al.* [2012] (solid line). The electric field diffusion coefficients derived from the mapped ground PSDs and the in situ CRRES measurements are in good agreement with a similar L shell dependence for radiation belt electrons with M values of 500 and 2000 MeV G^{-1} [see Figure 3]. However, as discussed by *Ozeke et al.* [2012], the magnetic field diffusion coefficient is typically more than an order of magnitude lower than the electric field diffusion coefficient.

If the phase between the electric and magnetic fluctuations are random, then the diffusion coefficients shown in equations (2) and (3) can be added together to form the total diffusion coefficient D_{LL}^{Total} where

$$D_{LL}^{\text{Total}} = D_{LL}^{E} + D_{LL}^{B}. \tag{9}$$

Consequently, the diffusion coefficients D_{LL}^{E} and D_{LL}^{B} cannot be added together in this way if the azimuthal electric and the compressional magnetic field fluctuations used in equations (2) and (3) belong to the same wave. However, in reality, some portions of the in situ measured azimuthal electric field used in equation (2) will be related to the compressional magnetic field used in equation (3) by Faradays law [see, e.g., *Ozeke et al.*, 2012]. Consequently, when D_{LL}^{E} already contains a contribution from the compressional wave, adding D_{LL}^{B} to determine D_{LL}^{Total} may not be correct. For example, by calculating the trajectories of electrons in an analytic model of the ULF wavefields, *Perry et al.* [2005] present results indicating that the total diffusion coefficient derived by the combined electric and magnetic fields of a compressional wave is lower than the diffusion coefficient derived from just the electric field of the compressional wave.

One pragmatic approach to avoid this problem of the phase relationship between the compressional magnetic field fluctuations used to determine D_{LL}^{B} and the azimuthal electric field component resulting from the compressional magnetic field, which contributes to D_{LL}^{E}, is to simply neglect the D_{LL}^{B} term. This approach may be valid since the magnetic diffusion coefficient is typically over an order of magnitude less than D_{LL}^{E} (as illustrated in Figure 3). In section 4, we investigate what impact neglecting D_{LL}^{B} has on the flux of the electrons in a series of radiation belt simulations.

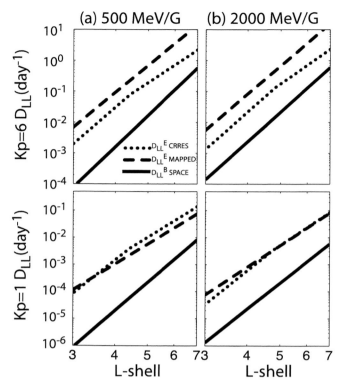

Figure 3. Diffusion coefficients as a function of L shell for $Kp = 6$ and $Kp = 1$ conditions derived from CRRES PSD electric field fits from *Brautigam et al.* [2005] (dotted line), from the mapped azimuthal electric field PSD presented by *Ozeke et al.* [2012] (dashed line), and from the GOES and Active Magnetospheric Particle Tracer Explorers compressional magnetic field PSD measurements presented by *Ozeke et al.* [2012] (solid line). Diffusion coefficients for electrons with first adiabatic invariants of (a) 500 MeV G^{-1} and (b) 2000 MeV G^{-1}.

It should be noted that in the work of *Brautigam and Albert* [2000] approach, the electrostatic diffusion coefficient (expressed as a function of the electric field PSD) is much smaller than the electromagnetic diffusion coefficient (expressed as a function of the compressional magnetic field PSD), as illustrated in Figure 4. Consequently, radial diffusion simulations such as that presented by *Shprits et al.* [2005] only include the electromagnetic term.

In Figure 5, the electric and magnetic diffusion coefficients presented in Figure 3 are compared with the electromagnetic and electrostatic diffusion coefficients presented in Figure 4. Figure 5 illustrates that the electromagnetic diffusion coefficient from *Brautigam and Albert* [2000] dominates over both the electric and magnetic diffusion coefficients from *Ozeke et al.* [2012]. The electrostatic diffusion coefficient also dominates over magnetic diffusion coefficient from *Ozeke et al.* [2012]. The results in Figure 5 indicate that the *Brautigam*

and Albert [2000] coefficients may overestimate the impact of radial diffusion on the dynamics and energization of radiation belt electrons [see also *Ozeke et al.*, 2012].

4. THE IMPACT OF D_{LL}^B AND D_{LL}^E ON THE ELECTRON FLUX

In the work of *Shprits et al.* [2005], the electron flux during 1990 was analyzed using the radial diffusion coefficients presented by *Brautigam and Albert* [2000]. To determine what impact the electric and magnetic diffusion coefficients presented by *Ozeke et al.* [2012] have on the radiation belt electrons, we have also simulated the same time interval using the *Ozeke et al.* [2012] diffusion coefficients. To model the electron flux variation, we solve equation (1) adopting the same inner and outer boundary conditions used in the work of *Shprits et al.* [2005] and an electron lifetime of $\tau = 10$ days.

The inner boundary has $f(L = 1) = 0$, and at the outer boundary, the phase space density $f(L = 7)$ is derived from a model of the differential electron flux, J, where $J =$

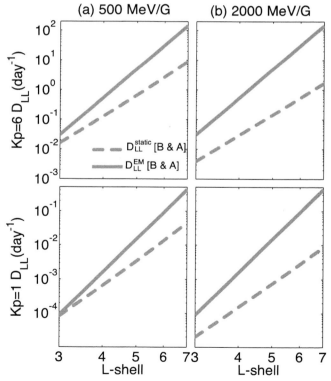

Figure 4. Electrostatic D_{LL}^{static} and electromagnetic D_{LL}^{EM} diffusion coefficients from *Brautigam and Albert* [2000] as a function of L shell for electrons with first adiabatic invariants of (a) 500 MeV G^{-1} and (b) 2000 MeV G^{-1}.

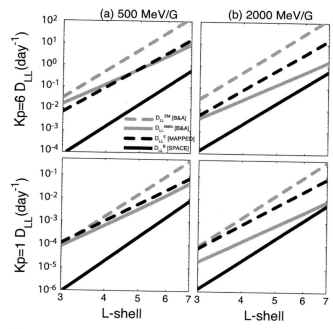

Figure 5. Comparison between the electrostatic and electromagnetic diffusion coefficients shown in Figure 3 and the electric and magnetic field diffusion coefficients shown in Figure 4 for electrons with first adiabatic invariants of (a) 500 MeV G^{-1} and (b) 2000 MeV G^{-1}.

8222.6exp($-$7.068) cm^{-2} sr^{-1} keV^{-1} s^{-1}, and K is the kinetic energy in MeV [see *Shprits et al.*, 2005]. The relationship between the phase space density, f, and the differential electron flux, J, is given by

$$f = \frac{J(L, E)}{p^2}, \qquad (10)$$

where E and p are the electrons' kinetic energy and momentum, respectively. Equation (10) is solved for a range of f values each with a different first adiabatic invariant, M, and equation (10) is then used to determine the differential electron flux at a constant energy. Figure 6 illustrates the change in the differential electron flux at 0.95 MeV during 1990 driven by the time series of Kp, which is shown in (a). Panels (b) and (c) show the radial diffusion response, which arises when both the electric and magnetic diffusion coefficients are included, and with only the electric diffusion coefficient is included, respectively. Panel (d) illustrates the ratio of the differential electron flux shown in (b) in units of cm^{-2} MeV^{-1} s^{-1} over the differential electron flux shown in (c) again in units of cm^{-2} MeV^{-1} s^{-1}.

The results in Figures 6b and 6c indicate that the electron flux response arising from diffusion, including both D_{LL}^E and D_{LL}^B, shows very similar characteristic electron enhance-

ments to the response, which occurs when using only D_{LL}^E. However, (d) shows that despite the similarity of the temporal dynamics of the flux enhancements in both simulations, the inclusion of D_{LL}^B in addition to D_{LL}^E enhances the electron flux by up to a factor of 1.2 on L shells between $L = 5$ and $L = 3$. This additional enhancement caused by the D_{LL}^B term is still relatively small compared with the total change in the electron flux, which can vary by a factor of ~100 along a fixed L shell over a timescale of a few days. Compared to the overall dynamic range, the effect of adding D_{LL}^B is relatively small, even in the heart of the outer zone radiation belt around $L \sim 3$–5.

Figure 7 illustrates what difference the electrostatic and electromagnetic diffusion coefficients, from *Brautigam and Albert* [2000], have on the electron flux compared to the electric and magnetic diffusion coefficients from *Ozeke et al.* [2012]. Figure 7b shows electron flux simulations during 1990 using the electrostatic and electromagnetic diffusion coefficients presented by *Brautigam and Albert* [2000] with the same boundary conditions and electron lifetime τ used in Figure 6.

Figure 7c illustrates the electron flux simulation run with electric and magnetic diffusion coefficients also presented in Figure 6b. Figure 7d illustrates the ratio of the differential electron flux shown in Figure 7b in units of cm^{-2} MeV^{-1} s^{-1} over the differential electron flux shown in (c) again in units of cm^{-2} MeV^{-1} s^{-1}. These results indicate that the electron flux simulated using the diffusion coefficient presented by *Brautigam and Albert* [2000] is enhanced on low L shells ($L < 3.5$) by a factor of 10,000 more than the electron flux simulated using the *Ozeke et al.* [2012] diffusion coefficients. For $L > 3.5$, this difference in the enhanced electron flux is less than a factor of 100, as illustrated in Figure 7d.

5. DISCUSSION AND CONCLUSIONS

Accurate determination of the ULF wave radial diffusion is critically important for simulations of the radiation belts. Currently, most radiation belt models use the radial diffusion coefficients presented by *Brautigam and Albert* [2000]. *Brautigam and Albert* [2000] separate the diffusion coefficients into a term resulting from diffusion due to electromagnetic wave fluctuations, D_{LL}^{EM}, and a term due to electrostatic fluctuations, D_{LL}^{static}. These diffusion coefficients D_{LL}^{EM} and D_{LL}^{static} are expressed in terms of the power spectral densities of the azimuthal symmetric compressional magnetic field and the azimuthal electric field. In order to estimate how D_{LL}^{EM} varies with time, *Brautigam and Albert* [2000] used a statistical database of compressional magnetic PSD values binned with Kp using 1 month of ATS-6 geosynchronous measurements and 18 days of ground magnetometer

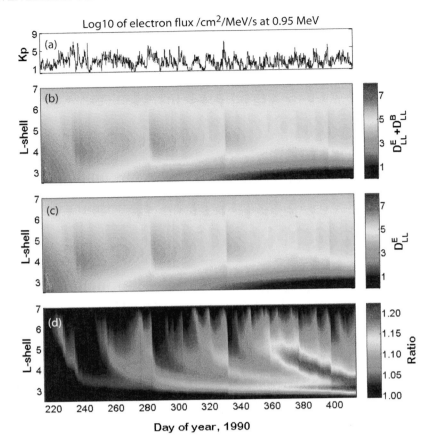

Figure 6. Impact of the mapped azimuthal electric field and in situ compressional magnetic field diffusion coefficients on the flux of 0.95 MeV electrons derived with a constant loss time of 10 days. (a) Time variation of *Kp*. (b) Electron flux variation derived with both the electric and magnetic diffusion coefficients summed together from *Ozeke et al.* [2012]. (c) Electron flux variation with only the electric field diffusion coefficient from *Ozeke et al.* [2012]. (d) Flux derived in Figure 6b over the flux derived in Figure 6c.

measurements made at $L \sim 4$. The D_{LL}^{static} diffusion coefficient in the work of *Brautigam and Albert* [2000] is derived using an analytic model for the azimuthal electric field PSD as a function of *Kp* resulting from substorm-driven convection electric field fluctuations. Alternative expressions for the diffusion coefficients were also derived by *Brizard and Chan* [2001] and *Fei et al.* [2006] in terms of the azimuthal electric field PSD and the compressional magnetic field PSD. In this paper, we compare the electromagnetic and electrostatic diffusion coefficients derived by *Brautigam and Albert* [2000] with the electric and magnetic diffusion coefficient derived by *Ozeke et al.* [2012] using the expressions presented by *Fei et al.* [2006]. *Ozeke et al.* [2012] determine the diffusion coefficients as a function of *Kp* using a database of over a solar cycle of in situ spacecraft and ground magnetometer measurements. The electromagnetic diffusion coefficients derived by *Brautigam and Albert* [2000] are

much larger than either the electric or magnetic diffusion coefficients presented by *Ozeke et al.* [2012] (see Figure 5). *Huang et al.* [2010] derive diffusion coefficients by following electron trajectories in the fields produced by the Lyon-Fedder-Mobbary MHD model. This approach also produces diffusion coefficients much lower than those derived by *Brautigam and Albert* [2000].

To investigate what impact the diffusion coefficients presented by *Ozeke et al.* [2012] have on the electron flux, we run a simulation of the year 1990 (see Figure 6). The results presented in Figure 6 indicate that the electron flux simulations driven by the electric and magnetic diffusion coefficients can be enhanced by a factor of ~100 in a few days. The electric field diffusion coefficient has the greatest impact on the electron flux; neglecting the magnetic diffusion term reduces the amplitude of the flux enhancements by only a factor ~1.2. The impact of a phase relationship

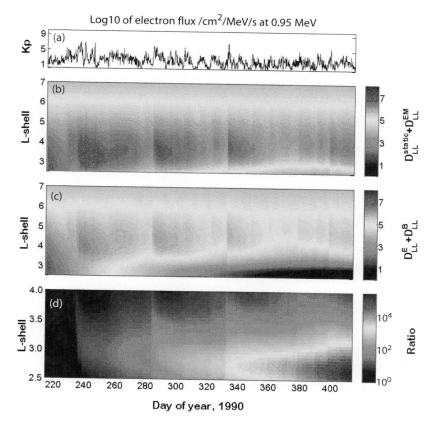

Figure 7. Impact of the mapped azimuthal electric field and in situ compressional magnetic field diffusion coefficients on the flux of 0.95 MeV electrons derived with a constant loss time of 10 days. (a) Time variation of *Kp*. (b) Electron flux variation derived with both the electrostatic and electromagnetic diffusion coefficients summed together from *Brautigam and Albert* [2000]. (c) Electron flux variation with the electric and magnetic field diffusion coefficient from *Ozeke et al.* [2012]. (d) Flux derived in Figure 7b over the flux derived in Figure 7c.

between the electric and magnetic PSDs may prevent the diffusion coefficients D_{LL}^B and D_{LL}^E from simply being added together to form a total diffusion coefficient (see discussion in section 4). However, the results illustrated in Figure 6 show that radiation belt simulations could be run using only the electric diffusion coefficient without affecting the magnitude of the electron flux by more than a factor of 1.2.

In Figure 7, we compared the electron flux simulated using the diffusion coefficients presented by *Brautigam and Albert* [2000] with the electron flux simulated using the diffusion coefficient presented by *Ozeke et al.* [2012]. These results indicate that the electron flux simulated using the diffusion coefficient presented by *Ozeke et al.* [2012] can be significantly lower (up to a factor of 10,000 lower) than the electron flux determined using the diffusion coefficients presented by *Brautigam and Albert* [2000], particularly on low *L* shells (*L* < 3.5).

Overall, we have demonstrated that in order to understand the relative importance of radial diffusion, it is critical to accurately specify the form of the diffusion coefficients. Close to the outer boundary of the simulation, the simulated fluxes remain largely controlled by the boundary condition imposed there; on lower *L* shells, a much larger difference in flux can develop due to the differences in the magnitude of the diffusion coefficients. As discussed, for example, by *Reeves et al.* [2003], the response of the radiation belts arises from a delicate balance between acceleration and loss. The new formalism for specifying the form of the radial diffusion coefficients presented by *Ozeke et al.* [2012], based on observed ULF wave power over a solar cycle, presents a new approach to this problem. Simulations run with the *Ozeke et al.* [2012] coefficients suggest a slower buildup of flux on lower *L* shells than that of *Brautigam and Albert* [2000]; however, any comparisons to data must also use realistic loss rates. When comparing to observed fluxes, slower diffusion rates can, of course, be somewhat compensated by the imposition of slower loss rates (and vice versa). Nonetheless, using data-driven diffusion coefficients, such as those presented by *Ozeke et al.* [2012], represents an important alternative basis for imposing the strength of radiation belt diffusion. This may be important for assessing the relative importance of radial diffusion compared to alternative acceleration processes, such as resonance with lower band chorus waves [cf. *O'Brien et al.*, 2003], for controlling the morphology and dynamics of the Earth's outer zone electron radiation belt.

Acknowledgments. IRM is supported by a Canadian NSERC Discovery Grant. K. R. M. is funded by Alberta Innovates and an NSERC Canadian Graduate Scholarship. I. J. R. and L. G. O. are funded by the Canadian Space Agency. CARISMA is operated by the University of Alberta and funded by the Canadian Space Agency. This work was supported, in part, by participation in the Monitoring, Analyzing and Assessing Radiation Belt Loss and Energization (MAARBLE) consortium. MAARBLE has received funding from the European Community's Seventh Framework Programme (FP7-SPACE-2010-1, SP1 Cooperation, Collaborative Project) under grant agreement 284520. This paper reflects only the authors' views, and the European Union is not liable for any use that may be made of the information contained herein.

REFERENCES

Beutier, T., and D. Boscher (1995), A three-dimensional analysis of the electron-radiation belt by the Salammbô Code, *J. Geophys. Res.*, *100*(A8), 14,853–14,861, doi:10.1029/94JA03066.

Brautigam, D. H., and J. M. Albert (2000), Radial diffusion analysis of outer radiation belt electrons during the October 9, 1990, magnetic storm, *J. Geophys. Res.*, *105*(A1), 291–309.

Brautigam, D. H., G. P. Ginet, J. M. Albert, J. R. Wygant, D. E. Rowland, A. Ling, and J. Bass (2005), CRRES electric field power spectra and radial diffusion coefficients, *J. Geophys. Res.*, *110*, A02214, doi:10.1029/2004JA010612.

Brizard, A. J., and A. A. Chan (2001), Relativistic bounce-averaged quasilinear diffusion equation for low-frequency electromagnetic fluctuations, *Phys. Plasmas*, *8*(11), 4762–4771.

Chisham, G., and I. R. Mann (1999), A Pc5 ULF wave with large azimuthal wavenumber observed within the morning sector plasmasphere by Sub-Auroral Magnetometer Network, *J. Geophys. Res.*, *104*(A7), 14,717–14,727, doi:10.1029/1999JA900147.

Cornwall, J. M. (1968), Diffusion processes influenced by conjugate-point wave phenomena, *Radio Sci.*, *3*(7), 740–744.

Fälthammar, C. G. (1968), Radial diffusion by violation of the third adiabatic invariant, in *Earth's Particles and Fields*, edited by B. M. McCormac, p. 157, Reinhold, New York.

Fei, Y., A. A. Chan, S. R. Elkington, and M. J. Wiltberger (2006), Radial diffusion and MHD particle simulations of relativistic electron transport by ULF waves in the September 1998 storm, *J. Geophys. Res.*, *111*, A12209, doi:10.1029/2005JA011211.

Holzworth, R. H., and F. S. Mozer (1979), Direct evaluation of the radial diffusion coefficient near L = 6 due to electric field fluctuations, *J. Geophys. Res.*, *84*(A6), 2559–2566, doi:10.1029/JA084iA06p02559.

Huang, C.-L., H. E. Spence, M. K. Hudson, and S. R. Elkington (2010), Modeling radiation belt radial diffusion in ULF wave fields: 2. Estimating rates of radial diffusion using combined MHD and particle codes, *J. Geophys. Res.*, *115*, A06216, doi:10.1029/2009JA014918.

Koller, J., Y. Chen, G. D. Reeves, R. H. W. Friedel, T. E. Cayton, and J. A. Vrugt (2007), Identifying the radiation belt source region by data assimilation, *J. Geophys. Res.*, *112*(A6), A06244, doi:10.1029/2006JA012196.

Lanzerotti, L. J., and C. G. Morgan (1973), ULF geomagnetic power near L = 4: 2. Temporal variation of the radial diffusion coefficient for relativistic electrons, *J. Geophys. Res.*, *78*(22), 4600–4610, doi:10.1029/JA078i022p04600.

Lanzerotti, L. J., D. C. Webb, and C. W. Arthur (1978), Geomagnetic field fluctuations at synchronous orbit 2. Radial diffusion, *J. Geophys. Res.*, *83*(A8), 3866–3870, doi:10.1029/JA083iA08p03866.

O'Brien, T. P., K. R. Lorentzen, I. R. Mann, N. P. Meredith, J. B. Blake, J. F. Fennell, M. D. Looper, D. K. Milling, and R. R. Anderson (2003), Energization of relativistic electrons in the presence of ULF power and MeV microbursts: Evidence for dual ULF and VLF acceleration, *J. Geophys. Res.*, *108*(A8), 1329, doi:10.1029/2002JA009784.

Ozeke, L. G., I. R. Mann, K. R. Murphy, I. J. Rae, D. K. Milling, S. R. Elkington, A. A. Chan, and H. J. Singer (2012), ULF wave derived radiation belt radial diffusion coefficients, *J. Geophys. Res.*, *117*, A04222, doi:10.1029/2011JA017463.

Pahud, D. M., I. J. Rae, I. R. Mann, K. R. Murphy, and V. Amalraj (2009), Ground-based Pc5 ULF wave power: Solar wind speed and MLT dependence, *J. Atmos. Sol. Terr. Phys.*, *71*(10–11), 1082–1092, doi:10.1016/j.jastp.2008.12.004.

Perry, K. L., M. K. Hudson, and S. R. Elkington (2005), Incorporating spectral characteristics of Pc5 waves into three-dimensional radiation belt modeling and the diffusion of relativistic electrons, *J. Geophys. Res.*, *110*, A03215, doi:10.1029/2004JA010760.

Rae, I. J., I. R. Mann, K. R. Murphy, L. G. Ozeke, D. K. Milling, A. A. Chan, S. R. Elkington, and F. Honary (2012), Ground-based magnetometer determination of in situ Pc4–5 ULF electric field wave spectra as a function of solar wind speed, *J. Geophys. Res.*, *117*, A04221, doi:10.1029/2011JA017335.

Reeves, G. D., K. L. McAdams, R. H. W. Friedel, and T. P. O'Brien (2003), Acceleration and loss of relativistic electrons during geomagnetic storms, *Geophys. Res. Lett.*, *30*(10), 1529, doi:10.1029/2002GL016513.

Schulz, M., and L. J. Lanzerotti (1974), *Particle Diffusion in the Radiation Belts*, *Phys. Chem. Space*, vol. 7, 215 pp., Springer, New York.

Shprits, Y. Y., R. M. Thorne, G. D. Reeves, and R. Friedel (2005), Radial diffusion modeling with empirical lifetimes: Comparison with CRRES observations, *Ann. Geophys.*, *23*(4), 1467–1471.

Southwood, D. J., and M. G. Kivelson (1982), Charged particle behavior in low-frequency geomagnetic pulsations, 2. Graphical approach, *J. Geophys. Res.*, *87*(A3), 1707–1710, doi:10.1029/JA 087iA03p01707.

Subbotin, D. A., and Y. Y. Shprits (2009), Three-dimensional modeling of the radiation belts using the Versatile Electron Radiation Belt (VERB) code, *Space Weather*, *7*, S10001, doi:10.1029/2008SW000452.

Tu, W., S. R. Elkington, X. Li, W. Liu, and J. Bonnell (2012), Quantifying radial diffusion coefficients of radiation belt electrons based on global MHD simulation and spacecraft measurements, *J. Geophys. Res.*, *117*, A10210, doi:10.1029/2012JA 017901.

A. A. Chan, Department of Physics and Astronomy, Rice University, Houston, TX 77251-1892, USA.

L. G. Ozeke, I. R. Mann, K. R. Murphy, and I. J. Rae, Department of Physics, University of Alberta, Edmonton, AB T6G 2E1, Canada. (lozeke@ualberta.ca)

Nonlinear Radial Transport in the Earth's Radiation Belts

B. T. Kress and M. K. Hudson

Department of Physics and Astronomy, Dartmouth College, Hanover, New Hampshire, USA

A. Y. Ukhorskiy

Applied Physics Laboratory, Johns Hopkins University, Laurel, Maryland, USA

H.-R. Mueller

Department of Physics and Astronomy, Dartmouth College, Hanover, New Hampshire, USA

Radial transport in the Earth's radiation belts is frequently modeled with a diffusion equation. In some cases, however, solar wind dynamic pressure fluctuations can cause significant changes in the phase space density radial profile that are not diffusive, e.g., shock-associated injections observed during storm sudden commencements. Three forms of radial transport driven by solar wind dynamic pressure fluctuations are briefly reviewed: (1) radial diffusion, (2) large changes in the phase space density radial profile due to a single or few ULF drift-resonant interactions, and (3) shock-associated injections of radiation belt electrons occurring in less than a single drift period. This is followed by a presentation of numerical results comparing radial diffusion with transport modeled by computing electron-guiding center trajectories in MHD magnetospheric model fields. Initial results suggest that radial transport of MeV electrons in the outer radiation belt due to moderate solar wind fluctuations is not well modeled by a diffusion equation.

1. INTRODUCTION

For over 40 years, the standard approach to understanding and modeling the dynamics of the Earth's radiation belts has been based on the solution of a diffusion equation solved in a space of one or several of the three adiabatic invariants associated with the motions of a charged particle in a dipole magnetic field. The creation of the Earth's outer electron belts is attributed to earthward transport and adiabatic acceleration of electrons through drift-resonant interactions with electromagnetic fluctuations in the magnetosphere [e.g., *Roederer*, 1970, pp. 112, 130; *Schulz and Lanzerotti*, 1974, pp. 47, 81]. Under conditions where the first and second adiabatic invariants are conserved, and individual drift-resonant interactions produce small random changes in the distribution function, the transport can be modeled with an L shell diffusion equation

$$\frac{\partial F}{\partial t} = L^2 \frac{\partial}{\partial L} \left\{ D_{LL} \frac{1}{L^2} \frac{\partial F}{\partial L} \right\}. \tag{1}$$

$F(L)$ is the number of particles per unit phase space volume, and D_{LL} is a diffusion coefficient that depends on the electromagnetic fluctuations that are driving the diffusion. Models based on equation (1) have been used extensively to study radiation belt dynamics [e.g., *Nakada and Mead*, 1965; *Schulz and Lanzerotti*, 1974; *Spjeldvik et al.*, 1998; *Brautigam and*

Dynamics of the Earth's Radiation Belts and Inner Magnetosphere
Geophysical Monograph Series 199
10.1029/2012GM001333

Albert, 2000; *Li et al.*, 2001]. Usually, a loss term is also included in equation (1) to model the effects of losses to the Earth's atmosphere by pitch angle scattering due to cyclotron-resonant particle interactions with VLF waves. A more comprehensive description of the radiation belts is provided by including the effects of pitch angle and energy scattering with a multidimensional diffusion equation, e.g., by simultaneously modeling diffusion in L, energy, and equatorial pitch angle. Recent development of comprehensive radiation belt models is mainly based on solving a multidimensional diffusion equation [e.g., *Bourdarie et al.*, 1996; *Brizard and Chan*, 2001; *Albert and Young*, 2005; *Shprits et al.*, 2008].

Equation (1) is usually obtained from a restricted form of the Fokker-Planck equation, and diffusion is assumed a priori [*Fälthammar*, 1968; *Walt*, 1994]. A *quasilinear* diffusion equation that describes the slow timescale evolution of a background distribution due to fast timescale perturbations can also be derived directly from first principles, beginning with the Vlasov equation. This approach appears in the work of *Kaufman* [1971, 1972] and is put into a formalism appropriate for 3-D radiation belt modeling by *Brizard and Chan* [2001]. By expanding the distribution function and Hamiltonian in terms of a small dimensionless parameter ε associated with the perturbing wave field, *Brizard and Chan* [2001] derived a 3-D radiation belt diffusion equation. The main requirement for the validity of the quasilinear treatment is that the perturbation to the distribution function εδF, due to a single drift-resonant interaction, is small with respect to the zeroth-order distribution function F_0. A fully nonlinear treatment is needed in cases where wavefields produce a perturbation of order F_0.

Recent observations and theoretical results suggest that fully nonlinear transport plays a significant role in radiation belt dynamics. *Albert* [2002] identifies three types of cyclotron-resonant interactions between electrons and VLF waves that lead to pitch angle and energy scattering: diffusion, phase bunching, and phase trapping. The latter two involve fully nonlinear interaction between the radiation belt particle distribution and VLF waves. Recently discovered, very efficient forms of electron acceleration known as relativistic turning acceleration [*Omura et al.*, 2007] and ultrarelativistic acceleration [*Summers and Omura*, 2007] are based on special forms of nonlinear phase trapping by VLF waves. *Bortnik et al.* [2008] describe the regimes where diffusion versus fully nonlinear interaction is in effect. *Bortnik et al.* [2008] also suggest that large-amplitude whistlers recently reported by *Cattell et al.* [2008] and *Cully et al.* [2008] may produce radiation belt transport not modeled with a diffusion equation.

Nonlinear wave-particle interaction also plays a role in *radial* transport in the radiation belts. A dramatic example of this are shock-drift injections, where outer zone electrons are transported radially inward and accelerated by an electric field pulse launched when an interplanetary shock impacts the magnetosphere during a storm sudden commencement (SSC) [*Li et al.*, 1993; *Hudson et al.*, 1997; *Kress et al.*, 2007]. The leading portion of the bipolar pulse can exceed ~100 mV m^{-1} [*Wygant et al.*, 1994] and is predominantly westward. Energetic electrons in drift resonance with the azimuthally propagating SSC electric field pulse may **E** × **B** drift inward several L shells undergoing significant energization in a fraction of a drift period. *Li et al.* [1993] modeled the formation of a new electron belt at $L \sim 2.5$ during the 24 March 1991 storm by following electron-guiding centers restricted to the equatorial plane in a pure dipole magnetic field traversed by an analytically modeled bipolar electric field pulse. The model was found to reproduce the observed electron drift echoes well.

A different form of radial transport was demonstrated by *Ukhorskiy et al.* [2006]. In that work, 10^4 electron-guiding center trajectories restricted to the equatorial plane were computed in fields from the *Tsyganenko and Sitnov* [2005] (TS05) empirical geomagnetic field model. The TS05 model was driven with time-dependent parameters derived from solar wind data, and the resulting $\Delta B / \Delta t$ was used in a Biot-Savart-type integral to obtain an inductive electric field. The resulting transport is dominated by a few drift-resonant interactions, each producing large coherent displacements of the distribution function in L. An initial distribution with guiding centers located on a single L shell produces a spiky irregular distribution function spanning a significant portion of the outer belt region. The conclusion is that, over the range of L spanned by the outer belt, the transport resulting from global magnetospheric compressions is inconsistent with radial diffusion. In a subsequent study, *Ukhorskiy and Sitnov* [2008] use a simplified model to show that if the magnetosphere is treated as boundless (i.e., extending to infinity with no magnetopause), the diffusion limit is obtained after ~100 h.

One characteristic of a solution of equation (1), with no source or loss terms and fixed boundary conditions, is that a peak in $F(L)$ will not form. The formation of phase space density peaks in L is sometimes interpreted as evidence of local heating, as opposed to radial transport, as a mechanism for radiation belt enhancements [*Green and Kivelson*, 2004]. However, *Degeling et al.* [2008] showed that a burst of narrow band ULF waves can produce strong localized peaks in the particle distribution with respect to L. *Degeling et al.* [2008] modeled an outer electron belt distribution in the $L = 4$–9 range by computing equatorial guiding center trajectories in an MHD wave field model. The drift-resonant interaction between a single $m = 3$ ULF wave mode and particles initially at $L = 6.6$ transported a significant portion of the particles over several L shells. After 5–10 wave periods, one or more localized peaks in electron phase space density were formed. The shock injections modeled by *Li et al.* [1993], *Hudson et al.* [1997], and *Kress et al.* [2007] differ from the radial transport demonstrated by *Ukhorskiy et al.* [2006] and *Degeling et al.* [2008] partly due to the different structure of the ULF fluctuations and also since, in the former case, the transport occurs in less than one drift period, while in the simulations by *Ukhorskiy et al.* [2006] and *Degeling et al.* [2008], the distribution undergoes large coherent displacements due to one or several drift-resonant interactions sustained for many drift periods.

For the numerical results presented in this paper, radial transport in the Earth's outer electron belts was modeled by computing guiding center trajectories in electric and magnetic fields from the Lyon-Feder-Mobarry (LFM) global MHD magnetospheric simulation code [*Lyon et al.*, 2004]. Trajectory computations and analysis were performed using radiation belt codes developed by the Center for Integrated Space Weather Modeling at Dartmouth College. Two initial case studies were performed to investigate radial transport under quiet and moderate solar wind driving conditions. Results from the run with moderate solar wind driving are mainly presented here. This study is focused on particles with 90° pitch angles; however, since the minimum B surface in the MHD model fields does not lie in a plane and is, in general, time-dependent, full 3-D test-particle dynamics are simulated. Guiding centers are initiated on a single drift shell with uniform, constant values of the first and second adiabatic invariants. The resulting radial transport is similar to that found in the works of *Ukhorskiy et al.* [2006] and *Ukhorskiy and Sitnov* [2008] supporting the conclusions of those studies.

2. GUIDING CENTER INTEGRATION IN MHD FIELDS

To model radial transport driven by ULF fluctuations in the Earth's magnetosphere, test-particle guiding center trajectories were computed in time-dependent MHD magnetospheric model fields using the noncovariant relativistic guiding center equations of *Cary and Brizard* [2009] (Equations 6.26 and 6.25 from *Cary and Brizard* [2009]). Since they are derived from a variational principle, these equations exactly conserve phase-space volume and energy in the static magnetic field case. The *Cary and Brizard* [2009] guiding center equations also exhibit better conservation of bounce and drift invariants in a general geomagnetic field model than do the *Northrop* [1963] guiding center equations, e.g., see the works of *Wan et al.* [2010] where the *Brizard and Chan* [1999] guiding center equations are used. Note that the

Brizard and Chan [1999] guiding center equations apply only to cases with a static magnetic field and no electric field, whereas the *Cary and Brizard* [2009] guiding center equations include phase space-preserving terms for guiding center dynamics in time-dependent magnetic and electric fields.

Test particle trajectories are integrated using a fourth-order Runge-Kutta integrator. An adaptive guiding center time step analogous to the one used by *Kress et al.* [2007, 2008] is used. An adiabaticity or "epsilon" parameter that characterizes the length scale of the magnetic field variations with respect to particle gyroradius is used to determine the validity of the guiding center approximation. Particles that are found not to conserve their first adiabatic invariant are removed from the simulation. Fast integration is achieved by linearly interpolating electric and perturbation magnetic fields from a 4-D space-time grid to the guiding center position at each time step. The grid spacing used for this study is 0.1 R_E, and the fields are linearly interpolated between ~1 min interval field snapshots from the MHD simulation. The domain for the test-particle trajectory computations is a $16 \times 16 \times 16$ R_E box centered on the Earth. All computations are performed in the solar-magnetic (SM) coordinate system. The inner boundary of the MHD magnetospheric model is at ~2 R_E, which does not affect the results, since this study mainly focuses on equatorially mirroring outer radiation belt particles.

Electric and magnetic fields were obtained from the LFM global MHD magnetospheric simulation code, which is driven at its sunward boundary by solar wind data. The two 10 h simulation periods used for this work are described in the next section. The MHD equations are solved on a distorted spherical grid. The grid is adapted to place higher resolution where needed, in the inner magnetosphere, magnetosheath, and magnetotail regions. The MHD simulation region spans 30 R_E upstream to 300 R_E downstream in the x direction and -100 to $+100$ R_E in the y and z directions. The LFM code uses a total variation diminishing advection scheme, which minimizes spurious oscillations without introducing excessive diffusion. The LFM has been shown to produce wave power at geosynchronous similar to observations, at all local times in all three magnetic field components [*Huang*, 2010a].

The MHD magnetospheric model fields can, in general, be quite complicated, and care has been taken to minimize particle scattering due to discontinuities introduced by interpolations in space and time. To characterize the properties of the transport in L, we wish to simulate a particle distribution with uniform, constant values of the first and second invariants and initially located on the same drift shell. In general, the minimum B surface in the MHD field model is time-dependent and does not coincide with the solar-magnetic equatorial plane; thus, the full 3-D guiding center dynamics have been simulated for this work.

The magnetic fields from the LFM simulation with moderate solar wind driving used for this study have double minima along dayside field lines at $L^* \gtrsim 6$. When particles gradient drift into this region, they typically undergo drift orbit bifurcation (DOB) leading to breaking of the second adiabatic invariant [*Öztürk and Wolf*, 2007; *Wan et al.*, 2010; *Ukhorskiy et al.*, 2011]. To illustrate, two example trajectories computed in static LFM field snapshots are shown in Figure 1. The trajectory shown in the top panel is initiated with $\alpha = 90°$ at the minimum B location along the field line passing through $x = -5$ R_E, $y = 0$, $z = 0$. After one drift period, the changes in L^* and pitch angle at the minimum B surface (henceforth α_o, analogous to equatorial pitch angle) are both less than 1%. The trajectory shown in the bottom panel is also initiated with $\alpha_o = 90°$, the minimum B surface, but in this case along the field line passing through $x = -6$ R_E, $y = 0$, $z = 0$. The orbit shown in the bottom panel is pitch angle scattered as it approaches the

dayside, when it encounters the transition to field lines with double minima. After one drift period, the trajectory shown in the bottom panel has $\alpha_o = 87°$. Since we are interested in following a population with uniform and constant values of the first and second invariants, only particles with α within 1% of 90° are included in the $f(L^*)$ distributions presented later in this paper. Also, this study is restricted to the region below $L = 6$.

The International Radiation Belt Environment Modeling Library (IRBEM-LIB) L^* calculator (formally Office National d'Etudes et de Recherche Aérospatiales-Département Environnement Spatial (ONERA-DESP)) (a modified version of calcul_Lstar_o.f (version 215) is used, from IRBEM-LIB at http://irbem.svn.sourceforge.net/; originally developed by D. Boscher et al., ONERA-DESP library, version Version 4.2, May 2008, at http://craterre.onecert.fr/support/user_guide.html) is used to bin the particles in L shell. The IRBEM-LIB L^* calculator returns an L^* value regardless of the presence of double minima along dayside field lines, by conserving the second invariant computed between mirror

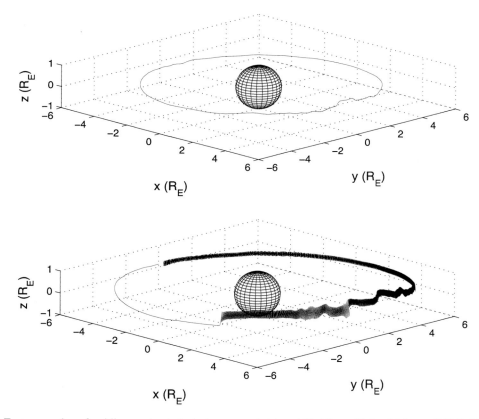

Figure 1. Two examples of guiding center trajectories computed in gridded Lyon-Feder-Mobarry (LFM) MHD model fields, initiated in the minimum B surface with $\alpha_o = 90°$. (top) Trajectory was initiated on the nightside near $x = -5$. (bottom) Trajectory was initiated on the nightside near $x = -6$ and undergoes pitch angle scattering as it enters the drift orbit bifurcation region on the dayside.

points that straddle the northern minimum and tracing guiding field lines to the northern polar cap where it performs the **B · ds** calculation. In the results that follow, $f(L^*)$ plots have abscissas that extend into the DOB region, but with the caveat that particle orbits in this region cannot necessarily be associated with an L^* value or drift shell. Since only particles with $\alpha \approx 90°$ are included in the distributions, and particles in the DOB region are rapidly scattered, $f(L^*)$ remains near zero above $L^* \sim 6$.

To numerically launch a distribution of guiding centers on a single drift shell, a single guiding center drift orbit is first computed in the $t = 0$ magnetic field snapshot, with the electric field set to zero. The guiding center location and pitch angle are saved to an array each time it passes through the minimum B surface. The distribution is then initialized by randomly choosing azimuthal coordinate values between 0 and 2π, and obtaining the minimum B surface location and pitch angle corresponding to that azimuthal coordinate by interpolating between the minimum B surface crossings obtained from the single guiding center drift orbit.

As a test of conservation of the second and third adiabatic invariants, 2000 guiding center trajectories with $\alpha = 90°$ were computed in a static magnetic field snapshot from the MHD simulations with the electric field set to zero. The guiding

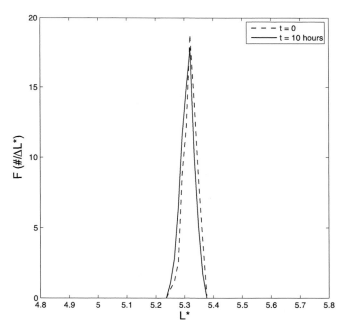

Figure 2. Initial and final distributions of 2000 guiding center trajectories computed in a single LFM static magnetic field snapshot with zero electric field. After 10 h, the mean and standard deviation are nearly identical with the initial conditions. The test demonstrates adequate conservation of the second and third adiabatic invariants for this study.

centers were initiated on a single L shell at approximately $L^* = 5.3$ with a uniform distribution in longitude and computed for 10 h of simulated time. The similarity between the initial and final distributions shown in Figure 2 is sufficient to substantiate the results presented in the following sections, where particles are transported across multiple L shells by time-dependent magnetic and electric fields.

3. SOLAR WIND INPUT FOR MHD MAGNETOSPHERIC MODEL SIMULATIONS

Observations show that magnetic field and velocity fluctuations in the solar wind are non-Gaussian and, thus, cannot be represented with a single spectral parameter [*Feynman and Ruzmaikin*, 1994; *Burlaga*, 1991, 1993; *Marsch et al.*, 1992; *Marsch and Lui*, 1993; *Viall et al.*, 2009]; e.g., *Viall et al.* [2009] analyzed 11 years of solar wind density data and found that certain discrete frequencies occur more often than others. Although statistically self similar periods of solar wind driving lasting many hours seldom occur, it is possible to select periods of driving that can be characterized qualitatively. Three examples of P_{dyn} time series from the Wind spacecraft that may be used as input to the LFM model are shown in Figure 3, obtained during quiet (top), moderate (middle), and strong (bottom) solar wind conditions. The mean value of P_{dyn} ranges from 1.4 nPa (top) to 13.9 nPa (bottom). In each case, a 10 h period is shown.

Data from the quiet and moderate periods of solar wind driving shown in the top two panels of Figure 3 were used as input to the MHD magnetospheric model for this study. ACE Solar wind density, velocity, temperature, and magnetic field data with a 1 min time resolution were used. The quiet period interval has solar wind speeds, densities, and interplanetary magnetic field (IMF) B_Z of approximately 450 km s^{-1}, 4–5 # cm^{-3}, and ± 4 nT. The moderate driving period has solar wind speeds, densities, and IMF B_Z ~550 km s^{-1}, 6–14 # cm^{-3} and ± 10 nT.

4. TEST PARTICLE RESULTS

The 10k guiding center trajectories with $\alpha_o = 90°$ were computed for 10 h of simulated time in time-dependent electric and magnetic fields from the 3 January 2003 LFM simulation. The guiding centers were initiated at random, uniformly chosen azimuthal locations in the minimum B surface on a single drift shell at $L_o^* = 5.3$. The drift shell is first located by computing a single guiding center drift trajectory as described in section 2. The initial values of the first and second adiabatic invariants are $\mu = 1560$ MeV G^{-1} and $J = 0$, respectively, corresponding to an energy of 1.6 MeV at $L^* = 5.3$. The drift period is ~1000 s. Particle data was saved

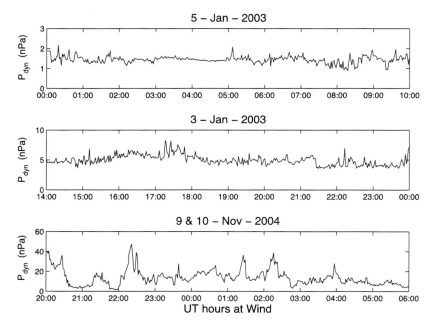

Figure 3. Three examples of P_{dyn} time series that may be used to drive the MHD magnetospheric model during (top) quiet, (middle) moderate, and (bottom) disturbed solar wind conditions. The mean values of P_{dyn} are (top) $<P_{dyn}> = 1.4$ nPa, (middle) $<P_{dyn}> = 5.0$ nPa, and (bottom) $<P_{dyn}> = 13.9$ nPa. The variances are (top) $\sigma^2 = 0.03$ nPa^2, (middle) $\sigma^2 = 0.72$ nPa^2, and (bottom) $\sigma^2 = 65$ nPa^2 (data sets provided by K. Ogilvie, NASA GSFC via CDAWeb).

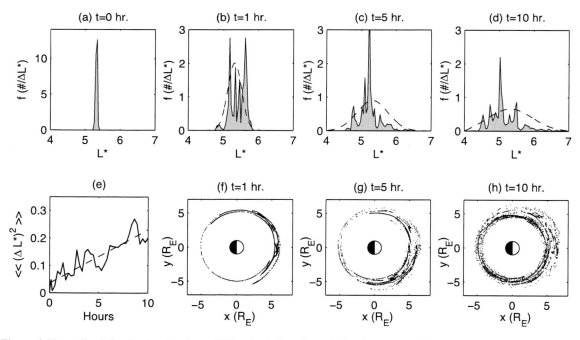

Figure 4. Normalized distribution functions $f(L^*)$ at (a–d) four times during the test-particle simulations $t = 0, 1, 5$ and 10 h and (f–h) corresponding positions in the SM equatorial plane. The time evolution of the variance of $f(L^*)$ is shown in Figure 4e. A linear fit to the variance is also included in Figure 4e (dashed line), and for comparison with the MHD test-particle results, a numerical solution of equation (1) is overlaid in the $f(L^*)$ plots using dashed lines.

to a file at regular time intervals throughout the simulation. The simulation took ~24 h to compute on a single 2.5 GHz CPU.

Time snapshots of the particle distribution at times $t = 0, 1, 5$, and 10 h are shown in Figure 4, which includes the normalized distribution function $f(L^*)$ in the top panels (a–d) and projections of the particle positions in the SM equatorial plane in the bottom three rightmost panels (f–h). Note that $f(L^*)$ *is the normalized density per unit L^**, to be distinguished from the phase space density $f(L^*)$ appearing in equation (1). The $t = 0$ particle positions in the SM equatorial plane, corresponding to the initial delta function appearing in (a), is simply a ring and is not shown. The time evolution of the distribution shows highly structured and irregular transport in L^* dominated by one or a few drift-resonant interactions that coherently displace particles in L^*. The macroscale distortion and folding of the initial ring of particles seen in (f–h) is typical of test-particle behavior in the LFM fields. The time evolution of the variance is shown in the lower left of Figure 4. The variance is computed,

$$\sigma^2 = \frac{1}{N_p} \sum_{i=1}^{N_p} (L_i^* - L_o^*)^2. \qquad (2)$$

A linear fit to the variance, yielding $\Delta\sigma/\Delta t = 0.020$ h^{-1}, provides an estimate of its growth rate. If the process governing the transport is diffusion, then $\sigma^2 = 2D_{LL}t$. For comparison with the results one would expect from diffusion, an FTCS finite difference method [e.g., *Press et al.*, 1992] is used to solve equation (1), using D_{LL} obtained from the linear fit to the variance growth rate and initial conditions $f(L^*, t = 0) = \delta(L_o^* - L^*)$. The solution is shown by the overlaid dashed line appearing in each of the $f(L^*)$ plots in Figure 4. The MHD test particle results clearly differ significantly from the diffusion model results.

Initial conditions identical with those described above were used for the test-particle simulation in MHD fields obtained with quiet solar wind driving conditions. The final time snapshot of the distribution at $t = 10$ h is shown in Figure 5. The evolution of $f(L^*)$ is irregular and shows similar structure to the case with moderate solar wind driving. The average rate of increase of the variance is $\Delta\sigma/\Delta t = 0.0013$ h^{-1}, which is approximately an order of magnitude smaller than $\Delta\sigma/\Delta t$ obtained with moderate solar wind driving.

5. DISCUSSION AND CONCLUSIONS

There is nothing unusual about the 3 January 2003 period chosen for the case study of radial transport under moderate solar wind driving. The solar wind speed and density during the interval are ~550 km s^{-1} and ~10 # cm^{-3}. The (GSM

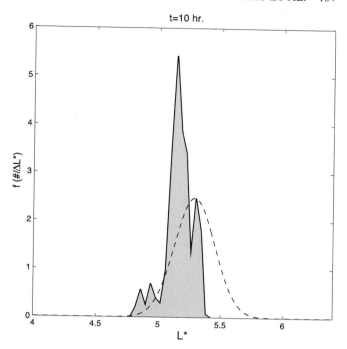

Figure 5. Final $f(L^*)$ distribution from test-particle run in MHD magnetospheric model fields driven by quiet solar wind conditions.

coordinates) IMF B_Z fluctuates between ±10 nT. The irregular and structured radial transport illustrated in Figure 4 is typical of test-particle simulations in time-dependent LFM fields and is easily reproduced. The simulations with quiet solar wind driving also produce transport in L^* that is irregular and structured (shown in Figure 5). If the solar wind P_{dyn} fluctuations are further reduced, approaching steady solar wind driving, it becomes difficult to distinguish between physical transport and transport driven by numerical effects associated with computing test-particle trajectories in gridded MHD fields, e.g., scattering due to field interpolations and discrete integration time steps. The dominant driver of radial transport of MeV outer belt electrons in the MHD test-particle model is drift-resonant interaction with ULF fluctuations driven by low mode number magnetospheric undulations that coherently displace large portions of the particle population over a significant fraction of the L^* range spanned by the outer belt region.

The results shown in Figure 4 are strikingly similar to those presented by *Ukhorskiy et al.* [2006], supporting the conclusion that radial transport in the outer electron belts driven by typical solar wind dynamic pressure fluctuations is not well modeled by an L shell diffusion equation. The equatorial plots in Figure 4 show drift phase bunching and folding similar to that shown in Figure 2 of *Ukhorskiy et al.* [2006], as particles displaced inward in L overtake those displaced outward. Given the similarity between the results presented here and those in the work of *Ukhorskiy et al.* [2006], one may conclude that the primary driver of radial transport in both cases is ULF fluctuations directly driven by the solar wind (breathing modes), since the time-dependently driven TS05 model used by *Ukhorskiy et al.* [2006] does not contain cavity mode and field line resonances or nightside Pc5 pulsations related to substorm activity.

The diffusion coefficients inferred from the average rate of increase of the $f(L^*)$ variances are similar to the outer electron belt radial diffusion coefficients found in earlier studies. For example, $D_{LL} = 0.031$ day^{-1} and $D_{LL} = 0.48$ day^{-1}, obtained from the quiet and moderately driven simulations, respectively, both fall within the range of theoretical and numerical D_{LL} values between 10^{-4} and 10^{1} at $L = 5$ shown in the Figure 6 of *Huang et al.* [2010b]. The diffusion coefficients numerically determined by *Huang et al.* [2010b], also in LFM fields, are considerably smaller than the diffusion coefficients found in this work. This discrepancy may, in part, be due to the fluctuating IMF B_Z contained in the MHD input used in this study, whereas in that work, idealized solar wind data with IMF $B_Z = +2$ nT

was used. In the diffusion limit, *Ukhorskiy et al.* [2006] obtain $D_{LL} = 0.028$ h^{-1} for 1 MeV equatorial guiding centers at $L = 5.8$, using solar wind input similar to the moderate solar wind driving conditions used in this work. The diffusion coefficient obtained by *Ukhorskiy et al.* [2006] is surprisingly close to $D_{LL} = 0.020$ h^{-1} obtained in this study, given the different magnetospheric field models used.

The dayside RMS electric field amplitude near $L = 5$ in the MHD field model is ~1 mV m^{-1}. This amplitude is sufficient to produce a single resonant island that transports particles over a significant portion of the outer electron belt. To illustrate this, Figure 6a shows a phase space portrait of the dynamics of equatorial guiding centers in a pure dipole magnetic field plus the simple ULF wave electric field model $E_\phi = A \sin(\phi - \omega t)$, with amplitude $A = 0.2$ mV m^{-1} and wave period $\tau = 540$ s. The abscissa value ($\psi = \phi - \omega t$) in Figure 6a is the angle between the azimuthally propagating wave crest and the azimuthal coordinate of the guiding center position. The width of the island is approximately one L^*, and the period corresponding to the phase space trajectories near the center of the island is ~2 h. Figure 6b shows the evolution of a particle distribution trapped in the resonant

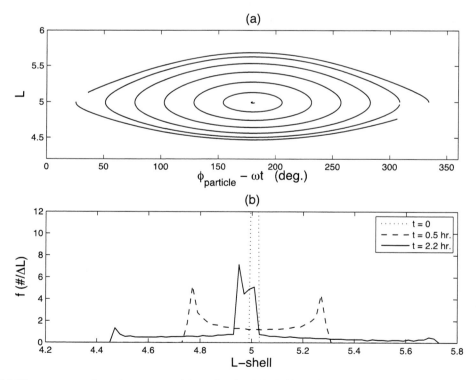

Figure 6. (a) Phase space trajectories computed in a simple ULF field model consisting of a pure dipole plus a wave electric field $E_\phi = A \sin(\phi - \omega t)$ with amplitude $A = 0.2$ mV m^{-1} and wave period $\tau = 540$ s. (b) Three time snapshots of $f(L^*)$ illustrating the evolution of the distribution in the simple field model, which is similar to the dynamics appearing in the MHD test-particle simulations.

island, computed with 2000 1 MeV guiding center trajectories initially at $L = 5$ drift resonant with the wave.

An animation of the particle distribution computed in the MHD field model reveals similar behavior to the dynamics of the distribution in the simple wave model illustrated in Figure 6. In the MHD test-particle simulations, an initial $f(L^*)$ peak splits into symmetric peaks that subsequently return to the initial L^*. The dynamics of the $f(L^*)$ distribution in the MHD field model is initially dominated by a single drift-resonant interaction, followed by subsequent transport due to adjacent resonances associated with new radial locations. *The single resonant island is large enough to coherently transport many particles over a significant portion of the outer belt.* Although the dynamics shown in Figure 4 are initially dominated by a resonance near $L^* = 5$, the time dependence of the spectral properties of the wavefields allow the distribution to spread and come under the influence of adjacent resonances, resulting in the irregular transport that is seen in the MHD test particle simulations.

An important question that remains unanswered is, what is the spectral and mode structure of the ULF fluctuations driving the transport illustrated in Figure 4? *Claudepierre et al.* [2008] find that at 0.5–3 mHz power is mainly in the low mode numbers. Additional work is needed to find the dominant source of radial transport in the simulations presented here. A second question is, under what circumstance is outer radiation belt transport well approximated by a diffusion equation? It may be speculated that a broad spectrum of ULF driving and a continuous radial distribution of particles will result in transport that is well modeled by diffusion. Alternatively, narrow frequency band ULF oscillations in the magnetosphere may produce a resonant island that dominates the transport and produces localized peaks in the phase space density radial profile (e.g., as shown in the work of *Degeling et al.* [2008]). These questions will be addressed by future studies.

Acknowledgments. This work was supported by NASA ROSES SR&T Grant NNX10AL87G.

REFERENCES

Albert, J. M. (2002), Nonlinear interaction of outer zone electrons with VLF waves, *Geophys. Res. Lett.*, *29*(8), 1275, doi:10.1029/2001GL013941.

Albert, J. M., and S. L. Young (2005), Multidimensional quasilinear diffusion of radiation belt electrons, *Geophys. Res. Lett.*, *32*, L14110, doi:10.1029/2005GL023191.

Bortnik, J., R. M. Thorne, and U. S. Inan (2008), Nonlinear interaction of energetic electrons with large amplitude chorus, *Geophys. Res. Lett.*, *35*, L21102, doi:10.1029/2008GL035500.

Bourdarie, S., D. Boscher, T. Beutier, J.-A. Sauvaud, and M. Blanc (1996), Magnetic storm modeling in the Earth's electron belt by the Salammbô code, *J. Geophys. Res.*, *101*(A12), 27,171–27,176, doi:10.1029/96JA02284.

Brautigam, D. H., and J. M. Albert (2000), Radial diffusion analysis of outer radiation belt electrons during the October 9, 1990, magnetic storm, *J. Geophys. Res.*, *105*(A1), 291–309, doi:10.1029/1999JA900344.

Brizard, A. J., and A. A. Chan (1999), Nonlinear relativistic gyrokinetic Vlasov-Maxwell equations, *Phys. Plasmas*, *6*, 4548, doi:10.1063/1.873742.

Brizard, A. J., and A. A. Chan (2001), Relativistic bounce-averaged quasilinear diffusion equation for low-frequency electromagnetic fluctuations, *Phys. Plasmas*, *8*(11), 4762–4771, doi:10.1063/1.1408623.

Burlaga, L. F. (1991), Intermittent turbulence in the solar wind, *J. Geophys. Res.*, *96*(A4), 5847–5851, doi:10.1029/91JA00087.

Burlaga, L. F. (1993), Intermittent turbulence in large-scale velocity fluctuations at 1 AU near solar maximum, *J. Geophys. Res.*, *98*(A10), 17,467–17,473, doi:10.1029/93JA01630.

Cary, J. R., and A. J. Brizard (2009), Hamiltonian theory of guiding-center motion, *Rev. Mod. Phys.*, *81*, 693–738, doi:10.1103/RevModPhys.81.693.

Cattell, C., et al. (2008), Discovery of very large amplitude whistler-mode waves in Earth's radiation belts, *Geophys. Res. Lett.*, *35*, L01105, doi:10.1029/2007GL032009.

Claudepierre, S. G., S. R. Elkington, and M. Wiltberger (2008), Solar wind driving of magnetospheric ULF waves: Pulsations driven by velocity shear at the magnetopause, *J. Geophys. Res.*, *113*, A05218, doi:10.1029/2007JA012890.

Cully, C. M., J. W. Bonnell, and R. E. Ergun (2008), THEMIS observations of long-lived regions of large-amplitude whistler waves in the inner magnetosphere, *Geophys. Res. Lett.*, *35*, L17S16, doi:10.1029/2008GL033643.

Degeling, A. W., L. G. Ozeke, R. Rankin, I. R. Mann, and K. Kabin (2008), Drift resonant generation of peaked relativistic electron distributions by Pc 5 ULF waves, *J. Geophys. Res.*, *113*, A02208, doi:10.1029/2007JA012411.

Fälthammar, C. (1968), Radial diffusion by violation of the third adiabatic invariant, in *Earth's Particles and Fields*, edited by B. M. McCormac, p. 157, Reinhold, New York.

Feynman, J., and A. Ruzmaikin (1994), Distributions of the interplanetary magnetic field revisited, *J. Geophys. Res.*, *99*(A9), 17,645–17,651, doi:10.1029/94JA01098.

Green, J. C., and M. G. Kivelson (2004), Relativistic electrons in the outer radiation belt: Differentiating between acceleration mechanisms, *J. Geophys. Res.*, *109*, A03213, doi:10.1029/2003JA010153.

Huang, C.-L., H. E. Spence, H. J. Singer, and W. J. Hughes (2010a), Modeling radiation belt radial diffusion in ULF wave fields: 1. Quantifying ULF wave power at geosynchronous orbit in observations and in global MHD model, *J. Geophys. Res.*, *115*, A06215, doi:10.1029/2009JA014917.

Huang, C.-L., H. E. Spence, M. K. Hudson, and S. R. Elkington (2010b), Modeling radiation belt radial diffusion in ULF wave fields: 2. Estimating rates of radial diffusion using combined MHD and particle codes, *J. Geophys. Res.*, *115*, A06216, doi:10.1029/2009JA014918.

Hudson, M. K., S. R. Elkington, J. G. Lyon, V. A. Marchenko, I. Roth, M. Temerin, J. B. Blake, M. S. Gussenhoven, and J. R. Wygant (1997), Simulations of radiation belt formation during storm sudden commencements, *J. Geophys. Res.*, *102*(A7), 14,087–14,102, doi:10.1029/97JA03995.

Kaufman, A. N. (1971), Resonant interactions between particles and normal modes in a cylindrical plasma, *Phys. Fluids*, *14*, 387–397, doi:10.1063/1.1693439.

Kaufman, A. N. (1972), Quasilinear diffusion of an axisymmetric toroidal plasma, *Phys. Fluids*, *15*, 1063–1069, doi:10.1063/1.1694031.

Kress, B. T., M. K. Hudson, M. D. Looper, J. Albert, J. G. Lyon, and C. C. Goodrich (2007), Global MHD test particle simulations of >10 MeV radiation belt electrons during storm sudden commencement, *J. Geophys. Res.*, *112*, A09215, doi:10.1029/2006JA012218.

Kress, B. T., M. K. Hudson, M. D. Looper, J. Albert, J. G. Lyon, and C. C. Goodrich (2008), Global MHD test particle simulations of solar energetic electron trapping in the Earth's radiation belts, *J. Atmos. Sol. Terr. Phys.*, *70*(14), 1727–1737, doi:10.1016/j.jastp.2008.05.018.

Li, X., I. Roth, M. Temerin, J. R. Wygant, M. K. Hudson, and J. B. Blake (1993), Simulation of the prompt energization and transport of radiation belt particles during the March 24, 1991 SSC, *Geophys. Res. Lett.*, *20*(22), 2423–2426, doi:10.1029/93GL02701.

Li, X., M. Temerin, D. N. Baker, G. D. Reeves, and D. Larson (2001), Quantitative prediction of radiation belt electrons at geostationary orbit based on solar wind measurements, *Geophys. Res. Lett.*, *28*(9), 1887–1890, doi:10.1029/2000GL012681.

Lyon, J. G., J. A. Fedder, and C. M. Mobarry (2004), The Lyon-Fedder-Mobarry (LFM) global MHD magnetospheric simulation code, *J. Atmos. Sol. Terr. Phys.*, *66*(15–16), 1333–1350, doi:10.1016/j.jastp.2004.03.020.

Marsch, E., and S. Lui (1993), Structure functions and intermittency of velocity fluctuations in the inner solar wind, *Ann. Geophys.*, *11*, 227–238.

Marsch, E., S. Lui, C.-Y. Tu, and H. Rosenbauer (1992), Spectra and structure functions of temperature fluctuations in the inner solar wind, in *Proceedings of the 26th ESLAB Symposium, Study of the Solar Terrestrial System, Killarney, Ireland, 16–19 June, ESA SP-346*, pp. 315–320, ESA Publ. Div., Noordwijk, The Netherlands.

Nakada, M. P., and G. D. Mead (1965), Diffusion of protons in the outer radiation belt, *J. Geophys. Res.*, *70*(19), 4777–4791, doi:10.1029/JZ070i019p04777.

Northrop, T. G. (1963), *The Adiabatic Motion of Charged Particles*, Interscience, New York.

Omura, Y., N. Furuya, and D. Summers (2007), Relativistic turning acceleration of resonant electrons by coherent whistler mode waves in a dipole magnetic field, *J. Geophys. Res.*, *112*, A06236, doi:10.1029/2006JA012243.

Öztürk, M. K., and R. A. Wolf (2007), Bifurcation of drift shells near the dayside magnetopause, *J. Geophys. Res.*, *112*, A07207, doi:10.1029/2006JA012102.

Press, W. H., B. P. Flannery, S. A. Teukolsky, and W. T. Vetterling (1992), *Numerical Recipes in C: The Art of Scientific Computing*, 2nd ed., Cambridge Univ. Press, Cambridge, U. K.

Roederer, J. G. (1970), *Dynamics of Geomagnetically Trapped Radiation*, Springer, Berlin.

Schulz, M., and L. J. Lanzerotti (1974), *Particle Diffusion in the Radiation Belts*, Springer, Berlin.

Shprits, Y. Y., D. A. Subbotin, N. P. Meredith, and S. R. Elkington (2008), Review of modeling of losses and sources of relativistic electrons in the outer radiation belt II: Local acceleration and loss, *J. Atmos. Sol. Terr. Phys.*, *70*(14), 1694–1713, doi:10.1016/j.jastp.2008.06.014.

Spjeldvik, W. N., G. I. Pugacheva, A. A. Gusev, I. M. Martin, and N. M. Sobolevsky (1998), Sources of inner radiation zone energetic helium ions: Cross-field transport versus in-situ nuclear reactions, *Adv. Space Res.*, *21*(12), 1675–1678, doi:10.1016/S0273-1177(98)00013-1.

Summers, D., and Y. Omura (2007), Ultra-relativistic acceleration of electrons in planetary magnetospheres, *Geophys. Res. Lett.*, *34*, L24205, doi:10.1029/2007GL032226.

Tsyganenko, N. A., and M. I. Sitnov (2005), Modeling the dynamics of the inner magnetosphere during strong geomagnetic storms, *J. Geophys. Res.*, *110*, A03208, doi:10.1029/2004JA010798.

Ukhorskiy, A. Y., and M. I. Sitnov (2008), Radial transport in the outer radiation belt due to global magnetospheric compressions, *J. Atmos. Sol. Terr. Phys.*, *70*(14), 1714–1726, doi:10.1016/j.jastp.2008.07.018.

Ukhorskiy, A. Y., B. J. Anderson, K. Takahashi, and N. A. Tsyganenko (2006), Impact of ULF oscillations in solar wind dynamic pressure on the outer radiation belt electrons, *Geophys. Res. Lett.*, *33*, L06111, doi:10.1029/2005GL024380.

Ukhorskiy, A. Y., M. I. Sitnov, R. M. Millan, and B. T. Kress (2011), The role of drift orbit bifurcations in energization and loss of electrons in the outer radiation belt, *J. Geophys. Res.*, *116*, A09208, doi:10.1029/2011JA016623.

Viall, N. M., L. Kepko, and H. E. Spence (2009), Relative occurrence rates and connection of discrete frequency oscillations in the solar wind density and dayside magnetosphere, *J. Geophys. Res.*, *114*, A01201, doi:10.1029/2008JA013334.

Walt, M. (1994), *Introduction to Geomagnetically Trapped Radiation*, Cambridge Univ. Press, Cambridge, U. K.

Wan, Y., S. Sazykin, R. A. Wolf, and M. K. Öztürk (2010), Drift shell bifurcation near the dayside magnetopause in realistic magnetospheric magnetic fields, *J. Geophys. Res.*, *115*, A10205, doi:10.1029/2010JA015395.

Wygant, J., F. Mozer, M. Temerin, J. Blake, N. Maynard, H. Singer, and M. Smiddy (1994), Large amplitude electric and magnetic field signatures in the inner magnetosphere during injection of 15 MeV electron drift echoes, *Geophys. Res. Lett.*, *21*(16), 1739–1742, doi:10.1029/94GL00375.

M. K. Hudson, B. T. Kress, and H.-R. Mueller, Department of Physics and Astronomy, Dartmouth College, Hanover, NH 03755-3528, USA. (bkress@northstar.dartmouth.edu)

A. Y. Ukhorskiy, Applied Physics Laboratory, Johns Hopkins University, MS MP3-E128, Laurel, MD 20723-6099, USA.

Time Scales for Localized Radiation Belt Injections to Become a Thin Shell

M. W. Liemohn, S. Xu, and S. Yan

Department of Atmospheric, Oceanic, and Space Sciences, University of Michigan, Ann Arbor, Michigan, USA

M.-C. Fok

NASA Goddard Space Flight Center, Greenbelt, Maryland, USA

Q. Zheng

Department of Astronomy, University of Maryland, College Park, Maryland, USA

Time scales for localized radiation belt injections to spread into a thin shell are presented. Near the slot region, where the numerical experiments are conducted, this transition from a localized injection into a thin shell is driven by scattering with plasmaspheric hiss, shifting the energy and pitch angle of the particles, which changes the drift speed of the particles. This mixing is energy dependent, taking much longer at the lower energies. It is shown that, at $L = 3.1$ (slot region) during quiet driving conditions, it takes 30–60 min for a narrow magnetic local time initial distribution of MeV energy electrons to transform into a uniformly distributed ring, but takes more than 6 h for ≤ 300 keV electrons to achieve a thin shell state. In addition, several numerical tests are conducted to demonstrate that the radiation belt environment model (RBE), the main computational tool for this study, is very appropriate for addressing the LT distribution of relativistic electrons in near-Earth space.

1. INTRODUCTION

When considering radiation belt enhancements, two processes are often discussed: radial diffusion and local acceleration [e.g., *Green and Kivelson*, 2001, 2004; *Li and Temerin*, 2001; *Friedel et al.*, 2002; *Summers et al.*, 2007]. Radial diffusion by drift resonance with ULF waves is possible when the magnetosphere is quiet enough to have trapped orbits of energetic electrons in the 6–10 R_E range but active enough to have sufficient wave power to push the

particles. Several studies have considered this source and found it to be sufficient to create the outer radiation belt after some magnetic storms [e.g., *Schulz and Lanzerotti*, 1974; *Elkington et al.*, 1999; *Iles et al.*, 2006]. The competing mechanism requires a seed population of hundreds of keV electrons already present in the 3–7 R_E inner magnetosphere, along with ample plasma wave activity (especially chorus waves on the nightside) to foster significant energy diffusion of these particles up to MeV energies [e.g., *Horne*, 2002; *Chen et al.*, 2006; *Shprits et al.*, 2006].

Another mechanism exists, however, that can lead to the enhancement of the outer zone radiation belt: localized injections. There is clear evidence for rapid increases in the outer radiation belt in a number of data sets, such as quick changes in the radio absorption as measured by ground-based riometers [e.g., *Keppler et al.*, 1962], jumps in the

Dynamics of the Earth's Radiation Belts and Inner Magnetosphere
Geophysical Monograph Series 199
10.1029/2012GM001335

precipitating flux as seen by the SAMPEX satellite [e.g., *Baker et al.*, 1994, 2004], strong rises in flux from one orbit to the next in CRRES data [e.g., *Hudson et al.*, 1996], sudden increases in geosynchronous satellite energetic particle data [e.g., *Birn et al.*, 1997a, 1997b], and step-like changes between the inbound and outbound passes of the Polar satellite [e.g., *Baker et al.*, 1998]. Recently, *Glocer et al.* [2011] used a number of different data sets to quantify a rapid rebuilding of the radiation belts during the magnetic storms on 4 September 2008 and 22 July 2009. These rises in phase space density (or flux) happen too quickly to be attributed to the two "standard" source processes mentioned above.

This process can be divided into two distinct classes of driving physical process. The first is a shock front passing through the inner magnetosphere, allowing certain particles to ride the inward wave and quickly become energized by an order of magnitude or more [e.g., *Li et al.*, 1993; *Hudson et al.*, 1996, 1997]. These are rare and easily identifiable because they are associated with an instigating solar wind pressure pulse. The second process is an injection from the magnetotail. While there is ample observational and numerical evidence that such injections can occur [e.g., *Birn et al.*, 1998; *Li et al.*, 1998; *Kim et al.*, 2000; *Fok et al.*, 2001; *Khazanov et al.*, 2004; *Elkington et al.*, 2004; *Glocer et al.*, 2009, 2011], this process is unfortunately often omitted from the discussion of radiation belt sources [e.g., *Green and Kivelson*, 2001].

Another source of localized injection of relativistic electrons into the radiation belts is a high-altitude nuclear explosion (HANE). The 1962 Starfish detonation resulted in a hard spectrum of energetic ($E > 200$ keV) electron fluxes of more than 10^9 cm^{-2} s^{-1} near the equator inside of $L = 2$ [e.g., *Durney et al.*, 1962; *Brown et al.*, 1963; *Van Allen et al.*, 1963; *Hess*, 1963]. Over time, the energy-dependent and spatially dependent decay lifetimes resulted in a softening of the inner zone spectra inside of $L = 2$ [*Beall et al.*, 1967], but a hardening of the outer zone spectra beyond $L = 3$ [*Welch et al.*, 1963]. The exact spatiotemporal variation of the flux intensities is highly dependent on the source location, solar-geomagnetic conditions, and the loss mechanisms acting on these particles [e.g., *Walt*, 1964; *Walt and Newkirk*, 1966; *Beall et al.*, 1967; *Stassinopoulos and Verzariu*, 1971]. A related source of "artificially injected" (as opposed to injections caused by "natural" geospace dynamics) is relativistic electron beam injection from a satellite. Space-based linear accelerators capable of producing MeV energy electrons have been used in a few test flights [*Banks et al.*, 1987, 1990], and numerical simulations of such injections have been conducted [e.g., *Neubert et al.*, 1996; *Habash-Krause*, 1998; *Khazanov et al.*, 1999a, 1999b, 2000; *Neubert and Gilchrist*, 2002].

The problem of understanding near-Earth relativistic electron enhancements is complicated because competing processes are driven by the same external factors. Some geospace storms enhance the radiation belts, while others do not [*Reeves*, 1998]. Relativistic electrons drift around the Earth in a matter of minutes, and regionally confined processes can directly influence the global radiation belt distribution [*Summers et al.*, 1998, 2007]. Depending on their energy and orbit, energetic electrons can experience severe losses or substantial energization.

An unresolved issue that is critical to the understanding of localized injections is the spread of these particles from a confined LT width into a symmetric (but perhaps thin) shell of particles around the Earth. *Khazanov et al.* [1999b] investigated this topic with a bounce-averaged, relativistic kinetic equation model, injecting an initial point source into a dipolar, quiet time magnetosphere. The point source spread into a thin shell of relatively uniform flux between 6 and 8 h after injection. However, this was for a static dipole magnetic field, not a nondipolar field or a time-varying one. *Lichtenberg and Lieberman* [2006] discuss the theoretical limits for using a diffusion model for describing a stochastic process, and *Degeling et al.* [2008, 2011] examined the magnetic local time (MLT)-symmetrization process of the outer radiation belt caused by steadily oscillating ULF waves. The present study reconsiders this issue with a similar numerical model to the *Khazanov et al.* [1999b] study, again for quiet activity levels but now with an empirically defined magnetic field configuration. The result has relevance to all four possible localized injection mechanisms mentioned above: traveling shock-front energization, magnetotail dipolarization, a HANE source population, and electron beam injection from a satellite-borne accelerator.

2. METHODOLOGY

To investigate the time for a localized injection to spread into a thin shell around the Earth, simulations will be conducted with the radiation belt environment (RBE) model [*Fok et al.*, 2001, 2008]. First, a test of the numerical conservation of RBE is conducted, followed by the presentation of a set of full-physics simulations with different initial and boundary conditions.

2.1. The Radiation Belt Environment Model

The RBE code is a relativistic version of the Fok ring current model [*Fok et al.*, 1996; *Fok and Moore*, 1997]. It solves the gyration and bounce averaged kinetic equation for electrons in the range from a few keV to a few MeV. It includes drift motion due to convective forces (corotative, potential, and inductive electric fields) as well as magnetic drift process (magnetic field gradient and curvature-induced particle motion), energy and pitch angle scattering due to various plasma wave processes (specifically for these runs, plasmaspheric hiss and VLF chorus waves), and loss to the atmosphere (a bounce-related attenuation factor applied to those equatorial pitch angles that map to thermospheric altitudes). *Zheng et al.* [2003] implemented a real-time version of the RBE model for radiation belt forecasting applications, and *Glocer et al.* [2009] coupled the RBE model into the Space Weather Modeling Framework (SWMF) [*Tóth et al.*, 2005, 2012] to allow for the use of MHD-derived magnetic and electric fields rather than the empirical models originally implemented in the code (specifically, a Tsyganenko magnetic field model [*Tsyganenko*, 1989, 1995] and a Weimer electric potential model [*Weimer*, 1996, 2001]). The model also requires a thermal plasma distribution throughout the simulation domain, given by the dynamic global core plasma model [*Ober et al.*, 1997].

Because the RBE code solves the kinetic equation with two spatial dimensions (radial distance and MLT), it is capable of resolving the injection and spreading of a localized source of relativistic electrons. In fact, this was the initial use by *Fok et al.* [2001], who examined the possibility of direct capture of MeV electrons from the magnetotail during substorm dipolarizations. *Glocer et al.* [2011] conclusively demonstrated that the rapid rise of relativistic electrons during the two events they studied were due to the capture of localized injections of MeV electrons from magnetotail dipolarizations. This model is ideally suited for investigations of this process.

This study will use the typical numerical setup for RBE. The grid configuration uses 48 evenly spaced cells in MLT and 51 cells in invariant latitude, plus 29 grid steps in the M and K invariant velocity space variables. Note that it is not a particle-tracking code, but rather employs fluid dynamic numerical techniques to obtain the phase space density at each of these 4-D grid cells. The time step is set to a maximum of 3 s, but the step is reduced according to 30% of the limiting advection stability criterion (the Courant-Friedrichs-Levy (CFL) condition) across all of phase space at each time during the simulation. The RBE model employs a second-order Lax-Wendroff numerical scheme with a Superbee flux limiter. By invoking the operator splitting numerical technique, advection and/or diffusion in each phase space variable is implemented separately, solving the operator processes first in one order and then in reverse order to maintain second-order accuracy of the solution. Because the spatial grid for the RBE model is in the ionosphere, inductive motion of the field lines by a time-varying magnetic field are implicitly included when the field line volume

integrals are calculated and when results are mapped to the equatorial plane. Therefore, only potential electric fields need to be included in the drift terms.

2.2. Numerical Tests of RBE

Before conducting numerical experiments with this model, it is useful to pause and explore the validity of applying this code to localized injection simulations. This will be done with a particle and energy conservation test. The test will be developed to determine how much a localized initial condition spreads in one variable as the distribution is advected in that direction. To do this, the RBE code is simplified to only azimuthal drift, turning off all other operators on the phase space density (i.e., no radial or velocity space drift, no wave-particle scattering terms, and no loss to the upper atmosphere). In addition, the azimuthal drift is set to a constant value, the same for all phase space grid cells. If the numerical scheme is a perfect representation of the physics, then the distribution should maintain its initial configuration.

These tests will be conducted at $L = 2.8$, a typical slot region radial distance examined with this code by *Zheng et al.* [2006]. Quiet time solar and geophysical settings will be used so that the magnetic field is nearly dipolar. Several different initial conditions were used, namely, a point source injection with a nonzero phase space density initially in only MLT bin, and a Gaussian distribution in MLT with the phase space density proportional to $\exp(-w^2/(2\delta w^2))$ where w is the MLT in hours. Three different values of δw were used for these tests: $\delta w = 1.5$, 3, and 4.5 h. A fast azimuthal drift speed is applied, resulting in a drift period of just under 160 s in order to reduce the simulation run time yet still yield a test with numerous circumdrifts around the Earth (very close to 23 periods are completed). Even though the time step is initialized to 3 s, the RBE code actually takes time steps of just under 1 s for this numerical test setup because it enforces a maximum step of 30% of the limiting CFL condition.

Figure 1 shows the MLT distribution of number flux for the point source initial condition. The top row shows the first few seconds of the simulation, while the second row shows the flux after one drift period around the Earth (Figure 1e, at $t = 180$ s), two drift periods (Figure 1f), 20 min into the simulation (Figure 1g), and 1h (Figure 1h). The y axis scale is linear, indicating that the pulse height drops to roughly 10% of its initial, single-bin value. It is seen that after an initial spread of the point source across several MLT bins, it then reaches a fairly stable configuration and maintains this shape for the rest of the simulation. No velocity space coordinate is given for these results because all of M–K space was given the same, artificially fast, azimuthal velocity. An alternative

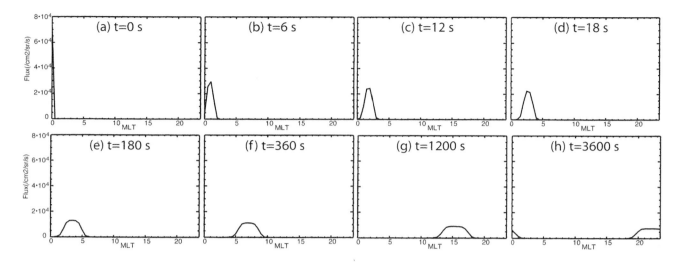

Figure 1. Point source conservation test simulation results.

independent variable to time, then, is drift periods (T_{drift} = 156.5 s).

Figure 2 is the same type of plot as Figure 1 except for the Gaussian initial condition with δw = 3 h. Again, the y axis scale is linear, revealing a drop in peak flux value of less than 5% over the course of the hour-long test simulation. In addition, it is seen that this initial condition maintains its shape very well throughout the 23 laps around the Earth, with only a bit of flattening of the top of the Gaussian distribution but very little spread in MLT.

To better see the spread of the initial condition with time, Figure 3 shows a superposition of the results for each of the four test simulations, aligning the peak of the flux distribution at MLT = 12 on the plot. Each line is at a different time throughout each simulation, plotted on a 6 s cadence with arbitrary coloring. In each of the four simulations, the distribution flattens across a small MLT range, and the automated routine of selecting the peak, therefore, sometimes shifts the distribution to the left or right, depending on each grid cell of the peak. While there is some time evolution in the point source simulation (Figure 3a) and the δw = 1.5 h Gaussian initial condition run (Figure 3b), the other two Gaussian initial conditions do not show much temporal variation at all. Much of the apparent spread in this plot is actually just a shift of the peak relative to $x = 0$ because the top of the distributions are so flat that the exact location of the peak value slightly changes as a function of time.

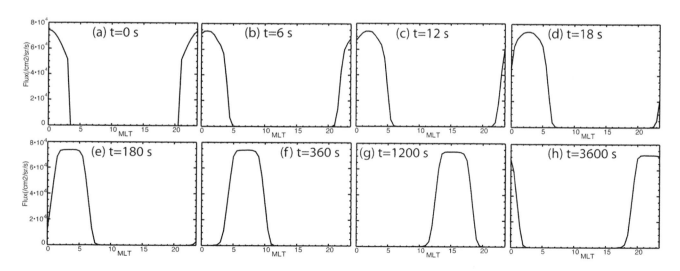

Figure 2. Conservation test results with a Gaussian initial condition with δw_{MLT} = 3 h.

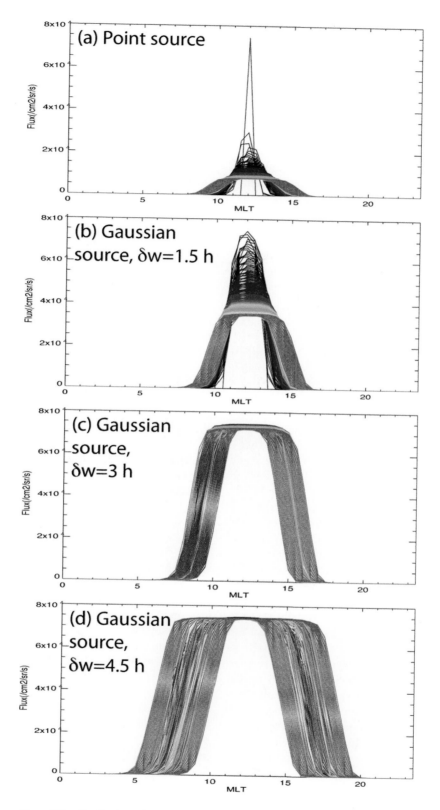

Figure 3. Superposition of the distribution via centering the peak flux value at MLT = 12, for initial conditions of (a) a point source, (b) a Gaussian source with $\delta w = 1.5$ h in LT, (c) a Gaussian with $\delta w = 3$ h, and (d) a Gaussian with $\delta w = 4.5$ h.

Because of the slight left-to-right wobble in the line plots of Figure 3, a better method of quantifying the temporal spread of the distribution is to examine the full width at half maximum (FWHM) value for each MLT distribution. Time series plots of this quantity for each initial condition are shown in Figure 4. For the Gaussian initial distributions, the FWHM value at $t = 0$ is close to double the δw value for that run. As in Figure 3, the point source and $\delta w = 1.5$ h simulations show a growth in the width of the distribution as a function of time. However, the $\delta w = 3$ and 4.5 h initial conditions show no growth at all in the FWHM value; in fact, there is a bit of a decrease in this spread measure as the peak flattens into a plateau during the first few seconds of the simulation.

This growth in the width of the distribution can be further quantified by considering the rate of change of the FWHM value with respect to time, shown in Figure 5. For the point source and $\delta w = 1.5$ Gaussian source, the rate of spread is initially large but drops below 0.04 MLT s^{-1} by the end of the first drift period (which equates to roughly 0.6 h of MLT per drift period) and is nearly zero (but still slightly positive) by 30 min into the simulation (i.e., after 11.5 drift periods). For the other two larger Gaussian initial conditions, the rate is initially negative for some part of the first drift period and then hovers around zero for the rest of the simulation.

To understand the periodic oscillation of the spreading rates in Figure 5, an integral of the total flux around the MLT ring is shown in Figure 6 for the point source (Figure 6a) and the $\delta w = 3$ h Gaussian. The two horizontal dashed lines in these two figures give the maximum and minimum values during the entire simulation. Also note that the y axis scale is truncated, not starting at zero but rather focusing in on the ~1% variation in the MLT-integrated flux. While a systematic oscillation is seen in this quantity, the overall value is exceptionally constant throughout the many drift periods during the simulation. The variation can be explained by considering the equatorial magnetic field strength around this invariant latitude, plotted in Figure 6c. The RBE model is set up to use a Tsyganenko or MHD magnetic field, and these test simulations used a very quiet T96 field configuration. However, the field at $L = 2.8$ is not quite symmetrically dipolar, but instead has a slight increase in field strength on the dayside relative to the local midnight. This variation introduces a systematic oscillation in the flux intensity, which causes a repeated pattern in the FWHM rate of change and a slight increase-decrease periodicity to the MLT-integrated flux. If one removes this systematic oscillation from consideration in Figures 6a and 6b, then the change in total flux versus time is extremely small; for the point source initial condition, the total variation in integrated flux is 10^{-4} over these 23 drift periods, and it is much smaller still for the Gaussian initial condition.

The rest of the simulation results were also examined in detail. There was no spread of the phase space density to other L values, and there was no systematic drift in M and K space either (not shown). This is expected because these drift and diffusion coefficients were set to zero. The fact that they did not influence the test simulation results is a sign that there is not a random or systematic leak of phase space density in these other variables. Additional simulations were conducted with full drift terms included, in both a dipole and a stretched magnetic field configuration. The stretched field test run had a slight increase in total energy of 0.05% during the first drift orbit and was fairly constant after that, and the dipole field yielded a total energy variation of less than 0.01% throughout the entire test simulation.

The conclusion to be drawn from these test simulations is that the RBE code can handle a localized injection and maintain the initial distribution with little or no numerical diffusion. For a localized source with a FWHM of 6 h or more in MLT, the spread is essentially zero. For a localized source with a smaller MLT extent, numerical diffusion causes a slight spread of the distribution across a few hours of MLT (over a time span of two dozen drift periods). Even a point source spread to a FWHM of only 5 h within a few drift periods and then exhibited very little spread over many more drift periods. Furthermore, the MLT-integrated flux was constant with time, implying excellent number and energy conservation. Therefore, the RBE code can be used with confidence for exploring the spatiotemporal evolution of localized radiation belt injections.

3. RESULTS

This study is focused on the time required for a localized injection to symmetrize in MLT and transform from an asymmetric distribution into a uniform thin shell. The initial results presented here are for quiet time solar and geophysical conditions. Specifically, the model is run with static solar wind, interplanetary magnetic field (IMF), and ground-based magnetometer indices needed to drive the T96 and Weimer-2000 models used by RBE. The solar wind density is set to 3.3 cm^{-3}, the solar wind speed to 450 km s^{-1}, the IMF vector is set to (+5, −1, +2.5) nT in GSM coordinates, the *SYM-H* index is set to −1 nT, and the *Kp* index is set to 1. These are the conditions at 00 UT on 23 October 2002, the beginning of a quiet interval with nearly constant solar wind parameters, a toward-sector Parker spiral IMF with B_z values hovering near zero.

The simulations shown here use an initial condition with a Gaussian distribution in MLT with $\delta w = 3$ h. The distribution in L value is a delta function at $L^* = 3.1$ (an arbitrarily chosen radial distance near the slot region), and the distribution in

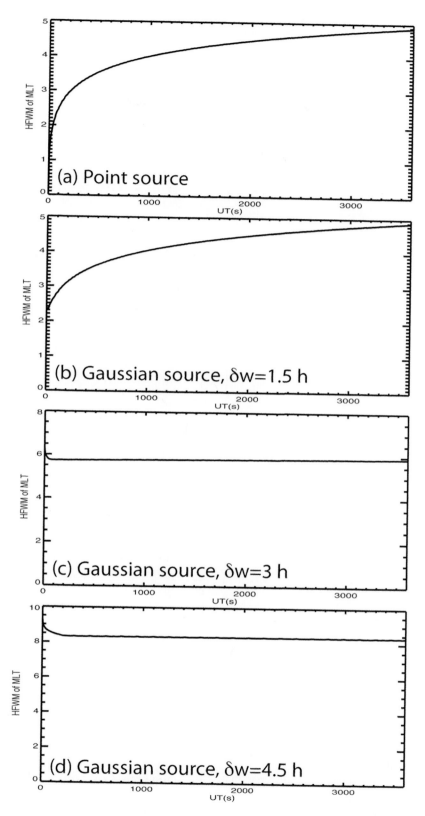

Figure 4. The FWHM value as a function of time for initial conditions of (a) a point source, (b) a Gaussian source with δw = 1.5 h in LT, (c) a Gaussian with δw = 3 h, and (d) a Gaussian with δw = 4.5 h.

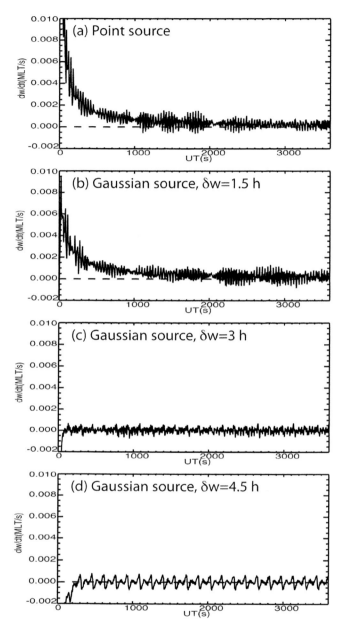

Figure 5. The time rate of change in the FWHM value throughout the simulation for initial conditions of (a) a point source, (b) a Gaussian source with $\delta w = 1.5$ h in LT, (c) a Gaussian with $\delta w = 3$ h, and (d) a Gaussian with $\delta w = 4.5$ h.

M and K space is a constant phase space density setting (arbitrarily chosen magnitude). This is slightly different than the test calculations presented above because, at present, the RBE model does not include wave interactions below $L^* = 3$. Another change between the results presented here and previous RBE simulations is the radial/latitudinal grid structure.

Typically, 51 grids are used between $L^* = 1.1$ and $L \geq 10$ (the outermost L shell changes with the magnetic field topology). For this run, the grid is set to a compact L^* grid from 2.0 to 4.0, keeping the same number of grid steps. Therefore, the radial grid spacing near an L^* of 3 is $\Delta L = 0.04$, which is roughly five times better than standard RBE simulations at this location. The initial condition will be placed at local midnight. A critical difference between these simulations and those in the numerical tests above is that now the full set of real, spatially and temporally varying drift velocities will be used, along with velocity space diffusion (from plasma wave scattering) and loss to the upper atmosphere.

In particular, these simulations include wave-particle interactions. Both plasmaspheric hiss and VLF chorus are included in the simulation. Figure 7 shows the bounce-averaged equatorial pitch angle diffusion coefficients, in arbitrary units, for four MLT locations around $L^* = 3.1$ for three different energies. Note that the y axis is on a logarithmic scale, with the diffusion coefficients near the loss cone (high K values) being significantly smaller than the values at higher pitch angles (low K values). At present, the RBE model includes wave-particle interactions up to a 5 MeV upper energy limit. While the true wave interactions continue above this energy, this cutoff is fortuitous because it allows us to examine the numerical diffusion in the code at the MeV energy range. This issue will be discussed in detail below.

Also, note that the calculation is conducted in M-K invariant velocity space, and the values are mapped to energy-pitch angle (E-α_0) velocity space with a high-order interpolation scheme for the wave-particle scattering. The M and K coordinates are given in grid cell index numbers used in the simulation and the color scale gives the corresponding energy (in keV) for that coordinate location. Note that K maps directly to equatorial pitch angle. The conversion between these two sets of variables is shown in Figure 8. The red and black lines indicate the lower and upper limits of the simulation domain computed in the model. The annotated text around the plot helps guide the conversion from M-K to E-α_0 space. Plasmaspheric hiss dominates this diffusion coefficient at this L value, and VLF chorus is hardly a factor. In addition, this coefficient is larger than the energy-scattering diffusion coefficient.

Figure 9 shows phase space density versus M and MLT for $K = 16$. These results were extracted at the L value of injection. The first row in Figure 9 shows the M-MLT distributions every minute very early in the simulation, while the second row shows every 10 min at the beginning of the simulation. The y axis scale is logarithmic and spans 4 orders of magnitude.

There are several interesting features to note in Figure 9. The drift motion periodicity of these electrons can be seen in

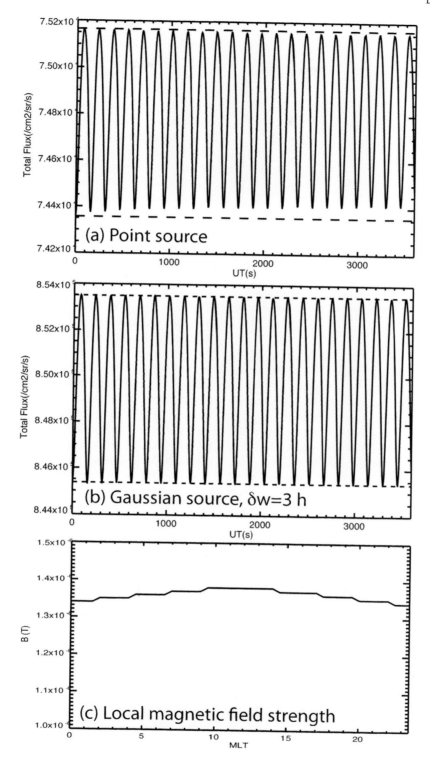

Figure 6. The total flux (integrated over all magnetic local time (MLT) bins) as a function of time for initial conditions of (a) a point source and (b) a Gaussian source with δw = 3 h, and (c) the equatorial magnetic field strength as a function of MLT around the invariant latitude of the conservation test simulations.

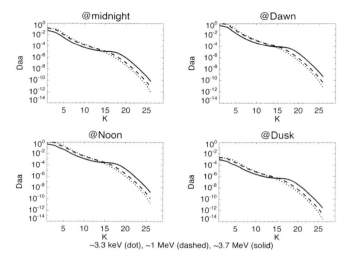

Figure 7. Pitch angle diffusion coefficients used in the radiation belt environment model for a few magnetic local times (the four panels) and energies (the lines within each panel). All of the panels are on the same logarithmic y axis scale.

these plots, with the high-energy electrons (high M values) circling the Earth in just a few minutes, while the lower-energy electrons (low M values) shifting about an hour in MLT during the 30 min of simulation time presented here. In addition, the high-energy distribution starts to fill in away from the peak sooner than the other two distributions. This is from velocity-space advection and scattering, and (inferred from the tests above) not from numerical diffusion. There is a clear break at the 5 MeV energy level in the amount of LT symmetrization occurring, with much more filling occurring

below this cutoff and far less above it. This is the energy extent of the wave interactions and gives an indication of the remaining amount of numerical diffusion (scattering above this M value) and physically based symmetrization.

To quantify the transition from a localized peak in MLT into a shell of electrons uniformly distributed around the Earth, it is convenient to consider the time progression of a ratio of phase space density values. In particular, Figure 10 shows the ratio of the MLT-averaged phase space density to the maximum phase space density value in this ring of numbers. The x and y axes are the M and K index numbers across velocity space, and the color scale is linear from zero to one. The ratio at $t = 0$ (Figure 10a) is uniformly blue at 0.22 because every M and K coordinate is given the same Gaussian distribution with $\delta w = 3$ h in LT. The regions of the plot with dotted stripes through it are M-K coordinates not used in the simulation, corresponding to energies beyond those of interest. In Figure 10b (at $t = 10$ min), the top of the K index values ($K = 23$ and higher) are shown in black because these are all in the loss cone where the electrons are depleted with a time scale of half a bounce period. Therefore, this ratio quantity is rather meaningless for these K indices, and the color scale results have been removed. At lower K indices, however, the ratio approaches unity for the 1–5 MeV energies first (M indices in the range of 17–21), even within the first 20 min. The situation is different for the ultrarelativistic electrons at the top of the scale, where the wave-particle interactions were not included in the simulation. At the highest M values, which are dominated by numerical diffusion, the ratios remain low during the first few hours. The physically based, wave-dominated scattering and MLT

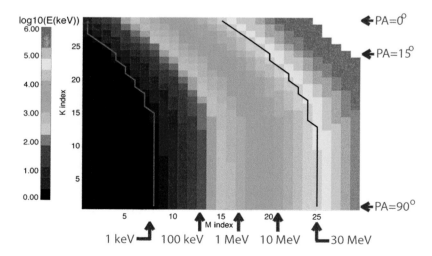

Figure 8. Conversion chart between M-K space and energy-equatorial pitch angle space for $L = 3$, averaged in LT. The red line indicates the lowest M value used in the simulation (near 1 keV), and the black line indicates the highest M value computed in the simulation (near 30 MeV).

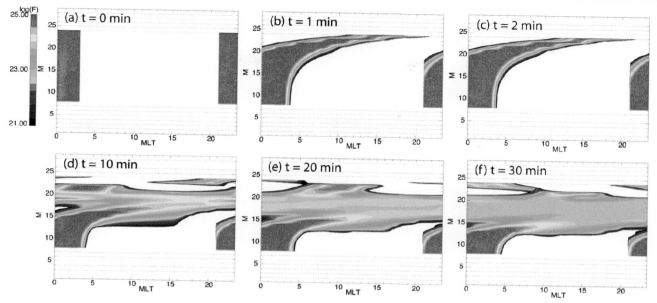

Figure 9. Phase space density versus M and MLT at the L value of injection for $K = 16$. Six times are shown during the first 30 min of the simulation.

symmetrization is faster than the numerical diffusion. The lower-M boundary of the red color systematically marches leftward as time progresses through the simulation. There is a fairly narrow range of M indices with ratios between 0.2 (blue) and 0.9 (red), with lower M values still near the initial condition ratio and higher M values with a ratio close to unity.

A final analysis of these results is shown in Figure 11. Presented here is the time needed for a particular M and K velocity space coordinate to reach a mean-to-max phase

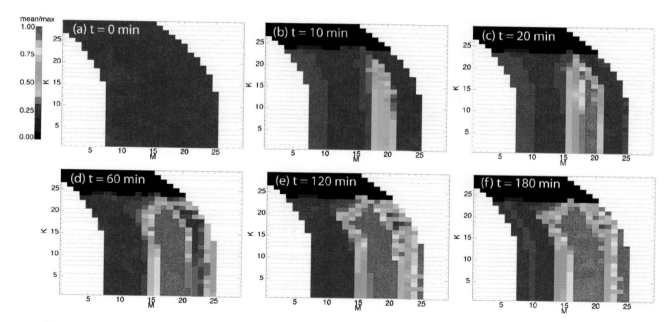

Figure 10. The ratio of the MLT-averaged mean phase space density value to the MLT maximum value at six times during the simulation. The color scale is linear from 0 to 1.

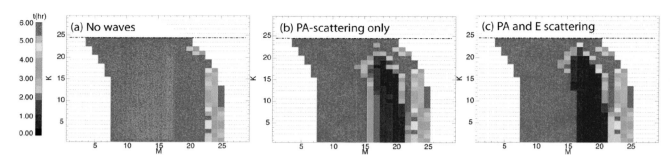

Figure 11. Time for the MLT distribution to reach a mean-to-maximum ratio of 0.9 at the L value of injection as a function of M and K for three different 6 h interval simulations: (a) without wave-particle interactions, (b) with pitch angle scattering only, and (c) with pitch angle and energy scattering included. The color scale is linear from 0 to $\geq$6 h.

space density ratio of 0.9. This corresponds to the transition to the red contour in Figure 10. The color scale for Figure 11 is linear with a range from 0 to $\geq$6 h. The three panels of Figure 11 show the results for different inclusions of the wave-particle interaction scattering terms: Figure 11a shows results from a simulation with no wave scattering included, Figure 11b has only pitch angle diffusion included in the calculation, and Figure 11c includes both pitch angle and energy scattering by the plasma waves.

The three panels of Figure 11 have some characteristic similarities and notable differences. First, the lower-M part of the figures is entirely red, indicating that it takes at least 6 h to reach symmetrization at these energies. For Figure 11a (no waves), this red section of the plot extends to around 15 MeV. Above this energy, some of the velocity space grid cells are green or even red. This shows the influence of numerical diffusion on the symmetrization process. For this compacted radial grid configuration in the RBE model (but standard MLT, M, and K grid spacings), the numerical diffusion is significantly less than the physical diffusion from the plasma waves (in particular, plasmaspheric hiss).

With the inclusion of waves (Figures 11b and 11c), the time to reach symmetrization looks similar between the two panels. That is, pitch angle scattering is the dominant diffusion term, and the addition of energy scattering changes the results only slightly. The energy range that takes longer than 6 h (i.e., the red part of each figure) extends over to M = 16 and M = 15 (at K = 0) in Figures 11b and 11c, respectively. These M indices correspond to roughly 500 and 300 keV. In the MeV-energy range, it takes an hour or less to reach MLT symmetrization, but at slightly lower energies, there is a sharp transition to taking longer than 6 h to reach the status of thin shell.

4. DISCUSSION AND CONCLUSION

This result, that it takes perhaps an hour for a localized injection of MeV electrons into the slot region to become a

thin shell that is fairly uniform in its MLT distribution, but takes many hours for hundreds of keV electrons to reach thin shell status, has several implications for radiation belt modeling. It implies that an assumption of instant MLT symmetrization of localized injections is incorrect. It takes time for these particles to spread out in MLT, during which the dynamics of the geospace system and the processes acting on the relativistic electrons can dramatically change. The time scale for plasmaspheric plume formation is on the order of an hour, as is the time scale for ring current particle injection. The longer time for the sub-MeV electrons to symmetrize is similar to the main phase of storms and longer than an entire isolated substorm growth-expansion-recovery phase sequence. The timing of the injection relative to variations in the geospace system will cause significant changes in the evolution of the localized phase space density pulse.

In addition, it has implications for numerical modeling. In particular, it means that results from drift-averaged radiation belt simulations should only be used to consider electron dynamics longer than this MLT-symmetrization time scale. Within such models, the coefficients for localized scattering and loss processes are often prorated according to their MLT extent. While this is reasonable for long time scale simulations spanning several days or more, it might not be appropriate for investigating the influences of short-lived processes that last a few hours or less.

Note that this study only focused on the MLT spread of a localized injection into the slot region during quiet times. This is because the MLT drift is, by far, the fastest of the advection operators and a quiet time outer boundary condition will yield the longest time scales for symmetrization. The spread in L value will be slower than the time to transform the distribution into a thin shell, while the use of disturbed-time driving conditions or an outer zone L value will yield a shorter time scale. These issues will be investigated in a future study.

A caveat to the work is that it used empirical models for the electric and magnetic fields and only included VLF chorus and plasmaspheric hiss. ULF wave influences are included via the spatiotemporal variation of the E and B fields, and the use of the T96 and Weimer-2000 models probably underestimates the amplitude of these waves. However, a quiet time choice at $L = 3.1$ should be within the plasmasphere, and therefore, ULF oscillations and other plasma wave influences are most likely small. The use of a global MHD model for the field configurations and variations is a logical next step to this study to improve the realism of the ULF wave distribution. Furthermore, the inclusion of additional plasma waves, in particular those that are amplified during storm times, is also a useful extension of this study to active driving conditions. Should numerical diffusion prove to be a problem, a refined grid spacing in MLT, M, and/or K space would help to reduce this unwanted, artificial scattering process.

Finally, it should be noted that the RBE model performed exceptionally well on the numerical tests conducted to verify its usage for localized injection studies. While very narrow localized injections are numerically spread by a few hours in MLT, the spread asymptotically slows, and an injection of 6 h in MLT width does not spread at all. Furthermore, the number and energy conservation of the code are nearly perfect.

Acknowledgments. The authors would like to thank the sponsors of this work, in particular NASA under grants NNX08AQ15G, NNX09AF45G, and NNX11AO60G, NSF under grant ATM-0802705, and the Defense Threat Reduction Agency under grant HDTRA1-10-1-0121.

REFERENCES

Baker, D. N., J. B. Blake, L. B. Callis, J. R. Cummings, D. Hovestadt, S. Kanekal, B. Klecker, R. A. Mewaldt, and R. D. Zwickl (1994), Relativistic electron acceleration and decay time scales in the inner and outer radiation belts: SAMPEX, *Geophys. Res. Lett.*, *21*(6), 409–412.

Baker, D. N., X. Li, J. B. Blake, and S. Kanekal (1998), Strong electron acceleration in the Earth's magnetosphere, *Adv. Space Res.*, *21*(4), 609–613.

Baker, D. N., S. G. Kanekal, X. Li, S. P. Monk, J. Goldstein, and J. L. Burch (2004), An extreme distortion of the Van Allen Belt arising from the 'Halloween' solar storm in 2003, *Nature*, *432*, 878–881.

Banks, P. M., A. C. Fraser-Smith, B. E. Gilchrist, K. J. Harker, L. R. O. Storey, and P. R. Williamson (1987), New concepts in ionosphere modification, *AFGL-TR-88-0133*, Air Force Geophys. Lab., Bedford, Mass.

Banks, P. M., A. C. Fraser-Smith, and B. E. Gilchrist (1990), Ionospheric modification using relativistic electron beams, *AGARD Conf. Proc.*, *485*, 22-1–22-18.

Beall, D. S., C. O. Bostrom, and D. J. Williams (1967), Structure and decay of the Starfish radiation belt, October 1963 to December 1965, *J. Geophys. Res.*, *72*(13), 3403–3424.

Birn, J., M. F. Thomsen, J. E. Borovsky, G. D. Reeves, D. J. McComas, R. D. Belian, and M. Hesse (1997a), Substorm ion injections: Geosynchronous observations and test particle orbits in three-dimensional dynamic MHD fields, *J. Geophys. Res.*, *102*(A2), 2325–2341.

Birn, J., M. F. Thomsen, J. E. Borovsky, G. D. Reeves, D. J. McComas, and R. D. Belian (1997b), Characteristic plasma properties during dispersionless substorm injections at geosynchronous orbit, *J. Geophys. Res.*, *102*(A2), 2309–2324.

Birn, J., M. F. Thomsen, J. E. Borovsky, G. D. Reeves, D. J. McComas, R. D. Belian, and M. Hesse (1998), Substorm electron injections: Geosynchronous observations and test particle simulations, *J. Geophys. Res.*, *103*(A5), 9235–9248.

Brown, W. L., W. N. Hess, and J. A. VanAllen (1963), Introduction, *J. Geophys. Res.*, *68*(3), 605–606.

Chen, Y., R. H. W. Friedel, and G. D. Reeves (2006), Phase space density distributions of energetic electrons in the outer radiation belt during two Geospace Environment Modeling Inner Magnetosphere/Storms selected storms, *J. Geophys. Res.*, *111*, A11S04, doi:10.1029/2006JA011703.

Degeling, A. W., L. G. Ozeke, R. Rankin, I. R. Mann, and K. Kabin (2008), Drift resonant generation of peaked relativistic electron distributions by Pc 5 ULF waves, *J. Geophys. Res.*, *113*, A02208, doi:10.1029/2007JA012411.

Degeling, A. W., R. Rankin, and S. R. Elkington (2011), Convective and diffusive ULF wave driven radiation belt electron transport, *J. Geophys. Res.*, *116*, A12217, doi:10.1029/2011JA 016896.

Durney, A. C., H. Elliot, R. J. Hynds, and J. J. Quenby (1962), Satellite observations of the energetic particle flux produced by the high-altitude nuclear explosion of July 9, 1962, *Nature*, *195*, 1245–1248.

Elkington, S. R., M. K. Hudson, and A. A. Chan (1999), Acceleration of relativistic electrons via drift-resonant interaction with toroidal-mode Pc-5 ULF oscillations, *Geophys. Res. Lett.*, *26*(21), 3273–3276.

Elkington, S. R., M. Wiltberger, A. A. Chan, and D. N. Baker (2004), Physical models of the geospace radiation environment, *J. Atmos. Sol. Terr. Phys.*, *66*(15–16), 1371–1387.

Fok, M.-C., and T. E. Moore (1997), Ring current modeling in a realistic magnetic field configuration, *Geophys. Res. Lett.*, *24*(14), 1775–1778.

Fok, M.-C., T. E. Moore, and M. E. Greenspan (1996), Ring current development during storm main phase, *J. Geophys. Res.*, *101*(A7), 15,311–15,322.

Fok, M.-C., T. E. Moore, and W. N. Spjeldvik (2001), Rapid enhancement of radiation belt electron fluxes due to substorm dipolarization of the geomagnetic field, *J. Geophys. Res.*, *106*(A3), 3873–3881.

Fok, M.-C., R. B. Horne, N. P. Meredith, and S. A. Glauert (2008), Radiation Belt Environment model: Application to space weather nowcasting, *J. Geophys. Res.*, *113*, A03S08, doi:10.1029/2007JA 012558.

Friedel, R. H. W., G. D. Reeves, and T. Obara (2002), Relativistic electron dynamics in the inner magnetosphere—A review, *J. Atmos. Sol. Terr. Phys.*, *64*, 265–282.

Glocer, A., G. Toth, M.-C. Fok, T. I. Gombosi, and M. W. Liemohn (2009), Integration of the radiation belt environment model into the Space Weather Modeling Framework, *J. Atmos. Sol. Terr. Phys.*, *71*, 1653–1663.

Glocer, A., M.-C. Fok, T. Nagai, G. Tóth, T. Guild, and J. Blake (2011), Rapid rebuilding of the outer radiation belt, *J. Geophys. Res.*, *116*, A09213, doi:10.1029/2011JA016516.

Green, J. C., and M. G. Kivelson (2001), A tale of two theories: How the adiabatic response and ULF waves affect relativistic electrons, *J. Geophys. Res.*, *106*(A11), 25,777–25,791.

Green, J. C., and M. G. Kivelson (2004), Relativistic electrons in the outer radiation belt: Differentiating between acceleration mechanisms, *J. Geophys. Res.*, *109*, A03213, doi:10.1029/ 2003JA010153.

Habash-Krause, L. (1998), The relativistic beam-interaction, Ph.D. thesis, Univ. of Mich., Ann Arbor.

Hess, W. N. (1963), The artificial radiation belt made on July 9, 1962, *J. Geophys. Res.*, *68*(3), 667–683.

Horne, R. B. (2002), The contribution of wave-particle interactions to electron loss and acceleration in the Earth's radiation belts during geomagnetic storms, in *Review of Radio Science 1999–2002*, edited by W. R. Stone, chap. 33, pp. 801–828, John Wiley, Hoboken, N. J.

Hudson, M. K., S. R. Elkington, J. G. Lyon, V. A. Marchenko, I. Roth, M. Temerin, and M. S. Gussenhoven (1996), MHD/particle simulations of radiation belt formation during a storm sudden commencement, in *Radiation Belts: Models and Standards*, *Geophys. Monogr. Ser.*, vol. 97, edited by J. F. Lemaire, D. Heynderickx and D. N. Baker, pp. 57–62, AGU, Washington, D. C., doi:10.1029/GM097p0057.

Hudson, M. K., S. R. Elkington, J. G. Lyon, V. A. Marchenko, I. Roth, M. Temerin, J. B. Blake, M. S. Gussenhoven, and J. R. Wygant (1997), Simulations of radiation belt formation during storm sudden commencements, *J. Geophys. Res.*, *102*(A7), 14,087–14,102.

Iles, R. H. A., N. P. Meredith, A. N. Fazakerley, and R. B. Horne (2006), Phase space density analysis of the outer radiation belt energetic electron dynamics, *J. Geophys. Res.*, *111*, A03204, doi:10.1029/2005JA011206.

Keppler, E., A. Ehmert, G. Pfotzer, and J. Ortner (1962), Sudden increase of radiation intensity coinciding with a geomagnetic storm sudden commencement, *J. Geophys. Res.*, *67*(13), 5343–5346.

Khazanov, G. V., M. W. Liemohn, E. N. Krivorutsky, J. U. Kozyra, and B. E. Gilchrist (1999a), Interhemispheric transport of relativistic electron beams, *Geophys. Res. Lett.*, *26*(5), 581–584.

Khazanov, G. V., M. W. Liemohn, E. N. Krivorutsky, J. M. Albert, J. U. Kozyra, and B. E. Gilchrist (1999b), Relativistic electron beam propagation in the Earth's magnetosphere, *J. Geophys. Res.*, *104*(A12), 28,587–28,599.

Khazanov, G. V., M. W. Liemohn, E. N. Krivorutsky, J. U. Kozyra, J. M. Albert, and B. E. Gilchrist (2000), On the influence of the initial pitch angle distribution on relativistic electron beam dynamics, *J. Geophys. Res.*, *105*(A7), 16,093–16,094.

Khazanov, G. V., M. W. Liemohn, M.-C. Fok, T. S. Newman, and A. J. Ridley (2004), Stormtime particle energization with high temporal resolution AMIE potentials, *J. Geophys. Res.*, *109*, A05209, doi:10.1029/2003JA010186.

Kim, H.-J., A. A. Chan, R. A. Wolf, and J. Birn (2000), Can substorms produce relativistic outer belt electrons?, *J. Geophys. Res.*, *105*(A4), 7721–7735.

Li, X., and M. Temerin (2001), The electron radiation belt, *Space Sci. Rev.*, *95*, 569–580.

Li, X., I. Roth, M. Temerin, J. R. Wygant, M. K. Hudson, and J. B. Blake (1993), Simulation of the prompt energization and transport of radiation belt particles during the March 24, 1991 SSC, *Geophys. Res. Lett.*, *20*(22), 2423–2426.

Li, X., D. N. Baker, M. Temerin, G. D. Reeves, and R. D. Belian (1998), Simulation of dispersionless injections and drift echoes of energetic electrons associated with substorms, *Geophys. Res. Lett.*, *25*(20), 3763–3766.

Lichtenberg, A. J., and M. A. Lieberman (2006), *Regular and Stochastic Motion*, Springer, Heidelberg, The Netherlands, doi:10.1002/zamm.19840640221.

Neubert, T., and B. E. Gilchrist (2002), Particle simulations of relativistic electron beam injection from spacecraft, *J. Geophys. Res.*, *107*(A8), 1167, doi:10.1029/2001JA900102.

Neubert, T., B. Gilchrist, S. Wilderman, L. Habash, and H. J. Wang (1996), Relativistic electron beam propagation in the Earth's atmosphere: Modeling results, *Geophys. Res. Lett.*, *23*(9), 1009–1012.

Ober, D. M., J. L. Horwitz, and D. L. Gallagher (1997), Formation of density troughs embedded in the outer plasmasphere by sub-auroral ion drift events, *J. Geophys. Res.*, *102*(A7), 14,595–14,602.

Reeves, G. D. (1998), Relativistic electrons and magnetic storms: 1992–1995, *Geophys. Res. Lett.*, *25*(11), 1817–1820.

Schulz, M., and L. J. Lanzerotti (1974), *Particle Diffusion in the Radiation Belts*, *Phys. Chem. Space*, vol. 7, Springer, New York.

Shprits, Y. Y., R. M. Thorne, R. B. Horne, S. A. Glauert, M. Cartwright, C. T. Russell, D. N. Baker, and S. G. Kanekal (2006), Acceleration mechanism responsible for the formation of the new radiation belt during the 2003 Halloween solar storm, *Geophys. Res. Lett.*, *33*, L05104, doi:10.1029/2005GL024256.

Stassinopoulos, E. G., and P. Verzariu (1971), General formula for decay lifetimes of starfish electrons, *J. Geophys. Res.*, *76*(7), 1841–1844.

Summers, D., R. M. Thorne, and F. Xiao (1998), Relativistic theory of wave-particle resonant diffusion with application to electron acceleration in the magnetosphere, *J. Geophys. Res.*, *103*(A9), 20,487–20,500.

Summers, D., B. Ni, and N. P. Meredith (2007), Timescales for radiation belt electron acceleration and loss due to resonant wave-particle interactions: 2. Evaluation for VLF chorus, ELF hiss, and electromagnetic ion cyclotron waves, *J. Geophys. Res.*, *112*, A04207, doi:10.1029/2006JA011993.

Tóth, G., et al. (2005), Space Weather Modeling Framework: A new tool for the space science community, *J. Geophys. Res.*, *110*, A12226, doi:10.1029/2005JA011126.

Tóth, G., et al. (2012), Adaptive numerical algorithms in space weather modeling, *J. Comput. Phys.*, *231*, 870–903.

Tsyganenko, N. A. (1989), A magnetospheric magnetic field model with a warped tail current sheet, *Planet. Space Sci.*, *37*, 5–20.

Tsyganenko, N. A. (1995), Modeling the Earth's magnetospheric magnetic field confined within a realistic magnetopause, *J. Geophys. Res.*, *100*(A4), 5599–5612.

Van Allen, J. A., L. A. Frank, and B. J. O'Brien (1963), Satellite observations of the artificial radiation belt of July 1962, *J. Geophys. Res.*, *68*(3), 619–627.

Walt, M. (1964), The effects of atmospheric collisions on geomagnetically trapped electrons, *J. Geophys. Res.*, *69*(19), 3947–3958.

Walt, M., and L. L. Newkirk (1966), Addition to investigation of the decay of the Starfish Radiation Belt, *J. Geophys. Res.*, *71*(13), 3265–3266.

Weimer, D. R. (1996), A flexible, IMF dependent model of high-latitude electric potentials having "Space Weather" applications, *Geophys. Res. Lett.*, *23*(18), 2549–2552, doi:10.1029/96GL02255.

Weimer, D. R. (2001), An improved model of ionospheric electric potentials including substorm perturbations and application to the Geospace Environment Modeling November 24, 1996, event, *J. Geophys. Res.*, *106*(A1), 407–416.

Welch, J. A., Jr., R. L. Kaufmann, and W. N. Hess (1963), Trapped electron time histories for $L = 1.18$ to $L = 1.30$, *J. Geophys. Res.*, *68*(3), 685–699.

Zheng, Y., M.-C. Fok, and G. V. Khazanov (2003), A radiation belt-ring current forecasting model, *Space Weather*, *1*(3), 1013, doi:10.1029/2003SW000007.

Zheng, Y., A. T. Y. Lui, X. Li, and M.-C. Fok (2006), Characteristics of 2–6 MeV electrons in the slot region and inner radiation belt, *J. Geophys. Res.*, *111*, A10204, doi:10.1029/2006JA011748.

M.-C. Fok, NASA Goddard Space Flight Center, Greenbelt, MD 20771, USA.

M. W. Liemohn, S. Xu, and S. Yan, Department of Atmospheric, Oceanic, and Space Sciences, University of Michigan, Ann Arbor, MI 48109-2143, USA. (liemohn@umich.edu)

Q. Zheng, Department of Astronomy, University of Maryland, College Park, MD 20742, USA.

Rebuilding Process of the Outer Electron Radiation Belt: The Spacecraft Akebono Observations

T. Nagai

Department of Earth and Planetary Sciences, Tokyo Institute of Technology, Tokyo, Japan

The spacecraft Akebono has made observations of relativistic electrons in the Earth's radiation belts with the radiation monitor. The highly elliptical, polar orbit with a period of 2.5 h can provide repeated observations of the outer belt. The unique observations by Akebono have revealed rapid rebuilding of the core part (around $L = 4$) of the electron outer belt with a time scale of less than a few hours during the Dst minimum period for several storms. The time scale of this process seems to be shorter than the time interval needed for radial diffusion or wave-particle interaction. This process has a significant contrast with the rebuilding process in the outer part (beyond $L = 5.5$) of the outer belt, where the electron flux increases with a time scale of 1 day or longer. The time scale in the outer part of the outer belt is consistent with the scenario that radial diffusion and/or wave particle interaction may play a central role in electron acceleration processes. The rapid rebuilding of the core part of the outer belt is associated with a storm time substorm, so that there is a possibility that the intense electric field caused by a substorm-associated large-scale dipolarization in the magnetic field results in transport and energization of the electrons in the outer radiation belt. The simulation studies supporting this mechanism are presented.

1. INTRODUCTION

It has been recognized that the flux of relativistic (>MeV) electrons in the Earth's outer radiation belt shows high variability in response to geomagnetic activity and solar wind conditions[e.g., *Williams et al.*, 1968; *Paulikas and Blake*, 1979]. Figure 1 shows a superposed epoch analysis of >2 MeV electron flux at geosynchronous orbit (6.6 R_E) during storms [*Nagai*, 1988]. The data were obtained by Space Environment Monitor on the Geostationary Meteorological Satellite-1 (GMS-1) at 135°E in 1978–1981, and 30 geomagnetic storms were used for this analysis. The electron flux decreases rapidly in association with an enhancement of geomagnetic activity, and then, it recovers rather gradually after a peak of geomagnetic activity. The flux level exceeds the preactivity level. The increase of the electron flux implies that some acceleration processes work in the outer belt. Mechanisms of electron acceleration to energies of >MeV by interaction with various magnetospheric waves have been proposed (see the work of *Thorne* [2010] as a recent review).

At geosynchronous orbit, the electron flux reaches its peak with a time scale of a day or longer than a day. This characteristic provides a possibility of "space weather forecast". Indeed, empirical prediction schemes for the relativistic electron flux at geosynchronous orbit have been proposed [e.g., *Nagai*, 1988; *Baker et al.*, 1990; *Koons and Gorney*, 1991]. As an example, Figure 2 shows a linear prediction filter of the electron flux at geosynchronous orbit as an input of Kp [*Nagai*, 1988]. This linear prediction filter was constructed using the data from >2 MeV electron flux from GMS-3 in 1984–1985 (433 days). An input is the daily sum of Kp, and an output is the daily average of the electron flux (log value).

Dynamics of the Earth's Radiation Belts and Inner Magnetosphere
Geophysical Monograph Series 199
10.1029/2012GM001281

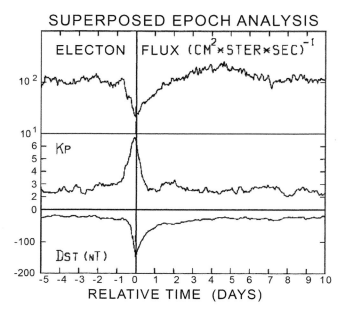

Figure 1. Superposed >2 MeV electron flux from GMS-1 at geosynchronous orbit, *Kp*, and *Dst* during 30 storms in 1978–1981. Hourly values are used. The zero epoch for each storm is taken to be the time of *Dst* minimum. From *Nagai* [1988].

An immediate response of enhanced activity is a flux decrease. A subsequent continuous activity produces an increase in the electron flux. This prediction filter and others can make a good skill for predicting a daily averaged flux level at geosynchronous orbit. This is mainly caused by the fact that the electron flux in the outer part of the outer belt increases with a time scale of a day or longer than a day.

The spacecraft Akebono was launched into a highly elliptical, polar orbit on 22 February 1989 [*Tsuruda and Oya*, 1991]. The initial apogee altitude of its orbit was 10,500 km. The radiation monitor (RDM) on Akebono has been observing electron fluxes in three energy ranges (0.3–0.95, 0.95–2.5, >2.5 MeV) [*Takagi et al.*, 1993]. The data since 1989 are presented at http://www.darts.isas.jaxa.jp/stp/akebono. RDM is not designed to be a scientific instrument so that its capability is very limited. The sensor view is perpendicular to the Sun-pointing spin axis, and the spin period is 8 s. The sensor has a wide view angle. Although basic time resolution for particle measurements is 0.5 s, 16 s averaged data have enough counts to be statistically meaningful. Isotropic flux distributions should be assumed, and no pitch angle information is available. Only real-time data are received at two ground stations in Japan and Sweden now, and data coverage is approximately 20%. On 1 March 2010, the orbit has its apogee at an altitude of 5140 km and its perigee at an altitude of 270 km. A great advantage of the Akebono/RDM observations is the capability of repeated observations crossing the

outer belt. The orbital period is approximately 2.5 h. Electron observations at low altitudes have significant flux variations caused by the Earth's nonuniform magnetic field, and this geographic effect provides a significant difficulty for interpreting temporal flux variations. When observations are made at altitudes of >3000 km, observed fluxes are free from the geographic effect.

The unique Akebono observations have revealed that the core part of the outer radiation belt forms within a few hours for several storms. *Nagai et al.* [2006] have examined >2.5 MeV electron flux variations during the 11 September 2005 and 24 August 2005 storms. The electron flux around *L* = 3 increases significantly with a time scale of a few hours near the peak of the *Dst* development. The typical time scale for local acceleration due to chorus waves to increase the 1 MeV electron flux by an order of magnitude at *L* = 4.5 is on the order of a day [e.g., *Horne et al.*, 2005]. This flux increase forms the core part of the outer belt. The simultaneous magnetic field observation at geosynchronous orbit shows a large-scale dipolarization caused by a storm time substorm near the timing of the electron flux increases inside *L* = 4. *Nagai et al.* [2006] have suggested that a large-scale dipolarization in the magnetic field can transport electrons earthward from the magnetotail and rebuilt the core part of the outer belt in a time scale of a few hours.

Currently, relevant observations for the inner part of the outer belt are limited. Further investigation is needed to clarify the rebuilding processes of the outer radiation belt and to improve space weather forecast. In this paper, we examine rebuilding processes of the outer belt with >2.5 MeV electron observations during the period from March to September in 2010 by Akebono. We compare MeV electron flux behavior in the core part of the outer belt (around *L* = 4)

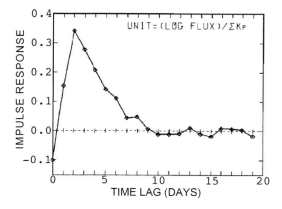

Figure 2. A linear prediction filter for >2 MeV electron flux at geosynchronous orbit constructed from the data from GMS-3 in 1984–1985. An input is daily sum of *Kp*, and an output is daily average of the electron flux (log value). From *Nagai* [1988].

to that in the outer part of the outer belt (beyond $L = 5.5$). Any magnetic field measurements by Akebono are not available now. We use McIlwain L values calculated from the spacecraft orbit information, and we discuss characteristics of flux variations, which do not depend significantly on the adopted L calculation method. *Glocer et al.* [2011] have made simulations to test the possibility that a large-scale dipolarization in the magnetic field can transfer electrons to form the core part of the outer belt. The simulation results are briefly reviewed. The present study shows that the core part of the outer belt forms much faster than the outer part of the outer belt. There is a possibility that the transport of electrons by a storm time substorm has a major contribution to the rebuilding of the core part of the outer belt.

2. OBSERVATIONS

Figure 3 shows the intensity of >2.5 MeV electron flux projected in the radial distance, geomagnetic latitude plane using all data taken by Akebono in March–August 2010. The data are averaged for 1 pixel (0.01 × 0.01 R_E) of the figure.

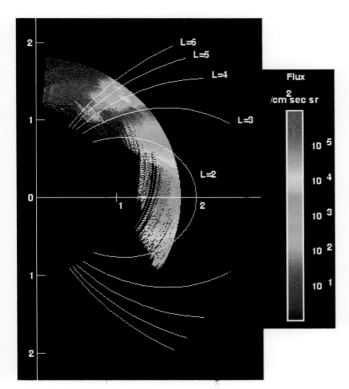

Figure 3. The >2.5 MeV electron flux (color-coded) observed by Akebono/RDM for the period of March–August 2010 in the geomagnetic latitude-radial distance format. The dipole magnetic field lines are drawn. Electron fluxes in this study are presented in units of cm^{-2} s^{-1} sr^{-1}. The unit length 1 corresponds to 1 R_E (6371.2 km).

During this period, Akebono has its apogee in the Northern Hemisphere and cuts the outer belt at high altitudes. Since particle counts are low at low altitudes in the outer belt ($L = 4$–6), we use the data taken only at altitudes of >3000 km in this study.

Figure 4a shows >2.5 MeV electron observations in the region of $L = 2$–8 in the period from March to August 2010. Electron flux values at $L = 4$ and $L = 6$ from each orbit are also plotted (Figures 4b and 4c). The flux of >2 MeV electrons from GOES 11 at 6.6 R_E is presented (Figure 4d). GOES 11 is located at 135°W, and its geomagnetic latitude is 5°, so that the flux from GOES 11 exhibits the equatorial flux at the outer edge of the outer belt. For the GOES 11 data, 1 h average values are used so that there is a diurnal variation (a noon peak is near 21 UT for each day). Gross features in flux variations in the region of $L > 5.5$ from Akebono (at low altitudes) can be found influx variations from GOES 11 near the equator. Kp and Dst values are presented to show geomagnetic activity (Figures 4e and 4f). The hourly averaged solar wind velocity data are also presented (Figure 4g). There are usually four or five data points for each L shell per 1 day in Akebono observations. The higher flux is observed at higher altitude (see Figure 3), so that this altitude effect produces a small diurnal variation in the observed flux. This altitude effect is small in comparison with flux variations examined here.

There are four major flux increases in the outer belt ($L = 4$–6) in association with four magnetic storms starting on 5 April, 2 May, 28 May, and 3 August 2010. An immediate response to these storms is a decrease in the electron flux in the almost whole outer belt. The flux near $L = 4$ shows an increase near the Dst minimum, and it tends to keep its level during the recovery phase. The flux beyond $L = 5$ tends to increase rather gradually, and the flux reaches its peak a few days after the Dst minimum. These characteristics will be examined for the individual storms in detail.

At geosynchronous orbit, the flux shows significant increases even for other periods starting on 11 March, 2 April, 25 June, 27 July, and 24 August 2010. The flux level after 27 July 2010 becomes higher than that after the 3 August 2010 storm at 6.6 R_E. Akebono observes similar flux variations at $L = 6$. However, the flux at $L = 4$ is almost unchanged after 27 July 2010, and it shows a decrease and stays in the lower level after 24 August 2010. For these periods, solar wind velocity is high, and high Kp values continue without any significant Dst development.

2.1. The 3–5 August 2010 Storm

Figure 5 shows fluxes of >2.5 MeV electrons from Akebono at $L = 4$ and $L = 5.5$ for the period of 3–6 August 2010. Representative flux profiles are presented in Figure 6. Before

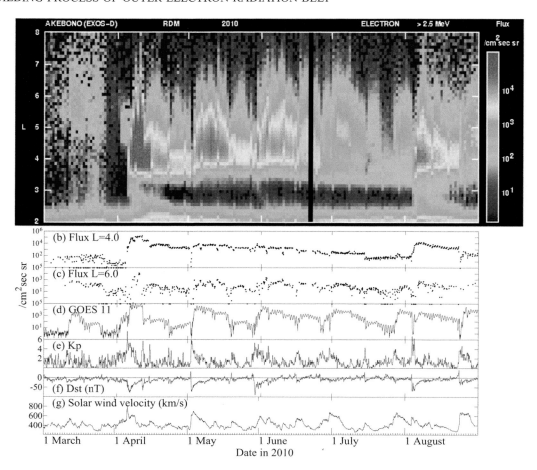

Figure 4. (a) Daily average intensity (color-coded) of >2.5 MeV electron flux observed by Akebono/RDM in the region of $L = 2–8$ at altitudes of >3000 km in the period from March to August 2010. (b) The >2.5 MeV electron flux values at $L = 4$ from Akebono. (c) The >2.5 MeV electron flux values at $L = 6$ from Akebono. (d) The >2 MeV electron flux at geosynchronous orbit from GOES 11. (e) Kp and (f) Dst geomagnetic indices. (g) Hourly solar wind velocity from ACE.

this storm, the outer belt forms with its center near $L = 5$ after 27 July 2010 (Figure 4). The ACE data are examined for monitoring solar wind conditions in this study. An interplanetary shock in the solar wind produces SSC at 17:40 UT (a spike in *SYM-H*, 1 min *Dst*), and *Dst* develops after 20:00 UT on 3 August 2010. The timing of SSC is determined with the 1 min ground magnetic field data, and the 1 s data from Kakioka, hereafter, in addition with the solar wind data from ACE, Wind, and Geotail. At 19:34 UT (Figure 6a) on 3 August 2010 (the time for the Akebono observation at $L = 4$), the flux beyond $L = 5$ decreases slightly, while the flux inside $L = 5$ is unchanged. At 21:56 UT (Figure 6b), the flux beyond $L = 3.5$ becomes low, indicating that the outer belt almost disappears. At 00:15 UT (Figure 6c) on 4 August 2010, an increase in the flux by an order of magnitude is observed only near $L = 4$. At 02:33 UT (Figure 6d), the flux

profile indicates formation of a new outer belt. The flux level at $L = 4$ is almost the same as that at 12:14 UT (Figure 6e). The flux beyond $L = 5$ stays in the low level, and it increases after 12 UT on 5 August 2010 (Figure 5b). Trapped >300 keV electrons (proton contamination is corrected) observed by the NOAA-POES satellites (NOAA 15, 16, 17, 18, 19, and METOP 02) also show an increase in the flux near 00:00 UT on 4 August 2010 (not shown here), like the events discussed by *Nagai et al.* [2006] and *Glocer et al.* [2011]. Hence, the >2.5 MeV electron flux increases with an order of magnitude in the core part of the outer belt (near $L = 4$) near the *Dst* minimum (see Figure 6f) with a time scale of less than 4 h, while the outer part of the outer belt (beyond $L = 5$) forms in the recovery phase of the storm.

Figure 6g shows the *H* component (northward) of the magnetic field at GOES 12 (60°W, its local midnight is

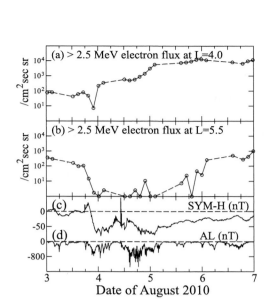

Figure 5. (a) The >2.5 MeV electron flux values at $L = 4$ observed by Akebono/RDM for the period of 3–6 August 2010. (b) The >2.5 MeV electron flux values at $L = 5.5$. (c) Geomagnetic index *SYM-H* (1 min *Dst*). (d) *AL* index (1 min values).

04 UT). A sharp increase in H at 17:40 UT on 3 August 2010 is an effect of SSC. There are several increases in H for substorm onsets on the nightside. A large-scale dipolarization in the magnetic field takes place near 00:00 UT on 4 August 2010, which is coincident with the increase in the electron flux in the core part of the outer belt ($L = 4$). An increase of the H magnetic field component is usually a recovery to the quiet-day level for nonstorm time substorms [e.g., *Nagai*, 1982a, 1982b]. For this event, the maximum H value (100 nT) exceeds the quiet-day level (80 nT). This substorm takes place near the *Dst* minimum (Figure 6f). Although the IMF B_z is continuously southward, *Dst* stays almost in the same level, suggesting that this substorm produces a partial collapse of the tail current. Hence, this dipolarization in the magnetic field is caused by a representative one for a storm time substorm.

There are sharp SSC-like spikes in the magnetic field on the ground (seen in *SYM-H* in Figure 6f) and at geosynchronous orbit (Figure 6g) near 10:20 UT on 4 August 2010. The electron flux does not show any significant change in the outer belt ($L = 3$–8). The dynamic pressure of the solar wind shows an increase after 18:00 UT on 3 August 2010 (1 h shifted). Although there are two spikes in the dynamic pressure at 04:30 and 08:00 UT on 4 August 2010, there are no changes in the dynamic pressure near 00:00 UT on 4 August 2010. Hence, any change in the solar wind dynamic pressure does not produce any significant increase in the electron flux in the outer belt during this storm.

2.2. The 28–31 May 2010 and 2–4 May 2010 Storms

Figure 7 shows fluxes of >2.5 MeV electrons from Akebono at $L = 4.2$ and $L = 5.5$ for the period of 28–31 May 2010. The flux stays in the prestorm level until 19:45 UT on 28 May 2010 and appears to decrease with the *Dst* development. This flux

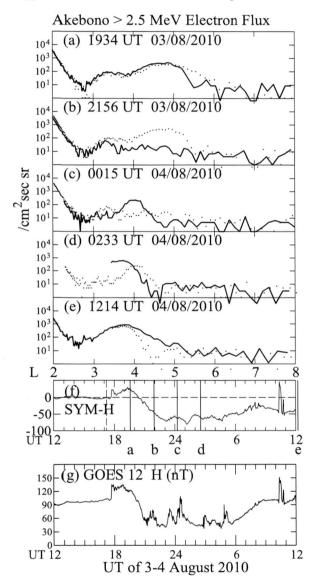

Figure 6. (a)–(e) Selected L profiles of >2.5 MeV electron flux observed by Akebono/RDM on 3–4 August 2010. The observation time at $L = 4$ is presented in each plot. Profiles presented by points are those obtained in the previous paths. (f) Geomagnetic index *SYM-H* for the 24 h period starting at 12:00 UT on 3 August 2010. The Akebono observation times for Figures 6a–6e are indicated by vertical lines. (g) The H (northward) component of the magnetic field from GOES 12 (its local midnight is 04 UT) at geosynchronous orbit.

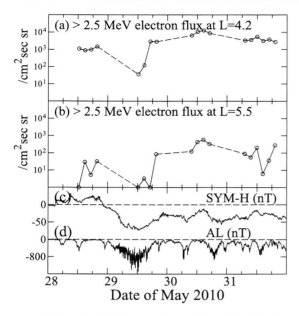

Figure 7. (a) The >2.5 MeV electron flux values at $L = 4.2$ observed by Akebono/RDM for the period of 28–31 May 2010. (b) The >2.5 MeV electron flux values at $L = 5.5$. (c) Geomagnetic index *SYM-H* (1 min *Dst*). (d) *AL* index (1 min values).

decrease is more evident at $L = 4.2$ than at $L = 4.0$ (Figure 4). An increase (more than one order of magnitude) in the flux at $L = 4.2$ takes place between 14:44 and 17:09 UT on 29 May 2010. The flux of trapped >300 keV electrons at $L = 4$ from the NOAA satellites shows the similar behavior (not shown here). A large substorm starts near 15:30 UT in the course of the *Dst* recovery. There is no relevant magnetic field observation at geosynchronous orbit for this event. The flux near $L = 4$ is almost constant after 29 May 2010. A flux increase in the outer part of the outer belt (beyond $L = 5.5$) is delayed, and it reaches a peak value after 1 June 2010 (see Figure 4).

Figure 8 shows fluxes of >2.5 MeV electrons from Akebono at $L = 4$ and $L = 5.5$ for the period of 1–4 May 2010. The outer belt with its center at $L = 3.8$ almost disappears with the *Dst* development (Figure 4). The flux is already in the low level at 14:03 UT on 2 May 2010, when the *Dst* is developing. The flux at $L = 4$ increases after 16:28 UT and reaches its peak value near 21:19 UT on 2 May 2010, when *Dst* stays in its minimum level. The flux of trapped >300 keV electrons at $L = 4$ from the NOAA satellites shows the similar behavior (not shown here). *AL* index indicates a large substorm staring at 18:00 UT. There is no relevant magnetic field observation at geosynchronous orbit for this event. An increase in the flux in the outer part of the outer belt (beyond $L = 5.5$) is delayed approximately 1 day, and the flux in this region reaches its peak on 4 May 2010.

2.3. The 5–6 April 2010 Storm

Figure 9 shows fluxes of >2.5 MeV electrons from Akebono at $L = 4$ and $L = 5.5$ for the period of 4–7 April 2010. Representative flux profiles are presented in Figure 10. Before this storm, the outer belt forms after 2 April 2010 with its peak around $L = 5.2$ (Figure 4). The flux at $L = 4$ is in the background level after 26 March 2010. Figure 10a shows the outer belt profile (at 01:04 UT) before this storm. At 12:55 UT on 5 April 2010, a new outer belt exists in the region of $L = 2.8$–4.5. Although the outer belt is slightly depressed at 15:18 UT, the outer belt becomes firm after 17:41 UT. The flux in the outer part of the outer belt is in the background level until the end of 6 April 2010.

Figure 10g presents the flux of trapped >300 keV electrons observed by the NOAA-POES satellites for the 24 h period on 5 April 2010. The interplanetary shock produces a large SSC (see *SYM-H* in Figure 9c) at 08:25 UT, and a large substorm starts at 08:56 UT (see *AL* in Figure 9d). The difference between these two timings can be identified definitely in the ground magnetic field data. There is a sudden increase in the flux of >300 keV electrons near 09:00 UT (a 30 min delay relative to the SSC event is evident), and the flux level continues to be high. The flux of >30 keV electrons also shows an increase at 09:00 UT (not shown here). Figure 10h shows the *H* component of the magnetic field at GOES 11 (135°W, its local midnight is 09 UT). The *H* component of

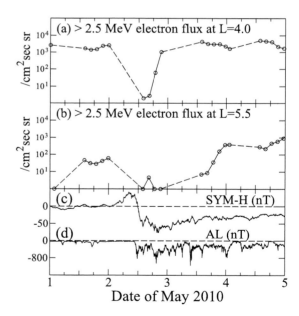

Figure 8. (a) The >2.5 MeV electron flux values at $L = 4$ observed by Akebono/RDM for the period of 1–4 May 2010. (b) The >2.5 MeV electron flux values at $L = 5.5$. (c) Geomagnetic index *SYM-H* (1 min *Dst*). (d) *AL* index (1 min values).

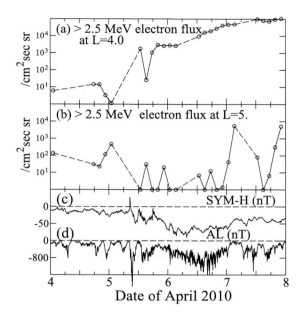

Figure 9. (a) The >2.5 MeV electron flux values at $L = 4$ observed by Akebono/RDM for the period of 4–7 April 2010. (b) The >2.5 MeV electron flux values at $L = 5.5$. (c) Geomagnetic index *SYM-H* (1 min *Dst*). (d) *AL* index (1 min values).

the magnetic field shows a large (unusual) overshooting of a dipolarization in the magnetic field for the 08:56 UT substorm. Hence, it is likely that the new outer belt observed by Akebono forms near 09:00 UT in association with the >300 keV electron flux increase at NOAA altitudes, although there is a data gap in the Akebono observations. The interplanetary shock produces another SSC at 17:45 UT. There is no evident change in the electron flux in the outer belt for this (the observations are made at 17:41 and 20:06 UT by Akebono, as seen in Figure 9).

2.4. Summary of Observations

We have examined the four storms in 2010 here. In the three storms in May and August, >2.5 MeV electron flux increases with an order of magnitude in the core part of the outer belt near $L = 4$ during the *Dst* minimum, and the time scale of the flux increase is less than a few hours. A large storm time substorm can be found in association with the flux increase. For the 5–6 April 2010 storm, the outer belt is probably reformed near $L = 4$ in association with the unusually large dipolarization caused by the large substorm. In these events, the outer part ($L > 5.5$) of the outer belt forms rather gradually in the recovery phase of storms. Hence, although the whole outer belt forms with a time scale of a day or more, the core part of the outer belt forms near $L = 4$ rapidly. Any significant changes in interplanetary conditions

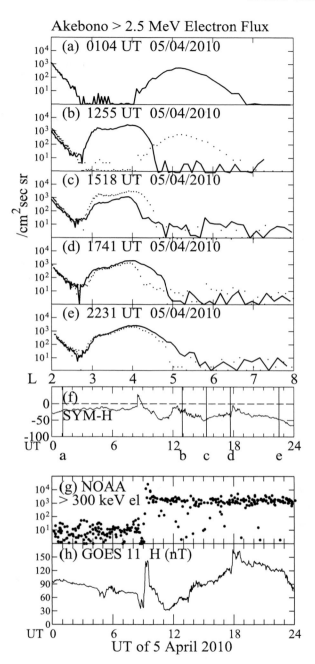

Figure 10. (a)–(e) Selected L profiles of >2.5 MeV electron flux observed by Akebono/RDM on 5 April 2010. The observation time at $L = 4$ is presented in each plot. Profiles presented by points are those obtained in the previous paths. (f) Geomagnetic index *SYM-H* on 5 April 2010. The Akebono observation times for Figures 10a–10e are indicated by vertical lines. (g) The flux of trapped >300 keV electrons observed by the NOAA-POES (NOAA 15, 16, 17, 18, 19, and METOP 02) satellites on 5 April 2010. (h) The H (northward) component of the magnetic field from GOES 11 (its local midnight is 09 UT) at geosynchronous orbit on 5 April 2010.

are not observed during the rapid rebuilding of the core part of the outer belt.

3. SIMULATION OF THE 3–4 SEPTEMBER 2008 STORM

Glocer et al. [2011] have made simulations of the rapid rebuilding process in the outer belt with the Radiation Belt Environment (RBE) model coupled with the Block-Adaptive-Tree Solar wind Roe-type Upwind Scheme (BATS-R-US) model [see also *Fok et al.*, 2011]. The main target of the simulations is to test whether or not MeV electrons can be transported and energized into the core part of the outer belt in the appropriate conditions. Details of this model are given in the work of *Glocer et al.* [2009, 2011]. The electron dynamics in the radiation belt is derived from the RBE model [*Fok et al.*, 2011], in which the bounce-averaged Boltzmann equation is solved under the given electric and magnetic field models with and without wave-particle interaction. The NASA AE8MAX model is used as the initial electron distribution. Although two events (on 4 September 2008 and 22 July 2009) are studied in the work of *Glocer et al.* [2011], the model results for the 4 September 2008 event are presented here.

Figure 11 shows variations of the outer belt in the 48 h period starting at 12:00 UT on 3 September 2008. The outer belt with its center at $L = 4.8$ exits at 15:10 UT on 3 September 2008, prior to SSC of this storm at 15:42 UT (Figures 11f and 11h). The outer belt is basically unchanged at least until 17:31 UT. *Dst* develops after 00:00 UT on 4 September 2008 (Figure 11f), and the outer belt almost disappears at 0321 UT near the *Dst* minimum. A new outer belt with higher fluxes appears at 05:46 UT (see Figures 11d and 11g). The peak flux at $L = 4.5$ is almost unchanged until the end of 7 September 2008. The flux of trapped >30 keV electrons at $L = 4.5$ from the NOAA satellites gradually increases after 00:00 UT on 4 September 2008 (not shown) with the development of *Dst*, and the flux of trapped >300 keV electrons at $L = 4.5$ from the NOAA satellites shows a sudden increase near 04:00 UT [Figure 8 of *Glocer et al.*, 2011]. Furthermore, GOES 10 (60°W, the local midnight is 04 UT) shows a large dipolarization at 04:00 UT (Figure 11h). For the period of 03:20–05:50 UT on 4 September 2008, any significant change is not observed in the solar wind dynamic pressure, and no SSC-like variation is detected on the ground magnetic field data. Hence, a rapid rebuilding of the outer belt with a time scale of less than 2.5 h is attributable to the storm time substorm at 04:00 UT in this event.

Glocer et al. [2011] have run the RBE model under different electric field and magnetic field conditions, with and

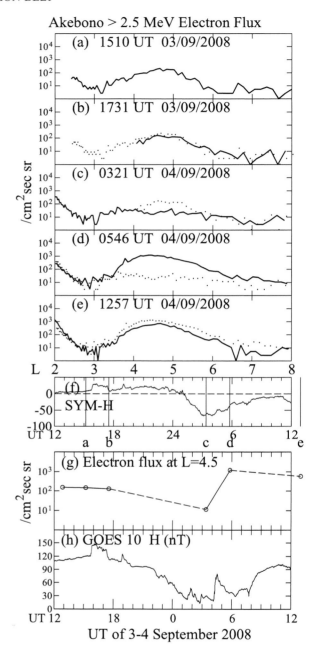

Figure 11. (a)–(e) Selected L profiles of >2.5 MeV electron flux observed by Akebono/RDM on 3–4 September 2008. The observation time at $L = 4$ is presented in each plot. Profiles presented by points are those obtained in the previous paths. (f) Geomagnetic index *SYM-H* for the 24 h period starting at 1200 UT on 3 September 2008. The Akebono observation times for Figures 11a–11e are indicated by vertical lines. (g) The >2.5 MeV electron flux values at $L = 4.5$ observed by Akebono/RDM for the 24 h period starting at 12:00 UT on 3 September 2008. (h) The H (northward) component of the magnetic field from GOES 10 (its local midnight is 04 UT) at geosynchronous orbit for the 24 h period starting at 12:00 UT on 3 September 2008.

without wave-particle interaction model. Figure 12 is adopted from Figure 6 of *Glocer et al.* [2011]. In case A, the radiation belt model is driven by the magnetic field and electric field model from the global MHD simulation. The global MHD simulation is run by observed solar wind conditions after 3 September 2008. A large-scale dipolarization in the magnetic field is reproduced in the simulation, although its magnitude is smaller than the observed one. No wave-particle interaction model is included in order to test the effect of the magnetic field changes. Case A can simulate a rapid flux enhancement with an appropriate time scale near $L = 4$. In case B, the Akebono/RDM observations are presented in the format similar to the model output for comparison. In cases C and D, the magnetic field model is the time-variable T04 model [*Tsyganenko and Sitnov*, 2005] according to the solar wind conditions, so that any significant dipolarization in the magnetic field is not reproduced. Case C does not include any wave effect, while case D has chorus-wave particle interactions. Radial diffusion is implicitly included due to magnetic field changes according to the solar wind conditions. Any

significant flux enhancement within an appropriate time scale does not take place in the course of these simulations. The simulation results suggest that the dipolarization in the magnetic field is important for the rapid flux increase around $L = 4$ in the outer belt.

4. DISCUSSION

It can be found in other observations that the outer belt does not grow uniformly in the region of $L = 3$–6. *Baker et al.* [1994] made a superposed epoch analysis of >400 keV electron increases for several L values using the SAMPEX data. Relative to the peak flux at $L = 3$, the flux at $L = 4$ reaches its peak 1 day later, and the flux at $L = 5$ reaches its peak 3 days later. The >1 MeV electron flux variations observed by CRRES during the October 1990 storm were presented [e.g., *Meredith et al.*, 2002; *Thorne et al.*, 2007]. It appears that the core part of the >MeV electron outer belt formed near the *Dst* minimum and that the >MeV electron flux in the outer part of the outer belt grew mostly in the recovery phase of the storm.

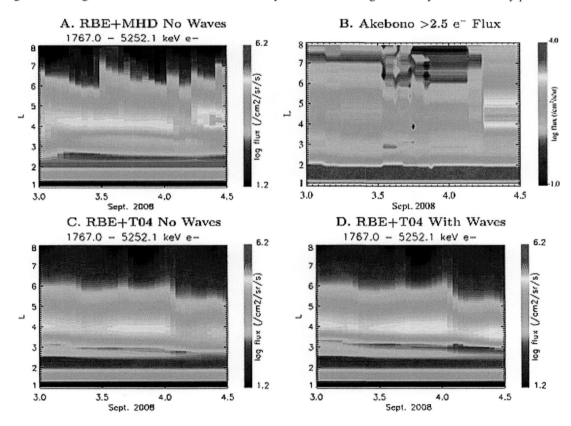

Figure 12. Color-coded energetic electron fluxes in the L time formats in the simulations by *Glocer et al.* [2011]. (a) The result from the radiation belt model driven by the global MHD model without chorus waves. (b) Akebono >2.5 MeV electron flux observations. The results from the radiation belt model driven by the T04 model (c) without chorus waves and (d) with chorus waves. From *Glocer et al.* [2011].

Since outbound CRRES trajectories did not cross the same L shell at the same magnetic field value, it is difficult to compare radial flux profiles quantitatively. *Vassiliadis et al.* [2002] examined 2–6 MeV electron flux variations by SAMPEX in the region of $L = 1.1$–10 using a linear filtering technique. In their statistical study, the outer belt at $L = 4$–8 responses to solar wind speed rather coherently. They insist that the flux at $L = 3$ shows a response to the solar wind density (dynamic pressure). These analyses are based on the daily averaged data. The time scale discussed by *Vassiliadis et al.* [2002] is different from that examined in this paper.

The present study shows that the core part (usually near $L = 4$) of the outer belt forms within a few hours during the *Dst* minimum period for several storms. The formation of this part of the outer belt appears to coincide with an onset of storm time substorms. In the past, observations of a rapid rebuilding of the outer belt were rarely reported. Probably, this is because repeated observations of the outer belt in the same L-B conditions with a short time scale cannot be made by a single satellite. *Li et al.* [1998] reported a rapid enhancement of >0.2 MeV electrons across $L = 4.2$–6 observed by three GPS satellites on 10 January 1997. They attributed the flux increases after 10:30 UT to the solar wind density pulse near 10:55 UT. Fortunately, Akebono made successive observations of the outer belt ($L = 3$–7) until 12:00 UT on 10 January 1997. The observed flux profiles were unchanged until 07:43 UT. The >2.5 MeV electron flux at $L = 4$ at 10:44 UT became more than 1 order of magnitude higher than that at 07:43 UT on 10 January 1997, and it is evident that the flux increase near $L = 4$ started before the solar wind density pulse at 10:55 UT. GOES 9 near midnight detected a large-scale dipolarization in the magnetic field after 10:40 UT (see Figure 5 of *Li et al.* [1998]). Hence, the Akebono observations do not show that an increase in the MeV electron flux is caused by the solar wind density pulse. In the events examined in the present study, any changes in the solar wind dynamic pressure do not produce any significant effects in the electron flux. It is likely that only exceptionally large interplanetary shocks (e.g., the 24 March 1991 event) can produce injection of energetic electrons [e.g., *Blake et al.*, 1992].

In the present study, we have examined the flux variations in the >2.5 MeV electrons in the outer belt by the Akebono/RDM observations for the four storms in 2010 with other relevant observations. The simulation studies by *Glocer et al.* [2011] have been examined. The present study suggests a scenario that the rebuilding of the outer belt has two steps for these storms. The outer belt can disappear in the early phase of storms. The core part of the outer belt (around $L = 4$) forms near the *Dst* minimum. It is likely associated with a large-scale dipolarization in the magnetic field caused by a storm time substorm. Electrons are transported with an intense electric field. It can be assumed that only low-energy (>30 keV) electrons are injected in the course of *Dst* development, and then, these electrons are accelerated to MeV electrons. In the 5–6 April 2010 storm, >30 and >300 keV electrons increase simultaneously at $L = 4$, suggesting that both >30 keV electrons and >300 keV electrons are transported in the same process. This rebuilding can be attributable only to radial transport of electrons by the intense electric field. No wave-particle interaction is needed for this process. However, *Omura et al.* [2007] have proposed a possibility of very rapid electron acceleration with wave-particle interaction. It is not known that the process proposed by *Omura et al.* [2007] can work in the highly variable conditions like the 5–6 April 2010 storm. During the recovery phase of storms, the outer part of the outer belt (mostly beyond $L = 5$) forms rather gradually in the course of continuous substorm activity. The wave-particle interaction may play a central role in this process, as proposed by many past studies [see *Thorne*, 2010]. There are some modeling studies on the substorm effect of the outer belt electron flux [e.g., *Kim et al.*, 2000; *Fok et al.*, 2001]. They show that electrons, which can produce a part of the outer belt, are transported by the substorm-associated electric field even for nonstorm time substorms.

5. CONCLUSIONS

The Akebono observations have shown that >2.5 MeV electron flux increases in the core part (around $L = 4$) of the electron radiation belt with a time scale of less than a few hours during the *Dst* minimum period for several storms. The time scale for the rebuilding of the core part of the outer belt seems to be much shorter than that required by electron acceleration mechanisms with wave-particle interactions. It is likely that an intense electric field caused by storm time substorms transports electrons into the outer belt. The outer part of the outer belt grows rather gradually in the recovery phase of storms, indicating that other mechanisms are working there. There are a number of limitations in the Akebono observations. Akebono does not provide any continuous observational data. We have examined flux variations with adequate timings near the *Dst* minimum period only for the limited number of events. The observations are made at low latitudes, off the equatorial plane, and information on electrons with various energies and their pitch angles are not obtained. The magnetic field and electric field measurements in the inner magnetosphere are not fully available. Hence, various verifications are needed for completing the theory of the rebuilding of the electron outer belt.

Acknowledgments. This work is supported by the JSPS Grant-in-Aid for Scientific Research (C) (225404588). The NOAA-POES

data were obtained from NGDC/NOAA (http://www.ngdc.noaa
.gov). The method for correction of proton contamination was
guided by J. C. Green. The ACE and GOES data were obtained
from the CDAWeb (http://cdaweb.gsfc.nasa.gov). The ground
magnetic field data and geomagnetic indices were provided by the
World Data Center for Geomagnetism, Kyoto. T. N. thanks M.-C.
Fok and Y. Ebihara for the useful comments.

REFERENCES

Baker, D. N., R. L. McPherron, T. E. Cayton, and R. W. Klebesadel
    (1990), Linear prediction filter analysis of relativistic electron
    properties at 6.6 R_E, J. Geophys. Res., 95(A9), 15,133–15,140.

Baker, D. N., J. B. Blake, L. B. Callis, J. R. Cummings, D. Hoves-
    tadt, S. Kanekal, B. Klecker, R. A. Mewaldt, and R. D. Zwickl
    (1994), Relativistic electron acceleration and decay time scales in
    the inner and outer radiation belts: SAMPEX, Geophys. Res.
    Lett., 21(6), 409–412.

Blake, J. B., W. A. Kolasinski, R. W. Fillius, and E. G. Mullen
    (1992), Injection of electrons and protons with energies of tens of
    MeV into L < 3 on 24 March 1991, Geophys. Res. Lett., 19(8),
    821–824.

Fok, M.-C., T. E. Moore, and W. N. Spjeldvik (2001), Rapid
    enhancement of radiation belt electron fluxes due to substorm
    dipolarization of the geomagnetic field, J. Geophys. Res.,
    106(A3), 3873–3881.

Fok, M.-C., A. Glocer, Q. Zheng, R. B. Horne, N. P. Meredith, J. M.
    Albert, and T. Nagai (2011), Recent developments in the radia-
    tion belt environment model, J. Atmos. Sol. Terr. Phys., 73,
    1435–1443, doi:10.1016/j.jastp.2010.09.033.

Glocer, A., G. Tóth, M.-C.Fok, T. Gombosi, and M. Liemohn
    (2009), Integration of the radiation belt environment model into
    the space weather modeling framework, J. Atmos. Sol. Terr.
    Phys., 71, 1653–1663, doi:10.1016/j.jastp.2009.01.003.

Glocer, A., M.-C. Fok, T. Nagai, G. Tóth, T. Guild, and J. Blake
    (2011), Rapid rebuilding of the outer radiation belt, J. Geophys.
    Res., 116, A09213, doi:10.1029/2011JA016516.

Horne, R. B., R. M. Thorne, S. A. Glauert, J. M. Albert, N. P.
    Meredith, and R. R. Anderson (2005), Timescale for radiation
    belt electron acceleration by whistler mode chorus waves,
    J. Geophys. Res., 110, A03225, doi:10.1029/2004JA010811.

Kim, H.-J., A. A. Chan, R. A. Wolf, and J. Birn (2000), Can
    substorms produce relativistic outer belt electrons?, J. Geophys.
    Res., 105(A4), 7721–7735.

Koons, H. C., and D. J. Gorney (1991), A neural network model of
    the relativistic electron flux at geosynchronous orbit, J. Geophys.
    Res., 96(A4), 5549–5556.

Li, X., D. N. Baker, M. Temerin, T. Cayton, G. D. Reeves, T. Araki,
    H. Singer, D. Larson, R. P. Lin, and S. G. Kanekal (1998),
    Energetic electron injections into the inner magnetosphere during
    the Jan. 10–11, 1997 magnetic storm, Geophys. Res. Lett.,
    25(14), 2561–2564.

Meredith, N. P., R. B. Horne, R. H. A. Iles, R. M. Thorne, D.
    Heynderickx, and R. R. Anderson (2002), Outer zone relativistic
    electron acceleration associated with substorm-enhanced whistler
    mode chorus, J. Geophys. Res., 107(A7), 1144, doi:10.1029/
    2001JA900146.

Nagai, T. (1982a), Local time dependence of electron flux changes
    during substorms derived from multi-satellite observation at syn-
    chronous orbit, J. Geophys. Res., 87(A5), 3456–3468.

Nagai, T. (1982b), Observed magnetic substorm signatures at syn-
    chronous altitude, J. Geophys. Res., 87(A6), 4405–4417.

Nagai, T. (1988), "Space weather forecast": Prediction of relativistic
    electron intensity at synchronous orbit, Geophys. Res. Lett.,
    15(5), 425–428.

Nagai, T., A. S. Yukimatu, A. Matsuoka, K. T. Asai, J. C. Green,
    T. G. Onsager, and H. J. Singer (2006), Timescales of relativistic
    electron enhancements in the slot region, J. Geophys. Res., 111,
    A11205, doi:10.1029/2006JA011837.

Omura, Y., N. Furuya, and D. Summers (2007), Relativistic turning
    acceleration of resonant electrons by coherent whistler mode
    waves in a dipole magnetic field, J. Geophys. Res., 112,
    A06236, doi:10.1029/2006JA012243.

Paulikas, G. A., and J. B. Blake (1979), Effects of the solar wind on
    magnetospheric dynamics: Energetic electrons at the synchronous
    orbit, in Quantitative Modeling of Magnetospheric Processes,
    Geophys. Monogr. Ser., vol. 21, edited by W. P. Olson, pp. 180–
    202, AGU, Washington, D. C., doi:10.1029/GM021p0180.

Takagi, S., T. Nakamura, T. Kohno, N. Shiono, and F. Makino
    (1993), Observations of space radiation environment with
    EXOS-D, IEEE Trans. Nucl. Sci., 40, 1491–1497.

Thorne, R. M. (2010), Radiation belt dynamics: The importance of
    wave-particle interactions, Geophys. Res. Lett., 37, L22107,
    doi:10.1029/2010GL044990.

Thorne, R. M., Y. Y. Shprits, N. P. Meredith, R. B. Horne, W. Li, and
    L. R. Lyons (2007), Refilling of the slot region between the inner
    and outer electron radiation belts during geomagnetic storms,
    J. Geophys. Res., 112, A06203, doi:10.1029/2006JA012176.

Tsuruda, K., and H. Oya (1991), Introduction to the EXOS-D
    (Akebono) Project, Geophys. Res. Lett., 18(2), 293–295.

Tsyganenko, N. A., and M. I. Sitnov (2005), Modeling the dynamics
    of the inner magnetosphere during strong geomagnetic storms,
    J. Geophys. Res., 110, A03208, doi:10.1029/2004JA010798.

Vassiliadis, D., A. J. Klimas, S. G. Kanekal, D. N. Baker, and R. S.
    Weigel (2002), Long-term-average, solar cycle, and seasonal re-
    sponse of magnetospheric energetic electrons to the solar wind speed,
    J. Geophys. Res., 107(A11), 1383, doi:10.1029/2001JA000506.

Williams, D. J., J. F. Arens, and L. J. Lanzerotti (1968), Observa-
    tions of trapped electrons at low and high altitudes, J. Geophys.
    Res., 73(17), 5673–5696.

T. Nagai, Department of Earth and Planetary Sciences, Tokyo
Institute of Technology, Tokyo152-8551, Japan. (nagai@geo.
titech.ac.jp)

The Shock Injection of 24 March 1991: Another Look

J. B. Blake

Space Science Department, The Aerospace Corporation, Los Angeles, California, USA

A strong shock that arrived at Earth on 24 March 1991 created a new radiation belt in the slot region consisting of energetic protons and relativistic electrons. The data from the CRRES satellite allowed for a prompt theoretical explanation of the injection event. Some CRRES observational details of the first minutes of the prompt injection were not published. This chapter presents some of these details in the hope that they might prove useful in the analysis of data from the Radiation Belt Storm Probes mission.

1. INTRODUCTION

The realization that the shock that arrived at Earth early on 24 March 1991 had created a new radiation belt in the slot region consisting of protons and electrons with energies of tens of MeV came as a big surprise. It engendered a renewed interest in radiation belt science that has persisted to the present day. One might argue that this injection event had a significant role in leading to the upcoming NASA Radiation Belt Storm Probe (RBSP) mission.

In space science, as in many aspects of life, reference often is made to Murphy's law: if something can go wrong, it will. However, in the case of the CRRES observations of the shock injection event, extraordinary good luck placed CRRES in just the right spot at just the right time. CRRES had entered the slot inbound, in the postmidnight sector when the shock struck. The probability of being so well placed was extremely small.

Furthermore, CRRES carried a Cherenkov counter, an instrument rarely used in radiation belt studies and one very well suited to measuring relativistic electrons. CRRES was able to observe multiple passages of longitudinally bunched drifting electrons and protons as the spacecraft proceeded through the slot. The Earth's magnetic field acted as a giant magnetic spectrometer, enabling the instruments to provide

information that would not have otherwise been revealed. This information enabled *Li et al.* [1993] to quickly model the physics and explain how the injection process worked.

A short description of the CRRES observations was published [*Blake*, 1992; *Blake et al.*, 1992] that was used by *Li et al.* [1993] in their model. Perhaps because of their success in modeling the injection event, some of the details of the CRRES observations were not published. However, in looking forward to the RBSP mission, it is interesting to examine the CRRES observations in light of what might be observed with the capable complement of instruments aboard RBSP if a similar injection event should occur during the mission.

The discussion in this chapter is confined mainly to the inbound pass of CRRES from just prior to the first detection of newly injected particles until CRRES passed below the stable trapping region. During this time interval, the injected particles were periodic visitors to the position of CRRES without other widespread disturbances.

2. OBSERVATIONS

Some of the salient features of the Cherenkov counter aboard CRRES are important in the interpretation of the observations. The CRRES Cherenkov counter was fabricated using spare parts from the University of California at San Diego (UCSD) investigation aboard Pioneer 10 and Pioneer 11 missions to Jupiter and beyond [*Fillius and McIlwain*, 1974]. The counter itself was a glass cylinder containing water and methanol (index of refraction of 1.33), which was viewed end on. The three discriminator thresholds provided three integral electron channels of $E > 6$ MeV, $E > 9$ MeV,

Dynamics of the Earth's Radiation Belts and Inner Magnetosphere
Geophysical Monograph Series 199
10.1029/2012GM001311

and $E > 13$ MeV (protons require more than 485 MeV to be detected). The angular response of the sensor was 120°, 90°, and 65° full width for the three channels.

The injection event, as seen in three channels of the Cherenkov counter, is shown in Figure 1 as CRRES traversed the slot, then the inner zone with its CRAND protons, and finally went below the trapping region. The relativistic electrons first appeared when CRRES was at $\sim L = 2.54$, and by $L = 2$, no further response due to newly injected particles could be seen. CRRES was near the geomagnetic equator at this time; the dipole latitude was around $-10°$.

The multiple peaks at a fixed spacing immediately suggested drift echoes, but the shape of the peaks and the very similar shape for all channels seemed strange. All three-channel count rates ramp up together and then turn over at the same time, falling more rapidly. The crucial realization was that the simultaneous turnover meant that all of the electrons in the echoes had energies above 13 MeV! Otherwise, the count rate of the >13 MeV channel would turn over first, while the two lower-energy channels would continue to rise as slower-drifting, lower-energy electrons reached CRRES. The simultaneous rollover in all three channels means there were no electrons drifting more slowly than 13 MeV electrons and, thus, no electrons with energies below 13 MeV. One can think of the magnetosphere as acting here as a giant drift-frequency spectrometer.

This conclusion, although straightforward, seemed hard to believe based upon previous radiation belt measurements of

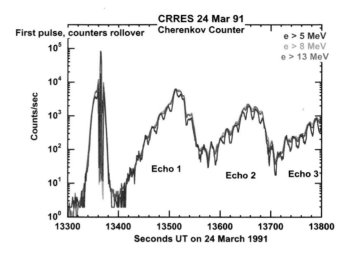

Figure 2. Count rates from the three Cherenkov channels converted to electron flux using the geometric factors given by *Fillius and McIlwain* [1974] and plotted on an expanded scale. All channels give the same value for the integral flux, revealing that there are no electrons greater than 6 MeV but less than 13 MeV. Large pitch angle modulation is seen in the >13 MeV channel only because the other two channels have a broad angular response.

energetic electrons. It seemed prudent to examine the data further. *Fillius and McIlwain* [1974] give geometric factors for the three Cherenkov counter channels, which were used with the count rates displayed in Figure 1 to create a similar time history using electron flux rather than count rate. The results are shown in Figure 2. It is satisfying to see that the time histories of the three channels lie on top of one another. The >13 MeV channel observed the same electron flux as the >6 MeV channel because there were no injected electrons between 6 and 13 MeV. The substantially larger pitch angle modulation seen in the >13 MeV channel simply is due to its substantially more narrow angular response. The spin modulation indicates that the newly injected electrons are highly peaked at a local pitch angle of 90°. The steep pitch angle distributions will be discussed further using measurements of newly injected protons. (The spin period of CRRES was approximately 2 rpm.)

A third approach to energy determination is shown in Figure 3, from which the drift frequency is estimated by timing the return of the first echo at the maximum energy where the Cherenkov counter has a detectable response, and the time of the count rate rollover, which will be the minimum energy. These frequencies correspond to 31 and 15 MeV, respectively. This estimate of the minimum energy in the injection of 15 MeV is (just) above the minimum energy estimated by the rollover in the count rates, as expected. The energy spectrum can be seen to be nicely exponential; the

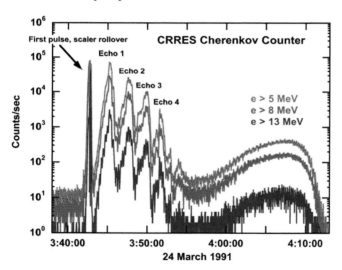

Figure 1. Time history of the count rate in the three Cherenkov counter channels plotted from prior to the first arrival of injected electrons until CRRES moved below the radiation belts. The L value of the s/c is plotted at the top of the plot. The count rate profiles of the three channels in each echo are alike, with the rollover at the peak intensity in a given echo occurring at the same time.

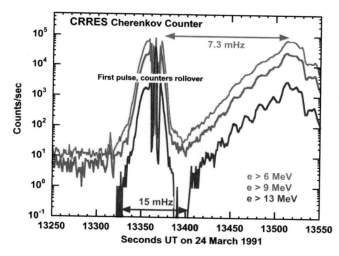

Figure 3. The drift frequency of the newly injected electrons can be estimated by timing returning echoes. Because the Cherenkov counter makes integral measurements, the timing is done by measuring the interval between onset-to-onset (the highest electron energy) and the interval between rollover-to-rollover (the lowest electron energy).

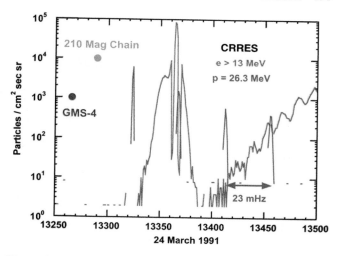

Figure 4. Time history of newly injected protons of 26.3 MeV plotted with that of the >13 MeV electrons. The differential energy channels and narrow angular response of the proton detector result in very sharp echoes. The 26.3 MeV channel was selected because, at that energy, the echo returned when the instrument was again viewing normal to the magnetic field. The shift in time between the flux maxima in the two channels is due to the fact that the instruments were not coaligned on the CRRES satellite. The time that the shock was observed near noon, at GEO (GMS-4), and on the ground (210 Mag Chain) is shown also.

rising count rate is a straight line on the semilog plot (Figure 3) of intensity versus time. Conversion of this time history to the energy dependence of the drift frequency gives an e-fold energy of 2.2 MeV.

Up to this point, the discussion has been confined to the highly relativistic electrons. However, protons were also injected into a new radiation belt. The PROTEL instruments aboard CRRES measured protons from 1 to 100 MeV [*Violet et al.*, 1993]. Figure 4 shows proton drift echoes as seen by PROTEL at 26.3 MeV along with the >13 MeV channel from the Cherenkov counter. The proton sensors had a relatively narrow field of view so it is necessary to select an energy channel such that PROTEL was viewing normal to the magnetic field at the return of each echo. The proton timing is clearer because, instead of being integral channels, they are quite narrow differential channels.

The proton injection also was modeled, by *Hudson et al.* [1995], based upon the same model of *Li et al.* [1993] but using a different source population for the protons.

The angular distribution of the newly injected particles can be seen much more clearly after the particles have spread out around the Earth over many drifts. Figure 5 shows the local pitch angle distribution of three differential proton channels over two CRRES spins. These data were taken approximately 1.4 h after the shock injection. CRRES had passed perigee and was outbound through the new radiation belt region. The newly injected protons are extremely sharply peaked in local pitch angle.

The injection disturbance was seen by GEO spacecraft and ground-based measurements. These non-CRRES data have been discussed in detail by *Araki et al.* [1997]. Figure 6 shows just two of these other timing markers: one from the Japanese METSAT GMS-4, and one of the 210 Meridian

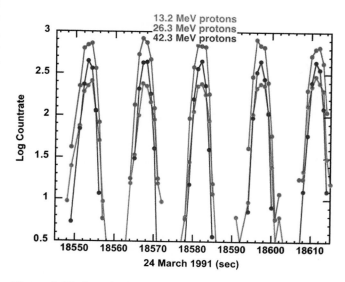

Figure 5. Pitch-angle distribution of energetic protons plotted at 1.4 hours after the shock injection, showing the strong peaking around 90° local pitch angle.

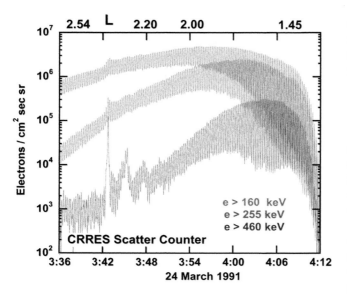

Figure 6. Time history of the relatively low energy electrons plotted over the same time period as the relativistic electrons shown in Figure 1. The highly penetrating relativistic electrons create significant background in the e > 460 keV channel but only a very modest perturbation in the two low-energy channels. This is a typical slot/inner zone traversal, indicating that the shock had not as yet modified the electron spectrum in the post-midnight sector.

magnetometer stations. GMS-4 was just postnoon, and the magnetometer station was at noon.

Li et al. [1993] and *Blake et al.* [1992] suggested that the magnetic pulse started propagating from the magnetopause midafternoon based upon the electron timing. Therefore, both electrons drifting eastward and protons drifting westward had to travel approximately 180° to reach CRRES. Figure 4 shows that the 26.3 MeV proton pulse arrived first, as expected for the injection position and relevant drift frequencies.

This analysis says that all of the injected electrons have energies above 13 MeV and that there are no electrons between 6 and 13 MeV. But what was the electron population at much lower energies? We know that the slot usually contains large electron fluxes at energies below 1 MeV. CRRES also carried the UCSD Scatter Counter from their Pioneer 10 and Pioneer 11 complement. This sensor had three integral electron channels: >160, >255, and >460 keV. Figure 6 shows data from the scatter counter. The slot region is seen to contain the expected large electron fluxes with a steep energy spectrum. The newly injected relativistic electrons create background events in the scatter counter, a large number in the >460 keV channel, but a relatively modest number in the two lower-energy channels because the electron fluxes in those channels was so high. The lower-energy

channels show smooth pitch angle modulation and a rising intensity as CRRES moved to lower *L*. Clearly, the typical slot electron population in the postmidnight sector has not yet been modified by the shock.

3. DISCUSSION

The chance location of CRRES at the time of the shock arrival, indeed, provided unique information, while other assets also gave valuable information [*Araki et al.*, 1997]. Sadly, there was no solar wind monitor to provide an observational understanding of the conditions in the solar wind, which must have been highly unusual.

Looking to the RBSP and what dramatic events might be observed during the mission, one wonders how unusual the event of 24 March 1991 might be. Was this event the space version of a 100 year storm? Has anything even roughly similar ever been seen? *McIlwain* [1963] published a paper giving results from Explorer 15. In this paper, he presents plots of constant intensity in magnetic-dipole coordinates for protons from 40 to 110 MeV. In addition to an intensity peak at *L* = 1.5, there is a second one at *L* = 2.2. This second peak does not fit current models of the Earth's radiation belts, but based upon the CRRES event, one could imagine that sometime in the year or so before the Explorer 15 observations, a similar shock event created a new electron and proton radiation belt. The relativistic electron situation was confused by the Starfish high-altitude nuclear test that injected fission electrons into the magnetosphere that proved to have long lifetimes.

Acknowledgments. The author is indebted to many colleagues for countless discussions over the last two decades about this marvelous geophysical event. Hopefully, we will share a similar situation in the near future using RBSP. This work was supported under NASA contract NA55-01072.

REFERENCES

Araki, T., et al. (1997), Anomalous sudden commencement on March 24, 1991, *J. Geophys. Res.*, *102*(A7), 14,075–14,086, doi:10.1029/96JA03637.

Blake, J. B. (1992), CRRES records creation of new radiation belt, *STEP Newsl.*, *2*(8), 1.

Blake, J. B., W. A. Kolasinski, R. W. Fillius, and E. G. Mullen (1992), Injection of electrons and protons with energies of tens of MeV into L < 3 on 24 March 1991, *Geophys. Res. Lett.*, *19*(8), 821–824, doi:10.1029/92GL00624.

Fillius, R. W., and C. E. McIlwain (1974), Measurements of the Jovian radiation belts, *J. Geophys. Res.*, *79*(25), 3589–3599, doi:10.1029/JA079i025p03589.

Hudson, M. K., A. D. Kotelnikov, X. Li, I. Roth, M. Temerin, J. Wygant, J. B. Blake, and M. S. Gussenhoven (1995), Simulation of proton radiation belt formation during the March 24, 1991 SSC, *Geophys. Res. Lett.*, *22*(3), 291–294, doi:10.1029/95GL00009.

Li, X., I. Roth, M. Temerin, J. R. Wygant, M. K. Hudson, and J. B. Blake (1993), Simulation of the prompt energization and transport of radiation belt particles during the March 24, 1991 SSC, *Geophys. Res. Lett.*, *20*(22), 2423–2426, doi:10.1029/93GL02701.

McIlwain, C. E. (1963), The radiation belts, natural and artificial, *Science*, *142*, 355–361.

Violet, M. D., K. Lynch, R. Redus, K. Riehl, E. Boughan, and C. Hein (1993), The Proton Telescope (PROTEL) on the CRRES spacecraft, *IEEE Trans. Nucl. Sci.*, *40*, 242–245.

J. B. Blake, Space Science Department, The Aerospace Corporation, M2/259, 2350 El Segundo Blvd., Los Angeles, CA 90245-4691, USA. (jbernard.blake@aero.org)

Outer Radiation Belt Flux Dropouts: Current Understanding and Unresolved Questions

D. L. Turner,[1] S. K. Morley,[2] Y. Miyoshi,[3] B. Ni,[4] and C.-L. Huang[5]

One of the most drastic events that occur in the Earth's outer radiation belt is the sudden depletion of relativistic electron flux, known as a flux dropout. Dropouts are characterized by electron fluxes dropping by up to several orders of magnitude over a broad range of energies, L shells, and equatorial pitch angles in only a few hours. Here we provide a review of the historical understanding of these events and the most recent research that has shed light on their true nature. Originally thought to be the result of entirely reversible, adiabatic effects, it is now understood from multi-spacecraft analysis that, in addition to adiabatic variations, dropouts also include a significant amount of true loss of electrons from the system. We provide brief discussions of the current theories explaining true loss during dropouts including loss to the atmosphere due to wave-particle interactions or current sheet scattering, loss by magnetopause shadowing and/or enhanced outward radial transport, and sudden local deceleration due to nonlinear wave-particle interactions. With a detailed review of the most recent studies, we discuss evidence in favor of each of these theories and detail some of the outstanding questions concerning flux dropouts. We finish with a brief discussion of the importance of multispacecraft observations, providing pitch angle–resolved differential energy flux and magnetic field observations from both the equatorial region and low-Earth orbit, for addressing those outstanding questions.

1. INTRODUCTION

Van Allen and Frank [1959] and *Vernov and Chudakov* [1960] independently reported the first in situ observations of relativistic electrons trapped in near-Earth space by the geomagnetic field. Since those preliminary observations, much attention has been devoted to understanding the electron radiation belts, which consist of an inner zone between ~1.5 and ~2 Earth radii (R_E) and an outer zone between ~3 and ~7 R_E around the Earth in the equatorial plane and extending in latitude along magnetic field lines to the Earth's atmosphere (~100 km altitude). Relative to the outer electron radiation belt, the inner belt is stable, with boundaries and intensities changing drastically only over periods of several years or during intense geomagnetic activity [e.g., *Williams and Smith*, 1965; *Pfitzer and Winkler*, 1968]. The outer electron belt, however, is highly dynamic, with flux intensities changing by orders of magnitude and boundaries shifting

[1]Department of Earth and Space Sciences, University of California, Los Angeles, California, USA.

[2]Los Alamos National Laboratory, Los Alamos, New Mexico, USA.

[3]Solar-Terrestrial Environment Laboratory, Nagoya University, Nagoya, Japan.

[4]Department of Atmospheric and Oceanic Sciences, University of California, Los Angeles, California, USA.

[5]Institute for the Study of Earth, Oceans, and Space, University of New Hampshire, Durham, New Hampshire, USA.

Dynamics of the Earth's Radiation Belts and Inner Magnetosphere
Geophysical Monograph Series 199
10.1029/2012GM001310

by thousands of kilometers (in the equatorial plane) over a variety of timescales, from minutes to years (see reviews by *Friedel et al.* [2002] and *Shprits et al.* [2008a, 2008b]). Understanding what drives these drastic variations is of increasing societal importance since the relativistic electrons that make up the outer belt pose a threat to both astronauts and spacecraft [e.g., *Baker et al.*, 2002].

One of the most drastic and sudden variations in the outer radiation belt electron population is the rapid depletion of flux by one or more orders of magnitude over a broad range of energy (i.e., tens of keV up to several MeV) and L shells (often $L > \sim 4$) in only a few hours. These events are known as flux dropouts. Dropouts are often associated with the main phase of geomagnetic storms, though they can occur independent of storm activity as well [e.g., *Morley et al.*, 2010b]. Different criteria have been used to identify flux dropouts in various data sets, many of which are specific to certain orbits or storm time conditions. Here we propose a more general definition of *outer belt flux dropout*: an event in which the flux of trapped electrons decreases by at least a factor of 50 (or by less than $50\times$ if the flux level drops from some significant level to the instrumental background/noise level) as measured at approximately the same L shell, equatorial pitch angle, and magnetic local time (MLT) by the same spacecraft in a period less than or equal to 24 h. This definition is independent of observational location, since it just as easily applies to a spacecraft in a low-altitude, high-inclination low-Earth orbit (LEO) as it does to one in a very high altitude, near-equatorial geosynchronous Earth orbit (GEO). For spacecraft that only provide omnidirectional flux observations, where the equatorial pitch angle is not resolved, an equivalent pitch angle can be assumed (based on some knowledge of the instrument functionality), or the observed fluxes can be coupled to a model pitch angle distribution. This definition is also independent of the mechanism responsible for the dropout, i.e., adiabatic flux modulation or nonadiabatic electron loss from the system (see below). The timescale of 24 h is longer than most dropout observations, which normally occur in only a few hours [e.g., *Morley et al.*, 2010b]; however, 24 h provides sufficient time for a spacecraft to make two observations from approximately the same L shell, equatorial pitch angle, and MLT. The factor of 50 (i.e., 98% reduction of the original flux) is arbitrary, though it is in the range for dropouts observed in previous studies [e.g., *Kim and Chan*, 1997; *Ohtani et al.*, 2009; *Morley et al.*, 2010b]. Also, being general, it is straightforward to apply to a systematic, data-mining search in a variety of electron flux data sets and can be further refined based on the requirements of a specific study. We note, however, that this definition will not capture all rapid loss events, but is instead a simple mechanistic way to identify clear dropout events.

Here we focus entirely on flux dropout events with a review of recent progress in understanding this important outer radiation belt phenomenon. For a more general review of outer belt loss processes, see the work of *Millan and Thorne* [2007]. This paper is laid out as follows. First, we provide a brief background and discussion of the historical understanding of dropout events. Next, we review the works that have drawn conclusions on one or more of the mechanisms thought to result in true loss during dropouts: loss to the atmosphere, loss due to magnetopause shadowing and/or enhanced outward radial transport, and sudden local deceleration. We finish with a summary of the current understanding, outstanding questions, the critical importance of multipoint observations, and some suggestions for future research.

2. HISTORICAL UNDERSTANDING

To understand the dynamics of the Earth's radiation belt particles, and in particular the previous and current work on outer belt dropouts, it is first important to describe the three characteristic, periodic motions that result in trapping and the corresponding adiabatic invariants associated with each (for more detailed explanations, see the works of *Northrop and Teller* [1960], *Roederer* [1970], and *Schulz and Lanzerotti* [1974]). A charged particle moving through the Earth's quasidipolar, inner magnetosphere undergoes three periodic motions resulting from the field geometry: gyromotion, i.e., spiraling around field lines, bounce motion along field lines between mirror points, and drift motion perpendicular to field lines around the Earth. Associated with the gyromotion, the first adiabatic invariant, M, is the relativistic magnetic moment of the particle:

$$M = \frac{(p\sin(\alpha))^2}{2m_0 B} = \frac{E_k(E_k + 2m_0 c^2)\sin^2(\alpha)}{2m_0 c^2 B}, \quad (1)$$

where p is the relativistic momentum, α is the local pitch angle, m_0 is particle rest mass, and B is the local magnetic field strength. The rightmost side of the equation uses the relativistic momentum-energy relation to show M as a function of a particle's kinetic energy, E_k, which is a quantity measured by energetic particle detectors. M is the first adiabatic invariant, since it remains constant so long as no changes to the system, particularly in the form of varying electric and/or magnetic fields, occur on timescales similar to or shorter than the particle's gyroperiod or length scales similar to or smaller than the gyroradius [e.g., *Sergeev and Tsyganenko*, 1982]. The second invariant, associated with the bounce motion, can be expressed as the path integral of parallel momentum between mirror points, and it remains conserved so long as the system does not change faster than the bounce period. To decouple the first and second invariants, it is most useful to work with the following form of the second invariant, K:

$$K = \frac{\oint p_{\parallel} ds}{2\sqrt{2m_0 M}} = \int_{S_m}^{S'_m} \sqrt{B_m - B(s)}\ ds = I\sqrt{B_m}, \quad (2)$$

where B_m is the field strength at the mirror points (S_m and S_m'), and $B(s)$ is the field strength as a function of distance along a field line, and I is another commonly used form of the second invariant. Finally, the third invariant, Φ, is defined as the magnetic flux enclosed by a drift shell, though a more useful quantity is L^*, a dimensionless quantity that is inversely proportional to the third invariant (and thus, also an invariant quantity for a given planetary dipole moment). L^* approximately defines the equatorial crossing point of the drift shell in R_E from the center of the Earth if the field were to relax to a dipolar configuration [Roederer, 1970]:

$$L^* = \frac{2\pi\mu_E}{\Phi R_E}, \quad (3)$$

where μ_E is the dipole moment of the Earth's magnetic field. If the system changes slower than the drift period, then any variations that occur are considered fully adiabatic and are reversible.

Flux dropouts were first thought to occur due to purely adiabatic, reversible effects, as theorized by Dessler and Karplus [1961]. McIlwain [1966] presented the first observational evidence in support of this adiabatic response theory to explain the drastic flux dropouts observed in the outer radiation belt. The concept is that during the main phase of a geomagnetic storm, the ring current becomes enhanced, resulting in a decrease in the total magnetic field strength within it, since the induced field is opposite to that of the Earth. In response to this weaker field, the drift shells of radiation belt particles expand to conserve the third adiabatic invariant. As their drift shells expand away from the Earth into regions of weaker field strength, particles lose energy to conserve their first adiabatic invariant, and since there are exponentially fewer particles at higher energies, the flux of particles observed by a spacecraft over some fixed energy range should drop precipitously during such an event. Afterward, during the recovery phase of a storm when the ring current relaxes to prestorm levels, the drift shells compress inward again, reversing the process and returning observed fluxes to their prestorm level.

Adiabatic effects can also explain some dropout observations at LEO. Tu and Li [2011] showed how adiabatic effects at low altitude are not analogous to those closer to the magnetic equator for the range of L shells spanning the radiation belts. At low altitudes, mirror points move up (i.e., away from the Earth) along field lines in response to the drift shell expansion that occurs during storm main phase. This occurs to conserve the second adiabatic invariant; basically, since the field lines become longer for a drift shell of larger circumference, mirror points move closer together to conserve the path integral of parallel momentum between the mirror points. For a satellite at LEO, this motion of the mirror points brings the spacecraft closer to or fully into the drift and/or bounce loss cones, resulting in a significant reduction of particle flux.

To accurately determine the cause of observed dynamics, it is important to study the electron phase space density (PSD) as a function of the three adiabatic invariants, invoking Liouville's theorem that states that PSD should remain constant along any trajectory in phase space in the absence of external sources or losses. If a dropout was entirely due to adiabatic effects, then the PSD for fixed invariants should remain constant during the entire event. Kim and Chan [1997] used PSD for fixed invariants derived from CRRES [e.g., Vampola et al., 1992] satellite data and studied a storm (minimum Dst of −100 nT) on 2–5 November 1993. They found that while adiabatic effects could account for a significant percentage of the observed flux variation, nonadiabatic effects also played an important role. Li et al. [1997] coined the term "Dst effect" for describing the adiabatic response of electrons during storm time main phase. In their study, they used data from Solar, Anomalous, and Magnetospheric Particle Explorer (SAMPEX) [Baker et al., 1993] in LEO, GPS [Distel et al., 1999], and Los Alamos geosynchronous (LANL-GEO) [Reeves et al., 1997] spacecraft during the same November 1993 event. They found that the flux variations at lower L shells ($L < {\sim}4$) result from combined adiabatic effects and loss to the atmosphere, while the variation at $L \geq 4$ is dominated by true loss, either via precipitation to the atmosphere or drift loss into the magnetopause (i.e., "magnetopause shadowing"). Onsager et al. [2002] examined a series of dropouts in which the flux of >2 MeV electrons at GEO did not recover for up to 1 week following the dropout, while the magnetic field recovered as normal within a couple of days, indicating that nonadiabatic loss must have occurred for these events.

To illustrate this, Figure 1 shows a flux dropout observed by GOES 13 (for details on NOAA-Geostationary Operational Environment Satellites and instrumentation, see <http://www.swpc.noaa.gov/Data/goes.html>) at geosynchronous orbit (GEO) on 6 January 2011. Provided that there are only observations of electron flux and magnetic field strength with no additional analysis, this flux dropout could be interpreted as a result of fully adiabatic effects. Storm

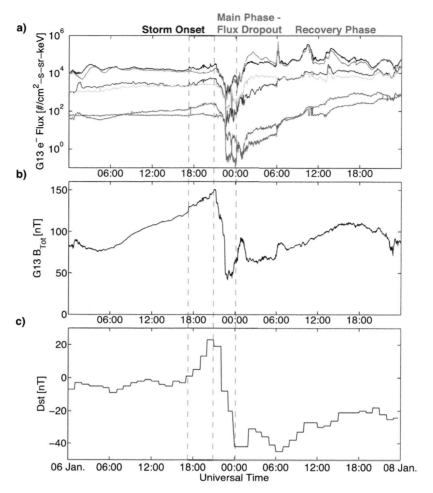

Figure 1. An example flux dropout event. (a) GOES-13 Magnetosphere Electron Detector (MAGED) electron flux observations for 50–60 keV (black: $\alpha_{loc} = 90°$; gray: $\alpha_{loc} = 45°$), 100–200 keV (blue: $\alpha_{loc} = 90°$; cyan: $\alpha_{loc} = 45°$), and 350–600 keV (red: $\alpha_{loc} = 90°$; magenta: $\alpha_{loc} = 45°$). (b) GOES-13 magnetic field strength. (c) Hourly *Dst* index. The dashed, vertical lines indicate different phases of the storm, as labeled at the top of the plots.

onset occurs around 18:00 UT on 6 January, visible in Figure 1c from the positive enhancement in the *Dst* index used to identify geomagnetic storms. Storm main phase begins around 21:00 UT, when the *Dst* falls rapidly to <−40 nT in only a few hours (Figure 1c). During the main phase, note how the magnetic field strength observed by GOES 13 also drops by a factor of ~3 (Figure 1b). Also at this time, the fluxes of energetic electrons observed by GOES 13 drop over a broad range of energy (tens to hundreds of keV) and pitch angle (curves for local $\alpha \approx 45°$ and $\alpha \approx 90°$ are shown). The tens to low hundreds of keV electrons are replenished soon after the main phase due to substorm injections and enhanced convection from a source in the magnetotail, but the relativistic, 350–600 keV electrons seem to return to prestorm levels only as the magnetic field strength recovers, just as

Dessler and Karplus [1961] predicted. So based on these observations alone, one could explain these storm time dynamics as having resulted from purely adiabatic dynamics. However, this assumes that there is an absence of both sources and losses of relativistic electrons, which is not the case during storm activity [e.g., *Reeves et al.*, 2003].

When the fluxes from the 6 January 2011 storm are converted to PSD as a function of the three adiabatic invariants, they reveal a different picture. *Turner et al.* [2012] studied this event and used PSD derived from both GOES and Time History of Events and Macroscale Interactions during Substorms (THEMIS) [*Angelopoulos*, 2008] spacecraft to show that the dropout was nonadiabatic: the PSD for fixed invariants, *M*, *K*, and *L**, decreased by an order of magnitude or more during the main phase when compared to the prestorm levels. As shown by this and many other studies reviewed in this paper, outer radiation belt dropout events most often include true loss of electrons. Thus, the question remains, what is the dominant loss process responsible for nonadiabatic loss during flux dropouts?

3. ATMOSPHERIC LOSS

When radiation belt electrons are able to mirror at altitudes where collisions with atmospheric particles become likely (normally below ~100 km altitude) [*Kennel*, 1969], they are considered to be in the bounce loss cone since collisions should result in loss to the atmosphere within one bounce period. Owing to the offset in the Earth's magnetic dipole, particles have easiest access to the bounce loss cone near the South Atlantic Anomaly (SAA). Quasitrapped particles in a specific range of pitch angle, such that over the course of one drift period they are lost to the atmosphere near the SAA, are said to be in the drift loss cone (see discussion of the bounce and drift loss cones in the work of *Tu et al.* [2010]). *Selesnick* [2006] used a numerical model to evolve the electron distribution as a function of equatorial pitch angle, longitude, and time fit to observations of two storm time dropouts at LEO and found that atmospheric losses due to pitch angle diffusion might be able to deplete the prestorm electron population in only ~1 h. Several mechanisms can result in loss of radiation belt electrons into the Earth's atmosphere via transport of previously trapped electrons into the drift or bounce loss cones, including pitch angle diffusion [e.g., *Selesnick*, 2006; *Shprits et al.*, 2006a; *Albert and Shprits*, 2009], rapid (i.e., in a nondiffusive manner on a timescale much less than the bounce period) scattering by wave-particle interactions [e.g., *Bortnik et al.*, 2008a, 2008b; *Tao et al.*, 2012], and current sheet scattering in the near-Earth magnetotail [e.g., *Sergeev and Tsyganenko*, 1982].

Various plasma wave modes have been identified as important for scattering radiation belt electrons into the atmospheric loss cones. Whistler mode chorus can preferentially accelerate hundreds of keV to MeV electrons [e.g., *Summers et al.*, 1998, 2007a, 2007b; *Horne and Thorne*, 1998; *Subbotin et al.*, 2010; *Xiao et al.*, 2010], and pitch angle scatter, tens to low hundreds of keV electrons into the loss cone [e.g., *Horne et al.*, 2005]. *Horne and Thorne* [2003], showed that at higher latitudes, chorus might also be responsible for scattering MeV electrons into the loss cone, which they speculated might result in the MeV microbursts of precipitation observed by SAMPEX. Individual interactions occur in only milliseconds, though it takes hours to a couple of days for the effects to systematically modulate the entire electron population. Electromagnetic ion cyclotron (EMIC) waves [e.g., *Kangas et al.*, 1986; *Anderson et al.*, 1992; *Anderson*

and Hamilton, 1993; *Summers and Thorne*, 2003; *Usanova et al.*, 2008; *McCollough et al.*, 2010; *Ukhorskiy et al.*, 2010] have also been identified as a potentially important wave for rapidly scattering relativistic electrons during main phase dropouts. EMIC wave growth is strongest in regions with large density gradients, such as the plasmapause and plasmaspheric plumes [*Kozyra et al.*, 1984], and they can rapidly scatter electrons with energy >~500 keV into the loss cones [*Summers and Thorne*, 2003; *Albert*, 2003; *Meredith et al.*, 2003]. Plasmaspheric hiss [e.g., *Bortnik et al.*, 2008a, 2008b] is also important for scattering radiation belt electrons and is thought to be responsible for the slot region between the inner and outer belts [*Lyons and Thorne*, 1973]. However, hiss is not thought to result in dropouts since it only occurs inside the plasmasphere, which becomes highly compressed during storm main phase, and scattering times are too slow [e.g., *Meredith et al.*, 2006; *Summers et al.*, 2007b].

Like the example shown in Figure 2, the SAMPEX spacecraft often observed short-duration (subsecond), high-intensity "microbursts" of precipitating, relativistic electron flux during geomagnetic storms [e.g., *Lorentzen et al.*, 2001a, 2001b; *O'Brien et al.*, 2004]. *Lorentzen et al.* [2001a, 2001b] found that these microbursts generally increased in intensity and moved to lower L shells during the recovery phases of three storms and that the amount of electrons lost during the events represented a significant fraction of the total radiation belt population. *O'Brien et al.* [2004] expanded upon the Lorentzen et al. study and found that the microburst precipitation might be sufficient to empty the prestorm outer

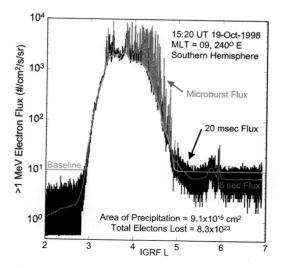

Figure 2. SAMPEX observations of microbursts. The >1 MeV electron fluxes are shown from a single SAMPEX pass through the outer belt on 19 October 1998. The 20 ms fluxes indicated in red were identified as microbursts using an automated algorithm. From *O'Brien et al.* [2004, Figure 1].

radiation belt in approximately 1 day based on globally extrapolating the SAMPEX observations. They also found that the flux intensity of microbursts was 10–100 times stronger during the main phase than during the recovery phase. The occurrence rates of microbursts are much higher during storm main and recovery phases (-100 nT $<$ Dst $<$ -50 nT) in the prenoon MLT sector [*O'Brien et al.*, 2003] and outside the plasmapause [*Johnston and Anderson*, 2010], which is consistent with resonant interactions of particles with whistler mode chorus waves (see statistics of *Meredith et al.* [2001], *Li et al.* [2009]). *Lam et al.* [2010] studied NOAA-Polar Orbiting Environmental Satellites (POES) [*Evans and Greer*, 2004] precipitating fluxes at LEO and found that the >30 keV electron precipitation was most enhanced during periods of substorm activity in the recovery phase of storms, primarily outside of the plasmapause in the dawn MLT sector, which is further evidence that these electrons are scattered by chorus waves. Interestingly, chorus waves are also thought to be a dominant acceleration mechanism of relativistic electrons in the outer radiation belt [e.g., *Chen et al.*, 2007], and thus, it is necessary to quantify both the loss due to microbursts and slower pitch angle diffusion and source due to local acceleration simultaneously to disambiguate the offsetting effects of each.

Occurrence rates of EMIC waves are higher during geomagnetic storms than during quiet times and peak during the main phase [*Erlandson and Ukhorskiy*, 2001; *Halford et al.*, 2010]. Using coincident observations from a NOAA-POES spacecraft and a ground-based observatory at Athabasca, Canada, *Miyoshi et al.* [2008] presented evidence of dual precipitation loss of tens of keV ions and relativistic electrons by EMIC waves that resulted in an isolated proton aurora. Using Antarctic balloon observations of MeV X-rays, *Millan et al.* [2002] showed that the MeV events occurred entirely in the premidnight and dusk MLT sectors, consistent with relativistic electrons being lost to the atmosphere due to interactions with EMIC waves. They concluded that scattering by EMIC waves can serve as a dominant loss mechanism since $\sim 5 \times 10^{25}$ electrons were calculated to be lost from the outer belt to the atmosphere over a period of 8 days compared to $\sim 2 \times 10^{25}$ stably trapped electrons being present over the same range, as estimated from GPS dosimeters. *Meredith et al.* [2003] conducted a statistical study of wave-particle interactions between EMIC waves and outer belt electrons using observations from the CRRES spacecraft. They found that for about 90% of cases, the minimum resonant electron energy was >2 MeV, and the spatial range for lower energies to be resonant was limited to between 13:00 and 18:00 MLT in regions of high plasma density and/or low magnetic field strength. They calculated drift-averaged scattering lifetimes for these electrons to be several hours to a day. *Ukhorskiy et*

al. [2010] showed an event in which the minimum resonant energy dropped to as low as 400 keV due to significant EMIC wave power near the helium ion gyrofrequency. They calculated that these waves could scatter electrons over a wide range of equatorial pitch angles on timescales of only seconds, and they discussed statistical results that suggested the results were not anomalous. *Meredith et al.* [2003] and *Ukhorskiy et al.* [2010] both concluded that rapid loss by EMIC waves might be a dominant loss mechanism during the main phases of geomagnetic storms.

Another mechanism that can cause rapid loss of trapped particles to the atmosphere is current sheet scattering [*Sergeev and Tsyganenko*, 1982]. When the radius of curvature of the field lines in the stretched magnetotail become comparable to the gyroradius of a trapped particle, μ is no longer conserved, and the particle's pitch angle can change such that it is lost to the atmosphere. *Sergeev et al.* [1983] reported observations of isotropic ion precipitation bands on the nightside, which were consistent with the expected results of this mechanism. Concerning radiation belt electrons, *Millan et al.* [2007] compared indirect observations of precipitating electrons from balloon X-ray instrument to trapped fluxes observed in situ with GOES and GPS spacecraft during a flux dropout. They found that the precipitation for electrons with energy of at least 400 keV occurred during a magnetotail field-stretching event and was observed between 19:20 and 22:40 MLT, consistent with electrons drifting into the stretched field region. They found that even though the loss cone population represented $\sim 1\%$ of the equatorial flux, the 2–3 h range in MLT was sufficient to account for the rate of flux decrease observed during the dropout. However, *Green et al.* [2004] argued that this mechanism should be most effective at high L shells and cannot explain dropouts observed down to $L^* \approx 4$.

Two statistical studies of dropout events have also concluded that precipitation to the atmosphere is the dominant loss mechanism. *Green et al.* [2004] examined 52 events identified from GOES observations. They performed a superposed epoch analysis and found that the flux decreases were observed first in the dusk local time sector associated with tail-stretching periods resulting from a partial ring current. By examining the innermost location of the magnetopause, they concluded that this alone could not be responsible for the dropouts at lower L shells, though additional transport mechanisms were not considered (as discussed below). Based on concurrent observations of enhanced electron flux in the loss cones from SAMPEX, they concluded that precipitation to the atmosphere due to wave-particle interactions was likely the cause of the observed nonadiabatic electron losses, though they were unable to identify which type of wave was most likely responsible. *Borovsky and Denton* [2009] studied

124 high-speed stream-driven storms over two solar cycles using 1.1–1.5 MeV flux observations from the LANL-GEO spacecraft. They found that the onset of relativistic electron dropouts coincides with the formation of dense plasmaspheric plumes in the afternoon local time sector. Though they did not include any observations of precipitation into the atmosphere, they concluded that the correlation between the presence of the plumes, which stretch across the outer belt from the plasmapause to the magnetopause in the afternoon sector, and the lack of relativistic electrons at GEO suggests that electron scattering by EMIC waves is the dominant loss mechanism during dropouts.

Interestingly, when statistics of particle precipitation into the atmosphere from NOAA-POES spacecraft are examined for geomagnetic storms, the results do not agree with the conclusions stating that atmospheric loss is the dominant loss mechanism during main phase dropouts. This is an interesting inconsistency, since several of the conflicting results using SAMPEX data were also statistical studies [e.g., *Lorentzen et al.*, 2001a, 2001b; *O'Brien et al.*, 2003]; see more on this in the conclusion. *Horne et al.* [2009] conducted a study of NOAA-POES observations of precipitating and trapped electron populations at LEO during nonhigh-speed stream-driven storms. They found that the precipitation of >300 keV electrons is strongest during the main phase of storms at all MLT in both the Northern and Southern Hemispheres, but the precipitation of >1 MeV is strongest during the recovery phase primarily near the SAA (see Figure 3). They suggested that wave-particle interactions were able to scatter >300 keV electrons directly into the bounce loss cone, but >1 MeV electrons only into the drift loss cone, and the >1 MeV electron precipitation is most enhanced during the recovery phase because it takes 1–2 days for the MeV electron population to replenish themselves and be precipitated by EMIC waves. *Meredith et al.* [2011] conducted a complementary study using the same data set for high-speed stream-driven storms, and they also found that relativistic precipitation was strongest during the recovery phases of the storms. They stressed that there was no evidence of enhanced precipitation of relativistic electrons during the flux dropouts in the main phases of the storms. Consistent with the work of *Horne et al.* [2009], the precipitation of relativistic electrons actually decreased during the main phase compared to that prior to the storm. Recent analysis by *Hendry et al.* [this volume] using POES and AARDDVARK observations also showed that high-speed stream-driven dropouts are not a result of energetic electron precipitation; rather, increased precipitation appears to accompany the recovery of trapped electron fluxes. These results suggest that though atmospheric loss is surely important to outer belt flux variability, some other process may be the dominant loss mechanism effective during main phase flux dropouts.

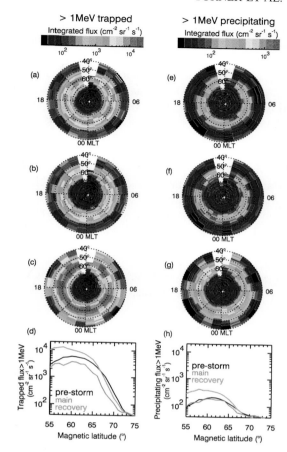

Figure 3. NOAA-Polar-Orbiting Environmental Satellites (POES) (left) trapped and (right) precipitating >1 MeV electron fluxes during different phases of geomagnetic storms. Color maps show Southern Hemisphere averaged fluxes binned by magnetic local time (MLT) and magnetic latitude during (a and e) prestorm, (b and f) main, and (c and g) recovery phases. (d and h) The same fluxes averaged over MLT as a function of magnetic latitude for the three different phases of storms. From *Horne et al.* [2009, Figure 1].

4. LOSS DUE TO OUTWARD RADIAL TRANSPORT

Both the inner and outer boundaries of the radiation belts are sinks that result in loss of particles from the system. Atmospheric losses are the sink at the inner boundary, but the magnetopause ultimately serves as the sink at the outer boundary. If electrons drifting eastward into the dayside magnetosphere are on an open drift shell, i.e., L^* is undefined, they will encounter the magnetopause and be lost into the magnetosheath. When this loss occurs due to the sudden, inward motion of the magnetopause in response to increased dynamic pressure in the solar wind, it is known as magnetopause shadowing. Several studies have proposed theories explaining flux dropouts of outer belt electrons via loss through the magnetopause.

Using test-particle simulations in realistic, time-dependent magnetic fields, *Ukhorskiy et al.* [2006] showed that diamagnetic effects from a partial ring current during the main phase of a storm resulted in nonadiabatic loss of electrons. The diamagnetic effects from the partial ring current caused electron drift shells to expand in such a way that electrons throughout much of the belt ($L > 5$) encountered the magnetopause and were lost from the system. They showed that this process occurred in only a couple of hours after the storm onset. This timescale, range in L, and independence on energy are consistent with observed flux dropout characteristics. *Nishimura et al.* [2007] also used test-particle simulations including electric field and potential distributions determined using Akebono observations. They found that the dawn-dusk electric field was enhanced (up to 6 mV m^{-1} amplitude) during the main phase of a storm, particularly in the range $2 < L < 5$. From the simulation results, they showed how this enhanced electric field could result in rapid outward radial transport and loss of outer belt relativistic electrons through the magnetopause.

Diffusion theory can also explain how rapid outward radial transport can result in a flux dropout throughout the outer radiation belt following magnetopause-shadowing events.

Using a 1-D, radial diffusion model with a *Kp*-driven diffusion coefficient and a time-dependent outer boundary condition at $L^* = 6$, *Brautigam and Albert* [2000] were able to reproduce a main phase dropout of relativistic electron PSD throughout the outer belt. Subsequent studies by *Miyoshi et al.* [2003, 2006] were also able to successfully reproduce main phase dropouts of two different storms using a radial diffusion model with a variable outer boundary condition. *Shprits et al.* [2006b] conducted a detailed study of the radial diffusion model's success at reproducing dropouts. They described how a sudden decrease in the PSD at the outer boundary, which simulates enhanced loss at higher L shells due to magnetopause shadowing, produces a sharp gradient in the radial PSD distribution. Since the time rate of change of PSD due to radial diffusion is dependent on both the diffusion coefficient and the magnitude of the PSD gradient, *Shprits et al.* [2006b] showed how magnetopause shadowing could drive subsequent outward radial diffusion that effectively reduces the PSD throughout most of the outer radiation belt ($L^* > 4$) in only a few hours. Figure 4 shows the results of a simple 1-D radial diffusion simulation with no explicit source or loss terms. The top two plots show the results when the distribution is allowed to diffuse for 5 h using the

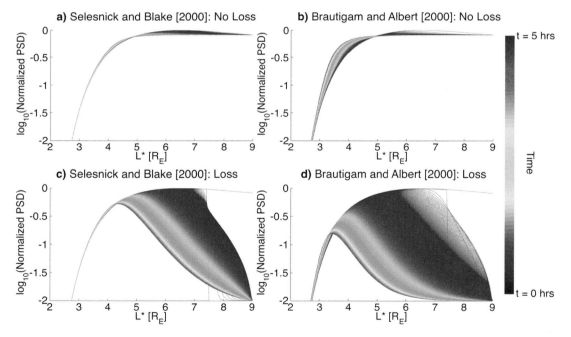

Figure 4. Phase space density (PSD) distributions generated by solving a simple 1-D radial diffusion equation. The diffusion coefficients used are those of (a and c) *Selesnick and Blake* [2000]: $D_{LL} = 1e^{-8} \times L^*10$ (1 day^{-1}) and (b and d) *Brautigam and Albert* [2000]: DLL = $10^{(0.506*Kp-9.325)} \times L^*10$ (1 day^{-1}) with $Kp = 5.0$. To simulate loss due to magnetopause shadowing, the normalized PSD at $7.5 \leq L^* \leq 9$ is dropped to 0.01 shortly after the model starts running (10th time step). This shows that enhanced diffusion can indeed occur in response to a magnetopause-shadowing event and results in a dropout of electrons throughout much of the outer belt in only a few hours.

Selesnick and Blake [2000] and *Brautigam and Albert* [2000] diffusion coefficients, and the bottom two plots show the results when the PSD at $L^* \geq 7.5$ is reduced to 1% of the peak value early in the simulation. This simulates magnetopause shadowing at higher L shells, and the results show how the PSD is reduced throughout much of the outer belt (down to L ~4 or 5) in only a few hours using either form of the diffusion coefficient. It is important to note that, although not all electrons are lost to the outer boundary due to this mechanism, by diffusing out, away from the Earth into weaker field, the particles are decelerated due to conservation of the first and second invariants. *Jordanova et al.* [2008] came to a similar conclusion when modeling the 21 October 2001 storm using the kinetic RAM model [*Jordanova and Miyoshi*, 2005]; they found that although EMIC wave-particle interactions were included in their simulation, the main phase dropout of relativistic electrons at $L > 5$ was best explained due to enhanced outward radial diffusion driven by the PSD decrease observed at higher L.

With these theories introduced, their expected results can be compared to in situ observations. *Morley et al.* [2010a] reported a case in which the GPS constellation observed a dropout of radiation belt electrons over the range $4 < L^* < 6$ in only 2 h. The dropout followed after the passage of a stream interface region, in which high dynamic pressure in the solar wind moved the subsolar magnetopause to $L = 8$. In a follow-up study, *Morley et al.* [2010b] compiled statistical results from 67 stream interfaces using GPS observations of relativistic electron counts at $L^* > 4$. By their nature, stream interface regions are accompanied by high solar wind pressures that result in magnetopause compression events. At all L^*, *Morley et al.* [2010b] found that dropouts of 0.4–1.8 orders of magnitude occurred with a median timescale of ~7 h. Of the 67 events examined, only 3 were not associated with a radiation belt dropout. Based on riometer data, they concluded that microburst loss to the atmosphere due to interactions with chorus may play a role, but only several hours after the dropouts had already occurred. Using inter-calibrated THEMIS and LANL-GEO flux data, *Ni et al.* [2011] analyzed the response of radiation belt electrons to a sudden solar wind pressure enhancement event and reported a clear correlation between observed PSD dropouts and the solar wind pressure pulse, which they attributed to the combination of magnetopause shadowing and outward radial diffusion. *Shprits et al.* [2012] examined radiation belt electron PSD reanalysis results during 200 days in 1990–1991 compiled from CRRES, Akebono, GPS, and LANL-GEO flux observations. They found that 73% of the 25 flux dropouts that occurred during the period were associated with sudden increases in solar wind dynamic pressure. The remaining 27% was split between 15% that were associated

with small or gradual increases in dynamic pressure and 12% (3 of the 25 dropouts) that occurred during relatively steady periods. By reconstructing the radial profiles of the outer belt electron PSD for the entire year for 2002 (solar maximum) based on a combination of a 1-D radial diffusion model and data assimilation of six-satellite, sparse observations, *Ni et al.* [2012] found 59 electron PSD dropout events. Of these, 81% were associated with solar wind dynamic pressure enhancements or pulses. They also identified 41 pressure pulses, of which 68% were associated with PSD dropouts, and 49 pressure enhancements, of which 41% were associated with PSD dropouts. These results indicate that the suddenness of solar wind dynamic pressure increases is an important variable in producing dropouts. However, it should be noted that previous studies [e.g., *Kangas et al.*, 1986; *Anderson and Hamilton*, 1993; *Usanova et al.*, 2008; *McCollough et al.*, 2010] showed that EMIC waves could also be generated during magnetospheric compressions resulting from high solar wind dynamic pressure, so the relationship between pressure pulses and dropouts does not only implicate loss to the magnetopause but may also result in loss to the atmosphere, especially for high-energy (>500 keV) electrons.

Using magnetospheric configurations predicted by the *Tsyganenko and Sitnov* [2005] model (TS05), C.-L. Huang et al. (Magnetopause shadowing signature during electron dropout events, manuscript in preparation, 2012, hereinafter referred to as Huang et al., manuscript in preparation, 2012) found that ~80% of geosynchronous flux dropouts are on open drift paths, a signature of magnetopause shadowing. The dropout events with magnetopause shadowing signatures displayed faster flux decreases (<6 h to flux minimum) compared to events without evidence of magnetopause shadowing. They concluded that magnetopause shadowing could be a major loss mechanism for outer belt electrons, even for small and moderate storms. Huang et al. (manuscript in preparation, 201) also explained the local time dependence of the dropout onsets observed by *Onsager et al.* [2002]. Dropouts at GEO were first observed at duskside MLTs because of the partial ring current effect and particle drift paths being connected to the magnetopause. Subsequently, GEO satellites on the nightside measured dropouts when the nightside field lines stretched and became open contours. Finally, measurements at dayside GEO showed flux depletion after particles experience outward radial transport while still well inside the magnetopause.

Ohtani et al. [2009] also examined flux dropout events at GEO statistically. They found that the day-night asymmetry of the geosynchronous H component of the magnetic field was pronounced during electron loss events, suggesting that MeV electrons on the nightside are very often on open drift paths during dropouts. They concluded that magnetopause shadowing is a plausible loss process of MeV electrons at GEO. *Matsumura et al.* [2011] examined the relationship between the outer edge of the outer radiation belt, as observed by THEMIS and the magnetopause standoff distance during electron loss events observed at GEO. They found that the outer edge of the outer belt moves earthward during the electron loss events at GEO, and the L^* of the outer edge is correlated to the magnetopause standoff distance. Using 3-D test-particle simulations (GEMSIS-RB) [*Saito et al.*, 2010], *Matsumura et al.* [2011] studied this relationship between the outer edge of the outer belt and the magnetopause and concluded that it results from magnetopause shadowing.

Two recent event studies have also concluded that magnetopause shadowing and subsequent rapid outward radial transport is the dominant mechanism for dropouts based on multisatellite observations. *Loto'aniu et al.* [2010] used THEMIS and GOES spacecraft observations and magnetic field measurements from the CARISMA ground network during a dropout event associated with magnetopause shadowing. They showed that the dropout was nonadiabatic and that the ULF wave activity was sufficient to result in diffusion timescales throughout the belt of only a few hours. *Turner et al.* [2012] also used GOES and THEMIS spacecraft

observations and a network of ground magnetometers, though they also included observations of trapped and precipitating electron fluxes from 6 NOAA-POES spacecraft in LEO (shown in Figure 5). For the 6 January 2011 flux dropout, they found that (1) the dropout was nonadiabatic based on both the GOES and THEMIS observations, (2) the dropout started immediately after sudden inward motion of the magnetopause (see Figure 5) and occurred for both ions and electrons within only a few hours over all pitch angles and a broad range in L ($L > 4$), (3) the recovery time was energy dependent, with electrons at tens to low hundreds of keV being replenished quickly after the dropout by injections from the tail and the relativistic population taking around a day or more to recover, and (4) there was insufficient precipitation of relativistic electrons into the atmosphere to explain the loss. Using NOAA-POES data for other dropouts, they were able to show that this lack of precipitation, particularly at higher L shells, is typical. At >800 km POES altitudes, adiabatic effects from a changing mirror point [*Tu and Li*, 2011] are negligible and cannot explain the several orders of magnitude drop in fluxes. Furthermore, using the THEMIS satellite and ground magnetometer network, they directly observed the magnetopause motion into around 7.5 R_E and showed that the ULF activity during the subsequent dropout was globally enhanced over a broad range in L ($L > 4$) over a

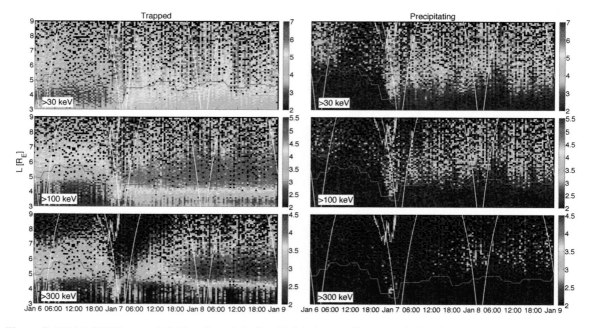

Figure 5. NOAA-POES trapped (left) and precipitating (right) electron flux maps during the 6 January 2011 storm. The logarithm of integral electron fluxes for (top) >30 keV, (middle) >100 keV, and (bottom) >300 keV are binned by L and time and shown in color. The white traces indicate the THEMIS-A orbit track, while the green and gray traces are for model plasmapause L shell and radial distance of the magnetopause, respectively. From *Turner et al.* [2012, Figure 2].

range in frequency that would break the relativistic electrons' third adiabatic invariants and allow for radial diffusion [e.g., *Mathie and Mann*, 2000; *Elkington*, 2006].

5. ALTERNATIVE EXPLANATIONS FOR FLUX DROPOUTS

Other than loss either to the atmosphere or the magnetopause with outward transport, outer radiation belt electron loss may also occur via local, nonadiabatic deceleration due to nonlinear wave-particle interactions with large-amplitude whistler waves. *Albert* [2002] derived analytical expressions for different types of cyclotron-resonant interactions between outer belt electrons and VLF waves; results showed that nonlinear phase bunching could lead to localized electron deceleration. *Bortnik et al.* [2008a, 2008b] conducted relativistic test-particle simulations of electrons interacting with individual chorus elements of fixed amplitudes. They found that low-amplitude waves at low latitudes symmetrically scatter electrons in pitch angle and energy, which agrees with quasilinear diffusion theory. However, when large-amplitude waves were used in simulations, electrons at low latitudes consistently experienced decreases in their equatorial pitch angles and energies, and at high latitudes, a small portion of the electrons became phase-trapped in the wave and experienced nonlinear acceleration in only ~80 ms, which is on the same timescale as a typical interaction with a chorus element. These results are summarized in Figure 6. Large-amplitude whistler waves have been observed in situ [e.g., *Cattell et al.*, 2008; *Cully et al.*, 2008] and display fine structure to their waveforms [e.g., *Santolík et al.*, 2004; *Tao et al.*, 2012]. *Tao et al.* [2012] extended the *Bortnik et al.* [2008a, 2008b] study

to include variable-amplitude waveforms in the test-particle simulations. They found that the chorus subpackets (i.e., waveform) play an important role in the wave-particle interactions, but the effect of the interactions can also result in the majority of electrons being rapidly decelerated. *Hikishima et al.* [2010] showed with self-consistent PIC simulations that the microbursts of tens of keV of electrons are caused by chorus waves.

Being a complex system, it is quite possible that for many flux dropouts, losses occur due to some combination of the above-mentioned processes, as speculated by *Li et al.* [1997]. *Bortnik et al.* [2006] studied a dropout during a large storm on 20 November 2003 using multiple spacecraft including SAMPEX, HEO, ACE, POES, and FAST. They concluded that during the dropout, loss at $L > 5$ occurred primarily to the magnetopause and outward radial transport, while at lower L shells, the loss was primarily to the atmosphere. *Millan et al.* [2010] examined observations of precipitating flux from a NOAA-POES spacecraft and found evidence for strong precipitation only inside of GEO. They concluded that loss to the magnetopause could then explain the flux dropout observed by the GOES spacecraft. *Turner et al.* [2012] also presented NOAA-POES precipitation maps during several dropout events (see supplementary material), and the select cases that exhibited any significant precipitation of >300 keV electrons also revealed that the precipitation was limited to $L < \sim 5$.

6. SUMMARY AND FUTURE WORK

Flux dropouts of outer radiation belt electrons are sudden (occurring in only a few hours), and drastic depletions of the

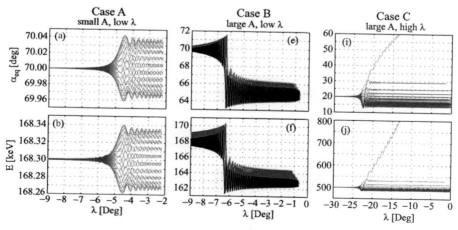

Figure 6. Example results from test-particle simulations of electrons interacting with large-amplitude whistler waves. Case A demonstrates the diffusive nature of interactions with small-amplitude (A) whistlers at low geomagnetic latitude (λ). Cases B and C show the advective nature of results for interactions with large-amplitude waves at low and high latitudes, respectively. From *Bortnik et al.* [2008a, Figure 2].

population of trapped electrons throughout the outer radiation belt over a broad range of energy (from tens of keV to several MeV), pitch angle (from equatorially mirroring to near the bounce loss cone), and L shells (throughout most of the outer belt at $L > \sim 4$). Here we have presented a general definition for dropout events and have reviewed the recent literature on potential mechanisms that can result in nonadiabatic loss during flux dropouts. Previous work has focused primarily on the effects of adiabatic variability and nonadiabatic loss processes such as precipitation to the atmosphere and/or loss to the magnetopause and outward radial transport. The relative importance of sudden electron deceleration by nonlinear wave-particle interactions is still not well quantified or understood.

Both losses to the atmosphere and loss to the outer boundary are seemingly important at different times and locations. Based on many of the studies reviewed here, precipitation loss to the atmosphere can consist of a significant percentage of the trapped population. However, simultaneous, intercalibrated, pitch angle–resolved observations at the same L shell from both high altitudes in the equatorial plane and over a broad range of MLT at LEO are necessary to more accurately quantify the percentage of the trapped population lost to the atmosphere during precipitation events. Recent statistical results [e.g., *Horne et al.*, 2009; *Morley et al.*, 2010b; *Matsumura et al.*, 2011; *Meredith et al.*, 2011; *Ni et al.*, 2012; *Shprits et al.*, 2012; *Huang et al.*, manuscript in preparation, 2012] indicate that the majority (~70%–80%) of dropouts are most likely the result of magnetopause shadowing at higher L shells (beyond GEO for most cases) and subsequent rapid outward radial transport, as depicted in the cartoon in Figure 7. Interestingly, sudden loss at higher L shells due to current sheet scattering and/or ballistic transport due to Shabansky drift orbits on the dayside [e.g., *Ukhorskiy et al.*, 2011] may also perpetuate rapid outward radial transport during active conditions. This mechanism of sudden loss at high L and subsequent rapid outward radial transport can explain both the spatial extent ($L > 4$) and timescale (a few hours) of dropouts [e.g., *Shprits et al.*, 2006b; *Turner et al.*, 2012]. However, around 20%–30% of dropouts occur independent of magnetopause shadowing events [e.g., *Morley et al.*, 2010b; *Matsumura et al.*, 2011; *Ni et al.*, 2012; *Shprits et al.*, 2012; *Huang et al.*, manuscript in preparation, 2012], and it is still unclear what the dominant loss mechanism is for these events.

Many outstanding questions remain concerning flux dropouts in the outer radiation belt. First, for flux dropouts associated with loss due to outward radial transport, what are the comparative contributions of diamagnetic effects from a partial ring current, enhanced dawn-dusk electric fields, and outward radial diffusion in driving rapid outward

radial transport? Another question concerning outward radial transport is what the PSD distribution across the last closed drift shell boundary looks like under normal, steady conditions? Since this boundary on the dayside is ultimately a sink for the system, there must be sufficient source to balance the transport across it under stable conditions. For dropouts that are not associated with sudden losses at higher L shells, what causes the sudden loss of electrons over a broad range of energies and pitch angles throughout the outer belt in only a few hours? This is a particularly interesting and challenging question considering that most electron interactions with the various waves present in the inner magnetosphere are both energy and pitch angle dependent. Also, a major question remains concerning the inconsistency between SAMPEX observations of enhanced relativistic electron flux loss in microbursts during storm main phases [e.g., *Lorentzen et al.*, 2001a, 2001b; *O'Brien et al.*, 2004; *Green et al.*, 2004] and the insufficient relativistic electron loss to the atmosphere observed in global flux maps compiled from NOAA-POES spacecraft [e.g., *Horne et al.*, 2009; *Meredith et al.*, 2011; *Turner et al.*, 2012; *Hendry et al.*, this volume]. The energetic particle instruments on NOAA-POES do not have the temporal resolution to directly observe microbursts, but if microbursts occur during the instrument integration time, those additional particles in the loss cone should be counted and contribute to an individual data bin. To address this puzzling inconsistency, a study of events directly comparing SAMPEX and NOAA-POES data can and should be conducted. Concerning the importance of sudden, local deceleration to flux dropouts, many important questions remain including: what is the occurrence rate of large-amplitude whistlers? What are the spatial range and temporal extent of large-amplitude whistler activity? How does each of these depend on solar wind and geomagnetic conditions? Can these explain the ranges in L ($L > \sim 4$) and pitch angle and timescale (a few hours) of flux dropouts? Finally, dropouts may involve loss due to a combination of these processes, and if so, it is critical to quantify the percent loss due to each for different events as a function of electron energy, pitch angle, and L shell and to identify how these may change during dropouts under different solar wind and geomagnetic conditions.

Of critical importance for addressing all of these questions and the many more that are not listed here is the use of simultaneous, multipoint observations. Pitch angle resolved flux observations from both the equatorial region and LEO are required to quantify the trapped and precipitating populations. For example, to disambiguate offsetting source and loss effects of chorus (loss due to scattering at higher latitudes versus acceleration due to interactions at

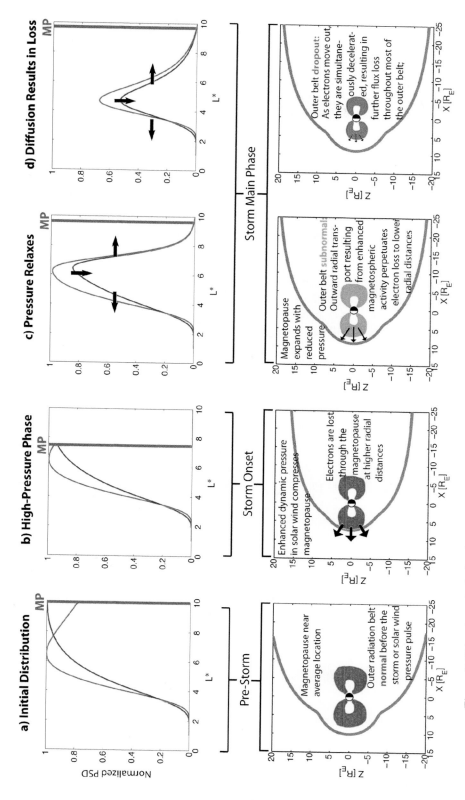

Figure 7. This schematic depicts how sudden loss at higher L due to magnetopause shadowing can result in rapid outward radial diffusion and additional flux loss throughout the majority of the outer belt over a broad range of energy in only a few hours. The results are the same regardless of the initial PSD distribution in L^*, whether it be peaked (blue curve) or positive (red curve). Modified from *Turner et al.* [2012].

lower latitudes) in the dawn/prenoon MLT sector, equatorial PADs and simultaneous observations of the loss cone population at LEO are necessary. Near-equatorial observations are currently provided by THEMIS, LANL-GEO, GOES, and GPS, and they will be supplemented in the near future by the upcoming NASA Radiation Belt Storm Probes (RBSP) and Japanese Energization and Radiation in Geospace [*Miyoshi et al.*, this volume] missions. At LEO, the NOAA-POES spacecraft provide continual coverage of trapped and precipitating fluxes throughout the outer belt over a broad range in MLT, and SAMPEX provided more than two decades of flux observations capable of resolving microbursts (see release of postmission data described in the work of *Baker et al.* [2012]). Future CubeSat missions [e.g., *Li et al.*, this volume; *Spence et al.*, 2012] and the NASA Balloon Array for Radiation-belt Relativistic Electron Losses [*Millan et al.*, 2011] campaign will provide additional measurements of precipitating electrons during the RBSP era, though none of these LEO observations are able to provide pitch angle resolution *within* the bounce loss cone, which would be a major development for improved understanding in future missions. Overall, future work on flux dropouts should incorporate data from as many vantage points as are available to develop a more complete picture of the system and better understanding, ideally to the point of predictability, of which loss mechanisms are active during specific solar wind, geomagnetic, and inner-magnetospheric plasma conditions. With a better understanding of flux dropouts, we will be able to better identify and understand the subsequent source and transport processes responsible for rebuilding, and often enhancing, the relativistic electrons in the Earth's outer radiation belt.

Acknowledgments. The authors would like to thank Vassilis Angelopoulos and Yuri Shprits for insightful conversations and comments. D. L. Turner is thankful for funding support from NASA's THEMIS mission (NASA contract NAS5-02099) and a NASA grant (NNX12AJ55G).

REFERENCES

Albert, J. M. (2002), Nonlinear interaction of outer zone electrons with VLF waves, *Geophys. Res. Lett.*, 29(8), 1275, doi:10.1029/2001GL013941.

Albert, J. M. (2003), Evaluation of quasi-linear diffusion coefficients for EMIC waves in a multispecies plasma, *J. Geophys. Res.*, 108(A6), 1249, doi:10.1029/2002JA009792.

Albert, J. M., and Y. Y. Shprits (2009), Estimates of lifetimes against pitch angle diffusion, *J. Atmos. Sol. Terr. Phys.*, 71, 1647–1652, doi:10.1016/j.jastp.2008.07.004.

Anderson, B. J., and D. C. Hamilton (1993), Electromagnetic ion cyclotron waves stimulated by modest magnetospheric compressions, *J. Geophys. Res.*, 98(A7), 11,369–11,382, doi:10.1029/93JA00605.

Anderson, B. J., R. E. Erlandson, and L. J. Zanetti (1992), A statistical study of Pc 1–2 magnetic pulsations in the equatorial magnetosphere: 1. Equatorial occurrence distributions, *J. Geophys. Res.*, 97(A3), 3075–3088.

Angelopoulos, V. (2008), The THEMIS mission, *Space Sci. Rev.*, 141, 5–34, doi:10.1007/s11214-008-9336-1.

Baker, D. N. (2002), How to cope with space weather, *Science*, 297, 1486–1487.

Baker, D. N., G. M. Mason, O. Figueroa, G. Colon, J. G. Watzin, and R. M. Aleman (1993), An overview of the Solar, Anomalous, and Magnetospheric Particle EXplorer (SAMPEX) mission, *IEEE Trans. Geosci. Remote Sens.*, 31, 531–541.

Baker, D. N., J. E. Mazur, and G. M. Mason (2012), SAMPEX to reenter atmosphere: Twenty-year mission will end, *Space Weather*, 10, S05006, doi:10.1029/2012SW000804.

Borovsky, J. E., and M. H. Denton (2009), Relativistic-electron dropouts and recovery: A superposed epoch study of the magnetosphere and the solar wind, *J. Geophys. Res.*, 114, A02201, doi:10.1029/2008JA013128.

Bortnik, J., R. M. Thorne, T. P. O'Brien, J. C. Green, R. J. Strangeway, Y. Y. Shprits, and D. N. Baker (2006), Observation of two distinct, rapid loss mechanisms during the 20 November 2003 radiation belt dropout event, *J. Geophys. Res.*, 111, A12216, doi:10.1029/2006JA011802.

Bortnik, J., R. M. Thorne, and U. S. Inan (2008a), Nonlinear interaction of energetic electrons with large amplitude chorus, *Geophys. Res. Lett.*, 35, L21102, doi:10.1029/2008GL035500.

Bortnik, J., R. M. Thorne, and N. P. Meredith (2008b), The unexpected origin of plasmaspheric hiss from discrete chorus emissions, *Nature*, 452, 62–66, doi:10.1038/nature06741.

Brautigam, D. H., and J. M. Albert (2000), Radial diffusion analysis of outer radiation belt electrons during the October 9, 1990, magnetic storm, *J. Geophys. Res.*, 105(A1), 291–309.

Cattell, C., et al. (2008), Discovery of very large amplitude whistler-mode waves in Earth's radiation belts, *Geophys. Res. Lett.*, 35, L01105, doi:10.1029/2007GL032009.

Chen, Y., G. D. Reeves, and R. H. W. Friedel (2007), The energization of relativistic electrons in the outer Van Allen radiation belt, *Nat. Phys.*, 3, 614–617, doi:10.1038/nphys655.

Cully, C. M., J. W. Bonnell, and R. E. Ergun (2008), THEMIS observations of long-lived regions of large-amplitude whistler waves in the inner magnetosphere, *Geophys. Res. Lett.*, 35, L17S16, doi:10.1029/2008GL033643.

Dessler, A. J., and R. Karplus (1961), Some effects of diamagnetic ring currents on Van Allen radiation, *J. Geophys. Res.*, 66(8), 2289–2295.

Distel, J. R., S. G. Blair, T. E. Cayton, R. Dingler, F. Guyker, J. C. Ingraham, E. Noveroske, R. C. Reedy, K. M. Spencer, and T. J. Wehne (1999), The combined x-ray dosimeter (CXD) on GPS Block IIR satellites, *Tech. Rep. LA-UR-99-2280*, Los Alamos Natl. Lab., Los Alamos, N. M.

Elkington, S. R. (2006), A review of ULF interactions with radiation belt electrons, in *Magnetospheric ULF Waves: Synthesis and New Directions*, *Geophys. Monogr. Ser.*, vol. 169, edited by K. Takahashi et al., pp. 177–193, AGU, Washington, D. C., doi:10.1029/169GM12.

Erlandson, R. E., and A. J. Ukhorskiy (2001), Observations of electromagnetic ion cyclotron waves during geomagnetic storms: Wave occurrence and pitch angle scattering, *J. Geophys. Res.*, *106*(A3), 3883–3895.

Evans, D. S., and M. S. Greer (2004), Polar orbiting environmental satellite space environment monitor – 2: Instrument descriptions and archive data documentation, *NOAA Tech. Memo. OAR SEC-93*, version 1.4, Space Environ. Lab., Boulder, Colo.

Friedel, R. H. W., G. D. Reeves, and T. Obara (2002), Relativistic electron dynamics in the inner magnetosphere – A review, *J. Atmos. Sol. Terr. Phys.*, *64*, 265–282.

Green, J. C., T. G. Onsager, T. P. O'Brien, and D. N. Baker (2004), Testing loss mechanisms capable of rapidly depleting relativistic electron flux in the Earth's outer radiation belt, *J. Geophys. Res.*, *109*, A12211, doi:10.1029/2004JA010579.

Halford, A. J., B. J. Fraser, and S. K. Morley (2010), EMIC wave activity during geomagnetic storm and nonstorm periods: CRRES results, *J. Geophys. Res.*, *115*, A12248, doi:10.1029/2010JA015716.

Hendry, A. T., C. J. Rodger, M. A. Clilverd, N. R. Thomson, S. K. Morley, and T. Raita (2012), Rapid radiation belt losses occurring during high-speed solar wind stream–driven storms: Importance of energetic electron precipitation, in *Dynamics of the Earth's Radiation Belts and Inner Magnetosphere*, *Geophys. Monogr. Ser.*, doi:10.1029/2012GM001299, this volume.

Hikishima, M., Y. Omura, and D. Summers (2010), Microburst precipitation of energetic electrons associated with chorus wave generation, *Geophys. Res. Lett.*, *37*, L07103, doi:10.1029/2010GL042678.

Horne, R. B., and R. M. Thorne (1998), Potential waves for relativistic electron scattering and stochastic acceleration during magnetic storms, *Geophys. Res. Lett.*, *25*(15), 3011–3014.

Horne, R. B., and R. M. Thorne (2003), Relativistic electron acceleration and precipitation during resonant interactions with whistler-mode chorus, *Geophys. Res. Lett.*, *30*(10), 1527, doi:10.1029/2003GL016973.

Horne, R. B., R. M. Thorne, S. A. Glauert, J. M. Albert, N. P. Meredith, and R. R. Anderson (2005), Timescale for radiation belt electron acceleration by whistler mode chorus waves, *J. Geophys. Res.*, *110*, A03225, doi:10.1029/2004JA010811.

Horne, R. B., M. M. Lam, and J. C. Green (2009), Energetic electron precipitation from the outer radiation belt during geomagnetic storms, *Geophys. Res. Lett.*, *36*, L19104, doi:10.1029/2009GL040236.

Johnston, W. R., and P. C. Anderson (2010), Storm time occurrence of relativistic electron microbursts in relation to the plasmapause, *J. Geophys. Res.*, *115*, A02205, doi:10.1029/2009JA014328.

Jordanova, V. K., and Y. Miyoshi (2005), Relativistic model of ring current and radiation belt ions and electrons: Initial results, *Geophys. Res. Lett.*, *32*, L14104, doi:10.1029/2005GL023020.

Jordanova, V. K., J. Albert, and Y. Miyoshi (2008), Relativistic electron precipitation by EMIC waves from self-consistent global simulations, *J. Geophys. Res.*, *113*, A00A10, doi:10.1029/2008JA013239. [Printed 114(A3), 2009.]

Kangas, J., A. Aiko, and J. V. Olson (1986), Multistation correlation of ULF pulsation spectra associated with sudden impulses, *Planet. Space Sci.*, *34*, 543–553.

Kennel, C. F. (1969), Consequences of a magnetospheric plasma, *Rev. Geophys.*, *7*(1–2), 379–419.

Kim, H.-J., and A. A. Chan (1997), Fully adiabatic changes in storm time relativistic electron fluxes, *J. Geophys. Res.*, *102*(A10), 22,107–22,116.

Kozyra, J. U., T. E. Cravens, A. F. Nagy, E. G. Fontheim, and R. S. B. Ong (1984), Effects of energetic heavy ions on electromagnetic ion cyclotron wave generation in the plasmapause region, *J. Geophys. Res.*, *89*(A4), 2217–2233.

Lam, M. M., R. B. Horne, N. P. Meredith, S. A. Glauert, T. Moffat-Griffin, and J. C. Green (2010), Origin of energetic electron precipitation >30 keV into the atmosphere, *J. Geophys. Res.*, *115*, A00F08, doi:10.1029/2009JA014619.

Li, W., R. M. Thorne, V. Angelopoulos, J. Bortnik, C. M. Cully, B. Ni, O. LeContel, A. Roux, U. Auster, and W. Magnes (2009), Global distribution of whistler-mode chorus waves observed on the THEMIS spacecraft, *Geophys. Res. Lett.*, *36*, L09104, doi:10.1029/2009GL037595.

Li, X., D. N. Baker, M. Temerin, T. E. Cayton, E. G. D. Reeves, R. A. Christensen, J. B. Blake, M. D. Looper, R. Nakamura, and S. G. Kanekal (1997), Multisatellite observations of the outer zone electron variation during the November 3–4, 1993, magnetic storm, *J. Geophys. Res.*, *102*(A7), 14,123–14,140.

Li, X., S. Palo, R. Kohnert, D. Gerhardt, L. Blum, Q. Schiller, D. Turner, W. Tu, N. Sheiko, and C. ShearerCooper (2012), Colorado Student Space Weather Experiment: Differential flux measurements of energetic particles in a highly inclined low Earth orbit, in *Dynamics of the Earth's Radiation Belts and Inner Magnetosphere*, *Geophys. Monogr. Ser.*, doi:10.1029/2012GM001313, this volume.

Lorentzen, K. R., M. D. Looper, and J. B. Blake (2001a), Relativistic electron microbursts during the GEM storms, *Geophys. Res. Lett.*, *28*(13), 2573–2576.

Lorentzen, K. R., J. B. Blake, U. S. Inan, and J. Bortnik (2001b), Observations of relativistic electron microbursts in association with VLF chorus, *J. Geophys. Res.*, *106*(A4), 6017–6027.

Loto'aniu, T. M., H. J. Singer, C. L. Waters, V. Angelopoulos, I. R. Mann, S. R. Elkington, and J. W. Bonnell (2010), Relativistic electron loss due to ultralow frequency waves and enhanced outward radial diffusion, *J. Geophys. Res.*, *115*, A12245, doi:10.1029/2010JA015755.

Lyons, L. R., and R. M. Thorne (1973), Equilibrium structure of radiation belt electrons, *J. Geophys. Res.*, *78*(13), 2142–2149.

Mathie, R. A., and I. R. Mann (2000), A correlation between extended intervals of Ulf wave power and storm-time geosynchronous relativistic electron flux enhancements, *Geophys. Res. Lett.*, *27*(20), 3261–3264.

Matsumura, C., Y. Miyoshi, K. Seki, S. Saito, V. Angelopoulos, and J. Koller (2011), Outer radiation belt boundary location relative to the magnetopause: Implications for magnetopause shadowing, *J. Geophys. Res.*, *116*, A06212, doi:10.1029/2011JA016575.

McCollough, J. P., S. R. Elkington, M. E. Usanova, I. R. Mann, D. N. Baker, and Z. C. Kale (2010), Physical mechanisms of compressional EMIC wave growth, *J. Geophys. Res.*, *115*, A10214, doi:10.1029/2010JA015393.

McIlwain, C. E. (1966), Ring current effects on trapped particles, *J. Geophys. Res.*, *71*(15), 3623–3628.

Meredith, N. P., R. B. Horne, and R. R. Anderson (2001), Substorm dependence of chorus amplitudes: Implications for the acceleration of electrons to relativistic energies, *J. Geophys. Res.*, *106*(A7), 13,165–13,178.

Meredith, N. P., R. M. Thorne, R. B. Horne, D. Summers, B. J. Fraser, and R. R. Anderson (2003), Statistical analysis of relativistic electron energies for cyclotron resonance with EMIC waves observed on CRRES, *J. Geophys. Res.*, *108*(A6), 1250, doi:10.1029/2002JA009700.

Meredith, N. P., R. B. Horne, S. A. Glauert, R. M. Thorne, D. Summers, J. M. Albert, and R. R. Anderson (2006), Energetic outer zone electron loss timescales during low geomagnetic activity, *J. Geophys. Res.*, *111*, A05212, doi:10.1029/2005JA011516.

Meredith, N. P., R. B. Horne, M. M. Lam, M. H. Denton, J. E. Borovsky, and J. C. Green (2011), Energetic electron precipitation during high-speed solar wind stream driven storms, *J. Geophys. Res.*, *116*, A05223, doi:10.1029/2010JA016293.

Millan, R. M., and R. M. Thorne (2007), Review of radiation belt relativistic electron losses, *J. Atmos. Sol. Terr. Phys.*, *69*, 362–377, doi:10.1016/j.jastp.2006.06.019.

Millan, R. M., R. P. Lin, D. M. Smith, K. R. Lorentzen, and M. P. McCarthy (2002), X-ray observations of MeV electron precipitation with a balloon-borne germanium spectrometer, *Geophys. Res. Lett.*, *29*(24), 2194, doi:10.1029/2002GL015922.

Millan, R. M., R. P. Lin, D. M. Smith, and M. P. McCarthy (2007), Observation of relativistic electron precipitation during a rapid decrease of trapped relativistic electron flux, *Geophys. Res. Lett.*, *34*, L10101, doi:10.1029/2006GL028653.

Millan, R. M., K. B. Yando, J. C. Green, and A. Y. Ukhorskiy (2010), Spatial distribution of relativistic electron precipitation during a radiation belt depletion event, *Geophys. Res. Lett.*, *37*, L20103, doi:10.1029/2010GL044919.

Millan, R. M., et al. (2011), Understanding relativistic electron losses with BARREL, *J. Atmos. Sol. Terr. Phys.*, *73*, 1425–1434, doi:10.1016/j.jastp.2011.01.006.

Miyoshi, Y., A. Morioka, H. Misawa, T. Obara, T. Nagai, and Y. Kasahara (2003), Rebuilding process of the outer radiation belt during the 3 November 1993 magnetic storm: NOAA and Exos-D observations, *J. Geophys. Res.*, *108*(A1), 1004, doi:10.1029/2001JA007542.

Miyoshi, Y. S., V. K. Jordanova, A. Morioka, M. F. Thomsen, G. D. Reeves, D. S. Evans, and J. C. Green (2006), Observations and modeling of energetic electron dynamics during the October 2001 storm, *J. Geophys. Res.*, *111*, A11S02, doi:10.1029/2005JA011351.

Miyoshi, Y., K. Sakaguchi, K. Shiokawa, D. Evans, J. Albert, M. Connors, and V. Jordanova (2008), Precipitation of radiation belt electrons by EMIC waves, observed from ground and space, *Geophys. Res. Lett.*, *35*, L23101, doi:10.1029/2008GL035727.

Miyoshi, Y., et al. (2012), The Energization and Radiation in Geospace (ERG) project, in *Dynamics of the Earth's Radiation Belts and Inner Magnetosphere, Geophys. Monogr. Ser.*, doi:10.1029/2012GM001304, this volume.

Morley, S. K., R. H. W. Friedel, T. E. Cayton, and E. Noveroske (2010a), A rapid, global and prolonged electron radiation belt dropout observed with the Global Positioning System constellation, *Geophys. Res. Lett.*, *37*, L06102, doi:10.1029/2010GL042772.

Morley, S. K., R. H. W. Friedel, E. L. Spanswick, G. D. Reeves, J. T. Steinberg, J. Koller, T. E. Cayton, and E. Noveroske (2010b), Dropouts of the outer electron radiation belt in response to solar wind stream interfaces: Global Positioning System observations, *Proc. R. Soc. A*, *466*, 3329–3350, doi:10.1098/rspa.2010.0078.

Ni, B., Y. Shprits, M. Hartinger, V. Angelopoulos, X. Gu, and D. Larson (2011), Analysis of radiation belt energetic electron phase space density using THEMIS SST measurements: Cross-satellite calibration and a case study, *J. Geophys. Res.*, *116*, A03208, doi:10.1029/2010JA016104.

Ni, B., Y. Y. Shprits, R. H. W. Friedel, R. M. Thorne, and M. Daae (2012), Potential correlation between radiation belt electron phase space density distributions and solar wind conditions: Survey of reanalysis results in 2002, paper presented at the Response of the Magnetosphere to High Speed Streams Workshop, Santa Fe, N. M.

Nishimura, Y., A. Shinbori, T. Ono, M. Iizima, and A. Kumamoto (2007), Evolution of ring current and radiation belt particles under the influence of storm-time electric fields, *J. Geophys. Res.*, *112*, A06241, doi:10.1029/2006JA012177.

Northrop, T. G., and E. Teller (1960), Stability of the adiabatic motion of charged particles in the Earth's field, *Phys. Rev.*, *117*, 215–225.

O'Brien, T. P., K. R. Lorentzen, I. R. Mann, N. P. Meredith, J. B. Blake, J. F. Fennell, M. D. Looper, D. K. Milling, and R. R. Anderson (2003), Energization of relativistic electrons in the presence of ULF power and MeV microbursts: Evidence for dual ULF and VLF acceleration, *J. Geophys. Res.*, *108*(A8), 1329, doi:10.1029/2002JA009784.

O'Brien, T. P., M. D. Looper, and J. B. Blake (2004), Quantification of relativistic electron microburst losses during the GEM storms, *Geophys. Res. Lett.*, *31*, L04802, doi:10.1029/2003GL018621.

Ohtani, S., Y. Miyoshi, H. J. Singer, and J. M. Weygand (2009), On the loss of relativistic electrons at geosynchronous altitude: Its dependence on magnetic configurations and external conditions, *J. Geophys. Res.*, *114*, A01202, doi:10.1029/2008JA013391.

Onsager, T. G., G. Rostoker, H.-J. Kim, G. D. Reeves, T. Obara, H. J. Singer, and C. Smithtro (2002), Radiation belt electron flux dropouts: Local time, radial, and particle-energy dependence, *J. Geophys. Res.*, *107*(A11), 1382, doi:10.1029/2001JA000187.

Pfitzer, K. A., and J. R. Winckler (1968), Experimental observation of a large addition to the electron inner radiation belt after a solar flare event, *J. Geophys. Res.*, *73*(17), 5792–5797.

Reeves, G. D., R. Belian, T. Cayton, M. Henderson, R. Christensen, P. McLachlan, and J. Ingraham (1997), Using Los Alamos geosynchronous energetic particle data in support of other missions, in *Satellite-Ground Based Coordination Source Book*, edited by M. Lockwood and H. Opgenoorth, pp. 263–272, Eur. Space Agency, Noordwijk, The Netherlands.

Reeves, G. D., K. L. McAdams, R. H. W. Friedel, and T. P. O'Brien (2003), Acceleration and loss of relativistic electrons during geomagnetic storms, *Geophys. Res. Lett.*, *30*(10), 1529, doi:10. 1029/2002GL016513.

Roederer, J. G. (1970), *Dynamics of Geomagnetically Trapped Radiation*, Springer, New York.

Saito, S., Y. Miyoshi, and K. Seki (2010), A split in the outer radiation belt by magnetopause shadowing: Test particle simulations, *J. Geophys. Res.*, *115*, A08210, doi:10.1029/2009JA014738.

Santolík, O., D. A. Gurnett, J. S. Pickett, M. Parrot, and N. Cornilleau-Wehrlin (2004), A microscopic and nanoscopic view of storm-time chorus on 31 March 2001, *Geophys. Res. Lett.*, *31*, L02801, doi:10. 1029/2003GL018757.

Schulz, M., and L. J. Lanzerotti (1974), *Particle Diffusion in the Radiation Belts*, Springer, New York.

Selesnick, R. S. (2006), Source and loss rates of radiation belt relativistic electrons during magnetic storms, *J. Geophys. Res.*, *111*, A04210, doi:10.1029/2005JA011473.

Selesnick, R. S., and J. B. Blake (2000), On the source location of radiation belt relativistic electrons, *J. Geophys. Res.*, *105*(A2), 2607–2624.

Sergeev, V. A., and N. A. Tsyganenko (1982), Energetic particle losses and trapping boundaries as deduced from calculations with a realistic magnetic field model, *Planet. Space Sci.*, *30*(10), 999–1006.

Sergeev, V. A., E. M. Sazhina, N. A. Tsyganenko, J. A. Lundblad, and F. Soraas (1983), Pitch-angle scattering of energetic electrons in the magnetotail current sheet as the dominant source of their isotropic precipitation into the nightside ionosphere, *Planet. Space Sci.*, *31*(10), 1147–1155.

Shprits, Y. Y., W. Li, and R. M. Thorne (2006a), Controlling effect of the pitch angle scattering rates near the edge of the loss cone on electron lifetimes, *J. Geophys. Res.*, *111*, A12206, doi:10.1029/ 2006JA011758.

Shprits, Y. Y., R. M. Thorne, R. Friedel, G. D. Reeves, J. Fennell, D. N. Baker, and S. G. Kanekal (2006b), Outward radial diffusion driven by losses at magnetopause, *J. Geophys. Res.*, *111*, A11214, doi:10.1029/2006JA011657.

Shprits, Y. Y., S. R. Elkington, N. P. Meredith, and D. Subbotin (2008a), Review of modeling losses and sources of relativistic electrons in the outer radiation belt I: Radial transport, *J. Atmos. Sol. Terr. Phys.*, *70*, 1679–1693, doi:10.1016/j.jastp. 2008.06.008.

Shprits, Y. Y., D. Subbotin, N. P. Meredith, and S. R. Elkington (2008b), Review of modeling of losses and sources of relativistic electrons in the outer radiation belt II: Local acceleration and loss, *J. Atmos. Sol. Terr. Phys.*, *70*, 1694–1713, doi:10.1016/j. jastp.2008.06.014.

Shprits, Y., M. Daae, and B. Ni (2012), Statistical analysis of phase space density buildups and dropouts, *J. Geophys. Res.*, *117*, A01219, doi:10.1029/2011JA016939.

Spence, H. E., et al. (2012), Focusing on size and energy dependence of electron microbursts from the Van Allen radiation belts, *Space Weather*, *10*, S11004, doi:10.1029/2012SW000869.

Subbotin, D., Y. Shprits, and B. Ni (2010), Three-dimensional VERB radiation belt simulations including mixed diffusion, *J. Geophys. Res.*, *115*, A03205, doi:10.1029/2009JA015070.

Summers, D., and R. M. Thorne (2003), Relativistic electron pitch-angle scattering by electromagnetic ion cyclotron waves during geomagnetic storms, *J. Geophys. Res.*, *108*(A4), 1143, doi:10. 1029/2002JA009489.

Summers, D., R. M. Thorne, and F. Xiao (1998), Relativistic theory of wave-particle resonant diffusion with application to electron acceleration in the magnetosphere, *J. Geophys. Res.*, *103*(A9), 20,487–20,500.

Summers, D., B. Ni, and N. P. Meredith (2007a), Timescales for radiation belt electron acceleration and loss due to resonant wave-particle interactions: 1. Theory, *J. Geophys. Res.*, *112*, A04206, doi:10.1029/2006JA011801.

Summers, D., B. Ni, and N. P. Meredith (2007b), Timescales for radiation belt electron acceleration and loss due to resonant wave-particle interactions: 2. Evaluation for VLF chorus, ELF hiss, and electromagnetic ion cyclotron waves, *J. Geophys. Res.*, *112*, A04207, doi:10.1029/2006JA011993.

Tao, X., J. Bortnik, R. M. Thorne, J. M. Albert, and W. Li (2012), Effects of amplitude modulation on nonlinear interactions between electrons and chorus waves, *Geophys. Res. Lett.*, *39*, L06102, doi:10.1029/2012GL051202.

Tsyganenko, N. A., and M. I. Sitnov (2005), Modeling the dynamics of the inner magnetosphere during strong geomagnetic storms, *J. Geophys. Res.*, *110*, A03208, doi:10.1029/2004JA010798.

Tu, W., and X. Li (2011), Adiabatic effects on radiation belt electrons at low altitude, *J. Geophys. Res.*, *116*, A09201, doi:10. 1029/2011JA016468.

Tu, W., R. Selesnick, X. Li, and M. Looper (2010), Quantification of the precipitation loss of radiation belt electrons observed by SAMPEX, *J. Geophys. Res.*, *115*, A07210, doi:10.1029/2009JA014949.

Turner, D. L., Y. Shprits, M. Hartinger, and V. Angelopoulos (2012), Explaining sudden losses of outer radiation belt electrons during geomagnetic storms, *Nat. Phys.*, *8*, 208–212, doi:10.1038/ nphys2185.

Ukhorskiy, A. Y., B. J. Anderson, P. C. Brandt, and N. A. Tsyganenko (2006), Storm time evolution of the outer radiation belt: Transport and losses, *J. Geophys. Res.*, *111*, A11S03, doi:10. 1029/2006JA011690.

Ukhorskiy, A. Y., Y. Y. Shprits, B. J. Anderson, K. Takahashi, and R. M. Thorne (2010), Rapid scattering of radiation belt electrons by storm-time EMIC waves, *Geophys. Res. Lett.*, *37*, L09101, doi:10.1029/2010GL042906.

Ukhorskiy, A. Y., M. I. Sitnov, R. M. Millan, and B. T. Kress (2011), The role of drift orbit bifurcations in energization and loss of electrons in the outer radiation belt, *J. Geophys. Res.*, *116*, A09208, doi:10.1029/2011JA016623.

Usanova, M. E., I. R. Mann, I. J. Rae, Z. C. Kale, V. Angelopoulos, J. W. Bonnell, K.-H. Glassmeier, H. U. Auster, and H. J. Singer (2008), Multipoint observations of magnetospheric compression-related EMIC Pc1 waves by THEMIS and CARISMA, *Geophys. Res. Lett.*, *35*, L17S25, doi:10.1029/2008GL034458.

Vampola, A. L., J. V. Osborne, and B. M. Johnson (1992), CRRES magnetic electron spectrometer AFGL-701-5A (MEA), *J. Spacecr. Rockets*, *29*, 592–594.

Van Allen, J. A., and L. A. Frank (1959), Radiation around the Earth to a radial distance of 107,400 km, *Nature*, *183*, 430–434.

Vernov, S. N., and A. E. Chudukov (1960), Investigation of radiation in outer space, in *Proceedings of the Moscow Cosmic Ray Conference*, vol. III, edited by S. I. Syrovatsky, p. 19, Int. Union of Pure and Appl. Phys., Moscow, Russia.

Williams, D. J., and A. M. Smith (1965), Daytime trapped electron intensities at high latitudes at 1100 kilometers, *J. Geophys. Res.*, *70*(3), 541–556.

Xiao, F., Z. Su, H. Zheng, and S. Wang (2010), Three-dimensional simulations of outer radiation belt electron dynamics including cross-diffusion terms, *J. Geophys. Res.*, *115*, A05216, doi:10.1029/2009JA014541.

C.-L. Huang, Institute for the Study of Earth, Oceans, and Space, University of New Hampshire, Durham, NH 03824, USA.

Y. Miyoshi, Solar-Terrestrial Environment Laboratory, Nagoya University, Nagoya 464-8601, Japan.

S. K. Morley, Los Alamos National Laboratory, Los Alamos, NM 87545, USA.

B. Ni, Department of Atmospheric and Oceanic Sciences, University of California, Los Angeles, CA 90095-1565, USA.

D. L. Turner, Department of Earth and Space Sciences, University of California, Los Angeles, CA 90095-1567, USA. (drew.lawson.turner@gmail.com)

Rapid Radiation Belt Losses Occurring During High-Speed Solar Wind Stream–Driven Storms: Importance of Energetic Electron Precipitation

Aaron T. Hendry,[1] Craig J. Rodger,[1] Mark A. Clilverd,[2] Neil R. Thomson,[1] Steven K. Morley,[3] and Tero Raita[4]

Recent studies have shown how trapped energetic radiation belt electron fluxes rapidly "drop out" during small geomagnetic disturbances triggered by the arrival of a solar wind stream interface (SWSI). In the current study, we use satellite and ground-based observations to describe the significance of energetic electron precipitation (EEP) and direct magnetopause shadowing loss mechanisms, both of which have been suggested as possible causes of the dropouts. Superposed epoch analysis of low-Earth-orbiting Polar-Orbiting Environmental Satellites (POES) spacecraft observations indicate that neither "classic" magnetopause shadowing nor EEP appear able to explain the dropouts. However, SWSI-triggered dropouts in trapped flux are followed ~3 h later by large increases of EEP, which start as the trapped electron fluxes begin to recover and may be signatures of the acceleration process, which rebuilds the trapped fluxes. Ground-based observations indicate typical >30 keV EEP flux magnitudes of ~8 × 10^5 electrons cm^{-2} sr^{-1} s^{-1}. While these are ~10 times larger than the equivalent precipitating fluxes measured by POES, this is consistent with the small viewing window of the POES telescopes.

1. INTRODUCTION

The basic structure of the Van Allen radiation belts was recognized from shortly after their discovery following the International Geophysical Year [*Van Allen and Frank*, 1959; *Hess*, 1968]. Despite being discovered at the dawn of the space age, there are still fundamental questions concerning the acceleration and loss of highly energetic radiation belt electrons [*Thorne*, 2010]; energetic electron fluxes can increase or decrease by several orders of magnitude on timescales of less

than a day. The coupling of the Van Allen radiation belts to the Earth's atmosphere through precipitating particles is an area of intense scientific interest, principally due to two differing research activities. One of these concerns the physics of the radiation belts and primarily the evolution of energetic electron fluxes during and after geomagnetic storms [e.g., *Reeves et al.*, 2003]. The other focuses on the response of the atmosphere to precipitating particles, with a possible linkage to climate variability [e.g., *Turunen et al.*, 2009; *Seppälä et al.*, 2009]. Both scientific areas require increased understanding of the nature of the precipitation, particularly as to the precipitation drivers, as well as the variation of the fluxes and energy spectrum for electrons lost from the outer radiation belts. One area of interest has been the link between the weak geomagnetic storms triggered by the arrival of a high-speed solar wind stream interface (SWSI) and associated "dropouts" in energetic electron fluxes [e.g., *O'Brien et al.*, 2001; *Miyoshi and Kataoka*, 2008; *Morley et al.*, 2010a]. These events highlight the dynamic nature of the outer radiation belt electron fluxes and are the subject of a review in the current monograph [*Turner et al.*, this volume].

The combination of observations from a large number of spacecraft provides a much higher time resolution than

---

[1]Department of Physics, University of Otago, Dunedin, New Zealand.

[2]British Antarctic Survey, Cambridge, UK.

[3]Los Alamos National Laboratory, Los Alamos, New Mexico, USA.

[4]Sodankylä Geophysical Observatory, University of Oulu, Sodankylä, Finland.

Dynamics of the Earth's Radiation Belts and Inner Magnetosphere
Geophysical Monograph Series 199
10.1029/2012GM001299

possible from a single spacecraft, and this has recently provided new understanding into the SWSI-linked dropout events. A statistical study utilizing nine GPS-borne particle detectors and superposed epoch analysis (SEA) around the arrival of 67 SWSIs showed a strong repeatable "signal" of a rapid electron flux dropout [*Morley et al.*, 2010b]. Dropouts occurred in a median time scale of ~7 h, with median electron counts falling by 0.4–1.8 orders of magnitude for all $L*$ (where $L*$ is a magnetic drift invariant) [*Roederer*, 1970]. The SWSI triggered geomagnetic storms with small *Dst* excursions (−40 nT) and small *Kp* increases ($Kp \approx 4$). Indeed, while these events show a storm-like evolution in *Dst* and *Kp*, the majority have maximum *Dst* excursions less than −30 nT and thus are not storms by the "traditional" definitions [e.g., *Loewe and Prölss*, 1997], although we will refer to them as such for want of a better label. The storms started ~6 h before the epoch defined by the expected arrival of the SWSI at the Earth's bow shock nose. While the radiation belt dropouts and recoveries depended on both $L*$ and energy, only 3 of 67 SWSIs did not have an associated dropout in the electron data.

In the current study, we reconsider satellite and ground-based observations to describe the significance of energetic electron precipitation (EEP) during SWSI-driven geomagnetic storms. We make use of the *Morley et al.* [2010b] epochs to allow "like with like" comparisons with the earlier GPS study. Here we show that the EEP occurs well after the dropout has started and confirm the EEP energy dependence reported earlier. From the existing literature, it appears possible that the dropout is caused by magnetopause shadowing. However, this study shows that the SWSI also triggers a geomagnetic storm some hours after the dropout, which enhances wave-particle interactions leading to EEP, as well as the recovery and enhancement of the trapped electron fluxes. We go on to use ground-based EEP observations to determine the likely precipitation flux into the atmosphere. SWSI-driven events are highly repeatable in form and lead to order of magnitude enhancements in EEP up to very high L shells. As such, the EEP will couple efficiently into the polar vortex and may influence the chemistry and dynamics of the polar neutral atmosphere. Recent work has demonstrated that geomagnetic storms produce levels of EEP that are significant in the lower ionosphere [e.g., *Rodger et al.*, 2007] and can significantly alter mesospheric neutral chemistry [*Newnham et al.*, 2011].

2. POES OBSERVATIONS

2.1. SWSI Event Epochs

As noted above, we make use of the epochs given by *Morley et al.* [2010b, Table A.1]. The experimental data we

use in this study, Polar-Orbiting Environmental Satellites (POES) electron counts and subionospheric VLF propagation, are both strongly affected by high-energy protons, which are likely to dominate over any electron response. We, therefore, removed two of the Morley epochs (7 May 2005 and 28 July 2005) from our list as these occurred in the declining phase of solar proton events. We, therefore, have 65 epochs in total from Table 2 of *Morley et al.* [2010b].

In our investigation of the POES spacecraft data described below, we follow the approach of earlier authors and undertake SEA. We explicitly follow the approach of *Morley et al.* [2010b].

2.2. POES SEM-2 Observations

We make use of measurements from the Space Environment Monitor (SEM-2) instrument package onboard the POES, which are in Sun-synchronous orbits at ~800–850 km altitudes. SEM-2 includes the Medium Energy Proton and Electron Detector (MEPED). For a detailed description of the SEM-2 instruments, see the works of *Evans and Greer* [2004]. We use SEM-2 observations from the NOAA-15 through 19 satellites plus the METOP-2 satellite, which also carries an SEM-2. All POES data is available from http://poes.ngdc.noaa.gov/data/ with the full-resolution data having 2 s time resolution. NOAA has developed new techniques to remove the significant low-energy proton contamination from the POES SEM-2 electron observations [e.g., *Rodger et al.*, 2010a], which has been described in Appendix A of *Lam et al.* [2010]. This algorithm is available for download through the Virtual Radiation Belt Observatory (http://virbo.org).

The SEM-2 detectors include integral electron telescopes with energies of >30 keV (e1), >100 keV (e2), and >300 keV (e3), pointed in two directions. Modeling work has established that the 0° telescopes monitor particles in the atmospheric bounce loss cone that will enter the Earth's atmosphere below the satellite when the spacecraft is poleward of $L < 1.5$–1.6 [*Rodger et al.*, 2010b, Appendix A]. Note however, that the 0° telescopes only observe a fraction of the bounce loss cone even when they are directed such that they only measure bounce loss cone fluxes; building on the *Rodger et al.* [2010b] modeling, we find that, in practice, at best 10% of the total bounce loss cone area is sampled, a value that can drop to less than ~2.5% depending on the location. In contrast, the 90° directed MEPED telescope tends to detect electrons with higher pitch angles, i.e., the drift loss cone and trapped electron populations. In practice, once even a small fraction of trapped electron fluxes are visible to the instrument, these will strongly dominate over any fluxes inside a loss cone. This occurs from roughly $L = 4$–5 and above, depending on the location.

In addition to the electron telescopes, the MEPED instrument also includes a number of proton telescopes. The SEM-2 proton detectors also suffer from contamination, falsely responding to electrons with relativistic energies, which can be useful for radiation belt studies [e.g., *Sandanger et al.*, 2007; *Yando et al.*, 2011] outside of solar proton events when significant energetic proton fluxes are present. In particular, the P6 telescope detectors, which are designed to measure >6.9 MeV protons, also respond to either trapped or bounce loss electrons (depending on L shell) with energies in the relativistic range [*Yando et al.*, 2011]. As shown in Figure 8 of *Yando et al.* [2011], the P6 channel plays a complementary role to the e1–e3 channels for detection of relativistic electrons and is sensitive to electrons of energy larger than roughly 1000 keV.

2.3. Superposed Epoch Analysis of MEPED Electrons

Before undertaking SEA, we first combine the POES-reported particle fluxes varying with L and time, using 0.25 L and 1 h time resolution. Observations from inside and around the South Atlantic Magnetic Anomaly are excluded before the measurements are combined. From this dataset, SEA is undertaken using the 65 epochs from *Morley et al.* [2010b]; in addition, another SEA is undertaken with a set of 65 epochs, which are randomly selected from the period January 2004 to December 2008, after having been filtered for solar proton events. This allows an additional check of the significance of any changes observed in the Morley epoch SEA.

2.3.1. POES observations of trapped flux changes. Figure 1 shows the results of this analysis on the 90° directed telescopes, i.e., those primarily showing the effect of SWSI on trapped fluxes. The left-hand panels show the results of analysis using the Morley epochs, while the right-hand side are for the random epochs. The upper panels are the integral flux observations from the >100 keV 90° telescope, the middle panels show the relativistic electron flux variation from the P6 90° telescope, and the lower panels the differential proton flux at 346 keV from the P3 90° telescope.

As a guide, all of the left-hand panels include the result of the SEA applied to GOES >600 keV trapped flux observations for the Morley epochs (green line). The SEA of the >600 keV trapped electrons from geostationary orbits at $L \approx$ 6.6 shows a very similar timing to the dropouts in trapped electron fluxes from the GPS spacecraft, which were also made near the geomagnetic equator (i.e., around geostationary orbit). The GOES SEA has been scaled and shifted to fit on this plot, but involves a flux drop of ~1.5 orders of magnitude, with a recovery to a flux level that is ~50% larger

than the initial levels. The rapid dropout starts at −0.7 day (relative to the epoch time), reaching the deepest point at +0.2 day, with the fluxes having returned to the same level by about +1 day.

The POES data shown in Figure 1 indicates that the observations of the trapped electrons and protons near the bottom of the geomagnetic field lines are very different from that near the geomagnetic equator and different from one another. While there is some evidence for a dropout in the >100 keV electrons, this is only true for L greater than about 6 and starts just around the zero epoch. The >300 keV 90° electron fluxes also include some evidence of a dropout (from L greater than about 5.5; not shown, although similar plots have been produced by *Miyoshi and Kataoka* [2008, Figure 3]), while the >30 keV 90° do not show a clear dropout (not shown). The relativistic electron observations provided by the P6 90° telescope do show a dropout, but this seems to start well after the dropout occurring near the geomagnetic equator. In contrast to all of the electron observations, the trapped differential 346 keV proton fluxes from P3 increase around the time the dropout begins in the electron fluxes near the geomagnetic equator. The same behavior is seen in the 90° directed P1, P2, and P4 detectors (not shown). The significance of the variation shown in the left-hand panels is particularly clear when contrasted for the random epoch SEA results presented in the right-hand panels.

In order to clarify the differences between the electron responses, Figure 2 presents line plots of the changing 90° electron observations from the >100 keV and P6 telescopes at $L = 5.4$. Following the format of *Morley et al.* [2010b], we show the superposed epoch median of the quantity by a black line. The 95% confidence interval for the median is given by the dark gray band. The inner bands mark the interquartile range (medium gray) and the 95% confidence interval about it (light gray). Figure 2 demonstrates the strong differences between the responses of the >100 keV electrons and the relativistic electrons from the P6 channel. During the quiet period before the start of the SWSI-triggered geomagnetic storm, the >100 keV trapped electron fluxes steadily drop. This is reversed at the zero epoch, very close to the time when the electron flux dropouts observed near the geomagnetic equator by GOES and GPS reach their "deepest" extent. In contrast to the >100 keV trapped fluxes, the relativistic electrons exhibit a well-defined dropout, which starts around the same time as seen in the GOES SEA, recovers after 1–1.5 days, and climbs to a slightly higher flux level.

2.3.2. POES observations of precipitating flux changes. Figure 3 shows SEA applied to two of the 0° directed

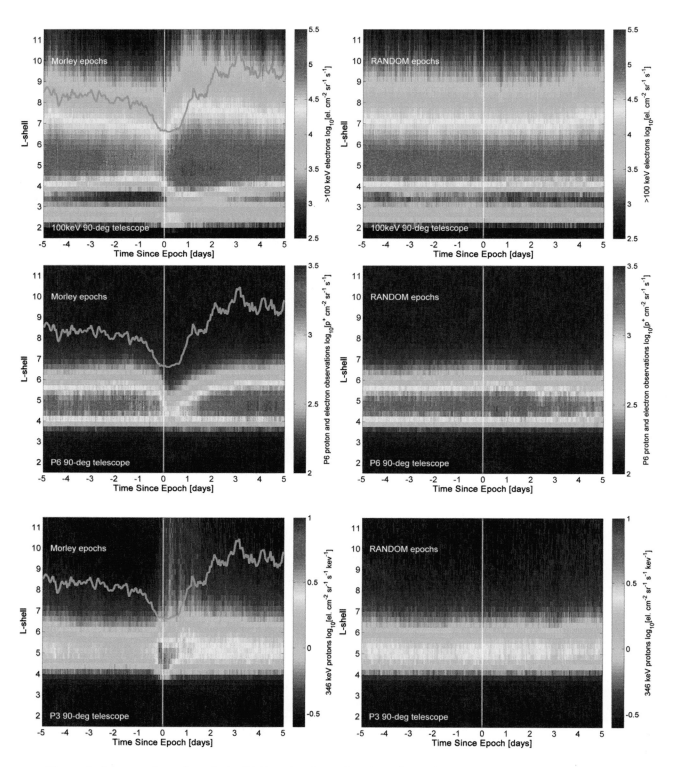

Figure 1. Superposed epoch analysis (SEA) undertaken using the Polar-Orbiting Environmental Satellites (POES) Medium Energy Proton and Electron Detector (MEPED) data. (left) Analysis applied to 65 of the *Morley et al.* [2010b] epochs. (right) A set of random epochs. (top) The integral flux observations from the >100 keV 90° telescope. (middle) Counts from the P6 90° telescope (which responds to relativistic electrons). (bottom) Differential proton flux at 346 keV from the P3 90° telescope. As a guide to the eye, all of the left-hand panels have the result of the Morley SEA applied to GOES >600 keV trapped flux observations (green line).

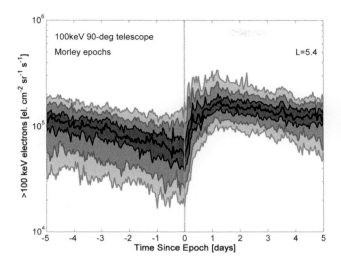
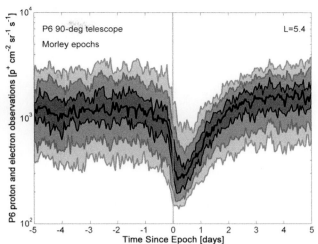

Figure 2. SEA of the POES 90° telescope >100 keV and P6-measured relativistic trapped electron fluxes at $L = 5.4$. The superposed epoch median of the quantity is given by a black line. The 95% confidence interval for the median is given by the dark gray band. The medium gray bands mark the interquartile range and the 95% confidence interval about it (light gray).

telescopes, i.e., those telescopes that detect a portion of the electrons that precipitate into the atmosphere. The format of Figure 3 is otherwise the same as Figure 1. The top panels of this figure show the variation of the >100 keV 0° telescope. By comparison with the random epoch analysis shown on the right-hand side, it is apparent that 4–5 days before the SWSI arrives, the magnitude of >100 keV precipitation is "normal" and then steadily decreases by ~0.5 order. This is likely to be linked to the "calm before the storm," intervals of unusually calm geomagnetic activity, which have been previously reported [e.g., *Clilverd et al.*, 1993]. Very shortly before the zero epoch, the >100 keV flux begins to increase by nearly 2 orders of magnitude (but only slightly more than 1 order of magnitude larger than normal conditions). This peak of precipitation is ~+0.3 days after the zero epoch, around the time the GPS- and GOES-observed electron dropouts are at their deepest. The most significant EEP stretches from $L = 5$ to $L = 8.5$, although there is a clear increase to at least $L = 14$. The EEP decays slowly over the 5 days after the epoch to roughly normal levels. Similar patterns occur with the >30 and >300 keV EEP (not shown) [*Meredith et al.*, 2011, Figure1]. In contrast, however, the relativistic electron flux from the 0° P6 telescope does not display a decrease before the SWSI arrives (i.e., no "calm" in relativistic EEP) and exhibits a small decrease in precipitation magnitude during the peak timing of the >100 keV EEP, lasting perhaps 1–2 days.

3. CONSISTENCY WITH LOSS MECHANISMS

As noted in the introduction, the existing literature has identified three possible causal mechanisms to explain the GPS-

observed dropout in trapped electron fluxes: (1) magnetopause shadowing, (2) EEP into the atmosphere due to wave-particle interactions, and (3) outward diffusion through the magnetopause. The POES SEA described in the previous section allows us to make conclusions as to the validity of the first two of these loss mechanisms. Note that the monograph in which this paper appears also contains a review covering the major loss mechanisms associated with dropouts [*Turner at al.*, this volume].

3.1. Mechanism 1: Magnetopause Shadowing

As previously noted by *Morley et al.* [2010a], our existing understanding is that the loss timescales possible from the EEP or outward diffusion are not fast enough to explain the dropouts as observed. As such, magnetopause shadowing, sometimes termed "magnetopause encounters" may appear the more likely candidate. This mechanism involves radiation belt particles drifting around the Earth, encountering the magnetopause boundary and being swept away by the solar wind and permanently lost. Characteristics of particle losses by magnetopause shadowing are (1) pitch angle independence of losses of particles on a given drift shell, such that losses would be expected for both high pitch angle particles, which spend most of their time near the geomagnetic equator, and low pitch angle particles, which mirror near the top of the atmosphere and (2) independence of particle charge, mass, or energy, such that electrons or protons that are drifting around the Earth on the same L shell (but in opposite directions) will encounter the magnetopause and, hence, be lost.

On this basis, one would expect the dropouts of electrons observed in the trapped electron fluxes near the geomagnetic

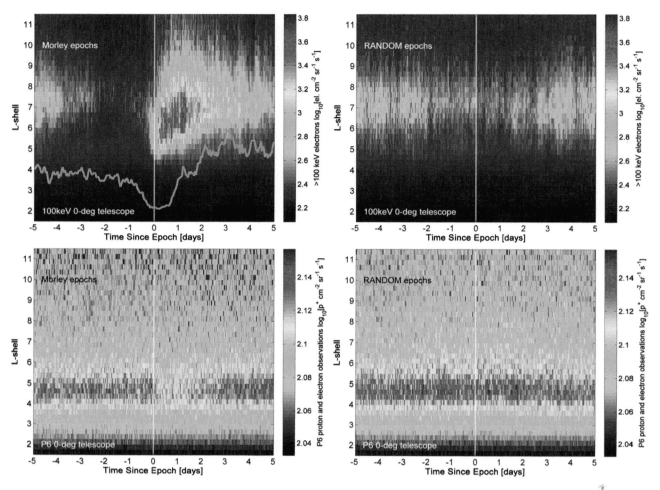

Figure 3. SEA of 0° directed MEPED telescopes in the same format as Figure 1. (top) Integral flux observations from the >100 keV 0° telescope. (bottom) Counts from the P6 0° telescope (which responds to relativistic electrons).

equator by the GPS spacecraft to also be seen in both the trapped electron and proton fluxes measured by the POES low-Earth-orbiting spacecraft. As reported in section 2.3.1, neither of these conditions hold, as the >30 and >100 keV trapped electron fluxes do not show the dropouts reported by satellites near the geomagnetic equator. In addition, the trapped proton fluxes increase rather than decrease during the dropouts. This suggests that direct magnetopause encounters cannot be used to explain the electron flux dropouts. Note that a similar argument was previously employed by *Green et al.* [2004] who contrasted the observed losses of protons and electrons to exclude magnetopause encounters as the dominant loss mechanism.

3.2. Mechanism 2: EEP Into the Atmosphere

As shown in Figure 3 and discussed in section 2 above, the >30, >100, and >300 keV 0° electron telescopes (which

measure part of the bounce loss cone) do show significant increases in EEP, but starting at the point that the dropout is at its deepest point and beginning to recover. In contrast, the relativistic electron fluxes measured by the P6 0° telescope show a small decrease in EEP at the same time. Clearly, these observations are not consistent with EEP as the primary mechanism to explain the dropouts. Indeed, it is possible that the opposite is true, that the EEP is the signature of wave-particle-driven acceleration processes, which serve to reverse the electron flux dropouts [e.g., *Thorne*, 2010, and references within].

4. AARDDVARK OBSERVATIONS

Subionospheric VLF propagation detects precipitation due to changes in the ionization number density at altitudes around the lower D region boundary. As the VLF waves propagate beneath the ionosphere in the Earth-ionosphere waveguide, the EEP-induced ionization produces changes in the received amplitude and phase. Owing to the low attenuation of VLF subionospheric propagation, the EEP-modified ionospheric region may be far from the transmitter or the receiver. As the received subionospheric amplitude is the sum of multiple propagation modes, the response to changes in the waveguide is often complex, and both increases and decreases in amplitude are possible when increased ionization occurs in the waveguide (see, for example, Figure 4 of *Rodger et al.* [2012]). The response also depends on the solar zenith angle along the path. As a result of these factors, subionospheric VLF is not particularly suitable for analysis through superposed epoch. We have, therefore, checked individual paths for a set of specific events to confirm the occurrence of significant EEP.

In this study, we make use of narrow-band subionospheric VLF data received at Churchill (CHUR, 58.75°N, 265.1°E, $L = 7.6$) and Sodankylä, Finland (SGO, 67.4°N, 26.4°E, $L = 5.3$). Both these receivers are part of the Antarctic-Arctic Radiation-belt Dynamic Deposition VLF Atmospheric Research Konsortia (AARDDVARK) [*Clilverd et al.*, 2009]. While the AARDDVARK observations have subsecond time resolution, we will restrict ourselves to 1 min median values to describe the overall transmitter operations. Figure 4 shows the transmitter-receiver great circle paths (GCP), which have been monitored by the Churchill and Sodankylä receivers for at least some part of the time period considered.

The AARDDVARK data was manually examined for evidence of EEP around the time of the Morley epochs. The process was as follows: for each Morley epoch, AARDDVARK data plots were made for all the transmitters monitored by the Churchill and Sodankylä receivers. Data for 4 days both before and after the epoch day were plotted. The days before the epoch were included primarily to construct a quiet-day curve (QDC) to provide comparisons with the epoch day. POES observations show that EEP levels are low immediately before the SWSI epoch, which should allow a good AARDDVARK QDC to contrast with the epoch day. Figure 5 shows examples of the data examined in this way. The upper panel of this figure presents the received amplitude of the GVT 22.1 kHz transmitter received at Sodankylä ($2.5 < L < 5.3$) around the Morley epoch at 06:30 UT on 28 May 2008. Observations on the days before the epoch day are plotted in gray, the epoch day in black. At this time of year, most of the GVT-SGO GCP is sunlit throughout the day, although the transmitter end of the path will have a nighttime ionosphere from ~20 to 04 UT. In general, subionospheric VLF propagation is more sensitive to EEP for nighttime rather than daytime ionospheric conditions [*Rodger et al.*, 2012], due to the extremely large D region energy input from the Sun during the day. On this day, there is a very clear precipitation-induced decrease in the received amplitude starting about 1 h after the epoch (the epoch being marked by the vertical line) continuing through to ~11 UT where the amplitude clearly returns to the QDC defined by the previous days. In this time, the amplitude initially

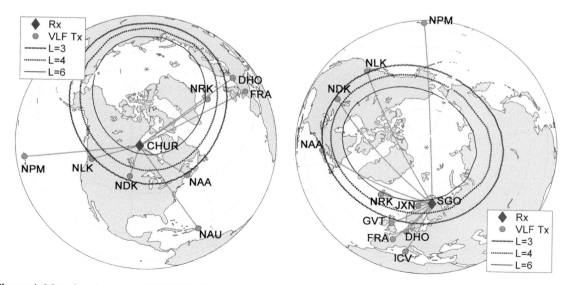

Figure 4. Map showing the AARDDVARK receivers at Churchill and Sodankylä (diamonds) and the VLF transmitters monitored by these receivers (circles). This map also indicates the great circle propagation paths between the transmitter and receiver, as well as a number of fixed L shell contours evaluated at 100 km altitude.

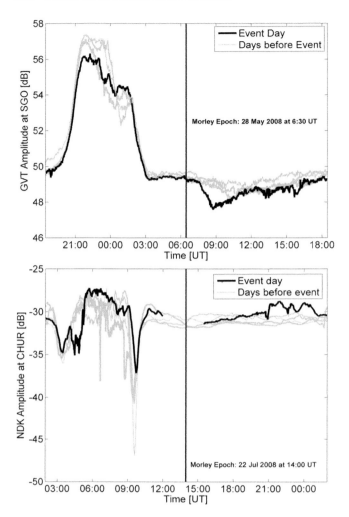

Figure 5. Examples of AARDDVARK observations made around the time of the Morley epochs (the epoch time is marked by the vertical lines in the center of the plots. Observations on the days before the epoch day are plotted in gray; the epoch day is in black. (top) GVT-Sodankylä amplitudes for the epoch at 6:30 UT on 28 May 2008. (bottom) NDK-Churchill amplitudes for the epoch at 14:00 UT on 22 July 2008.

decreases by ~1.3 dB, after which it returns to near-QDC levels. In the hours following, there are several subsequent examples of likely precipitation periods (i.e., ~13.5 and 15.5 UT), both of which have quite small amplitude changes in comparison with the first precipitation period.

The lower panel of Figure 5 presents the received amplitude of the NDK 25.2 kHz transmitter received at Churchill ($2.8 < L < 7.4$) around the Morley epoch at 14:00 UT on 22 July 2008. Once again, there is a long period when the entire path is sunlit, from ~12 to 02 UT. Unfortunately, the transmitter was not operating for a few hours around the time

of the epoch. However, the amplitude on the event day is well behaved from ~2 h after the epoch, showing a steady rise from 15.5 to 20.75 UT, followed by three broad bursts of precipitation at 21.5, 23, and 0.25 UT on the following day.

Of the 65 Morley epochs that we studied, there were 7 epochs for which there were no Sodankylä AARDDVARK observations either on the epoch day or on one of the days immediately before the epoch, leaving 58 epochs to examine. The Churchill AARDDVARK receiver was not installed until May 2007, halfway through the period containing the Morley epochs and was also not operating from December 2007 through May 2008. As a result, only 14 epochs were able to be examined in the Churchill data, although most of these epochs are also represented in Sodankylä observations. We classified data as showing evidence of EEP if an obvious deviation from the QDC could be seen concurrently on at least two different transmitter-receiver paths; this was to ensure the deviations we were seeing were, indeed, due to EEP, and not through random fluctuations in the AARDD-VARK data.

We performed the above analysis on both the aforementioned Morley epochs. For the 67 Morley epochs, 2 epochs were removed due to solar proton activity, 4 were removed as neither receiver was operating, and 15 epochs were removed as there was no transmitter-receiver path with a good QDC. Of the remaining 46 epochs, 34 of these showed clear signs of EEP across multiple paths (i.e., 74%). This confirms the riometer- and satellite-based observations of significant EEP occurring during the SWSI dropouts and also provides us with an additional data set in order to determine the magnitude of the EEP entering the atmosphere.

5. AARDDVARK MODELING

For the next step, we returned to the Morley epochs, focusing on the paths that had a well-behaved QDC, again concentrating on times when the path is sunlit. We then focused on only three subionospheric transmitter-receiver paths; NAA 24.0 kHz to Churchill and GVT to Sodankylä, both of which are relatively short paths, which span a limited magnetic local time range, and the rather long path from NAA to Sodankylä. There is a significant amount of variability in the observed amplitude changes; this is hardly surprising given the large variability in the magnitude of the EEP from event to event evidenced from the interquartile range (not shown). However, we are still in a position to establish "typical" amplitude changes for the subionospheric VLF-observed SWSI-associated precipitation. These are shown in Table 1.

In order to determine the typical magnitude of the EEP triggered by the SWSI, we follow the modeling approach

Table 1. Summary of Ground-Based Instrument Responses During the SWSI-Triggered Geomagnetic Storms[a]

	ΔVLF obs. (dB)	ΔVLF calc. (dB)	ΔCNA calc. (dB)	EEP Flux
NAA-CHUR	+2.0	+2.0	1.41	1×10^6
GVT-SGO	−1.5	−1.1	1.05	4×10^5
NAA-SGO	+2.5	+2.5	1.35	9×10^5
	ΔCNA obs. (dB)	ΔVLF calc. (dB)	ΔCNA calc. (dB)	EEP Flux
NAA-CHUR	1.25	+1.7	1.25	8×10^5
GVT-SGO	1.25	−0.9	1.25	8×10^5
NAA-SGO	1.25	+2.4	1.25	8×10^5

[a]The top half of the table examines the energetic electron precipitation (EEP) values necessary to best reproduce the subionospheric VLF observations from this study (ΔVLF obs.), while the lower half examines the EEP values necessary to best reproduce the *Morley et al.* [2010b]-reported riometer observations (ΔCNA obs.). In each case, the calculated change of the other ground-based instrument response is shown. The EEP values listed are >30 keV electron fluxes with units of electrons cm^{-2} sr^{-1} s^{-1}.

outlined in the work of *Rodger et al.* [2012]. Here our goal is to determine the fluxes, which will lead to the changes in VLF amplitude shown in Table 1. In addition, *Morley et al.* [2010b] reported that the SWSI-associated radiation belt dropouts were linked to increases in riometer-measured absorption of "cosmic noise," which is expected due to increases in the ionization number density in the ionospheric D and E regions caused by EEP. A SEA of riometer data found that the change in cosmic noise absorption (ΔCNA) in Canadian and European instruments peaked at ~1.25 dB in the period 3–6 h following the epoch [*Morley et al.*, 2010b]. Thus, our modeling goal is to reproduce both the subionospheric VLF changes as well as those from the riometer SEA.

For each VLF transmitter-receiver path, we take a modeling point midway along the path and use a combination of International Reference Ionosphere (IRI-2007) [online from http://omniweb.gsfc.nasa.gov/vitmo/iri_vitmo.html] and typical D region electron-density profiles determined for high latitudes at noon [*Thomson et al.*, 2011]. We model the SWSI-associated EEP signature in ground-based data using 10 keV to 2.0 MeV precipitating electrons with an energy spectra determined by the POES SEA observations. During the peak precipitation period, the >30, >100, and >300 keV precipitating fluxes are best fitted terms of a power law where the slope (scaling exponent, k) is −3.5. Otherwise, our modeling techniques follow that described by *Rodger et al.* [2012].

As shown in Table 1, a relatively small range of EEP flux magnitudes will reproduce the ground-based instrument responses observed during the SWSI-triggered geomagnetic storms. The top half of the table examines the EEP values necessary to best reproduce the subionospheric VLF amplitude change observations (ΔVLF obs.), and shows the predicted change in riometers absorption (ΔCNA calc.) predicted for that EEP flux striking the atmosphere at the midpoint of that path. Although the lower bound of the EEP

was assumed to be 10 keV (to more accurately capture the riometer responses), we report the >30 keV EEP flux magnitude to allow direct comparison with the POES 0° telescope observations, given below. Table 1 shows there is very good agreement between the modeled and predicted VLF responses (ΔVLF calc.) for the paths NAA-CHUR and NAA-SGO with >30 keV EEP flux magnitudes of 9–10 × 10^5 electrons cm^{-2} sr^{-1} s^{-1} and slightly lower quality matching for the GVT-SGO path, where −1.1 dB is the largest negative amplitude change we can produce (compare, a typical change of −1.5 dB observed) for a >30 keV EEP flux magnitude of 4 × 10^5 electrons cm^{-2} sr^{-1} s^{-1}. These EEP are calculated to produce riometer absorption changes, which are similar to those reported (1–1.4 dB). The lower half of Table 1 examines the EEP values necessary to best reproduce the *Morley et al.* [2010b] reported peak riometer observations (ΔCNA obs. of 1.25 dB), and contrasts the ΔVLF calc. predicted for these fluxes. In all cases, despite the different undistorted ionospheric electron density-height profiles and neutral atmospheric parameters, the typical observed ΔCNA is reproduced by an EEP flux magnitude of ~8 × 10^5 electrons cm^{-2} sr^{-1} s^{-1}, with relatively small differences in the ΔVLF calc. relative to those observed.

6. DISCUSSION

The >30 keV EEP flux magnitude determined from Table 1 should be contrasted with that found in the SEA of the POES precipitating electrons. The peak in the median >30 keV POES 0° telescope fluxes is ~7 × 10^4 electrons cm^{-2} sr^{-1} s^{-1}, with the 95% confidence interval for the median spanning ~4 × 10^4 to 1 × 10^5 electrons cm^{-2} sr^{-1} s^{-1}. Clearly, this is approximately one order of magnitude smaller than the EEP determined in section 5 from the ground-based measurements. The difference is significant; if the EEP flux was 7 × 10^4 electrons cm^{-2} sr^{-1} s^{-1}, the riometer absorption change would be only 0.27 dB. It is not unexpected that the POES-reported 0° telescope flux is a fraction of that in the bounce loss cone and striking the atmosphere. As noted in section 2.2, the POES SEM-2 0° telescope only samples a fraction of the loss cone, with 10% being a common "best case."

Note that the typical SWSI-triggered >30 keV electron precipitation flux of 8×10^5 electrons cm^{-2} sr^{-1} s^{-1} determined from the ground-based instruments should be considered a large electron precipitation event, although with a softer energy spectra than the $k = -2$ spectra reported as typical by *Clilverd et al.* [2010]. A >30 keV precipitation flux of 2.2×10^6 electrons cm^{-2} sr^{-1} s^{-1} would occur if the entire electron flux stored in a $L = 6.5$ flux tube was precipitated out in a 10 min period, with the population calculated using the European Space Agency Space Environmental Effects-1 radiation belt model [*Vampola*, 1996]. In practice, the POES observations indicate that SWSI-triggered geomagnetic storm have roughly constant precipitation fluxes with values similar to those of the peak level for ~1.5 days. We speculate that this is evidence that the acceleration process, which "rebuilds" the energetic electron fluxes after the dropout, also produces electron precipitation, with a significant fraction of the accelerated electrons being lost into the atmosphere.

7. SUMMARY AND CONCLUSIONS

In this study, we have examined satellite- and ground-based observations to describe the significance of EEP during SWSI-driven geomagnetic storms, focusing on the *Morley et al.* [2010b] epochs to allow "like with like" comparisons with the earlier study focused primarily upon GPS observations. The SEA of the low-Earth-orbiting POES satellite observations confirm that SWSI-driven geomagnetic storms are strongly associated with large EEP events. However, the EEP only becomes significant at the time that the dropout is at its deepest point and is starting to recover, such that EEP cannot be used to explain the observed dropouts in trapped energetic radiation belt electrons for any energy range. Our observations are more suggestive of the opposite phenomena, where the EEP is the signature of wave-particle-driven acceleration processes, which serve to reverse the electron flux dropouts.

Previous studies have suggested that magnetopause shadowing may be the primary reason for the rapid dropouts. Our SEA is, however, not consistent with a simple model of direct magnetopause shadowing causing the losses. In particular, we found that the trapped proton fluxes increased rather than decreased during the dropouts, while the classic direct magnetopause shadowing explanation that would predict this mechanism would be independent of particle charge, mass, or energy, such that electrons or protons, which are

drifting around the Earth on the same L shell (but in opposite directions), will encounter the magnetopause and, hence, be lost.

Ground-based observation of subionospheric VLF propagation from the AARDDVARK network has been used to confirm the POES observations of large EEP events generated by the SWSI-triggered storms. For the epochs for which there were data available and well-defined QDCs, 74% of the Morley epochs showed evidence of EEP occurring, producing amplitude changes of several decibels. The EEP was observed typically ~3 h after the Morley epochs. The AARDDVARK observations were combined with riometer measurements made for the *Morley et al.* [2010b] epochs in order to model the magnitude of the EEP occurring in these events. The very high levels of agreement in the modeling, which involved multiple instruments, and multiple transmitter-receiver paths, indicates a strong probability that the >30 keV EEP flux magnitude has a value close to 8×10^5 electrons cm^{-2} sr^{-1} s^{-1}. This is ~11 times larger than the >30 keV EEP flux reported by the 0° directed >30 keV electron telescope measurements made onboard POES, which is expected as the POES telescopes only view ~10% of the bounce loss cone.

Acknowledgments. A.T.H., C.J.R., M.A.C., and N.R.T. were supported by the New Zealand Marsden Fund. The research leading to these results has also received funding from the European Union Seventh Framework Programme (FP7/2007-2013) under grant agreement 263218.

REFERENCES

Clilverd, M. A., T. D. G. Clark, A. J. Smith, and N. R. Thomson (1993), Observation of a decrease in midlatitude whistler-mode signal occurrence prior to geomagnetic storms, *J. Atmos. Terr. Phys.*, *55*, 1479–1485, doi:10.1016/0021-9169(93)90113-D.

Clilverd, M. A., et al. (2009), Remote sensing space weather events: Antarctic-Arctic Radiation-belt (Dynamic) Deposition-VLF Atmospheric Research Konsortium network, *Space Weather*, *7*, S04001, doi:10.1029/2008SW000412.

Clilverd, M. A., C. J. Rodger, R. J. Gamble, T. Ulich, T. Raita, A. Seppälä, J. C. Green, N. R. Thomson, J.-A. Sauvaud, and M. Parrot (2010), Ground-based estimates of outer radiation belt energetic electron precipitation fluxes into the atmosphere, *J. Geophys. Res.*, *115*, A12304, doi:10.1029/2010JA015638.

Evans, D. S., and M. S. Greer (2004), Polar orbiting environmental satellite space environment monitor – 2: Instrument descriptions and archive data documentation, *NOAA Tech. Memo. OAR SEC-93*, version 1.4, Space Environ. Lab., Boulder, Colo.

Green, J. C., T. G. Onsager, T. P. O'Brien, and D. N. Baker (2004), Testing loss mechanisms capable of rapidly depleting relativistic electron flux in the Earth's outer radiation belt, *J. Geophys. Res.*, *109*, A12211, doi:10.1029/2004JA010579.

Hess, W. N. (1968), *The Radiation Belt and Magnetosphere*, Blaisdell, London, U. K.

Lam, M. M., R. B. Horne, N. P. Meredith, S. A. Glauert, T. Moffat-Griffin, and J. C. Green (2010), Origin of energetic electron precipitation >30 keV into the atmosphere, *J. Geophys. Res.*, 115, A00F08, doi:10.1029/2009JA014619.

Loewe, C. A., and G. W. Prölss (1997), Classification and mean behavior of magnetic storms, *J. Geophys. Res.*, 102(A7), 14,209–14,213, doi:10.1029/96JA04020.

Meredith, N. P., R. B. Horne, M. M. Lam, M. H. Denton, J. E. Borovsky, and J. C. Green (2011), Energetic electron precipitation during high-speed solar wind stream driven storms, *J. Geophys. Res.*, 116, A05223, doi:10.1029/2010JA016293.

Miyoshi, Y., and R. Kataoka (2008), Flux enhancement of the outer radiation belt electrons after the arrival of stream interaction regions, *J. Geophys. Res.*, 113, A03S09, doi:10.1029/2007JA012506.

Morley, S. K., R. H. W. Friedel, T. E. Cayton, and E. Noveroske (2010a), A rapid, global and prolonged electron radiation belt dropout observed with the Global Positioning System constellation, *Geophys. Res. Lett.*, 37, L06102, doi:10.1029/2010GL042772.

Morley, S. K., R. H. W. Friedel, E. L. Spanswick, G. D. Reeves, J. T. Steinberg, J. Koller, T. Cayton, and E. Noveroske (2010b), Dropouts of the outer electron radiation belt in response to solar wind stream interfaces: Global positioning system observations, *Proc. R. Soc. A*, 466(2123), 3329–3350, doi:10.1098/rspa.2010.0078.

Newnham, D. A., P. J. Espy, M. A. Clilverd, C. J. Rodger, A. Seppälä, D. J. Maxfield, P. Hartogh, K. Holmén, and R. B. Horne (2011), Direct observations of nitric oxide produced by energetic electron precipitation into the Antarctic middle atmosphere, *Geophys. Res. Lett.*, 38, L20104, doi:10.1029/2011GL048666.

O'Brien, T. P., R. L. McPherron, D. Sornette, G. D. Reeves, R. Friedel, and H. J. Singer (2001), Which magnetic storms produce relativistic electrons at geosynchronous orbit?, *J. Geophys. Res.*, 106(A8), 15,533–15,544.

Reeves, G. D., K. L. McAdams, R. H. W. Friedel, and T. P. O'Brien (2003), Acceleration and loss of relativistic electrons during geomagnetic storms, *Geophys. Res. Lett.*, 30(10), 1529, doi:10.1029/2002GL016513.

Rodger, C. J., M. A. Clilverd, N. R. Thomson, R. J. Gamble, A. Seppälä, E. Turunen, N. P. Meredith, M. Parrot, J.-A. Sauvaud, and J.-J. Berthelier (2007), Radiation belt electron precipitation into the atmosphere: Recovery from a geomagnetic storm, *J. Geophys. Res.*, 112, A11307, doi:10.1029/2007JA012383.

Rodger, C. J., M. A. Clilverd, J. C. Green, and M. M. Lam (2010a), Use of POES SEM-2 observations to examine radiation belt dynamics and energetic electron precipitation into the atmosphere, *J. Geophys. Res.*, 115, A04202, doi:10.1029/2008JA014023.

Rodger, C. J., B. R. Carson, S. A. Cummer, R. J. Gamble, M. A. Clilverd, J. C. Green, J.-A. Sauvaud, M. Parrot, and J.-J. Berthelier (2010b), Contrasting the efficiency of radiation belt losses caused by ducted and nonducted whistler-mode waves from ground-based transmitters, *J. Geophys. Res.*, 115, A12208, doi:10.1029/2010JA015880.

Rodger, C. J., M. A. Clilverd, A. J. Kavanagh, C. E. J. Watt, P. T. Verronen, and T. Raita (2012), Contrasting the responses of three different ground-based instruments to energetic electron precipitation, *Radio Sci.*, 47, RS2021, doi:10.1029/2011RS004971.

Roederer, J. G. (1970), *Dynamics of Geomagnetically Trapped Radiation*, Springer, New York.

Sandanger, M., F. Søraas, K. Aarsnes, K. Oksavik, and D. S. Evans (2007), Loss of relativistic electrons: Evidence for pitch angle scattering by electromagnetic ion cyclotron waves excited by unstable ring current protons, *J. Geophys. Res.*, 112, A12213, doi:10.1029/2006JA012138.

Seppälä, A., C. E. Randall, M. A. Clilverd, E. Rozanov, and C. J. Rodger (2009), Geomagnetic activity and polar surface air temperature variability, *J. Geophys. Res.*, 114, A10312, doi:10.1029/2008JA014029.

Thomson, N. R., C. J. Rodger, and M. A. Clilverd (2011), Daytime *D* region parameters from long-path VLF phase and amplitude, *J. Geophys. Res.*, 116, A11305, doi:10.1029/2011JA016910.

Thorne, R. M. (2010), Radiation belt dynamics: The importance of wave-particle interactions, *Geophys. Res. Lett.*, 37, L22107, doi:10.1029/2010GL044990.

Turner, D. L., S. K. Morley, Y. Miyoshi, B. Ni, and C.-L. Huang (2012), Outer radiation belt flux dropouts: Current understanding and unresolved questions, in *Dynamics of the Earth's Radiation Belts and Inner Magnetosphere*, Geophys. Monogr. Ser., doi:10.1029/2012GM001310, this volume.

Turunen, E., P. T. Verronen, A. Seppälä, C. J. Rodger, M. A. Clilverd, J. Tamminen, C. F. Enell, and T. Ulich (2009), Impact of different energies of precipitating particles on NOx generation in the middle and upper atmosphere during geomagnetic storms, *J. Atmos. Sol. Terr. Phys.*, 71, 1176–1189, doi:10.1016/j.jastp.2008.07.005.

Vampola, A. L. (1996), Outer zone energetic electron environment update, *P.O. #151351*, Eur. Space Agency, Noordwijk, The Netherlands.

Van Allen, J. A., and L. A. Frank (1959), Radiation measurements to 658,300 kilometers with Pioneer IV, *Nature*, 184, 219–224.

Yando, K., R. M. Millan, J. C. Green, and D. S. Evans (2011), A Monte Carlo simulation of the NOAA POES Medium Energy Proton and Electron Detector instrument, *J. Geophys. Res.*, 116, A10231, doi:10.1029/2011JA016671.

M. A. Clilverd, British Antarctic Survey, High Cross, Madingley Road, Cambridge CB3 0ET, England, UK. (macl@bas.ac.uk)

A. T. Hendry, C. J. Rodger, and N. R. Thomson, Department of Physics, University of Otago, P.O. Box 56, Dunedin, New Zealand. (ahendry@physics.otago.ac.nz; crodger@physics.otago.ac.nz; n_thomson@physics.otago.ac.nz)

S. K. Morley, Los Alamos National Laboratory, Bikini Atoll Road, SM 30, Los Alamos, NM 87545, USA. (smorley@lanl.gov)

T. Raita, Sodankylä Geophysical Observatory, University of Oulu, Sodankylä, Finland. (tero.raita@sgo.fi)

Background Magnetospheric Variability as Inferred From Long Time Series of GOES Data

David J. Thomson

Department of Mathematics and Statistics, Queen's University, Kingston, Ontario, Canada

Examination of the background state of the equatorial magnetosphere in multi-year time series of magnetic fields, relativistic electrons, protons, and solar X-rays from the GOES 10 satellite shows that fluctuations in these series are dominated by the normal modes of the Sun. The magnetic fields at the ACE spacecraft and GOES are coherent. The power spectrum contains many narrow line components superimposed on a smooth background. Many of the lines are split by the Earth's rotation and orbit. At p mode, or Pc5, frequencies, these can explain wave packets. Between 100 and 500 μHz, the lower ULF band, the line components usually contain more than one half the power and are strong enough that all 11 singlets of the $l = 5$, $n = -1$ g mode are detected above the 99.9% significance level in GOES > 2 MeV electrons. The uncorrected center frequency is estimated to be 383.812 μHz with an average synodic splitting of 908 ± 18 nHz, implying that the Sun has a rapidly rotating core.

1. INTRODUCTION

In this paper, I analyze data from the GOES 10 geostationary satellite with emphasis on the long-term, low-frequency background processes, that is, on what is present in the magnetosphere most of the time. The analysis concentrates on extremely low frequencies, ≤500 μHz or periods of 2000 s and longer. This is the lower part of the magnetospheric ULF band, approximately "Pc6." This frequency band includes about 98% of the power in the magnetic fields. Flares, coronal mass ejections (CMEs), substorms, and similar disruptive transient events are of minor interest except as a motivating factor. A large part of this background consists of high-Q discrete modes that appear to be forced by the normal modes of the Sun, specifically solar pressure or p modes, and gravity (buoyancy) or g modes.

Most of my previous work has been on data collected either below the magnetosphere or above it. Coherence, between these disparate kinds of data [*Thomson et al.*, 2007, hereinafter referred to as P1; *Thomson*, 2007], imply that the common parts of these widely separated data sources must pass through the magnetosphere and, consequently, should be seen there as well. My expertise is in spectrum estimation, and what I see in the power spectra of magnetospheric data raises serious doubts about much of what I see in papers on the magnetosphere. Thus, I hope that this work will give a different way to look at the dynamics of the magnetosphere that may help explain some of its many puzzling features.

Canton's observations in the 1750s and subsequent work has confirmed that geomagnetic data are nonstationary. When analyzing nonstationary time series, there is always a conflict between time and frequency resolution. One desires good time resolution to isolate discrete events that can have serious implications for technology and the general field of space weather. However, when the source of the data is the Sun, with its 10 million normal modes [*Harvey*, 1995], one also needs exceptional frequency resolution. Here in choosing frequency resolution as a primary objective, I am looking

Dynamics of the Earth's Radiation Belts and Inner Magnetosphere
Geophysical Monograph Series 199
Published in 2012 by the American Geophysical Union.
10.1029/2012GM001318

at something closer to "space climate" as opposed to "space weather." The two are obviously not independent.

A similar conflict lies at the heart of our understanding. One commonly describes the underlying physics as partial differential equations, a form where everything is primarily a function of time and position, but then one invokes boundaries and symmetries to describe the solution of these PDEs as normal mode expansions, which are often a form of frequency decomposition. Normal mode seismology has given most of our detailed knowledge about the interiors of both the Earth and the Sun. Terrestrial normal mode seismology developed during the 1950s, and following the great Chilean earthquake in May 1960, many normal modes were identified. The differences between the predicted and observed mode frequencies were resolved by Backus and Gilbert's invention of inverse theory (see chapter 1 of *Dahlen and Tromp* [1998]). Helioseismology began with *Leighton's* [1960] discovery of solar p modes, but there was little development before about 1975 (see chapter 5.5 of *Stix* [2004]). A similarly slow process, compounded by even greater observational difficulties, appears to be taking place in magnetospheric physics. *Chapman and Bartels'* [1940] book, the magnum opus of geomagnetism before Dungey's work, does not appear to include "mode" in its extensive index, whereas the analysis of "modes" occupies much of chapter 5 of *Dungey* [1958]. Use of modal expansions has continued, and the predecessor to the present volume, *Magnetospheric ULF Waves* (the work of *Takahashi et al.* [2006] contains two chapters whose titles include "eigenmodes") [*Wright and Mann*, 2006; *Lee and Takahashi*, 2006]. The review [*Menk*, 2011] continues this spirit, but notes some recent work where previously accepted explanations are now deemed inadequate. Even more recently, section 5 of *Hartinger et al.* [2012] excludes several mechanisms because they are inconsistent with Time History of Events and Macroscale Interactions during Substorms (THEMIS) observations. A major purpose of this paper is to emphasize that analysis, identification, and refinement of magnetospheric modes cannot be done without consideration of helioseismology.

Our previous work showed that numerous frequencies obtained by spectrum analysis of time series of interplanetary magnetic fields and particle densities agree closely with those predicted for solar gravity, g, and pressure, p mode oscillations (see the work of *Thomson et al.* [1995], hereinafter referred to as TML). These observations were made at 1 AU by various spacecraft and satellites and beyond 1 AU by Ulysses and Voyager II. *Roberts et al.* [1996] immediately objected that the line spectral features of these modes, which we call discrete modes, would be destroyed by turbulence both in the corona and also by the additional dynamical processing expected in the solar wind during its transit from the source regions to the point of observation. We responded to these and other objections given by *Thomson et al.* [1996, 2001]. Further, in the work of P1, we showed (1) an example, Figure 28, where the magnitude-squared-coherence (MSC) between magnetic fields at ACE and GOES 10 exceeded 0.95 in the solar p mode band (the probability of observing such a high MSC at a given frequency by chance is less than $\sim 4 \times 10^{-10}$, and there are three peaks of this size in the MSC, plus many more with MSC $\gtrsim 0.9$) and (2) high canonical coherences between the vector magnetic fields at ACE and Ulysses in Figure 18, and between ACE and the South Pole magnetometer in Figure 30. Further information on the ACE-Ulysses coherences is given by *Thomson et al.* [2000]. Thus, as mentioned earlier, one sees coherent solar signals both above and below the magnetosphere implying that they should be seen in the magnetosphere as well.

Furthermore, simulations of propagation through the corona [*Dmitruk et al.*, 2004] and interplanetary space [*Ghosh et al.*, 2009] show that discrete modes and turbulence can coexist. Simulations by *Fujita et al.* [2011] show that periodic ULF disturbances in the solar wind couple into the magnetosphere. The Ghosh et al. simulation was done using three space dimensions in contrast to that of *Roberts et al.* [1996], where only a single space dimension was used. The conditions for turbulence to exist [*Kim and Powers*, 1978] are that waves exist at three frequencies related by $f_1 + f_2 = f_3$, with the corresponding wavenumbers simultaneously satisfying $k_1 + k_2 = k_3$. With a single space dimension, these are always true, and such simulations must result in turbulence.

Following this introduction, there is a brief description of the data used plus an example showing why filtering must be done carefully to prevent aliasing, which is followed by section 3 on the background. This section sets out two hypotheses on the nature of the solar wind and its interaction with the magnetosphere: the conventional one, where the solar wind is assumed to be turbulent, and an alternative one, where the solar wind is assumed to be commonly dominated by the normal modes of the Sun, and the turbulent background is a fossil relic of the solar convection zone. This does not change the presence of the various cavity and field line resonances in the magnetosphere itself [see, e.g., *Kivelson*, 2006], but does make their identification more difficult. Section 4 gives a concise summary of multitaper spectrum estimates with emphasis on their statistical properties. Section 5 describes coherence estimates and applies them between the magnetic fields on ACE and GOES. Section 6 gives evidence for mode splitting by ± 1 cycle d^{-1} and ± 1 cycle yr^{-1} in GOES data. Section 7 addresses mode identification. It is shown that the same frequencies detected by the GOLF helioseismology instrument on SOHO are found in GOES data. The signal-to-noise ratio is better in much of the

GOES data than it is in GOLF data, and this permits *identification* of solar gravity modes. As an example, the $l = 5$, $n = -1$ solar *g* mode at 383.812 µHz is shown where *all* 11 singlets are detected above the 99.9% level. The synodic splitting of this mode is 904 ± 18 nHz, implying that the Sun has a rapidly rotating core. Finally, discussion and conclusions are given in section 8.

2. DATA AND EXPLORATORY DATA ANALYSIS

Most of the data analyzed here are from GOES 10, spanning the time interval 1999 to 2004 and concentrating on a few series: the three components of the magnetic fields, the >2 MeV electrons; the 15 to 40 MeV protons (P4); and, to a minor extent, the 1 to 8 Å solar X-rays (XL). Magnetic field data from the ACE spacecraft are also used. The first problem is to determine a suitable sampling rate.

As an exploratory study, the sum of the spectra of the three magnetic field components from GOES 10 for the years 1999, 2000, and 2001 using 1 min resolution was computed. This was done on 100 20 day blocks and 2136 1 day blocks. Both sets used a nominal 50% overlap, omitting some blocks having extensive missing data. Multitaper spectrum estimates, described in section 4, with time-bandwidths of four and six tapers were used with the results summarized in Figure 1. The average spectra (the black and green curves) agree well. The long dashed magenta line has an arbitrary amplitude and a slope of $-5/3$. It is approximately followed by the average over $\sim 10\ \mu\text{Hz} \le f \le \sim 5000\ \mu\text{Hz}$, approximately the acoustic cutoff frequency of the Sun [*Jiminéz*, 2006]. The prominent peak centered at 11.57 µHz is the 1 cycle d^{-1} component and dominates the spectrum except for occasional very low frequency effects. It is not strictly periodic and, as will be seen later in section 6, modulates the statistical structure of the data. Its apparent width, ± 2.32 µHz, is the product of the time-bandwidth product of the estimate and the Rayleigh resolution for 20 day blocks, 0.579 µHz. The dotted red curve at the bottom of Figure 1 shows that, on average, ~98% of the power is restricted to frequencies below 500 µHz. Even on the maximum of the 1 day spectra, one finds about 90% of the power in this band. Consequently, the emphasis in this paper is on these very low frequencies. The ULF electron data has been found to be surprisingly predictable [*Li et al.*, 2011], relative to a purely nondeterministic process with a similar spectrum, suggesting that systematic components are present.

The maxima of the 100 spectra computed on 20 day blocks are plotted in Figure 2, and it can be seen that what appears as a blur in Figure 1 resolves into sharp peaks mostly concentrated around the frequencies of low-degree solar *p* modes. These cannot be resolved with a 20 day block; however, the *Q* of the

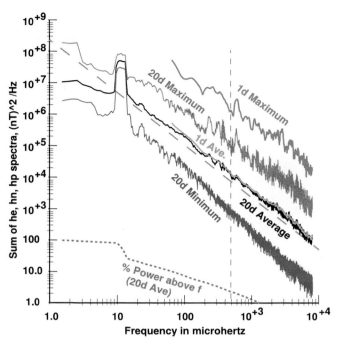

Figure 1. Several statistics of the total power spectrum of GOES 10 magnetic fields, $S_{he}(f) + S_{hn}(f) + S_{hp}(f)$. Two sets of estimates are summarized: *first*, the curves that begin at ~1.5 µHz were estimated on 100 20-day blocks offset by 50%; *second*, the two curves beginning at ~80 µHz were computed from 2136 one-day blocks. Both sets used data sampled at one minute resolution covering 1999 to 2001. The dashed red curve at the bottom represents the percentage of power at frequencies above *f*, showing that frequencies below 500 µHz include approximately 98% of the total power. The remainder of the curves are in units of $(nT)^2/Hz$. The top red and magenta curves show the maxima of both sets of spectra. Note that the apparent "roughness" of this red curve is an artifact of the plotting scale and resolves into discrete *p*–mode peaks in Figure 2.

~3647 µHz line (possibly the $p_{6,23}$ mode despite being marked "1,25") is at least 1200. In contrast, the measured *Q*s of magnetospheric cavities are low, and the observations reported by *Crowley et al.* [1987] imply $Q \sim 6$ near 1.6 mHz. In the 1 day spectra, these cannot be even approximately resolved, but peaks in the maxima of the 1 day spectra tend to occur where the mode density is high. This suggests that large peaks occur when several modes are aligned "in phase," as in section 7.1.

Given the evidence in TML and related works for the influence of solar *g* modes on the solar wind together with the density of *g* modes at low frequencies, it is desirable to use very long blocks to get good frequency resolution. This is most easily done by low-pass filtering and decimating the 1 min raw data to a 16 min sampling rate except the proton data that was available as 5 min samples have been filtered and decimated to 15 min. These result in Nyquist frequencies of

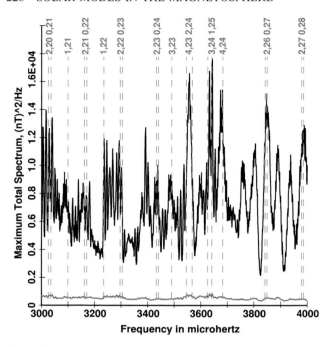

Figure 2. An expanded view of the 3 to 4 mHz part of the maximum 20-day spectrum shown in Figure 1. The frequencies of many of the larger peaks correspond to those of low order solar p modes, and a selection of such modes are marked with dashed red lines and the l, n mode identification. The largest (double) peak is marked "3,24 1,25" that indicates the $P_{3,24}$ and $P_{1,25}$ modes at frequencies of 3626.078 and 3640.475 µHz, respectively, *Broomhall et al.* [2009]. P modes of degree l have $2l + 1$ singlets spaced ~450 nHz, approximately the Sun's rotational frequency, so that a small frequency range around these two peaks can be expected to contain ten singlets and possibly more from higher l modes. These are not resolved with 20-day blocks and can interfere so the apparent amplitude of such modes varies as the unresolved singlets drift in and out of phase.

520 and 555 µHz, respectively, and most of my results are for frequencies ≤500 µHz. These zero-phase filters reject out-of-band signals by >80 dB and are flat within the passband to better than 1 part in 10^5. The 80 dB cutoff effectively keeps p modes from aliasing into the low-frequency band. The Qs of low-frequency solar p modes are expected to be very high, perhaps $>10^4$, so lifetimes may exceed a year, and it is necessary to reduce the amplitudes of possible aliases to very small amplitudes to avoid confusion with the elusive g modes.

3. BACKGROUND AND OBSERVATIONS

Despite early observations such as those of *Belcher et al.* [1969, p. 2302] who, in an analysis of Mariner 5 data, commented "They definitely establish the existence of aperiodic Alfvén waves, propagating outward from the Sun . . . ", it appears that many space scientists believed that

the solar wind is primarily turbulent with the view summarized by *Ghosh et al.* [2009, p. 1]: "Solar wind fluctuations are usually regarded as a more or less featureless random superposition of fluctuations. These may have an origin as coronal waves and turbulence, or as *in situ* generated turbulence or as some mixture of these *Tu and Marsch* [1995]. Regardless of the origin, the prevalent view is that the baseline description is a spectrum that is of the broad band rather than of the point spectral type." There *were* good reasons why this conventional view was adopted:

3.1. The Conventional View

C1 The partial differential equations describing waves in plasma are nonlinear, and analytic solutions are nonexistent or, at best, intractable. Consequently, much of the insight was based on simulations. Because these are computationally demanding, almost all of these were done, until recently, for time and one or two space dimensions as in, e.g., the work of *Roberts et al.* [1996]. These simulations implied only turbulence.

C2 Power spectra of interplanetary data estimated using procedures known at the beginning of the space age generally gave a spectrum with a $\sim f^{-5/3}$ shape, the "signature" of Kolmogorov turbulence. Thus, both theory and observation supported the idea that the interplanetary medium was turbulent with two subsequent implications:

C3 If it were possible to inject periodic components into the solar wind at the Sun, they would be largely destroyed before reaching the Earth. Long-period components such as solar rotation where the propagation time from Sun to Earth is more than about one half the period is excepted.

C4 Power spectra of interplanetary data should be generally smooth with their slope and evolution with distance from the Sun the main quantities of interest.

C5 These imply that most of the oscillations seen in the near-Earth environment have to be *locally* generated, possibly by solar wind turbulence, and the detritus of flares or CMEs exciting various resonances and nonlinearities in the magnetosphere and related structures.

3.2. An Alternative View

*The premise of this paper is that **C1–C5** are largely incorrect because they are contradicted by observations. This failure has been compounded by poor data analysis methods. This does not imply that the cavity and field line resonances attributed to the magnetosphere do not exist, but it does mean that much greater care must be exercised in identifying and characterizing them than has been common. Specifically, the characteristics of the solar source must be considered when analyzing magnetospheric data.*

At low frequencies, turbulence in the solar wind is a minor component, and the spectra are dominated by discrete lines. These are amplified by the bow shock and related structures so that magnetospheric data are primarily discrete lines. These lines originate in the normal modes of the Sun. To elaborate on these points in the same order as before, an alternative to **C1–C5** is:

A1 Simulations of the solar wind done in time and *all three* space dimensions [*Ghosh et al.*, 2009] show that discrete modes and turbulence can coexist and that high-Q modes propagate better than low-Q modes. The Qs of low-frequency solar modes are predicted to range from 10^4 to 10^{11}, so individual modes persist from weeks to millenia.

A2 The spectrum analysis procedures known at the beginning of the space age were unreliable (see section 4.1). There were occasional peaks in these spectra, but given the volume of data being processed combined with the terrible statistical properties of the spectrum estimates, these peaks were usually within the range expected. Moreover, the frequencies and amplitudes of most of these peaks varied erratically between data sets and times, reinforcing the opinion that they were just random fluctuations.

The multitaper method of spectrum estimation, outlined in section 4, allows consistent, efficient, and approximately maximum likelihood estimates. They have well-defined statistical properties. Multitaper estimates of spectra of interplanetary data *exclude* a purely turbulent spectrum.

A3 Discrete solar modes are easily observed in Ulysses data at 5 AU. It was shown in TML that even the 5 min solar p modes were detectable in Ulysses electron data, and with a small Doppler correction, their frequencies matched those of direct optical helioseismology. Figure 2 shows p modes in the spectrum of GOES magnetic field data. Persistent frequencies in the solar wind and the dayside magnetosphere have been reported by *Viall et al.* [2009]. Although details of the process may not be understood, solar modes are obviously being implanted into the solar wind.

A4 The power spectrum of interplanetary quantities consists of a smooth $f^{-5/3}$ background with discrete mode signatures superimposed. I hypothesize that this turbulent background is largely a fossil signature of turbulence in the solar convection zone. The fact that it is measured in space does not imply that it originates in space, any more than the fact that the high-Q modes that are measured in space implies that they originate there. Most studies of the spectral index and its evolution with distance from the Sun are unreliable because of selection bias. The simplest explanation is that a fossil turbulence spectrum is imprinted onto the solar wind by the same mechanism that imprints the modes. Figure 1 shows that the average spectrum of the GOES 10 magnetic fields has a $f^{-5/3}$ dependence out to ~5000 µHz and also that

the maximum spectrum is much further above the average than one would usually expect.

A5 As is shown in section 5, fluctuations in GOES data are coherent with data at the ACE spacecraft at L_1, and discrete modes dominate the continuous part of the spectrum.

Ideas in space physics, such as the radial flow of the solar wind [*Nolte and Roelof*, 1973] imply mode splittings that are observed in this analysis. Also, the idea of propagation paths leading to "footprints" on the Sun implies phenomena similar to multipath in communications systems and may help explain some of the observed amplitude and phase variations. This may also explain why low l modes appear most commonly. The exact propagation mechanism between the Sun and the Earth is unclear, but the recent observations of "pulsed" Alfvénic waves in the solar wind [*Gosling et al.*, 2011], may help explain many puzzles.

The hypothesis that solar modes drive many of the magnetospheric fluctuations implies that solar physics must be considered when planning analysis of magnetospheric data. In particular, there are *many* solar modes: at least 2000 g modes and possibly millions of p modes. Solar g modes are approximately equally spaced in period, becoming dense at low frequencies, while p modes are approximately equally spaced in frequency and become dense at frequencies above ~1000 µHz. G modes all occur below ~400 µHz and p modes above ~250 µHz, so there is some overlap. The frequency range ~100 to 500 µHz has a low-enough mode density that identification of individual modes may be possible. Thus, I concentrate on this frequency range. However, even with fewer modes to contend with, one has the problem that the consensus in the work of *Appourchaux et al.* [2010] is that g modes have not been identified. There are even relatively few detections of p modes below 1 mHz [*Garcia et al.*, 2001]. One must also remember that the Sun is not a laboratory signal source, and in the p mode band [see *Stix*, 2004, p. 183], "one-third oscillates." At ULF, much less is known.

4. SPECTRUM ESTIMATION

Almost all the inferences made about the magnetosphere come from analyzing time series, that is, measurements made sequentially in time. The most reliable way to make inferences about time series is by estimating their power spectrum by *multitaper methods*. These were introduced by *Thomson* [1982], and the theory has been further developed in numerous papers and books in the intervening 30 years, with recent developments and references in the work of *Haykin et al.* [2009] and, of particular relevance here that of P1. Multitaper analysis begins by assuming that one has a representative sample of a stationary process, $x(t)$, of duration N samples at $t = 0, 1, \ldots, N - 1$. Such processes have a spectral representation, that is, they may be written as a Fourier transform

$$x(t) = \int_{-\frac{1}{2}}^{\frac{1}{2}} e^{i2\pi\xi\left(t-\frac{N-1}{2}\right)} \, dX(\xi), \tag{1}$$

where the spectrum is *defined* by the expected value $S(f)df = \mathbf{E}\{|dX(f)|^2\}$ [*Doob*, 1953; *Loève*, 1963]. However, space physics data are rarely, if ever, Gaussian so *Kolmogorov's* [1960] cautions should be heeded. Given the comments of *He and Thomson* [2009a, 2009b], we use the restricted representation

$$x(t) = \int_{-\frac{1}{2}}^{\frac{1}{2}} e^{i2\pi\xi\left(t-\frac{N-1}{2}\right)} X(\xi)d\xi \tag{2}$$

with the origin, $t = 0$, taken at the center of the observation interval. This representation is assumed to be valid for all reasonable times, t, whether included in the sample or not. Taking the Fourier transform of the sample, one obtains a convolution of $X(\xi)$ with a Dirichlet kernel.

$$y(f) = \sum_{t=0}^{N-1} x(t)e^{-i2\pi\left(t-\frac{N-1}{2}\right)}$$
$$= \int_{-\frac{1}{2}}^{\frac{1}{2}} \frac{\sin N\pi(f-\xi)}{\sin \pi(f-\xi)} X(\xi)d\xi, \tag{3}$$

If, instead of thinking of equation (3) as a convolution, one calls it an integral equation, the implication is that one should try to solve it. This is what the multitaper method does. In spectrum estimation work, the most important parameter is the Rayleigh resolution bandwidth $\mathcal{R} = 1/N$ at $\delta t = 1$ or more generally, $\mathcal{R} = 1/T$ where $T = N\delta t$ is the total time duration of the analyzed sample. In multitaper analysis, one chooses a bandwidth, W, typically a few Rayleigh resolutions, $W = C_{\mathcal{R}}/N = C_{\mathcal{R}}\mathcal{R}$ wide, then constructs a local least-squares solution of equation (3) in a band $(f - W, f + W)$. The key to doing this is the set of *discrete prolate spheroidal wave functions*. These functions define the uncertainty principle for the Fourier transforms of finite sequences [*Slepian*, 1978]. The first $K \lesssim \lfloor 2NW \rfloor$ of these *define* the dimensionality of the time-frequency space $[0, N-1] \times (f - W, f + W)$ and are now known as *Slepian functions*, and the corresponding sequences, denoted $v_t^{(k)}$, are called *Slepian sequences*. The corresponding eigenvalues, λ_k, give the fraction of energy in the interval $(-W, W)$ of their Fourier transforms. The first K of the λ_k are close to one, then drop rapidly to zero for $k > 2NW$. Details of the derivation are given by *Thomson* [1982] and depend on the *eigencoefficients*

$$x_k(f) = \sum_{t=0}^{N-1} x(t)v_t^{(k)}e^{-i2\pi ft}, \tag{4}$$

for $k = 0, 1, \ldots, K - 1$. This calculation is usually done with a fast Fourier transform (FFT) zero-padded to M points, with

$M \geq N$. The frequency increment, or bin, is $1/M$ in standard units or $1/(M\delta t)$ in physical units. A simple multitaper spectrum estimate is then

$$\hat{S}(f) = \frac{1}{2NW} \sum_{k=0}^{K} \lambda_k |x_k(f)|^2. \tag{5}$$

Each term in this equation is a conventional direct spectrum estimate and, because the Slepian sequences are orthonormal, they are approximately uncorrelated. Each term thus contributes two degrees-of-freedom (d.f.), so the overall estimate has $\approx 2K$ d.f. This improvement from 2 to $2K$ d.f. is obtained *without* increasing the bandwidth.

Adaptive weights for more complicated spectra are given in the references and are generally used in practice. Numerous studies, listed in section V-c of the work of P1, have shown that multitaper estimates generally make more efficient use of data than conventional spectrum estimates. For Gaussian data, the multitaper spectrum estimate is approximately maximum likelihood [*Stoica and Sundin*, 1999]. The estimate has an approximately χ^2 distribution with $2K$ d.f., χ^2_{2K}, and a locally smooth spectrum. This is rigorously true for data with a Gaussian distribution. Heuristically [*Blackman and Tukey*, 1959; *Brillinger*, 1975, 1984], for the large sample sizes used to compute the spectra in this study, the central limit theorem suggests that the windowed transforms, the $x_k(f)$, equation (4), should be Gaussian. (This is proven in *Mallows* [1967] for time-series data under reasonable assumptions.) Deterministic signals (such as modes) can cause the distribution of $\hat{S}(f)$ in equation (5) to be *noncentral* χ^2.

Let me emphasize that this method *does not* use sample autocovariances: such methods are obsolete and should not be used. Indeed, *McWhorter and Scharf* [1998] have shown that the *definition* of autocovariances implies that multiple taper methods should be used.

4.1. Statistics of Spectrum Estimates

Many of the problems with spectra of space physics data have begun, as mentioned in C2 and A2, through the use of the periodogram, that is $P(f) = |y(f)|^2/N$, where $y(f)$ is just the discrete Fourier transform (usually computed with a FFT) of the data, equation (3). The periodogram was named by Schuster in 1898 and shown in 1903 by Lord Rayleigh to be inconsistent, that is, its variance does not decrease with sample size [*Rayleigh*, 1903]. Its bias is even worse. For example, Figure 7 of P1 shows a spectrum of barometric pressure data where the periodogram is biased by a factor of $\approx 10^8$, and the physically significant features of the spectrum are buried under this mountain of bias. In this example, the estimated spectral index is -2.0 with a periodogram, instead of -5.5 with a multitaper. Consequently, physics inferred from periodograms *cannot* be trusted.

A further problem shared by periodograms and single-window direct estimates is their poor sampling properties. If one has a set of J such spectrum estimates, their minimum is a factor of J too low, and their maximum a factor of $\sim \ln 1.78\,J$ too large. For example, if one has a sample of size $N = 100,000$ and computes a single-taper spectrum estimate, its range may easily exceed the range of the true spectrum by a factor of a million. This is in addition to the bias. This has left the impression that spectrum estimates are unreliable. It should be an embarrassment that papers still appear using an estimate that Lord Rayleigh proved to be unreliable over a century ago.

The situation with multitapers is quite different, and if one uses $K = 10$ tapers to get an estimate with 20 d.f. with again $N = 100,000$, the lower extreme is about 0.17 and, the upper extreme, 2.95 times the true spectrum. False peaks in such a spectrum that are a factor of five larger than the mean should only occur once in $\sim 10^{12}$ cases.

More formally, a multitaper estimate with K tapers will have a distribution proportional to a χ_ν^2 and $\nu \le 2K$. If ξ_j is a set of ν independent, real-valued, Gaussian random variables with means μ_j and unit variances, then their sum of squares

$$S = \sum_{j=1}^{\nu} \xi_j^2, \qquad \lambda = \sum_{j=1}^{\nu} \mu_j^2 \qquad (6)$$

has, by definition [*Lancaster*, 1969], a noncentral χ^2 distribution with ν d.f. and a noncentrality parameter, λ, equal to the sum of the squares of the means. If the μ_j are all zero, as one expects for the Fourier transforms of random noise, $\lambda = 0$, and the distribution is *central* χ^2. With the unit variance, or power, of the ξ_j in the definition (6), the expected value $\mathbf{E}\{S\} = \nu + \lambda$ means that ν also represents the noise power and, λ, the signal power. The signal-to-noise power ratio is $\rho = \lambda/\nu$.

The standard form of the central χ_ν^2 distribution has expected value ν and to be used has to be scaled to match the process. I have found the scaling

$$z(f) = \frac{\hat{S}(f)}{\mathbf{E}\{\hat{S}_c(f)\}} \qquad (7)$$

to be particularly convenient for spectrum estimation work. Here the subscript c is used to emphasize that one is dealing with the central, or noise, part of the spectrum, not the peaks. Clearly, $z(f)$ has an expected value of one in noise. Defining the d.f. parameter $\nu = 2\alpha$, so $\alpha \approx K$, but keeping α distinct from K to allow for weighting, z has the scaled χ^2 or gamma distribution

$$p(z;\alpha) = \frac{\alpha^\alpha}{\Gamma(\alpha)} z^{\alpha-1} e^{-\alpha z} \qquad (8)$$

that has a unit mean and variance $1/\alpha$. The CDF, the probability that an estimate is less than z, is

$$P(z;\alpha) = \int_0^z p(t;\alpha)\,dt, \qquad (9)$$

and the probability that the estimate exceeds z is $1 - P(z;\alpha)$. It is convenient to define an inverse, or quantile, function $\mathcal{Q}(\gamma;\alpha)$ where γ is a probability so $0 \le \gamma \le 1$. The quantile $\mathcal{Q}$ is defined implicitly by $\gamma = P(\mathcal{Q}(\gamma;\alpha);\alpha)$ and here is used for two distinct purposes: *first*, to set approximate confidence levels and, *second*, in estimation of the "baseline," or noise level, of spectrum estimates.

4.2. Estimation of the Baseline

Most of the spectrum estimates, made from interplanetary and GOES data, *do not* have a central χ^2 distribution, but a mixture of central and noncentral χ^2 distributions. Specifically, the distribution is noncentral at modal frequencies and central when one is between modes in a noise-like part of the spectrum. It is possible, but tedious, to fit a mixture of the two distributions [see *Thomson et al.*, 2001, Section 7]. It is simpler to use a robust procedure to estimate the noise spectrum. The procedure is as follows: (1) Postwhiten the raw spectrum estimate by multiplying by $f^{+5/3}$ so that it is reasonably flat. (2) Take a band many times wider than the bandwidth parameter W, around frequencies of interest, say J frequencies (or FFT bins), sort the estimates in the band into increasing order, and find the lower 100γ percent point, usually 5% or 10%, denoted $\bar{S}_{(\gamma)}(f)$. (3) Scale this estimate by a gain, $\mathcal{G}(\gamma;\alpha) = 1/\mathcal{Q}(\gamma;\alpha)$, to obtain an unbiased estimate of the noise spectrum,

$$\hat{S}_{N;\gamma}(f) = \mathcal{G}(\gamma;\alpha)\bar{S}_{(\gamma)}(f). \qquad (10)$$

Within reason, this estimate should not depend critically on the choice of γ. Taking the spectrum of the GOES 10 electrons over 200 to 500 μHz as an example, the average baseline shifts by 0.0, 0.026, 0.059, and 0.156 as γ ranges through 5%, 10%, 20%, and 50%, respectively. The 5% to 20% estimates are close, and for those whose only experience with spectra is with single taper or periodograms, such changes will appear insignificant. With multitaper estimates, however, the fact that the median is biased high by more than one standard deviation relative to the 20% estimate is a strong indication that the spectrum is not noise-like. The explanation is that in some bands, more than half the frequencies are modal. The lesson is that one must examine the distribution of spectrum estimates, not just blindly take their average. The average shown in Figure 1 would lead one to believe in turbulence, but the fact that the maxima were not as expected leads to new insight.

The variance of this estimate can be found using standard methods [see *Kendall and Stuart*, 1963, p. 237, (10.29)], and is given by

$$\mathbf{Var}\{\hat{S}_{N;\gamma}(f)\} = \frac{\mathcal{G}^2(\gamma;\alpha)\gamma(1-\gamma)}{Jp^2(\mathcal{Q}(\gamma;\alpha);\alpha)}. \qquad (11)$$

Figure 3 shows the gain $G(\gamma;\alpha)$ and scaled variance, $J \cdot \mathbf{Var}\{S_{N;\gamma}(f)\}$, for a sample of $J = 100$ independent χ^2_{20} random variates as a function of $\gamma_j = (2j-1)/2J$ for $j = 1,\ldots, J$. If, for example, one takes the 10th smallest sample of a set of 100 as a robust estimate, it will be biased below the average. Multiplying by $\mathcal{G} = 1.6247$ will correct this bias. Similarly, the variance of such an estimate will be 0.004126. For an uncontaminated sample, the simple average would have a relative variance of $1/(\alpha J)$ or, in these units, 0.1, so the lower 10% point only has an efficiency of $0.1/0.41 \approx 24\%$. Note that equation (11) applies to *independent* random variates; multitaper spectrum estimates have an approximately triangular correlation function that is almost zero for frequency separations $>\pm 2W$. Thus, one must reduce the value of J used in equation (11) by a factor $\delta f/(2W)$ where δf is the frequency step used in the spectrum estimate. If, as before, one zero pads a sample of size N and uses an M point FFT, the effective sample size is

$$J_{eff} \approx J \cdot \frac{N}{M} \cdot \frac{1}{K}. \qquad (12)$$

For most of the spectra used here the baseline was estimated from the $\gamma = 5\%$ level in 1.6 μHz bands offset by 50%.

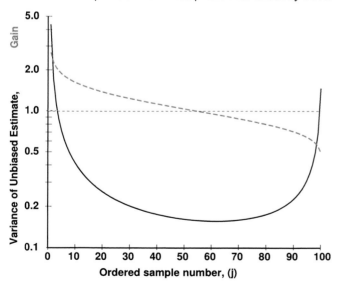

Figure 3. The red dashed curve shows the gain, $(\mathcal{G}(\gamma;\alpha)$ for $\alpha = 10$ (20 d.f.) and $J = 100$ as a function of quantile required to produce unbiased spectrum estimates from their order statistics. The solid black curve shows the variance (11) of the unbiased estimate scaled by J.

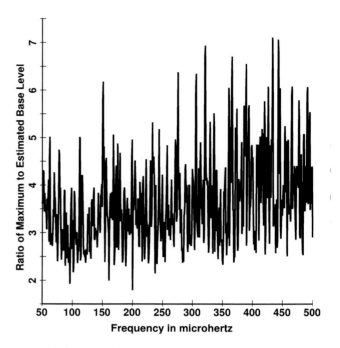

Figure 4. The ratio of the maximum to the baseline spectrum. The raw spectrum was multiplied by $f^{+5/3}$, effectively flattening it. Next, the lower 5% point in blocks of width ~1.6 μHz (± 200 FFT bins) was found and scaled to give an unbiased estimate. The ratio of the maximum in each block to the baseline is plotted.

(Multiply the raw spectrum by $f^{5/3}$, then do the baseline estimation. See Appendices C and D of *Thomson et al.* [2001] for details.) Figure 4 shows the ratio of the maximum to baseline estimates. A ratio of peak to "average" of six in single-window estimates is common, but the estimates shown here have 16 d.f., so such an event only happens with a probability of $\sim 1.9 \times 10^{-13}$ (Table 1). One such event would be remarkable, but this happens in 19 of the 567 bands, and even allowing for overlap, there are 13 independent events. Even less probable, there are 34 independent frequencies where the ratio exceeds five. The probability of this, $\sim 10^{-330}$, is so low that the hypothesis that these peaks arise from sampling variations in a basically featureless spectrum from simple Kolmogorov turbulence must be absolutely rejected. Thus, there are discrete line components. Whether these are locally generated or externally forced is discussed in the next section.

To conclude this section, I must emphasize the necessity of taking long-enough data sections to resolve the possible line components. The GONG helioseismology network routinely works with 60 day "months," and most of the series analyzed for low-frequency modes [see, e.g., *Salabert et al.*, 2009] are longer than 2000 days. The danger of working with short data sections is that when two or more modes fall within the

Table 1. Number of Peaks Above Level z for the 2 MeV Electron Spectrum Shown in Figure 4[a]

z	$Q = 1 - P$	Bands	Peaks
3	4.75×10^{-5}	409	143
4	1.09×10^{-7}	174	87
5	1.66×10^{-10}	56	34
6	1.94×10^{-13}	19	13
7	1.87×10^{-16}	2	2

[a]Here Q is the probability of a *central* χ_{16}^2 scaled to have an expected value of 1.0 exceeding z. The penultimate column gives the number of bands (of 567) where z is exceeded, and the last gives the number of local maxima (in the 567 bands) above z.

analysis bandwidth, $(f - W, f + W)$, spectrum estimates are erratic. Consider the case of just two modes with good signal-to-noise ratios and similar amplitude within the analysis band. When they happen to be "in phase," one will see a peak in the spectrum near the "average" frequency. Conversely, if the two modes are out of phase, one will not. The result is that spectra will appear unnecessarily variable. Moreover, because there are many modes, the odds are good that in a given data block, there will be some frequencies where modes are in phase and others where they are not, with the result that peak frequencies will vary erratically when comparing results from different data sections. However, because the frequencies of solar p modes are known to vary with the solar cycle [*Woodard and Noyes*, 1985; *Jain et al.*, 2009], and it appears probable that g mode frequencies do as well, one must be cautious with exceptionally long blocks because the bandwidth decreases with the length, T, of the series, $B = C_R/T$. When the bandwidth becomes narrower than the frequency shift of the mode, detections become sporadic.

5. COUPLING

There are many papers devoted to solar-terrestrial coupling and, here I consider the MSC computed between the magnetic fields on ACE and GOES, noting the following complicating factors:

1. There are three components on ACE and three on GOES giving nine coherence estimates.

2. Daily variation in geomagnetism has been known since it was discovered by Graham in 1724, and the inclination of the Earth's dipole axis [see section IV of P1] implies that, in addition to direct coupling, modes coming from the Sun will be split by multiples of a cycle per day. That is, in the magnetosphere, data is *cyclostationary* or *periodically correlated*. In magnetospheric work, cyclostationarity is usually described in terms of local solar time ("dayside," "dawn

sector," and similar terms) and because the dependence can be strong, data segments of a few hours or less are commonly used in a quasistationary approximation. An alternative is given in section 6.

3. It has been known since Canton's work in 1759 that geomagnetic activity is greater in summer than in winter. So, again, one has a periodically correlated process, this time with a period of 1 year. The whole sequence of geometrically induced correlations is obviously driven by the nonstationary solar cycle.

4. Magnetospheric processes are nonlinear, and consequently, one also sees their sum and difference frequencies [*He and Thomson*, 2009b], in addition to the "direct" solar modes.

5. As we show in section 7, frequencies measured in GOES electrons and protons are different; this may be a simple Doppler shift or, because the electron energies exceed 2 MeV (four times the rest mass of an electron), possibly a relativistic effect, or both.

The fundamental tool for analyzing such effects is *coherence* or correlation in the frequency domain. If $\{x_k(f_1)\}$ and $\{y_k(f_2)\}$ represent two sets of eigencoefficients, equation (4), for compatible series $x(t)$ and $y(t)$, respectively, the general complex coherency is

$$\hat{C}_{x,y}(f_1, f_2) = \frac{\sum_{k=0}^{K-1} x_k^*(f_1) y_k(f_2)}{[\sum_{k=0}^{K-1} |x_k(f_1)|^2 \sum_{k=0}^{K-1} |y_k(f_2)|^2]^{\frac{1}{2}}}. \quad (13)$$

For the most part, one uses the stationary form with $f_1 = f_2$ expressed as MSC,

$$\text{MSC}_{x,y}(f) = |\hat{C}_{x,y}(f, f)|^2 \quad (14)$$

An important feature of coherence estimates is that their statistical properties are extremely well understood [*Carter*, 1993; *Thomson and Chave*, 1991] and are particularly simple under the null hypothesis that the two series are independent. In principle, one should use canonical coherences (CCs) between ACE and GOES as described in section XII of P1 as this typically gives two or three functions to be examined instead of nine. The drawback is that the statistical distributions of CCs are far more difficult, and demonstrations of significant coupling between the interplanetary magnetic field and magnetospheric processes might possibly be taken less seriously than with the simpler, but well-calibrated MSC.

5.1. ACE to GOES Coherences

To allow for the nonstationarity and the Doppler shifts noted in section 7, the MSC was computed between the magnetic fields on ACE and GOES on 10 day blocks offset

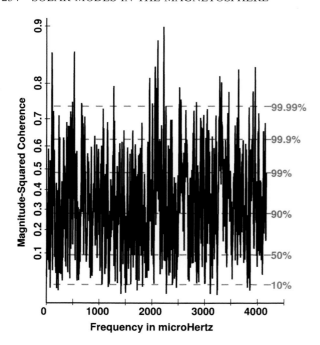

Figure 5. Magnitude-squared-coherence (MSC) between the GSE-Z magnetic field at ACE and the HE component on GOES 10 for July 25 to Aug. 3, 1999. Only the lower half of the frequency range is shown. Significance levels are indicated by the dashed red lines. Note that only 2.3% of the estimates are below the 10% level, while 56.8% of the estimates are above the 90% level.

by 5 days for 1999. Figure 5 shows MSC between the GSE B_z component on ACE and the **HE** component of the magnetic fields on GOES 10 between days 206 and 216 of 1999 using 1 min resolution data. A time-bandwidth NW = 5 with $K = 8$ tapers were used. Such figures are graphically demanding even though this one only includes the lower half of the Nyquist band. Note that there are more than 20 peaks above the 99.99% significance level, although on average, only 0.72 such peaks are expected. To avoid a proliferation of such plots, Table 2 gives the average and maximum percentage of estimates above the nominal 90% significance line. (In Figure 5, this line is at an MSC of 0.2803, and 56.8% of the estimates exceed it.) It may be seen that, on average, *none* of the nine coherences between the magnetic fields on ACE and GOES can be considered independent.

This calculation was repeated with 60 day data blocks, and on plotting the fraction of the coherence estimates above given levels with time, one finds the maximum coherence in summer. Although the coherence fluctuates, there are many frequencies where the *minimum* coherence never drops to negligible values. In section 7, we turn to the identification of these frequencies.

5.2. Local Resonance Versus Forcing

The problem of identifying local resonances as opposed to external forcing may be partly resolved by considerations of the Q of the resonator. The Q of a system is the ratio of 2π times the energy stored in the resonator to that dissipated per cycle. If the dissipation is balanced by incoming power, the ratio of the spectra in the resonance to that outside should reflect this. For a simple resonance, Q is also the ratio of the center frequency to the full width of the peak at the half-power point. In the magnetosphere, Q are not expected to be high, possibly between 1 and 20, but the width can decrease with time [*Mann et al.*, 1995]. The Q of low frequency solar p modes are expected to be $\gtrsim 10^4$ and possibly $\sim 10^{11}$ for g modes. The line width of the p modes shown in Figure 2 implied $Q \gtrsim 1000$ even though with 20 day blocks, the singlets of the modes are not resolved. For such modes, the Q implies that the modes are externally forced. The envelope of the peaks in Figure 2 may have a peak centered at ~3600 µHz. This is slightly above the 3.2–3.4 mHz predicted in the work of *Samson et al.* [1992], but the frequency of this resonance may vary with time.

6. CYCLOSTATIONARITY

One of the indicators of external forcing, as opposed to local generation of modes, is the fact that GOES data are cyclostationary, a phenomenon also referred to as periodically correlated. It is a useful intermediate case between stationarity and fully nonstationary processes that arises in climate and communications problems and, as we show here, in magnetospheric data. There are, of course, the simple daily and annual additive terms, but cyclostationary implies correlation between different frequencies. The daily and annual periodic terms are best considered to be part of the average, but in periodically correlated data, the variance and autocorrelations also vary periodically. To put this into a magnetospheric context, parameters such as conductivity vary on both annual and diurnal rates [*Hurtaud et al.*, 2007]. Now imagine that the local solar time dependence of the coupling

Table 2. Average (Left) and Maximum (Right) Percentage of MSC Estimates Above the Nominal 90% Level Between Individual GSE Components at ACE and GOES 10 during 1999[a]

ACE	Average			Maximum		
	He	Hn	Hp	He	Hn	Hp
X	22.34	22.16	20.33	66.7	72.6	74.8
Y	24.22	23.70	21.39	66.3	72.9	82.4
Z	26.43	24.35	21.39	68.2	63.9	64.1

[a]MSC computed on 70 10-day blocks with a 50% overlap.

between the incoming solar signals and the response seen at GOES is written as a Fourier series with a period of 1 day. Thus, the output of the system, the GOES measurements, will be a convolution of the solar wind input and the periodically varying impulse response. In the frequency domain, this gives replicas of the input signal translated in frequency by multiples of a cycle d^{-1}. At polar latitudes, these copies are often obvious in a spectrum.

Stationarity is an assumption that can be checked by testing for nonstationarity using the fact that in nonstationary processes, signal elements at different frequencies are correlated. This fact underlies Loève's [1963] fundamental theory of nonstationary processes. This correlation can be measured by computing coherence between signal elements at frequency f and $f + 1$ cycle/day (c/d) or f and $f + 1$ cycle/year (c/y), ($f - 1$ is similar) by an offset MSC (OMSC) adapting equation (13),

$$\text{OMSC}_x(f) = |C_{x,x}(f, \ f + 1)|^2 \qquad (15)$$

where the "1" denotes either 1 cycle per day or per year. Mathematical details and examples are given in section XV of P1 and sections V to VIII of *Haykin et al.* [2009].

Table 3 compares offset MSC between ACE solar wind protons and GOES electrons. Both sets have $N \approx 200,000$ samples at $\delta t = 16$ min, and the calculations use the same time-bandwidth parameter. In *random* Gaussian data, one expects 10% of the estimates and about 20,000 peaks above the 90% significance level and approximately 20 peaks above the 99.99% level. However, there are many modes in both data sets, and if f is at a modal frequency and $f + 1$ is not, the false detection rate increases. Thus, in the ACE data, 15.8% of the estimates are above the 90% level instead of the

10% that one would expect. In the GOES data, in contrast, 45% of the 1 c/d OMSC's exceed the 90% level. Turning to the peaks, in a random set of $N = 200,000$, one expects an average of ~20 peaks above the $1 - P = 10^{-4}$ (or 99.99%) significance level, that is, OMSC ≥ 0.627. In the ACE data, there are 108 peaks (0.1% of the bandwidth), while in the GOES data, there are over 2400.

One finds 84 peaks at ACE above the 99.99% level at an offset of 1 c/y, and 192 at a 2 c/y offset. This is reasonable as one expects some splitting by 1 c/y from the eccentricity of the orbit and the fact that the 7.5° inclination of the ecliptic on the solar equator splits odd-parity modes by ±1 c/y. In the GOES data, one sees 2649 peaks with a 1 c/y splitting. Many of these are triplets with $f - 1$, f, and $f + 1$ c/y peaks, so there is some overcounting; however, one is faced with an average of 3 or 4 peaks per µHz.

Finally, one may also compute coherences between frequency f at ACE and $f \pm 1$ at GOES. This confirms the splitting. These coherences are typically as large, or larger, than the direct coherences. These observations have several implications. First, if the solar wind were truly turbulent, one would be surprised if the ±7.5° annual change in Earth's heliographic latitude would obviously split modes. Second, the change in the fraction of OMSCs above the 90% level from 15.8% at ACE to 45% at GOES implies systematic strong coupling. Third, this is confirmed by the direct measurement of offset coherences between ACE and GOES. Thus, the daily splitting occurs between ACE and GOES.

7. MODE IDENTIFICATIONS

Notwithstanding the consensus mentioned above in the work of *Appourchaux et al.* [2010] that g modes have not been detected, *Turck-Chièze et al.* [2004a, 2004b] and *Mathur et al.* [2007] have identified some "candidate" singlets from long series of observations made with the GOLF instrument on the SOHO spacecraft. SOHO is also at L_1, and much of the SOHO data overlap that from ACE and GOES-10, so comparisons are useful. Several of these "candidates" can be seen in GOES data. In the 1 to 8 Å XL, one finds three lines, 220.576, 220.136, and 220.728 µHz, whose average difference from the corresponding "case 1" lines is 3.3 ± 12.7 nHz. The GOLF frequencies are given to the nearest 10 nHz, while my estimates of frequencies from GOES data are accurate to ~±4 nHz. The GOES X-ray data should not depend on magnetospheric processes, so this comparison shows that these lines are real and not an artifact of either spacecraft.

In contrast to the GOES X-rays, the electrons and protons are influenced by magnetospheric processes, and the frequencies differ systematically from the GOLF frequencies.

Table 3. Summary Statistics for the Cyclostationary Coherence of the >2 MeV Electrons[a]

Theory		ACE, 1c/d		GOES, 1c/d		GOES, 1c/y	
$1 - P$	Level	Over	Peaks	Over	Peaks	Over	Peaks
10^{-1}	0.280	15.8	15375	45.3	34754	43.8	32894
10^{-2}	0.482	2.5	3105	18.2	16103	18.5	15607
10^{-3}	0.627	0.4	565	6.7	6669	7.2	6739
10^{-4}	0.732	0.1	108	2.2	2411	2.5	2649
10^{-5}	0.807	0	12	0.7	821	0.8	924
10^{-6}	0.861	0	1	0.2	260	0.3	305

[a]The first column gives the significance, $1 - P$, the second the level. The third and fourth columns are the percent of frequencies above the nominal level (marked "Over") and the number of peaks for one cycle per day offsets in ACE protons. The fifth and sixth columns are GOES at one cycle per day offset, and the last two are GOES at one cycle per year. There are 197,150 data samples in the GOES data, and 200,000 in the ACE.

Being cautioned that there are only six "case 1" frequencies to match, the electron lines appear to be shifted *down* by 28.7 ± 16 nHz, while the P4 proton (15–40 MeV) lines are shifted *up* by 56.9 ± 6.2 nHz relative to the published "case 1" frequencies. These may be Doppler shifts with different signs because electrons drift eastward and protons drift westward.

It must be emphasized, however, that while these "candidates" are believed to be *g* mode singlets, they appear to come from several modes. Recall that a *g* mode of spherical harmonic degree *l* is split by solar rotation into $2l + 1$ singlets. We have also noted [*Thomson et al.*, 2001] that magnetic fields on the surface of the Sun are frozen in and so will respond more to horizontal motions than to vertical. At a given frequency, the horizontal power of a *g* mode at the Sun's surface is a factor of $l(l + 1)$ larger than the vertical [*Christensen-Dalsgaard*, 2002, eqn. (3)], a possible explanation of the high peaks seen in ACE and GOES data.

Figure 6 shows a candidate for the $l = 5$, $n = -1$ solar gravity mode. In Figure 6, all 11 singlets appear at levels

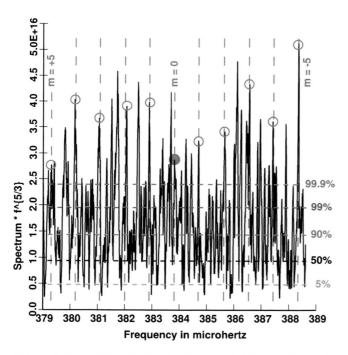

Figure 6. A candidate for the $l = 5$, $n = -1$ solar gravity mode measured in GOES 10 2 MeV electrons. Here there are 11 singlets marked by the dashed vertical lines with *all* 11 significant above the 99.9% level. The average splitting between singlets is 904 ± 18 nHz with a range 884 to 930 nHz. The center $m = 0$ peak is marked with a red dot and has a frequency of 383.812 µHz. The peaks on the other 10 singlets are marked with open circles. The other peaks may be from higher *lg* modes or the $l = 2$, $n = 1$ *p* mode, predicted at 384.54 µHz [*Provost et al.*, 2000].

above the 99.9% significance level. The center frequency, that of the $m = 0$ singlet, is 383.812 µHz. If one corrects by the same relative factor, 130 ppm, found by comparing with the GOLF "case 1" frequencies, this becomes 383.862 µHz. Two predicted frequencies for this mode are 386.39 µHz [*Provost et al.*, 2000] and 385.55 µHz [*Couvidat et al.*, 2003], both slightly above that seen here. It is also possible that this is a partial detection of some mode with $l \geq 7$, but a search for singlets with $|m| \geq 6$ failed. The splitting between individual singlets ranges between 884 and 930 nHz with an average 904 ± 18 nHz or 12.80 days (12.365 sidereal). For reference, the 11 singlet synodic frequencies, beginning with $m = +5$, are: 379.296, 380.184, 381.076, 382.004, 382.888, 383.812, 384.696, 385.626, 386.540, 387.428, and 388.336 µHz, where positive m represents motion in the prograde direction. The differences between these spacings may be systematic. Moreover, these are all solid detections with $-\log_{10}(1 - P)$ values being 3.37, 5.79, 4.38, 4.10, 5.48, 3.47, 3.74, 4.17, 5.50, 3.81, and 7.58, respectively, where P is the peak probability. These were computed from the lower 5% point of the distribution using equation (10) and are stable because the baseline was estimated from a bandwidth in excess of 1800 Rayleigh resolutions.

In addition to the 11 main singlets, there are many other peaks in this spectrum. Some of these are expected as the $l = 2$, $n = 1$ *p* mode may fall within this band. Predictions of this mode's frequency vary considerably: *Guenther et al.* [1992] gives 379.326 and 382.616 µHz for the "standard" and "combined" solar models, respectively, and *Provost et al.* [2000] gives 384.54 µHz. There are several other *g* modes identifiable in GOES data that space limitations dictate must be presented elsewhere.

7.1. Observational Implications

Detections of discrete mode singlets imply that one should see "wave packets" in time. Suppose one has $P = 2l + 1$ singlets spaced γ Hz apart around some center frequency f_0 and that these singlets are phase aligned at some time, say $t = 0$. Each singlet has the form $\mu \cos 2\pi (f + m\gamma)t$ and sum to the wave packet, $\mu \cos 2\pi f_0 t \dfrac{\sin \pi P \gamma t}{\sin \pi \gamma t}$, where μ is the amplitude of the individual singlets. The numerator implies that the envelope will have its first zeroes at $t_z = \pm 1/(P\gamma)$. The denominator implies that packets should recur with a period of $1/\gamma$. There are two cases of interest: For the $l = 5$ *g* mode just discussed, $\gamma \sim 930$ nHz and $P = 11$, the first zeros of the envelope occur at $\pm 1/(P\gamma) \sim \pm 1.1$ days, so one expects low-frequency packets that last a few days to maybe 2 weeks for an $l = 1$ *g* mode. *P* mode frequencies are approximately $\nu_{l,m} \sim \nu_0 (n + l/2 + \cdots)$ where $\nu_0 \approx 136$ µHz, so if *P* p-modes are aligned in *l*, one will see roughly symmetric packets of roughly quantized duration $\sim\!\pm 2/P$ h and twice that if the alignment is in n. These idealized packet shapes will obviously be modified by magnetospheric coupling effects that depend on local time, the exact singlet frequencies, differences in singlet amplitudes and phases, missing singlets, and similar effects. Both *Villante et al.* [2007] and *Piersanti et al.* [2012] show such packet shapes.

7.2. Peak Density as a Function of Frequency

Plotting even the spectra available from GOES data at sufficient resolution to be useful requires in excess of 1000 pages, clearly impractical. As an alternative, Figure 7 shows the number of peaks in 10 µHz bands for the electrons and P4 protons above 99.99% and 99.9% significance levels, respectively. These, in common with any number of other statistics from GOES data, show that the probability of these peak counts arising from random data is essentially zero. If the data were random, one would expect one peak above the 99.99% in these spectra about every 44 µHz; however, in the electrons, one observes 360 times the expected rate.

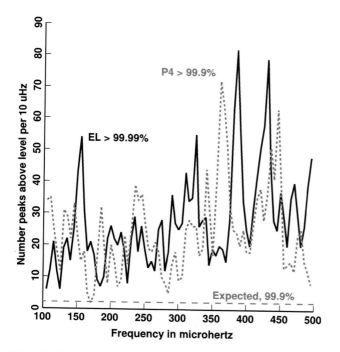

Figure 7. The number of peaks in the spectra of GOES-10 electrons (solid black line) and P4 protons (dotted red line). These were computed in 10 µHz bands offset 5 µHz with a threshold at the 99.99% significance level for the electrons and the 99.9% level for the proton spectrum. For the electrons, the expected peak count is 0.228 and the observed maximum was 82. The P4 protons had a lower threshold of 99.9% giving an expected background rate of 1.97, reached in 1 of 80 bands with a maximum of 73. The high concentration of peaks near 370 µHz is a potential source of aliases.

Speculations on the cause of this are: Are the peaks in this plot from Samson resonances [see, e.g., *Kivelson*, 2006, section 4.2], and if so, why are the electron and proton peaks at different frequencies? Since these peaks are close to the Sun's Brunt-Väisälä frequency, $\lesssim\!450$ µHz [*Provost et al.*, 2000], could they be from an accumulation of high-l g modes or from some other condition on the Sun? Or could they be solar effects magnified by the Samson resonance? In general, it is easier to excite a resonant structure than a nonresonant one, and for example, *Wright and Mann* [2006] have characterized the magnetosphere's response to a pressure pulse. *Walker* [2005] has analyzed periodic forcing and the coupling between the solar wind and the magnetosphere. The coupling is complicated and, given the variability of particle density in the magnetosphere over times much longer than the periods of the modes, clearly requires more study.

Figure 7 also indicates a potential inferential trap. If the processes at the altitudes of low Earth orbit (LEO) satellites are similar to those observed at geostationary orbit, there is a danger of aliasing. A 90 min orbit corresponds to a sampling frequency of 185 µHz for a given location, and this aliases to 370 µHz, close enough to the maximum peak density to be of concern. That is, if one sees an apparently static or slowly varying phenomenon in LEO data at a given ground location (Studies where ground and satellite data are combined are common.), there is no guarantee that the LEO observations represent what they appear to be, rather than actually being an alias of the peaks around 370 µHz. These modes can be quite energetic as occasionally one can have several of them "in phase," which causes people to comment on the large oscillations visible in plots of raw data. As discussed in section 7.1, the total width of many of these peaks such as that of $g_{5,1}$ shown above in Figure 6, is $\sim\!10$ µHz implying that when the singlets become phase aligned, the condition will persist for a day or more depending on the mode or modes. Are solar g modes the origin of Pc6 pulsations? One also cannot help but note that the Pc5 band corresponds almost exactly with the main solar p mode band. *Parkinson* [1983, p. 300] gives the Pc5 band as 150 to 600 s or 1.7 to 6.7 mHz. Optical detection of p modes is almost impossible below 1 mHz, and there is an acoustic cutoff at 5100 µHz [*Jiminéz*, 2006]. However, as shown in Figure 32 of P1, p modes may also be seen at higher frequencies in magnetic fields. The peaks shown in this figure, from induced voltages on an ocean cable, can be easily detected in 1 s magnetic field data from ACE, and in geomagnetic data; so to get from ACE to the bottom of the Pacific Ocean, it clearly must pass through the magnetosphere. The peaks near 20 mHz have the 136 µHz spacing that is characteristic of p modes and match the Pc3 band. Thus, many of the differences between geomagnetic and solar phenomena may be simply that they have been given different names.

8. DISCUSSION AND CONCLUSIONS

This paper has sketched an idiosyncratic view of magnetospheric data concentrating on the long-term background processes. Some differences between the conventional picture of a fully turbulent solar wind exciting internal resonances of the magnetosphere and the hypothesis here, namely, that the solar wind consists of a mixture of imprinted normal modes of the Sun superimposed on fossil turbulence from the solar convection zone was sketched in section 3. I chose GOES data for this study because their geostationary orbit is simple. Satellites with orbital periods other than 1 day should produce additional splittings of spectral lines in an already complicated spectrum and may be aliased.

Section 4 briefly reviewed multitaper spectrum estimates. It was emphasized that the periodogram (or squared FFT) is still commonly used despite the fact that Lord Rayleigh proved that it was inconsistent. The relation between multitaper estimates and periodograms is similar to that between modern spacecraft instrumentation and a brass and mahogany ballistic galvanometer. The number and size of spurious peaks in multitaper spectrum estimates of "random" data can be predicted, and departures from these usually imply the presence of nonrandom components.

It was demonstrated that much of the power in the spectrum of solar wind components at ACE, usually more than one-half of the total, originates in the normal modes of the Sun. In section 5, it was shown that the magnetic field components at GOES are coherent with those at ACE. This coherence has a bimodal distribution, high at modes and low between them. The incoming modes are modulated by the structures associated with the bow shock and magnetopause. The geometry of the Earth's rotation and orbit forces the bow shock and magnetopause to vary periodically at daily and annual rates, splitting the incoming modes. It was shown in section 6 that the data at GOES are periodically correlated with many modes appearing at their original solar frequency, f, plus sidebands at $f \pm 1$ cycle d^{-1} and $f \pm 1$ cycle yr^{-1}.

Interplanetary and magnetospheric data are *not* simple and contain thousands of approximately periodic components driven by the normal modes of the Sun. This implies that plans for data analysis must include both solar physics and aspects of the Earth's orbit with the consequence that one needs at least 5 years of data to resolve fine splitting.

Many solar modes are detectable in magnetospheric data. In section 7, it was shown that some of the "case 1 candidates" singlets of solar gravity modes [*Turck-Chièze et al.*, 2004a, 2004b], were detectable in GOES data. The frequencies were identical in X-ray data, those from electrons shifted down, and protons shifted up. The solar X-rays should not be affected by magnetospheric processes so the agreement with the optical data is expected. Magnetospheric electrons drift eastward, while protons drift westward; however, as waves in the solar wind propagate down the windward side of the magnetopause, the coupling [see *Hasegawa et al.*, 2004, Figure 1] should produce opposite signs in the Doppler shifts of the frequencies. The signal-to-noise ratio of solar modes is reasonably high in GOES electrons, making detection easier and allowing several solar gravity modes to be detected. All 11 singlets of what is probably the $l = 5$, $n = -1$ solar gravity mode are detected above the 99.9% significance level. This gives an estimated center frequency of 383.812 μHz and a synodic splitting of 904 nHz, or ≈12.80 days. This implies that the Sun's core rotates considerably faster than the surface rate. It is possible that these 11 singlets are part of a mode with $l \geq 7$ or, improbably, an $l = 10$ mode with most of the odd-parity singlets missing. These modes remain coherent for the duration of the data, over 5 years.

Finally, why does this matter? Foremost is the simple idea that one should get the physics right. Unlike laboratory physics, one cannot put the magnetosphere in a shielded box, but must consider the driver of its dynamics, the Sun. If one assumes that the driving signal has a featureless spectrum, as opposed to a line spectrum, it is my belief that the probability of arriving at the correct physics is exceedingly low, maybe zero. Thus, for example, one is forced to invent various mechanisms to account for the various observed pulsations and similar phenomena.

Similarly, there have been many papers devoted to coupling between the solar wind, particularly the interplanetary magnetic field, and the terrestrial environment. These have had some, but not remarkable, success. Although not discussed here, the modes all have distinct polarization properties and orientations, and given the usual assumptions when working with transfer functions, the coupling calculations should be done on a mode-by-mode basis.

Currently, there is much interest in space weather and related phenomena. These are immensely important practical problems. It appears that, once a flare or CME occurs and is observed, its trajectory can be computed with moderate accuracy. The larger problem of what causes the flare in the first instance does not seem to have received as much attention. Sometime after the "Bastille Day" flare, I computed a harmonic expansion for a year of data from Ulysses and was surprised to see that the sequence of flares was reproduced with reasonable accuracy, although background details were not. This harmonic expansion only included the frequencies that appeared to be periodic in a multitaper F test, so were mostly modal frequencies. The implication, however, is that flares may be triggered when several modes become aligned with the proper phase, suggesting that the flare process may be somewhat predictable.

To summarize, the premise of this paper is that the dominant component of the variability of the solar wind, the driver of the magnetosphere, comes from the discrete normal modes of the Sun. This is superimposed on a background that is a fossil of Kolmogorov turbulence in the Sun's convection zone. The signatures of both this fossil turbulence and the normal modes are imprinted onto the solar wind. When this mixture encounters the Earth's bow shock and related structures, the modal parts are coherent over the whole structure and are enhanced when viewed inside the magnetosphere. Thus, solar gravity modes are detectable in GOES data. If one is attempting to measure a transfer function in the laboratory, common practice is to use a sinusoidal signal for linear cases and a pair of sinusoids for nonlinearities. Studies of the magnetosphere are essentially attempts to assess a large-scale nonlinear transfer function, and nature has built in the sinusoidal test signals. The next job is to exploit them.

Acknowledgments. This work was supported by the Canada Research Chairs program, the Killam Foundation, and NSERC. The ACE data were obtained from NASA's National Space Science Data Center and Space Physics Data Facility and the GOES data from NOAA's Space Environment Laboratory via the National Geophysical Data Center. I thank Wesley Burr, Charlotte Haley, and particularly Maja-Lisa Thomson for editorial assistance and the referees for helpful comments.

REFERENCES

Appourchaux, T., et al. (2010), The quest for the solar g modes, *Astron. Astrophys Rev.*, *18*, 197–277.

Belcher, J. W., L. Davis Jr., and E. J. Smith (1969), Large-amplitude Alfvén waves in the interplanetary medium: Mariner 5, *J. Geophys. Res.*, *74*(9), 2302–2308.

Blackman, R. B., and J. W. Tukey (1959), *The Measurement of Power Spectra*, Dover, New York.

Brillinger, D. R. (1975), *Time Series, Data Analysis and Theory*, Holt, Rinehart, and Winston, New York.

Brillinger, D. R. (1984), *The Collected Works of John W. Tukey*, vol. I, *Time Series: 1949-1964*, Wadsworth, Belmont, Calif.

Broomhall, A.-, W. J. Chaplin, C. R. Davies, Y. Elsworth, and S. T. Fletcher (2009), Definitive Sun-as-a-star p-mode frequencies: 23 years of BISON observations, *Mon. Not. R. Astron. Soc.*, *396*, L100–L104.

Carter, G. C. (1993), *Coherence and Time Delay Estimation*, IEEE Press, New York.

Chapman, S., and J. Bartels (1940), *Geomagnetism*, Oxford Univ. Press, Oxford, U. K.

Christensen-Dalsgaard, J. (2002), Solar g-mode oscillations: Experimental detection efforts and theoretical estimates, *Int. J. Mod. Phys. D*, *11*, 995–1009.

Couvidat, S., S. Turck-Chièze, and A. G. Kosovichev (2003), Solar seismic models and the neutrino predictions, *Astrophys. J.*, *599*, 1434–1438.

Crowley, G., W. J. Hughes, and T. B. Jones (1987), Observational evidence of cavity modes in the Earth's magnetosphere, *J. Geophys. Res.*, *92*(A11), 12,233–12,240, doi:10.1029/JA092iA11p12233.

Dahlen, F. A., and J. Tromp (1998), *Theoretical Global Seismology*, Princeton Univ. Press, Princeton, N. J.

Dmitruk, P., W. H. Matthaeus, and L. J. Lanzerotti (2004), Discrete modes and turbulence in a wave-driven strongly magnetized plasma, *Geophys. Res. Lett.*, *31*, L21805, doi:10.1029/2004GL021119.

Doob, J. L. (1953), *Stochastic Processes*, John Wiley, New York.

Dungey, J. W. (1958), *Cosmic Electrodynamics*, Cambridge Univ. Press, Cambridge, U. K.

Fujita, S., T. Tanaka, and T. Motoba (2011), Long-period ULF waves driven by periodic solar wind disturbances, in *The Dynamic Magnetosphere*, edited by W. Liu and M. Fujimoto, pp. 39–45, Springer, Dordrecht, The Netherlands.

García, R. A., et al. (2001), Low-degree low-order solar p modes as seen by GOLF on board SOHO, *Sol. Phys.*, *200*, 361–379.

Ghosh, S., D. J. Thomson, W. H. Matthaeus, and L. J. Lanzerotti (2009), Coexistence of turbulence and discrete modes in the solar wind, *J. Geophys. Res.*, *114*, A08106, doi:10.1029/2009JA014092.

Gosling, J. T., H. Tian, and T. D. Phan (2011), Pulsed Alfvénic waves in the solar wind, *Astrophys. J. Lett.*, *737*, L35, doi:10.1088/2041-8205/737/2/L35.

Guenther, D. B., P. Demarque, Y.-C. Kim, and M. H. Pinsonneault (1992), Standard solar models, *Astrophys. J.*, *387*, 372–393.

Hartinger, M., V. Angelopoulos, M. B. Moldwin, Y. Nishimura, D. L. Turner, K.-H. Glassmeier, M. G. Kivelson, J. Matzka, and C. Stolle (2012), Observations of a Pc5 global (cavity/waveguide) mode outside the plasmasphere by THEMIS, *J. Geophys. Res.*, *117*, A06202, doi:10.1029/2011JA017266.

Harvey, J. W. (1995), Helioseismology, *Phys. Today*, *48*, 32–38.

Hasegawa, H., M. Fujimoto, T. Phan, H. Reème, A. Balogh, M. W. Dunlop, C. Hashimoto, and R. TanDokoro (2004), Transport of solar wind into Earth's magnetosphere through rolled-up Kelvin-Helmholtz vortices, *Nature*, *430*, 755–758.

Haykin, S., D. J. Thomson, and J. Reed (2009), Spectrum sensing for cognitive radio, *Proc. IEEE*, *97*, 849–877.

He, H., and D. J. Thomson (2009a), The canonical bicoherence: Part I – Definitions, properties, and multitaper estimates, *IEEE Trans. Signal Process.*, *57*, 1273–1284.

He, H., and D. J. Thomson (2009b), The canonical bicoherence: Part II – Statistics, and application in geomagnetic data analysis, *IEEE Trans. Signal Process.*, *57*, 1285–1292.

Hurtaud, Y., C. Peymirat, and A. D. Richmond (2007), Modeling seasonal and diurnal effects on ionospheric conductances, region-2 currents, and plasma convection in the inner magnetosphere, *J. Geophys. Res.*, *112*, A09217, doi:10.1029/2007JA012257.

Jain, K., S. C. Tripathy, and F. Hill (2009), Solar activity phases and intermediate-degree mode frequencies, *Astrophys. J.*, *695*, 1567–1576.

Jiménez, A. (2006), An estimation of the acoustic cutoff frequency of the Sun based on the properties of the low-degree pseudo-modes, *Astrophys. J.*, *646*, 1398–1404.

Kendall, M. G., and A. Stuart (1963), *The Advanced Theory of Statistics*, Hafner, New York.

Kim, Y. C., and E. J. Powers (1978), Digital bispectral analysis of self-excited fluctuation spectra, *Phys. Fluids*, *21*(8), 1452–1453.

Kivelson, M. G. (2006), ULF waves from the ionosphere to the outer planets, in *Magnetospheric ULF Waves: Synthesis and New Directions, Geophys. Monogr. Ser.*, vol. 169, edited by K. Takahashi et al., pp. 11–30, AGU, Washington, D. C., doi:10.1029/169GM04.

Kolmogorov, A. N. (1960), On the $\Phi^{(n)}$ classes of Fortet and Blanc-Lapierre, *Theory Probab. Appl.*, *5*, 337.

Lancaster, H. O. (1969), *The Chi-Squared Distribution*, John Wiley, New York.

Lee, D.-H., and K. Takahashi (2006), MHD eigenmodes in the inner magnetosphere, in *Magnetospheric ULF Waves: Synthesis and New Directions, Geophys. Monogr. Ser.*, vol. 169, edited by K. Takahashi et al., pp. 73–89, AGU, Washington, D. C., doi:10.1029/169GM07.

Leighton, R. B. (1960), Transcript of talk, *Proc. IAU Symp.*, *12*, 321–325.

Li, X., M. Temerin, D. N. Baker, and G. D. Reeves (2011), Behavior of MeV electrons at geosynchronous orbit during last two solar cycles, *J. Geophys. Res.*, *116*, A11207, doi:10.1029/2011JA016934.

Loève, M. (1963), *Probability Theory*, D. Van Nostrand, Princeton, N. J.

Mallows, C. L. (1967), Linear processes are nearly Gaussian, *J. Appl. Probab.*, *4*, 313–329.

Mann, I. R., A. N. Wright, and P. S. Cally (1995), Coupling of magnetospheric cavity modes to field line resonances: A study of resonance widths, *J. Geophys. Res.*, *100*(A10), 19,441–19,456.

Mathur, S., S. Turck-Chièze, S. Couvidat, and R. A. García (2007), On the characteristics of the solar gravity mode frequencies, *Astrophys. J.*, *668*, 594–602.

McWhorter, L. T., and L. L. Scharf (1998), Multiwindow estimators of correlation, *IEEE Trans. Signal Process.*, *46*, 440–448.

Menk, F. W. (2011), Magnetospheric ULF waves: A review, in *The Dynamic Magnetosphere*, edited by W. Liu and M. Fujimoto, pp. 223–256, Springer, Dordrecht, The Netherlands.

Nolte, J. T., and E. C. Roelof (1973), Large-scale structure of the interplanetary medium I: High coronal source longitude of the quiet-time solar wind, *Sol. Phys.*, *33*, 241–257.

Parkinson, W. D. (1983), *Introduction to Geomagnetism*, Elsevier, New York.

Piersanti, M., U. Villante, C. Waters, and I. Coco (2012), The 8 June 2000 ULF wave activity: A case study, *J. Geophys. Res.*, *117*, A02204, doi:10.1029/2011JA016857.

Provost, J., G. Berthomieu, and P. Morel (2000), Low-frequency *p*- and *g*-mode solar oscillations, *Astron. Astrophys.*, *353*, 775–785.

Rayleigh, L. (1903), On the spectrum of an irregular disturbance, *Philos. Mag.*, *41*, 238–243. [Reprinted in *Scientific Papers by Lord Rayleigh*, vol. V, Article 285, pp. 98–102, Dover, New York, 1964.]

Roberts, D. A., K. W. Ogilvie, and M. L. Goldstein (1996), The nature of the solar wind, *Nature*, *381*, 31–32.

Salabert, D., J. Leibacker, T. Appourchaux, and F. Hill (2009), Measurement of low signal-to-noise ratio solar *p*-modes in spatially resolved helioseismic data, *Astrophys. J.*, *696*, 653–667.

Samson, J. C., B. G. Harrold, J. M. Ruohoniemi, R. A. Greenwald, and A. D. M. Walker (1992), Field line resonances associated with MHD waveguides in the magnetosphere, *Geophys. Res. Lett.*, *19*(5), 441–444.

Slepian, D. (1978), Prolate spheroidal wave functions, Fourier analysis, and uncertainty V: The discrete case, *Bell Syst. Tech. J.*, *57*, 1371–1429.

Stix, M. (2004), *The Sun*, 2nd ed., Springer, Berlin.

Stoica, P., and T. Sundin (1999), On nonparametric spectral estimation, *Circuits Syst. Signal Process.*, *18*, 169–181.

Takahashi, K., P. J. Chi, R. E. Denton, and R. L. Lysak (Eds.) (2006), *Magnetospheric ULF Waves: Synthesis and New Directions, Geophys. Monogr. Ser.*, vol. 169, 359 pp., AGU, Washington, D. C., doi:10.1029/GM169.

Thomson, D. J. (1982), Spectrum estimation and harmonic analysis, *Proc. IEEE*, *70*, 1055–1096.

Thomson, D. J. (2007), Jackknifing multitaper spectrum estimates, *IEEE Signal Process. Mag.*, *24*(7), 20–30.

Thomson, D. J., and A. D. Chave (1991), Jackknifed error estimates for spectra, coherences, and transfer functions, in *Advances in Spectrum Analysis and Array Processing*, vol. 1, edited by S. Haykin, chap. 2, pp. 58–113, Prentice Hall, Upper Saddle River, N. J.

Thomson, D. J., C. G. Maclennan, and L. J. Lanzerotti (1995), Propagation of solar oscillations through the interplanetary medium, *Nature*, *376*, 139–144.

Thomson, D. J., C. G. Maclennan, and L. J. Lanzerotti (1996), The nature of the solar wind, *Nature*, *381*, 32, doi:10.1038/381032a0.

Thomson, D. J., L. J. Lanzerotti, and C. G. Maclennan (2000), Coherent frequency variations in electron fluxes at 1 and 5 AU in the inner heliosphere, in *Acceleration and Transport of Energetic Particles Observed in the Heliosphere*, edited by R. A. Mewaldt, *AIP Conf. Proc.*, *528*, 278–281.

Thomson, D. J., L. J. Lanzerotti, and C. G. Maclennan (2001), Interplanetary magnetic field: Statistical properties and discrete modes, *J. Geophys. Res.*, *106*(A8), 15,941–15,962.

Thomson, D. J., L. J. Lanzerotti, F. L. Vernon III, M. R. Lessard, and L. T. P. Smith (2007), Solar modal structure of the engineering environment, *Proc. IEEE*, *95*, 1085–1132.

Tu, C., and E. Marsch (1995), MHD structures, waves, and turbulence in the solar wind, *Space Sci. Rev.*, *73*, 1–210.

Turck-Chièze, S., et al. (2004a), Looking for gravity-mode multiplets with the GOLF experiment aboard SOHO, *Astrophys. J.*, *604*, 455.

Turck-Chièze, S., et al. (2004b), Erratum: "Looking for gravity-mode multiplets with the GOLF experiment aboard SOHO", *Astrophys. J.*, *608*, 610.

Viall, N. M., L. Kepko, and H. E. Spence (2009), Relative occurrence rates and connection of discrete frequency oscillations in the solar wind density and dayside magnetosphere, *J. Geophys. Res.*, *114*, A01201, doi:10.1029/2008JA013334.

Villante, U., P. Francia, M. Vellante, P. Di Giuseppe, A. Nubile, and M. Piersanti (2007), Long-period oscillations at discrete frequencies: A comparative analysis of ground, magnetospheric, and interplanetary observations, *J. Geophys. Res.*, *112*, A04210, doi:10.1029/2006JA011896.

Walker, A. D. M. (2005), Excitation of field line resonances by sources outside the magnetosphere, *Ann. Geophys.*, *23*, 3375–3388.

Woodard, M. F., and R. W. Noyes (1985), Change in solar oscillation eigenfrequencies with the solar cycle, *Nature*, *318*, 449–450.

Wright, A. N., and I. R. Mann (2006), Global MHD eigenmodes of the outer magnetosphere, in *Magnetospheric ULF Waves: Synthesis and New Directions*, *Geophys. Monogr. Ser.*, vol. 169, edited by K. Takahashi et al., pp. 51–72, AGU, Washington, D. C., doi:10.1029/169GM06.

D. J. Thomson, Department of Mathematics and Statistics, Queen's University, Kingston, ON K7L 3N6, Canada. (djt@mast.queensu.ca)

Generation Processes of Whistler Mode Chorus Emissions: Current Status of Nonlinear Wave Growth Theory

Yoshiharu Omura

Research Institute for Sustainable Humanosphere, Kyoto University, Kyoto, Japan

David Nunn

Research Institute for Sustainable Humanosphere, Kyoto University, Kyoto, Japan
School of Electronics and Computer Science, University of Southampton, Southampton, UK

Danny Summers

Department of Mathematics and Statistics, Memorial University of Newfoundland, St. John's, Newfoundland, Canada
School of Space Research, Kyung Hee University, Yongin, South Korea

There has been significant progress in understanding the generation mechanism of whistler mode chorus emissions in recent years. This is partly due to the successful reproduction of chorus emissions by computer simulations and partly because of precise observations of chorus emissions by spacecraft. We give a brief review of the nonlinear processes related to the generation mechanism of chorus emissions that have been revealed by the simulations or observations. We describe the nonlinear dynamics of resonant electrons and the formation of electromagnetic electron holes or hills that result in resonant currents generating rising-tone emissions or falling-tone emissions, respectively. We also describe the mechanism of nonlinear wave damping due to quasi-oblique propagation, which results in the formation of a gap at half the electron cyclotron frequency.

1. INTRODUCTION

Coherent electromagnetic waves called chorus emissions have been frequently observed in the inner magnetosphere [e.g., *Tsurutani and Smith*, 1974; *Meredith et al.*, 2003; *Santolík et al.*, 2003; *Santolík*, 2008; *Sigsbee et al.*, 2010; *Kasashara et al.*, 2009]. Whistler mode chorus emissions typically consist of a series of rising tones. They are generated near the

Dynamics of the Earth's Radiation Belts and Inner Magnetosphere
Geophysical Monograph Series 199
10.1029/2012GM001347

magnetic equator through interaction with energetic electrons from several keV to tens of keV injected into the inner magnetosphere at the time of a geomagnetic disturbance. In recent years, chorus emissions have been studied extensively because of their role as a viable mechanism for accelerating radiation belt electrons [*Summers et al.*, 1998, 2002; *Albert*, 2002; *Omura et al.*, 2007; *Katoh and Omura*, 2007a; *Summers and Omura*, 2007; *Katoh et al.*, 2008].

Numerical modeling of chorus emissions have been performed using a Vlasov-hybrid simulation (VHS) based on simplified field equations derived from Maxwell's equations under the assumption of a coherent whistler mode wave [*Nunn et al.*, 1997, 2009]. The initial wave amplitude and the wave phase are specified in such simulations. The VHS code has a finite bandwidth ~100 Hz and so is unable to simulate the physics whereby one chorus element is triggered by the previous one, which requires a full broadband code. The VHS also requires a trigger signal within the simulation bandwidth. On the other hand, chorus emissions with rising tones were reproduced successfully in an electron-hybrid electromagnetic code starting from broadband thermal noise. Here Maxwell's equations are solved directly together with the electron fluid equation for the cold dense electrons and with the equations of motion for the hot resonant electrons [Katoh and Omura, 2007a, 2007b]. The mechanism of rising chorus emissions has been analyzed theoretically in terms of nonlinear wave growth due to the formation of an electromagnetic electron hole in velocity phase space [Omura et al., 2008]. The relation between the wave amplitude and the frequency sweep rate in the generation region of chorus emissions has been derived [Omura et al., 2008, Equation (50)], which matches the rates from observations [Cully et al., 2011]. The validity of this relation has been demonstrated in a full-particle electromagnetic simulation [Hikishima et al., 2009] as well as in an electron-hybrid simulation [Katoh and Omura, 2007a]. These simulations show that seeds of chorus emissions with rising tones are formed in a localized region near the magnetic equator. The seed emissions grow as a result of the formation of a resonant current arising from nonlinear trajectories of resonant untrapped electrons.

2. DYNAMICS OF RESONANT ELECTRONS

We assume a coherent electromagnetic wave propagating parallel to a static magnetic field $\boldsymbol{B}_0$ directed along the h axis, and h is the distance along the magnetic field line from the magnetic equator. The wavefields are in the transverse plane containing x and y axes. We express the electric and magnetic field vectors of the wave in the transverse plane by the complex forms $\overrightarrow{E_w} = E_w\exp(i\psi_E)$ and $\overrightarrow{B_w} = B_w\exp(i\psi_B)$, respectively. We assume an electron with a parallel velocity $v_\parallel$, a perpendicular velocity $v_\perp$, and a phase ϕ of $v_\perp$ in a transverse plane to the static magnetic field $\boldsymbol{B}_0$. We define the relative phase angle ζ between the perpendicular velocity and the wave magnetic field $\overrightarrow{B_w}$ defined by

$$\zeta = \phi - \psi_B. \tag{1}$$

We also define the difference between the parallel velocity $v_\parallel$ and the resonance velocity V_R as

$$\theta = k(v_\parallel - V_R), \tag{2}$$

where k is the wavenumber given by $k = -\partial\psi_B/\partial h$, and V_R is the resonance velocity given by $V_R = (\omega - \Omega_e/\gamma)/k$. The parameters Ω_e and γ are the electron cyclotron frequency

and the Lorenz factor given by $\gamma = [1 - (v/c)^2]^{-1/2}$, respectively, where c is the speed of light. The frequency ω is considered to be variable and given by $\omega = \partial\psi_B/\partial t$. Using the relativistic equations of motion for a relativistic electron interacting with the whistler mode wave, we derive simplified equations of motion with the average perpendicular velocity $V_{\perp 0}$ [Omura et al., 2008].

$$\frac{d\zeta}{dt} = \theta, \tag{3}$$

and

$$\frac{d\theta}{dt} = \omega_{tr}^2(\sin\zeta + S), \tag{4}$$

where we find the relativistic trapping frequency $\omega_{tr} = \omega_t\chi\gamma^{-1/2}$, and we have assumed that $v_\parallel \sim V_R$, i.e., $\theta \sim 0$. Here ω_t is the nonrelativistic trapping frequency given by $\omega_t = \sqrt{kV_{\perp 0}\Omega_w}$ [Omura et al., 1991], and Ω_w is the normalized wave amplitude defined by $\Omega_w = eB_w/m_0$, where $-e$ and m_0 are the charge and rest mass of an electron, respectively. The condition for nonlinear trapping of a resonant electron is given by $d\theta/dt = d^2\zeta/dt^2 = 0$, which is called the second-order resonance condition. The shape of the electron hole is determined by the inhomogeneity factor S given by

$$S = -\frac{1}{s_0\omega\Omega_w}\left(s_1\frac{\partial\omega}{\partial t} + cs_2\frac{\partial\Omega_e}{\partial h}\right), \tag{5}$$

where

$$s_0 = \frac{\chi}{\xi}\frac{V_{\perp 0}}{c}, \tag{6}$$

$$s_1 = \gamma\left(1 - \frac{V_R}{V_g}\right)^2, \tag{7}$$

and

$$s_2 = \frac{1}{2\xi\chi}\left\{\frac{\gamma\omega}{\Omega_e}\left(\frac{V_{\perp 0}}{c}\right)^2 - \left[2 + \Lambda\frac{\chi^2(\Omega_e - \gamma\omega)}{\Omega_e - \omega}\right]\frac{V_R V_p}{c^2}\right\}, \tag{8}$$

where $\xi^2 = \omega(\Omega_e - \omega)/\omega_{pe}^2$ and $\chi^2 = 1 - (\omega/ck)^2$ [Omura et al., 2008]. The parameter ω_{pe} is the electron plasma frequency. When we evaluate γ in the equations above, we substitute $v_\parallel = V_R$ and $v_\perp = V_{\perp 0}$. We have incorporated the variation of the cold electron density $N_e(h)$ along the magnetic field line as $N_e(h) = N_{e0}\Omega_e(h)/\Omega_{e0}$, where N_{e0} and Ω_{e0} are, respectively, the cold electron density and the electron cyclotron frequency at the equator. We have $\Lambda = \omega/\Omega_e$ for this inhomogeneous electron density model, while $\Lambda = 1$ for the constant electron density model [Omura et al., 2009]. We have plotted the trajectories of resonant electrons in the $\zeta - \theta$

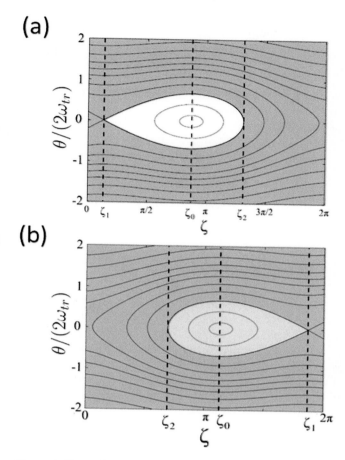

(a)

(b)

Figure 1. Illustration of the wave-trapping potential in the velocity phase space ($\theta - \zeta$). The contours correspond to trajectories of resonant electrons. Because of the difference in the numbers of trapped electrons and untrapped resonant electrons, there arises either (a) an electron hole with $S = -0.4$ or (b) an electron hill with $S = 0.4$. In both cases, the phase-bunched resonant electrons in the phase range $\pi < \zeta < 2\pi$ result in a negative J_E causing wave growth. The electron hole results in a negative J_B inducing a rising-tone emission, while the electron hill causes a positive J_B inducing a falling-tone emission.

phase space with $S = -0.4$ and $S = 0.4$ in Figures 1a and 1b, respectively.

For the case of constant-frequency field-aligned waves, the inhomogeneity factor S given by equation (5) is identically equal to the parameter R used by *Albert* [2000] in a Hamiltonian formulation of the resonant interaction of a charged particle with a monochromatic electromagnetic wave.

3. RESONANT CURRENTS

A coherent whistler mode wave is excited by a linear instability driven by temperature anisotropy near the equator.

It can also be artificially excited by an antenna of a spacecraft, a virtual antenna of an ionospheric current modulated by high-frequency waves from the ground, or an antenna of VLF transmitters on the ground. A coherent wave with a finite amplitude forms a wave potential around the resonance velocity that can trap some of the energetic resonant electrons or distort the adiabatic trajectories of untrapped resonant electrons. As a result of the inhomogeneous magnetic field, trajectories of trapped and untrapped electrons become significantly different, and there arises a hole or a hill in the velocity phase space, as shown in Figure 1. The deformation of the velocity distribution function of energetic electrons around the resonance velocity results in currents formed by resonant electrons that modify the coherent wave.

From Maxwell's equations, we obtain the following equation for the amplitude B_w of the wave magnetic field in the form [*Nunn*, 1974; *Omura et al.*, 2008],

$$\frac{\partial B_w}{\partial t} + V_g \frac{\partial B_w}{\partial h} = -\frac{\mu_0 V_g}{2} J_E, \tag{9}$$

where μ_0 and J_E are the vacuum permeability and the component of the resonant current parallel to the wave electric field, respectively. Under the assumption that the growth rate ω_i is much smaller than the wave frequency ω, i.e., $\omega_i \ll \omega$, the resonant current parallel to the wave magnetic field J_B is neglected. This ensures that the frequency ω is constant in the frame of reference moving with the group velocity V_g as expressed by the equation,

$$\frac{\partial \omega}{\partial t} + V_g \frac{\partial \omega}{\partial h} = 0. \tag{10}$$

The frequency ω and wave number k satisfy the cold plasma dispersion relation for the whistler mode wave, which we write as

$$\chi^2 = \frac{1}{1 + \xi^2}. \tag{11}$$

Using the parameters ξ and χ, we express the phase velocity and group velocity of the whistler mode wave as [*Omura et al.*, 2008]

$$V_p = \frac{\omega}{k} = c\chi\xi \tag{12}$$

and

$$V_g = \frac{c\xi}{\chi} \left[\xi^2 + \frac{\Omega_e}{2(\Omega_e - \omega)} \right]^{-1}. \tag{13}$$

The electron resonance velocity for an electron with a speed v is then

$$V_R = c\chi\xi \left(1 - \frac{\Omega_e}{\gamma\omega} \right). \tag{14}$$

Solving equations (3) and (4), we find that the first-order resonance condition $d\zeta/dt = 0$ and the second-order resonance condition $d^2\zeta/dt^2 = d\theta/dt = 0$ are satisfied at two different phases ζ_0 and ζ_1 that satisfy $\sin \zeta + S = 0$. At $\theta = 0$, the phase ζ_0 represents a stable equilibrium point around which trapped resonant electrons rotate in time. The phase ζ_1 represents an unstable saddle equilibrium point. The separatrix of the trapping region that originates from this saddle point ($\theta = 0$ and $\zeta = \zeta_1$) is given by

$$\theta_s(\zeta) = \pm\omega_{tr}\sqrt{2[\cos\zeta_1 - \cos\zeta + S(\zeta - \zeta_1)]}, \quad (15)$$

for $\zeta_1 < \zeta < \zeta_2$, where ζ_2 is the phase of the separatrix crossing $\theta = 0$. The separatrix boundary of the trapping region plays an important role in the formation of the resonant current J_E, as reported by *Omura and Summers* [2006]. Under the action of a slowly varying whistler mode wave packet, there develops an electromagnetic electron hole in the phase space ($v_\parallel$, ζ) as shown in Figure 1a for $S = -0.4$. (trajectories of resonant electrons with different values of S are found in the work of *Omura and Matsumoto* [1982]) This is because the initially trapped resonant electrons inside the trapping region are guided along the resonance velocity, while the untrapped electrons outside the trapping region cannot cross the separatrix so long as the structure of the trapping region remains constant.

The currents J_E and J_B are components of the resonant current, parallel to the wave electric field $\overrightarrow{E}_w$ and wave magnetic field $\overrightarrow{B}_w$, respectively. The resonant current is formed by the interaction of resonant electrons with the wave. Expressions for J_E and J_B are

$$J_E = \int_0^\infty \int_0^{2\pi} \int_{-\infty}^\infty [ev_\perp \sin \zeta] f(v_\parallel, \zeta, v_\perp) v_\perp dv_\parallel d\zeta dv_\perp \quad (16)$$

and

$$J_B = \int_0^\infty \int_0^{2\pi} \int_{-\infty}^\infty [-ev_\perp \cos \zeta] f(v_\parallel, \zeta, v_\perp) v_\perp dv_\parallel d\zeta dv_\perp, \quad (17)$$

where $f(v_\parallel, \zeta, v_\perp)$ is the distribution function of energetic electrons. It should be noted that the cold electrons supporting the whistler mode waves are not included in the distribution function. It will be seen in the following section that the component current J_B occurs in the nonlinear correction term to the wave dispersion relation.

From the analysis of trajectories of resonant electrons as described by equations (3) and (4), it is found that the maximum value of J_E is realized when $S = -0.4$ [*Omura et al.*, 2008]. The magnitude of J_E is calculated by assuming a distribution function in the velocity phase space in the presence of a coherent whistler mode wave as

$$g(v_\parallel, \zeta) = g_0(v_\parallel) - Qg_t(v_\parallel, \zeta), \quad (18)$$

and we have

$$J_E = -eQV_{\perp 0}^2 \int_0^{2\pi} \int_{-\infty}^\infty g_t(v_\parallel, \zeta) \sin \zeta dv_\parallel d\zeta, \quad (19)$$

where we have assumed a Dirac delta function $\delta(v_\perp - V_{\perp 0})$ for the perpendicular velocity $v_\perp$. The functions $g_0(v_\parallel)$ and $g_t(v_\parallel, \zeta)$ are the unperturbed velocity distribution function and the part of g_0 that corresponds to trapping by the wave.

The resonant currents J_E and J_B are expressed, respectively, as

$$J_E = -J_0 \int_{\zeta_1}^{\zeta_2} [\cos \zeta_1 - \cos \zeta + S(\zeta - \zeta_1)]^{1/2} \sin \zeta d\zeta, \quad (20)$$

and

$$J_B = J_0 \int_{\zeta_1}^{\zeta_2} [\cos \zeta_1 - \cos \zeta + S(\zeta - \zeta_1)]^{1/2} \cos \zeta d\zeta, \quad (21)$$

where $J_0 = (2e)^{3/2}(m_0 k\gamma)^{-1/2} V_{\perp 0}^{5/2} \chi QGB_w^{1/2}$, and e and m_0 are the charge and rest mass of an electron. The factor Q represents the depth of the electron hole [*Omura et al.*, 2009]. The phase angles ζ_1 and ζ_2 define the boundary of the trapping wave potential as described by *Omura et al.* [2009]. The parameter G is the value of the velocity distribution function $g(v_\parallel, \zeta)$ in the trapping region around the resonance velocity.

4. GENERATION PROCESS OF CHORUS EMISSIONS

From Maxwell's equations, we obtain the wave equation (9) describing the evolution of the wave amplitude, and the dispersion relation with a nonlinear correction term,

$$c^2 k^2 - \omega^2 - \frac{\omega \omega_{pe}^2}{\Omega_e - \omega} = \mu_0 c^2 k \frac{J_B}{B_w}, \quad (22)$$

where μ_0 is the magnetic permittivity in vacuum. Details of the derivation of equations (9) and (22) are provided in Appendix A of *Omura et al.* [2008]. The currents J_E and J_B are components of the resonant current, parallel to the wave electric field $\overrightarrow{E}_w$ and wave magnetic field $\overrightarrow{B}_w$, respectively.

While the resonant current J_E modifies the wave amplitude B_w, the quantity J_B/B_w changes the frequency ω of the triggered wave. It is noted that the wave number k or the wavelength does not change in space and time because it is imposed by the triggering wave with the constant frequency ω_0 in the present situation. Denoting the frequency deviation from ω_0 as ω_1 (where $\omega = \omega_0 + \omega_1$) and assuming $\omega_1 \ll \omega_0$, we expand equation (22) around ω_0 to obtain

$$\omega_1 = -\frac{\mu_0 V_g}{2} \frac{J_B}{B_w}, \quad (23)$$

where V_g is given by equation (13).

We assume that the velocity distribution function f of hot energetic electrons is given in terms of the relativistic momentum per unit mass $u = \gamma v$; u has components $u_\parallel = \gamma v_\parallel$ and $u_\perp = \gamma v_\perp$, respectively, parallel and perpendicular to the ambient magnetic field. We specify f as

$$f(u_\parallel, u_\perp) = \frac{N_h}{(2\pi)^{3/2} U_{t\parallel} U_{\perp 0}} \exp\left(-\frac{u_\parallel^2}{2 U_{t\parallel}^2}\right) \delta(u_\perp - U_{\perp 0}), \quad (24)$$

where $U_{\perp 0} = \gamma V_{\perp 0}$, $U_{t\parallel}$ is the thermal momentum in the parallel direction, and δ is the Dirac delta function, and we have normalized f to the density of hot electrons N_h. Integrating over $u_\perp$ and taking an average over ζ, we obtain the magnitude G of the unperturbed distribution function $g(v_\parallel, \zeta)$ at the resonance velocity V_R as

$$G = \frac{N_h}{(2\pi)^{3/2} U_{t\parallel} U_{\perp 0}} \exp\left(-\frac{\gamma^2 V_R^2}{2 U_{t\parallel}^2}\right). \quad (25)$$

Omura et al. [2008] found that the maximum value of $-J_E$ takes place when $S = -0.413$. Solving equation (21) for $S = -0.413$, we obtain $J_B = -1.3 J_0$, which we rewrite as

$$J_B = -1.3(2e)^{3/2} \left(\frac{B_w}{m_0 k \gamma}\right)^{1/2} V_{\perp 0}^{5/2} \chi Q G. \quad (26)$$

The nonlinear transition time T_N for formation of the nonlinear resonant current is roughly estimated by the nonlinear trapping period T_{tr} given by

$$T_{tr} = \frac{2\pi}{\omega_{tr}} = \frac{2\pi}{\chi} \left(\frac{m_0 \gamma}{k V_{\perp 0} e B_w}\right)^{1/2}. \quad (27)$$

We define a ratio $\tau = T_N / T_{tr}$, which is to be determined by numerical simulations and observations.

During a time interval equal to the nonlinear transition time T_N, an electron hole is gradually formed. Along with the formation of J_B, the frequency of the triggered wave gradually deviates from ω_0 to $\omega_0 + \omega_1$. From equations (23), (25), (26), and (27), we obtain the frequency sweep rate over the nonlinear trapping period as

$$\frac{\omega_1}{T_N} = \frac{1.3}{4} \pi^{-5/2} \frac{Q}{\tau} \left(\frac{\chi \omega_{ph} V_{\perp 0}}{\gamma c}\right)^2 \frac{V_g}{U_{t\parallel}} \exp\left(-\frac{\gamma^2 V_R^2}{2 U_{t\parallel}^2}\right), \quad (28)$$

where ω_{ph} is the plasma frequency of the hot energetic electrons defined by $\omega_{ph}^2 = \mu_0 c^2 N_h e^2 / m_0$.

At the equator, the inhomogeneity of the magnetic field is zero, and the second term on the right-hand side of equation (5) vanishes. Since the maximum nonlinear wave growth takes place when $S = -0.4$ [*Omura et al.*, 2008], we can derive from equation (5) the relation between the frequency sweep rate and the normalized wave amplitude at the equator $\Omega_{w0} = e B_{w0} / m_0$ in the form,

$$\frac{\partial \omega}{\partial t} = \frac{0.4 s_0 \omega_0}{s_1} \Omega_{w0}. \quad (29)$$

Equating the left-hand sides of equations (28) and (29), we obtain an optimum wave amplitude Ω_{opt} that can trigger the rising-tone chorus element as

$$\tilde{\Omega}_{opt} = 0.81 \pi^{-5/2} \frac{Q}{\tau} \frac{s_1 \tilde{V}_g}{s_0 \tilde{\omega} \tilde{U}_{t\parallel}} \left(\frac{\chi \omega_{ph} \tilde{V}_{\perp 0}}{\gamma}\right)^2 \exp\left(-\frac{\gamma^2 \tilde{V}_R^2}{2 \tilde{U}_{t\parallel}^2}\right), \quad (30)$$

where $\tilde{\Omega}_{opt} = \Omega_{opt} / \Omega_{e0}$, $\tilde{\omega}_{ph} = \omega_{ph} / \Omega_{e0}$, and $\tilde{U}_{t\parallel} = U_{t\parallel} / c$. In Figure 2a, we have plotted the optimum wave amplitude given by equation (30) in solid lines. The numbers attached to the lines are the time scale factors τ. We also plotted the threshold for the nonlinear wave growth, which is the subject of the next section. We can calculate the nonlinear transition time T_N using the wave amplitude obtained above. Using equation (12), we rewrite equation (27) as

$$T_N \Omega_{e0} = 2\pi \tau \left(\frac{\gamma \xi}{\chi \tilde{\omega} \tilde{V}_{\perp 0} \tilde{\Omega}_{w0}}\right)^{1/2}, \quad (31)$$

which is plotted in Figure 2b for different values of τ. The nonlinear transition time is roughly the time between the onset of the triggering wave and the start of the nonlinear wave growth. The value can be obtained from simulations and observations, and it can be used in determining the time scale factor τ used in the theory.

5. NONLINEAR WAVE GROWTH

We consider the wave evolution in time in a frame of reference moving with the group velocity V_g as described by equation (9). The variation of the wave amplitude is due to the resonant current $-J_E$, which is a function of S and maximizes at $S = -0.4$. The maximum value is given by $-J_E / J_0 = 0.975 \sim 1$. We, thus, have

$$J_E = -(2e)^{3/2} (m_0 k \gamma)^{-1/2} V_{\perp 0}^{5/2} \chi Q G B_w^{1/2}. \quad (32)$$

Writing the right-hand side of equation (9) as dB_w/dt, we obtain

$$\frac{dB_w}{dt} = \Gamma_N B_w, \quad (33)$$

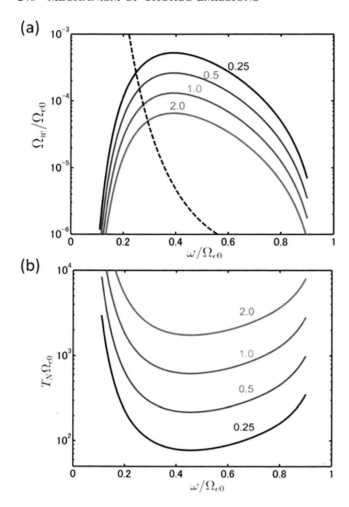

Figure 2. (a) The optimum wave amplitudes (solid lines) for triggering rising tone emissions with different values of the time scale factor τ (attached numbers) and the threshold of the wave amplitude for nonlinear wave growth (dashed line) with the parameters used in the Vlasov-hybrid simulation run. (b) The corresponding nonlinear transition time T_N for formation of the nonlinear resonant current $-J_B$ with different values of the time scale factor τ. From *Omura and Nunn* [2011].

where we define

$$\Gamma_N = \frac{Q\omega_{ph}^2}{2}\left(\frac{\xi}{\omega\Omega_w}\right)^{1/2}\frac{V_g}{U_{t\parallel}}\left(\frac{\chi V_{\perp 0}}{c\pi\gamma}\right)^{3/2}\exp\left(-\frac{\gamma^2 V_R^2}{2U_{t\parallel}^2}\right) \quad (34)$$

as the nonlinear growth rate. The parameter ω_{ph} is the plasma frequency of hot electrons given by $\omega_{ph}^2 = N_h e^2/(\varepsilon_0 m_0)$, where ε_0 is the vacuum permittivity. Figure 3a shows the variation of Γ_N as a function of the wave frequency for different plasma frequencies indicated by the attached numbers. *Sum-*

mers et al. [2012a] generalize the expression (34) for Γ_N to an arbitrary particle distribution function.

As described by equation (10), the wave frequency is constant in the frame of reference moving with the group velocity. For a frequency variation $\partial\omega/\partial t \neq 0$, the wave should grow in time at a fixed position. Expressing the derivative dB_w/dt in equation (33) in terms of temporal and spatial derivatives, and normalizing the wave amplitude, we obtain

$$\frac{\partial\Omega_w}{\partial t} + V_g\frac{\partial\Omega_w}{\partial h} = \Gamma_N\Omega_w. \quad (35)$$

For chorus emissions to grow at the equator, the temporal growth rate should be positive, namely, $\partial\Omega_w/\partial t > 0$. From equation (35), we therefore obtain

$$\frac{\partial\Omega_w}{\partial h} < \frac{\Gamma_N}{V_g}\Omega_w, \quad (36)$$

where we have assumed that the chorus waves propagate in the positive direction, i.e., $V_g > 0$.

Near the magnetic equator, we assume a parabolic variation along the magnetic field line, which is specified by the L value and the Earth's radius R_E, as expressed by $\Omega_e = \Omega_{e0}(1 + ah^2)$ with $a = 4.5/(LR_E)^2$. Noting that $\partial\Omega_e/\partial h = 2a\Omega_{e0}h$, we consider the distance h_c at which the first and second terms of the right-hand side of equation (5) become equal. Equating the two terms and using equation (29), we obtain the critical distance h_c as

$$h_c = \frac{s_0\omega\Omega_{w0}}{5cas_2\Omega_{e0}}. \quad (37)$$

The distance h_c is used in identifying the dominant terms of the inhomogeneity factor S as we discuss in the following.

As the chorus emission propagates further from the equator to the distance h ($\gg h_c$), the second term of the inhomogeneity factor (equation (5)) becomes much greater than the first term. For the chorus element to maintain maximum growth at this distance, a negative resonant current J_E must be formed with $S = -0.4$. Neglecting the first term on the right-hand side of equation (5) and setting $S = -0.4$, we obtain

$$\Omega_w = \frac{cs_2}{0.4s_0\omega}\frac{\partial\Omega_e}{\partial h}. \quad (38)$$

Taking the spatial derivative of equation (38), we obtain

$$\frac{\partial\Omega_w}{\partial h} = \frac{cs_2}{0.4s_0\omega}\frac{\partial^2\Omega_e}{\partial h^2} = \frac{5cas_2\Omega_{e0}}{s_0\omega}. \quad (39)$$

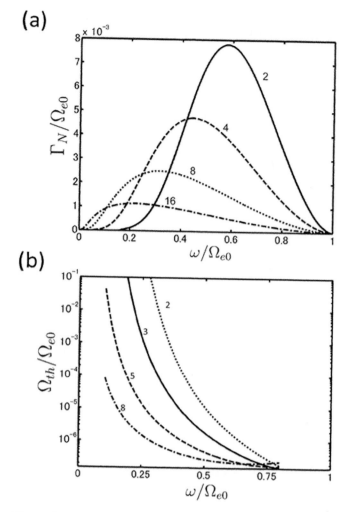

(a)

(b)

Figure 3. (a) Nonlinear growth rate Γ_N as a function of wave frequency ω for the plasma frequencies $\omega_{pe}/\Omega_{eo} = 2, 4, 8$, and 16 (indicated by the attached numbers) and the parameters $U_{t\parallel}/c = 0.18$, $V_{\perp 0}/c = 0.21$, $\omega_{ph}/\Omega_{e0} = 0.2$, $Q = 0.5$, and $\Omega_w/\Omega_{e0} = 0.0001$. (b) The threshold amplitude Ω_{th} for the generation of self-sustaining chorus emissions for the plasma frequencies $\tilde{\omega}_{pe} = 2, 3, 5$, and 8 (indicated by the attached numbers) and for the parameters $\tilde{U}_{t\parallel} = 0.18$, $\tilde{V}_{\perp 0} = 0.21$, $\tilde{a} = 2 \times 10^{-7}$, $\tilde{\omega}_{ph} = 0.2$, and $Q = 0.5$. From *Omura et al.* [2009].

Self-sustaining nonlinear wave growth during propagation near the equator, where the dipole magnetic field is approximated by the parabolic function, requires that the spatial gradient of the wave amplitude $\partial \Omega_w/\partial h$ is constant as shown in equation (39). It should be noted that the spatial gradient of the wave amplitude does not depend on the wave amplitude itself. When the optimum self-sustaining wave growth is realized in the initial generation process of a chorus element,

the gradient of the wave amplitude should be close to the value given by equation (39).

Inserting equation (39) into equation (36), we obtain the inequality,

$$\tilde{\Omega}_{w0} > \tilde{\Omega}_{th}, \qquad (40)$$

where

$$\tilde{\Omega}_{th} = \frac{100\pi^3 \gamma^3 \xi}{\tilde{\omega}\,\tilde{\omega}_{ph}^4 \tilde{V}_{\perp 0}^5 \chi^5} \left(\frac{\tilde{a} s_2 \tilde{U}_{t\parallel}}{Q} \right)^2 \exp\left(\frac{\gamma^2 \tilde{V}_R^2}{\tilde{U}_{t\parallel}^2} \right). \qquad (41)$$

We have used the normalized parameters, $\tilde{V}_{\perp 0} = V_{\perp 0}/c$, $\tilde{\omega} = \omega/\Omega_{e0}$, $\tilde{a} = ac^2/\Omega_{e0}^2$, $\tilde{U}_{t\parallel} = U_{t\parallel}/c$, $\tilde{\omega}_{ph} = \omega_{ph}/\Omega_{e0}$, $\tilde{\Omega}_{w0} = \Omega_{w0}/\Omega_{e0}$, and $\tilde{\Omega}_{th} = \Omega_{th}/\Omega_{e0}$.

It is clear from equation (38) that the self-sustaining mechanism only works for $h > 0$ with the positive gradient of the magnetic field. That is, nonlinear wave growth takes place only when the wave propagates away from the equator with an amplitude greater than the threshold amplitude Ω_{th} given by equation (41). In Figure 3b, we plot the threshold amplitude for typical parameters at Earth ($L = 4.4$) and for the normalized electron plasma frequencies $\tilde{\omega}_{pe} = 2, 3, 5, 8$. The threshold amplitude is higher for a lower wave frequency $\tilde{\omega}$ and for a smaller plasma frequency $\tilde{\omega}_{pe}$. Since the linear wave growth rate usually maximizes in the lower frequency range [e.g., *Omura and Summers*, 2004], the threshold amplitude becomes especially important for smaller plasma frequencies.

6. TEMPORAL AND SPATIAL EVOLUTION OF RISING-TONE EMISSIONS

Nonlinear wave growth is due to the formation of a resonant current as described by the second-order resonance condition, while linear wave growth is due to particle diffusion at the resonance velocity determined by the first-order resonance condition. In the linear growth phase starting from incoherent thermal noise, there arises a coherency at a frequency corresponding to the maximum linear growth rate. Once the amplitude of a coherent wave exceeds the threshold value for self-sustaining emissions, nonlinear wave growth sets in, driven by the second-order phase variation $\partial \omega/\partial t$ corresponding to the maximum value of the resonant current J_E.

We evaluate the temporal variation of the wave amplitude by assuming that the spatial derivative of the wave amplitude in equation (35) takes the threshold value for self-sustaining wave growth given by equation (41). Assuming the minimum spatial gradient of the growing wave amplitude in equation (39), and inserting this into equation (35), we derive the equation,

$$\frac{\partial \tilde{\Omega}_{w0}}{\partial \tilde{t}} = \tilde{V}_g \left[\frac{Q \tilde{\omega}_{ph}^2}{2 \tilde{U}_{t\parallel}} \left(\frac{\chi \tilde{V}_{\perp 0}}{\pi \gamma} \right)^{3/2} \left(\frac{\xi \tilde{\Omega}_{w0}}{\tilde{\omega}} \right)^{1/2} \right. \tag{42}$$

$$\left. \times \exp \left(-\frac{\gamma^2 \tilde{V}_R^2}{2 \tilde{U}_{t\parallel}^2} \right) - \frac{5 s_2 \tilde{a}}{s_0 \tilde{\omega}} \right].$$

We now rewrite equation (29) in the form,

$$\frac{\partial \tilde{\omega}}{\partial \tilde{t}} = \frac{2 s_0}{5 s_1} \tilde{\omega} \tilde{\Omega}_{w0}. \tag{43}$$

The temporal evolution of a chorus element at the equator is determined by the pair of coupled differential equations (42) and (43) referred to as "chorus equations." *Summers et al.* [2012a] generalized the chorus equations to an arbitrary distribution, and obtained numerical solutions to these equations for three specified particle distributions, and various sets of parameters. The chorus equations were used by *Summers et al.* [2011] to determine the effects of nonlinear growth of whistler mode waves on extreme radiation belt fluxes.

The spatial evolution of the chorus element is obtained by solving the wave equations (9) and (10) in space and time along with appropriate evaluation of the resonant current J_E. In calculating $J_E(h)$ at a position h, we need to approximate the distribution function of energetic electrons at h by the same form as equation (24). Assuming the adiabatic motion of energetic electrons, we can approximate the off-equatorial distribution function by

$$f(u_\parallel, u_\perp, h)$$
$$= \frac{W(h) N_h}{(2\pi)^{3/2} U_{t\parallel} U_{\perp 0}} \exp \left(-\frac{u_\parallel^2}{2 U_{t\parallel}^2} \right) \delta(u_\perp - W(h) U_{\perp 0}), \tag{44}$$

where $U_{t\parallel}$ and $U_{\perp 0} (= \gamma V_{\perp 0})$ and N_h are values at the magnetic equator [*Summers et al.*, 2012b]. The variation of the density and the average perpendicular velocity are modified by a function

$$W(h) = \left(1 + \frac{A_{eq} a h^2}{1 + a h^2} \right)^{-1/2}, \tag{45}$$

where $A_{eq} = (2/\pi)(U_{\perp 0}/U_{t\parallel})^2 - 1$. In evaluating $J_E(h)$, we incorporate the spatial variations of the density and the average perpendicular velocity in equation (44) as

$$J_E(h) = -(2e)^{3/2} (m_0 k \gamma)^{-1/2} (W(h) V_{\perp 0})^{5/2} \chi Q G(h) B_w^{1/2}, \tag{46}$$

where

$$G(h) = \frac{W(h) N_h}{(2\pi)^{3/2} U_{t\parallel} U_{\perp 0}} \exp \left(-\frac{\gamma^2 V_R^2}{2 U_{t\parallel}^2} \right). \tag{47}$$

7. SATURATION MECHANISM OF RISING-TONE EMISSIONS

The chorus equations describe the time evolution of the wave amplitude and frequency at the equator. However, the evolution of the wave amplitude does not follow the prediction of the chorus equation, and the wave amplitude saturates at a certain level. The wave amplitude exhibits intermittent growth on a continual basis. This suggests that nonlinear wave growth only takes place over a short period of time as the process of the triggering with an optimum wave amplitude discussed in the previous section. The triggering is repeated several times in forming a rising-tone emission. Therefore, there arise subpacket structures in the chorus element.

The saturation takes place because of the decrease of the Q value corresponding to the depth of the electron hole. More exactly, there arises a substantial number of electrons trapped in the wave potential as the wave packet propagates toward higher latitudes. The trapped electrons appear at different locations in the velocity phase space as demonstrated by *Hikishima et al.* [2010], shown in Figure 4.

The waves propagate away from the equator undergoing convective nonlinear growth as expressed by equation (33). The nonlinear growth rate (equation (34) decreases rapidly as the absolute value of the resonance velocity V_R increases. Both adiabatic variation of the distribution function and the increasing resonance velocity result in the saturation of the nonlinear wave growth [*Summers et al.*, 2012b].

8. NONLINEAR DAMPING AT HALF THE GYROFREQUENCY

Chorus emissions with a rising tone are generated near the magnetic equator. As they propagate away from the equator, they are amplified by the nonlinear growth mechanism. The wave packet propagates with the group velocity V_g given by equation (13), while its phase varies with the phase velocity given by equation (12). By inserting $\omega = 0.5 \Omega_e$ into equation (13), we find $V_g = V_p$. In the frame of reference moving with the group velocity V_g, the phase of the wave becomes stationary. In this frame of reference, the frequency ω is constant as expressed by equation (10). The amplitude of the wave is a slowly varying function modified by the resonant

(a)

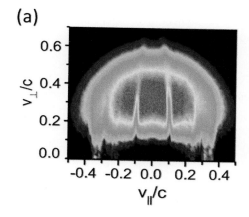

(b)

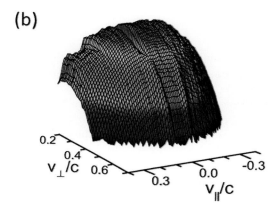

Figure 4. Electron velocity distribution function $f(v_\parallel, v_\perp)$ at $t = 5980\Omega_{e0}^{-1}$ for a simulation of triggered emissions: (a) $f(v_\parallel, v_\perp)$ plotted by color contours on a logarithmic scale and (b) perspective view of the distribution from a high pitch angle. Enhancement of electrons (a hill) is formed at higher pitch angles due to entrapping of resonant electrons, while depletion of electrons (a hole) is found at lower pitch angles [from *Hikishima et al.*, 2010]. The hill cancels the resonant current formed by the electron hole, resulting in the saturation of the nonlinear wave growth.

current given by equation (9). Taking into account the spatial inhomogeneity of the magnetic field and the plasma density of the inner magnetosphere, we assume the wave normal angle deviates gradually from the parallel direction.

We assume quasiparallel propagation in which the wave normal angle Ψ satisfies $\sin^2\Psi \ll 1$, while at the same time, we retain the term involving $\sin\Psi$. Under the assumption of quasiparallel propagation, the polarization of the transverse electromagnetic field remains circular (see Appendix C of *Omura et al.* [2009]). Therefore, we can assume a constant wave amplitude B_w in the plane perpendicular to the static magnetic field. In addition, there appears a longitudinal wave electric field $E_{w\parallel}$ parallel to the static magnetic field $\boldsymbol{B}_0$ which we express as

$$E_{w\parallel} = \frac{\omega \sin\Psi}{\chi^2 \Omega_e - \omega} E_w. \tag{48}$$

The equation of motion of energetic electrons interacting with the quasiparallel whistler mode wave is given by

$$\frac{d(\gamma v_\parallel)}{dt} = -\frac{eE_{w\parallel}}{m_0}\sin\phi + \frac{ev_\perp B_w}{m_0}\sin\zeta - \frac{\gamma v_\perp^2}{2\Omega_e}\frac{\partial\Omega_e}{\partial h}, \tag{49}$$

where $\phi = \int(\omega - kv_\parallel)dt$ and $\zeta = \int(\Omega - \omega + kv_\parallel)dt$.

We consider energetic particles with velocities near the wave phase velocity, i.e., $v_\parallel \sim \omega/k$. Denoting $\bar{v}_\parallel = v_\parallel - \omega/k$, we find that $\phi = -\int k\bar{v}_\parallel dt$. Since the phase of the second term on the right-hand side of equation (49) changes very quickly with frequencies close to Ω_e, we can neglect the contribution of this term to the variation of $v_\parallel$. Solving for the time derivative of $\bar{v}_\parallel$ in equation (49), we obtain a pair of coupled differential equations of $\bar{v}_\parallel$ and ϕ

$$\frac{d\bar{v}_\parallel}{dt} = -\frac{eE_{w\parallel}}{\gamma m_0}\left(1 - \frac{v_\parallel^2}{c^2}\right)\sin\phi - \frac{v_\perp^2}{2\Omega_e}\frac{\partial\Omega_e}{\partial h} \tag{50}$$

and

$$\frac{d\phi}{dt} = -k\bar{v}_\parallel. \tag{51}$$

Assuming that $\bar{v}_\parallel \sim 0$, and calculating the second-order derivative of ϕ, we obtain from equations (50) and (51)

$$\frac{d^2\phi}{dt^2} = \omega_{t\parallel}^2(\sin\phi + S_\parallel), \tag{52}$$

where

$$\omega_{t\parallel}^2 = \frac{ekE_{w\parallel}\chi^2}{\gamma m_0} \tag{53}$$

and

$$S_\parallel = \frac{kv_\perp^2}{2\omega_{t\parallel}^2\Omega_e}\frac{\partial\Omega_e}{\partial h}. \tag{54}$$

If the condition $|S_\parallel| < 1$ is satisfied, the parallel electric field of the whistler mode wave packet can trap some of the energetic electrons that satisfy $v_\parallel \sim V_p$. Trapped electrons move with the phase velocity, which increases at higher latitude. Since the density of the untrapped energetic electrons decreases at higher latitude because of reflection at the mirror points, the electrons trapped by the parallel electric field become isolated in the phase space, thus forming the resonant current $J_\parallel$. The center of the trapping potential (V_p, ϕ_c) is given by the second-order resonance condition $d^2\phi/dt^2 = 0$. From equation (52), we obtain the condition $\sin \phi_c + S_\parallel = 0$. Since we assume that the chorus element propagates in the positive h region, i.e., moves away from the equator, we find that $S_\parallel > 0$ and $\sin \phi_c < 0$. Taking the average over the wave phase from $\phi = 0$ to $\phi = 2\pi$, we obtain

$$\overline{E_{w\parallel}J_\parallel} = -\frac{e}{2\pi}\int_0^{2\pi}E_{w\parallel}\int_{-\infty}^{\infty}v_\parallel g_t(v_\parallel, \phi)\sin \phi\, dv_\parallel d\phi > 0, \quad (55)$$

where $g_t(v_\parallel, \phi)$ is the distribution function of resonant electrons trapped by the wave potential. Thus, trapped electrons moving with the phase velocity of the wave are accelerated, while they are trapped by the longitudinal wave potential. In the dipole magnetic field, both the phase velocity and group velocity increase as the distance from the equator increases. The increase of the phase velocity corresponds to an increase in kinetic energy of the trapped electrons. This is a further interpretation of the process, whereby the trapped electrons are accelerated.

Since we assume the quasiparallel wave propagation, we have $E_w \sim V_p B_w$ and the parallel wave electric field is given by

$$E_{w\parallel} = \frac{\omega}{\chi^2\Omega_e - \omega}V_p B_w \sin \Psi. \quad (56)$$

Substituting equations (53), (54), and (56) into the trapping condition $S_\parallel < 1$, we thereby express the necessary condition for effective nonlinear damping as $h < h_N$ where

$$h_N = \frac{\xi\chi^3 c\Omega_w\omega}{\gamma aV_{\perp 0}^2(\chi^2\Omega_e - \omega)}\sin \Psi \sim \frac{V_p\Omega_w}{\gamma aV_{\perp 0}^2}\sin \Psi. \quad (57)$$

Here we have assumed that $\omega_{pe} \gg \Omega_e$, i.e., $\chi^2 \sim 1$, and that $\omega \sim 0.5\, \Omega_e$.

In the course of the generation of a rising-tone chorus element, waves with frequencies near half the electron cyclotron frequency can also be generated near the magnetic equator during the process of nonlinear wave growth. Before leaving the equatorial region ($h < h_N$), however, the waves lose a substantial amount of energy to the Landau resonant electrons due to the deviation of the wave number vector from the parallel direction of the geomagnetic field. Since the magnitude of the resonant current depends on the width of the trapping potential (which is itself proportional to the trapping velocity), the rate of the nonlinear damping is proportional to $\sqrt{B_w}\sin \Psi$. As waves grow with a rising frequency at the equator, wave amplitudes can be larger at higher frequencies near half the gyrofrequency. However, the larger-amplitude waves, with frequencies close to half the gyrofrequency, are subject to stronger nonlinear damping as they propagate along the magnetic field line.

9. FALLING-TONE EMISSIONS

Nunn and Omura [2012] successfully reproduced the falling-tone emissions by confining the simulation system to be the region upstream from the magnetic equator. Effective entrapping of resonant electrons at the equator induces an enhanced distribution function or "hill" in the resonant particle trap, whereas risers have a "hole." To maintain the nonlinear wave growth near the equator where $\partial\Omega_e/\partial h \sim 0$, the following relation must be satisfied.

$$S = -\frac{s_1}{s_0\omega\Omega_w}\frac{\partial\omega}{\partial t}. \quad (58)$$

To generate a falling tone $\partial\omega/\partial t < 0$, we need $S > 0$. Since the second-order resonance condition for the stably trapped electrons gives $\sin \zeta + S = 0$, we have $\sin \zeta < 0$ for $S > 0$.

In the theory of rising-tone emissions, we assumed an electron hole expressed by equation (18) and the resonant current J_E given by equation (19). Since the trapped electrons forming $g_t(v_\parallel, \zeta)$ take the phase range $\sin \zeta < 0$ for $S > 0$, Q should be negative to yield negative J_E, which causes wave growth. The negative Q corresponds to an electron hill formed by trapped electrons as shown in Figure 1b. Assuming that $g_t(v_\parallel, \zeta) = G$ inside the trapping region and $g_t(v_\parallel, \zeta) = 0$ outside it, we obtain

$$J_E = -J_0\int_{\zeta_2}^{\zeta_1}[\cos \zeta_1 - \cos \zeta + S(\zeta - \zeta_1)]^{1/2}\sin \zeta d\zeta \quad (59)$$

and

$$J_B = J_0\int_{\zeta_2}^{\zeta_1}[\cos \zeta_1 - \cos \zeta + S(\zeta - \zeta_1)]^{1/2}\cos \zeta d\zeta. \quad (60)$$

The phase angles ζ_1 and ζ_2 define the boundary of the trapping wave potential as described by *Omura et al.* [2008]. The resonant current J_B takes a positive value, inducing a negative frequency offset given by equation (23).

The maximum value of the resonant current J_E for the nonlinear wave growth is attained with $S = 0.4$ [*Omura et al.*, 2008], namely,

$$J_{E,\max} = (2e)^{3/2}(m_0k\gamma)^{-1/2}(WV_{\perp 0})^{5/2}\chi QGB_w^{1/2}, \quad (61)$$

where we have incorporated the adiabatic variation of the equatorial distribution function (24) [*Summers et al.*, 2012b]. We then obtain the nonlinear growth rate at a generation point h as

$$\Gamma_N = -\frac{Q\omega_{ph}^2}{2}\left(\frac{\xi}{\omega\Omega_w}\right)^{1/2}\frac{V_g}{U_{t\parallel}}\left(\frac{\chi W V_{\perp 0}}{c\pi\gamma}\right)^{3/2}\exp\left(-\frac{\gamma^2 V_R^2}{2U_{t\parallel}^2}\right).$$

(62)

The parameters other than $U_{t\parallel}$ and $V_{\perp 0}$ are to be evaluated at the generation point h. It is noted that $-1 < Q < 0$, which corresponds to an electron hill formed by resonant electrons trapped by the wave potential. When a whistler mode wave packet propagates toward the equator, effective entrapping of resonant particles occurs at the wavefront in the upstream region from the magnetic equator. It is necessary to have an absolute instability to generate a falling tone. The value of Γ_N has to be large enough to cancel the effect of the convective term in equation (35).

10. DISCUSSION

We have reviewed recent progress of the theoretical analyses of the generation mechanism of chorus emissions. Electron dynamics related to the formation of an electromagnetic electron hole in the velocity phase space has been confirmed in particle code simulations [*Hikishima and Omura*, 2012]. The generation mechanism of rising tone emissions is due to the formation of an electromagnetic electron hole in the velocity phase space. The relation between frequency sweep rates of rising tones and their wave amplitudes expressed by equation (29) are confirmed by simulations [*Hikishima et al.*, 2009; *Katoh and Omura*, 2011] and observations by THEMIS spacecraft [*Cully et al.*, 2011]. The optimum amplitude given by equation (30) and the threshold for rising-tone emissions given by equation (41) are also confirmed by recent simulations.

The generation mechanism of falling-tone emissions is due to the formation of an electron hill in the velocity phase space. We express the electron hill by a negative value of Q in equation (18). The electron hill is mostly likely to be formed in the region upstream from the magnetic equator, where the hill produces a negative J_E value due to the gradient of the magnetic field as well as the negative frequency offset induced by the positive J_B expressed by equation (23). This mechanism has been confirmed by a Vlasov-Hybrid simulation [*Nunn and Omura*, 2012], but is yet to be confirmed by particle simulations and observations.

Acknowledgments. This work was supported by Grant-in-Aid 23340147 of the Ministry of Education, Culture, Sports, Science and Technology in Japan. D. S. holds a Discovery Grant of the Natural Sciences and Engineering Research Council of Canada and acknowledges additional support from WCU grant (R31-10016) funded by the Korean Ministry of Education, Science and Technology.

REFERENCES

Albert, J. M. (2000), Gyroresonant interactions of radiation belt particles with a monochromatic electromagnetic wave, *J. Geophys. Res.*, *105*(A9), 21,191–21,209, doi:10.1029/2000JA000008.

Albert, J. M. (2002), Nonlinear interaction of outer zone electrons with VLF waves, *Geophys. Res. Lett.*, *29*(8), 1275, doi:10.1029/2001GL013941.

Cully, C. M., V. Angelopoulos, U. Auster, J. Bonnell, and O. Le Contel (2011), Observational evidence of the generation mechanism for rising-tone chorus, *Geophys. Res. Lett.*, *38*, L01106, doi:10.1029/2010GL045793.

Hikishima, M., and Y. Omura (2012), Particle simulations of whistler-mode rising-tone emissions triggered by waves with different amplitudes, *J. Geophys. Res.*, *117*, A04226, doi:10.1029/2011JA017428.

Hikishima, M., S. Yagitani, Y. Omura, and I. Nagano (2009), Full particle simulation of whistler-mode rising chorus emissions in the magnetosphere, *J. Geophys. Res.*, *114*, A01203, doi:10.1029/2008JA013625.

Hikishima, M., Y. Omura, and D. Summers (2010), Self-consistent particle simulation of whistler mode triggered emissions, *J. Geophys. Res.*, *115*, A12246, doi:10.1029/2010JA015860.

Kasahara, Y., Y. Miyoshi, Y. Omura, O. P. Verkhoglyadova, I. Nagano, I. Kimura, and B. T. Tsurutani (2009), Simultaneous satellite observations of VLF chorus, hot and relativistic electrons in a magnetic storm "recovery" phase, *Geophys. Res. Lett.*, *36*, L01106, doi:10.1029/2008GL036454.

Katoh, Y., and Y. Omura (2007a), Computer simulation of chorus wave generation in the Earth's inner magnetosphere, *Geophys. Res. Lett.*, *34*, L03102, doi:10.1029/2006GL028594.

Katoh, Y., and Y. Omura (2007b), Relativistic particle acceleration in the process of whistler-mode chorus wave generation, *Geophys. Res. Lett.*, *34*, L13102, doi:10.1029/2007GL029758.

Katoh, Y., and Y. Omura (2011), Amplitude dependence of frequency sweep rates of whistler mode chorus emissions, *J. Geophys. Res.*, *116*, A07201, doi:10.1029/2011JA016496.

Katoh, Y., Y. Omura, and D. Summers (2008), Rapid energization of radiation belt electrons by nonlinear wave trapping, *Ann. Geophys.*, *26*, 3451–3456.

Meredith, N. P., R. B. Horne, R. M. Thorne, and R. R. Anderson (2003), Favored regions for chorus-driven electron acceleration to relativistic energies in the Earth's outer radiation belt, *Geophys. Res. Lett.*, *30*(16), 1871, doi:10.1029/2003GL017698.

Nunn, D. (1974), A self-consistent theory of triggered VLF emissions, *Planet. Space Sci.*, *22*, 349–378.

Nunn, D., and Y. Omura (2012), A computational and theoretical analysis of falling frequency VLF emissions, *J. Geophys. Res.*, *117*, A08228, doi:10.1029/2012JA017557.

Nunn, D., Y. Omura, H. Matsumoto, I. Nagano, and S. Yagitani (1997), The numerical simulation of VLF chorus and discrete emissions observed on the Geotail satellite using a Vlasov code, *J. Geophys. Res.*, *102*(A12), 27,083–27,097.

Nunn, D., O. Santolik, M. Rycroft, and V. Trakhtengerts (2009), On the numerical modelling of VLF chorus dynamical spectra, *Ann. Geophys.*, *27*, 2341–2359.

Omura, Y., and H. Matsumoto (1982), Computer simulations of basic processes of coherent whistler wave-particle interactions in the magnetosphere, *J. Geophys. Res.*, *87*(A6), 4435–4444.

Omura, Y., and D. Nunn (2011), Triggering process of whistler mode chorus emissions in the magnetosphere, *J. Geophys. Res.*, *116*, A05205, doi:10.1029/2010JA016280.

Omura, Y., and D. Summers (2004), Computer simulations of relativistic whistler-mode wave-particle interactions, *Phys. Plasmas*, *11*, 3530–3534.

Omura, Y., and D. Summers (2006), Dynamics of high-energy electrons interacting with whistler mode chorus emissions in the magnetosphere, *J. Geophys. Res.*, *111*, A09222, doi:10.1029/2006JA011600.

Omura, Y., D. Nunn, H. Matsumoto, and M. J. Rycroft (1991), A review of observational, theoretical and numerical studies of VLF triggered emissions, *J. Atmos. Terr. Phys.*, *53*, 351–368.

Omura, Y., N. Furuya, and D. Summers (2007), Relativistic turning acceleration of resonant electrons by coherent whistler mode waves in a dipole magnetic field, *J. Geophys. Res.*, *112*, A06236, doi:10.1029/2006JA012243.

Omura, Y., Y. Katoh, and D. Summers (2008), Theory and simulation of the generation of whistler-mode chorus, *J. Geophys. Res.*, *113*, A04223, doi:10.1029/2007JA012622.

Omura, Y., M. Hikishima, Y. Katoh, D. Summers, and S. Yagitani (2009), Nonlinear mechanisms of lower-band and upper-band VLF chorus emissions in the magnetosphere, *J. Geophys. Res.*, *114*, A07217, doi:10.1029/2009JA014206.

Santolík, O. (2008), New results of investigations of whistler-mode chorus emissions, *Nonlin. Processes Geophys.*, *15*, 621–630.

Santolík, O., D. A. Gurnett, J. S. Pickett, M. Parrot, and N. Cornilleau-Wehrlin (2003), Spatio-temporal structure of storm-time chorus, *J. Geophys. Res.*, *108*(A7), 1278, doi:10.1029/2002JA009791.

Sigsbee, K., J. D. Menietti, O. Santolík, and J. S. Pickett (2010), Locations of chorus emissions observed by the Polar Plasma Wave Instrument, *J. Geophys. Res.*, *115*, A00F12, doi:10.1029/2009JA014579.

Summers, D., and Y. Omura (2007), Ultra-relativistic acceleration of electrons in planetary magnetospheres, *Geophys. Res. Lett.*, *34*, L24205, doi:10.1029/2007GL032226.

Summers, D., R. M. Thorne, and F. Xiao (1998), Relativistic theory of wave-particle resonant diffusion with application to electron acceleration in the magnetosphere, *J. Geophys. Res.*, *103*(A9), 20,487–20,500.

Summers, D., C. Ma, N. P. Meredith, R. B. Horne, R. M. Thorne, D. Heynderickx, and R. R. Anderson (2002), Model of the energization of outer-zone electrons by whistler-mode chorus during the October 9, 1990 geomagnetic storm, *Geophys. Res. Lett.*, *29*(24), 2174, doi:10.1029/2002GL016039.

Summers, D., R. Tang, and Y. Omura (2011), Effects of nonlinear wave growth on extreme radiation belt electron fluxes, *J. Geophys. Res.*, *116*, A10226, doi:10.1029/2011JA016602.

Summers, D., R. Tang, and Y. Omura (2012a), Linear and nonlinear growth of magnetospheric whistler mode waves, in *Dynamics of the Earth's Radiation Belts and Inner Magnetosphere, Geophys. Monogr. Ser.*, doi:10.1029/2012GM001298, this volume.

Summers, D., Y. Omura, Y. Miyashita, and D.-H. Lee (2012b), Nonlinear spatiotemporal evolution of whistler mode chorus waves in Earth's inner magnetosphere, *J. Geophys. Res.*, *117*, A09206, doi:10.1029/2012JA017842.

Tsurutani, B. T., and E. J. Smith (1974), Postmidnight chorus: A substorm phenomenon, *J. Geophys. Res.*, *79*(1), 118–127.

D. Nunn, School of Electronics and Computer Science, University of Southampton, Southampton, U.K. (dn@ecs.soton.ac.uk)

Y. Omura, Research Institute for Sustainable Humanosphere, Kyoto University, Uji, Kyoto 611-0011, Japan. (omura@rish.kyoto-u.ac.jp)

D. Summers, Department of Mathematics and Statistics, Memorial University of Newfoundland, St. John's, Newfoundland A1C 5S7, Canada. (dsummers@mun.ca)

Aspects of Nonlinear Wave-Particle Interactions

Jay M. Albert

Space Vehicles Directorate, Air Force Research Laboratory, Kirtland Air Force Base, New Mexico, USA

Xin Tao and Jacob Bortnik

Department of Atmospheric and Oceanic Sciences, University of California, Los Angeles, California, USA

Gyro-averaged equations, valid near an arbitrary resonance, are developed both directly from the Lorentz equations and from a Hamiltonian approach. The Hamiltonian development, with several significant approximations, gives a reduction to a much lower dimensional system, which can be approximated analytically. Previous treatments of diffusion, phase bunching, and phase trapping are reviewed and extended in various ways. These results, and future work along these lines, may allow practical, long-term modeling of nonlinear wave-particle interactions in the radiation belts.

1. INTRODUCTION

Wave-particle interactions are widely recognized as a crucial ingredient in radiation belt electron behavior. Quasi-linear theory, with waves specified empirically rather than self-consistently, has been widely used to model these processes [*Lyons*, 1974], often with considerable qualitative success [e.g., *Albert et al.*, 2009]. While this seems appropriate for broadband waves, such as plasmaspheric hiss, it seems dubious for the coherent, quasi-monochromatic waves that seem to be a dominant component of chorus and which can occur with unexpectedly large amplitudes. In recent years, such waves have come to the fore, both observationally [*Cattell et al.*, 2008; *Cully et al.*, 2008] and in large-scale simulations [*Omura et al.*, 2009]. The full, self-consistent story is undoubtedly complex, but a simplified, non-self-consistent description of coherent interactions may prove fruitful for radiation belt modeling, as it has in the broadband context. In this paper, we review some of the ideas and tools that are available for understanding test particle behavior, deliberately ignoring many complications. In particular, frequency sweeping and amplitude modulation are not treated, although they have been considered by others [e.g., *Chang et al.*, 1983; *Demekhov et al.*, 2006; *Tao et al.*, 2012a, 2012b].

In section 2, gyro-averaged equations, valid near an arbitrary resonance, are developed directly from the Lorentz equations of motion. Relativistic particles and oblique waves are both included, generalizing previous results. Next, the Hamiltonian treatment is presented in section 3, starting with the straightforward equations of motion and progressing, through variable transformations, resonance averaging, and identification of constants of motion, to a reduced "1½-dimensional" system. At this stage, the difference between "linear" and "nonlinear" behavior becomes apparent, and delineated by a so-called inhomogeneity parameter. It also becomes possible to develop analytical treatments of the resonant changes in adiabatic invariants, energy, and equatorial pitch angle in the different regimes of diffusion, phase bunching, and phase trapping, and this is reviewed in section 4. For diffusion, a previous result is generalized for the case of resonance right at the equator. For phase bunching, the spread often seen in numerical results is discussed in terms of the range of resonance phase. For phase trapping, a closed-form solution is given for the particle energy as a function of position, which could also serve as a simple model for relativistic turning acceleration (RTA) [*Omura et al.*, 2007] and ultrarelativistic acceleration (URA) [*Summers and Omura*, 2007].

Dynamics of the Earth's Radiation Belts and Inner Magnetosphere
Geophysical Monograph Series 199
10.1029/2012GM001324

2. LORENTZ EQUATIONS OF MOTION

From the Lorentz force law for a charged particle in a background magnetic field in the presence of a field-aligned wave,

$$\frac{d\mathbf{p}}{dt} = q[\mathbf{E_w} + \mathbf{v} \times (\mathbf{B_0} + \mathbf{B_w})], \qquad \frac{d\mathbf{x}}{dt} = \mathbf{v}, \qquad (1)$$

it is straightforward to derive the gyro-averaged equations of motion as given by *Chang and Inan* [1983] for an electron with $q = -e$:

$$\frac{dp_\parallel}{dt} = -\frac{p_\perp^2}{2m\gamma\Omega}\frac{\partial\Omega}{\partial z} + eB_w\frac{p_\perp}{m\gamma}\sin\xi,$$

$$\frac{dp_\perp}{dt} = \frac{p_\perp p_\parallel}{2m\gamma\Omega}\frac{\partial\Omega}{\partial z} - eB_w\left(\frac{p_\parallel}{m\gamma} + \frac{\omega}{k}\right)\sin\xi,$$

$$\frac{d\xi}{dt} = \frac{\Omega}{\gamma} - \omega - k\frac{p_\parallel}{m\gamma} - \frac{eB_w}{p_\perp}\left(\frac{p_\parallel}{m\gamma} + \frac{\omega}{k}\right)\cos\xi,$$

$$\frac{dz}{dt} = \frac{p_\parallel}{m\gamma}. \qquad (2)$$

Here ω is the wave frequency, Ω is the electron gyrofrequency eB_0/m, $\mathbf{p} = m\mathbf{v}\gamma$, $\gamma = [1 + (p_\perp/mc)^2 + (p_\parallel/mc)^2]^{1/2}$, and ξ is the angle between $p_\perp$ and $-B_w$. Note that for field-aligned waves, and in the SI units used above, the combination $eB_w(p_\parallel/m\gamma + \omega/k)$ is just $e(E_w + v_\parallel B_w)$. Heavy use has been made of the approximations that the wave amplitude is small compared to the background magnetic field and that the dependence of wave quantities on position is slow compared to the timescales set by ω and Ω.

For an oblique wave, the calculations are a bit more involved. *Tao and Bortnik* [2010] generalized the work of *Bell* [1984] to account for relativistic particles and arbitrary resonance number ℓ, but did not gyro-average the results. Also, neither paper retained terms proportional to the wavefields in the phase equation. Incorporating both extensions yields the set of equations

$$\frac{dp_\parallel}{dt} = -\frac{p_\perp^2}{2m\gamma\Omega}\frac{\partial\Omega}{\partial z} + |q|\frac{k_\perp v_\perp}{\omega}\left\{\left[\frac{E_1+E_2}{2}J_{\ell+1} + \frac{E_1-E_2}{2}J_{\ell-1}\right]\right.$$

$$\left. - s\frac{v_\parallel}{v_\perp}E_3 J_\ell\right\}\cos\xi,$$

$$\frac{dp_\perp}{dt} = \frac{p_\parallel p_\perp}{2m\gamma\Omega}\frac{\partial\Omega}{\partial z} + |q|\frac{s\ell\Omega}{\omega\gamma}\left\{\left[\frac{E_1+E_2}{2}J_{\ell+1} + \frac{E_1-E_2}{2}J_{\ell-1}\right]\right.$$

$$\left. - s\frac{v_\parallel}{v_\perp}E_3 J_\ell\right\}\cos\xi,$$

$$\frac{d\xi}{dt} = k_\parallel v_\parallel + \frac{s\ell\Omega}{\gamma} - \omega - \frac{q}{p_\perp}\left\{\frac{s\Omega}{\omega\gamma}\left[\frac{E_1+E_2}{2}J_{\ell+1} + \frac{E_1-E_2}{2}J_{\ell-1}\right]\right.$$

$$+ s\frac{k_\perp v_\perp}{\omega}\left[\frac{E_1+E_2}{2}J'_{\ell+1} + \frac{E_1-E_2}{2}J'_{\ell-1}\right]$$

$$\left. - \frac{k_\perp v_\parallel}{\omega}E_3 J'_\ell\right\}\sin\xi,$$

$$\frac{dz}{dt} = \frac{p_\parallel}{m\gamma}, \qquad (3)$$

where $\xi = k_\perp x_0 + k_\parallel z - \omega t + s\ell\phi$, $\tan\phi = -p_y/sp_x$, $s = q/|q| = \pm 1$, and x_0 refers to the guiding center. Here and from now on, cgs units are used. The Bessel functions J have argument $k_\perp\rho$, and the combinations of wavefield components in the equations for $p_\parallel$ and $p_\perp$ are familiar from quasi-linear theory [*Lyons*, 1974]. A detailed derivation will be given in a forthcoming publication. The notation of *Ginet and Heinemann* [1990] has been used, so $\Omega = |q|B_0/mc$, $\rho = mv_\perp\gamma/|q|B_0$,

$$\mathbf{E_w} = E_1\cos\Phi\hat{\mathbf{e}}_\mathbf{x} - E_2\sin\Phi\hat{\mathbf{e}}_\mathbf{y} - E_3\cos\Phi\hat{\mathbf{e}}_\mathbf{z}, \qquad (4)$$

$$\mathbf{B_w} = B_1\sin\Phi\hat{\mathbf{e}}_\mathbf{x} + B_2\cos\Phi\hat{\mathbf{e}}_\mathbf{y} - B_3\sin\Phi\hat{\mathbf{e}}_\mathbf{z},$$

and $\Phi = \mathbf{k}\cdot\mathbf{x} - \omega t$. Faraday's law gives $(B_1, B_2, B_3) = (c/\omega)$ $(k_\parallel E_2, k_\parallel E_1 + k_\perp E_3, k_\perp E_2)$. Both $p_\parallel$ and $k_\parallel$ are positive when $\mathbf{p}$ and $\mathbf{k}$, respectively, point in the $+z$ direction. Thus, for electrons, with $p_\parallel < 0$ but $k_\parallel > 0$, the primary resonance has $\ell = -1$. Furthermore, if the wave is field aligned (i.e., $k_\perp = 0$), this is the only resonance for which the wave terms in equation (3) do not vanish.

These equations may be used to derive the so-called inhomogeneity parameter and frequency associated with nonlinear particle behavior, namely, phase bunching and phase trapping. This is illustrated most clearly in the work of *Inan et al.* [1978]. Using the nonrelativistic version of equation (2), and neglecting the wave term in the equation for $d\xi/dt$, gives

$$\frac{d^2\xi}{dt^2} = -k\frac{dv_\parallel}{dt} + v_\parallel\frac{\partial}{\partial z}(\Omega - kv_\parallel). \qquad (5)$$

Using the full (nonrelativistic) form of $dv_\parallel/dt$ from equation (2) yields

$$\frac{d^2\xi}{dt^2} + kv_\perp\frac{eB_w}{m}\sin\xi = \left(v_\parallel + \frac{kv_\perp^2}{2\Omega}\right)\frac{\partial\Omega}{\partial z} - v_\parallel^2\frac{\partial k}{\partial z}, \qquad (6)$$

where $\partial k/\partial z$ is evaluated using the dispersion relation for the wave. Relativistically, one has to consider the equation for $dp_\perp/dt$ as well as $dp_\parallel/dt$ because of the γ factors, but the procedure is the same. *Omura et al.* [2008] write equation (6) (relativistically) in a form similar to

$$\frac{d^2\xi}{dt^2} + \omega_t^2(\sin\xi + S) = 0, \qquad (7)$$

which defines the trapping (or nonlinear, or oscillation) frequency ω_t and the inhomogeneity parameter S. *Bell* [1984, 1986] gives a much more detailed, but less transparent, treatment of S for an oblique wave.

According to equation (3), and using the relation $\gamma^2 = 1 + (p/mc)^2$, the rate of change of energy is given by

$$\gamma\frac{d\gamma}{dt} = \frac{|q|}{mc}\left\{\frac{p_\perp}{mc}\left[\frac{E_1+E_2}{2}J_{\ell+1} + \frac{E_1-E_2}{2}J_{\ell-1}\right]\right.$$

$$\left. - s\frac{p_\parallel}{mc}E_z J_\ell\right\}\cos\xi. \qquad (8)$$

The overall sign of $d\gamma/dt$ depends on the detailed behavior of the phase angle, as discussed in section 4.

3. HAMILTONIAN EQUATIONS OF MOTION

3.1. Three-Dimensional Formulation

The Hamiltonian formulation of the equations of motion, exactly equivalent to equation (1), is

$$\frac{d\mathbf{x}}{dt} = \frac{\partial H}{\partial \mathbf{P}}, \quad \frac{d\mathbf{P}}{dt} = -\frac{\partial H}{\partial \mathbf{x}}, \tag{9}$$

where

$$H = mc^2 \sqrt{1 + \left(\frac{\mathbf{P} - q\mathbf{A}/c}{mc}\right)^2}. \tag{10}$$

Here $\mathbf{P}$ is the canonical momentum, related to the usual mechanical momentum $\mathbf{p}$ by $\mathbf{P} = \mathbf{p} + q\mathbf{A}/c$, and $\mathbf{A} = \mathbf{A}_0 + \mathbf{A_w}$ is the vector potential of the electromagnetic fields. Thus, $\mathbf{B}_0 = \nabla \times \mathbf{A}_0$ is the background magnetic field, and the wavefields are $\mathbf{B_w} = \nabla \times \mathbf{A_w}$ and $\mathbf{E_w} = (-1/c)\partial\mathbf{A_w}/\partial t$. Consistent with, and in the same spirit as, the slow-variation assumptions mentioned in connection with equation (2), it is permissible to write

$$\mathbf{A_w} = A_1 \sin \Phi \hat{\mathbf{e}}_x + A_2 \cos \Phi \hat{\mathbf{e}}_y - A_3 \sin \Phi \hat{\mathbf{e}}_z, \tag{11}$$

and only consider the variation of Φ when differentiating $\mathbf{A_w}$ to obtain the wavefields. This leads to $(A_1, A_2, A_3) = (c/\omega)(E_1, E_2, E_3)$ [Ginet and Heinemann, 1990].

Within the framework of Hamiltonian mechanics, changes of variable can be made using so-called canonical transformations [e.g., Goldstein et al., 2001; Lichtenberg and Lieberman, 1992]. A fruitful approach, when possible, is to obtain sets of action-angle variables, with the lowest-order Hamiltonian a function of only the action variables $\mathcal{J}$, and dependence on the angle variables θ only occurring in higher-order terms. In this case, the angle variables will be periodic, and the action variables will be adiabatic invariants, robust to all but resonant periodic perturbations. Since any dependence on angular variables can be expanded in a Fourier series, the Hamiltonian will have the generic form

$$H(\mathcal{J}, \theta) = H_0(\mathcal{J}) + \sum_{n=-\infty}^{\infty} H_{1,n}(\mathcal{J})\sin n\theta. \tag{12}$$

The $H_{1,n}$ coefficients are proportional to the wave amplitude and are considered small and slowly varying, while the purely adiabatic (no-wave) motion is completely described

by H_0. Then $d\mathcal{J}/dt = -\sum n H_{1,n} \cos n\theta$ will rapidly oscillate about zero unless $d\theta/dt \approx \partial H_0/\partial \mathcal{J} \approx 0$, which is the resonance condition in this description.

Essentially, the advantage of this approach is that by working with action-angle variables, and making various approximations on the single function H, physically important properties like adiabatic invariance are well maintained. The corresponding relationships between "direct" variables like $\mathbf{x}$ and $\mathbf{p}$ can be delicate and easily lost if approximations are made separately and inconsistently to each of the six components of equations like equation (1).

Assuming a constant background magnetic field, Ginet and Heinemann [1990] followed these standard procedures [Lichtenberg and Lieberman, 1992] to transform to modified guiding center variables, which can be written as

$$I = \frac{(p_x + qA_x^w/c)^2 + (p_y + qA_y^w/c)^2}{2m\Omega},$$
$$\tan\phi = \frac{-(p_y + qA_y^w/c)}{s(p_x + qA_x^w/c)}. \tag{13}$$

In this notation, I is approximately the standard first adiabatic invariant $\mathcal{J}_1$ [Schulz, 1991], and ϕ is essentially the gyrophase, but both are modified by the wave. Albert [1993], following Shklyar [1986], generalized this to account for spatial variation of the background magnetic field. In terms of the new variables, H near an isolated resonance $n = \ell$ takes the form

$$H = H_0(I, P_Z, Z) + H_1(I, P_Z, Z)\sin \xi, \tag{14}$$

where ξ means the combination $k_x X + k_z Z - \omega t + s\ell\phi$. H_0 and H_1 have the forms

$$H_0 = mc^2 \Upsilon, \quad H_1 = mc^2 \frac{a_\ell}{2\Upsilon}, \tag{15}$$

where

$$\Upsilon = \sqrt{1 + \frac{2\Omega I}{mc^2} + \left(\frac{P_Z}{mc}\right)^2},$$

$$a_\ell = -2\frac{|q|}{mc\omega}\left\{\sqrt{\frac{2\Omega I}{mc^2}}\left[\frac{E_1 - E_2}{2}J_{\ell-1} + \frac{E_1 + E_2}{2}J_{\ell+1}\right]\right.$$
$$\left. - s\frac{P_Z}{mc}E_z J_\ell\right\} \tag{16}$$

[Albert, 1993, 2000]. As before, the Bessel functions J have argument $k_\perp\rho$. In the absence of the wave, Υ is just the usual relativistic factor γ, and $2\Omega I/mc^2$ becomes $(p_\perp/mc)^2$.

Using $dP_Z/dt = -\partial H/\partial Z \approx -k_z H_1 \cos \xi$ and $dI/dt = -\partial H/\partial\phi \approx -s\ell H_1 \cos \xi$, it can be seen that both $d\Upsilon/dt$ and dI/dt have a dependence on the wave components that is consistent with equation (8). In fact, it can be shown that the equations for I, P_z, Z, and ξ are consistent with equation (3) to first order in the wave amplitude. Furthermore, the Hamiltonian-derived system can be reduced to a single pair of action-angle variables, whose behavior can be visualized and analyzed in two dimensions. This leads to very useful simplifications and analytical estimates, as will be discussed in the next section.

Because ξ will assume a primary role below, its rate of change is calculated as

$$\frac{d\xi}{dt} = k_z \frac{\partial H}{\partial P_Z} + s\ell \frac{\partial H}{\partial I} - \omega$$

$$= \left[\frac{k_z P_Z}{m\Upsilon} + \frac{s\ell\Omega}{\Upsilon} - \omega\right] + \left[k_z \frac{\partial H_1}{\partial P_Z} + s\ell \frac{\partial H_1}{\partial I}\right]\sin\xi. \quad (17)$$

From the lowest-order, wave-independent terms, the usual resonance condition corresponds to $d\xi/dt = 0$.

3.2. Reduction to One Dimension

This development closely follows the work of *Shklyar* [1986]. From the equations

$$\frac{dP_X}{dt} = -\frac{\partial H}{\partial X} = -k_x H_1 \cos\xi,$$

$$\frac{dI}{dt} = -\frac{\partial H}{\partial\phi} = -s\ell H_1 \cos\xi, \quad (18)$$

$$\frac{dH}{dt} = \frac{\partial H}{\partial t} = -\omega H_1 \cos\xi,$$

and assuming constant ω, it is seen that both $s\ell P_X - k_x I \equiv c_1$ and $\omega I - s\ell H \equiv c_2$ are constant in time. The second relation is especially useful because for $\ell \neq 0$, it allows both Υ and P_z to be expressed as functions of only I and Z:

$$\Upsilon \approx \frac{H}{mc^2} = \frac{\omega I - c_2}{s\ell mc^2},$$

$$\left(\frac{P_Z}{mc}\right)^2 \approx \left(\frac{\omega I - c_2}{s\ell mc^2}\right)^2 - 1 - \frac{2\Omega I}{mc^2}. \quad (19)$$

Thus, the equations

$$\frac{dI}{dt} = -s\ell H_1 \cos\xi,$$

$$\frac{d\xi}{dt} = \left[\frac{k_z P_Z}{m\Upsilon} + \frac{s\ell\Omega}{\Upsilon} - \omega\right] + \left[k_z \frac{\partial H_1}{\partial P_Z} + s\ell \frac{\partial H_1}{\partial I}\right]\sin\xi, \quad (20)$$

$$\frac{dZ}{dt} = \frac{P_Z}{m\Upsilon},$$

constitute a closed set. The system can be condensed even further by dividing the first two equations by the third, resulting in

$$\frac{dI}{dZ} = -s\ell(mc)^2 \frac{a_\ell}{2P_Z}\cos\xi,$$

$$\frac{d\xi}{dZ} = \left[\frac{k_z P_Z/m + s\ell\Omega - \omega\Upsilon}{P_Z/m}\right] + (\ldots)\sin\xi. \quad (21)$$

Thus, from here on, Z will be considered the independent variable. Finally, we try to fit these equations into the form of a 1-D Hamiltonian system with

$$K(I, \xi, Z) = K_0(I, Z) + K_1(I, Z)\sin\xi \quad (22)$$

so that

$$\frac{dI}{dZ} = -K_1\cos\xi,$$

$$\frac{d\xi}{dZ} = \frac{\partial K_0}{\partial I} + \frac{\partial K_1}{\partial I}\sin\xi. \quad (23)$$

This readily gives

$$K_1 = s\ell(mc)^2 \frac{a_\ell}{2P_Z}, \quad (24)$$

and it can be checked that

$$K_0 = k_z I - s\ell P_Z(I, Z) \quad (25)$$

gives back the lowest-order, resonance condition terms of $d\xi/dZ$. Reproduction of the wave part of $d\xi/dZ$ in equation (21) has been sacrificed in order to obtain a 1-D Hamiltonian system. Actually, since K_0 and K_1 explicitly depend on Z, this is a "1½-dimensional" system.

For the case $\ell = 0$, I is constant. *Albert* [2000] used equation (15) and the resonance condition to write the third line of equation (18) as

$$\frac{d(\Upsilon^2)}{dZ} = -k_z a_0 \cos\xi \equiv -\frac{\partial K}{\partial\xi}. \quad (26)$$

Choosing $\Gamma = \Upsilon^2$ to be the action conjugate to ξ in a 1-D system gives $d\Gamma/dz = -K_1\cos\xi$, so

$$K_1(\Gamma, Z) = k_z a_0, \quad (27)$$

while the lowest-order part of the angle equation is

$$\frac{d\xi}{dZ} = k_z - \frac{\omega}{P_Z/m\Upsilon} \equiv \frac{\partial K_0}{\partial\Gamma}. \quad (28)$$

Expressing $P_Z/m\Upsilon$ in terms of Γ and integrating with respect to Γ, as in Appendix A of *Albert* [2000], gives

$$K_0(\Gamma, Z) = k_z\Gamma - \frac{\omega}{c}\left\{\sqrt{\Gamma(\Gamma - \Gamma_0)} - \Gamma_0 \log\left[\sqrt{\Gamma} + \sqrt{\Gamma - \Gamma_0}\right]\right\}, \quad (29)$$

where Γ_0 is $1 + 2\Omega I/mc^2$.

4. ONE-DIMENSIONAL HAMILTONIAN ANALYSIS

Appendix C of *Albert* [2000] shows how the previously discussed constants of motion may be used to determine changes in energy E corresponding to changes in I or Γ:

$$E = \begin{cases} \omega \Delta I / s\ell, & \ell \neq 0, \\ \Delta \Gamma / 2\gamma, & \ell = 0. \end{cases} \quad (30)$$

For any ℓ, the corresponding change in equatorial pitch angle α_0 is

$$\Delta \alpha_0 = \frac{-\sin^2 \alpha_0 + s\ell\Omega_{\text{eq}}/\omega}{\sin \alpha_0 \cos \alpha_0} \frac{\Delta p}{p}, \quad (31)$$

where $\Delta p = (m\gamma/p)\Delta E$.

4.1. Inhomogeneity Parameter

Expanding the reduced Hamiltonian near a resonance, where $\partial K_0 / \partial I \approx 0$, gives

$$K - K_{\text{res}} = \frac{\partial^2 K_0}{\partial I^2} \frac{(I - I_{\text{res}})^2}{2} + K_1 \sin \xi, \quad (32)$$

where I_{res} is approximately determined by the resonance condition $\partial K_0 / \partial I = 0$. At each value of the time-like variable Z, this has the same form as that of a plane pendulum, with an associated phase portrait in the (I, ξ) plane as shown in Figure 1. The width of the island, W, and the oscillation frequency, ω_0, of a trajectory on a curve near the center of the island, are given by

$$W = 4\sqrt{\left| \frac{K_1}{\partial^2 K_0 / \partial I^2} \right|}, \quad \omega_0 = \sqrt{\left| K_1 \frac{\partial^2 K_0}{\partial I^2} \right|}. \quad (33)$$

In a strictly 1-D system, trajectories would stay on one of the constant-K contours shown, but because of the explicit Z dependence, $dK/dZ = \partial K/\partial Z \neq 0$. One can visualize the particles "trying" to follow a contour while the contours themselves are changing in "time" Z.

The island center moves at a rate that can be estimated from the resonance condition:

$$\frac{\partial^2 K_0}{\partial I^2} dI + \frac{\partial^2 K_0}{\partial Z \partial I} dZ = 0$$
$$\Rightarrow \quad \frac{dI_{\text{res}}}{dZ} = -\frac{\partial^2 K_0 / \partial Z \partial I}{\partial^2 K_0 / \partial I^2}. \quad (34)$$

The timescale (in Z) for this is estimated by $W/(dI_{\text{res}}/dZ)$, which can be compared to ω_0:

$$\mathcal{R}^{-1} \approx \omega_0 \frac{W}{dI_{\text{res}}/dZ}; \quad \mathcal{R} \equiv \frac{\partial^2 K_0 / \partial Z \partial I}{K_1 (\partial^2 K_0 / \partial I^2)}. \quad (35)$$

For small values of $|\mathcal{R}|$, particles move along the contours quickly, while the contours themselves change slowly, but for large values of $|\mathcal{R}|$, the island and contours shift in less than an oscillation period. $\mathcal{R}$ is proportional to the background inhomogeneity, indicated by $\partial/\partial Z$, and inversely proportional to

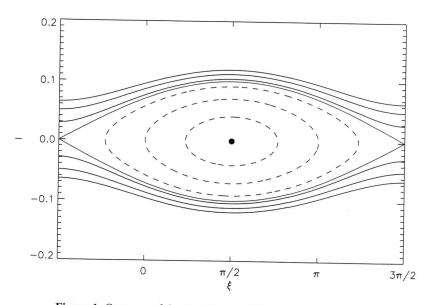

Figure 1. Contours of the Hamiltonian $K(I, \xi)$ and fixed "time" Z.

the wave amplitude, through K_1. In fact, as noted by *Albert and Bortnik* [2009], this definition of R, for parallel-propagating, constant-frequency waves, coincides exactly with the inhomogeneity parameter S of *Omura et al.* [2008].

4.2. Diffusion

As discussed in *Albert* [1993], when $|\mathcal{R}| \gg 1$, the behavior of ξ near resonance is dominated by its Z dependence. Thus,

$$\frac{d\xi}{dZ} \approx \frac{\partial}{\partial Z}\left(\frac{d\xi}{dZ}\right)(Z - Z_{\text{res}}) \equiv a(Z - Z_{\text{res}});$$
$$\xi \approx \xi_{\text{res}} + \frac{a}{2}(Z - Z_{\text{res}})^2. \qquad (36)$$

Using equation (21), a is given by

$$a \approx \frac{\partial^2 K_0}{\partial Z \partial I} = \frac{\partial}{\partial Z}\left[\frac{k_z P_z}{m\gamma} + \frac{s\ell\Omega}{\gamma} - \omega\right]. \qquad (37)$$

The change in I across the resonance is calculated by integrating $dI/dZ = -K_1 \cos \xi$ using the stationary phase method:

$$\Delta I = -K_1 \int_{-\infty}^{\infty} dZ \cos\left[\xi_{\text{res}} + \frac{a}{2}(Z - Z_{\text{res}})^2\right]$$
$$= -K_1 \sqrt{\frac{2\pi}{|a|}} \cos\left(\xi_{\text{res}} + \sigma_a \frac{\pi}{4}\right), \qquad (38)$$

where σ_a is the sign of a. Averaging over a uniform distribution of ξ_{res} gives a zero mean, but squaring before averaging gives a diffusion coefficient in I, which can be converted to diffusion in E and α_0 using equations (30) and (31). *Albert* [2010] showed that these results are identical to the single-wave limit of bounce-averaged quasi-linear theory, or conversely, averaging these results over a distribution of waves reproduces broadband quasi-linear theory.

This treatment breaks down for resonance right at the equator or anywhere that the Z derivative leads to $a = 0$. The simple fix is to Taylor expand $d\xi/dZ$ to one order higher in $Z - Z_{\text{res}}$:

$$\frac{d\xi}{dZ} \approx \frac{\partial^2}{\partial Z^2}\left(\frac{d\xi}{dZ}\right)\frac{(Z - Z_{\text{res}})^2}{2} \equiv \frac{b}{2}(Z - Z_{\text{res}})^2;$$
$$\xi \approx \xi_{\text{res}} + \frac{b}{6}(Z - Z_{\text{res}})^3, \qquad (39)$$

where b can be written as

$$b = \frac{\partial^3 K_0}{\partial^2 Z \partial I} = \frac{\partial^2}{\partial Z^2}\left[\frac{k_z P_z}{m\gamma} + \frac{s\ell\Omega}{\gamma} - \omega\right]. \qquad (40)$$

Then, the stationary phase method gives

$$\Delta I = -K_1 \int_{-\infty}^{\infty} dZ \cos\left[\xi_{\text{res}} + \frac{b}{6}(Z - Z_{\text{res}})^3\right]$$
$$= -K_1 \left\{\frac{\Gamma(1/3)}{\sqrt{3}}\left(\frac{6}{|b|}\right)^{1/3}\right\}\cos \xi_{\text{res}}. \qquad (41)$$

Squaring and averaging over ξ_{res} again gives diffusion coefficients.

The "interaction length" associated with near-stationary phase in equation (36) is approximately given by $(a/2)(\delta Z)^2 \sim \pi$, or $\delta Z = (2\pi/|a|)^{1/2}$, while equation (39) gives $(b/6)(\delta Z)^3 \sim \pi$, or $\delta Z = (6\pi/|b|)^{1/3}$. In either case, the diffusion coefficient constructed from $(\Delta I)^2$ is proportional to δZ. Whichever value of δZ is smaller should give a better estimate of the true interaction length. Similar concepts and calculations have a long history [e.g., *Helliwell*, 1967; *Howard*, 1981] and were also incorporated into self-consistent analyses by *Bespalov and Trakhtengerts* [1986] and *Trakhtengerts et al.* [1999].

4.3. Phase Bunching

Albert [1993, 2000] also considered the limit $|\mathcal{R}| \ll 1$. In this limit, particles approximately follow the K contours outside the island in Figure 1 while gradually approaching it. When they reach it, they typically move along the separatrix until reaching the x point (where they "bunch"), cross to the other side of the island, and gradually leave it, as shown in Figure 2 of *Albert* [1993] and Figure 8 of *Albert* [2000]. Adiabatic invariant considerations lead to the estimate

$$\Delta I = -\sigma_Z \frac{8}{\pi}\sqrt{\left|\frac{K_1}{\partial^2 K_0/\partial I^2}\right|}, \qquad (42)$$

where σ_Z is the sign of P_Z at the resonance. Direct numerical simulations [*Albert*, 1993, 2000; *Albert and Bortnik*, 2009; *Tao et al.*, 2012a, 2012b] show that most particles in a distribution will experience resonant changes in I with the same sign and that equation (42) correctly predicts the sign and maximum value. However, there is a spread in the value of ΔI, which increases as $|\mathcal{R}|$ increases toward 1; this behavior is not captured by equation (42).

To understand this, a more detailed treatment of ξ near resonance is needed. *Albert* [2000] derived the following equation for ξ and $d\xi/dZ$:

$$\frac{1}{2}\left(\frac{d\xi}{dZ}\right)^2 + \omega_0^2(\sin \xi + R\xi) = N, \qquad (43)$$

where N is constant. Contours are shown in Figure 2, where trajectories on dotted curves cross resonance ($d\xi/dZ = 0$) within a range of ξ values. The range of ξ_{res} is small when $|\mathcal{R}|$ is small, but increases as $|\mathcal{R}|$ becomes comparable to 1. The range of ξ_{res} can be found analytically and must be accounted for in an improved formulation of ΔI. (Such work will be reported in an upcoming publication.)

4.4. Phase Trapping

Under favorable conditions, a particle may end up on a trajectory deep inside the island, shown as dashed curves in Figures 1 and 2. In this case, the particle does not pass quickly through resonance, with an abrupt but isolated change in I, but experiences sustained resonance and continual change. A good estimate of the rate of change is given by movement of the island itself, described by equation (34). In fact, simply combining the resonance condition,

$$\frac{k_z P_Z}{m} + s\ell\Omega - \omega\Upsilon = 0, \tag{44}$$

with the definition

$$\Upsilon = \sqrt{1 + \frac{2\Omega I}{mc^2} + \left(\frac{P_Z}{mc}\right)^2} \tag{45}$$

and the constant of motion

$$\gamma \approx \Upsilon \approx \frac{\omega I - c_2}{s\ell mc^2} \tag{46}$$

gives, to lowest order, a quadratic equation for $\gamma(z)$:

$$\left(\frac{k_z^2 c^2}{\omega^2} - 1\right)\gamma^2 - 2\left(\frac{k_z^2 c^2}{\omega^2} - 1\right)\frac{s\ell\Omega}{\omega}\gamma$$
$$- \left[\frac{k_z^2 c^2}{\omega^2}\left(1 + 2\frac{\Omega}{\omega}\frac{c_2}{mc^2}\right) + \left(\frac{s\ell\Omega}{\omega}\right)^2\right] = 0. \tag{47}$$

The reliability of this estimate has been verified by direct numerical simulation [*Tao et al.*, 2012b]. A constant-frequency wave has been assumed, but it is possible to generalize this approach to varying frequencies [*Demekhov et al.*, 2006].

It has been found that for certain particle and wave parameters, solutions $\gamma(Z)$ only exist for Z above a certain value. Thus, particles trapped in resonance and with Z decreasing are eventually required to have Z increase. This is nothing other than an instance of RTA [*Omura et al.*, 2007]. Mathematically, equation (47) admits two roots γ at each value of Z, but only one is continuously linked to the initial value at resonance (and trapping); at turning, the roots coalesce, and thereafter, the larger (increasing) value applies. Figure 3 shows solutions evaluated for an electron at $L = 4$, with $\gamma = 1.8$ and $\alpha_0 = 60°$, interacting with a field-aligned whistler wave with $B_w = 120$ pT. These parameters, as well as the values $\omega_{pe}/\Omega_0 = 2$ and $\omega/\Omega_0 = 0.4$, where Ω_0 is the equatorial value of Ω, follow the work of *Omura et al.* [2007]. The reversal of direction of the electron, and continual energy gain, are similar to their Figure 1. Also shown in Figure 3 are the local pitch angle and the inhomogeneity parameter $\mathcal{R}$.

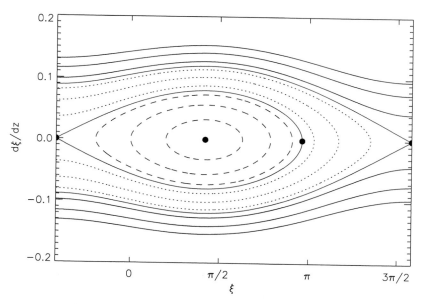

Figure 2. Contours of the constant of motion $N(\xi, d\xi/dZ)$.

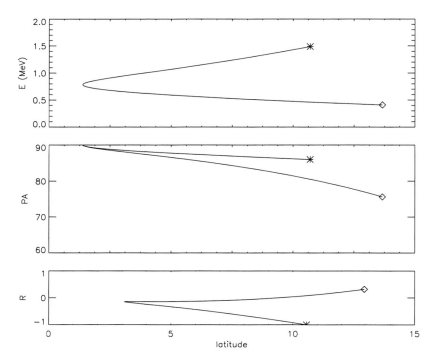

Figure 3. Analytical estimate of the behavior of a phase-trapped electron, with wave and particle parameters given in the text. The open symbol indicates initial trapping, and the asterisk shows the final values. The particle exhibits relativistic turning acceleration. Shown are (top) kinetic energy, which increases, (middle) the local pitch angle, which goes through 90°, and (bottom) the inhomogeneity parameter $\mathcal{R}$, which is less than 1 in absolute value while the particle is trapped.

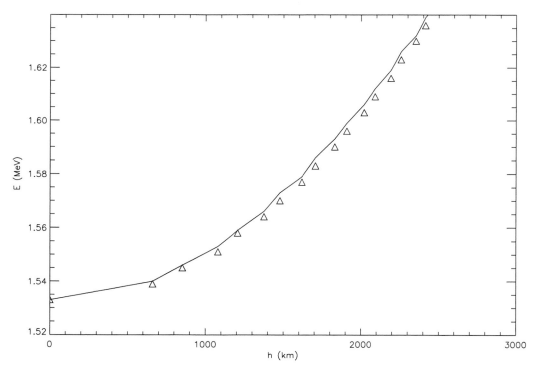

Figure 4. Analytical estimate of the behavior of a phase-trapped electron undergoing ultrarelativistic acceleration, with wave and particle parameters given in the text. The solid curve is from equation (47), while the symbols are from *Summers and Omura* [2007].

Figure 4 applies equation (47) to a phase-trapped electron with $\gamma > \Omega/\omega$, a situation which *Summers and Omura* [2007] refer to as URA. Here an electron at $L = 4$ with $\gamma = 4$ is used, with α_0 calculated for resonance at the equator. The energy gain as analytically estimated by *Summers and Omura* [2007] (their Equation 17) is also shown; the agreement is quite good.

5. SUMMARY

Gyro-averaged equations, valid near an arbitrary resonance, have been developed both directly from the Lorentz equations and from a Hamiltonian approach. The Hamiltonian development made several significant approximations, whose fidelity ultimately has to be checked against the original equations. The benefit is a reduction to a much smaller dimensional system, which can be approximately treated analytically. This is advantageous both conceptually and for large-scale modeling. Regimes of behavior include diffusion, phase bunching, and phase trapping, for each of which the previous treatment was extended. It is expected that further improvements and generalizations, including accounting for time-varying wave frequencies and amplitudes, will soon be developed and incorporated into realistic models of particle behavior in the presence of resonant waves.

Acknowledgments. This work was supported by the Space Vehicles Directorate of the Air Force Research Laboratory, by the NASA LWS TR&T program, and by NSF grant 0903802, which was awarded through the NSF/DOE Plasma Partnership program.

REFERENCES

Albert, J. M. (1993), Cyclotron resonance in an inhomogeneous magnetic field, *Phys. Fluids B, 5*, 2744–2750, doi:10.1063/1.860715.

Albert, J. M. (2000), Gyroresonant interactions of radiation belt particles with a monochromatic electromagnetic wave, *J. Geophys. Res., 105*(A9), 21,191–21,209, doi:10.1029/2000JA000008.

Albert, J. M. (2010), Diffusion by one wave and by many waves, *J. Geophys. Res., 115*, A00F05, doi:10.1029/2009JA014732.

Albert, J. M., and J. Bortnik (2009), Nonlinear interaction of radiation belt electrons with electromagnetic ion cyclotron waves, *Geophys. Res. Lett., 36*, L12110, doi:10.1029/2009GL038904.

Albert, J. M., N. P. Meredith, and R. B. Horne (2009), Three-dimensional diffusion simulation of outer radiation belt electrons during the 9 October 1990 magnetic storm, *J. Geophys. Res., 114*, A09214, doi:10.1029/2009JA014336.

Bell, T. F. (1984), The nonlinear gyroresonance interaction between energetic electrons and coherent VLF waves propagating at an arbitrary angle with respect to the Earth's magnetic field, *J. Geophys. Res., 89*(A2), 905–918.

Bell, T. F. (1986), The wave magnetic field amplitude threshold for nonlinear trapping of energetic gyroresonant and Landau resonant electrons by nonducted VLF waves in the magnetosphere, *J. Geophys. Res., 91*(A4), 4365–4379.

Bespalov, P. A., and V. Y. Trakhtengerts (1986), The cyclotron instability in the Earth radiation belts, in *Reviews of Plasma Physics*, vol. 10, edited by M. A. Leontovich, pp. 155–192, Springer, New York.

Cattell, C., et al. (2008), Discovery of very large amplitude whistler-mode waves in Earth's radiation belts, *Geophys. Res. Lett., 35*, L01105, doi:10.1029/2007GL032009.

Chang, H. C., and U. S. Inan (1983), Quasi-relativistic electron precipitation due to interactions with coherent VLF waves in the magnetosphere, *J. Geophys. Res., 88*(A1), 318–328.

Chang, H. C., U. S. Inan, and T. F. Bell (1983), Energetic electron precipitation due to gyroresonant interactions in the magnetosphere involving coherent VLF waves with slowly varying frequency, *J. Geophys. Res., 88*(A9), 7037–7050.

Cully, C. M., J. W. Bonnell, and R. E. Ergun (2008), THEMIS observations of long-lived regions of large-amplitude whistler waves in the inner magnetosphere, *Geophys. Res. Lett., 35*, L17S16, doi:10.1029/2008GL033643.

Demekhov, A. G., V. Y. Trakhtengerts, M. J. Rycroft, and D. Nunn (2006), Electron acceleration in the magnetosphere by whistler-mode waves of varying frequency, *Geomagn. Aeron., 46*, 711–716.

Ginet, G. P., and M. A. Heinemann (1990), Test particle acceleration by small amplitude electromagnetic waves in a uniform magnetic field, *Phys. Fluids B, 2*, 700–714.

Goldstein, H., C. P. Poole, and J. L. Safko (2001), *Classical Mechanics*, 3rd ed., Addison Wesley, Boston, Mass.

Helliwell, R. A. (1967), A theory of discrete VLF emissions from the magnetosphere, *J. Geophys. Res., 72*(19), 4773–4790, doi:10.1029/JZ072i019p04773.

Howard, J. E. (1981), Effective time and resonance width in cyclotron resonance heating, *Plasma Phys., 23*, 597, doi:10.1088/0032-1028/23/7/001.

Inan, U. S., T. F. Bell, and R. A. Helliwell (1978), Nonlinear pitch angle scattering of energetic electrons by coherent VLF waves in the magnetosphere, *J. Geophys. Res., 83*(A7), 3235–3253, doi:10.1029/JA083iA07p03235.

Lichtenberg, A. J., and M. A. Lieberman (1992), *Regular and Chaotic Dynamics*, 2nd ed., Springer, New York.

Lyons, L. R. (1974), Pitch angle and energy diffusion coefficients from resonant interactions with ion-cyclotron and whistler waves, *J. Plasma Phys., 12*, 417–432.

Omura, Y., N. Furuya, and D. Summers (2007), Relativistic turning acceleration of resonant electrons by coherent whistler mode waves in a dipole magnetic field, *J. Geophys. Res., 112*, A06236, doi:10.1029/2006JA012243.

Omura, Y., Y. Katoh, and D. Summers (2008), Theory and simulation of the generation of whistler-mode chorus, *J. Geophys. Res., 113*, A04223, doi:10.1029/2007JA012622.

Omura, Y., M. Hikishima, Y. Katoh, D. Summers, and S. Yagitani (2009), Nonlinear mechanisms of lower-band and upper-band

VLF chorus emissions in the magnetosphere, *J. Geophys. Res.*, *114*, A07217, doi:10.1029/2009JA014206.

Schulz, M. (1991), The magnetosphere, *in Geomagnetism*, vol. 4, edited by J. A. Jacobs, pp. 87–293, Academic, San Diego, Calif.

Shklyar, D. R. (1986), Particle interaction with an electrostatic VLF wave in the magnetosphere with an application to proton precipitation, *Planet. Space Sci.*, *34*, 1091–1099.

Summers, D., and Y. Omura (2007), Ultra-relativistic acceleration of electrons in planetary magnetospheres, *Geophys. Res. Lett.*, *34*, L24205, doi:10.1029/2007GL032226.

Tao, X., and J. Bortnik (2010), Nonlinear interactions between relativistic radiation belt electrons and oblique whistler mode waves, *Nonlin. Processes Geophys.*, *17*, 599–604, doi:10.5194/npg-17-599-2010.

Tao, X., J. Bortnik, R. M. Thorne, J. M. Albert, and W. Li (2012a), Effects of amplitude modulation on nonlinear interactions between electrons and chorus waves, *Geophys. Res. Lett.*, *39*, L06102, doi:10.1029/2012GL051202.

Tao, X., J. Bortnik, J. M. Albert, R. M. Thorne, and W. Li (2012b), The importance of amplitude modulation in nonlinear interactions between electrons and large amplitude whistler waves, *J. Atmos. Sol. Terr. Phys.*, doi:10.1016/j.jastp.2012.05.012, in press.

Trakhtengerts, V. Y., Y. Hobara, A. G. Demekhov, and M. Hayakawa (1999), Beam-plasma instability in inhomogeneous magnetic field and second order cyclotron resonance effects, *Phys. Plasmas*, *6*, 692–698, doi:10.1063/1.873305.

J. M. Albert, Air Force Research Laboratory/RVBX, 3550 Aberdeen Avenue SE, Kirtland AFB, NM 87117, USA. (jay.albert@us.af.mil)

J. Bortnik and X. Tao, Department of Atmospheric and Oceanic Sciences, University of California, Los Angeles, CA 90095-1565, USA. (jbortnik@gmail.com; xtao@atmos.ucla.edu)

Linear and Nonlinear Growth of Magnetospheric Whistler Mode Waves

Danny Summers

Department of Mathematics and Statistics, Memorial University of Newfoundland, St. John's, Newfoundland, Canada

School of Space Research, Kyung Hee University, Yongin, South Korea

Rongxin Tang

Department of Mathematics and Statistics, Memorial University of Newfoundland, St. John's, Newfoundland, Canada

Academy of Space Technology, Nanchang University, Nanchang, China

Yoshiharu Omura

Research Institute for Sustainable Humanosphere, Kyoto University, Kyoto, Japan

According to recently developed nonlinear cyclotron resonance theory, the generation of a whistler mode rising-tone chorus element is determined by a pair of coupled nonlinear ordinary differential equations referred to as "chorus equations." We generalize the chorus equations to an arbitrary energetic electron distribution and calculate the associated threshold wave amplitude for sustained nonlinear growth. For three distinct particle distributions and various sets of parameters, from the chorus equations, we obtain solutions for the wave magnetic field amplitude, wave frequency, total nonlinear growth rate, and temporal nonlinear growth rate. Further, we obtain complete time profiles for the wave amplitude that smoothly match at the interface of the linear and nonlinear growth phases. The matching process and, consequently, nonlinear wave growth can only take place over restricted regions of parameter space.

1. INTRODUCTION

Whistler mode chorus is a coherent electromagnetic wave with time-changing frequency, naturally produced in the Earth's magnetosphere close to the equatorial plane in two bands: 0.1–$0.5|\Omega_e|$ and 0.5–$0.7|\Omega_e|$, where $|\Omega_e|$ is the local electron gyrofrequency. Recent observations of chorus have been made, for instance, by *Santolík et al.* [2010], *Sigsbee et*

al. [2010], *Spasojevic and Inan* [2010], *Tsurutani et al.* [2011], and *Li et al.* [2011]. Chorus waves play an important role in controlling electron radiation belt dynamics. Stochastic acceleration due to electron gyroresonance with whistler mode chorus can generate relativistic (>1 MeV) electrons in the Earth's outer radiation belt [*Summers et al.*, 1998, 2002, 2007a, 2007b; *Horne et al.*, 2005; *Varotsou et al.*, 2008; *Xiao et al.*, 2010]. Gyroresonant interaction with whistler mode chorus can also cause electron losses from the inner magnetosphere via pitch angle scattering loss to the atmosphere [*Thorne et al.*, 2005; *Summers et al.*, 2007a, 2007b; *Ni et al.*, 2008; *Hikishima et al.*, 2010]. Whistler mode instability driven by injected seed electrons can serve to suppress the radiation belt electron flux below the *Kennel-Petschek* [1966]

Dynamics of the Earth's Radiation Belts and Inner Magnetosphere
Geophysical Monograph Series 199
10.1029/2012GM001298

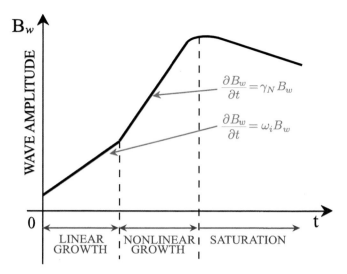

Figure 1. Schematic representation of the time profile of the magnetic field amplitude B_w of a whistler mode chorus wave; ω_i is the temporal linear growth rate and γ_N is the temporal nonlinear growth rate.

limit [*Summers et al.*, 2009, 2011; *Mauk and Fox*, 2010; *Tang and Summers*, 2012].

The generation process of whistler mode chorus comprises a linear growth phase followed by a nonlinear growth phase (see Figure 1). It is well accepted that linear wave growth is described by the cyclotron resonance theory of *Kennel and Petschek* [1966], whereby waves are excited by injected anisotropic electrons of energies ~tens of keV. *Omura et al.* [2008, 2009] have recently developed a nonlinear cyclotron resonance theory to describe the nonlinear growth phase of whistler mode waves. Nonlinear phase trapping, the generation of an electromagnetic hole in phase space, and the associated nonlinear resonant current are found to be instrumental in the nonlinear wave growth process. *Omura et al.* [2012] provide a compact review of the nonlinear processes related to the generation mechanism of chorus emissions. Special forms of nonlinear phase trapping by coherent whistler mode waves provide the basis for efficient electron energization mechanisms known as relativistic turning acceleration [*Omura et al.*, 2007] and ultrarelativistic acceleration [*Summers and Omura*, 2007; see also *Katoh et al.*, 2008].

In this paper, we further examine the theory of linear and nonlinear growth of whistler mode chorus waves. In section 2, we calculate the relativistic linear whistler mode wave growth rates for the bi-Maxwellian, bi-Lorentzian (kappa), and Dory-Guest-Harris [*Dory et al.*, 1965] loss cone particle distributions. In sections 3 and 4, we extend the nonlinear wave growth theory of *Omura et al.* [2008, 2009]. Specifi-

cally, we generalize to an arbitrary energetic electron distribution the "chorus equations" [*Omura et al.*, 2009], which are nonlinear ordinary differential equations that describe the generation of a whistler mode chorus element at the equator, and we calculate the associated threshold wave amplitude for sustained nonlinear growth. We solve the chorus equations for the wave magnetic field amplitude and frequency and, hence, obtain solutions for the total nonlinear growth rate and the temporal nonlinear growth rate. In section 5, we present numerical solutions to the chorus equations for the bi-Maxwellian, bi-Lorentzian, and Dory-Guest-Harris loss cone distributions, for a range of parameters. As well, for each of these distributions, we construct complete time profiles for the wave magnetic field amplitude that smoothly match at the interface of the linear and nonlinear growth phases. Finally, in section 6, we summarize our results.

2. LINEAR GROWTH RATE

We consider the linear growth rate for R-mode electromagnetic waves propagating parallel to a uniform magnetic field in a spatially homogeneous plasma, for a hot anisotropic relativistic electron population with a distribution function $f(p_\parallel, p_\perp)$ in the presence of a dominant cold electron population; $p_\parallel = \gamma m_e v_\parallel$ and $p_\perp = \gamma m_e v_\perp$ are the components of the relativistic momentum $\boldsymbol{p} = \gamma m_e \boldsymbol{v}$, where m_e is the electron rest mass, $\boldsymbol{v}$ is the electron velocity with components $v_\parallel$ and $v_\perp$, respectively, parallel and perpendicular to the ambient magnetic field, $\gamma = (1 - v^2/c^2)^{-1/2} = (1 + p^2/(m_e c)^2)^{1/2}$, with $v^2 = v_\parallel^2 + v_\perp^2$ and $p^2 = p_\parallel^2 + p_\perp^2$, and c is the speed of light. We require that the distribution satisfies $\int f(p_\parallel, p_\perp) d^3 p = N_h$ where N_h is the hot electron number density, with $d^3 p = 2\pi p_\perp dp_\perp dp_\parallel$. We assume that $\omega_{pe} = (4\pi N_0 e^2/m_e)^{1/2}$ is the plasma frequency, and $|\Omega_e| = eB_0/(m_e c)$ is the electron gyrofrequency, where $-e$ is the electron charge, N_0 is the cold electron number density, and B_0 is the magnitude of the zeroth-order magnetic field. Then, following *Xiao et al.* [1998] and considering the Fourier components of the wavefield ($e^{i(kx-\omega t)}$) with real wave number k (>0) and complex wave frequency $\omega = \omega_r + i\omega_i$, we write the relativistic linear growth/damping rate in the form,

$$\omega_i = \frac{\pi \omega_{pe}^2 \tilde{\eta}_{\text{rel}} \{\tilde{A}_{\text{rel}} - A_c\}}{[2\omega_r + \omega_{pe}^2 |\Omega_e|/(\omega_r - |\Omega_e|)^2]}, \qquad (1)$$

where

$$\tilde{\eta}_{\text{rel}} = \frac{\pi m_e}{N_0} \frac{(\omega_r - |\Omega_e|)}{k} \int_0^\infty \frac{dp_\perp}{\Delta_R} p_\perp^2 \left[\frac{\partial f}{\partial p_\perp}\right]_{p_\parallel = p_R} \qquad (2)$$

is the fraction of the relativistic particle distribution near resonance,

$$\tilde{A}_{\text{rel}} = \frac{\frac{k}{(\omega_r - |\Omega_e|)} \frac{1}{m_e} \int_0^\infty \frac{dp_\perp p_\perp^2}{\Delta_R} \frac{1}{\gamma_R} \left[p_\perp \frac{\partial f}{\partial p_\parallel} - p_\parallel \frac{\partial f}{\partial p_\perp} \right]_{p_\parallel = p_R}}{\int_0^\infty \frac{dp_\perp}{\Delta_R} p_\perp^2 \left[\frac{\partial f}{\partial p_\perp} \right]_{p_\parallel = p_R}} \quad (3)$$

is the relativistic pitch angle anisotropy of the resonant particles,

$$\gamma_R = \frac{-1 + \left(\frac{ck}{\omega_r} \right) \left[\left\{ \left(\frac{ck}{\omega_r} \right)^2 - 1 \right\} \left(1 + \frac{p_\perp^2}{(m_e c)^2} \right) \left(\frac{\omega_r}{|\Omega_e|} \right)^2 + 1 \right]^{1/2}}{\left\{ \left(\frac{ck}{\omega_r} \right)^2 - 1 \right\} \left(\frac{\omega_r}{|\Omega_e|} \right)} \quad (4)$$

is the resonant value of the Lorentz factor,

$$p_R = \frac{m_e}{k} (\gamma_R \omega_r - |\Omega_e|) \quad (5)$$

is the resonant value of the electron parallel momentum,

$$\Delta_R = 1 - \frac{\omega_r p_R}{m_e c^2 k \gamma_R} \quad (6)$$

is the resonant denominator in expression (2),

$$A_c = \frac{\omega_r}{(|\Omega_e| - \omega_r)} \quad (7)$$

is the minimum anisotropy required for wave instability ($\omega_i > 0$), and

$$\frac{c^2 k^2}{\omega_r^2} = 1 + \frac{\omega_{pe}^2}{\omega_r(|\Omega_e| - \omega_r)} \quad (8)$$

is the cold plasma R-mode dispersion relation. The nonrelativistic version of equation (1) is recovered by setting $\gamma = 1$, $\gamma_R = 1$, and $\Delta_R = 1$.

We define the thermal anisotropy A_T as

$$A_T = \frac{\langle p_\perp^2 \rangle}{2\langle p_\parallel^2 \rangle} - 1, \quad (9)$$

where

$$\langle p_\perp^2 \rangle = \int p_\perp^2 f(p_\parallel, p_\perp) \mathrm{d}^3 p, \langle p_\parallel^2 \rangle = \int p_\parallel^2 f(p_\parallel, p_\perp) \mathrm{d}^3 p. \quad (10)$$

In the nonrelativistic limit, relations (10) are equivalent to $T_\perp = \langle p_\perp^2 \rangle/(2m_e)$, $T_\parallel = \langle p_\parallel^2 \rangle/m_e$, and equation (9) then becomes $A_T = T_\perp/T_\parallel - 1$, where $T_\perp$ and $T_\parallel$ are the perpendicular and parallel electron temperatures, respectively. In Table 1, for three specific distributions, namely, the bi-Maxwellian, bi-Lorentzian (kappa), and Dory-Guest-Harris [Dory et al., 1965] loss cone distribution, we provide expressions for $T_\parallel$, $T_\perp$, and A_T.

Throughout this paper, for the cold plasma, we set $\omega_{pe}/|\Omega_e| = 4$. Also, for the bi-Lorentzian distribution, we fix $\kappa = 2$, and for the Dory-Guest-Harris distribution, we fix $\sigma = 4$. In general, N_h, $\theta_\parallel$, $\theta_\perp$, κ, and σ are fitting parameters for the chosen distributions.

In Figure 2, we plot the relativistic and nonrelativistic linear growth rates for the bi-Maxwellian, bi-Lorentzian (kappa), and Dory-Guest-Harris [Dory et al., 1965] loss cone distributions, in each case for thermal anisotropy values $A_T = 0.32$, 1.25, and 3.99, and for the various indicated values of the hot electron number density N_h. For each distribution, we set $T_\parallel/(mc^2) = 0.123$; $\theta_\parallel$ is then obtained using the appropriate expressions in Table 1. For each chosen value of A_T, the corresponding value of $\theta_\perp$ (and, hence, $T_\perp$) can then be found, likewise using Table 1. Table 2 contains the complete

Table 1. Electron Temperatures $T_\parallel$ and $T_\perp$ and Thermal Anisotropy A_T for the Given Hot Distributions f

$f(p_\parallel, p_\perp)$	$T_\parallel$	$T_\perp$	A_T
1. Bi-Maxwellian $\dfrac{N_h}{\pi^{3/2}\theta_\parallel \theta_\perp^2} \exp\left(-\dfrac{p_\parallel^2}{\theta_\parallel^2} - \dfrac{p_\perp^2}{\theta_\perp^2} \right)$	$\theta_\parallel^2/(2m)$	$\theta_\perp^2/(2m)$	$\theta_\perp^2/\theta_\parallel^2 - 1$
2. Bi-Lorentzian (kappa) $\dfrac{N_h}{\pi^{3/2}\theta_\parallel \theta_\perp^2} \dfrac{\Gamma(\kappa+1)}{\kappa^{3/2}\Gamma\left(\kappa - \frac{1}{2}\right)} \left(1 + \dfrac{p_\parallel^2}{\kappa\theta_\parallel^2} + \dfrac{p_\perp^2}{\kappa\theta_\perp^2} \right)^{-(\kappa+1)}$	$[\kappa/(2\kappa - 3)]\theta_\parallel^2/m$	$[\kappa/(2\kappa - 3)]\theta_\perp^2/m$	$\theta_\perp^2/\theta_\parallel^2 - 1$
3. Dory-Guest-Harris (1965) loss cone $\dfrac{N_h}{\pi^{3/2}\theta_\parallel \theta_\perp^2} \dfrac{1}{\Gamma(\sigma+1)} \left(\dfrac{p_\perp}{\theta_\perp} \right)^{2\sigma} \exp\left(-\dfrac{p_\parallel^2}{\theta_\parallel^2} - \dfrac{p_\perp^2}{\theta_\perp^2} \right)$	$\theta_\parallel^2/(2m)$	$(\sigma + 1)\theta_\perp^2/(2m)$	$(\sigma + 1)\theta_\perp^2/\theta_\parallel^2 - 1$

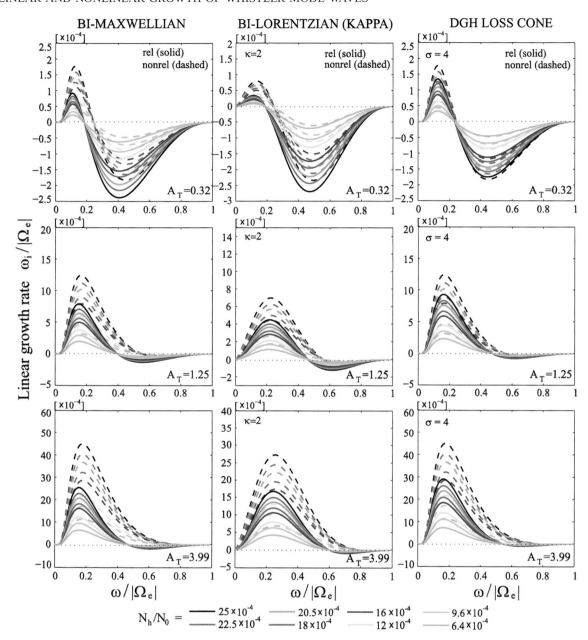

Figure 2. Linear relativistic and nonrelativistic whistler mode wave growth rates ω_i for the bi-Maxwellian, bi-Lorentzian (kappa), and Dory-Guest-Harris [*Dory et al.*, 1965] loss cone distributions, for the specified values of the hot (energetic) electron number density N_h and thermal anisotropy A_T.

set of values of A_T, $\theta_\parallel$, $\theta_\perp$ used in Figure 2. In Figure 2, we see that for each distribution, the maximum growth rate increases as the anisotropy A_T increases and as the hot number density N_h increases, as is obviously expected. As A_T increases, the bandwidth for wave growth ($\omega_i > 0$) and the frequency at which the maximum growth rate occurs both increase for each distribution. Relativistic growth rates are

less than the corresponding nonrelativistic growth rates for each distribution. Both relativistic and nonrelativistic growth rates for the bi-Lorentzian are less than those of the bi-Maxwellian, but the frequencies at which the growth rates of the bi-Lorentzian maximize are larger than those of the bi-Maxwellian. The relativistic growth rates of the Dory-Guest-Harris [*Dory et al.*, 1965] loss cone distribution exceed those

Table 2. Parameter Values Used in Figures 2, 4–9

	A_T	$\theta_\parallel/(m_e c)$	$\theta_\perp/(m_e c)$	$V_{\perp 0}/c$
Bi-Maxwellian	0.32	0.495	0.569	0.450
	1.25	0.495	0.743	0.550
	3.99	0.495	1.106	0.700
Bi-Lorentzian	0.32	0.248	0.284	0.373
	1.25	0.248	0.372	0.465
	3.99	0.248	0.553	0.616
DGH loss cone	0.32	0.495	0.254	0.485
	1.25	0.495	0.332	0.587
	3.99	0.495	0.495	0.733

of the bi-Maxwellian, while the nonrelativistic growth rates of these two distributions are identical.

3. NONLINEAR GROWTH RATE

The generation process of magnetospheric whistler mode chorus emissions has recently been analyzed by *Omura et al.* [2008, 2009] by both theory and simulation. Here we adopt the nonlinear wave growth theory of *Omura et al.* [2009] to apply to an arbitrary distribution of hot electrons. We express the nonlinear growth rate Γ_N for field-aligned whistler mode (R-mode) waves of frequency $\omega(t)$ and wave magnetic field amplitude $B_w(t)$ by the relation,

$$\frac{dB_w}{dt} = \Gamma_N B_w, \tag{11}$$

with

$$\frac{\Gamma_N}{|\Omega_e|} = \sqrt{2} Q \chi^{3/2} \left(\frac{\xi}{\tilde{\gamma}_R}\right)^{1/2} \left(\frac{B_0}{B_w}\right)^{1/2} \left(\frac{|\Omega_e|}{\omega}\right)^{1/2} \left(\frac{V_g}{c}\right)$$
$$\times \left(\frac{V_{\perp 0}}{c}\right)^{5/2} \frac{(m_e^2 c^2 \tilde{G})}{a N_0}, \tag{12}$$

and

$$a = |\Omega_e|^2 / \omega_{pe}^2 \tag{13}$$

is the cold plasma parameter,

$$\xi^2 = \frac{\omega(|\Omega_e| - \omega)}{\omega_{pe}^2}, \chi^2 = \frac{1}{1+\xi^2}, \tag{14}$$

$$\frac{V_g}{c} = \frac{\xi}{\chi} \left[\xi^2 + \frac{\Omega_e}{2(|\Omega_e|-\omega)}\right]^{-1}, \tag{15}$$

$$\tilde{\gamma}_R = \left[1 - \left(\frac{\tilde{V}_R}{c}\right)^2 - \left(\frac{V_{\perp 0}}{c}\right)^2\right]^{-1/2},$$

$$\frac{\tilde{V}_R}{c} = \chi\xi\left(1 - \frac{|\Omega_e|}{\tilde{\gamma}_R\omega}\right), \tag{16}$$

where $V_g(t)$ is the wave group speed, $\tilde{\gamma}_R(t)$ is the resonant Lorentz factor, $\tilde{V}_R(t)$ is the resonant parallel particle velocity, $V_{\perp 0}$ (= constant) is the average perpendicular particle velocity, and Q is the dimensionless factor that represents the depth of the electromagnetic electron hole within which nonlinear particle trapping takes place. The quantity $\tilde{G}$ in equation (12) is a measure of the average value of the assumed hot electron distribution F_t trapped by the wave. The trapped distribution F_t is expressed as the electron ring distribution,

$$F_t(p_\parallel, p_\perp) = \Phi(p_\parallel)\delta(p_\perp - p_{\perp 0}) \tag{17}$$

with

$$p_{\perp 0} = \gamma_0 m_e V_{\perp 0}, \gamma_0 = \left(1 - \frac{V_{\perp 0}^2}{c^2}\right)^{-1/2} \tag{18}$$

where δ is the Dirac delta function, and Φ is a function of parallel particle momentum $p_\parallel$ only; $\tilde{G}$ is given by

$$\tilde{G} = \left[\int F_t dp_\perp\right]_{p_\parallel = \tilde{p}_R} = \Phi(\tilde{p}_R) \tag{19}$$

where $\tilde{p}_R = \tilde{\gamma}_R m_e \tilde{V}_R$.

Generalizing the "chorus equations" (equations (40) and (41) given by *Omura et al.* [2009]), we find that, corresponding to the nonlinear growth rate Γ_N given here by equation (12) for the arbitrary distribution (17), the normalized wave amplitude $\tilde{B}_w = B_w(t)/B_0$ and normalized frequency $\tilde{\omega} = \omega(t)/|\Omega_e|$ satisfy the equations,

$$\frac{\partial \tilde{B}_w}{\partial \tilde{t}} = \frac{\Gamma_N}{|\Omega_e|} \tilde{B}_w - 5\frac{s_2}{s_0}\frac{V_g}{c}\frac{\tilde{a}}{\tilde{\omega}}, \tag{20}$$

$$\frac{\partial \tilde{\omega}}{\partial \tilde{t}} = \frac{2s_0}{5s_1} \tilde{\omega}\tilde{B}_w, \tag{21}$$

with $\tilde{t} = |\Omega_e|t$, and

$$s_0 = \frac{\chi}{\xi}\frac{V_{\perp 0}}{c}, \tag{22}$$

$$s_1 = \tilde{\gamma}_R\left(1 - \frac{\tilde{V}_R}{V_g}\right)^2, \tag{23}$$

$$s_2 = \frac{1}{2\xi\chi}\left\{\tilde{\gamma}_R\tilde{\omega}\left(\frac{V_{\perp 0}}{c}\right)^2 - \left[2 + \chi^2\frac{(1-\tilde{\gamma}_R\tilde{\omega})}{(1-\tilde{\omega})}\right]\frac{\tilde{V}_R V_P}{c^2}\right\},$$

(24)

$$\tilde{a} = \frac{4.5}{(LR_E)^2}\left(\frac{c}{\Omega_e}\right)^2, \quad \frac{V_P}{c} = \chi\xi$$

(25)

where $V_P(t)$ is the wave phase speed. Corresponding to the wave frequency ω as given by equations (20) and (21), the wave number k is a function of time and is given by the cold plasma dispersion relation.

The wave equations (20) and (21) hold at the magnetic equator of an assumed dipole field and, in general, for wave frequencies in the range $0.1 \leq \tilde{\omega} \leq 0.5$. Further, in order for self-sustaining emissions to exist and for waves to be amplified during propagation from the equator to higher latitudes, a threshold condition must be satisfied. Adapting section 4 of *Omura et al.* [2009] to a general distribution, we find that wave amplitudes must satisfy $\tilde{B}_w > \tilde{B}_{th}$, where $\tilde{B}_{th} = B_{th}/B_0$ is the normalized threshold amplitude given by

$$\tilde{B}_{th} = \frac{25}{2}\frac{\xi\tilde{\gamma}_R}{\chi^5\tilde{\omega}}\left(\frac{\tilde{a}s_2}{Q}\right)^2\left(\frac{c}{V_{\perp 0}}\right)^7\frac{a^2 N_0^2}{(m_e^2 c^2 \tilde{G})^2}.$$

(26)

To determine the nonlinear wave growth rate, we proceed as follows. We choose an initial wave frequency $\tilde{\omega}_0$ and calculate the corresponding wave amplitude threshold

$\tilde{B}_{th}$ at $\tilde{\omega} = \tilde{\omega}_0$ using equation (26). We then solve equations (20) and (21) for $\tilde{B}_w(\tilde{t})$, $\tilde{\omega}(\tilde{t})$ subject to the condition $\tilde{\omega}(0) = \tilde{\omega}_0$, and we choose $\tilde{B}_w(0) = B_w(0)/B_0$ to be slightly greater than the threshold value. The solutions for $\tilde{B}_w(\tilde{t})$ and $\tilde{\omega}(\tilde{t})$ can then be substituted into equation (12) to provide the nonlinear growth rate Γ_N as a function of time $\tilde{t}$ or frequency $\tilde{\omega}$.

The nonlinear growth rate Γ_N as expressed by equation (11) can be regarded as a "total" growth rate, since $d/dt \equiv \partial/\partial t + V_g\,\partial/\partial h$, where h is the distance measured along the magnetic field line from the magnetic equator [*Omura et al.*, 2009]. We may also define a "temporal" nonlinear growth rate γ_N given by

$$\frac{\partial B_w}{\partial t} = \gamma_N B_w.$$

(27)

Hence, from equation (20), we see that

$$\gamma_N = \Gamma_N - 5\frac{|\Omega_e|}{\tilde{B}_w}\frac{s_2}{s_0}\frac{V_g}{c}\frac{\tilde{a}}{\tilde{\omega}}.$$

(28)

Result (27) is analogous to the corresponding relation describing temporal linear wave growth, namely,

$$\frac{\partial B_w}{\partial t} = \omega_i B_w$$

(29)

where ω_i is the linear growth rate introduced in section 2. Figure 1 schematically illustrates the linear and nonlinear growth phases.

Table 3. Trapped Distribution $F_t(p_\parallel, p_\perp)$ and Value of $m_e^2 c^2 \tilde{G}$ Corresponding to Each Hot Electron Distribution Specified in Table 1

	$F_t(p_\parallel, p_\perp)$	$p_{\perp 0} = \gamma_0 m_e V_{\perp 0}$	$m_e^2 c^2 \tilde{G}$
1. Bi-Maxwellian	$\dfrac{N_h}{\pi^2\theta_\parallel\theta_\perp}\delta(p_\perp - p_{\perp 0})\exp\left(-\dfrac{p_\parallel^2}{\theta_\parallel^2}\right)$	$\dfrac{\sqrt{\pi}}{2}\theta_\perp$	$\dfrac{N_h}{\pi^2}\left(\dfrac{m_e c}{\theta_\parallel}\right)\left(\dfrac{m_e c}{\theta_\perp}\right)\exp\left(-\dfrac{\tilde{p}_R^2}{\theta_\parallel^2}\right)$
2. Bi-Lorentzian	$\dfrac{N_h}{\pi^2\theta_\parallel\theta_\perp}\dfrac{\Gamma(\kappa+1)\Gamma\left(\kappa-\frac{1}{2}\right)}{\kappa\Gamma\left(\kappa+\frac{1}{2}\right)\Gamma(\kappa-1)}\delta(p_\perp - p_{\perp 0})$ $\times\left(1+\dfrac{p_\parallel^2}{\kappa\theta_\parallel^2}\right)^{-(\kappa+1)}$	$\dfrac{\sqrt{\pi}}{2}\dfrac{\kappa^{1/2}\Gamma(\kappa-1)}{\Gamma\left(\kappa-\frac{1}{2}\right)}\theta_\perp$	$\dfrac{N_h}{\pi^2}\left(\dfrac{m_e c}{\theta_\parallel}\right)\left(\dfrac{m_e c}{\theta_\perp}\right)\dfrac{\Gamma(\kappa+1)\Gamma\left(\kappa-\frac{1}{2}\right)}{\kappa\Gamma\left(\kappa+\frac{1}{2}\right)\Gamma(\kappa-1)}$ $\times\left(1+\dfrac{\tilde{p}_R^2}{\kappa\theta_\parallel^2}\right)^{-(\kappa+1)}$
3. DGH loss cone	$\dfrac{N_h}{2\pi^{3/2}\theta_\parallel\theta_\perp}\dfrac{\Gamma(\sigma+1)}{\Gamma\left(\sigma+\frac{3}{2}\right)}\delta(p_\perp - p_{\perp 0})$ $\times\exp\left(-\dfrac{p_\parallel^2}{\theta_\parallel^2}\right)$	$\dfrac{\Gamma\left(\sigma+\frac{3}{2}\right)}{\Gamma(\sigma+1)}\theta_\perp$	$\dfrac{N_h}{2\pi^{3/2}}\left(\dfrac{m_e c}{\theta_\parallel}\right)\left(\dfrac{m_e c}{\theta_\perp}\right)\dfrac{\Gamma(\sigma+1)}{\Gamma\left(\sigma+\frac{3}{2}\right)}$ $\times\exp\left(-\dfrac{\tilde{p}_R^2}{\theta_\parallel^2}\right)$

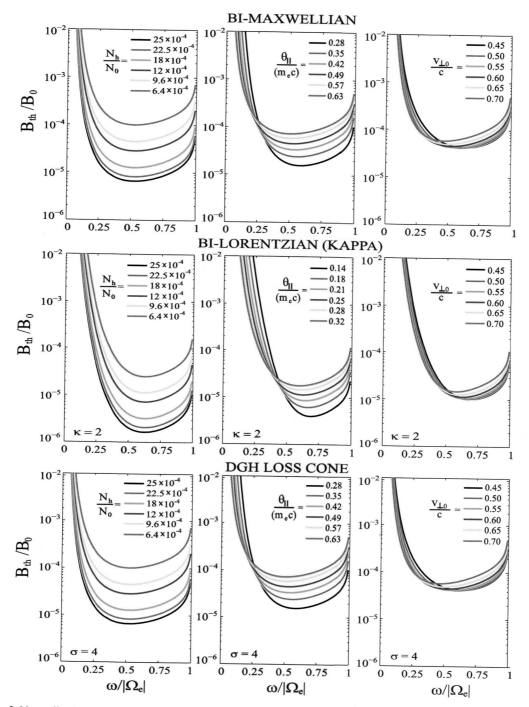

Figure 3. Normalized threshold wave amplitude, given by equation (26) as a function of frequency for the bi-Maxwellian, bi-Lorentzian, and Dory-Guest-Harris [*Dory et al.*, 1965] loss cone distributions, for the specified values of the hot electron number density N_h, the parallel thermal speed $\theta_{\parallel}$, and the perpendicular electron velocity $V_{\perp 0}$.

4. DETERMINATION OF THE TRAPPED DISTRIBUTION F_t

In order to determine the trapped distribution $F_t(p_\parallel, p_\perp)$ given by equation (17), corresponding to a given anisotropic hot electron distribution $f(p_\parallel, p_\perp)$, we apply the conditions

$$\int f(p_\parallel, p_\perp) d^3 p = \int F_t(p_\parallel, p_\perp) d^3 p \tag{30}$$

and

$$\int f(p_\parallel, p_\perp) p_\perp d^3 p = \int F_t(p_\parallel, p_\perp) p_\perp d^3 p, \tag{31}$$

where, in equations (30) and (31), we adopt a suitable trial form for F_t. From equation (17), this involves choosing a trial form for $\Phi(p_\parallel)$. For instance, corresponding to the bi-Maxwellian distribution,

$$f(p_\parallel, p_\perp) = \frac{N_h}{\pi^{3/2} \theta_\parallel \theta_\perp^2} \exp\left(-\frac{p_\parallel^2}{\theta_\parallel^2} - \frac{p_\perp^2}{\theta_\perp^2}\right), \tag{32}$$

we set $\Phi(p_\parallel) = A \exp(-p_\parallel^2/\theta_\parallel^2)$, with $A =$ constant, so that the trial form for F_t is

$$F_t(p_\parallel, p_\perp) = A \exp\left(-\frac{p_\parallel^2}{\theta_\parallel^2}\right) \delta(p_\perp - p_{\perp 0}), \tag{33}$$

where A and $p_{\perp 0}$ are to be determined. Substitution of distributions (32) and (33) into equations (30)–(31) gives

$$N_h = 2\pi^{3/2} A \theta_\parallel p_{\perp 0}, \tag{34}$$

$$\frac{\sqrt{\pi}}{2} \theta_\perp N_h = 2\pi^{3/2} A \theta_\parallel p_{\perp 0}^2. \tag{35}$$

From equations (34) and (35), we obtain

$$A = \frac{N_h}{\pi^2 \theta_\parallel \theta_\perp}, p_{\perp 0} = \frac{\sqrt{\pi}}{2} \theta_\perp, \tag{36}$$

and hence, from equation (19), we find

$$\tilde{G} = \frac{N_h}{\pi^2 \theta_\parallel \theta_\perp} \exp\left(-\frac{\tilde{p}_R^2}{\theta_\parallel^2}\right). \tag{37}$$

In line 1 of Table 3, corresponding to the bi-Maxwellian distribution, we give the trapped distribution F_t and the corresponding values of $p_{\perp 0}$ and $m_e^2 c^2 \tilde{G}$. We carried out a similar procedure for finding the trapped distribution associated with the bi-Lorentzian (kappa) distribution and the Dory-Guest-Harris [*Dory et al.*, 1965] loss cone distribution. The results for these distributions are given, respectively, in lines 2 and 3 of Table 3. Values of the quantity $m_e^2 c^2 \tilde{G}$, which occurs in expression (12) for the nonlinear growth rate Γ_N and also in result (26) for $\tilde{B}_{th}$, are thus provided for the three distributions.

5. NUMERICAL RESULTS

As noted above, we set $\omega_{pe}/|\Omega_e| = 4$. Hence, from equation (13), it follows that $a = 0.0625$, which in the Earth's dipole magnetic field corresponds to $N_0 = 44$ cm^{-3} at $L = 3.9$, for which $\tilde{a} = 9.8 \times 10^{-7}$. A plot of the parameter a as a function of L for a dipole field is given in Figure 2 of *Summers et al.* [2007a]. For the parameter Q, which relates to the depth of the electron hole, we fix $Q = 0.5$. Before obtaining numerical solutions to the chorus equations (20) and (21), we find it useful to examine the behavior of the threshold wave amplitude $\tilde{B}_{th} = B_{th}/B_0$, as given by formula (26). Noting that $V_{\perp 0}$ and $\theta_\perp$ are connected by relations given in Table 3, then for fixed values of a, $\tilde{a}$, and Q, we can regard $\tilde{B}_{th}$ as a function of $\tilde{\omega}$, N_h, $\theta_\parallel$, and $V_{\perp 0}$. In Figure 3, we plot $\tilde{B}_{th}$ as a function of frequency $\tilde{\omega}$ for the bi-Maxwellian, bi-Lorentzian, and Dory-Guest-Harris loss cone distributions, each for the specified ranges of values of N_h, $\theta_\parallel$, and $V_{\perp 0}$. The values of the remaining parameters used to construct Figure 3 are given in Table 4. The profiles in Figure 3 for each distribution in all panels possess a minimum value of $\tilde{B}_{th}$ at a

Table 4. Parameter Values Used in Figure 3

	Left Panel	Middle Panel	Right Panel
Bi-Maxwellian	$\theta_\parallel/(m_e c) = 0.495$	$N_h/N_0 = 9.6 \times 10^{-4}$	$N_h/N_0 = 9.6 \times 10^{-4}$
	$V_{\perp 0}/c = 0.605$	$V_{\perp 0}/c = 0.605$	$\theta_\parallel/(m_e c) = 0.495$
	$\theta_\perp/(m_e c) = 0.857$	$\theta_\perp/(m_e c) = 0.857$	$\theta_\perp/(m_e c) = 0.57, 0.65, 0.74, 0.86, 0.97, 1.11$
Bi-Lorentzian	$\theta_\parallel/(m_e c) = 0.247$	$N_h/N_0 = 9.6 \times 10^{-4}$	$N_h/N_0 = 9.6 \times 10^{-4}$
	$V_{\perp 0}/c = 0.605$	$V_{\perp 0}/c = 0.605$	$\theta_\parallel/(m_e c) = 0.248$
	$\theta_\perp/(m_e c) = 0.537$	$\theta_\perp/(m_e c) = 0.537$	$\theta_\perp/(m_e c) = 0.36, 0.41, 0.47, 0.54, 0.60, 0.69$
DGH loss cone	$\theta_\parallel/(m_e c) = 0.495$	$N_h/N_0 = 9.6 \times 10^{-4}$	$N_h/N_0 = 9.6 \times 10^{-4}$
	$V_{\perp 0}/c = 0.605$	$V_{\perp 0}/c = 0.605$	$\theta_\parallel/(m_e c) = 0.495$
	$\theta_\perp/(m_e c) = 0.348$	$\theta_\perp/(m_e c) = 0.348$	$\theta_\perp/(m_e c) = 0.23, 0.26, 0.30, 0.35, 0.39, 0.45$

"midrange" value of frequency, approximately. Since the frequency at which $\tilde{B}_{th}$ is typically evaluated lies in the range $0 < \tilde{\omega} < 0.3$, then, in practice, $\tilde{B}_{th}$ is a decreasing function of frequency for each distribution. In addition, over this frequency range, the magnitude of $\tilde{B}_{th}$ decreases as N_h increases, for each distribution (left panels). It is also useful to observe that in the frequency range $0 < \tilde{\omega} < 0.25$, $\tilde{B}_{th}$ decreases as $V_{\perp 0}$ (or $\theta_\perp$) increases, for each distribution (right panels).

In Figure 4, for the bi-Maxwellian distribution, we plot time profiles, obtained by solving the chorus equations (20)–(21), for the wave amplitude B_w, wave frequency ω, total

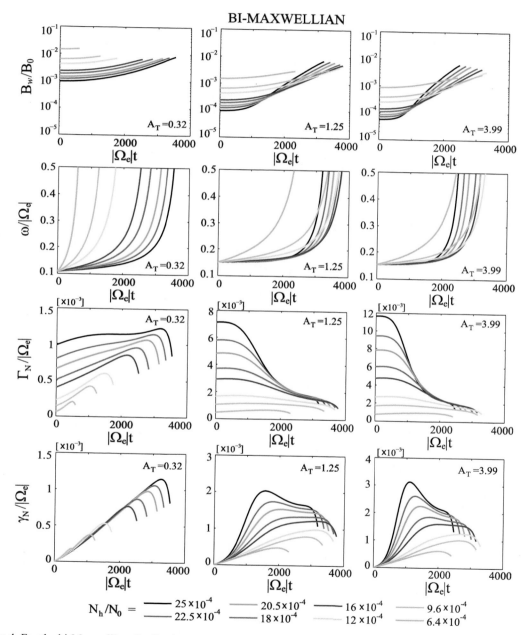

Figure 4. For the bi-Maxwellian distribution, solutions of the chorus equations (20) and (21) for the wave amplitude B_w, the frequency ω, the total nonlinear growth rate Γ_N, and the temporal nonlinear growth rate γ_N, for the specified values of the hot electron number density N_h and thermal anisotropy A_T.

nonlinear growth rate Γ_N, and temporal nonlinear growth rate γ_N. As in Figure 2, in Figure 4, we set $T_\parallel/(mc^2) = 0.123$, and we find $\theta_\parallel$ and $\theta_\perp$ (for each of the specified values of $A_T = 0.32$, 1.25, 3.99) using Table 1. Parameter values used in Figure 4 are given in Table 2. To solve the chorus equations, here we adopt the initial conditions $B_w(0) = B_{w0}$, $\omega(0) = \omega_m$, where ω_m is the frequency at which the linear growth rate

maximizes (for a given value of A_T), and B_{w0} slightly exceeds the threshold amplitude B_{th} corresponding to $\omega = \omega_m$. From equations (26) and (37), we observe that B_{th} also depends on N_h, so B_{th} must be calculated separately for each of the eight solutions (corresponding to the eight selected N_h values) for each value of A_T adopted in Figure 4. In general, as we have seen in Figure 2, as A_T increases, ω_m increases. Further, B_{th}

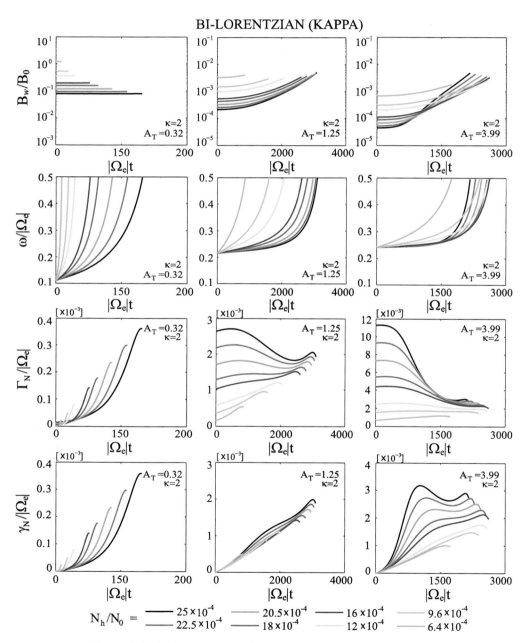

Figure 5. As in Figure 4, except here for the bi-Lorentzian distribution.

decreases as A_T increases. The latter follows since, as $\theta_\perp$ increases, B_{th} decreases (from our discussion of Figure 3) and A_T increases (from the formulae for A_T in Table 1). Thus, as expected, we see from the top three panels of Figure 4 that for each chosen value of N_h, the values of $B_w(0)$ decrease with increasing values of A_T. Figure 4 captures the typical properties of whistler mode chorus, namely, a rising tone (increasing frequency) with increasing wave amplitude. At lower anisotropies, Γ_N increases to a maximum value before

decreasing, while for larger anisotropies, Γ_N monotonically decreases with time. For all anisotropies, γ_N rises to a maximum value before falling. Figure 4 also provides evidence (see the solutions for B_w in the top row of panels) that for sustained nonlinear wave growth, sufficiently large values of A_T and N_h are required.

In Figures 5 and 6, we show solutions for B_w, ω, Γ_N, γ_N corresponding to the bi-Lorentzian and Dory-Guest-Harris loss cone distributions, respectively. The solutions in

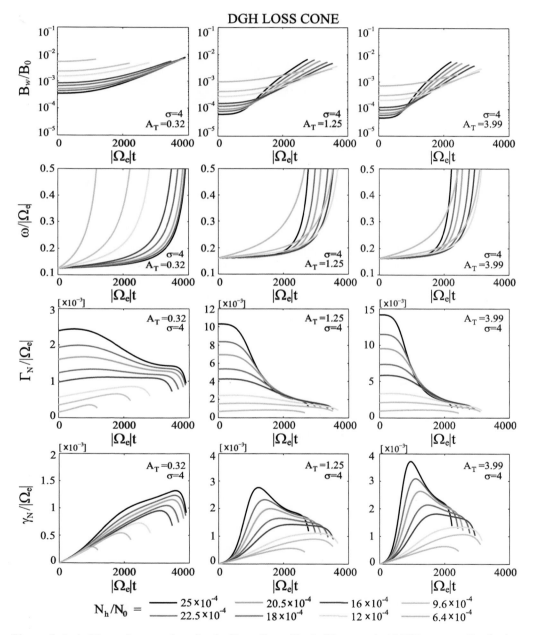

Figure 6. As in Figure 4, except here for the Dory-Guest-Harris [*Dory et al.*, 1965] loss cone distribution.

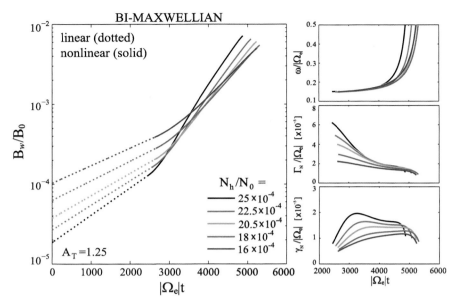

Figure 7. (left) For the bi-Maxwellian distribution, time variation of the whistler mode wave magnetic field consisting of smoothly matching linear and nonlinear profiles, for the given values of the electron number density N_h. (right) Corresponding to the left panel in the nonlinear phase, profiles of the wave frequency ω, the total nonlinear growth rate Γ_N, and the temporal nonlinear growth rate γ_N.

Figures 5 and 6 were obtained by similar techniques to those in Figure 4 and, generally, are similar to the latter solutions. As an exception, for the bi-Lorentzian distribution, the profiles for Γ_N and γ_N can possess two local maxima (Figure 5, third row center panel, and fourth row right panel). Further evidence that sustained nonlinear growth requires sufficiently large values of A_T, N_h is provided in the solutions for B_w in the top row of Figure 5.

Hitherto, we have solved the chorus equations to obtain typical solutions for the wave amplitude B_w during the

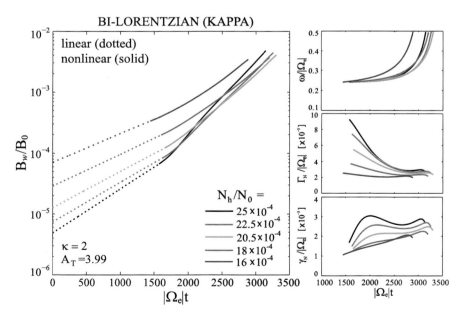

Figure 8. As in Figure 7, except here for the bi-Lorentzian distribution.

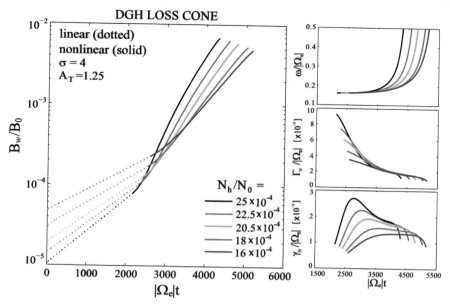

Figure 9. As in Figure 7, except here for the Dory-Guest-Harris [*Dory et al.*, 1965] loss cone distribution.

nonlinear growth phase. We now refine our study by constructing time profiles of B_w that are valid throughout the wave growth process comprising both the linear and nonlinear phases. Equations (29) and (27) describe wave growth during the linear and nonlinear phases, respectively. In order to construct linear and nonlinear solutions that smoothly match at some "matching" amplitude B_M (at which linear growth ends and nonlinear growth begins), we chose B_M so that $\gamma_N = (\omega_i)_{max}$, where γ_N is given by equation (28), and $(\omega_i)_{max}$ is the maximum value of the linear growth rate, which occurs at frequency $\omega = \omega_m$. Having determined B_M, we then solve the chorus equations (20)–(21) subject to the condition $B_w = B_M$ at $\omega = \omega_m$ to obtain the solution for $B_w(t)$ during the nonlinear phase. The matching solution for $B_w(t)$ during the linear phase is simply obtained from equation (29), with $\omega_i = (\omega_i)_{max}$, and we assume the time period over which linear growth takes place is of the order of the time period for nonlinear growth (the latter being specified as the time required for the frequency to increase from $\omega = \omega_m$ to $\omega = 0.5|\Omega_e|$). The results of this matching process are shown for the bi-Maxwellian distribution in Figure 7, for $A_T = 1.25$, and the specified values of N_h. The left panel shows the smoothly matched profiles for B_w, and the right panels show the corresponding profiles for ω, Γ_N, and γ_N during the nonlinear phase. It should be noted that the matching process described can only be carried out over limited regions of parameter space, for instance, for sufficiently large values of A_T, N_h. Detailed analysis of such regions, which can be expected to shed further light on the process of nonlinear wave growth, is left as a future project.

In Figures 8 and 9, we show further examples of matched profiles for B_w for the bi-Lorentzian and Dory-Guest-Harris loss cone distributions, respectively, for the indicated values of A_T, N_h.

6. SUMMARY

We have examined the linear and nonlinear growth phases of coherent whistler mode waves generated at the magnetic equator. We first calculated relativistic linear growth rates corresponding to the bi-Maxwellian, bi-Lorentzian (kappa), and Dory-Guest-Harris [*Dory et al.*, 1965] loss cone distributions for a range of values of the thermal anisotropy A_T and the hot electron number density N_h. We then generalized the nonlinear wave growth theory of *Omura et al.* [2008, 2009] to arbitrary energetic electron distributions. Specifically, we expressed in terms of an arbitrary electron distribution (1) the total nonlinear growth rate Γ_N, (2) the temporal nonlinear growth rate γ_N, (3) the chorus equations, which describe the nonlinear evolution of the wave magnetic field B_w and frequency ω, and (4) the threshold wave amplitude B_{th} required for sustained nonlinear growth.

For the bi-Maxwellian, bi-Lorentzian, and Dory-Guest-Harris loss cone distributions, we solved the chorus equations to obtain profiles for $B_w(t)$, $\omega(t)$, $\Gamma_N(t)$, and $\gamma_N(t)$ in the nonlinear growth phase, for various sets of parameters. For each of these distributions for selected A_T values and a range of N_h values, we constructed complete profiles for $B_w(t)$ by smoothly matching solutions in the linear and nonlinear regimes. Further work is required to determine the parameter space over which sustained nonlinear wave growth is possible.

This paper has been concerned with the temporal growth of whistler mode waves at the equator. A recent study by *Summers et al.* [2012] addresses the spatiotemporal evolution of whistler mode waves as they evolve along a field line from their equatorial source.

Acknowledgments. This work is supported by a Discovery Grant of the Natural Sciences and Engineering Research Council of Canada to D.S. and by the WCU grant R31-10016 funded by the Korean Ministry of Education, Science and Technology. Additional support is acknowledged from grant-in-aid 23340147 of the Ministry of Education, Science, Sports and Culture of Japan.

REFERENCES

Dory, R. A., G. E. Guest, and E. G. Harris (1965), Unstable electrostatic plasma waves propagating perpendicular to a magnetic field, *Phys. Rev. Lett.*, *14*, 131–133.

Hikishima, M., Y. Omura, and D. Summers (2010), Microburst precipitation of energetic electrons associated with chorus wave generation, *Geophys. Res. Lett.*, *37*, L07103, doi:10.1029/2010GL042678.

Horne, R. B., R. M. Thorne, S. A. Glauert, J. M. Albert, N. P. Meredith, and R. R. Anderson (2005), Timescale for radiation belt electron acceleration by whistler mode chorus waves, *J. Geophys. Res.*, *110*, A03225, doi:10.1029/2004JA010811.

Katoh, Y., Y. Omura, and D. Summers (2008), Rapid energization of radiation belt electrons by nonlinear wave trapping, *Ann. Geophys.*, *26*, 3451–3456.

Kennel, C. F., and H. E. Petschek (1966), Limit on stably trapped particle fluxes, *J. Geophys. Res.*, *71*(1), 1–28.

Li, W., R. M. Thorne, J. Bortnik, Y. Y. Shprits, Y. Nishimura, V. Angelopoulos, C. Chaston, O. Le Contel, and J. W. Bonnell (2011), Typical properties of rising and falling tone chorus waves, *Geophys. Res. Lett.*, *38*, L14103, doi:10.1029/2011GL047925.

Mauk, B. H., and N. J. Fox (2010), Electron radiation belts of the solar system, *J. Geophys. Res.*, *115*, A12220, doi:10.1029/2010JA015660.

Ni, B., R. M. Thorne, Y. Y. Shprits, and J. Bortnik (2008), Resonant scattering of plasma sheet electrons by whistler-mode chorus: Contribution to diffuse auroral precipitation, *Geophys. Res. Lett.*, *35*, L11106, doi:10.1029/2008GL034032.

Omura, Y., N. Furuya, and D. Summers (2007), Relativistic turning acceleration of resonant electrons by coherent whistler mode waves in a dipole magnetic field, *J. Geophys. Res.*, *112*, A06236, doi:10.1029/2006JA012243.

Omura, Y., Y. Katoh, and D. Summers (2008), Theory and simulation of the generation of whistler-mode chorus, *J. Geophys. Res.*, *113*, A04223, doi:10.1029/2007JA012622.

Omura, Y., M. Hikishima, Y. Katoh, D. Summers, and S. Yagitani (2009), Nonlinear mechanisms of lower-band and upper-band VLF chorus emissions in the magnetosphere, *J. Geophys. Res.*, *114*, A07217, doi:10.1029/2009JA014206.

Omura, Y., D. Nunn, and D. Summers (2012), Generation processes of whistler mode chorus emissions: Current status of nonlinear wave growth theory, in *Dynamics of the Earth's Radiation Belts and Inner Magnetosphere, Geophys. Monogr. Ser.*, doi:10.1029/2012GM001347, this volume.

Santolík, O., J. S. Pickett, D. A. Gurnett, J. D. Menietti, B. T. Tsurutani, and O. Verkhoglyadova (2010), Survey of Poynting flux of whistler mode chorus in the outer zone, *J. Geophys. Res.*, *115*, A00F13, doi:10.1029/2009JA014925.

Sigsbee, K., J. D. Menietti, O. Santolík, and J. S. Pickett (2010), Locations of chorus emissions observed by the Polar Plasma Wave Instrument, *J. Geophys. Res.*, *115*, A00F12, doi:10.1029/2009JA014579.

Spasojevic, M., and U. S. Inan (2010), Drivers of chorus in the outer dayside magnetosphere, *J. Geophys. Res.*, *115*, A00F09, doi:10.1029/2009JA014452.

Summers, D., and Y. Omura (2007), Ultra-relativistic acceleration of electrons in planetary magnetospheres, *Geophys. Res. Lett.*, *34*, L24205, doi:10.1029/2007GL032226.

Summers, D., R. M. Thorne, and F. Xiao (1998), Relativistic theory of wave-particle resonant diffusion with application to electron acceleration in the magnetosphere, *J. Geophys. Res.*, *103*(A9), 20,487–20,500.

Summers, D., C. Ma, N. P. Meredith, R. B. Horne, R. M. Thorne, D. Heynderickx, and R. R. Anderson (2002), Model of the energization of outer-zone electrons by whistler-mode chorus during the October 9, 1990 geomagnetic storm, *Geophys. Res. Lett.*, *29*(24), 2174, doi:10.1029/2002GL016039.

Summers, D., B. Ni, and N. P. Meredith (2007a), Timescales for radiation belt electron acceleration and loss due to resonant wave-particle interactions: 1. Theory, *J. Geophys. Res.*, *112*, A04206, doi:10.1029/2006JA011801.

Summers, D., B. Ni, and N. P. Meredith (2007b), Timescales for radiation belt electron acceleration and loss due to resonant wave-particle interactions: 2. Evaluation for VLF chorus, ELF hiss, and electromagnetic ion cyclotron waves, *J. Geophys. Res.*, *112*, A04207, doi:10.1029/2006JA011993.

Summers, D., R. Tang, and R. M. Thorne (2009), Limit on stably trapped particle fluxes in planetary magnetospheres, *J. Geophys. Res.*, *114*, A10210, doi:10.1029/2009JA014428.

Summers, D., R. Tang, and Y. Omura (2011), Effects of nonlinear wave growth on extreme radiation belt electron fluxes, *J. Geophys. Res.*, *116*, A10226, doi:10.1029/2011JA016602.

Summers, D., Y. Omura, Y. Miyashita, and D.-H. Lee (2012), Nonlinear spatiotemporal evolution of whistler mode chorus waves in Earth's inner magnetosphere, *J. Geophys. Res.*, *117*, A09206, doi:10.1029/2012JA017842.

Tang, R., and D. Summers (2012), Energetic electron fluxes at Saturn from Cassini observations, *J. Geophys. Res.*, *117*, A06221, doi:10.1029/2011JA017394.

Thorne, R. M., T. P. O'Brien, Y. Y. Shprits, D. Summers, and R. B. Horne (2005), Timescale for MeV electron microburst loss during geomagnetic storms, *J. Geophys. Res.*, *110*, A09202, doi:10.1029/2004JA010882.

Tsurutani, B. T., B. J. Falkowski, O. P. Verkhoglyadova, J. S. Pickett, O. Santolík, and G. S. Lakhina (2011), Quasi-coherent chorus properties: 1. Implications for wave-particle interactions, *J. Geophys. Res.*, *116*, A09210, doi:10.1029/2010JA016237.

Varotsou, A., D. Boscher, S. Bourdarie, R. B. Horne, N. P. Meredith, S. A. Glauert, and R. H. Friedel (2008), Three-dimensional test simulations of the outer radiation belt electron dynamics including electron-chorus resonant interactions, *J. Geophys. Res.*, *113*, A12212, doi:10.1029/2007JA012862.

Xiao, F., R. M. Thorne, and D. Summers (1998), Instability of electromagnetic R-mode waves in a relativistic plasma, *Phys. Plasmas*, *5*, 2489–2497.

Xiao, F., Z. Su, H. Zheng, and S. Wang (2010), Three-dimensional simulations of outer radiation belt electron dynamics including cross-diffusion terms, *J. Geophys. Res.*, *115*, A05216, doi:10.1029/2009JA014541.

Y. Omura, Research Institute for Sustainable Humanosphere 611-0011, Kyoto University, Kyoto, Japan. (omura@rish.kyoto-u.ac.jp)

D. Summers and R. Tang, Department of Mathematics and Statistics, Memorial University of Newfoundland, St. John's, Newfoundland A1C 5S7, Canada. (dsummers@mun.ca; tangrx@gmail.com)

High-Energy Electron Diffusion by Resonant Interactions With Whistler Mode Hiss

J.-F. Ripoll and D. Mourenas

CEA, DAM, DIF, Arpajon, France

The dynamics of trapped electrons in the Earth's radiation belts can be described by quasilinear diffusion theory. Computer simulation of the evolution of radiation belt electrons is, however, time-consuming, so it is useful to seek approximate models. In this chapter, the simple parallel propagation approximation model and the mean value approximation model are compared with the Lyons (1972) exact formulation. The models have been further simplified by using the low-frequency wave limit and the dense plasma limit, which apply to high-energy electrons interacting with hiss waves in the plasmasphere. Precipitation lifetimes are deduced directly from quasilinear diffusion coefficients. Recent analytical developments are used throughout to complement the analysis.

1. INTRODUCTION

Orbiting satellites are subject to the impact of relativistic electrons, which are trapped in the radiation belts or transit through the slot region. Geomagnetic activity and solar storms contribute to inject or accelerate these electrons in the magnetosphere. The knowledge of the electron dynamics and their lifetime is therefore crucial for the protection of onboard electronics [*Baker*, 2002; *Hughes and Benedetto*, 2003] as well as for predicting electron precipitations into the Earth atmosphere.

Most of the loss of trapped electrons occurs by pitch angle diffusion from resonant interactions with whistler waves [*Dungey*, 1963; *Andronov and Trakhtengerts*, 1964; *Kennel and Petschek*, 1966]. Plasmaspheric wave energy causes the electron pitch angle to decrease until precipitation into the atmosphere can eventually occur, while high pitch angle electrons bounce between the mirror points. Pitch angle diffusion coefficients have been derived by

quasilinear theory [*Roberts*, 1969; *Lyons et al.*, 1971, 1972; *Lyons*, 1974], among which the latter remains the most detailed reference model. Electron precipitation lifetimes can then be either computed from the electron distribution, solution of a Fokker-Planck diffusion transport equation [*Lyons et al.*, 1972], or be approximated directly from the diffusion coefficients [*Albert and Shprits*, 2009]. However, full simulations require a very large number of operations because six dimensions are involved, namely, the electron momentum p, the McIlwain parameter L, the latitude λ, the pitch angle α, and the wave normal angle θ for each harmonic n. Two approximations, the parallel propagation approximation (PPA) [*Summers*, 2005] and the mean value approximation (MVA) [*Albert*, 2007], have recently been derived for fast computations. They are based on the selection of a single wave normal angle, being null for the PPA and representative of the actual average over the whole distribution in the case of the MVA. Both have been compared by *Albert* [2008] to the full model [*Lyons*, 1974], showing a good accuracy of MVA and of PPA at small to moderate pitch angles.

In both the inner belt and the slot region, on which this article focuses, hiss, VLF, and lightning-generated waves are responsible for the main pitch angle diffusion [*Abel and Thorne*, 1998; *Summers et al.*, 2007a, 2007b; *Pokhotelov et al.*, 2008; *Meredith et al.*, 2007, 2009]. Because of wave

Dynamics of the Earth's Radiation Belts and Inner Magnetosphere
Geophysical Monograph Series 199
10.1029/2012GM001309

and plasma properties, the low-frequency (LF) wave limit and the dense plasma (LF) limit can be used [*Lyons et al.*, 1972; *Albert*, 1999]. They allow simplifying significantly all models, in particular, MVA and the determination of the mean angle, since the resonance frequencies become more tractable.

After simplifying all three models in both the LF wave and dense plasma limits and showing the validation of our quasilinear code (CEVA) [*Réveillé et al.*, 2001], the models' accuracy is assessed for high-energy electrons, which computation often requires a very high number of harmonics. The model will also be challenged with the use of complex and realistic wave conditions in the plasmasphere, based on CRRES measurements [*Meredith et al.*, 2007]. Approximated precipitation lifetimes are also deduced and discussed. Recent analytical developments [*Mourenas and Ripoll*, 2012] are used throughout and contribute to better explaining and understanding the numerical results.

2. DIFFUSION COEFFICIENTS FOR LOW-FREQUENCY WAVES AND DENSE PLASMAS

2.1. General Formulation

The local pitch angle quasilinear diffusion coefficient $D_{\alpha\alpha}(\alpha)$ of *Lyons* [1974] (with dimensions of p^2/t) has been conveniently rewritten by *Albert* [2005, 2007] as

$$
D_{\alpha\alpha} = \frac{D_{\alpha\alpha}}{p^2} = \frac{\Omega_c}{\gamma^2} \frac{B_{wave}^2}{B_0^2} \sum_{n=-\infty}^{+\infty} \sum_{\omega} D_{\alpha\alpha}^n, \quad D_{\alpha\alpha}^n = \int_{\theta_{min}}^{\theta_{max}} D_{\alpha\alpha}^{n,\theta} \mathrm{d}\theta
$$

$$
= \int_{\theta_{min}}^{\theta_{max}} \sin(\theta)\Delta_n G_1 G_2, \mathrm{d}\theta, \tag{1}
$$

with

$$
\Delta_n = \frac{\pi}{2\cos\theta}\left|\frac{c}{V_\parallel}\right|^3 \frac{(\sin^2(\alpha) + n\Omega_c/\gamma\omega)^2}{|1 - (\partial\omega/\partial k_\parallel)_\theta/V_\parallel|}\Phi_n^2, \tag{2}
$$

$$
G_1 = \frac{\Omega_c B^2(\omega)}{\langle B^2 \rangle}, \quad G_2(\omega,\theta) = \frac{g(\theta)}{N(\omega,\theta)}, \tag{3}
$$

$$
N(\omega,\theta) = \int_{\theta_{min}}^{\theta_{max}} \mathrm{d}\theta' \sin(\theta')\Gamma g(\theta'),
$$
$$
\Gamma(\theta',\omega(\theta)) = \mu^2|\mu + \omega\partial\mu/\partial\omega|, \tag{4}
$$

with $\mu(\omega, \theta) = kc/\omega$ is the wave refractive index, B_0 is the amplitude of the local dipolar magnetic field. Φ_n^2 contains Bessel functions $J_{n\pm1}$ with the argument $x = (-\omega\gamma/\Omega_c + n)\tan\alpha\tan\theta$. Ω_c is the local electron gyrofrequency, and γ is the relativistic factor. $B^2(\omega) = \exp[-(\omega - \omega_m)^2/\Delta\omega^2]$ is the wave power frequency distribution and $\langle B^2 \rangle = \int_{\omega_{LC}}^{\omega_{UC}} B^2(\omega')\mathrm{d}\omega'$. The distribution with wave normal angle θ is described by $g(\theta) = \exp[-(\tan\theta - \tan\theta_m)^2/\tan^2\Delta\theta]$, which is nonnull for θ between θ_{max} and θ_{min}. $D_{\alpha\alpha}$ has to be bounce averaged as described in the work of *Lyons et al.* [1972].

2.2. Low-Frequency Resonance Approximations

Both $G_2(\omega, \theta)$ and $D''(\omega, \theta)$ in equation (1) are evaluated at the resonant frequencies ω corresponding to a particular θ, n, and α and which is determined from the cyclotron (or Landau, $n = 0$) resonance condition

$$
\omega + n\Omega_c/\gamma = k\cos\theta \, V\cos\alpha, \tag{5}
$$

in which V denotes the electron velocity and k the wave vector. In both the LF wave and dense plasma limits, i.e., $\Omega_{ci} \ll \omega \ll \Omega_c$ and $\Omega_{pe}^2 \gg \omega\Omega_c$ with Ω_{ci}, Ω_c, and Ω_{pe}, the ion gyrofrequency, the electron gyrofrequency, and the plasma frequency [*Carpenter and Anderson*, 1992], the dispersion relation can be reduced to [*Lyons et al.*, 1972]

$$
\mu^2 = \Omega_{pe}^2/(\Omega_c\omega\cos\theta), \tag{6}
$$

in which the term $\omega_m/\Omega_c/\cos\theta$ has been neglected, which corresponds to considering θ values not too large. This expression has been shown to be a very good approximation for plasmaspheric hiss and lightning-generated whistlers for $L < 4.5$ [*Albert*, 1999]. Combining equation (5) for small ω_m/Ω_c and equation (6), one gets the following cyclotron and Landau ($n = 0$) resonant frequencies

$$
\omega_{res} = n^2\Omega_c^3/(\cos^2\alpha(\gamma^2 - 1)\Omega_{pe}^2\cos\theta), \quad \forall n \neq 0,
$$
$$
\omega_{res} = \Omega_c(\gamma^2 - 1)\Omega_{pe}^2\cos^2\alpha\cos\theta/(\gamma^2\Omega_c^2), \quad n = 0. \tag{7}
$$

If one assumes $\omega_{res} \sim \omega_m$ in equation (7), it gives $\cos\alpha \sim n/(p\varepsilon_m)$ with $\varepsilon_m = (\Omega_{pe}^2\omega_m/\Omega_c^3)^{1/2}$ and $p = (\gamma^2 - 1)^{1/2}$. This relationship will be used below as a model for determining analytically the value of the resonant pitch angle.

Moreover, the Bessel function term Φ_n^2 can be simplified in our limits [*Stix*, 1962; *Lyons et al.*, 1972] and reads

$$
\Phi_n^2 = ((1 + \cos\theta)J_{n+1} + (1 - \cos\theta)J_{n-1})^2/8\cos^2\theta, \quad \forall n \neq 0
$$
$$
\Phi_n^2 = \left(\cos\theta J_1 - {}^\omega\!/_{\Omega_c}\cot\alpha\sin\theta J_0\right)^2\Big/2\cos^2\theta, \quad n = 0. \tag{8}
$$

2.3. MVA and PPA Models for Fast Computations

Using equations (1)–(4) and (8), the diffusion coefficients can be rewritten as

$$
D_{\alpha\alpha}^{n,\theta} = \frac{\pi}{4}\frac{\Omega_c^2}{\gamma^2}\frac{B_{\text{wave}}^2}{B_0^2}\left|\frac{c^3}{V_{\parallel}^3}\right|\frac{1}{\cos^3\theta}\left|\frac{\Phi'^2_n}{1-V_g/V_{\parallel}}\right|\frac{g(\theta)}{N(\omega)}
$$
$$
\times\left(\sin(\alpha)^2 + n\Omega_c/\gamma\omega\right)^2\frac{B^2(\omega)}{\langle B^2\rangle},
\tag{9}
$$

with $\Phi'^2_n = 2\cos^2\theta\Phi_n^2$ for all n and the group velocity defined by

$$
V_g = \frac{\partial\omega}{\partial k_{\parallel}} = 2V_{\parallel}\frac{\omega}{\left(\omega + n\frac{\Omega_e}{\gamma}\right)} = 2c\frac{|n|}{n}\frac{\sqrt{\Omega_c\omega_{\text{res}}/\cos\theta}}{\omega_{pe}}.
\tag{10}
$$

Albert [2007] has showed that $\omega(\theta)$ is nearly constant for $\theta \leq \pi/3$, allowing taking $N(\omega)$ outside the first θ integral of equation (1). It leads to a new diffusion coefficient defined from a weighted average between θ_{LC} and θ_{UC}

$$
D_n \approx \int_{\theta_{LC}}^{\theta_{UC}}\sin(\theta)g(\theta)\Gamma d\theta\ \Delta_n G_1/\Gamma\ \bigg/\int_{\theta_{LC}}^{\theta_{UC}}\sin(\theta)g(\theta)\Gamma d\theta
$$
$$
\equiv \langle\Delta_n G_1/\Gamma\rangle_\theta.
\tag{11}
$$

This weighted average is expected to be rather insensitive to the exact form of the weighting function. When the averaged function is taken at a single "carefully chosen point," by using what is called the MVA [*Albert*, 2007, 2008], it becomes $D^n \sim (\Delta^n G_1/\Gamma)_{|\theta_{\text{MVA}}}$. Moreover, taking our limits, the term Γ can be simplified to

$$
\Gamma = \mu^2\left|\mu + \omega\frac{\partial\mu}{\partial\omega}\right| \rightarrow \frac{1}{2}|\mu^3| = \frac{1}{2}\frac{\left(\omega + \frac{n\Omega_c}{\gamma}\right)^3}{\omega^3\cos^3\theta}\frac{c^3}{V_{\parallel}^3}.
\tag{12}
$$

Using the previous expressions, the MVA diffusion coefficients are given by

$$
D_{\alpha\alpha}^{n,\text{MVA}}/p^2 = \frac{\pi}{4}\frac{\Omega_c^2}{\gamma^2}\frac{B_{\text{wave}}^2}{B_0^2}\frac{\Phi'^2_n}{|V_{\parallel}/V_g - 1|}\frac{\omega_{\text{res}}^2}{\left(\omega_{\text{res}} + n\Omega_c/\gamma\right)^2}
$$
$$
\times\left(\sin^2\alpha + n\Omega_c/\gamma\omega_{\text{res}}\right)^2\frac{B^2(\omega)}{\langle B^2\rangle}\bigg|_{\theta_{\text{MVA}}}.
\tag{13}
$$

The particular θ_{MVA} angle is determined by combining the resonant condition, the dispersion relation, the frequency cutoffs, and the wave normal angle cutoffs:

$$
\cos\theta_{LC}^{UC} = \frac{\Omega_c^2(\gamma\omega_{LC}^{UC} + n\Omega_c)^2}{\omega_{pe}^2\cos^2\alpha(\gamma^2-1)\Omega_c\omega_{LC}^{UC}}\quad\text{and}
$$
$$
\theta_{\min} < \theta_{\text{MVA}} < \theta_{\max},\ \forall n
\tag{14}
$$

with the MVA angle given by $\tan\theta_{\text{MVA}} = \frac{1}{2}(\tan\theta_{LC} + \tan\theta_{UC})$. A drawback of the model comes from the wave normal angle width $\Delta\theta$, which is not used anymore, except in the definition $\theta_{UC} = \max(\Delta\theta, \theta_{\max})$ used below.

The PPA, introduced by *Summers* [2005], can simply be deduced for LF waves from equation (13) by evaluating the diffusion coefficient at $\theta = 0$. In this case, the Bessel function terms simplify to $\Phi'^2_n = J_{n+1}^2$. Since the argument is $x = 0$, only the $n = -1$ harmonic remains. Using $\Phi'^2_{-1}(0) = J_0^2(0) = 1$ leads to

$$
D_{\alpha\alpha}^{\text{PPA}}/p^2 = \frac{\pi}{4}\frac{\Omega_c^2}{\gamma^2}\frac{B_{\text{wave}}^2}{B_0^2}\frac{1}{|V_{\parallel}/V_g - 1|}\frac{\omega_{\text{res}}^2}{\left(\omega_{\text{res}} - \Omega_c/\gamma\right)^2}
$$
$$
\times\left(-\sin(\alpha)^2 + \Omega_c/\gamma\omega_{\text{res}}\right)^2\frac{B^2(\omega)}{\langle B^2\rangle},
\tag{15}
$$

with $V_g = -2c(\Omega_c\omega_{\text{res}}/\omega_{pe}^2)^{1/2}$. In this paper, the LF resonance equation (7) is always used in equations (13) and (15) defining the LF PPA and LF MVA, sometimes abusively noted below MVA or PPA.

2.4. Fast Precipitation Lifetime Computations

Assuming that the bounce averaged $\langle D_{\alpha\alpha}(\alpha)\rangle$ has one or two deep minima, a simple expression of the electron lifetime τ is derived by *Albert and Shprits* [2009]

$$
\tau \sim \sigma\int_{\alpha_L}^{\pi/2}\frac{d\alpha_0}{2\langle D_{\alpha\alpha}\rangle_B\tan\alpha_0}.
\tag{16}
$$

The lifetime is therefore determined by the minimum values of $\langle D_{\alpha\alpha}\rangle\tan\alpha$ [*Lyons and Thorne*, 1973; *Albert and Shprits*, 2009]. The parameter σ is originally one as given by *Albert and Shprits* [2009], but could be lower (up to 0.5), as discussed by *Mourenas and Ripoll* [2012], when the lifetime is dominated by variations at small pitch angles.

2.5. Analytical Developments and Trapped Electron Lifetime Estimates

Analytical developments of diffusion coefficients and lifetimes are performed for LF waves in the work of *Mourenas and Ripoll* [2012]. It is necessary to assume the resonance maximum always occurs at $\omega \sim \omega_m$ with a wave model composed of a single Gaussian function of half-width

$\Delta\omega \leq \omega_m/2$, with lower and upper cutoffs at $\omega_{LC} \sim \omega_m/2$ and $\omega_{UC} \sim 2\omega_m$. Furthermore, $\theta_m \sim \theta_{min} \sim 0$ and $\Delta\theta \leq \theta_{max} \leq \pi/4$ are imposed as in most studies of hiss waves. At $\theta \sim \theta_{max}$, assuming the resonance occurs at the Bessel function maximum, the arguments $x = (-\omega\gamma/\Omega_c + n)\tan(\alpha)\tan(\theta) \sim n\tan(\alpha_{min})\tan(\theta_{max}) \sim n + 1$ gives $1 \sim \tan \alpha_{min} \tan \theta_{max}$. From both the latter and the simplified resonance relation, the maximum number of harmonics $N_r = p\varepsilon_m \cos\alpha_{R,min}$ can be expressed as $N_r = \cos[\tan^{-1}(1/(\tan\theta_{max})]p\varepsilon_m$, leading to

$$N_r = p\varepsilon_m \sin\theta_{max} = p\left(\omega_m\omega_{pe}^2/\Omega_c^3\right)^{1/2}\sin\theta_{max}, \quad (17)$$

which is valid for $p\varepsilon_m > 1$. Note that the last resonant pitch angle is then given by $\alpha_{R,min} = \pi/2 - \theta_{max}$. Greater are θ_{max} (leading to $\alpha_{R,min}$ closer to the loss cone) and N_r (through $p\varepsilon_m$), greater will be the diffusion coefficient at small pitch angles implying therefore a smaller lifetime. An analytical estimate of the electron lifetime τ has also been derived in the work of *Mourenas and Ripoll* [2012] after many simplifications of both $D_{\alpha\alpha}$ and its bounce averaging (not recalled here). When diffusion coefficient minima occur at small pitch angles, the lifetime can be approximated by

$$\tau \sim \frac{B_0^2\gamma}{B_{wave}^2 G_1}\left(\frac{p\omega_m^{1/2}\Omega_{pe}}{\Omega_c^{03/2}}\right)^{13/9}$$

$$\times \frac{T(\alpha^*)(\cos^2\alpha_+ - \cos^2\alpha_{LC} + 2\ln(\sin\alpha_+) - 2\ln(\sin\alpha_{LC}))}{10(1 + 3\sin^2\lambda_R)^{7/12}\Delta\lambda_R\omega_m},$$

$$(18)$$

in which Ω_c^0 denotes the gyrofrequency taken at the equator, α_{LC} is the loss cone pitch angle value, $G_1 = \Omega_c^0/(\pi^{1/2}erf(1)\Delta\omega)$, $T(\alpha) = 1.38 - 0.64\sin^{3/4}\alpha$ and $\alpha^* = (\alpha_+ - \alpha_{LC})/2$. Moreover, the domain $\mathcal{D} = [\alpha_{LC}, \alpha_+]$ with $\alpha_+ = \max(7\alpha_{R,min}/10, 2.2\alpha_{LC})$ represents the high $1/(D_{\alpha\alpha}\tan\alpha)$ values at low pitch angle, with a definition improved compared to the original one [*Mourenas and Ripoll*, 2012]. The value of α_+ cannot exceed (70% of) $\alpha_{R,min}$, the value at which the large pitch angle $D_{\alpha\alpha}$ become predominant. The domain $\mathcal{D}$ has then been restricted to not too large pitch angle values, which are imposed by the validity domain of equation (18). Such a model for α_+ is crucial and can change τ by a factor of 1.5 up to 2 if wrongly chosen. Finally, the estimation of both resonant latitude λ_R and width $\Delta\lambda_R$, defined as the latitude location and domain at and over which resonance occurs, are given by $\lambda_R = 2(1 - 1/(p\varepsilon_m)^{1/9})^{1/2}/3^{1/2}$ and $\Delta\lambda_R = (\Delta\omega/\omega_m)/(3^{5/2}(p\varepsilon_m)^{1/9}(1 - 1/(p\varepsilon_m)^{1/9})^{1/2})$.

3. NUMERICAL RESULTS

3.1. On the Large Number of Harmonics for MeV Electrons

Hiss waves (B_{wave} = 35 pT with a mean frequency f_m = 600 Hz and frequency width Δf = 300 Hz) acting with a broad wave normal angle distribution ($\Delta\theta$ = 78.7°, θ_{max} = 45°) in a dense plasma are considered using the LF formulation, as given by *Lyons et al.* [1972] and by *Albert* [1994]. The plasma density model is given by N_e (cm^{-3}) = 1000(4/L)^4. A validation of the diffusion coefficient $D_{\alpha\alpha}/p^2$ is shown in Figure 1 for 0.2 MeV electrons at $L = 4$. The theoretical maximal harmonic number given by equation (17) is $N_r = 4$ and $N_r = 6$ if equation (39) of *Albert* [1999] is used instead (with $\omega_{UC} = \omega_m + 2\Delta\omega_m$), while three are numerically found to be enough. Equation (17) is validated numerically in Figure 2 for similar wave conditions, but in the slot region at $L = 3$. The lowest harmonic number that contributes to less than 1% of $D_{\alpha\alpha}$ agrees to about one or two harmonics with equation (17) for all energies. If Eq. (34) of *Albert* [1999] was used instead for the calculation of N_r, its value would be about 35% higher. The difference is still minor for this particular case because of the use of $\cos\alpha_{R,min} \sim 1$ in the derivation of equation (39), which is correct here because of the broad wave normal angle width assumed in the wave model ($\alpha_{R,min} \sim \pi/2 - \max(\theta_{max} = 45°, \Delta\theta = 79°) \sim 0$). But this expression would overestimate N_r much more if a more recent wave model was used because θ_{max} or $\Delta\theta$ are now considered as being smaller [*Meredith et al.*, 2007, 2009]. Both our and *Albert's* [1999] expressions show anyhow

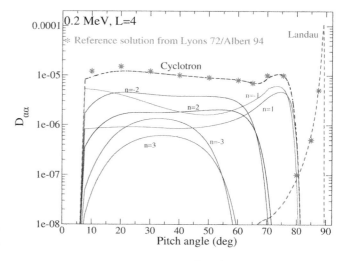

Figure 1. Diffusion coefficients $D_{\alpha\alpha}/p^2$ (s^{-1}) at $L = 4$ for 0.2 MeV electrons. Reference (stars) and CEVA (dashed line) solutions agree well. The three positive and negative contributing harmonics are also plotted (plain line).

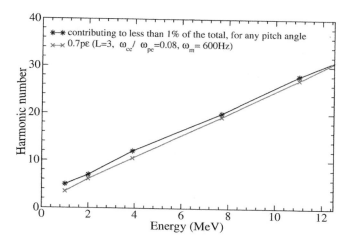

Figure 2. Theoretical (equation (17), crosses) and numerical (stars) harmonic number needed for the computation of $D_{\alpha\alpha}$ at $L = 3$ with hiss waves versus the electron energy.

many harmonics are needed for computing diffusion coefficients of highly energetic electrons [cf. *Albert*, 1994, 1999].

3.2. The 3 MeV Electrons at L = 4.5 Under Storm Conditions

The interaction of strong hiss waves ($B_w = 100$ pT, $f_m = 573$ Hz, $\Delta f = 286$ Hz, $\Delta\theta = 30°$, and $\theta_{max} = 45°$) occurring during the main phase of a storm with 3.18 MeV electrons at $L = 4.5$ ($\Omega_{pe} = 15\Omega_c$) is treated similarly to those of *Li et al.*

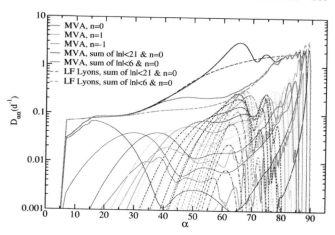

Figure 4. $D_{\alpha\alpha}/p^2$ (d^{-1}) in the hiss case of *Li et al.* [2007] for 3.18 MeV electrons as a function of pitch angle, computed with the LF mean value approximation (MVA). All harmonics are shown as well as the sum over ±5 (red) and over ±20 (blue) harmonics. Individual harmonics are plotted with solid line for $|n| < 6$, dashed line for $6 \le |n| < 11$, dash-dotted line for $11 \le |n| < 16$ and dotted line for $16 \le |n| < 21$.

[2007] and *Albert* [2008]. However, the latitude cutoff at $\lambda = 40°$ has not been accounted for in our CEVA/Lyons simulations, leading to a quite constant diffusion at small pitch angles. In contrast, it decreases quickly below 15° in our MVA and PPA computations (Figures 4, 5, and 6), for which the latitude cutoff has been implemented, similarly to *Albert* [2008] simulations. According to Figures 3 and 4, between

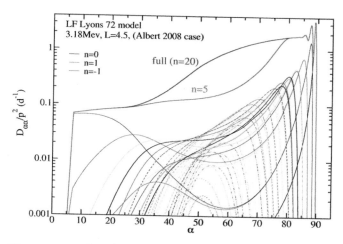

Figure 3. $D_{\alpha\alpha}/p^2$ (d^{-1}) in the hiss case of *Li et al.* [2007] for 3.18 MeV electrons as a function of pitch angle, computed with the low-frequency (LF) Lyons 72 model. All harmonics are shown as well as the sum over ±5 (red) and over ±20 (blue) harmonics. Individual harmonics are plotted with solid line for $|n| < 6$, dashed line for $6 \le |n| < 11$, dash-dotted line for $11 \le |n| < 16$ and dotted line for $16 \le |n| < 21$.

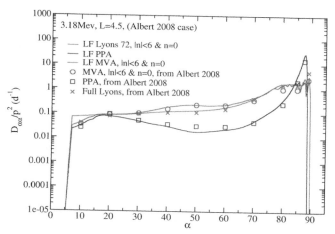

Figure 5. $D_{\alpha\alpha}/p^2$ (d^{-1}) due to hiss [*Li et al.*, 2007] of 3.18 MeV electrons, as a function of pitch angle, computed using LF Lyons, MVA, and PPV models. Reference solutions for each of this model but obtained using the full dispersion relation are reported from *Albert* [2008] (denoted with crosses, circles, and squares, respectively).

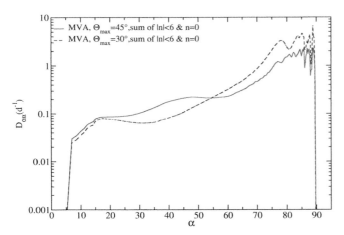

Figure 6. The influence of the wave normal angle cutoff on the LF MVA solution using 5 harmonics.

15 and 17 harmonics are numerically needed. Equation (17) gives a maximum of 18 harmonics, while only 5 are used in the work of *Albert* [2008], which explains the double bump profile of Figure 5 (dashed curve) and leads to a 1 order of magnitude underestimation of $D_{\alpha\alpha}$ at intermediate pitch angles. Overall, our results agree well with the computations shown by *Albert* [2008] (Figure 5), signifying the LF limit is applicable for this case since it compares well with the full dispersion used by *Albert* [2008]. In Figures 4 and 5, the accuracy of the MVA is found to be acceptable because deviations are very localized, being around 45° with 5 harmonics (65° with 20) and not exceeding 60% (a factor of 2.5 with 20 harmonics). The sensitivity to the MVA mean angle, itself, defined from the frequency and wave cutoffs, is shown in Figure 6, and profiles can become wiggly. The plot of all harmonics in Figure 3 allows enlightening that the maximum of the last Bessel function, assimilated to the lowest resonant pitch angle, is around $\alpha_{R,min} \sim \pi/2 - \theta_{max} \sim 45°$. Assuming the resonant pitch angle is located at each coefficient maximum, the bounce averaged resonant wave normal angle can be deduced from $\theta_R \sim \pi/2 - \alpha_R$, showing for instance when it differs from zero. Moreover, the equatorial resonant pitch angle can be estimated from the bounce averaged one (or vice versa) by using $\sin^3\alpha_{R,eq} \sim |n|/(p\varepsilon_{m,eq})(\sin^3\alpha_R/\cos\alpha_R)$ and similarly $\theta_{R,eq} \sim \pi/2 - \alpha_{R,eq}$ can be deduced [*Mourenas and Ripoll*, 2012]. On the other hand, PPA is found to be accurate at low/moderate pitch angle, which is usually critical for lifetime computations. But the present case is rather not favorable to advertising PPA since many harmonics are needed between 30° and 80°. Finally, both PPA and MVA are very efficient methods allowing quasi-instantaneous computations, which is useful for testing parameters or models and evaluating their influence. As a matter of fact, they

are 120 times faster than the Lyons 72 model, with 15 s for a single harmonic (compared to 30 min) for computations made on our TERA 100 supercomputer, using a single processor, 256 p, 256 μ, 200 latitudes (without optimizing).

3.3. The 2 MeV Electrons Trapped in the Inner Belt and Slot Region

A recent realistic model of waves acting below $L = 3$ has been built with two (to three) Gaussian functions fitted from CRRES measurements by *Meredith et al.* [2007]. The 0.1 to 2 kHz waves are generated from plasma turbulence in space, while lightning-generated waves found in the range 2–5 kHz are not here accounted for. Pitch angle diffusion is solved at $L = 2, 2.5, 3$ for quiet conditions ($\omega_{pe}/\Omega_c = 5.6, 7.3, 9.3$). Pitch Angle Diffusion of Ions and Electrons (PADIE) solutions [*Meredith et al.*, 2009] computed by using the full dispersion relation are reported in Figure 7 (left). Compared to the results obtained with the Lyons 72 model, the disagreement does not exceed 10%, on average, except at the intersection of the Landau and the first harmonic where it reaches 30%–35% at all L. It could be attributed to the use of different plasma densities or to the LF approximation. Figure 7 (right) shows PPA is accurate at small pitch angles (below 20% for $\alpha < 65°$), but the solution deviates from the reference one, up to a factor of 2 at $L = 3$ for larger pitch angle. However, the accuracy of MVA is found to be very good for all pitch angles and L, for instance below 10% for all $\alpha < 60°$ and below 20% at $L = 3$ for $60° < \alpha < 78°$. Only the maximum of the Landau harmonic can be overestimated until a factor of 1.8 at $L = 3$. MVA has to be used with the wave normal angle upper bound defined by $\max(\theta_{max}, \Delta\theta) = 30°$ here. If $\Delta\theta = 20°$ was used instead, MVA solutions would become comparable or barely better than the PPA ones at large pitch angle.

3.4. Electron Precipitation Lifetimes

Lifetimes are computed by using equation (16) for all diffusion coefficients shown in Figure 7 and compared in Table 1 to the PADIE computations [*Meredith et al.*, 2009] (first line). At $L = 2$, the lifetime is controlled by Landau resonance, which explains why PPA and equation (18) are not used. But one way to always obtain lifetimes with PPA is to complement it with the MVA Landau in order to build a coefficient defined over all pitch angles. Using the Landau resonance of both Gaussians, it would give 475 days at $L = 2$, which is almost identical to the MVA result. The lifetime expression equation (36) given by *Mourenas and Ripoll* [2012] is also applicable when the intersection of the Landau and the first resonance determines the lifetime, but not directly when more than one Gaussian function is used for the

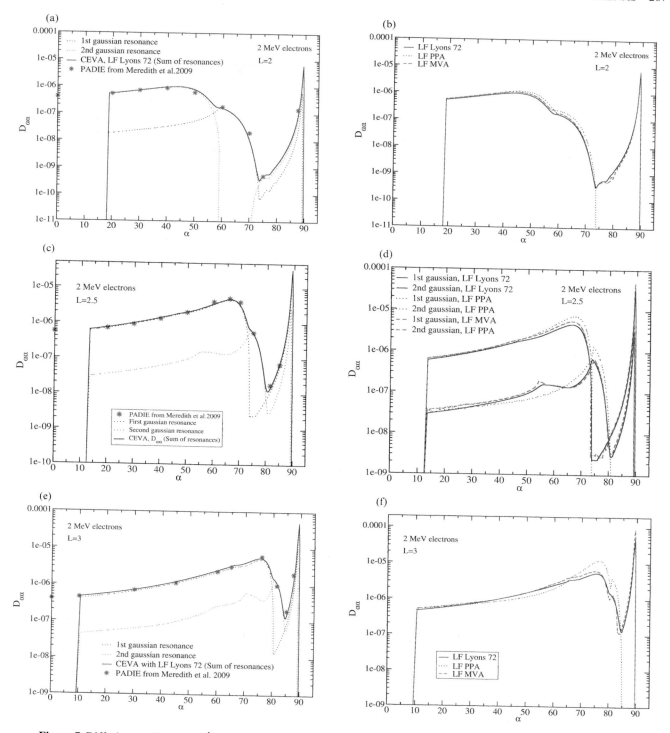

Figure 7. Diffusion coefficients (s^{-1}) at (a and b) $L = 2$, (c and d) $L = 2.5$, and (e and f) $L = 3$ of 2 MeV electrons interacting with hiss waves. (left) CEVA solutions are compared with reference points reported from *Meredith et al.* [2009]. Also shown with dotted lines are the contributions of each of the two Gaussian functions to the resonant diffusion coefficient (denoted first and second Gaussian resonance). (right) PPA and MVA are compared to the Lyons 72 model. Each coefficient is shown at $L = 2.5$, while the sum is directly plotted at $L = 2$ and $L = 3$.

Table 1. Lifetimes in Days of 2 MeV Electrons Interacting With Hiss Waves, Showing Large Resident Times in the Inner Belt and Much Shorter Ones in the Slot Region

Models/McIlwain Numbers	$L = 2$	$L = 2.5$	$L = 3$
PADIE	278	7.5	12.4
Lyons 72	381	13.6	14.8
Low-frequency (LF) mean value approximation (MVA)	423	12	13.4
LF parallel propagation approximation (PPA)	(475)	(17.4)	15.5
Analytic	/	(6.5)	11.3

wave model. This explains the absence of an analytical value at $L = 2$. At $L = 2.5$, the lifetime results from the competition between the small pitch angle values and the Landau minimum, explaining why both PPA and analytic lifetimes are written within brackets. Overall, the few tens of percent of difference found in the diffusion coefficients added to the loss of accuracy due to the use of equation (16) have direct repercussions to the lifetimes. The maximal relative error found is 50% with LF MVA at $L = 2$, 80% with LF Lyons at $L = 2.5$, and 25% at $L = 3$ with either LF PPA or the analytical estimate, comparing with PADIE. However, MVA/PPA agrees to about 10%/20% with the Lyons 72 exact lifetime predictions. At $L = 2.5$, where waves act strongly, removing particles, the discrepancy between the LF Lyons and PADIE lifetimes is more than the 30% difference noticed between each minima, which has to be explained by the use of equation (16). The analytic lifetime is found to be more accurate with the newly proposed definition of α_+, but it should be understood that a different choice of α_+ could lead to lower (or higher) lifetimes. In Table 2, lifetimes are presented for the 3.18 MeV case, showing a mean time around 15 days. Accounting for the wave influence at latitudes higher than 30° shortens the electron lifetime by 25%. It would increase to about 170 days during storm recovery phases, reaching as much as 50 months under minimal geomagnetic activity. An overestimation of only 10% is caused by the lack of harmonics in Lyons and MVA computations, which use

only 5 harmonics. It can be explained by the fact that most harmonics are confined to quite large pitch angles (only the seven first harmonics do not vanish for $\alpha < 30°$, Figure 3), which does not affect $D_{\alpha\alpha}$ minimal values close to the loss cone and, therefore, the lifetime value. With a greater θ_{max}, there would be more harmonics reaching α_{LC} (because $\alpha_{R,min}$ would be higher), and it would lower the accuracy of the lifetime prediction. However, taking only the $n = -1$ harmonic represents here a strong approximation leading to an overestimation of the lifetime by a factor of two. Finally, the analytic expression gives an estimation which is 1.7 times too low compared to the exact one. Overall, we believe a precision better than a factor of two cannot be expected with this model.

4. CONCLUSIONS

Interactions of energetic electrons with hiss waves in both the inner belt and the slot region can be described by quasi-linear pitch angle diffusion coefficients, simplified by using the LF wave limit and the dense plasma limit. The accuracy of fast computational methods has been assessed for high-energy electrons and realistic wave models. Differences between the exact and the approximate solutions do not exceed a few tens of percent on average, reaching a factor of two on localized domains, which we regard as satisfactory. The advantage is a reduction of the computational time by a factor 120 using MVA or PPA, which makes these models very attractive. The electron lifetimes calculated with MVA/PPA are also in good agreement with those computed using the exact analysis of *Lyons et al.* [1972]. Analytical lifetimes (equation (18)) are found to be quite accurate, within about a factor of 2. An extension of these approximation techniques to LF chorus waves has recently been carried out [*Mourenas et al.*, 2012].

Table 2. Lifetimes (day) of 3.18 MeV Electrons Interacting With Hiss Waves During the Main Phase of a Storm

Models/McIlwain Numbers	$L = 4.5$	Without λ Cutoff
Lyons 72 ($n < 21$)	14.5	11
Lyons 72 ($n < 6$)	15.5	12
LF MVA ($n < 21$)	14.3	
LF MVA ($n < 6$)	15	
LF PPA		30
Analytic		8.75

REFERENCES

Abel, B., and R. M. Thorne (1998), Electron scattering and loss in Earth's inner magnetosphere, 1: Dominant physical processes, *J. Geophys. Res.*, *103*(A2), 2385–2396, doi:10.1029/97JA02919.

Andronov, A. A., and V. Y. Trakhtengerts (1964), Kinetic instability of the Earth's outer radiation belt, *Geomagn. Aeron.*, *4*, 181–188.

Albert, J. M. (1994), Quasi-linear pitch angle diffusion coefficients: Retaining high harmonics, *J. Geophys. Res.*, *99*(A12), 23,741–23,745, doi:10.1029/94JA02345.

Albert, J. M. (1999), Analysis of quasi-linear diffusion coefficients, *J. Geophys. Res.*, *104*(A2), 2429–2441, doi:10.1029/1998JA 900113.

Albert, J. M. (2005), Evaluation of quasi-linear diffusion coefficients for whistler mode waves in a plasma with arbitrary density ratio, *J. Geophys. Res.*, *110*, A03218, doi:10.1029/2004JA 010844.

Albert, J. M. (2007), Simple approximations of quasi-linear diffusion coefficients, *J. Geophys. Res.*, *112*, A12202, doi:10.1029/ 2007JA012551.

Albert, J. M. (2008), Efficient approximations of quasi-linear diffusion coefficients in the radiation belts, *J. Geophys. Res.*, *113*, A06208, doi:10.1029/2007JA012936.

Albert, J. M., and Y. Y. Shprits (2009), Estimates of lifetimes against pitch angle diffusion, *J. Atmos. Sol. Terr. Phys.*, *71*(16), 1647–1652, doi:10.1016/j.jastp.2008.07.004.

Baker, D. N. (2002), How to cope with space weather, *Science*, *297* (5586), 1486–1487, doi:10.1126/science.1074956.

Carpenter, D. L., and R. R. Anderson (1992), An ISEE/Whistler model of equatorial electron density in the magnetosphere, *J. Geophys. Res.*, *97*(A2), 1097–1108, doi:10.1029/91JA01548.

Dungey, J. W. (1963), The loss of Van Allen electrons due to whistlers, *Planet. Space Sci.*, *11*(6), 591–595, doi:10.1016/ 0032-0633(63)90166-1.

Hughes, H. L., and J. M. Benedetto (2003), Radiation effects and hardening of MOS technology: Devices and circuits, *IEEE Trans. Nucl. Sci.*, *50*(3), 500–521, doi:10.1109/TNS.2003.812928.

Kennel, C. F., and H. E. Petschek (1966), Limit on stably trapped particle fluxes, *J. Geophys. Res.*, *71*(1), 1–28, doi:10.1029/ JZ071i001p00001.

Li, W., Y. Y. Shprits, and R. M. Thorne (2007), Dynamic evolution of energetic outer zone electrons due to wave-particle interactions during storms, *J. Geophys. Res.*, *112*, A10220, doi:10.1029/ 2007JA012368.

Lyons, L. R. (1974), Pitch-angle and energy diffusion coefficients from resonant interactions with ion-cyclotron and whistler waves, *J. Plasma Phys.*, *12*(3), 417–432, doi:10.1017/S002237780002537X.

Lyons, L. R., and R. M. Thorne (1973), Equilibrium structure of radiation belt electrons, *J. Geophys. Res.*, *78*(13), 2142–2149, doi:10.1029/JA078i013p02142.

Lyons, L. R., R. M. Thorne, and C. F. Kennel (1971), Electron pitch-angle diffusion driven by oblique whistler-mode turbulence, *J. Plasma Phys.*, *6*(3), 589–606, doi:10.1017/S0022377800006310.

Lyons, L. R., R. M. Thorne, and C. F. Kennel (1972), Pitch-angle diffusion of radiation belt electrons within the plasmasphere, *J. Geophys. Res.*, *77*(19), 3455–3474.

Meredith, N. P., R. M. Horne, S. A. Glauert, and R. R. Anderson (2007), Slot region electron loss timescales due to plasmaspheric hiss and lightning-generated whistlers, *J. Geophys. Res.*, *112*, A08214, doi:10.1029/2007JA012413.

Meredith, N. P., R. M. Horne, S. A. Glauert, D. N. Baker, S. G. Kanekal, and J. M. Albert (2009), Relativistic electron loss timescales in the slot region, *J. Geophys. Res.*, *114*, A03222, doi:10. 1029/2008JA013889.

Mourenas, D., and J.-F. Ripoll (2012), Analytical estimates of quasi-linear diffusion coefficients and electron lifetimes in the inner radiation belt, *J. Geophys. Res.*, *117*, A01204, doi:10.1029/ 2011JA016985.

Mourenas, D., A. V. Artemyev, J.-F. Ripoll, O. V. Agapitov, and V. V. Krasnoselskikh (2012), Timescales for electron quasi-linear diffusion by parallel and oblique lower-band chorus waves, *J. Geophys. Res.*, *117*, A06234, doi:10.1029/2012JA017717.

Pokhotelov, D., F. Lefeuvre, R. B. Horne, and N. Cornilleau-Wehrlin (2008), Survey of ELF-VLF plasma waves in outer radiation belt observed by Cluster STAFF-SA experiment, *Ann. Geophys.*, *26*, 3269–3277, doi:10.5194/angeo-26-3269-2008.

Réveillé, T., P. Bertrand, A. Ghizzo, F. Simonet, and N. Baussart (2001), Dynamic evolution of relativistic electrons in the radiation belts, *J. Geophys. Res.*, *106*(A9), 18,883–18,894, doi:10. 1029/2000JA900177.

Roberts, C. S. (1969), Pitch angle diffusion of electrons in the magnetosphere, *Rev. Geophys.*, *7*(1,2), 305–337, doi:10.1029/ RG007i001p00305.

Stix, T. H. (1962), *The Theory Of Plasma Waves*, 283 pp., McGraw-Hill, New York.

Summers, D. (2005), Quasi-linear diffusion coefficients for field-aligned electromagnetic waves with applications to the magnetosphere, *J. Geophys. Res.*, *110*, A08213, doi:10.1029/2005JA011159.

Summers, D., B. Ni, and N. P. Meredith (2007a), Timescales for radiation belt electron acceleration and loss due to resonant wave-particle interactions: 1. Theory, *J. Geophys. Res.*, *112*, A04206, doi:10.1029/2006JA011801.

Summers, D., B. Ni, and N. P. Meredith (2007b), Timescales for radiation belt electron acceleration and loss due to resonant wave-particle interactions: 2. Evaluation for VLF chorus, ELF hiss, and electromagnetic ion cyclotron waves, *J. Geophys. Res.*, *112*, A04207, doi:10.1029/2006JA011993.

D. Mourenas and J.-F. Ripoll, CEA, DAM, DIF, F-91297, Arpajon, France. (didier.mourenas@cea.fr; jean-francois.ripoll@cea.fr)

Recent Advances in Understanding the Diffuse Auroral Precipitation: The Role of Resonant Wave-Particle Interactions

Binbin Ni and Richard M. Thorne

Department of Atmospheric and Oceanic Sciences, University of California, Los Angeles, California, USA

Diffuse auroral precipitation provides the major source of energy input into the nightside upper atmosphere and strong magnetosphere-ionosphere coupling. Resonant wave-particle interactions play a dominant role in the scattering of injected plasma sheet electrons leading to the diffuse auroral precipitation. Here we review the recent advances in understanding the origin of the diffuse aurora and quantifying the exact role of various magnetospheric waves in producing the global distribution of diffuse auroral precipitation and its variability with geomagnetic activity. A combination of upper and lower-band chorus (LBC) scattering has been identified as the major contributor to the most intense inner magnetospheric diffuse auroral precipitation on the nightside, while electrostatic electron cyclotron harmonic waves are an important or even dominant cause for nightside diffuse auroral precipitation beyond $\sim 8\ R_E$. Dayside chorus could be responsible for the weaker dayside diffuse auroral precipitation. LBC can also modulate the intensity of quasiperiodic pulsating auroral emissions. We conclude the review with a summary of current understanding, outstanding questions, and some suggestions for future research.

1. INTRODUCTION

As a broad belt of emissions extending around the entire auroral oval, the diffuse aurora typically comprises 80% of the total auroral energy input into the polar region at solar maximum and 50% at solar minimum [*Sandford*, 1968]. Figure 1 shows the pattern for electron diffuse aurora, which is believed to be the dominant contributor to the global precipitation budget, based on POLAR PIXIE X-ray observations (top) [*Petrinec et al.*, 1999] and DMSP particle observations (bottom) [*Newell et al.*, 2009]. The diffuse aurora extends over a broad latitude range of 5° to 10° and maps along field lines from the outer radiation belts to the central plasma sheet (CPS). Owing to the predominant east-ward transport of electrons as a result of a combination of **E × B** and gradient drifting from the nightside plasma sheet, the diffuse aurora is most intense in the magnetic local time (MLT) sector from premidnight to dawn. Rapid precipitation loss leads to greatly reduced energy deposition on the dayside and relatively insignificant input from postnoon through dusk. Compared to plasma sheet ions, electron precipitation plays a dominant role in driving the diffuse auroral activity [*Hardy et al.*, 1985, 1989], which will be the major focus of this review.

It is generally accepted that ~0.1–10 keV CPS electrons are the dominant source population for the diffuse aurora [e.g., *Lui et al.*, 1977] and that the occurrence of the diffuse aurora is a result of electron pitch angle scattering into the loss cone by resonant wave-particle interactions [e.g., *Swift*, 1981]. It has been long proposed that electrostatic electron cyclotron harmonic (ECH) waves and chorus can both resonate with hundreds of eV to 10 keV electrons, but which of them is more influential in the formation of the diffuse aurora has remained a subject of controversy for over 40 years [see

Dynamics of the Earth's Radiation Belts and Inner Magnetosphere
Geophysical Monograph Series 199
10.1029/2012GM001337

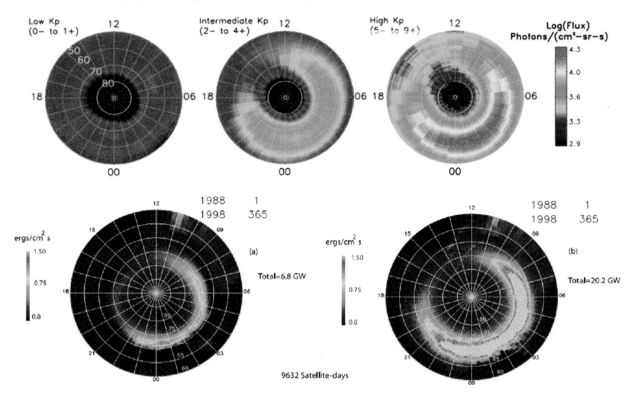

Figure 1. (top) Average statistical X-ray aurora (northern hemisphere) observed onboard POLAR/PIXIE. From *Petrinec et al.* [1999, Figure 2]. (bottom) Electron diffuse aurora hemispheric energy fluxes for low and high solar wind driving, obtained using the DMSP 11 year particle database. Adapted from *Newell et al.* [2009, Figure 5].

Thorne et al., 2010; *Ni et al.*, 2011a, 2011b, and references therein]. Electrostatic ECH emissions occur within bands between the harmonics of electron gyrofrequency, f_{ce}, with the dominant frequency often located around $(n + 1/2) f_{ce}$ [e.g., *Kennel et al.*, 1970]. These waves propagate at very large angles with respect to the ambient magnetic field. *Kennel et al.* [1970] first suggested that ECH waves could cause intense pitch angle diffusion of auroral zone electrons based on OGO 5 observations of large-amplitude (1–10 mV m^{-1} and up to 100 mV m^{-1}) ECH emissions, which was later quantified by *Lyons* [1974] but challenged by *Belmont et al.* [1983]. The latter pointed out that the stronger ECH events (>1 mV m^{-1}) occur less than 2% of the time compared to 88% occurrence of much weaker electric fields (<0.1 mV m^{-1}) based on the GEOS 2 data, similar to the findings of a number of following studies using other satellite databases [e.g., *Roeder and Koons*, 1989; *Meredith et al.*, 2009; *Ni et al.*, 2011c]. Electromagnetic chorus waves occur in the low-density plasmatrough region characteristically in a lower band (0.1–0.5 f_{ce}) and an upper band (0.5–0.8 f_{ce}) [*Burtis and Helliwell*, 1976]. Typically, chorus amplitudes lie in the range 1–100 pT [e.g., *Meredith et al.*, 2001; *Li*

et al., 2009]. Observations also show distinct differences in the characteristics of nightside (22–06 MLT) and dayside (06–13 MLT) chorus. Nightside chorus is predominantly confined to magnetic latitudes below 15°, while dayside chorus can propagate to much higher latitudes [e.g., *Meredith et al.*, 2001; *Li et al.*, 2009]. In addition, while nightside chorus occurs predominantly during disturbed periods, confined to $L <$ ~8 R_E, dayside chorus can persist even for relatively quiet conditions with the highest occurrence at $L > 7$ [*Li et al.*, 2009]. The interactions between chorus and energetic electrons (> ~100 keV) have been well investigated, indicating that chorus can play an efficient dual role in driving their stochastic acceleration and scattering loss [e.g., *Summers et al.*, 2007; *Thorne*, 2010]. Chorus waves are also able to resonate with the bulk of the plasma sheet electron population and have a spatial (MLT, L) distribution similar to that of the diffuse aurora [*Meredith et al.*, 2009].

In the present paper, we will review recent advances in understanding the origin of diffuse auroral precipitation and its global distribution in which resonant wave-particle interactions play a fundamental role. The outline of this paper is as follows. Recent understanding of the origin of nightside diffuse aurora in the inner magnetosphere is described in section 2. We describe in section 3 the possibly dominant mechanism for nightside diffuse auroral precipitation at higher L shells ($L > \sim 8$) in the CPS. Some recent efforts to understand the occurrence of the dayside diffuse aurora are reviewed in section 4. We conclude in section 5 with a summary of current understanding, outstanding questions, and some suggestions for future research.

2. THE INNER MAGNETOSPHERIC NIGHTSIDE DIFFUSE AURORAL PRECIPITATION

Here "nightside" means the coverage of ~23–6 MLT; the inner magnetosphere refers to the region with the equatorial crossings from ~4 R_E to ~8 R_E, corresponding to the magnetic latitudes (mapped to the Earth's surface) ~60°–~67°. The diffuse aurora is most intense within this spatial coverage (Figure 1). For simplicity, we use "N-I-DAP" to denote the Nightside, Inner-magnetospheric Diffuse Auroral Precipitation. Interestingly, the N-I-DAP has a distribution and geomagnetic activity dependence similar to that of both chorus and ECH waves, as shown in Figure 2 (from the work of *Thorne et al.* [2010]). Their results show that ECH waves and chorus emissions (observed by CRRES) both increase with geomagnetic activity, accompanying the intensification of nightside average diffuse auroral 2–12 keV X-ray emissions (observed by POLAR/PIXIE). Consequently, it is natural to connect the two wave modes with the diffuse aurora on the nightside where both phenomena are most intense. Both ECH and chorus waves can resonate with plasma sheet electrons whenever the Doppler-shifted frequency of the wave (viewed from the particle rest frame) is a multiple of f_{ce}. While the early work by *Lyons* [1974] regarding ECH wave scattering was disputed due to the extremely large adopted ECH wave amplitude that is rarely observed, a careful analysis of CRRES wave data by *Meredith et al.* [2000] led to a resurgence in interest of ECH scattering [e.g., *Horne and Thorne*, 2000], when they established that wave amplitudes following substorm activity were typically above 1 mV m^{-1} near the magnetic equator. On the other hand, *Inan et al.* [1992] suggested that lower-band chorus (LBC) could cause scattering loss of 10–50 keV electrons often related to the pulsating aurora, while upper-band chorus (UBC) could scatter the 1–10 keV electrons responsible for the diffuse aurora. *Ni et al.* [2008] quantified LBC and UBC scattering of plasma sheet electrons at $L = 6$ using statistical wave information, showing that UBC is the dominant scattering cause for electrons below ~5 keV, while LBC is more effective at higher energies, especially near the loss cone.

However, which wave mode is more influential in driving the most intense N-I-DAP required further investigation. *Thorne et al.* [2010] performed comprehensive theoretical and modeling studies using CRRES observations to compute the bounce-averaged quasilinear diffusion coefficients for scattering by ECH waves, UBC, and LBC and to simulate the temporal evolution of plasma sheet electron pitch angle distribution (Figure 3) at $L = 5$ (a representative location where the N-I-DAP and the waves are strongest). Their analysis confirmed that, although ECH waves alone can lead to scattering loss of electrons within 15° of the loss cone, only the combination of UBC and LBC can drastically modify the entire population of electrons injected into the inner magnetosphere, leading to the formation of pancake distributions at energies below a few keV. The dominant features of Figure 3c are very similar to the CRRES observations of electron distributions a couple of hours after low-energy electron injection from the plasma sheet, as well as similar to electron distributions observed by Time History of Events and Macroscale Interactions during Substorms (THEMIS) satellites [*Tao et al.*, 2011]. Significantly, the study of *Thorne et al.* [2010] provided a resolution to two major aspects of N-I-DAP: (1) nightside ECH waves and chorus waves can drive efficient loss of plasma sheet electrons into the atmosphere to cause the occurrence of N-I-DAP, and chorus waves are the dominant contributor; and (2) chorus waves and only chorus waves can explain the remnant pancake electron distribution left behind in space and the pronounced depletion of trapped electrons on the dayside. Conclusively, *Thorne et al.* [2010] have revealed that scattering by chorus is the dominant cause of the most intense diffuse auroral precipitation on the nightside in the inner magnetosphere ($<\sim 8$ R_E). A following study by *Tao et al.* [2011] reported that the formation of observed electron pitch angle distribution is mainly due to resonant interactions with a combination of UBC and LBC. They also pointed out that the pancake distributions at lower energies <2 keV, the flattened pitch angle distributions at medium energies of 2–3 keV, and the distributions with enhanced pitch angle anisotropy at high energies >3 keV, can all be explained using the banded chorus wave structure with a power minimum at ½ f_{ce}. Evaluations of ECH and chorus wave-induced resonant scattering rates of plasma sheet electrons under different geomagnetic conditions were performed by *Ni et al.* [2011a, 2011b] using CRRES wave data. They found that chorus scattering varies from above the strong diffusion limit (approximately less than hours) during active times to weak scattering (~day) during quiet times, accounting for the formation of observed electron pancake distribution. In contrast, ECH wave scattering coefficients are at least 1 order of magnitude smaller and are

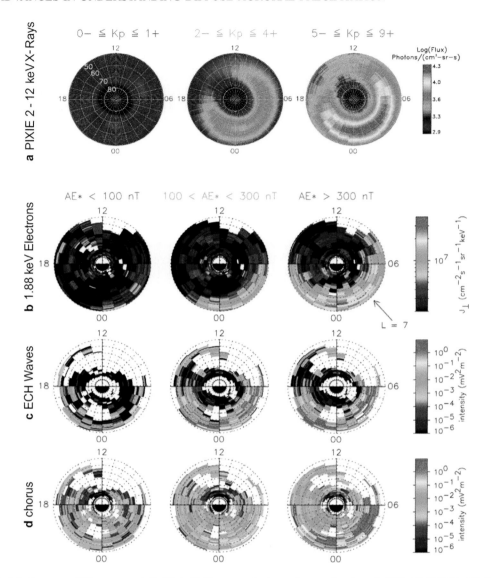

Figure 2. (a) Average diffuse auroral 2–12 keV X-ray emissions observed onboard POLAR/PIXIE. (b) Statistical distribution of 1.88 keV plasma sheet electrons measured by CRRES. Statistical distribution of (c) electron cyclotron harmonic (ECH) waves and (d) chorus observed on CRRES. From *Thorne et al.* [2010, Figure 2].

negligible under all geomagnetic conditions in the inner magnetosphere.

As the more dynamic structure embedded in the diffuse aurora, the pulsating aurora has also received intensive investigations in recent years. Using the in situ and ground observations of THEMIS mission, *Nishimura et al.* [2010] provided direct evidence that a modulation of LBC can drive the pulsating aurora, which is known to be excited by modulated, downward streaming electrons. Figure 4 shows the simultaneous observations of the pulsating aurora and chorus waves on 15 February 2009. The auroral pulsations had an almost one-to-one correspondence with each burst of chorus (Figure 4d), supporting the idea that intensity-modulated LBC can be the driver of the pulsating aurora, which was further supported by a THEMIS multievent study of *Nishimura et al.* [2011] and by a REIMEI time-of-flight analysis of *Miyoshi et al.* [2010]. However, there are other explanations on the driver(s) of the pulsating aurora, e.g., temporal variations of the field-aligned acceleration [*Sato et al.*, 2004], ECH wave scattering [*Liang et al.*, 2010], and Fermi-type

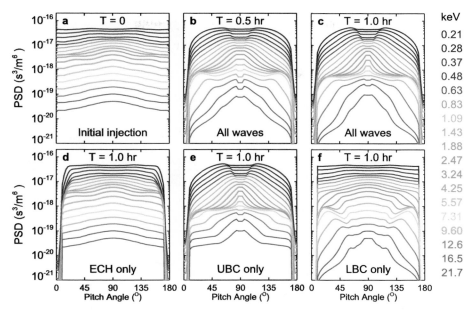

Figure 3. (a) Fokker-Planck diffusion simulations of the evolution of the electron phase space density pitch angle distribution after 1 h due to scattering by (d) ECH waves alone, (e) upper-band chorus alone, (f) lower-band chorus (LBC) alone and a combination of all three types of waves at (b) 0.5 h and (c) 1 h, for the color-coded 18 electron energies from 0.21 to 21.7 keV. From *Thorne et al.* [2010, Figure 4].

acceleration associated with midtail earthward plasma flow [*Nakajima et al.*, 2012]. Obviously, further careful and detailed studies are required to pursue a full understanding of the driver(s) of the pulsating aurora and its dynamics.

3. THE OUTER MAGNETOSPHERIC NIGHTSIDE DIFFUSE AURORAL PRECIPITATION

The mechanism(s) for nightside diffuse auroral precipitation from the outer magnetosphere (N-O-DAP) remains unclear. Here the outer magnetosphere covers the spatial range with equatorial crossings from ~8 R_E up to the CPS outer boundary. A THEMIS survey by *Li et al.* [2009] showed that chorus waves are statistically weak (less than a few pT) above ~8 R_E in the night-to-dawn MLT sector and, thus, insufficient to cause efficient diffuse auroral scattering at higher L shells. On the other hand, moderately strong ECH waves (>0.1 mV m^{-1}) can extend to L > ~12 [*Roeder and Koons*, 1989; *Ni et al.*, 2011c], suggesting that ECH waves could be potentially important for understanding the origin of the diffuse aurora and the evolution of plasma sheet electron distribution at L > 8 (or λ_{inv} > ~67°). While strong ECH waves ≥1 mV m^{-1} are most likely to occur within L = 5–10 [*Ni et al.*, 2011c], moderate ECH waves of 0.1–1 mV m^{-1} are still capable of scattering plasma sheet electrons at a rate comparable to the higher L shell strong diffusion limit

that is lower due to the smaller loss cone and a longer bounce period.

The scenario that ECH waves might play a leading role in driving the N-O-DAP has been implied implicitly by some recent studies. Following the study of *Liang et al.* [2011] that performed a detailed event study using the simultaneous in situ wave and particle observations by THEMIS probes and ground-based NORSTAR optical auroral observations during 08–09 UT on 5 February 2009, *Ni et al.* [2012] carried out a more comprehensive theoretical and numerical analysis for this event, through a systematic combination of quasi-linear theory, realistic nondipolar magnetic field mapping, and the concept of strong and weak diffusion. They found that the observed ECH wave activity could cause intense pitch angle scattering of plasma sheet electrons of 0.1–5 keV at a rate of >10^{-4} s^{-1} for equatorial pitch angles α_{eq} < 30°, approaching the strong diffusion limit. Use of the auroral electron transport model [*Lummerzheim*, 1987] produced an intensity of ~2.3 kR for the green-line diffuse aurora, and separately Maxwellian fitting to the electron precipitation flux produced a green-line auroral intensity of ~2.6 kR, both in good agreement with the observed ~2.4 kR green-line auroral intensity (as shown in Figure 5b). *Ni et al.* [2012] was the first attempt to quantify ECH wave scattering CPS electrons and to simulate the ionospheric precipitation flux and resultant auroral brightness to study the magnetospheric

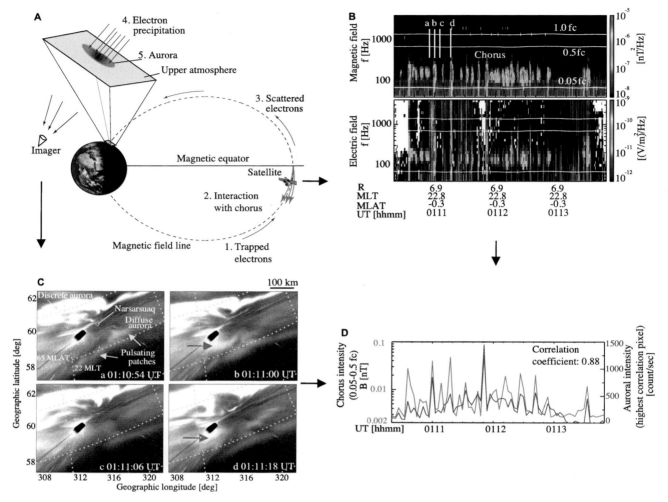

Figure 4. (a) Schematic diagram showing the geometry of chorus wave propagation (red arrows), electron precipitation (blue arrows), and pulsating aurora. (b) Time History of Events and Macroscale Interactions during Substorms-A observation of LBC bursts shown in electromagnetic field spectra. (c) Snapshots of imager data projected onto the geographic coordinates at 110 km altitude. (d) Correlation of LBC-integrated magnetic field intensity over 0.05–0.5 f_{ce} (red) and auroral intensity (blue) at the highest cross-correlation pixel. From *Nishimura et al.* [2010, Figure 1]. Reprinted with permission from AAAS.

cause of the diffuse aurora in the outer CPS. Their results support that enhanced ECH waves can be an important or even dominant driver of N-O-DAP. But a statistical investigation of ECH waves and N-O-DAP under different geomagnetic conditions is required to put a closure to the cause of N-O-DAP.

4. THE DAYSIDE DIFFUSE AURORAL PRECIPITATION

Compared to the intensity of the nightside diffuse aurora, the dayside diffuse auroral activity is generally much weaker,

which is easily understood due to the rapid loss of plasma sheet electrons during the eastward drift after the injection from the nightside plasma sheet. However, the dayside diffuse aurora can become very intense occasionally, and its effect on the dayside magnetosphere-ionosphere coupling is also important. *Hu et al.* [2009, 2012], using the ground-based all-sky imager (ASI) data at the Chinese Yellow River Station (YRS) in Ny-Ålesund, Svalbard, surveyed the synoptic distribution of dayside aurora emissions and their potential correlation with interplanetary magnetic field (IMF), indicating a prenoon (07:30–09:30 MLT) "warm spot" featured uniquely by an increase of 557.7 nm emissions and a

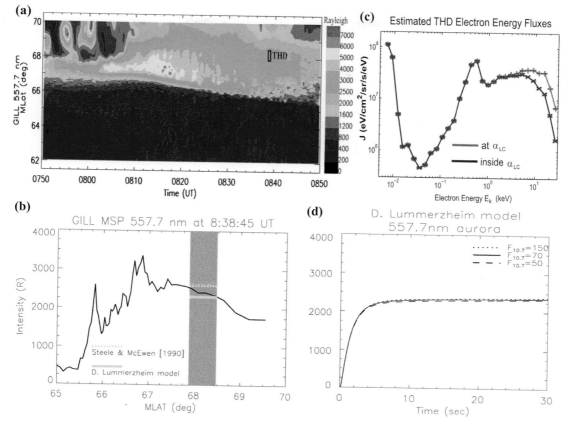

Figure 5. (a) NORSTAR GILL MSP green-line (557.7 nm) auroral observations for 07:50–08:50 UT on 5 February 2009. (b) The latitudinal variation of MSP green-line auroral intensity for the time stamp centered at 08:38:45 UT. The gray band shows the THEMIS-D (THD) footprint in magnetic latitude, ±0.3°. The solid and dotted yellow horizontal lines represent, respectively, the modeled green-line auroral intensities. (c) Electron energy flux at the loss cone (red) and estimated electron precipitation flux inside the loss cone (blue). (d) *Lummerzheim* [1987] model results of the 557.7 nm auroral brightness for three different profiles of the neutral atmosphere with the solar F10.7 flux of 150, 70, and 50. From *Ni et al.* [2012, Figure 6].

midday (09:30–13:00 MLT) gap of relatively weak green-line emissions, both for the discrete and diffuse aurora.

The mechanism(s) responsible for the dayside diffuse aurora remains poorly understood. *Li et al.* [2009] showed that moderately strong (~10 pT) dayside chorus can be persistently present up to $L \sim 10$, peaking around $L \sim 8$, therefore possibly acting as an important cause to the dayside diffuse aurora. Most recently, *Shi et al.* [2012] presented a representative event of intense dayside diffuse aurora observed by the Chinese YRS ASI near local noon at $L \sim 9.5$. The observed intensification of YRS ASI green-line diffuse aurora was related to solar wind changes manifested by fluctuations in the three components of the IMF and noticeable increases in solar wind velocity and solar wind dynamic pressure. Such change of the solar wind conditions can facilitate the excitation and amplification of magnetospheric whistler mode

chorus waves [e.g., *Li et al.*, 2010, 2011], which consequently favors resonant wave-particle interactions. Using a statistical model of dayside chorus waves at high L shells based upon the THEMIS survey [*Li et al.*, 2009], *Shi et al.* [2012] further computed dayside chorus-driven bounce-averaged pitch angle diffusion coefficients for plasma sheet electrons in realistic magnetic field models. The scattering rate results (top, Figure 6) demonstrated that dayside chorus scattering can produce intense precipitation losses of plasma sheet electrons on timescales of hours (approaching the strong diffusion limit) over a broad range of energy and pitch angle. The resultant loss cone filling for >500 eV electrons by dayside chorus pitch angle scattering (bottom, Figure 6) can reasonably explain the YRS ASI-observed green-line diffuse auroral intensification. However, the lack of simultaneous conjugate wave measurements in space for this particular

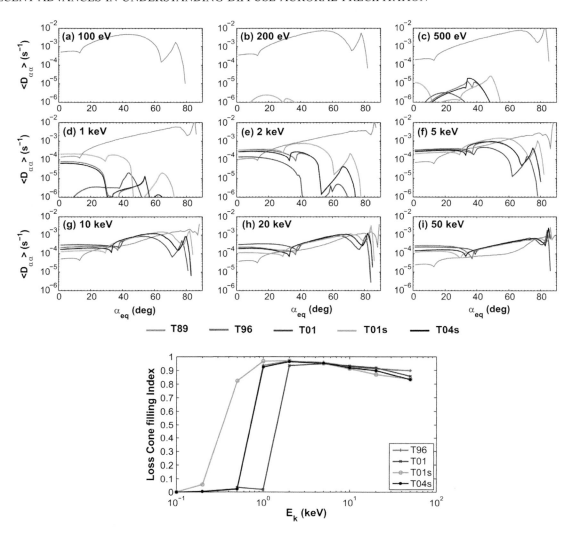

Figure 6. (top) $\langle D_{\alpha\alpha} \rangle$ for the adoption of different Tsyganenko magnetic field models. (bottom) Estimated loss cone filling index for the color-coded Tsyganenko models. From *Shi et al.* [2012, Figures 5 and 6].

case study raises questions on the general applicability of the dayside scattering model, which needs improvements in future investigations.

5. SUMMARY AND FUTURE WORK

The major recent findings regarding the diffuse aurora are summarized as follows:

1. Chorus waves provide the dominant scattering of plasma sheet electrons in the inner magnetosphere, being the major cause for the strongest nightside diffuse aurora ($<\sim 67°$).

2. Intense ECH waves can extend to >12 R_E on the nightside and, therefore, can be an important or even dominant driver of nightside diffuse auroral precipitation in the outer magnetosphere, which has been confirmed through a detailed conjunction case study but requires further multievent investigation and statistical studies.

3. Moderate chorus (>10 pT) is persistently present on the dayside even during quiet times, supporting the idea that dayside chorus could play a major role in the production of the dayside diffuse aurora, which needs future improvements due to current lack of simultaneous, conjugate wave measurements in space and ground auroral observations.

4. The intensity modulation of LBC is found to be well correlated with quasiperiodic pulsating aurora near the spacecraft magnetic footprint. The pulsating aurora can be caused by LBC modulation due to macroscopic density variations in

Resonant Wave-Particle Interactions Play a Critical Role!

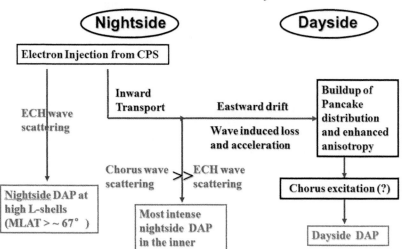

Figure 7. Schematic diagram summarizing our current understanding of the origin of the diffuse aurora and the roles of resonant wave-particle interactions. CPS, central plasma sheet; DAP, diffuse auroral precipitation.

the magnetosphere [*Li et al.*, 2011]; however, controversy remains regarding its periodicity and its dynamics.

5. Injected plasma sheet electrons are lost to the atmosphere via wave-induced pitch angle scattering. The resultant electron loss timescales can be faster than the electron transport to the dayside, leading to a pronounced MLT dependence on both the trapped and precipitation flux. Scattering by UBC causes pancake distribution for ~1 keV electrons. A combination of precipitation loss by LBC and stochastic acceleration by UBC drives enhanced anisotropy for ~10 keV electrons, which can provide a source for chorus excitation on the dayside. The dual-band chorus wave structure with a power minimum at $\frac{1}{2} f_{ce}$ can explain the flattened pitch angle distributions at medium energies of 2–3 keV. The temporal development of the plasma sheet source electron distribution can also lead to the excitation and the global distribution of a number of plasma waves. Figure 7 shows schematically our recent understanding of the origin of the diffuse aurora.

A number of outstanding questions remain concerning the origin of the diffuse aurora and the contribution of resonant wave-particle interactions. First, what is the exact role of ECH waves in driving nightside diffuse aurora at $L > 8$ and modifying the global morphology of plasma sheet electrons in a realistic magnetosphere under different geomagnetic conditions? Second, what is the role of dayside chorus in driving dayside diffuse auroral precipitation? Third, can we model the observed global distribution of the diffuse aurora and plasma sheet electron distribution by considering the effect of resonant

wave-particle interactions as a function of L shell, MLT, and geomagnetic activity level? Finally, what is the relationship between the pulsating aurora and substorms? What is the total energy involved in pulsating aurora events? What is the spatial extent of the pulsating aurora? To answer these questions, more data from space and ground, including wave and particle observations and ASI measurements, are required. A better understanding of the origin of the diffuse aurora and the role of resonant wave-particle interactions will enable us to better simulate the electron precipitation into the atmosphere and the magnetosphere-ionosphere coupling process for integrated incorporation into the development of self-consistent global models such as the Geospace General Circulation Models and associated wave excitation processes.

Acknowledgments. This research was supported by the NSF grant ATM-0802843 and NASA grant NNX12AD12G. The authors would like to thank J. Liang, V. Angelopoulos, Y.Y. Shprits, R.B. Horne, N.P. Meredith, E. Donovan, X. Zhang, W. Li, E. Spanswick, M. Kubyshkina, D. Lummerzheim, R. Shi, Z.-J. Hu, and D. Han for valuable help and useful discussions. B.N. thanks Chen Zhou for help in preparing the figures.

REFERENCES

Belmont, G., D. Fontaine, and P. Canu (1983), Are equatorial electron cyclotron waves responsible for diffuse auroral electron precipitation?, *J. Geophys. Res.*, *88*(A11), 9163–9170.

Burtis, W. J., and R. A. Helliwell (1976), Magnetospheric chorus: Occurrence patterns and normalized frequency, *Planet. Space Sci.*, *24*, 1007–1010, doi:10.1016/0032-0633(76)90119-7.

Hardy, D. A., M. S. Gussenhoven, and E. Holeman (1985), A statistical model of auroral electron precipitation, *J. Geophys. Res.*, *90*(A5), 4229–4248.

Hardy, D. A., M. S. Gussenhoven, and D. Brautigam (1989), A statistical model of auroral ion precipitation, *J. Geophys. Res.*, *94*(A1), 370–392.

Horne, R. B., and R. M. Thorne (2000), Electron pitch angle diffusion by electrostatic electron cyclotron harmonic waves: The origin of pancake distributions, *J. Geophys. Res.*, *105*(A3), 5391–5402.

Hu, Z.-J., et al. (2009), Synoptic distribution of dayside aurora: Multiple-wavelength all-sky observation at Yellow River Station in Ny-Ålesund, Svalbard, *J. Atmos. Sol. Terr. Phys.*, *71*, 794–804, doi:10.1016/j.jastp.2009.02.010.

Hu, Z.-J., H.-G. Yang, D.-S. Han, D.-H. Huang, B.-C. Zhang, H.-Q. Hu, and R.-Y. Liu (2012), Dayside auroral emissions controlled by IMF: A survey for dayside auroral excitation at 557.7 and 630.0 nm in Ny-Ålesund, Svalbard, *J. Geophys. Res.*, *117*, A02201, doi:10.1029/2011JA017188.

Inan, U. S., Y. T. Chiu, and G. T. Davidson (1992), Whistler-mode chorus and morningside aurorae, *Geophys. Res. Lett.*, *19*(7), 653–656.

Kennel, C. F., F. L. Scarf, R. W. Fredricks, J. H. McGehee, and F. V. Coroniti (1970), VLF electric field observations in the magnetosphere, *J. Geophys. Res.*, *75*(31), 6136–6152.

Li, W., R. M. Thorne, V. Angelopoulos, J. Bortnik, C. M. Cully, B. Ni, O. LeContel, A. Roux, U. Auster, and W. Magnes (2009), Global distribution of whistler-mode chorus waves observed on the THEMIS spacecraft, *Geophys. Res. Lett.*, *36*, L09104, doi:10.1029/2009GL037595.

Li, W., et al. (2010), THEMIS analysis of observed equatorial electron distributions responsible for the chorus excitation, *J. Geophys. Res.*, *115*, A00F11, doi:10.1029/2009JA014845.

Li, W., J. Bortnik, R. M. Thorne, Y. Nishimura, V. Angelopoulos, and L. Chen (2011), Modulation of whistler mode chorus waves: 2. Role of density variations, *J. Geophys. Res.*, *116*, A06206, doi:10.1029/2010JA016313.

Liang, J., V. Uritsky, E. Donovan, B. Ni, E. Spanswick, T. Trondsen, J. Bonnell, A. Roux, U. Auster, and D. Larson (2010), THEMIS observations of electron cyclotron harmonic emissions, ULF waves, and pulsating auroras, *J. Geophys. Res.*, *115*, A10235, doi:10.1029/2009JA015148.

Liang, J., B. Ni, E. Spanswick, M. Kubyshkina, E. F. Donovan, V. M. Uritsky, R. M. Thorne, and V. Angelopoulos (2011), Fast earthward flows, electron cyclotron harmonic waves, and diffuse auroras: Conjunctive observations and a synthesized scenario, *J. Geophys. Res.*, *116*, A12220, doi:10.1029/2011JA017094.

Lui, A. T. Y., D. Venkatesan, C. D. Anger, S.-I. Akasofu, W. J. Heikkila, J. D. Winningham, and J. R. Burrows (1977), Simultaneous observations of particle precipitations and auroral emissions by the Isis 2 satellite in the 19-24 MLT sector, *J. Geophys. Res.*, *82*(16), 2210–2226.

Lummerzheim, D. (1987), Electron transport and optical emissions in the aurora, Ph.D. thesis, Univ. of Alaska, Fairbanks.

Lyons, L. R. (1974), Electron diffusion driven by magnetospheric electrostatic waves, *J. Geophys. Res.*, *79*(4), 575–580.

Meredith, N. P., R. B. Horne, A. D. Johnstone, and R. R. Anderson (2000), The temporal evolution of electron distributions and associated wave activity following substorm injections in the inner magnetosphere, *J. Geophys. Res.*, *105*(A6), 12,907–12,917.

Meredith, N. P., R. B. Horne, and R. R. Anderson (2001), Substorm dependence of chorus amplitudes: Implications for the acceleration of electrons to relativistic energies, *J. Geophys. Res.*, *106*(A7), 13,165–13,178.

Meredith, N. P., R. B. Horne, R. M. Thorne, and R. R. Anderson (2009), Survey of upper band chorus and ECH waves: Implications for the diffuse aurora, *J. Geophys. Res.*, *114*, A07218, doi:10.1029/2009JA014230.

Miyoshi, Y., Y. Katoh, T. Nishiyama, T. Sakanoi, K. Asamura, and M. Hirahara (2010), Time of flight analysis of pulsating aurora electrons, considering wave-particle interactions with propagating whistler mode waves, *J. Geophys. Res.*, *115*, A10312, doi:10.1029/2009JA015127.

Nakajima, A., et al. (2012), Electron and wave characteristics observed by the THEMIS satellites near the magnetic equator during a pulsating aurora, *J. Geophys. Res.*, *117*, A03219, doi:10.1029/2011JA017066.

Newell, P. T., T. Sotirelis, and S. Wing (2009), Diffuse, monoenergetic, and broadband aurora: The global precipitation budget, *J. Geophys. Res.*, *114*, A09207, doi:10.1029/2009JA014326.

Ni, B., R. M. Thorne, Y. Y. Shprits, and J. Bortnik (2008), Resonant scattering of plasma sheet electrons by whistler-mode chorus: Contribution to diffuse auroral precipitation, *Geophys. Res. Lett.*, *35*, L11106, doi:10.1029/2008GL034032.

Ni, B., R. M. Thorne, R. B. Horne, N. P. Meredith, Y. Y. Shprits, L. Chen, and W. Li (2011a), Resonant scattering of plasma sheet electrons leading to diffuse auroral precipitation: 1. Evaluation for electrostatic electron cyclotron harmonic waves, *J. Geophys. Res.*, *116*, A04218, doi:10.1029/2010JA016232.

Ni, B., R. M. Thorne, N. P. Meredith, R. B. Horne, and Y. Y. Shprits (2011b), Resonant scattering of plasma sheet electrons leading to diffuse auroral precipitation: 2. Evaluation for whistler mode chorus waves, *J. Geophys. Res.*, *116*, A04219, doi:10.1029/2010JA016233.

Ni, B., R. Thorne, J. Liang, V. Angelopoulos, C. Cully, W. Li, X. Zhang, M. Hartinger, O. Le Contel, and A. Roux (2011c), Global distribution of electrostatic electron cyclotron harmonic waves observed on THEMIS, *Geophys. Res. Lett.*, *38*, L17105, doi:10.1029/2011GL048793.

Ni, B., J. Liang, R. M. Thorne, V. Angelopoulos, R. B. Horne, M. Kubyshkina, E. Spanswick, E. F. Donovan, and D. Lummerzheim (2012), Efficient diffuse auroral electron scattering by electrostatic electron cyclotron harmonic waves in the outer magnetosphere: A detailed case study, *J. Geophys. Res.*, *117*, A01218, doi:10.1029/2011JA017095.

Nishimura, Y., et al. (2010), Identifying the driver of pulsating aurora, *Science*, *330*, 81–84, doi:10.1126/science.1193186.

Nishimura, Y., et al. (2011), Multievent study of the correlation between pulsating aurora and whistler mode chorus emissions, *J. Geophys. Res.*, *116*, A11221, doi:10.1029/2011JA016876.

Petrinec, S. M., D. L. Chenette, J. Mobilia, M. A. Rinaldi, and W. L. Imhof (1999), Statistical X ray auroral emissions – PIXIE observations, *Geophys. Res. Lett.*, *26*(11), 1565–1568.

Roeder, J. L., and H. C. Koons (1989), A survey of electron cyclotron waves in the magnetosphere and the diffuse auroral electron precipitation, *J. Geophys. Res.*, *94*(A3), 2529–2541.

Sandford, B. P. (1968), Variations of auroral emissions with time, magnetic activity and the solar cycle, *J. Atmos. Sol. Terr. Phys.*, *30*, 1921–1942.

Sato, N., D. M. Wright, C. W. Carlson, Y. Ebihara, M. Sato, T. Saemundsson, S. E. Milan, and M. Lester (2004), Generation region of pulsating aurora obtained simultaneously by the FAST satellite and a Syowa-Iceland conjugate pair of observatories, *J. Geophys. Res.*, *109*, A10201, doi:10.1029/2004JA010419.

Shi, R., D. Han, B. Ni, Z.-J. Hu, C. Zhou, and X. Gu (2012), Intensification of dayside diffuse auroral precipitation: Contribution of dayside whistler-mode chorus waves in realistic magnetic fields, *Ann. Geophys.*, *30*, 1297–1307.

Steele, D. P., and D. J. McEwen (1990), Electron auroral excitation efficiencies and intensity ratios, *J. Geophys. Res.*, *95*, 10,321–10,336, doi:10.1029/JA095iA07p10321.

Summers, D., B. Ni, and N. P. Meredith (2007), Timescales for radiation belt electron acceleration and loss due to resonant wave-particle interactions: 2. Evaluation for VLF chorus, ELF hiss, and electromagnetic ion cyclotron waves, *J. Geophys. Res.*, *112*, A04207, doi:10.1029/2006JA011993.

Swift, D. W. (1981), Mechanisms for auroral precipitation: A review, *Rev. Geophys.*, *19*(1), 185–211.

Tao, X., R. M. Thorne, W. Li, B. Ni, N. P. Meredith, and R. B. Horne (2011), Evolution of electron pitch angle distributions following injection from the plasma sheet, *J. Geophys. Res.*, *116*, A04229, doi:10.1029/2010JA016245.

Thorne, R. M. (2010), Radiation belt dynamics: The importance of wave-particle interactions, *Geophys. Res. Lett.*, *37*, L22107, doi:10.1029/2010GL044990.

Thorne, R. M., B. Ni, X. Tao, R. B. Horne, and N. P. Meredith (2010), Scattering by chorus waves as the dominant cause of diffuse auroral precipitation, *Nature*, *467*, 943–946, doi:10.1038/nature09467.

B. Ni and R. M. Thorne, Department of Atmospheric and Oceanic Sciences, University of California, Los Angeles, CA 90095-1565, USA. (bbni@atmos.ucla.edu)

The Role of the Earth's Ring Current in Radiation Belt Dynamics

Vania K. Jordanova

Space Science and Applications, Los Alamos National Laboratory, Los Alamos, New Mexico, USA

The dynamical coupling of the ring current and the radiation belt populations is investigated during geomagnetic storms, employing recent ring current modeling studies that include time-dependent transport in realistic nondipolar and self-consistently calculated magnetic fields. We present results from a ring current-atmosphere interactions model (RAM) that solves the kinetic equation for H^+, O^+, and He^+ ions and electrons and is two-way coupled with a 3-D equilibrium code (SCB) that calculates self-consistently the magnetic field in force balance with the anisotropic ring current plasma pressure. The RAM-SCB boundary conditions are specified by a plasma sheet source population at geosynchronous orbit that varies both in space and time. It is demonstrated that the storm time ring current development affects radiation belt dynamics in three significant ways: (1) it depresses the background magnetic field on the nightside, which affects the subsequent transport of radiation belt electrons, (2) its electron component represents a highly variable, asymmetric, low-energy seed population of the radiation belts, and (3) the unstable ring current ion and electron populations generate electromagnetic ion cyclotron, magnetosonic, and chorus waves (with different intensities and spatial distributions) that scatter radiation belt particles. Therefore, to understand radiation belt dynamics, we need to consider the coupling in the inner magnetosphere across broad spatial, temporal, and energy scales.

1. INTRODUCTION

The fast rise of radiation belt electron fluxes, over orders of magnitude during geomagnetically active periods, represents a serious threat to modern instrumentation placed in ground-based and satellite-borne systems. Space weather warnings of such relativistic electron enhancements will be even more important as we enter the solar maximum in ~2014 following an unusual solar minimum [*Love et al.*, 2012]. The high variability of the radiation belts, however, remains not adequately explained due to their complex dynamics including competing particle acceleration and loss processes. In this paper, we discuss how changes in the lower-energy ring current population affect the evolution of the radiation belts.

The majority of large geomagnetic storms are driven by interplanetary (IP) coronal mass ejections associated with huge eruptions from the Sun of plasma and magnetic flux [e.g., *Gosling et al.*, 1991; *Richardson et al.*, 2001]. The large-scale electric field across the magnetotail intensifies when solar wind plasma carrying a strong, southward interplanetary magnetic field (IMF) passes Earth. This causes an enhancement of the earthward flow of ions and electrons inside the magnetosphere and a significant increase of the ring current energy density, as demonstrated in earlier ring current ion modeling studies [e.g., *Chen et al.*, 1994; *Fok et al.*, 1996; *Jordanova et al.*, 1998]. As a result, the magnetic field measured on the Earth's surface decreases rapidly, and a geomagnetic storm occurs [e.g., *Tsurutani and Gonzalez*, 1997]. During the peak of large storms, the tail-like deformation of the magnetic field penetrates close to Earth and the quasidipolar approximation breaks down at distances as

Dynamics of the Earth's Radiation Belts and Inner Magnetosphere
Geophysical Monograph Series 199
10.1029/2012GM001330

small as 3–4 R_E [e.g., *Tsyganenko et al.*, 2003]. The response of radiation belt electron fluxes, however, is not well correlated with geomagnetic storm strength, i.e., the *Dst* index. For example, the radiation belt electrons move outward during the storm main phase to conserve the third adiabatic invariant; hence, their drift paths may intersect the magnetopause and cause radiation belt electron loss known as magnetopause shadowing [e.g., *Kim and Chan*, 1997; *Ohtani et al.*, 2009]. During the storm recovery phase, the electrons move inward as the magnetic field recovers, but the electron fluxes may not return to the prestorm values. To determine the storm time transport of both ring current and radiation belt particles, a detailed knowledge of the magnetic field in the inner magnetosphere is thus required.

Radiation belt enhancements are known to be associated with high-speed solar wind streams (HSS) [e.g., *Paulikas and Blake*, 1979; *Miyoshi and Kataoka*, 2005, 2008]. Previous radiation belt studies suggested that the acceleration processes at geosynchronous orbit are dominated by ULF wave-enhanced radial diffusion [e.g., *Li et al.*, 2001; *Elkington et al.*, 2003]. Closer to Earth, the efficiency of ULF wave acceleration is likely to be reduced because of the reduction of the ULF power, and local heating of ~100 keV ring current electrons to MeV energies by VLF chorus was proposed [*Horne and Thorne*, 1998; *Summers et al.*, 1998]. This process is expected to be most efficient in the low-density region just outside the plasmapause [e.g., *Miyoshi et al.*, 2003]. Besides the pitch angle scattering by ELF hiss inside the plasmasphere causing the slot region [*Lyons et al.*, 1972; *Albert*, 1994], other processes removing radiation belt electrons are magnetopause shadowing, the *Dst* effect, scattering by VLF chorus waves outside the plasmasphere [e.g., *Lorentzen et al.*, 2001; *Thorne et al.*, 2005], and resonance with electromagnetic ion cyclotron (EMIC) waves [e.g., *Thorne and Kennel*, 1971; *Summers and Thorne*, 2003; *Jordanova et al.*, 2008]. All of these plasma waves have been observed in the magnetosphere with different intensities, spatial distributions, and occurrence rates. The free energy for the excitation of these plasma waves is provided by the anisotropic ring current ion and electron distributions. Modeling of wave-particle interactions on a global scale that includes the combined effect of these waves on trapped particles as they drift around the Earth is needed to obtain the global evolution of radiation belt electrons and determine the penetration of electron fluxes to the altitudes of near-Earth orbiting satellites. Moreover, the precipitating electrons will affect the ionospheric conductance and, subsequently, the self-consistently calculated electric field [e.g., *Harel et al.*, 1981] and the global inner magnetosphere dynamics. It is evident that the full physical coupling between the plasma and the fields has to be addressed.

Finally, the ring current electrons represent the low-energy seed population that gets accelerated to form the radiation belts. To accurately model radiation belt dynamics, particle energies must span 5–6 orders of magnitude. Such a broad cross-energy coupling, combined with the range of timescales between global transport and local acceleration and loss processes, over the large spatial scale of the radiation belts poses significant computational challenges. Recent advancements on these topics in preparation for the launch of the Radiation Belt Storm Probes [*Kessel*, this volume], a spacecraft mission dedicated to study the radiation belts, are reviewed in this paper.

2. RING CURRENT ELECTRONS

The dynamics of ring current electrons have not received much attention in the past as most of the ring current energy density is carried by ions [*Daglis et al.*, 1999]. Early observations from the OGO 3 satellite [*Frank*, 1967] of low-energy (<50 keV) electron and proton fluxes showed that the electron component may provide about 25% of the ring current energy during storm times. A more recent study [*Liu et al.*, 2005] indicated that this may be an overestimation because the high-energy (~50–200 keV) ring current particles were not included. Analyzing Explorer 45 data spanning larger energy (1–200 keV) and spatial range (2–5 R_E) during the 17 December 1971 storm with minimum *Dst* = −171 nT, Liu et al. found that ring current electrons may contribute ~7.5% of the ring current energy during quiet time; the electron contribution may increase to 19% during storm time. *Jordanova and Miyoshi* [2005] performed an initial simulation of the large storm of 21 October 2001 using an analytical *Volland* [1973]-*Stern* [1975] electric field and dipolar magnetic field models and obtained that the electron contribution was about 2% during quiet time and about 10% near peak *Dst* = −187 nT. The simulated electron fluxes increased on the dawnside during the rapid main phase in agreement with NOAA observations [*Miyoshi et al.*, 2006].

The dynamics of ring current electrons during the 19 October 1998 storm (min *Dst* = −112 nT) in nondipolar magnetic field geometry were studied in more detail by *Chen et al.* [2006]. A convective electric field model including quiescent Volland-Stern convection and storm-associated enhancements in the convection was used. Tracing equatorially mirroring protons and electrons and solving the force-balance equation in the equatorial plane, Chen et al. found enhancements in both the proton and electron pressure, which tended to occur on the duskside for protons and on the dawnside for electrons, consistent with the corresponding direction of magnetic drift. On average, the equatorial perpendicular pressure within the ring current region was smaller in the self-consistent simulations than in the nonself-consistent simulations. There were localized regions, however, where the equatorial perpendicular pressure was more intense. Similar features were found in 3-D magnetically self-consistent simulations of the ring current ion dynamics during the 21–23 April 2001 storm by *Zaharia et al.* [2006].

Ring current development during the high-speed solar wind stream (HSS) storm of 23–25 October 2002 was studied

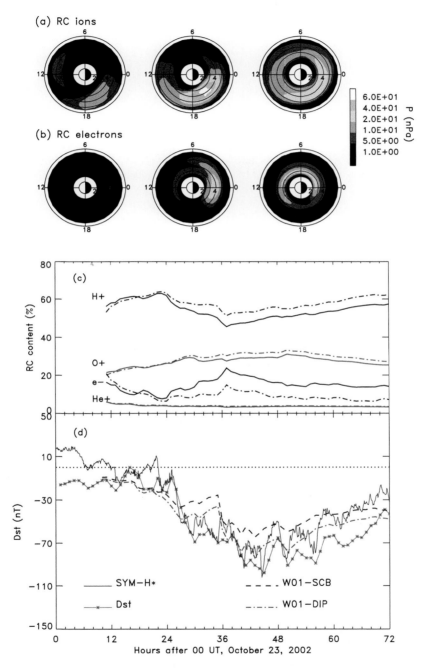

Figure 1. Ring current (a) ion and (b) electron pressure obtained with RAM-SCB at hours 24, 36, and 38 after 00 UT 23 October 2002. (c) Contribution to total ring current energy from various species using W01 potential and either dipolar (dash-dotted line) or self-consistent (solid line) magnetic field simulations. (d) Computed *Dst* index using different model formulations compared with measured *Dst* (starred line) and magnetopause current-corrected SYM-H (solid line).

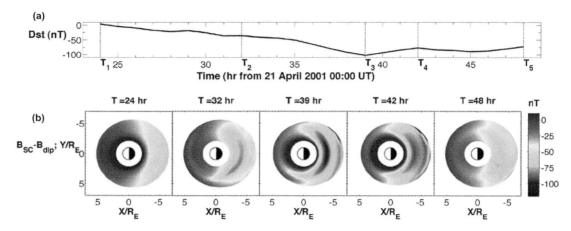

Figure 2. (a) *Dst* index during the April 2001 storm, time axis starts at 00 UT on 21 April. (b) Difference between the self-consistent magnetic field intensity obtained with RAM-SCB and the Earth dipolar field in the equatorial plane at hours 24, 32, 39, 42, and 48 after 00 UT 21 April 2001 [after *Zaharia et al.*, 2006].

by *Jordanova et al.* [2012]. This time period was one of the rare HSS passages at Earth that triggered significant geomagnetic activity: a magnetic storm with a step-like *Dst* profile and minimum $Dst \approx -100$ nT at 21 UT 24 October (Figure 1). The slow storm recovery lasted for several days due to the continuing fluctuations of the IMF B_z component in the high-speed flow behind the interaction region. In addition to ring current ion dynamics simulated in previous studies [*Jordanova et al.* 2006, 2010b; *Zaharia et al.*, 2006, 2010; *Welling et al.*, 2011], the effects of realistic nondipolar magnetic field configuration were investigated on both ring current ions and electrons using a ring current-atmosphere interactions model with self-consistent magnetic field (RAM-SCB). RAM solves the bounce-averaged kinetic equation for H^+, O^+, and He^+ ions and electrons in the solar magnetic (SM) equatorial plane and is two-way coupled with a 3-D equilibrium code (SCB) that calculates the magnetic field in force balance with the anisotropic ring current ion and electron distributions. To this end, the perpendicular and parallel pressures in the equatorial plane obtained from moments of the RAM distribution function are mapped along the 3-D magnetic field lines through energy and magnetic moment conservation. The electric field model in this study represented the gradient of the convection potential of W01 [*Weimer*, 2001] and a corotation potential. The W01 ionospheric potential was driven by IP data and the *AL* index and was mapped to the SM equatorial plane along SCB field lines. The nightside

boundary conditions were determined as they varied in space and time from plasma sheet flux measurements from the MPA and SOPA instruments on the Los Alamos National Laboratory geosynchronous spacecraft.

The total ring current ion and electron pressure calculated with RAM-SCB in the SM equatorial plane at hours 24, 36, and 38 from 00 UT 23 October 2002 is shown in Figures 1a and 1b, respectively. The pressure intensifies with storm development as the particles are injected from the plasma sheet, accelerated, and trapped. A strong asymmetry is seen during the main phase of the storm (hour 36) when the ion pressure peaks in the premidnight magnetic local time (MLT), while the electron pressure peaks near midnight. Both ions and electrons penetrate close to Earth reaching radial distances of ~2.5 R_E. The ring current pressure becomes more symmetric and extends in MLT as the storm develops (hour 38). The contribution to the total ring current energy from various species during this HSS-driven storm is shown in Figure 1c. The major contribution is from H^+, which is the dominant species, and it varies between 60% at the beginning of the storm at hour ~24 and ~50% during the HSS interface. The ring current O^+ content is about ~30%, while He^+ content remains at ~5% throughout the interval. The electron content is quite variable, being about 10% during quiet time and about 20% during disturbed time. Note that it is significantly larger in the self-consistent case (solid line) than in the dipolar case (dash-dotted line), indicating

Figure 3. (opposite) The B_x, B_y, and B_z components of the magnetic field computed with RAM-SCB model (dashed line) for the storm event on 31 August 2005 compared with TS05 model (dashed-dotted line) and in situ observations (solid line) from (left) Cluster and (right) Polar. (top) The satellite trajectories are shown for reference. (bottom) The normalized RMS errors for each field component and magnitude indicate the agreement between the model and the observations [after *Yu et al.*, 2012].

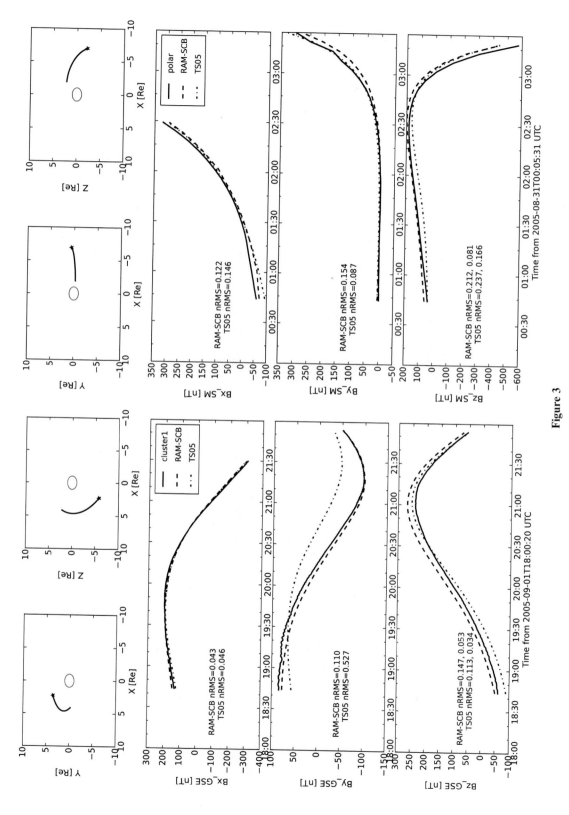

Figure 3

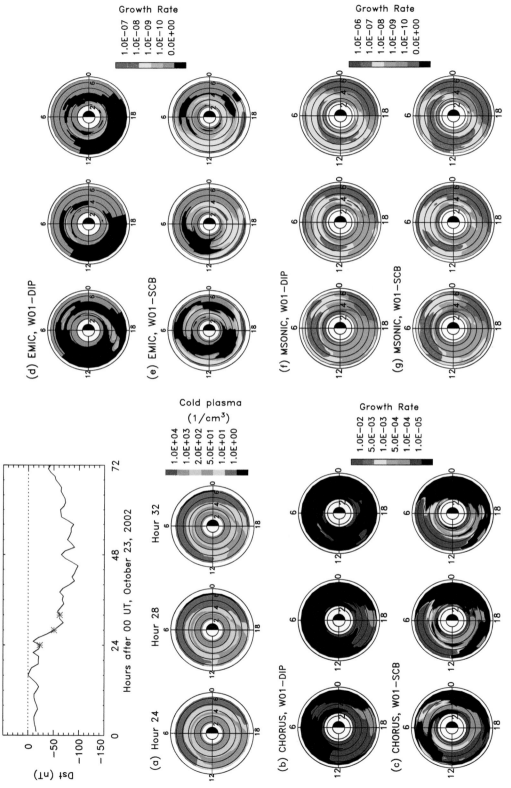

Figure 4

that simulations not taking into account the feedback of the ring current overestimate the actual proton and underestimate the actual electron contribution. The total ring current contribution to the *Dst* index is shown in Figure 1d. Results computed with the Dessler-Parker-Sckopke (DPS) relation [*Dessler and Parker*, 1959; *Sckopke*, 1966] are compared with the magnetopause current-corrected SYM-H index. *Jordanova et al.* [2012] calculated as well the ring current contribution to *Dst* using Biot-Savart integration of the currents in the whole domain and found that it differed little from the DPS relation except during quiet time. Overall, the simulated global evolution of the ring current reproduces very well the temporal variations of SYM-H. In agreement with previous studies, larger ring current injection and, correspondingly, SYM-H depression, is obtained using a dipolar (dashed-dotted line) rather than a self-consistently calculated **B** field (dashed line). This is caused by the increased drift velocities in the dusk to midnight MLT sector (where the SCB field is mostly depressed), combined with drift-shell splitting in realistic fields, which lead to enhanced magnetopause losses and significantly reduced nightside fluxes at larger distances from Earth. The RAM-SCB simulations underestimate |SYM-H| minimum by ~25% but reproduce well the storm recovery phase. Better agreement is expected once contributions from tail currents outside geosynchronous orbit are included in the simulations [e.g., *Ganushkina et al.*, 2012].

3. MAGNETIC FIELD CONFIGURATION

The global specification of the magnetic field in the inner magnetosphere affects strongly the radiation belt dynamics and is one of the critically needed inputs for radiation belt modeling efforts. A significant fraction of the observed decrease of relativistic electron fluxes at geosynchronous orbit during the main phase of a storm can be accounted by the fully adiabatic response (all three adiabatic invariants conserved) of the radiation belts to the magnetic field changes [*Kim and Chan*, 1997]. The contribution of prestorm electron fluxes to the geosynchronous flux increase during the storm recovery phase, however, is expected to be small because of losses to the dawnside magnetopause. *Green and Kivelson* [2004] demonstrated that calculations in a realistic magnetic field can change significantly the inferred radial profiles of the phase space density of radiation belt electrons. *Huang et al.* [2008] evaluated the performance of recent empirical

magnetic field models during active geomagnetic conditions and compared the model outputs with GOES observations. It was found that for major magnetic storms, the *Tsyganenko* [1996, 2002] models predicted anomalously strong negative B_z at geostationary orbit on the nightside due to input values exceeding the model limits. The more recent storm time model TS05 [*Tsyganenko and Sitnov*, 2005] had better overall performance across the entire range of local times, storm levels, and storm phases at geostationary orbit. Huang et al. showed as well that even small deviations between field models alter significantly the model-dependent invariants used to interpret radiation belt dynamics; the L^* parameter thus varied by ~13% using the *Tsyganenko* [1996] or *Tsyganenko and Sitnov* [2005] models at geosynchronous orbit.

The 3-D magnetic field during the 21–23 April 2001 storm (minimum $Dst \approx -100$ nT) was computed by *Zaharia et al.* [2006] in force balance with prescribed anisotropic ring current ion pressure using the RAM-SCB model. The simulations were driven by the *Weimer* [2001] electric field and geosynchronous plasma boundary conditions. Figure 2 shows the difference between the self-consistently calculated magnetic field intensity with RAM-SCB and the Earth dipolar field in the equatorial plane at selected hours during the storm. Large depressions in the equatorial magnetic field intensity were seen on the nightside during the storm main phase at hours 39 and 42. Significant differences were found in the self-consistent results compared to those obtained from the kinetic RAM model with a dipolar background field (using the same particle boundary conditions and electric fields). The computed pressure maximum was smaller in the self-consistent simulation, mostly due to enhanced drifts in regions of reduced magnetic intensity leading to lower particle densities. In addition, narrow pressure (and density) peaks appeared in certain locations, unlike the smooth picture obtained with a dipolar field. It was thus demonstrated that for storm modeling, the plasma and the magnetic field should represent a self-consistent configuration obeying the plasma momentum equation.

A validation study of the magnetically self-consistent inner magnetosphere model RAM-SCB was performed by *Yu et al.* [2012]. The validation was conducted by simulating two magnetic storm events and then comparing the model results against a variety of satellite observations, including the magnetic field from Cluster and Polar spacecraft, ion differential flux from the Cluster/CODIF analyzer, and the ground-based SYM-H index. The self-consistent magnetic field computed

Figure 4. (opposite) (a) Plasmaspheric electron density as a function of radial distance in the equatorial plane and MLT at selected hours during 23–25 October 2002 indicated with stars on the *Dst* plot. Linear growth rate of (b and c) chorus, (d and e) EMIC, and (f and g) magnetosonic waves calculated with RAM using W01 electric field and either a dipolar (DIP) or a self-consistently calculated (SCB) magnetic field.

along Cluster and Polar trajectories for the 31 August 2005 storm event is compared with TS05 model and in situ observations in Figure 3. The magnetic field at a given satellite location was obtained after interpolating the magnetic field values from nearby grids. It is clear that the RAM-SCB model performed equally well or better in reproducing the magnetic field components than the empirical TS05 model. The good agreement of RAM-SCB magnetic field prediction with observations demonstrated the model capability of representing well the inner magnetospheric field configuration. This provided confidence that RAM-SCB can be used for field line and drift shell tracing needed in radiation belt studies.

4. PLASMA WAVE DYNAMICS

The anisotropic ring current ion and electron populations generate various types of plasma waves in the inner magnetosphere with different occurrence rates, spatial distributions, and intensities. Whistler mode chorus emissions are usually observed outside the plasmasphere over a broad MLT range with peak storm time amplitudes in the range of 30–100 pT. Statistical surveys indicate that these VLF emissions intensify at distances larger than ~3 R_E on the dawnside, extending from premidnight to noon MLT [*Meredith et al.*, 2001; *Santolik and Gurnett*, 2003]. The nightside chorus is confined to ±15° of the equator, while the dayside chorus may propagate to higher latitudes. It is considered that chorus emissions are initially excited by cyclotron resonance with a highly anisotropic distribution of energetic (~10 keV) electrons [e.g., *Li et al.*, 2009; *Jordanova et al.*, 2010a], followed by nonlinear evolution into discrete chorus elements [*Nunn et al.*, 2003; *Omura et al.*, 2008]. Anisotropic distributions of energetic ring current ions excite EMIC waves [e.g., *Mauk and McPherron*, 1980; *Horne and Thorne*, 1993; *Jordanova et al.*, 1997]. The most intense EMIC waves occur during geomagnetic storms and are confined to the dusk MLT region [*Erlandson and Ukhorskiy*, 2001; *Meredith et al.*, 2003]. CRRES observations show a source region within ±11° of the equator [*Loto'aniu et al.*, 2005] in support of theoretical studies suggesting that the equatorial region is favored for EMIC wave generation. Finally, fast magnetosonic waves, also called equatorial noise [e.g., *Russell et al.*, 1970], occur at frequencies between the proton gyrofrequency and the lower hybrid frequency, and are primarily confined within several degrees of the geomagnetic equator both inside and outside the plasmapause. These waves are associated with ring-type (in the direction perpendicular to the magnetic field) proton velocity distributions [e.g., *Perraut et al.*, 1982; *Horne et al.*, 2000] at energies of the order of ~10 keV.

The linear growth rate of all dominant magnetospheric plasma waves driven by the free energy in the ring current distributions was simulated on a global scale by *Jordanova et al.* [2012] during the 23–25 October 2002 storm. Figure 4 displays results at hours 24, 28, and 32 after 00 UT 23 October obtained in the SM equatorial plane with RAM using W01 electric field and either dipolar or self-consistently calculated magnetic field. It is evident that the wave growth is significantly enhanced in the latter case (Figures 4c, 4e, and 4g) when transport in a realistic nondipolar magnetic field is considered. Whistler mode chorus emissions (Figure 4c) are excited in the dawnside MLT region outside the plasmasphere (Figure 4a) by the anisotropic ring current electron distributions that develop due to drift-shell splitting. After being injected from the plasma sheet, ring current electrons drift eastward from midnight to noon and populate different shells as higher pitch angle particles move outward. Electrons with near 90° pitch angle may intercept the magnetopause and will be depleted from the postnoon MLT region. The drift-shell splitting thus leads to the development of positive prenoon and negative postnoon anisotropy for ring current electrons at larger radial distances from Earth [*Jordanova et al.*, 2012]. By analogy, the ring current ion anisotropy increases in the postnoon region and decreases in the prenoon region as these particles drift westward from midnight to noon. Intense EMIC waves (Figure 4e) are thus generated at large radial distances (>5 R_E) in high-density plasmaspheric drainage plumes in the afternoon MLT and along the plasmapause. On the other hand, magnetosonic waves are generated by ion ring-type distributions [e.g. *Chen et al.*, 2010] that result from energy-dependent drifts, losses, and dispersed injections. The regions of magnetosonic instability cover broad MLT range both inside and outside the plasmasphere. The charge exchange losses are stronger on the dayside in a nondipolar magnetic field, leading to larger gradients in the phase space density and subsequently to larger magnetosonic wave growth (Figure 4g). As discussed in the introduction, these plasma waves are important for the acceleration and loss of radiation belt electrons.

5. SUMMARY AND CONCLUSIONS

Using recent magnetically self-consistent simulations, we investigated how the storm time ring current evolution affects radiation belt dynamics. We found the following:

1. As strong depressions in the self-consistent magnetic field develop at large radial distances (>4 R_E) on the nightside during the main phase of a storm, the particle's gradient-curvature drift velocity increases, the particle fluxes are reduced, and the ring current is confined close to Earth. The ring current electron component, which represents the

low-energy seed population of the radiation belts, is highly asymmetric with peaks near midnight to dawn. Its contribution increases with geomagnetic activity and reaches ~20% near minimum Dst.

2. Increased anisotropies develop in the ring current ion and electron velocity distributions as a result of particle injection, transport, and loss in a realistic nondipolar magnetic field. There is thus sufficient free energy in the equatorial magnetosphere to excite strong EMIC waves in the afternoon to midnight region, whistler mode chorus waves from postmidnight to noon, and magnetosonic waves both inside and outside the plasmasphere. These waves will scatter radiation belt particles in energy and pitch angle as they drift through the unstable regions, causing their further acceleration or loss.

3. Comparisons with satellite data during magnetic storm periods show that the RAM-SCB model represents better the magnetic field configuration in the inner magnetosphere than empirical models and is thus reliable to use in radiation belt studies. This indicates that global models should include self-consistent coupling of the plasma and the fields in order to accurately predict the dynamics of the radiation belts and the harmful enhancements of electron fluxes during disturbed times.

Acknowledgments. Work at Los Alamos was conducted under the auspices of the U.S. Department of Energy, with partial support from NASA and NSF programs. The *Dst* and SYM-H indices were provided by the World Data Center in Kyoto, Japan.

REFERENCES

Albert, J. M. (1994), Quasi-linear pitch angle diffusion coefficients: Retaining high harmonics, *J. Geophys. Res.*, *99*(A12), 23,741–23,745, doi:10.1029/94JA02345.

Chen, L., R. M. Thorne, V. K. Jordanova, and R. B. Horne (2010), Global simulation of magnetosonic wave instability in the storm time magnetosphere, *J. Geophys. Res.*, *115*, A11222, doi:10.1029/2010JA015707.

Chen, M. W., L. R. Lyons, and M. Schulz (1994), Simulations of phase space distributions of storm time proton ring current, *J. Geophys. Res.*, *99*(A4), 5745–5759, doi:10.1029/93JA02771.

Chen, M. W., S. Liu, M. Schulz, J. L. Roeder, and L. R. Lyons (2006), Magnetically self-consistent ring current simulations during the 19 October 1998 storm, *J. Geophys. Res.*, *111*, A11S15, doi:10.1029/2006JA011620.

Daglis, I. A., R. M. Thorne, W. Baumjohann, and S. Orsini (1999), The terrestrial ring current: Origin, formation, and decay, *Rev. Geophys.*, *37*(4), 407–438.

Dessler, A. J., and E. N. Parker (1959), Hydromagnetic theory of geomagnetic storms, *J. Geophys. Res.*, *64*(12), 2239–2252.

Elkington, S. R., M. K. Hudson, and A. A. Chan (2003), Resonant acceleration and diffusion of outer zone electrons in an asymmetric geomagnetic field, *J. Geophys. Res.*, *108*(A3), 1116, doi:10.1029/2001JA009202.

Erlandson, R. E., and A. J. Ukhorskiy (2001), Observations of electromagnetic ion cyclotron waves during geomagnetic storms: Wave occurrence and pitch angle scattering, *J. Geophys. Res.*, *106*(A3), 3883–3895, doi:10.1029/2000JA000083.

Fok, M.-C., T. E. Moore, and M. E. Greenspan (1996), Ring current development during storm main phase, *J. Geophys. Res.*, *101*(A7), 15,311–15,322, doi:10.1029/96JA01274.

Frank, L. A. (1967), On the extraterrestrial ring current during geomagnetic storms, *J. Geophys. Res.*, *72*(15), 3753–3767.

Ganushkina, N. Y., M. W. Liemohn, and T. I. Pulkkinen (2012), Storm-time ring current: Model-dependent results, *Ann. Geophys.*, *30*, 177–202.

Gosling, J. T., D. J. McComas, J. L. Phillips, and S. J. Bame (1991), Geomagnetic activity associated with Earth passage of interplanetary shock disturbances and coronal mass ejections, *J. Geophys. Res.*, *96*(A5), 7831–7839.

Green, J. C., and M. G. Kivelson (2004), Relativistic electrons in the outer radiation belt: Differentiating between acceleration mechanisms, *J. Geophys. Res.*, *109*, A03213, doi:10.1029/2003JA010153.

Harel, M., R. A. Wolf, P. H. Reiff, R. W. Spiro, W. J. Burke, F. J. Rich, and M. Smiddy (1981), Quantitative simulation of a magnetospheric substorm, 1. Model logic and overview, *J. Geophys. Res.*, *86*(A4), 2217–2241, doi:10.1029/JA086iA04p02217.

Horne, R. B., and R. M. Thorne (1993), On the preferred source location for the convective amplification of ion cyclotron waves, *J. Geophys. Res.*, *98*(A6), 9233–9247.

Horne, R. B., and R. M. Thorne (1998), Potential waves for relativistic electron scattering and stochastic acceleration during magnetic storms, *Geophys. Res. Lett.*, *25*(15), 3011–3014.

Horne, R. B., G. V. Wheeler, and H. S. C. K. Alleyne (2000), Proton and electron heating by radially propagating fast magnetosonic waves, *J. Geophys. Res.*, *105*(A12), 27,597–27,610.

Huang, C.-L., H. E. Spence, H. J. Singer, and N. A. Tsyganenko (2008), A quantitative assessment of empirical magnetic field models at geosynchronous orbit during magnetic storms, *J. Geophys. Res.*, *113*, A04208, doi:10.1029/2007JA012623.

Jordanova, V. K., and Y. Miyoshi (2005), Relativistic model of ring current and radiation belt ions and electrons: Initial results, *Geophys. Res. Lett.*, *32*, L14104, doi:10.1029/2005GL023020.

Jordanova, V. K., J. U. Kozyra, A. F. Nagy, and G. V. Khazanov (1997), Kinetic model of the ring current-atmosphere interactions, *J. Geophys. Res.*, *102*(A7), 14,279–14,292, doi:10.1029/96JA03699.

Jordanova, V. K., C. J. Farrugia, L. Janoo, J. M. Quinn, R. B. Torbert, K. W. Ogilvie, R. P. Lepping, J. T. Steinberg, D. J. McComas, and R. D. Belian (1998), October 1995 magnetic cloud and accompanying storm activity: Ring current evolution, *J. Geophys. Res.*, *103*(A1), 79–92, doi:10.1029/97JA02367.

Jordanova, V. K., Y. S. Miyoshi, S. Zaharia, M. F. Thomsen, G. D. Reeves, D. S. Evans, C. G. Mouikis, and J. F. Fennell (2006), Kinetic simulations of ring current evolution during the Geospace

Environment Modeling challenge events, *J. Geophys. Res.*, *111*, A11S10, doi:10.1029/2006JA011644.

Jordanova, V. K., J. Albert, and Y. Miyoshi (2008), Relativistic electron precipitation by EMIC waves from self-consistent global simulations, *J. Geophys. Res.*, *113*, A00A10, doi:10.1029/2008JA013239. [Printed 114(A3), 2009].

Jordanova, V. K., R. M. Thorne, W. Li, and Y. Miyoshi (2010a), Excitation of whistler mode chorus from global ring current simulations, *J. Geophys. Res.*, *115*, A00F10, doi:10.1029/2009JA014810.

Jordanova, V. K., S. Zaharia, and D. T. Welling (2010b), Comparative study of ring current development using empirical, dipolar, and self-consistent magnetic field simulations, *J. Geophys. Res.*, *115*, A00J11, doi:10.1029/2010JA015671.

Jordanova, V. K., D. T. Welling, S. G. Zaharia, L. Chen, and R. M. Thorne (2012), Modeling ring current ion and electron dynamics and plasma instabilities during a high-speed stream driven storm, *J. Geophys. Res.*, *117*, A00L08, doi:10.1029/2011JA017433.

Kessel, R. L. (2012), NASA's Radiation Belt Storm Probes mission: From concept to reality, in *Dynamics of the Earth's Radiation Belts and Inner Magnetosphsere, Geophys. Monogr. Ser.*, doi:10.1029/2012GM001312, this volume.

Kim, H.-J., and A. A. Chan (1997), Fully adiabatic changes in storm time relativistic electron fluxes, *J. Geophys. Res.*, *102*(A10), 22,107–22,116, doi:10.1029/97JA01814.

Li, W., R. M. Thorne, V. Angelopoulos, J. Bortnik, C. M. Cully, B. Ni, O. LeContel, A. Roux, U. Auster, and W. Magnes (2009), Global distribution of whistler-mode chorus waves observed on the THEMIS spacecraft, *Geophys. Res. Lett.*, *36*, L09104, doi:10.1029/2009GL037595.

Li, X., M. Temerin, D. N. Baker, G. D. Reeves, and D. Larson (2001), Quantitative prediction of radiation belt electrons at geostationary orbit based on solar wind measurements, *Geophys. Res. Lett.*, *28*(9), 1887–1890, doi:10.1029/2000GL012681.

Liu, S., M. W. Chen, J. L. Roeder, L. R. Lyons, and M. Schulz (2005), Relative contribution of electrons to the stormtime total ring current energy content, *Geophys. Res. Lett.*, *32*, L03110, doi:10.1029/2004GL021672.

Lorentzen, K. R., J. B. Blake, U. S. Inan, and J. Bortnik (2001), Observations of relativistic electron microbursts in association with VLF chorus, *J. Geophys. Res.*, *106*(A4), 6017–6027, doi:10.1029/2000JA003018.

Loto'aniu, T. M., B. J. Fraser, and C. L. Waters (2005), Propagation of electromagnetic ion cyclotron wave energy in the magnetosphere, *J. Geophys. Res.*, *110*, A07214, doi:10.1029/2004JA010816.

Love, J. J., E. Joshua Rigler, and S. E. Gibson (2012), Geomagnetic detection of the sectorial solar magnetic field and the historical peculiarity of minimum 23–24, *Geophys. Res. Lett.*, *39*, L04102, doi:10.1029/2011GL050702.

Lyons, L. R., R. M. Thorne, and C. F. Kennel (1972), Pitch angle diffusion of radiation belt electrons within the plasmasphere, *J. Geophys. Res.*, *77*(19), 3455–3474.

Mauk, B. H., and R. L. McPherron (1980), An experimental test of the electromagnetic ion cyclotron instability within the Earth's

magnetosphere, *Phys. Fluids*, *23*, 2111–2127, doi:10.1063/1.862873.

Meredith, N. P., R. B. Horne, and R. R. Anderson (2001), Substorm dependence of chorus amplitudes: Implications for the acceleration of electrons to relativistic energies, *J. Geophys. Res.*, *106*(A7), 13,165–13,178, doi:10.1029/2000JA900156.

Meredith, N. P., R. M. Thorne, R. B. Horne, D. Summers, B. J. Fraser, and R. R. Anderson (2003), Statistical analysis of relativistic electron energies for cyclotron resonance with EMIC waves observed on CRRES, *J. Geophys. Res.*, *108*(A6), 1250, doi:10.1029/2002JA009700.

Miyoshi, Y., and R. Kataoka (2005), Ring current ions and radiation belt electrons during geomagnetic storms driven by coronal mass ejections and corotating interaction regions, *Geophys. Res. Lett.*, *32*, L21105, doi:10.1029/2005GL024590.

Miyoshi, Y., and R. Kataoka (2008), Flux enhancement of the outer radiation belt electrons after the arrival of stream interaction regions, *J. Geophys. Res.*, *113*, A03S09, doi:10.1029/2007JA012506.

Miyoshi, Y., A. Morioka, H. Misawa, T. Obara, T. Nagai, and Y. Kasahara (2003), Rebuilding process of the outer radiation belt during the 3 November 1993 magnetic storm: NOAA and Exos-D observations, *J. Geophys. Res.*, *108*(A1), 1004, doi:10.1029/2001JA007542.

Miyoshi, Y. S., V. K. Jordanova, A. Morioka, M. F. Thomsen, G. D. Reeves, D. S. Evans, and J. C. Green (2006), Observations and modeling of energetic electron dynamics during the October 2001 storm, *J. Geophys. Res.*, *111*, A11S02, doi:10.1029/2005JA011351.

Nunn, D., A. Demekhov, V. Trakhtengerts, and M. J. Rycroft (2003), VLF emission triggering by a highly anisotropic electron plasma, *Ann. Geophys.*, *21*, 481–492.

Ohtani, S., Y. Miyoshi, H. J. Singer, and J. M. Weygand (2009), On the loss of relativistic electrons at geosynchronous altitude: Its dependence on magnetic configurations and external conditions, *J. Geophys. Res.*, *114*, A01202, doi:10.1029/2008JA013391.

Omura, Y., Y. Katoh, and D. Summers (2008), Theory and simulation of the generation of whistler-mode chorus, *J. Geophys. Res.*, *113*, A04223, doi:10.1029/2007JA012622.

Paulikas, G. A., and J. B. Blake (1979), Effects of the solar wind on magnetospheric dynamics: Energetic electrons at the synchronous orbit, in *Quantitative Modeling of Magnetospheric Processes, Geophys. Monogr. Ser.*, vol. 21, edited by W. P. Olson, pp. 180–202, AGU, Washington, D. C., doi:10.1029/GM021p0180.

Perraut, S., A. Roux, P. Robert, R. Gendrin, J.-A. Sauvaud, J.-M. Bosqued, G. Kremser, and A. Korth (1982), A systematic study of ULF waves above F_{H^+} from GEOS 1 and 2 measurements and their relationships with proton ring distributions, *J. Geophys. Res.*, *87*(A8), 6219–6236, doi:10.1029/JA087iA08p06219.

Richardson, I. G., E. W. Cliver, and H. V. Cane (2001), Sources of geomagnetic storms for solar minimum and maximum conditions during 1972–2000, *Geophys. Res. Lett.*, *28*(13), 2569–2572, doi:10.1029/2001GL013052.

Russell, C. T., R. E. Holzer, and E. J. Smith (1970), OGO 3 observations of ELF noise in the magnetosphere, 2. The nature of the equatorial noise, *J. Geophys. Res.*, *75*(4), 755–768, doi:10. 1029/JA075i004p00755.

Santolík, O., and D. A. Gurnett (2003), Transverse dimensions of chorus in the source region, *Geophys. Res. Lett.*, *30*(2), 1031, doi:10.1029/2002GL016178.

Sckopke, N. (1966), A general relation between the energy of trapped particles and the disturbance field near the Earth, *J. Geophys. Res.*, *71*(13), 3125–3130.

Stern, D. P. (1975), The motion of a proton in the equatorial magnetosphere, *J. Geophys. Res.*, *80*(4), 595–599.

Summers, D., and R. M. Thorne (2003), Relativistic electron pitch-angle scattering by electromagnetic ion cyclotron waves during geomagnetic storms, *J. Geophys. Res.*, *108*(A4), 1143, doi:10. 1029/2002JA009489.

Summers, D., R. M. Thorne, and F. Xiao (1998), Relativistic theory of wave-particle resonant diffusion with application to electron acceleration in the magnetosphere, *J. Geophys. Res.*, *103*(A9), 20,487–20,500.

Thorne, R. M., and C. F. Kennel (1971), Relativistic electron precipitation during magnetic storm main phase, *J. Geophys. Res.*, *76*(19), 4446–4453.

Thorne, R. M., T. P. O'Brien, Y. Y. Shprits, D. Summers, and R. B. Horne (2005), Timescale for MeV electron microburst loss during geomagnetic storms, *J. Geophys. Res.*, *110*, A09202, doi:10. 1029/2004JA010882.

Tsurutani, B. T., and W. D. Gonzalez (1997), The interplanetary causes of magnetic storms: A review, in *Magnetic Storms, Geophys. Monogr. Ser.*, vol. 98, edited by B. T. Tsurutani et al., pp. 77–89, AGU, Washington, D. C., doi:10.1029/GM098p0077.

Tsyganenko, N. A. (1996), Effects of the solar wind conditions on the global magnetospheric configuration as deduced from data-based field models, *ESA SP-389*, p. 181, Eur. Space Agency, Paris, France.

Tsyganenko, N. A. (2002), A model of the near magnetosphere with a dawn-dusk asymmetry 1. Mathematical structure, *J. Geophys. Res.*, *107*(A8), 1179, doi:10.1029/2001JA000219.

Tsyganenko, N. A., and M. I. Sitnov (2005), Modeling the dynamics of the inner magnetosphere during strong geomagnetic storms, *J. Geophys. Res.*, *110*, A03208, doi:10.1029/2004JA010798.

Tsyganenko, N. A., H. J. Singer, and J. C. Kasper (2003), Storm-time distortion of the inner magnetosphere: How severe can it get?, *J. Geophys. Res.*, *108*(A5), 1209, doi:10.1029/2002JA009808.

Volland, H. (1973), A semiempirical model of large-scale magnetospheric electric fields, *J. Geophys. Res.*, *78*(1), 171–180.

Weimer, D. R. (2001), An improved model of ionospheric electric potentials including substorm perturbations and application to the Geospace Environment Modeling November 24, 1996, event, *J. Geophys. Res.*, *106*(A1), 407–416, doi:10.1029/2000JA000604.

Welling, D. T., V. K. Jordanova, S. G. Zaharia, A. Glocer, and G. Toth (2011), The effects of dynamic ionospheric outflow on the ring current, *J. Geophys. Res.*, *116*, A00J19, doi:10.1029/2010JA 015642.

Yu, Y., V. Jordanova, S. Zaharia, J. Koller, J. Zhang, and L. M. Kistler (2012), Validation study of the magnetically self-consistent inner magnetosphere model RAM-SCB, *J. Geophys. Res.*, *117*, A03222, doi:10.1029/2011JA017321.

Zaharia, S., V. K. Jordanova, M. F. Thomsen, and G. D. Reeves (2006), Self-consistent modeling of magnetic fields and plasmas in the inner magnetosphere: Application to a geomagnetic storm, *J. Geophys. Res.*, *111*, A11S14, doi:10.1029/2006JA011619.

Zaharia, S., V. K. Jordanova, D. Welling, and G. Tóth (2010), Self-consistent inner magnetosphere simulation driven by a global MHD model, *J. Geophys. Res.*, *115*, A12228, doi:10.1029/2010JA015915.

V. K. Jordanova, Los Alamos National Laboratory, P.O. Box 1663, MS D466, Los Alamos, NM 87545, USA. (vania@lanl.gov)

Ring Current Asymmetry and the Love-Gannon Relation

G. L. Siscoe

Center for Space Physics, Boston University, Boston, Massachusetts, USA

M.-C. Fok

Geospace Physics Laboratory, NASA Goddard Space Flight Center, Greenbelt, Maryland, USA

The Love-Gannon (LG) relation connects the ring current to the dawn-dusk asymmetry in the low-latitude geomagnetic disturbance field. Although such a connection had been noticed earlier, Love and Gannon's treatment makes the connection between the asymmetry and the ring current appear as direct as the connection between *Dst* and the ring current. Indeed, this is precisely how Love and Gannon expressed the relation: the asymmetry is proportional to *Dst*. This way of understanding the LG result, direct, unmediated coupling, does not fit well with classical ideas of magnetospheric dynamics. For it to do so requires that magnetic storms be steady state phenomena (which they are not) or that the shielding time in magnetosphere-ionosphere coupling theory be an order of magnitude longer than observed, which is also inadmissible. In this paper, we explain why this is so, and we suggest that what needs to be added to the standard way of treating magnetosphere-ionosphere coupling is ring current dissipation as a cause of field-aligned current.

1. INTRODUCTION

This paper concerns the ring current's relation to the dawn-dusk asymmetry in the geomagnetic field observed on the ground at low latitudes. The asymmetry we refer to is the canonical one in which the H component is more depressed on the duskside than on the dawnside. This subject is not new, of course, since the mentioned asymmetry has long been known, and for about equally as long, it has been ascribed to the presence of a partial ring current. What we offer that is new to the subject is a reconsideration of the origin of the partial ring current that causes the mentioned asymmetry ascribing it not to the dawn-dusk asymmetry inherent in the creation of the ring current under magnetospheric advection, the usual idea, but rather to the ring current decay mechanism that modifies the advectively created ring current.

2. WHAT IS THE LOVE-GANNON RELATION?

The subject enjoys renewed interest because in 2009, Love and Gannon drew attention to an underappreciated feature of the relation between the subject dawn-dusk asymmetry (henceforth δ_{DD}) and *Dst*, the nominal measure of the intensity of the symmetric ring current: they are proportional. Specifically, they found

$$\delta_{DD}(\text{nT}/R_E) = -0.2\ Dst(\text{nT}). \qquad (1)$$

The dimension of δ_{DD} is nT/R_E, meaning the difference between the dawn and dusk values of H in nT divided by their separation, 2 in units of R_E. The Love-Gannon (LG) relation holds statistically; it is based on 50 years of observations. At

Dynamics of the Earth's Radiation Belts and Inner Magnetosphere
Geophysical Monograph Series 199
10.1029/2012GM001350

any time, departure from the LG relation can be large. Figure 1 illustrates both statements graphically. The left panel gives a 50 year average of the local time variation of the disturbance of the H component of the geomagnetic field (call it ΔH) normalized to the simultaneous determined Dst. Thus, the circle marked 1.0 corresponds to the ΔH field equaling Dst on a 50 year average. Note that this condition occurs very close to noon and midnight local time. That is, on a long time average, the ratio of ΔH to Dst at noon and midnight very nearly equals 1. But at dusk, the average value of that ratio is about 1.2 and, at dawn, about 0.8. Very nearly, the curve in the left panel is a circle of unit radius offset toward dusk by 0.2 units. That the radius is unity is no surprise since Dst is the average of the local time values of ΔH. However, the offset, its size and direction, and the circular shape of the contour are the significant points that Love and Gannon exposed and that need to be interpreted.

We note that *Newell and Gjerloev* [2012] find a similar but not identical result. A superposed epoch analysis of the storm-time profiles of 125 storms shows noon and midnight profiles nearly coinciding and roughly bisecting the space between the dawn and dusk profiles, which agrees with the LG noon-midnight finding. But the ΔH to Dst ratio seems bigger during the main phase than during the recovery phase. See their Figure 7, bottom. Nonetheless, the care that LG took to isolate the local time signal for all levels of Dst (see below in connection with *Weygand and McPherron* [2006]) justifies an attempt to find a cause for their result.

The right panel of Figure 1 simply illustrates the statement that at any given time, the actual local time variation of the ratio of ΔH to Dst can be far from the statistical, offset circle seen in the left panel. Thus, one can think about the local time behavior of the ΔH to Dst ratio by imagining the offset circle in Figure 1a to be an elastic band that is almost always being stretched and shifted away from the average shape by many processes, some of which are well known. For example, shock wave compression of the magnetosphere tends to shift the circle sunward (bigger ΔH on the dayside); both substorms and region 1 currents tend to shift the circle tailward (bigger ΔH on the nightside).

Overriding or underlying all these fluctuations, the transient and counteracting nature of which reduces their long-term average signals effectively to zero, is some unidirectional, ever-present process that acts to shift the center with respect to which the mentioned processes and others do their distorting and shifting work 0.2 units, the LG factor, duskward. Somehow, this process must be integrally connected with the ring current, must be part of its dynamical nature, because the offset is proportional to Dst. Just as Dst is a measure of ring current energy, δ_{DD} must be a measure of some related, but as yet, unknown dynamical property of the ring current. Said differently, some fundamental dynamical processes tied to the structure of the ring current set the asymmetric geometrical framework within which the various transient fluctuations in ΔH make their impressions like winds swaying the branches of a tree that is rooted 0.2 units duskward of the

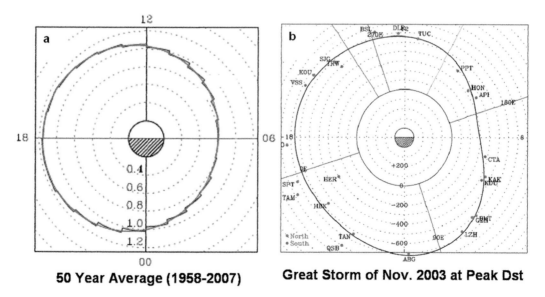

50 Year Average (1958-2007)   **Great Storm of Nov. 2003 at Peak Dst**

Figure 1. (a) Fifty year average of local time, H disturbance normalized to simultaneous Dst. The smooth line is a harmonic fit to the 24 separate, 1 h local time averages. From *Love and Gannon* [2009]. (b) A local time plot of the trace of the low-latitude disturbance field at the time of the peak Dst disturbance (-422 nT) of the great storm of November 2003. From *Love and Gannon* [2010].

center. The question arises: What is it in the structure of the ring current that organizes its contribution to the geomagnetic disturbance field in the fixed shape of a duskward displaced circle? Here we offer a suggestion for the mechanism that lies behind the LG asymmetry.

3. WHAT IS NEW?

A question that by now has probably occurred to the reader should be answered. Why are we treating the dawn-dusk asymmetry in the disturbance field as something new when it has been known since the 1960s and commented on frequently since then (even as recently as at the time of writing this paper by *Newell and Gjerloev* [2012])? What is new is the recognition of its significance for "classical" magnetospheric physics.

A question about the significance of something similar to the LG relation was brought up earlier by *Weygand and McPherron* [2006] in a study that also found the asymmetry to be proportional to *Dst* (albeit the asymmetry was defined as the local time range of *H* rather than as the dawn-dusk difference of *H* used by Love and Gannon). Weygand and McPherron regarded their result to be unphysical because it disagreed with what was expected from the classical model of magnetic storms as developed in a study of 74 storms by *Sugiura and Chapman* [1960]. According to it, the asymmetry should not track *Dst* but should dominate the early part of the storm and *Dst* the latter part. Weygand and McPherron found that they could recover the expected behavior of the relation between asymmetry and *Dst* by offsetting the zeros of the indices by about 20 nT in opposite directions. Then, the asymmetry-*Dst* tracking property, so conspicuous in the "uncorrected" data, went away. However, Love and Gannon meticulously cleansed their data with the precise intent to eliminate offsets, and their deliberate "clean room" care of the data leaves the underlying asymmetry-*Dst* proportionality of Figure 1a robustly present. More precisely, Weygand and McPherron used ratios that normalized out any measure of storm size (such as *Dst*), thus amplifying the effects of small nonzero residuals in *Dst*, which can affect ratios for small levels of disturbance, whereas Love and Gannon, rather than forming ratios, approach the analysis by fitting proportionality factors. This minimizes the impact of nonzero residuals and gives factors that are sensitive to the full range of disturbance (not just small levels).

Moreover, Figure 2 adds to the evidence that the standard expectation that asymmetry disappears during the recovery phase has a problem. Here are nine storms for which 1 min values of an asymmetry index are plotted against a symmetry index. In this case, the indices are determined by the Kyoto World Data Center. The symmetry index is basically *Dst*.

The asymmetry index is the same as Weygand and McPherron's, that is, the difference between the maximum and minimum local time values of *H*. We divide the value by 2 to give it the same units as δ_{DD} (as if the max and min always occurred on opposite sides of the earth, which is not true, of course).

The straight line in each plot of Figure 2 is the LG relation, equation (1). The data points in each plot arrange themselves into a high-asymmetry group and a low-asymmetry group, both of which span the full range of *Dst*. In each case, the low-asymmetry group tends to lie along the LG line. Now, the interesting thing is that low-asymmetry groups all correspond to the recovery phase of the storms. The high-asymmetry groups correspond to the main phase. Note that this is a third, independently computed data set that shows a correlation between asymmetry and *Dst* lasting throughout the storm. In this case, the LG relation is evident in the recovery phase points. The main phase points display the expected greater degree of asymmetry, which, because the asymmetry index here is defined as a local time range instead of dawn-dusk difference, probably owe their higher-than-LG values to strong noon-midnight contributions, not present during the recovery phase.

We take the convergence of evidence of a phenomenon that disagrees with expectations to be an opportunity to find something new about magnetospheric dynamics. Thus, we proceed under the assumption that, in fact, the LG relation describes the phenomena as it is. Then, the problem is to find a mechanism that ties the dawn-dusk asymmetry more-or-less fixedly to *Dst*.

4. STANDARD IDEAS AND WHY THEY DO NOT WORK

To begin the task, note first that the mere presence of more ring current pressures on the duskside than on the dawnside of the magnetosphere, the natural situation for the developing ring current, does not, by itself, guarantee an asymmetry in the geomagnetic disturbance field. This is because an asymmetry implies a force on the geomagnetic dipole that must be balanced by a force on the mass through which the current flows that produces the asymmetry, and the mass of the ring current is too small to sustain the force associated with the LG asymmetry [*Siscoe et al.*, 2012]. The force problem is solved by assuming that the ring current balances excess current on the duskside by shunting it along magnetic field lines where it closes through the ionosphere, which, because of its collisional coupling to the thermosphere, has sufficient effective mass to sustain the force. *Cummings* [1966] showed that an electrical circuit having the geometry just described, indeed, produces an asymmetry with the

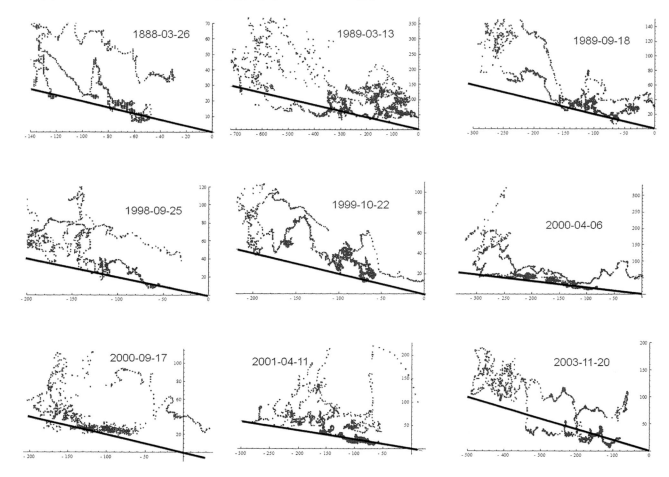

Figure 2. Nine magnetic storms as seen in plots of asymmetry (ordinate) versus *Dst* (abscissa). These are the ASY-H and SYM-H indices produced by the Kyoto World Data Center. The ASY-H values are divided by 2 to make them compatible with δ_{DD}. The main phase of each storm produces the high ASY-H group of points and the recovery phase, the low group. The solid lines show the Love-Gannon relation, equation (1).

observed properties. *Crooker and Siscoe* [1981] showed that most of the asymmetry comes from the field-aligned part of the current. They give the following formula for the asymmetry

$$\delta_{DD} = (\mu_o/4\,R_E)I_B\cos\lambda$$
$$= (\mu_o/4\,R_E)\Sigma_H\Phi\cos\lambda \approx 0.017\Sigma_H\Phi(\text{kV}), \quad (2)$$

where Σ_H is the auroral zone Hall conductance, and λ is the latitude at which the field-aligned currents (a.k.a. Birkeland currents) of strength I_B connected to the ring current enter and leave the ionosphere (I_B in each north-south hemispheres). The last term in equation (2) assumes that λ is 70°. The transpolar potential Φ measures the speed at which magnetic flux cycles around the magnetosphere and, hence,

also the flow of ring current particle through the magnetosphere. The asymmetry amplitude, δ_{DD}, is the dawn-dusk component of the gradient in the magnetic field. The relation between I_B and Φ in equation (2) is obtained from the magnetosphere-ionosphere (MI) coupling theory of *Vasyliunas* [1970, 1972] in which there is perfect shielding between auroral and subauroral ionospheres. In its operational version as subsequently elaborated by Wolf and coworkers, it forms the conceptual structure of the Rice Convection Model. But even in the simplest, idealized form represented by equation (1), it exposes a problem that the LG relation poses for the standard MI coupling theory. In equation (2), δ_{DD} is proportional to Φ, and in the LG relation, equation (1), it is proportional to *Dst*, which implies that *Dst* should be proportional to Φ. But the canonical dependence of *Dst* on Φ is, instead, the Burton-McPherron-Russell (BMR) relation [*Burton et al.*, 1975]

$$dDst/dt = -a\Phi - Dst/\tau_{sym}, \qquad (3)$$

where a and τ_{sym} are empirically determined positive constants. (Actually, instead of Φ, the BMR relation has VB_z, where V is the solar wind speed, B_s is the southward component of the IMF, zero if northward. But VB_z can be replaced by Φ through empirical formulas, e.g., 50 kV mV^{-1} m^{-1} [*Reiff and Luhmann*, 1986].) If we take the LG relation to be the empirical fact to which the other relations must conform, then either we must assume that under the circumstances when the LG relation holds, Dst has always reached a steady state value so that $dDst/dt = 0$, and equation (3) reduces to an equation like equation (2), or we must find a BMR-like relation between δ_{DD} and Φ to replace equation (2) with an equation like equation (3). In the first alternative, we can eliminate Φ between equations (2) and (3) to find

$$\begin{aligned} \delta_{DD}(\text{nT}/R_E) &= 0.017\Sigma_H Dst/a\tau_{sym} \\ &= -0.02\Sigma_H Dst(\text{nT}), \qquad (4) \end{aligned}$$

where we have used BMR values for a and τ_{sym}. Then, if we take Σ_H to be 10 S, a not unreasonable value for the auroral zones, we recover equation (1), the LG relation, precisely. So, does this not solve the problem? Unfortunately, no because during magnetic storms, when the LG relation is very much in effect (e.g., Figure 2), the $dDst/dt$ term in equation (3) is not small compared to the terms on the right hand side. For example, typically, Dst reaches its maximum storm-time depression in about 10 h, which is comparable to the value of the decay time, τ [e.g., *Weygand and McPherron*, 2006]. Thus, the assumption of steady state is invalid, and we must turn to option 2: find a BMR-like relation between δ_{DD} and Φ to replace equation (2) with an equation like equation (3).

5. A SUGGESTED SOLUTION

Let us cut the Gordian knot by simply writing down the desired result

$$d\delta_{DD}/dt = 0.017\Sigma\Phi/\tau_{asy} - \delta_{DD}/\tau_{asy}. \qquad (5)$$

This reduces to equation (2) in steady state but for the asymmetry index, δ_{DD}, it gives a time-dependent analog of the BMR relation for Dst. Here τ_{asy} is the timescale appropriate for asymmetry. *Siscoe et al.* [2012] derive equation (5) from MI coupling theory [*Vasyliunas*, 1970, 1972; *Siscoe*, 1982]. In this case, τ_{asy} turns out to be the shielding time, that is, the time ring current particles need to readjust to a change

in applied transpolar potential to reestablish the equipotential condition that in steady state characterizes the equatorial border of the region 2 current ring. The problem is that the shielding time scale is an order of magnitude smaller than the ring current time scale, τ_{sym} (1 h or less versus 10 h or more). Hence, Siscoe et al. conclude that MI coupling theory, which ties δ_{DD} to Dst through the applied potential, Φ, cannot account for the LG relation. They further conclude that the LG relation implies that $\tau_{asy} = \tau_{sym}$, which implies that δ_{DD} results directly from the ring current with no intermediary, such as Φ. That is, like Dst, δ_{DD} is a direct measure of some property of the ring current. They suggest that the property by which the ring current generates δ_{DD} is dissipative decay (e.g., charge exchange and precipitation loss) that distorts ring current pressure contours thereby disrupting normal current flow paths in the magnetosphere causing the current to close through the ionosphere, in at noon and out at midnight as needed to give a dawn-dusk asymmetry. This proposed solution to the LG problem has the virtue of operating during all phases of magnetic storms, especially during the recovery phase when the LG asymmetry is still evident. It also makes a prediction about the local time structure of the dissipative decay process.

To make an initial test of the idea that the ring current decay process is causing the LG asymmetry, we used the Fok Ring Current Model at the Community Coordinated Modeling Center (CCMC) [*Fok et al.*, 1995; *Fok and Moore*, 1997; *Fok et al.*, 2001], which has the desired feature of incorporating ring current dissipation processes, charge exchange, and loss-cone losses. The model was used to simulate the field-aligned currents for the storm of 10–11 April 1997. Figure 3 shows the computed region 2 currents at the peak of the storm in Dst. The important aspect to

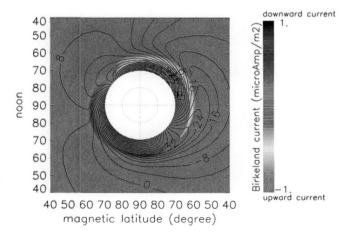

Figure 3. Region 2 field-aligned currents computed by the Fok ring current model at CCMC for the peak of the storm on 11 April 1997.

notice is the clockwise rotation of the current system relative to the noon-midnight axis. By contrast, region 1 currents are more symmetrical relative to the noon-midnight axis [*Iijima and Potemra*, 1978]. It is this rotation of the region 2 current system relative to the region 1 current system that can give a net flow of current in at noon and out at midnight (instead of all region 2 current crossing latitudinally to feed the region 1 ring) and, thus, produce a dawn-dusk asymmetry in *H* [*Crooker and Siscoe*, 1981]. Thus, to the extent that a model that incorporates ring current dissipation produces a rotation of the current system that connects the ring current to the ionosphere (i.e., the region 2 system) of an amount and in the right sense to generate an asymmetry of the LG type, it suggests that ring current dissipation might be the ring current property responsible for δ_{DD}. Then, we would have *Dst* measuring ring current energy and δ_{DD} measuring its rate of loss.

Acknowledgments. This work is supported, in part, by the National Science Foundation under grant ATM-0809307 and by NASA Science Mission Directorate, Heliophysics Division, Living With a Star Targeted Research and Technology Program.

REFERENCES

Burton, R. K., R. L. McPherron, and C. T. Russell (1975), The terrestrial magnetosphere: A half-wave rectifier of the interplanetary electric field, *Science*, *189*, 717–718.

Crooker, N. U., and G. L. Siscoe (1981), Birkeland currents as the cause of the low-latitude asymmetric disturbance field, *J. Geophys. Res.*, *86*(A13), 11,201–11,210.

Cummings, W. D. (1966), Asymmetric ring currents and the low-latitude disturbance daily variation, *J. Geophys. Res.*, *71*(19), 4495–4503.

Fok, M.-C., and T. E. Moore (1997), Ring current modeling in a realistic magnetic field configuration, *Geophys. Res. Lett.*, *24*(14), 1775–1778.

Fok, M.-C., T. E. Moore, J. U. Kozyra, G. C. Ho, and D. C. Hamilton (1995), Three-dimensional ring current decay model, *J. Geophys. Res.*, *100*(A6), 9619–9632.

Fok, M.-C., R. A. Wolf, R. W. Spiro, and T. E. Moore (2001), Comprehensive computational model of Earth's ring current, *J. Geophys. Res.*, *106*(A5), 8417–8424.

Iijima, T., and T. A. Potemra (1978), Large-scale characteristics of field-aligned currents associated with substorms, *J. Geophys. Res.*, *83*(A2), 599–615.

Love, J. J., and J. L. Gannon (2009), Revised Dst and the epicycles of magnetic disturbance: 1958–2007, *Ann. Geophys.*, *27*, 3101–3131.

Love, J. J., and J. L. Gannon (2010), Movie-maps of low-latitude magnetic storm disturbance, *Space Weather*, *8*, S06001, doi:10.1029/2009SW000518.

Newell, P. T., and J. W. Gjerloev (2012), SuperMAG-based partial ring current indices, *J. Geophys. Res.*, *117*, A05215, doi:10.1029/2012JA017586.

Reiff, P. H., and J. G. Luhmann (1986), Solar wind control of the polar-cap voltage, in *Solar Wind – Magnetosphere Coupling*, edited by Y. Kamide and J. A. Slavin, pp. 453–476, Terra Sci., Tokyo, Japan.

Siscoe, G. L. (1982), Energy coupling between regions 1 and 2 Birkeland current systems, *J. Geophys. Res.*, *87*(A7), 5124–5130.

Siscoe, G. L., J. J. Love, and J. L. Gannon (2012), Problem of the Love-Gannon relation between the asymmetric disturbance field and *Dst*, *J. Geophys. Res.*, *117*, A09216, doi:10.1029/2012JA017879.

Sugiura, M., and S. Chapman (1960), The average morphology of geomagnetic storms with sudden commencement, *Abhandl. Akad. Wiss. Goettingen Math. Physik. Kl.*, *4*, 51–53.

Vasyliunas, V. M. (1970), Mathematical models of magnetospheric convection and its coupling to the ionosphere, in *Particles and Fields in the Magnetosphere*, edited by B. M. McCormac, pp. 60–71, D. Reidel, Dordrecht, The Netherlands.

Vasyliunas, V. M. (1972), The interrelationship of magnetospheric processes, in *Earth's Magnetospheric Processes*, edited by B. M. McCormac, pp. 29–38, D. Reidel, Dordrecht, The Netherlands.

Weygand, J. M., and R. L. McPherron (2006), Dependence of ring current asymmetry on storm phase, *J. Geophys. Res.*, *111*, A11221, doi:10.1029/2006JA011808.

M.-C. Fok, Geospace Physics Laboratory, NASA Goddard Space Flight Center, Greenbelt, MD 20771, USA.

G. L. Siscoe, Center for Space Physics, Boston University, Boston, MA 02215, USA. (siscoe@bu.edu)

The Importance of the Plasmasphere Boundary Layer for Understanding Inner Magnetosphere Dynamics

Mark B. Moldwin and Shasha Zou

Department of Atmospheric, Oceanic and Space Sciences, University of Michigan, Ann Arbor, Michigan, USA

In the last decade, the role of the plasmapause in clearly separating different wave and particle environments in the inner magnetosphere has been demonstrated. In addition, a new conceptual framework has emerged that emphasizes that the plasmapause is most often highly structured and dynamic and more appropriately should be considered the plasmasphere boundary layer (PBL). The complex and dynamic PBL corresponds to clear density gradients observed in the ionosphere and overlaps the inner edge of the outer radiation belts. The local plasmapause acts as a clear ULF and higher-frequency plasma wave dividing line as well. This short review highlights recent observational studies describing the important role the PBL plays in separating a wide range of wave and particle populations and focuses on studies of the role of the PBL in modulating ULF waves as observed in space and on the ground.

1. INTRODUCTION

With the launch of the IMAGE satellite in 2000 and the capability of the EUV Imager Camera to take global images of the plasmasphere [e.g., *Sandel et al.*, 2001] (Figure 1), there has been renewed interest in the role of the plasmapause in organizing inner magnetospheric wave and plasma populations. For the most part, earlier studies used single point observations that clearly captured the local plasmapause structure and were often well explained by the classic MHD plasmapause picture [e.g., *Nishida*, 1966]. With the advent of global and multiinstrument and multipoint observations, significant shortcomings to the classic plasmapause model have been emphasized. The importance and complexity of the processes that occur at the plasmapause, as well as the observational disconnect from the classical zeroth-order picture of the plasmapause being well explained by ideal MHD, inspired the development of a new conceptual frame-work called the "plasmasphere boundary layer (PBL)" [*Carpenter and Lemaire*, 2004]. This conceptual picture emphasizes the physics of shear flows, sharp pressure gradients, field-aligned currents, and transient and localized electric fields that arise at the plasmasphere boundary. The classic MHD plasmapause model neglects the connection between the inner magnetosphere and ionosphere and, hence, has hindered our understanding and modeling efforts. Throughout the review, we will most often use PBL instead of the term plasmapause to emphasize the dynamics and structure of the boundary.

We know that there is a very close correspondence between the location of the PBL and other sharp boundaries in wave and particle locations including in the ionosphere [e.g., *Yizengaw et al.*, 2005], the inner edge of the outer radiation belt [e.g., *Li et al.*, 2006], the ULF waves [e.g., *Allan et al.*, 1986], and the higher-frequency plasma waves [e.g., *Summers et al.*, 2008; *Masson et al.*, 2009].

The PBL plays a very important role with regard to plasma waves, and considerable recent research has gone into understanding electromagnetic ion cyclotron, chorus, plasma-spheric hiss, and other high-frequency waves [e.g., *Golden et al.*, 2010; *Kalaee et al.*, 2010; *Pickett et al.*, 2010; *Usanova et al.*, 2010; *Agapitov et al.*, 2011; *Bortnik et al.*, 2011;

Dynamics of the Earth's Radiation Belts and Inner Magnetosphere
Geophysical Monograph Series 199

10.1029/2012GM001323

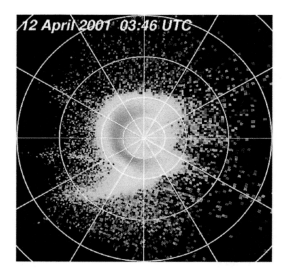

Figure 1. IMAGE EUV observations of the plasmasphere showing the complex and structured nature of the plasmapause including a clear drainage plume. From *Foster et al.* [2004].

Grimald et al., 2011; *K. Liu et al.*, 2011; *Wang et al.*, 2011]. Observational and theoretical studies of the role of the PBL in directly or indirectly modulating radiation belt particles has also been of considerable recent interest [e.g., *Chu et al.*, 2010; *Johnston and Anderson*, 2010; *Lam et al.*, 2010; *Miyoshi and Kataoka*, 2011; *Shprits et al.*, 2011, 2012].

In this short review, we synthesize the last 2 years (2010 to the first part of 2012) of research that has demonstrated the importance of the PBL in separating different plasma and wave environments and highlights the need to think of the plasmapause as a more complex boundary layer instead of a simple sharp density gradient. In particular, we emphasize recent studies of ionospheric signatures of the PBL and ULF waves modulated by the PBL. Though the present review does not focus on the detailed physics within the PBL itself, we provide a discussion on how the PBL concept can and should be used to understand density structure and ULF wave signatures at the plasmapause that are not easily explained using the traditional MHD plasmapause picture. A recent review by *Singh et al.* [2011] provides an excellent overview of processes specific to the plasmasphere itself such as refilling and plasma density structure and dynamics.

2. PBL AND IONOSPHERIC BOUNDARIES

Several recent studies have demonstrated the very close correspondence of the ionospheric midlatitude trough and the plasmapause. Though this relationship has been known for decades [e.g., *Grebowsky et al.*, 1976], only recently have global imaging, multisatellite, and multiinstrument studies

clearly showed this relationship and its behavior with local time (LT) and geomagnetic activity [e.g., *Obana et al.*, 2010].

Pedatella and Larson [2010] used the COSMIC constellation of Low-Earth Orbiting (LEO) satellites and the upward looking dual-frequency receivers to routinely identify the ionospheric midlatitude plasma trough. Following up the work of *Anderson et al.* [2008], *Pedatella and Larson* [2010] found that sharp gradients in total electron content (TEC) defined the ionospheric midlatitude trough over a wide variety of LTs and geomagnetic conditions. Figure 2 shows a vertical TEC profile from a single LEO satellite pass that clearly shows the trough region. Since there are a number of LEO satellites with upward looking dual-frequency GPS receivers, the study indicates that multipoint, relatively high time resolution determination of the PBL is possible. It also showed that the location of the trough follows general trends with enhanced convection, but that there is significant variability on the location of the trough as a function of LT and geomagnetic activity. This variability is suggestive of an azimuthally structured plasmapause boundary and is difficult to explain with slowly varying convective electric fields and a quasicircular last closed equipotential (MHD) picture.

Using both ground and space-based observations, *Zou et al.* [2011] examined the dynamics of the midlatitude trough region during substorms, suggesting that some of the changing structure within the PBL is driven by substorm-related phenomenon. Since structured substorm electric fields often have narrow LT extents on the nightside, and also since substorms occur on average many times per day, the results suggest that substorms could be responsible for much of the azimuthal and temporal structure observed in the PBL. Taken together with considerable work by *Goldstein et al.* [2002, 2003a, 2003b], among others [e.g., *Foster et al.*, 2002;

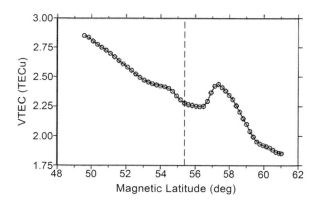

Figure 2. The vertical total electron content determined from a single COSMIC satellite pass with the location of the trough indicated by the vertical line. From *Pedatella and Larson* [2010].

Spasojević et al., 2003], the picture of a dynamic and highly structured plasmapause emerges emphasizing the need for the conceptual framework switch to PBL.

The recent work shows the clear connection between the topside ionosphere and PBL, and the role of both ionospheric and inner magnetospheric drivers on imposing structure within the PBL. This connection enables remote studies of the inner magnetosphere by using ground-based ionospheric observations such as radars [e.g., *Ribeiro et al.*, 2012], GPS TEC, and ground magnetometers to probe the PBL. The next section highlights several recent studies that used combinations of ground and space-based observations to understanding the source, propagation, and mode conversion of ULF waves across the PBL.

3. PBL AND ULF WAVES

Because of the often highly structured large-density gradients observed in the PBL, there are sharp Alfvén wave velocity gradients that greatly influence ULF wave propagation through the region [e.g., *Kelley et al.*, 2012]. As understanding of the role that ULF waves (especially at the low-frequency Pc5 band) have in the energization of radiation belts [e.g., *Elkington et al.*, 1999] increases, understanding the impact of the PBL on ULF waves (and on ULF waves on the PBL) increases in importance.

Hartinger et al. [2010] clearly demonstrated that the PBL plays a key role in damping ULF waves as they propagate from the trough into the plasmasphere, and the damping factor is LT dependent. Figure 3 shows the change in wave amplitude as the CRRES spacecraft travels across the PBL. This study showed that estimates of ULF Pc5 wave power that do not take into account the location where the observation is made (either inside or outside the plasmasphere) will greatly underestimate the power spectral density (PSD) of the ULF waves in the trough region and thus affect the radial diffusion coefficient calculation. Various empirical averages of ULF PSD are used in radiation belt diffusion modeling

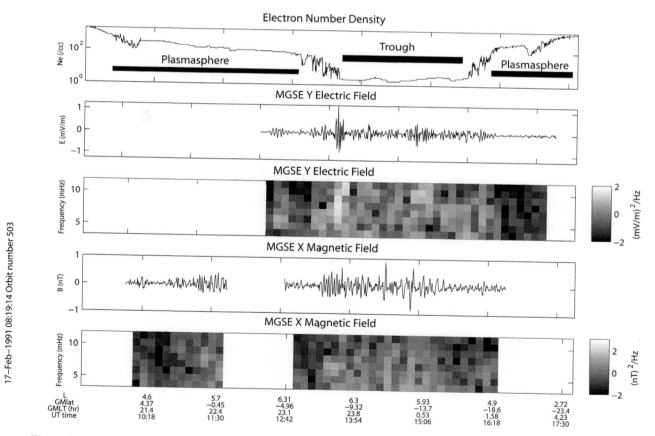

Figure 3. An example of CRRES data from 17 February 1991 showing (top to bottom) electron density with plasmasphere and trough intervals indicated, modified geocentric solar ecliptic (MGSE) *y* detrended electric field followed by corresponding dynamic power spectrum, MGSE *x* detrended magnetic field followed by corresponding dynamic power spectrum.

[e.g., *Huang et al.*, 2010], and therefore, it is essential to incorporate the local observations into the plasmaspheric context of the MHD model.

Kawano et al. [2011] used four Cluster satellites and multiple-ground magnetometer data to examine the propagation of a Pi2 wave in the inner magnetosphere. They found a clear phase shift pattern across the PBL consistent with the wave propagating from the outer magnetosphere at higher L shell, across the PBL and into the plasmasphere. Conversely, *Luo et al.* [2011] analyzed multiple Time History of Events and Macroscale Interactions during Substorms (THEMIS) satellites and ground magnetometer data to understand low-latitude Pi2 and found that observations inside and outside the plasmasphere are consistent with the picture that trapped fast-mode waves inside the plasmasphere are the source of low-latitude Pi2. Hence, the PBL can act as a cavity trapping ULF fast-mode waves inside as well as a damping or reflecting boundary for waves propagating into the inner magnetosphere. *Frissell et al.* [2011], using ground-based radars, magnetometers, and THEMIS satellite data, identified the source of Pi2 waves observed at the PBL as due to bursty bulk flows braking against the dipole field region in the inner magnetosphere.

Further evidence of instabilities due to shear flow within the PBL driving ULF waves was provided by *Henderson et al.* [2010]. They observed a giant auroral undulation event that was consistent with shear flow due to high-speed subauroral polarization stream flow at the plasmapause (see also *Kim et al.* [2010] and *Liléo et al.* [2010]). The giant undulation was associated with long period (Pc5) pulsations observed on the ground.

W. Liu et al. [2011] examined Pc5 waves on the dayside and found evidence of a plasmapause surface wave resonantly coupling to an adjacent field-line. This ULF wave power is confined to the PBL region and is argued to not be due to an external wave source. Dynamics of the PBL, hence, can give rise to localized ULF wave power. In addition, the study found that externally driven ULF waves outside the plasmasphere are damped by plasmaspheric dynamics consistent with observations of *Hartinger et al.* [2010] and *Kawano et al.* [2011].

Pc3–4 waves are predominantly observed on the dayside. They are due to upstream waves produced in the Earth's foreshock propagating into the magnetosphere. *Ponomarenko et al.* [2010] presented evidence that nightside Pc3–4 waves are also driven by upstream waves that propagate to the nightside in the closed field line region within the PBL. They used a combination of radar data and magnetometer data and observed similar frequency oscillations (Figure 4). They suggest that a resonance near the PBL helps amplify part of the upstream wave spectrum and that a cavity mode is responsible for the uniform frequency. Similarly, *Takahashi et al.* [2010] using THEMIS multipoint observations of

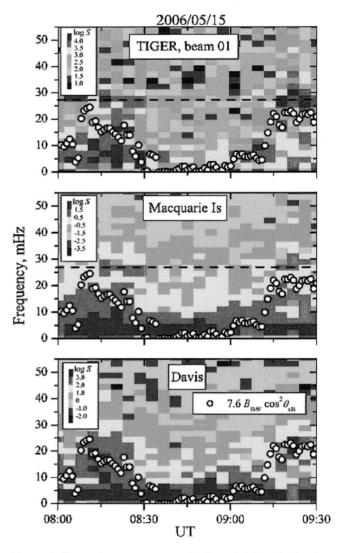

Figure 4. Dynamic power spectra for line-of-sight velocity data (top) for TIGER radar and trace spectra for ground magnetometer data from (middle) Macquarie Island and (bottom) Davis. The white circles represent theoretically predicted values of f_{UW} based on the work by *Takahashi et al.* [1984]. From *Ponomarenko et al.* [2010].

dayside Pc3–4 waves unambiguously presented observations of trapped fast-mode waves within the plasmasphere consistent with a cavity mode or virtual resonance [e.g., *Teramoto et al.*, 2011].

These studies have shown that ULF wave propagation is clearly modulated by the PBL. Pc 5 and Pi 2 waves have clear power spectral density changes and phase shifts across the boundary. This creates a relatively sharp gradient of ULF wave power from inside to outside the plasmasphere. In addition, plasmapause surface waves can resonantly couple to closely adjacent field lines creating enhanced ULF wave power at the PBL. Fast mode waves can also be trapped inside the plasmasphere cavity. The combination of damping, fast to Alfvén mode coupling via field line resonance, and surface waves and cavity modes make complicated ULF wave profiles as a function of L across the PBL and highlights the essential role that the PBL plays in organizing ULF wavefields. The role of shear flow in generating Pi2 [*Frissell et al.*, 2011] and giant undulations in the aurora at the plasmasphere boundary [*Henderson et al.*, 2010] again emphasizes the PBL conceptual picture.

4. SUMMARY AND CONCLUSIONS

Recent work has clearly demonstrated that the simple conceptual picture of the plasmapause being a nearly circular sharp density gradient has hindered our understanding of its important role in organizing and driving inner magnetospheric structure and dynamics. The plasmasphere boundary layer (PBL) concept more accurately captures the complexity of the plasmapause in both L shell and azimuth and highlights the important role that local dynamics of the PBL play. The PBL modulates ULF and plasma waves, which in turn modulate the higher-energy particle populations in the inner magnetosphere. The coupling to the ionosphere enables remote sensing of inner magnetospheric dynamics at the PBL using imaging, radar, GPS TEC, and ground magnetometers. In combination with multiple space-based observations, several studies have been able to unambiguously identify ULF wave sources and modes that now enable more realistic modeling of the coupled inner magnetosphere.

The next steps for the community will be coupling of global magnetospheric models, with inner magnetospheric ring current and radiation belt models and realistic ionospheric/plasmaspheric models to capture the important feedback of ion outflow, ionospheric conductivity, field-aligned currents, and the role of the PBL on organizing and modulating wave activity. Being able to capture the boundary layer physics is essential for understanding the highly interconnected fields and flows of the inner magnetosphere.

Acknowledgments. This work was supported by NASA LWS grant (NNH09ZDA001N-LWST), NASA Heliophysics Theory Program grant (NNX11AO60G), NASA Geospace grant (NNX09AI62G), and NSF grant (AGS 1111476).

REFERENCES

Agapitov, O., V. Krasnoselskikh, T. Dudok de Wit, Y. Khotyaintsev, J. S. Pickett, O. Santolík, and G. Rolland (2011), Multispacecraft observations of chorus emissions as a tool for the plasma density fluctuations' remote sensing, *J. Geophys. Res.*, *116*, A09222, doi:10.1029/2011JA016540.

Allan, W., E. M. Poulter, and S. P. White (1986), Hydromagnetic waves coupling in the magnetosphere-plasmapause effects on impulse-excited resonances, *Planet. Space Sci.*, *34*(12), 1189–1220.

Anderson, P. C., W. R. Johnston, and J. Goldstein (2008), Observations of the ionospheric projection of the plasmapause, *Geophys. Res. Lett.*, *35*, L15110, doi:10.1029/2008GL033978.

Bortnik, J., L. Chen, W. Li, R. M. Thorne, and R. B. Horne (2011), Modeling the evolution of chorus waves into plasmaspheric hiss, *J. Geophys. Res.*, *116*, A08221, doi:10.1029/2011JA016499.

Carpenter, D. L., and J. Lemaire (2004), The plasmasphere boundary layer, *Ann. Geophys.*, *22*, 4291–4298, doi:10.5194/angeo-22-4291-2004.

Chu, F., M. K. Hudson, P. Haines, and Y. Shprits (2010), Dynamic modeling of radiation belt electrons by radial diffusion simulation for a 2 month interval following the 24 March 1991 storm injection, *J. Geophys. Res.*, *115*, A03210, doi:10.1029/2009JA014409.

Elkington, S. R., M. K. Hudson, and A. A. Chan (1999), Acceleration of relativistic electrons via drift-resonant interaction with toroidal-mode Pc-5 ULF oscillations, *Geophys. Res. Lett.*, *26*(21), 3273–3276, doi:10.1029/1999GL003659.

Foster, J. C., P. J. Erickson, A. J. Coster, J. Goldstein, and F. J. Rich (2002), Ionospheric signatures of plasmaspheric tails, *Geophys. Res. Lett.*, *29*(13), 1623, doi:10.1029/2002GL015067.

Foster, J. C., A. J. Coster, P. J. Erickson, F. J. Rich, and B. R. Sandel (2004), Stormtime observations of the flux of plasmaspheric ions to the dayside cusp/magnetopause, *Geophys. Res. Lett.*, *31*, L08809, doi:10.1029/2004GL020082.

Frissell, N. A., J. B. H. Baker, J. M. Ruohoniemi, L. B. N. Clausen, Z. C. Kale, I. J. Rae, L. Kepko, K. Oksavik, R. A. Greenwald, and M. L. West (2011), First radar observations in the vicinity of the plasmapause of pulsed ionospheric flows generated by bursty bulk flows, *Geophys. Res. Lett.*, *38*, L01103, doi:10.1029/2010GL045857.

Golden, D. I., M. Spasojević, F. R. Foust, N. G. Lehtinen, N. P. Meredith, and U. S. Inan (2010), Role of the plasmapause in dictating the ground accessibility of ELF/VLF chorus, *J. Geophys. Res.*, *115*, A11211, doi:10.1029/2010JA015955.

Goldstein, J., R. W. Spiro, P. H. Reiff, R. A. Wolf, B. R. Sandel, J. W. Freeman, and R. L. Lambour (2002), IMF-driven overshielding electric field and the origin of the plasmaspheric shoulder of May 24, 2000, *Geophys. Res. Lett.*, *29*(16), 1819, doi:10.1029/2001GL014534.

Goldstein, J., B. R. Sandel, W. T. Forrester, and P. H. Reiff (2003a), IMF-driven plasmasphere erosion of 10 July 2000, *Geophys. Res. Lett.*, *30*(3), 1146, doi:10.1029/2002GL016478.

Goldstein, J., B. R. Sandel, M. R. Hairston, and P. H. Reiff (2003b), Control of plasmaspheric dynamics by both convection and subauroral polarization stream, *Geophys. Res. Lett.*, *30*(24), 2243, doi:10.1029/2003GL018390.

Grebowsky, J. M., N. C. Maynard, Y. K. Tulunay, and L. J. Lanzer-otti (1976), Coincident observations of ionospheric troughs and the equatorial plasmapause, *Planet. Space Sci.*, *24*(12), 1177–1185.

Grimald, S., F. El-Lemdani-Mazouz, C. Foullon, P. M. E. Décréau, S. A. Boardsen, and X. Vallières (2011), Study of nonthermal continuum patches: Wave propagation and plasmapause study, *J. Geophys. Res.*, *116*, A07219, doi:10.1029/2011JA016476.

Hartinger, M., M. B. Moldwin, V. Angelopoulos, K. Takahashi, H. J. Singer, R. R. Anderson, Y. Nishimura, and J. R. Wygant (2010), Pc5 wave power in the quiet-time plasmasphere and trough: CRRES observations, *Geophys. Res. Lett.*, *37*, L07107, doi:10.1029/2010GL042475.

Henderson, M. G., E. F. Donovan, J. C. Foster, I. R. Mann, T. J. Immel, S. B. Mende, and J. B. Sigwarth (2010), Start-to-end global imaging of a sunward propagating, SAPS-associated giant undulation event, *J. Geophys. Res.*, *115*, A04210, doi:10.1029/2009JA014106.

Huang, C.-L., H. E. Spence, M. K. Hudson, and S. R. Elkington (2010), Modeling radiation belt radial diffusion in ULF wave fields: 2. Estimating rates of radial diffusion using combined MHD and particle codes, *J. Geophys. Res.*, *115*, A06216, doi:10.1029/2009JA014918.

Johnston, W. R., and P. C. Anderson (2010), Storm time occurrence of relativistic electron microbursts in relation to the plasmapause, *J. Geophys. Res.*, *115*, A02205, doi:10.1029/2009JA014328.

Kalaee, M. J., Y. Katoh, A. Kumamoto, T. Ono, and Y. Nishimura (2010), Simulation of mode conversion process from upper-hybrid waves to LO-mode waves in the vicinity of the plasmapause, *Ann. Geophys.*, *28*, 1289–1297, doi:10.5194/angeo-28-1289-2010.

Kawano, H., S. Ohtani, T. Uozumi, T. Tokunaga, A. Yoshikawa, K. Yumoto, E. A. Lucek, M. André, and the CPMN group (2011), Pi 2 waves simultaneously observed by Cluster and CPMN ground-based magnetometers near the plasmapause, *Ann. Geophys.*, *29*, 1663–1672, doi:10.5194/angeo-29-1663-2011.

Kelley, M. C., J. Franz, and A. Jacobson (2012), On structuring of the plasmapause, *Geophys. Res. Lett.*, *39*, L01101, doi:10.1029/2011GL050048.

Kim, K.-H., F. S. Mozer, D.-H. Lee, and H. Jin (2010), Large electric field at the nightside plasmapause observed by the Polar spacecraft, *J. Geophys. Res.*, *115*, A07219, doi:10.1029/2010JA015439.

Lam, M. M., R. B. Horne, N. P. Meredith, S. A. Glauert, T. Moffat-Griffin, and J. C. Green (2010), Origin of energetic electron precipitation >30 keV into the atmosphere, *J. Geophys. Res.*, *115*, A00F08, doi:10.1029/2009JA014619.

Li, X., D. N. Baker, T. P. O'Brien, L. Xie, and Q. G. Zong (2006), Correlation between the inner edge of outer radiation belt electrons and the innermost plasmapause location, *Geophys. Res. Lett.*, *33*, L14107, doi:10.1029/2006GL026294.

Liléo, S., T. Karlsson, and G. T. Marklund (2010), Statistical study on the occurrence of ASAID electric fields, *Ann. Geophys.*, *28*, 439–448, doi:10.5194/angeo-28-439-2010.

Liu, K., S. P. Gary, and D. Winske (2011), Excitation of banded whistler waves in the magnetosphere, *Geophys. Res. Lett.*, *38*, L14108, doi:10.1029/2011GL048375.

Liu, W., T. E. Sarris, X. Li, Q.-G. Zong, R. Ergun, V. Angelopoulos, and K. H. Glassmeier (2011), Spatial structure and temporal evolution of a dayside poloidal ULF wave event, *Geophys. Res. Lett.*, *38*, L19104, doi:10.1029/2011GL049476.

Luo, H., G. X. Chen, A. M. Du, V. Angelopoulos, W. Y. Xu, X. D. Zhao, and Y. Wang (2011), THEMIS multipoint observations of Pi2 pulsations inside and outside the plasmasphere, *J. Geophys. Res.*, *116*, A12206, doi:10.1029/2011JA016746.

Masson, A., et al. (2009), Advances in plasmaspheric wave research with CLUSTER and IMAGE observations, *Space Sci. Rev.*, *145*, 137–191, doi:10.1007/s11214-009-9508-7.

Miyoshi, Y., and R. Kataoka (2011), Solar cycle variations of outer radiation belt and its relationship to solar wind structure dependences, *J. Atmos. Sol. Terr. Phys.*, *73*(1), 77–87, doi:10.1016/j.jastp.2010.09.031.

Nishida, A. (1966), Formation of plasmapause, or magnetospheric plasma knee, by the combined action of magnetospheric convection and plasma escape from the tail, *J. Geophys. Res.*, *71*(23), 5669–5679.

Obana, Y., G. Murakami, I. Yoshikawa, I. R. Mann, P. J. Chi, and M. B. Moldwin (2010), Conjunction study of plasmapause location using ground-based magnetometers, IMAGE-EUV, and Kaguya-TEX data, *J. Geophys. Res.*, *115*, A06208, doi:10.1029/2009JA014704.

Pedatella, N. M., and K. M. Larson (2010), Routine determination of the plasmapause based on COSMIC GPS total electron content observations of the midlatitude trough, *J. Geophys. Res.*, *115*, A09301, doi:10.1029/2010JA015265.

Pickett, J. S., et al. (2010), Cluster observations of EMIC triggered emissions in association with Pc1 waves near Earth's plasmapause, *Geophys. Res. Lett.*, *37*, L09104, doi:10.1029/2010GL042648.

Ponomarenko, P. V., C. L. Walters, and J.-P. St-Maurice (2010), Upstream Pc3-4 waves: Experimental evidence of propagation to the nightside plasmapause/plasmatrough, *Geophys. Res. Lett.*, *37*, L22102, doi:10.1029/2010GL045416.

Ribeiro, A. J., J. M. Ruohoniemi, J. B. H. Baker, L. B. N. Clausen, R. A. Greenwald, and M. Lester (2012), A survey of plasma irregularities as seen by the midlatitude Blackstone SuperDARN radar, *J. Geophys. Res.*, *117*, A02311, doi:10.1029/2011JA017207.

Sandel, B. R., R. A. King, W. T. Forrester, D. L. Gallagher, A. L. Broadfoot, and C. C. Curtis (2001), Initial results from the IMAGE Extreme Ultraviolet Imager, *Geophys. Res. Lett.*, *28*(8), 1439–1442, doi:10.1029/2001GL012885.

Shprits, Y., D. Subbotin, B. Ni, R. Horne, D. Baker, and P. Cruce (2011), Profound change of the near-Earth radiation environment caused by solar superstorms, *Space Weather*, *9*, S08007, doi:10.1029/2011SW000662.

Shprits, Y., M. Daae, and B. Ni (2012), Statistical analysis of phase space density buildups and dropouts, *J. Geophys. Res.*, *117*, A01219, doi:10.1029/2011JA016939.

Singh, A. K., R. P. Singh, and D. Siingh (2011), State studies of Earth's plasmasphere: A review, *Planet Space Sci.*, *59*(9), 810–834, doi:10.1016/j.pss.2011.03.013.

Spasojevic, M., J. Goldstein, D. L. Carpenter, U. S. Inan, B. R. Sandel, M. B. Moldwin, and B. W. Reinisch (2003), Global response of the plasmasphere to a geomagnetic disturbance, *J. Geophys. Res.*, *108*(A9), 1340, doi:10.1029/2003JA009987.

Summers, D., B. Ni, N. P. Meredith, R. B. Horne, R. M. Thorne, M. B. Moldwin, and R. R. Anderson (2008), Electron scattering by whistler-mode ELF hiss in plasmaspheric plumes, *J. Geophys. Res.*, *113*, A04219, doi:10.1029/2007JA012678.

Takahashi, K., R. L. McPherron, and T. Terasawa (1984), Dependence of the spectrum of Pc3-4 pulsations on the interplanetary magnetic field, *J. Geophys. Res.*, *89*, 2770–2780.

Takahashi, K., et al. (2010), Multipoint observation of fast mode waves trapped in the dayside plasmasphere, *J. Geophys. Res.*, *115*, A12247, doi:10.1029/2010JA015956.

Usanova, M. E., et al. (2010), Conjugate ground and multisatellite observations of compression-related EMIC Pc1 waves and associated proton precipitation, *J. Geophys. Res.*, *115*, A07208, doi:10.1029/2009JA014935.

Teramoto, M., K. Takahashi, M. Nosé, D.-H. Lee, and P. R. Sutcliffe (2011), Pi2 pulsations in the inner magnetosphere simultaneously observed by the Active Magnetospheric Particle Tracer Explorers/Charge Composition Explorer and Dynamics Explorer 1 satellites, *J. Geophys. Res.*, *116*, A07225, doi:10.1029/2010JA016199.

Wang, C., Q. Zong, F. Xiao, Z. Su, Y. Wang, and C. Yue (2011), The relations between magnetospheric chorus and hiss inside and outside the plasmasphere boundary layer: Cluster observation, *J. Geophys. Res.*, *116*, A07221, doi:10.1029/2010JA016240.

Yizengaw, E., H. Wei, M. B. Moldwin, D. Galvan, L. Mandrake, A. Mannucci, and X. Pi (2005), The correlation between mid-latitude trough and the plasmapause, *Geophys. Res. Lett.*, *32*, L10102, doi:10.1029/2005GL022954.

Zou, S., M. B. Moldwin, A. Coster, L. R. Lyons, and M. J. Nicolls (2011), GPS TEC observations of dynamics of the mid-latitude trough during substorms, *Geophys. Res. Lett.*, *38*, L14109, doi:10.1029/2011GL048178.

M. B. Moldwin and S. Zou, Atmospheric, Oceanic and Space Sciences, University of Michigan, Ann Arbor, MI 48109, USA. (mmoldwin@umich.edu)

The Role of Quiet Time Ionospheric Plasma in the Storm Time Inner Magnetosphere

Andrew W. Yau and Andrew Howarth

Department of Physics and Astronomy, University of Calgary, Calgary, Alberta, Canada

W. K. Peterson

Laboratory of Atmospheric and Space Physics, University of Colorado, Boulder, Colorado, USA

Takumi Abe

Institute of Space and Astronautical Science, Sagamihara, Japan

We investigate the role of quiet time ionospheric plasma in the storm time inner magnetosphere using single-particle trajectory simulation, specifically the influence of the plasma properties of polar wind O^+ ions on energetic plasma sheet O^+ ions preceding and during magnetic storms. Polar wind O^+ ions are frequently observed at ~1 R_E altitude in the polar cap, at temperatures of ~0.2–0.3 eV, flow velocities of a few km s^{-1}, and solar-maximum densities up to ~30 cm^{-3}. Owing to the centrifugal acceleration at altitudes above ~2–3 R_E, a small fraction (~10%–25%) of these ions are found to acquire sufficient energy to reach the plasma sheet with a typical transit time of ~5 ± 2 h. Such "in-transit" ions between the polar ionosphere and the inner magnetosphere can populate the plasma sheet to significant density levels over a period of several hours preceding a magnetic storm, and they could explain the prompt presence of energetic O^+ ions in the plasma sheet and ring current at the storm onset.

1. INTRODUCTION

The energetic O^+ ion population in the ring current is believed to originate from the ionosphere and enter the ring current primarily via the plasma sheet (see, for example, the work of *Hultqvist et al.* [1999]) and to play an important role in the particle and magnetic field pressure distributions during a geomagnetic storm [*Fok et al.*, 2001; *Jordanova*, 2003]. Thus, the nature of energization and transport of the seed ionospheric O^+ population is of interest to the ring current dynamics.

The thermal plasma in the polar ionosphere is dominated by the polar wind [*Chappell et al.*, 1987; *Yau and Andre*, 1997]. Its density is found to exhibit a significant variation with seasons and to transition from an exponential altitude dependence at low altitude (below ~0.4 R_E) to a power law relationship at high altitude up to at least ~1.6 R_E [*Kitamura et al.*, 2009]. Near solar maximum, the density (n_e) ranges from about 20 to 10^2 cm^{-3} near 1 R_E at equinox.

At Akebono altitudes (<1.6 R_E), the observed polar wind is comprised mainly of H^+, He^+, and O^+ ions, and electrons, and the ions have an average temperature of ~0.2–0.3 eV [*Drakou et al.*, 1997]. At a given altitude, the ion velocity is, on average, larger on the dayside than on the nightside and is

Dynamics of the Earth's Radiation Belts and Inner Magnetosphere
Geophysical Monograph Series 199
10.1029/2012GM001325

largest for H$^+$ and smallest for O$^+$ [*Abe et al.*, 1993]. The O$^+$ ions often constitute a significant component of the polar wind ion population and a potentially nonnegligible supply of escaping low-energy O$^+$ ions and may be characterized by a cold, drifting plasma with a drift velocity of ~2–4 km s^{-1} and a density of a few to ~30 cm^{-3} near 1 R_E. Observationally, it is not always possible to unambiguously separate these polar wind O$^+$ ions from other low-energy ion populations, such as low-energy ions from the "cleft ion fountain" that have convected into a polar wind flux tube, which are viable sources of ion outflows and plasma sheet ions.

The effect of polar wind ion acceleration is apparent in the Akebono and Polar data. The observed O$^+$ ion temperature was about 7.5 eV, and the drift velocity was 16.8 km s^{-1} at Polar apogee (~8 R_E) near solar minimum [*Su et al.*, 1998]. The observed density of ~0.03–0.05 cm^{-3} corresponds to a small fraction (20%–40%) of the observed density at Akebono altitude, assuming the latter to be ~10 cm^{-3} near solar minimum and scale with magnetic field density along a field line. This is indicative of acceleration between the two altitudes, which is required to lift a sufficient fraction of the ions from Akebono altitudes to the Polar apogee. Using both energetic (0.015–17 keV) and thermal (0–50 eV) ion composition data on Polar [*Lennartsson et al.*, 2004; *Su et al.*, 1998], *Peterson et al.* [2008] estimated the thermal component in the auroral region to be on the order of half of the quiet-time escaping O$^+$ ion flux near 1 R_E altitude and almost 100% in the polar cap.

As illustrated by *Yau et al.* [2012], an upward moving polar wind ion typically loses its kinetic energy initially, but it can gradually gain energy at higher altitude due to "centrifugal acceleration," as it experiences magnetic curvature drift through the convection electric field potential while convecting across the polar cap into the magnetospheric lobes; this is provided that it has sufficient initial kinetic energy to reach the centrifugal acceleration altitude region. Thereafter, as the ion traverses the highly curved magnetic field in the neutral sheet, it is further and abruptly accelerated to "plasma sheet" energies of hundreds of eV or beyond.

The classical Jeans escape velocity v_e of an ion in the polar ionosphere is independent of ion mass and inversely proportional to the square root of its geocentric distance r_e. At 1 R_E altitude ($z_e = 1$ R_E; $r_e = 2$ R_E), v_e is ~7.9 km s^{-1}. However, when an acceleration mechanism, such as centrifugal acceleration, is present starting at a higher altitude z_c, the minimum velocity v_{esc} for an upward moving ion to reach the magnetosphere corresponds to the velocity required to lift the ion to altitude z_c. In general, v_{esc} is smaller than the classical escape velocity v_e; the actual velocity required may be larger than v_{esc} and depends on the magnitude of the acceleration.

In the case of a drifting Maxwellian distribution of temperature T and drift velocity v_d, the maximum fraction of ions f_{esc} capable of reaching altitude z_c is given by

$$f_{esc} = \frac{1}{2}\left\{1 - \left[1 - \exp\left(-\frac{(v_{esc}-v_d)^2}{v_t^2}\right)\right]^{1/2}\right\} \quad ; \quad v_{esc} > v_d \tag{1a}$$

$$f_{esc} = \frac{1}{2}\left\{1 + \left[1 - \exp\left(-\frac{(v_{esc}-v_d)^2}{v_t^2}\right)\right]^{1/2}\right\} \quad ; \quad v_{esc} < v_d. \tag{1b}$$

The derivation of equation (1) and v_{esc} is given in the Appendix, and f_{esc} corresponds to the maximum fraction of escaping ions. Figure 1 shows the calculated values of v_{esc} and f_{esc} as a function of z_c for polar wind O$^+$ ions in the case of $z_e = 0.5$ and 1 R_E (short dash and solid trace), respectively. In Figure 1a, the value of v_{esc} decreases with decreasing z_c. In

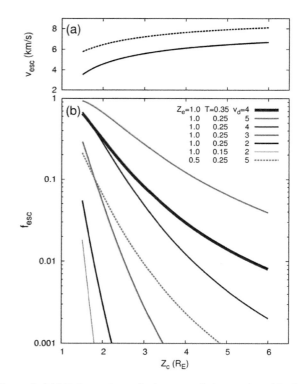

Figure 1. (a) Minimum ion velocity to reach the starting altitude of acceleration z_c from altitude z_e (v_{esc}). (b) Maximum fraction of escaping polar wind O$^+$ ions (f_{esc}) as a function of z_c: $z_e = 0.5$, 1.0 R_E (short dash, solid trace); ion temperature $T = 0.15$, 0.25, 0.35 eV (thin, medium, thick trace); parallel ion velocity $v_d = 2$, 3, 4, 5 km s^{-1} (black, green, blue, red traces).

the case of $z_e = 1\ R_E$, v_{esc} is about 4 km s^{-1}, which is about half of v_e, in the presence of ion acceleration starting at $z_c = 2\ R_E$; v_{esc} is about 6 km s^{-1} in the case of $z_c = 3\ R_E$. In comparison, in the case of $z_e = 0.5\ R_E$ and ion acceleration starting at $z_c = 2\ R_E$, v_{esc} is about 6 km s^{-1}, or ~50% larger. Since (as will be discussed below) centrifugal acceleration due to spatial and/or temporal magnetic field gradient typically occurs down to ~2–3 R_E, particularly during active times, the value of v_{esc} is often comparable to the parallel drift velocity of polar wind O$^+$ ions at 1 R_E, particularly on the dayside [Abe et al., 1993, 2004].

In Figure 1b, f_{esc} is shown for ion temperature T of 0.15, 0.25, and 0.35 eV (thin, medium, and thick traces) and for ion drift velocity v_d of 2, 3, 4, and 5 km s^{-1} (black, green, blue, and red), respectively. As z_c decreases, f_{esc} increases quasi-exponentially to a nonnegligible value exceeding 1%–10% in the case of $z_e = 1.0\ R_E$ (solid traces) and $z_c = 2$–$3\ R_E$ for typical polar wind ion temperature (~0.2–0.3 eV) and velocity (~2–4 km s^{-1}). In comparison, in the case of $z_e = 0.5\ R_E$ (short dash trace), f_{esc} is at least a factor of 5 smaller. Thus, centrifugal acceleration may well play a significant role in transporting a nonnegligible fraction of polar wind O$^+$ ions to the magnetosphere, particularly during magnetically active times when significant temporal magnetic field gradients may be present down to ~2 R_E.

A number of authors have used single-particle trajectory tracing to investigate the transport of ionospheric ions of various energies in the magnetosphere. Cladis [1986] used a simple 2-D model to demonstrate the importance of centrifugal ion acceleration in regions of curved or changing magnetic field. Delcourt et al. [1989, 1993, 1994] used both the guiding center and full equations of motion to study the 3-D dynamics of auroral, polar cap, cusp, and H$^+$ polar wind ions in different magnetospheric regions including the magnetotail. Cully et al. [2003] used the guiding center equations to address the question of auroral substorm trigger and the supply of various types of low-energy ions to the central plasma sheet under both northward and southward interplanetary magnetic field (IMF) conditions, including both light (H$^+$) and heavy (O$^+$) polar wind and "cleft ion fountain" ions.

In a follow-on study, Howarth and Yau [2008] examined the influence of the IMF and convection electric field on the trajectories and destinations of polar wind and other low-energy ions observed on Akebono in the case of steady IMF and ionospheric convection. Ions were found to preferentially feed the dusk sector of the plasma sheet over the dawn sector when the IMF is duskward ($B_Y > 0$) and to be more evenly distributed between the two sectors when the IMF is dawnward. In a similar study, Ebihara et al. [2006] [see also Kitamura et al., 2010] simulated the magnetospheric destination of O$^+$ ions above escape energy under an active-time

magnetic field and "midstrength" convection electric field, a source altitude of 1 R_E and using the full equation of motion and an empirical model of ion flux that is based on K_P and the sunspot number. The majority of the O$^+$ ions were found to reach the ring current.

Huddleston et al. [2005] extended the simulation code of Delcourt et al. [1993] to compute the full trajectories of H$^+$ and He$^+$ polar wind (0–3 eV) and cleft ion fountain (10–20 eV) ions, as well as O$^+$ cleft ion fountain and H$^+$ and O$^+$ auroral (500 eV) ions. The polar wind H$^+$ ions originating from or below auroral latitudes were found to be an important source of plasma sheet H$^+$ ions. In the case of the cleft ion fountain and auroral ions, the heavy O$^+$ ions were found to reach farthest down tail and the light H$^+$ ions least down tail.

Yau et al. [2012] extended the simulation model in the work of Howarth and Yau [2008] to the case of time-dependent magnetic and convection electric field to investigate the transport of polar wind O$^+$ ions in periods of variable IMF and ion convection, including periods of both quiet times and magnetic storms. The aim of the present study is to examine the dependence of this transport on the ambient plasma properties, particularly the energy evolution and transit time of each individual ion, and its destination in the magnetosphere. Our focus is to delineate a "lower limit" estimate of the possible role of quiet-time polar wind oxygen ions in ion outflows in the presence of centrifugal ion acceleration only, by excluding from our model the other ion populations and ion acceleration or heating mechanisms that are generally absent at quiet times.

2. PARTICLE TRAJECTORY SIMULATION

The details of the present simulation model were presented in Yau et al. [2012]. Briefly, the model consisted of a time-dependent model of the electric and magnetic field, respectively, over the region of $X_{GSM} = -70$ to $+10\ R_E$, $Y_{GSM} = -25$ to $+25\ R_E$, and $Z_{GSM} = -30$ to $+30\ R_E$. A nonuniformly spaced grid of $101 \times 101 \times 101$ grid points was used in which the grid spacing varied with the radial distance at a grid point from a minimum of ~1/3 R_E.

The magnetic field was a superposition of the Internal Geomagnetic Reference Field [Finlay et al., 2010] and the Tsyganenko 1996 (T96) external magnetic field model [Tsyganenko, 1995; Tsyganenko and Stern, 1996]; the T96 model was parameterized by the Dst index, the Earth's dipole tilt angle, and instantaneous values of solar wind dynamic pressure and IMF B_Y and B_Z derived from time-shifted solar wind measurements on the ACE.

The electric field was derived from SuperDARN ion convection velocity data [Howarth and Yau, 2008; Yau et al.,

2012] and supplemented by the Applied Physics Lab statistical model [*Ruohoniemi and Greenwald*, 1996; *Shepherd and Ruohoniemi*, 2000] in the case of data gaps, assuming the absence of parallel electric field (i.e., $E_\parallel = 0$) along the field lines. Both the magnetic and electric fields were updated every 10 min in response to the changing solar wind dynamic pressure and IMF.

The first-order guiding center equations of motion [*Northrop*, 1963] were used to solve for the trajectory of an ion starting at its source location $\mathbf{R}_0$ ($\equiv \mathbf{R}(t = t_0)$) and initial velocity $\mathbf{v}_0$ ($\equiv \mathbf{v}(t = t_0)$), to determine its evolving energy along the trajectory, its destination in the magnetosphere, and the transit time to reach its destination:

$$\frac{dv_\parallel}{dt} = G_\parallel + \frac{q}{m}E_\parallel - \frac{\mu}{m}\nabla_\parallel B + \overrightarrow{v}_{\overrightarrow{E}\times\overrightarrow{B}} \cdot \frac{d\hat{B}}{dt} \quad (2)$$

$$\overrightarrow{v}_\perp = \frac{\overrightarrow{E}\times\overrightarrow{B}}{B^2} + \frac{\mu}{qB^2}\overrightarrow{B}\times\nabla B + \frac{m}{qB^2}\overrightarrow{G}$$
$$\times\overrightarrow{B} - \frac{m}{qB^2}\left[v_\parallel\frac{d\hat{B}}{dt} + \frac{d\overrightarrow{v}_{\overrightarrow{E}\times\overrightarrow{B}}}{dt}\right]\times\overrightarrow{B}. \quad (3)$$

In equations (2) and (3), q is the ion charge, m is the ion mass, G is the gravitational acceleration, and μ is the magnetic moment. E is the electric field and B is the magnetic field: in the simulation, the E and B values at any position and time were obtained from the corresponding values at the nearest grid points in the nearest updates using linear interpolation in time and in space. The last term in equation (2) is the centrifugal acceleration term first discussed by *Cladis* [1986]. The convective derivatives depend on both time and space, i.e.,

$$\frac{d}{dt}A = \left[\frac{\partial}{\partial t} + v_\parallel\nabla_\parallel + (\overrightarrow{v}_{\overrightarrow{E}\times\overrightarrow{B}}\cdot\nabla)\right]A. \quad (4)$$

The guiding center equations of motion (equations (2) and (3)) are valid only in the adiabatic regime, where the electric and magnetic field do not vary significantly over a gyroperiod or gyroradius. In the nonadiabatic regime, the guiding center approximation is not applicable. Therefore, we confine our study to the adiabatic region beyond the plasma sheet boundary layer (PSBL), where the solution of the guiding center equations was found in *Yau et al.* [2012] to match the solution of the full equation of motion.

Figure 1b shows that over the range of observed ion temperature and velocity at Akebono altitudes, the maximum fraction of escaping ions f_{esc} varies strongly with ion temperature and velocity as well as the starting acceleration altitude in general. In the present study, one of three ion temperature cases was selected for each simulation run: low, medium, and high temperature (T_L, T_M, and T_H), corresponding to ion temperature of 0.15, 0.25, and 0.35 eV, respectively, to examine the temperature dependence of the overall polar wind O^+ transport. In *Yau et al.* [2012] (compare their Figure 1), a fixed ion temperature of 0.25 eV was assumed, and a K_P-dependent ion velocity selection scheme was used: one of two ion velocity distributions in invariant latitude and magnetic local time (MLT) "velocity maps", which correspond to $K_P \leq 3$ and ≥ 4, respectively, was dynamically selected in the simulation depending on the instantaneous K_P value. In this study, one of three velocity cases was selected for each simulation run: K_P-dependent velocity (V_K), following *Yau et al.* [2012] and low and high velocity (V_L and V_H), which correspond to the low- and high-K_P velocity distributions given by *Yau et al.* [2012], respectively, throughout a simulation run regardless of the actual variations of magnetic activity level.

A total of 10,000 ions were "launched" from a fixed altitude (7000 km) every 10 min, starting at 6 h before the onset of a magnetic storm. A Monte Carlo scheme was used to randomly select the initial velocity and location (invariant latitude between 70° and 86° and MLT) for each ion, by assuming uniform ion density at all locations, and a drifting Maxwellian velocity distribution in which the perpendicular ion velocity was derived from the SuperDARN data and the ion temperature and parallel ion drift velocity were specified as explained above.

The trajectory of each ion was followed until it reached one of nine possible final destinations, and the initial and final position and velocity and the transit time of each ion were captured for analysis. An ion was identified as "trapped" in the *atmosphere* if it was gravitationally trapped, descended below 600 km, or was below 1 R_E altitude after 2 h. An untrapped ion could convect to the *dayside* equatorial plane, reach the *magnetopause* or its vicinity where the field lines did not connect back to the ionosphere in the T96 model, flow to the *distant tail* beyond $X_{GSM} = -70$ R_E, cross the PSBL tailward of $X_{GSM} = -15$ R_E into the *dawn tail* or *dusk tail*, cross the PSBL earthward of -15 R_E into the *dawn plasma sheet* or *dusk plasma sheet*, or flow out of the simulation region altogether. The PSBL was defined as a function of the dipole tilt angle and serves as the cutoff where the ions begin to experience nonadiabatic energization in the plasma sheet. The value of -70 R_E was chosen for the limit for the distant tail in keeping with our previous studies.

3. DISTRIBUTION OF PARTICLE DESTINATIONS

Yau et al. [2012] determined that the percentage of untrapped ions reaching the different magnetospheric regions was as a function of time of ion launch for the periods preceding and during five large magnetic storms near the maximum and declining phase of solar cycle (SC) 23. The percentage of untrapped ions was used as a quantitative measure of the contribution of polar wind O^+ ions to the "in-transit" low-energy oxygen ion population between the ionosphere and the plasma sheet and ring current during the quiet-time period immediately preceding a magnetic storm. In this study, a series of five simulation runs were performed for the storm on 6 April 2000 (minimum $Dst = -320$ nT), in order to elucidate the dependence of this in-transit population on the polar wind plasma properties (ion temperature and velocity), and to facilitate data comparison with that of Yau et $al.$ [2012].

Figure 2 shows the geomagnetic and solar wind conditions preceding and during the 6 April 2000 storm, including (a) the 3 h K_P index (Figure 2a), 1 min symmetric disturbance field in the horizontal component SYM-H, and the time-shifted solar wind data from ACE (Figure 2b), including the solar wind dynamic pressure P_{dyn} (Figure 2c), IMF B_Y (Figure 2d), and B_Z (Figure 2e). The rapid decrease in SYM-H starting near 16:45 UT coincided with the sudden, order-of-magnitude increase in the solar wind dynamic pressure, and the southward turning of the IMF, which resulted in the decrease of B_Z from about -2 to -20 nT. The rapid decrease

signaled the onset of the main phase at this time. In the period preceding the onset, K_P increased from 2 at 09:00 UT to 3 at 12:00 UT and to 6 at 15:00 UT. The IMF was initially weakly southward ($B_Z \approx$ a few nT negative) and then turned northward around 14:30 UT, while B_Y was small and dawnward (negative). During the main phase of the storm, K_P reached a maximum value of 8, and SYM-H reached a minimum value of about -320 nT near 00:10 UT on 7 April 2000.

The first three of the five simulation runs correspond to the case of low, high, and K_P-dependent ion velocity, respectively, and medium temperature (0.25 eV), to examine the ion velocity dependence of the untrapped ion percentages. In other words, $T_1 = T_2 = T_3 = T_M = 0.25$ eV, and $V_1 = V_L$, $V_2 = V_H$, and $V_3 = V_K$; where T_i and V_i denote the temperature and velocity selection for the ith run (R_i), the third run (R_3) being identical to the corresponding simulation run in Yau et $al.$ [2012]. The two remaining runs correspond to the case of K_P-dependent ion velocity and of low and high ion temperature, respectively. In other words, $T_4 = T_L$ and $T_5 = T_H$, and $V_4 = V_5 = V_K$.

Figure 3 shows the distribution of percentage of ions at each destination (region of the magnetosphere) in the case of low and high ion velocity (V_L and V_H), respectively, as a function of the launch time of the ions before and after the

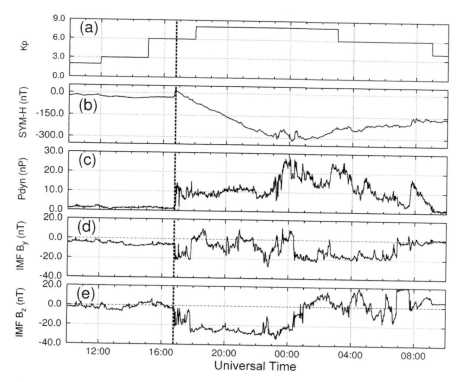

Figure 2. Geomagnetic and solar wind conditions preceding and during the 6 April 2000 magnetic storm: (a) K_P, (b) SYM-H, (c) solar wind dynamic pressure, (d) interplanetary magnetic field (IMF) B_Y and (e) IMF B_Z.

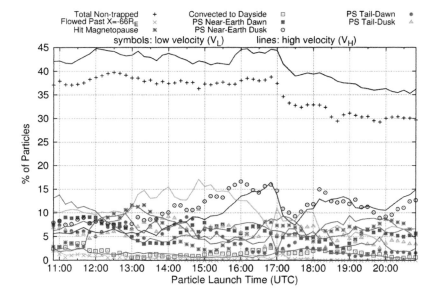

Figure 3. Distribution of percentage of ions at each destination (region of the magnetosphere) as a function of the launch time of the ions before and after the onset of the 6 April 2000 magnetic storm (onset at 16:45 UT): low parallel ion drift velocity (V_L, symbols), high ion velocity (V_H, lines); untrapped ions in the dayside (brown), magnetopause (green), distant tail ($X < -66\ R_E$, light blue), dawn tail (red), dusk tail (gold), dawn plasma sheet (purple), dusk plasma sheet (blue), and all regions of the magnetosphere (black); medium ion temperature ($T = 0.25$ eV).

onset at 16:45 UT. As noted above, the case of low ion velocity corresponds to the condition of $K_P \leq 3$, while the high-velocity case corresponds to $K_P \geq 4$. The distributions are shown for the untrapped ions in the dayside (brown), magnetopause (green), distant tail ($X < -66\ R_E$; light blue), dawn and dusk tail (red and gold), and dawn and dusk plasma sheet (purple and blue), respectively, as well as the total percentage of untrapped ions in all magnetospheric regions (black).

Figure 3 shows that the total percentage of untrapped ions in the atmosphere is about 5%–8% larger in the case of high ion velocity (black lines), compared with the low-velocity case (black crosses). The percentage of untrapped ions in the tail is also larger, while the percentage in the plasma sheet is at times smaller, resulting in a larger tail-to-plasma sheet ratio in general. The larger total untrapped percentage is attributed to the increase in f_{esc} with ion drift velocity (compare Figure 1b). As will be shown in Figure 5 below, the larger tail-to-plasma sheet percentage ratio is consistent with the fact that an ion with a larger initial parallel ion velocity will be centrifugally accelerated to a higher energy and will reach the PSBL at a greater geocentric distance. As noted above, K_P increased from 2 at 09:00 UT to 3 at 12:00 UT and to 6 at 15:00 UT. Not surprisingly, the combination of the low-velocity data before 15:00 UT from run R_1 with the high-velocity data thereafter from run R_2 reproduces the

result in simulation run R_3, in which K_P-dependent ion velocity was assumed, i.e., $T_3 = T_M$ and $V_3 = V_K$.

The last three simulation runs (R_3, R_4, and R_5) correspond to the case of medium, low, and high ion temperature, respectively, and K_P-dependent ion velocity, to examine the ion temperature dependence of the untrapped ion percentage. In other words, $V_3 = V_4 = V_5 = V_K$ and $T_3 = T_M$, $T_4 = T_L$ and $T_5 = T_H$. Figure 4 shows the distribution of percentage of ions at each magnetospheric destination in the case of low and high ion temperature (T_L and T_H), respectively, as a function of the launch time of the ions before and after the storm onset: the color coding for the different magnetospheric regions in this figure is the same as in Figure 3. In order to show the ion percentages with an expanded scale on the ordinate axis, the total percentage of untrapped ions is not shown in Figure 3.

In Figure 4, the low- and high-temperature distributions (symbols and lines) virtually overlap each other. The overlap is indicative of the negligible dependence of the untrapped ion percentage on the polar wind ion temperature, which is due to the relatively low starting acceleration altitude ($<3\ R_E$; compare Figure 1b) and to the fact that the parallel ion drift energy dominates over the thermal energy ($\frac{1}{2}\ mv_d^2 \gg \frac{1}{2}\ mv_t^2 = kT$) in the velocity phase space density and in the resulting maximum escaping fraction f_{esc} under typical polar wind velocity and temperature conditions.

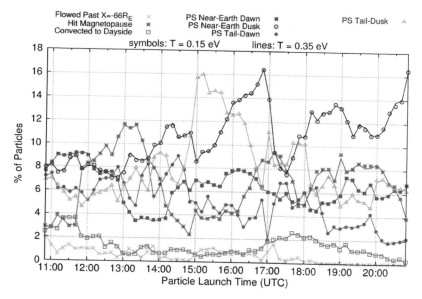

Figure 4. Distribution of percentage of ions at each destination (region of the magnetosphere) as a function of the launch time of the ions before and after the onset of the 6 April 2000 magnetic storm (onset at 16:45 UT): low ion temperature (T_L; $T = 0.15$ eV; symbols), high ion temperature (T_H; $T = 0.35$ eV; lines); untrapped ions in the dayside (brown), magnetopause (green), distant tail ($X < -66 R_E$; light blue), dawn tail (red), dusk tail (gold), dawn plasma sheet (purple), dusk plasma sheet (blue); K_P-dependent ion velocity.

Together, Figures 3 and 4 show that ~35%–40% of the ions launched before 15:00 UT was untrapped, when K_P was 3, and the percentage increased slightly to ~40%–45% for ions launched between 15:00 and 18:00 UT, when K_P increased to 6. About 7%–16% of the ions launched before the onset reached the dusk plasma sheet (blue circles and lines), while about 4%–9% of the ions reached the dawn plasma sheet (purple squares and lines). The corresponding fractions of ions reaching the dusk and dawn sectors of the tail (gold and red) were comparable, and the fractions reaching the dayside and the magnetopause (brown and green) were, in general, smaller in comparison. In other words, about 11% to 25% of the polar wind O^+ ions launched in the 6 h quiet-time period immediately preceding the start of this storm from 7000 km altitude were found to reach the plasma sheet.

Figure 5 shows how the energy of an ion evolves along its trajectory in the adiabatic region of the magnetosphere, for ions at initial energies of 2, 3, 5, 10, and 20 eV, respectively, launched at 13:00 UT. In this figure, the instantaneous ion energy in each case is color coded in logarithmic scale as the ion traversed from the starting location at 7000 km altitude, 75° invariant, and 12 MLT to the PSBL; in general, ions originating from this location are found to experience the largest ion acceleration [*Yau et al.*, 2012]. In the case of the 3 eV ion, the ion energy initially decreased to near 1 eV as the trajectory trace changed in color gradually from blue to black. However, the energy stopped decreasing and, instead,

started to increase as the ion reached higher altitude, to ~2.5 and >5 eV by the time the ion reached ~3 and 5 R_E altitude (Z_{GSM} ~4 and 6 R_E), respectively; correspondingly, the trajectory trace changed gradually from black at 3 R_E to blue at 5 R_E altitude. The ion energy eventually reached 40 eV (and the trace changed to gold) by the time the ion reached X_{GSM} = −20 R_E. In comparison, the 2 eV ion reached the PSBL at a closer distance (X_{GSM} ~ −15 R_E) and at a much lower energy (<5 eV). The difference between the two cases may be attributed to the dependence of the magnitude of ion acceleration on the parallel ion velocity in equation (4). Trajectories are also shown for ions at initial energies of 5, 10, and 20 eV, respectively, to provide a comparison with the trajectories of "cleft ion fountain" (CIF) ions, which typically have energies in the 5 to 20 eV range. In comparison, the CIF ions experience larger ion acceleration and reach the PSBL with higher energy and further down tail. In other words, the polar wind ions originating from the cleft region tend to arrive at the PSBL at lower energies and closer to Earth.

It is clear from the different trajectories in Figure 5 that ions launched at a given time and source altitude generally arrive at their respective destinations at different times, since the transit time of each ion from its source location to its destination varies with its initial ion position and velocity as well as the prevailing geomagnetic and solar wind conditions, which affect the overall convection path and the energy evolution of the ion.

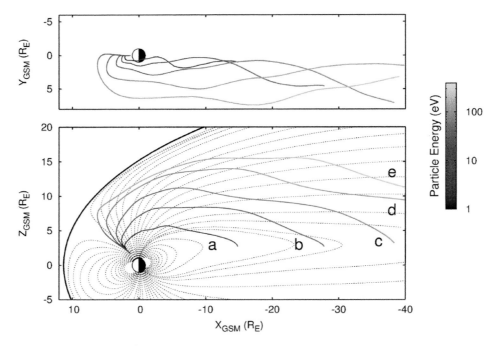

Figure 5. Ion energy of simulated O^+ ions originating at 7000 km altitude, 75° invariant and 12 MLT, with initial energies of (a) 2, (b) 3, (c) 5, (d) 10, and (e) 20 eV, respectively, along their respective trajectories in the adiabatic region.

Figure 6 shows the histogram of transit time of ions from their starting locations at 7000 km altitude to their destination in the magnetosphere in the case of K_P-dependent ion velocity and medium (0.25 eV) ion temperature, from simulation run R3. Only ions reaching the plasma sheet and distant tail are included in the histogram. This figure shows that the majority of the ions have a transit time of just under 5 ± 2 h between the topside ionosphere and the plasma sheet (boundary layer) or the tail, and a very small percentage of ions have a longer transit time of 10 h or greater. In comparison, the transit time of these ions through the plasma sheet is expected to be small, where they are likely to be accelerated to significantly higher energy and velocity. This implies that in the 5 to 10 h period immediately preceding the onset of the main phase of a magnetic storm, a small but substantial fraction of the polar wind O^+ ions can populate the plasma sheet as well as the "pipeline" between the plasma sheet and the topside ionosphere, and subsequently enter the ring current from the plasma sheet in the main phase of a storm where they may play a role in the inner magnetosphere dynamics.

4. DISCUSSION AND SUMMARY

Yau et al. [2012] used a time-dependent magnetic and convection electric field model to study the transport of thermal-energy polar wind O^+ ions in time periods of variable IMF preceding large magnetic storms. In this study, we use the same simulation model to investigate the dependence of this transport on the polar wind plasma properties and focus on the energy evolution and transit time of the individual ions and their destination in the magnetosphere. The simulation solves for the single-particle trajectories of

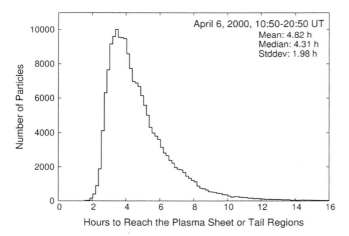

Figure 6. Histogram of simulated transit time of polar wind O^+ ions from their starting locations at 7000 km altitude to their destination in the magnetosphere (plasma sheet boundary layer or tail) in and preceding the magnetic storm of 6 April 2000, assuming K_P-dependent parallel ion velocity and medium (0.25 eV) ion temperature.

polar wind O^+ ions preceding and during a magnetic storm, to determine the fraction (percentage) of ions that is able to overcome gravitation and reach the different regions of the magnetosphere as a result of centrifugal ion acceleration at higher altitudes. A series of simulation runs were performed for a range of polar wind ion velocity and temperature for the storm of 6 April 2000, for comparison with one of the five storms studied by *Yau et al.* [2012].

Yau et al. [2012] found that under typical quiet-time geomagnetic and solar wind conditions, the higher-energy (>~2 eV) polar wind ions are able to reach the magnetosphere beyond ~10 R_E due to centrifugal acceleration as they traverse regions of magnetic field curvature. As a result, a variable and small percentage of the polar wind ions can reach the plasma sheet and magnetotail during the quiet-time period preceding a storm onset. A dawn-dusk asymmetry was found in the dusk plasma sheet and magnetotail, in that the dusk-to-dawn ratio of the percentage of untrapped ions is in the range of 1.7 to 7.3 and 3.1 on average. In addition, the percentage is strongly influenced by a number of factors, including the IMF B_Y and B_Z components, the rate of change of the magnetic field, the solar wind dynamic pressure, and the level of geomagnetic activity as characterized by the K_P index. In particular, an increase in both the convection strength due to increase in B_Y and B_Z and the rate of change of magnetic field results in a larger centrifugal ion acceleration and ultimately a larger percentage of untrapped ions as well as a larger magnetotail to plasma sheet percentage ratio.

The present study shows that for a given condition of solar wind and geomagnetic activity, the plasma properties of the polar wind O^+ ions affect the overall acceleration of the ions and their transport to the respective magnetospheric destinations, specifically as follows:

1. The parallel drift velocity of polar wind O^+ ions is well below the classical Jeans escape velocity in the 1 R_E altitude region (Figure 1a), but is often comparable to the minimum velocity required to reach the region of centrifugal ion acceleration, where it experiences the force of magnetic curvature as it drifts through the convection electric field potential across field lines in the polar cap.

2. In the case of typical polar wind ion velocity and temperature, a nonnegligible percentage (up to 10% or greater) of the O^+ ions can reach the centrifugal acceleration region where the magnetic field curvature (gradient) is sufficiently large so that the resulting ion acceleration is sufficient to overcome gravitation and thereby allows the ion to continue its upward motion (Figure 1b).

3. A larger polar wind ion velocity results in a significantly larger percentage of untrapped ions reaching the magnetosphere as well as proportionally more ions remaining at higher L shells and ultimately a larger magnetotail-to-plasma

sheet ion percentage ratio. In the case studied, the percentage of untrapped ions increased by as much as a third (Figure 3).

4. On the other hand, the percentage of untrapped ions is virtually independent of the polar wind ion temperature, due to the relatively low starting altitude of centrifugal ion acceleration (<3 R_E; Figure 1b) and to the fact that the parallel ion drift energy dominates over the thermal energy in the velocity phase space density and the resulting overall maximum escaping fraction (Figure 4).

5. The majority of the polar wind O^+ ions have a transit time of ~5 ± 2 h, and a very small percentage of ions have a much longer transit time of 10 h or greater, resulting in a small but substantial fraction of the O^+ ions populating the plasma sheet as well as the "pipeline" between the plasma sheet and the topside ionosphere in the 5 to 10 h period immediately preceding the onset of the main phase of a magnetic storm.

On Cluster, *Nilsson et al.* [2008] attributed a large portion of the observed O^+ ion velocity to centrifugal ion acceleration. Their analysis was confined to ion data above 40 eV and was focused on higher-energy ions on auroral or cleft/cusp field lines such as cleft ion fountain and upwelling ions that have initial ion energy of about 40 eV and typical transit times of 2 h or less as well as more tailward transport paths. The much longer transit time of the polar wind O^+ ions in the present study (~5 ± 2 h) may be attributed to their smaller initial parallel velocity: It can be seen in equations (2) and (3) above that the parallel ($v_{\parallel}$-dependent) component of the ion acceleration term depends on the initial ion velocity and that the transit time of an ion through an acceleration region varies inversely with its initial parallel velocity, while its overall energy increase through the region increases with the velocity.

On ISEE-1 [*Lennartsson and Shelley*, 1986], the density of O^+ ions above 100 eV was found to exhibit a dawn-dusk asymmetry in favor of the dusk sector, the quiet-time dusk-to-dawn density ratio being about 2. On Cluster, most of the observed O^+ ions were above 100 eV [*Kistler et al.*, 2005]. In our simulation, a large fraction of the untrapped O^+ ions reaching the plasma sheet boundary layer typically had energies of 100 eV or greater due to centrifugal ion acceleration [see, e.g., *Yau et al.* [2012, Figure 6] and the dusk-to-dawn percentage ratio of ions in the plasma sheet is ~1.5 to 5. These results are qualitatively consistent with those on Cluster and ISEE-1, respectively, given that the ions are expected to gain additional energies in the plasma sheet through processes such as current sheet acceleration and betatron acceleration, which are not included in our model.

As discussed by *Yau et al.* [2012], the O^+ mass content in the plasma sheet is estimated to be ~260 and >780 kg near solar minimum and maximum, respectively; the estimated O^+ mass content in the ring current is up to a factor of 5 smaller [*Peterson et al.*, 2009]. In comparison, the density of O^+ polar wind is seasonally dependent and is on the order of a few to ~30 cm^{-3} near solar maximum. Assuming a seasonally averaged density of 10 cm^{-3}, a K_P-averaged parallel ion velocity of 3 km s^{-1}, and an averaged untrapped ion percentage of 25% reaching the plasma sheet, the flux of untrapped O^+ ions reaching the plasma sheet is ~7.5 × 10⁵ cm^{-2} s^{-1}. Assuming regular occurrence of the polar wind poleward of 70° invariant and a rough estimate of ~3 × 10¹⁸ cm^2 for the total area of polar wind flux tubes at 7000 km altitude in both hemispheres, the total rate of untrapped ions into the plasma sheet is on the order of ~2 × 10²⁴ O^+ ions s^{-1} or ~200 kg h^{-1}. For an estimated O^+ plasma sheet mass content of ~780 kg, this suggests a "steady state" filling time of ~4 h, and a "refilling" time of ~9 ± 2 h when the averaged ion transit time is taken into account.

In conclusion, centrifugal acceleration at higher altitudes (above ~2.5–3 R_E altitude) energizes a significant fraction of thermal polar wind O^+ ions to sufficient velocity to reach the plasma sheet. The percentage of ions reaching the dusk sector of the plasma sheet is a factor of ~3 larger compared with the dawn sector, and the overall percentage in the two sectors depends on the parallel drift velocity of the polar wind ions for a given geomagnetic activity condition. Our results are consistent with the picture that low-energy ion flow constitutes a significant "in-transit" oxygen ion population over a period of several (~5–10) hours preceding a magnetic storm, which explains the presence of O^+ ions in the ring current shortly after the onset of the main phase of the storm, when the heavy ions could potentially modify the evolution of the ring current.

APPENDIX A

The minimum energy for an ion to reach a starting altitude of acceleration z_c (geocentric distance r_c) from an originating altitude of z_e (geocentric distance r_e; $z_c > z_e$) is

$$E_{esc}(r_e; \ r_c) = E_e(r_e) - E_e(r_c) = m \ g_0 (R_E^2/r_e - R_E^2/r_c). \quad (A1)$$

In equation (A1), $E_e(r)$ is the classical Jeans escape energy at r, m is the ion mass, g_0 is the acceleration of gravity at Earth's surface, and R_E is the Earth radius. The corresponding minimum velocity is

$$v_{esc}(r_e; \ r_c) = \sqrt{\frac{2E_{esc}}{m}} = \sqrt{\frac{2(E_e(r_e) - E_e(r_c))}{m}}$$
$$= \sqrt{[v_e(r_e)^2 - v_e(r_c)^2]}. \quad (A2)$$

Note that in equation (A2), the case of no acceleration is equivalent to the case of "infinite" starting acceleration altitude (i.e., $r_c = \infty$ in which case v_{esc} reduces to v_e).

Assuming a drifting Maxwellian of temperature T and drift velocity v_d (along vertical magnetic field line, for simplicity) for a polar wind ion population, the velocity phase space distribution may be written in Cartesian coordinates as:

$$f(v_x, v_y, v_z) = N \exp\left[-\frac{m}{2kT}(v_x^2 + v_y^2 + (v_z - v_d)^2)\right]$$
$$= N \exp\left[-\frac{1}{v_t^2}(v_x^2 + v_y^2 + (v_z - v_d)^2)\right], \quad (A3)$$

where z is in the vertical direction, N = normalization constant, and v_t = thermal speed. Thus, the maximum fraction f_{esc} of velocity phase space above velocity v_{esc} is given by equations (1a) and (1b) in the case of $v_{esc} > v_d$ and $v_{esc} < v_d$, respectively. In the latter case, as v_d increases for a constant v_t, the magnitude of the argument in the exponential term also increases, as does the square bracket term overall. On the other hand, as v_t increases for a constant v_d, the magnitude of the argument decreases, as does the square bracket term overall for a constant v_d, so that the escaping fraction becomes smaller. Physically, the larger temperature means a larger fraction of ions below the minimum escape velocity.

Acknowledgments. The authors gratefully acknowledge the support of both the Canadian Space Agency and the Natural Science and Engineering Research Council (NSERC) Industrial Research Chair Program for this research. W.K.P. was supported by NASA Grant NNX12AD25G.

REFERENCES

Abe, T., B. A. Whalen, A. W. Yau, R. E. Horita, S. Watanabe, and E. Sagawa (1993), EXOS D (Akebono) suprathermal mass spectrometer observations of the polar wind, J. Geophys. Res., 98(A7), 11,191–11,203.

Abe, T., A. W. Yau, S. Watanabe, M. Yamada, and E. Sagawa (2004), Long-term variation of the polar wind velocity and its implication for the ion acceleration process: Akebono/suprathermal ion mass spectrometer observations, J. Geophys. Res., 109, A09305, doi:10.1029/2003JA010223.

Chappell, C. R., T. E. Moore, and J. H. Waite Jr. (1987), The ionosphere as a fully adequate source of plasma for the Earth's magnetosphere, J. Geophys. Res., 92(A6), 5896–5910.

Cladis, J. B. (1986), Parallel acceleration and transport of ions from polar ionosphere to plasma sheet, Geophys. Res. Lett., 13(9), 893–896.

Cully, C. M., E. F. Donovan, A. W. Yau, and H. J. Opgenoorth (2003), Supply of thermal ionospheric ions to the central plasma sheet, J. Geophys. Res., 108(A2), 1092, doi:10.1029/2002JA 009457.

Delcourt, D. C., C. R. Chappell, T. E. Moore, and J. H. Waite Jr. (1989), A three-dimensional numerical model of ionospheric plasma in the magnetosphere, J. Geophys. Res., 94(A9), 11,893–11,920.

Delcourt, D. C., J. A. Sauvaud, and T. E. Moore (1993), Polar wind ion dynamics in the magnetotail, *J. Geophys. Res.*, *98*(A6), 9155–9169.

Delcourt, D. C., T. E. Moore, and C. R. Chappell (1994), Contribution of low-energy ionospheric protons to the plasma sheet, *J. Geophys. Res.*, *99*(A4), 5681–5689.

Drakou, E., A. W. Yau, and T. Abe (1997), Ion temperature measurements from the Akebono suprathermal mass spectrometer: Application to the polar wind, *J. Geophys. Res.*, *102*(A8), 17,523–17,539.

Ebihara, Y., M. Yamada, S. Watanabe, and M. Ejiri (2006), Fate of outflowing suprathermal oxygen ions that originate in the polar ionosphere, *J. Geophys. Res.*, *111*, A04219, doi:10.1029/2005JA011403.

Finlay, C. C., et al. (2010), International Geomagnetic Reference Field: The eleventh generation, *Geophys. J. Int.*, *183*, 1216–1230, doi:10.1111/j.1365-246X.2010.04804.x.

Fok, M.-C., R. A. Wolf, R. W. Spiro, and T. E. Moore (2001), Comprehensive computational model of Earth's ring current, *J. Geophys. Res.*, *106*(A5), 8417–8424.

Howarth, A., and A. W. Yau (2008), The effects of IMF and convection on thermal ion outflow in magnetosphere-ionosphere coupling, *J. Atmos. Sol. Terr. Phys.*, *70*, 2132–2143, doi:10.1016/j.jastp.2008.08.008.

Huddleston, M. M., C. R. Chappell, D. C. Delcourt, T. E. Moore, B. L. Giles, and M. O. Chandler (2005), An examination of the process and magnitude of ionospheric plasma supply to the magnetosphere, *J. Geophys. Res.*, *110*, A12202, doi:10.1029/2004JA010401.

Hultqvist, B., M. Øieroset, G. Paschmann, and R. Treumann (Eds.) (1999), *Magnetospheric Plasma Sources and Losses*, Kluwer Acad., Dordrecht, The Netherlands.

Jordanova, V. K. (2003), New insights on geomagnetic storms from model simulations using multi-spacecraft data, *Space Sci. Rev.*, *107*, 157–165.

Kistler, L. M., et al. (2005), Contribution of nonadiabatic ions to the cross-tail current in an O^+ dominated thin current sheet, *J. Geophys. Res.*, *110*, A06213, doi:10.1029/2004JA010653.

Kitamura, N., A. Shinbori, Y. Nishimura, T. Ono, M. Iizima, and A. Kumamoto (2009), Seasonal variations of the electron density distribution in the polar region during geomagnetically quiet periods near solar maximum, *J. Geophys. Res.*, *114*, A01206, doi:10.1029/2008JA013288.

Kitamura, N., et al. (2010), Observations of very-low-energy (<10 eV) ion outflows dominated by O^+ ions in the region of enhanced electron density in the polar cap magnetosphere during geomagnetic storms, *J. Geophys. Res.*, *115*, A00J06, doi:10.1029/2010JA015601.

Lennartsson, W., and E. G. Shelley (1986), Survey of 0.1- to 16-keV/*e* plasma sheet ion composition, *J. Geophys. Res.*, *91*(A3), 3061–3076.

Lennartsson, O. W., H. L. Collin, and W. K. Peterson (2004), Solar wind control of Earth's H^+ and O^+ outflow rates in the 15-eV to 33-keV energy range, *J. Geophys. Res.*, *109*, A12212, doi:10.1029/2004JA010690.

Nilsson, H., et al. (2008), An assessment of the role of the centrifugal acceleration mechanism in high altitude polar cap oxygen ion outflow, *Ann. Geophys.*, *26*, 145–157.

Northrop, T. G. (1963), *The Adiabatic Motion of Charged Particles*, Wiley-Interscience, New York.

Peterson, W. K., L. Andersson, B. C. Callahan, H. L. Collin, J. D. Scudder, and A. W. Yau (2008), Solar-minimum quiet time ion energization and outflow in dynamic boundary related coordinates, *J. Geophys. Res.*, *113*, A07222, doi:10.1029/2008JA013059.

Peterson, W. K., L. Andersson, B. Callahan, S. R. Elkington, R. W. Winglee, J. D. Scudder, and H. L. Collin (2009), Geomagnetic activity dependence of O^+ in transit from the ionosphere, *J. Atmos. Sol. Terr. Phys.*, *71*, 1623–1629.

Ruohoniemi, J. M., and R. A. Greenwald (1996), Statistical patterns of high-latitude convection obtained from Goose Bay HF radar observations, *J. Geophys. Res.*, *101*(A10), 21,743–21,763.

Shepherd, S. G., and J. M. Ruohoniemi (2000), Electrostatic potential patterns in the high-latitude ionosphere constrained by SuperDARN measurements, *J. Geophys. Res.*, *105*(A10), 23,005–23,014.

Su, Y.-J., J. L. Horwitz, T. E. Moore, B. L. Giles, M. O. Chandler, P. D. Craven, M. Hirahara, and C. J. Pollock (1998), Polar wind survey with the Thermal Ion Dynamics Experiment/Plasma Source Instrument suite aboard POLAR, *J. Geophys. Res.*, *103*(A12), 29,305–29,337.

Tsyganenko, N. A. (1995), Modeling the Earth's magnetospheric magnetic field confined within a realistic magnetopause, *J. Geophys. Res.*, *100*(A4), 5599–5612.

Tsyganenko, N. A., and D. P. Stern (1996), Modeling the global magnetic field of the large-scale Birkeland current systems, *J. Geophys. Res.*, *101*(A12), 27,187–27,198, doi:10.1029/96JA02735.

Yau, A. W., and M. Andre (1997), Source of ion outflow in the high latitude ionosphere, *Space Sci. Rev.*, *80*, 1–25.

Yau, A. W., A. Howarth, W. K. Peterson, and T. Abe (2012), Transport of thermal-energy ionospheric oxygen (O^+) ions between the ionosphere and the plasma sheet and ring current at quiet times preceding magnetic storms, *J. Geophys. Res.*, *117*, A07215, doi:10.1029/2012JA017803.

T. Abe, Institute of Space and Astronautical Science, Sagamihara, Kanagawa 229-8510, Japan.

A. Howarth and A. W. Yau, Department of Physics and Astronomy, University of Calgary, Calgary, AB T2N1N4, Canada. (yau@phys.ucalgary.ca)

W. K. Peterson, Laboratory of Atmospheric and Space Physics, University of Colorado, Boulder, CO 80303, USA.

Cold Ion Outflow as a Source of Plasma for the Magnetosphere

S. Haaland,[1,2] K. Li,[1,3] A. Eriksson,[4] M. André,[4] E. Engwall,[4] M. Förster,[5] C. Johnsen,[2] B. Lybekk,[6] H. Nilsson,[7] N. Østgaard,[2] A. Pedersen,[6] and K. Svenes[8]

The importance of ion outflow as a supplier of plasma to the terrestrial magnetosphere has been recognized for decades, and there are suggestions that the ionosphere alone is a sufficient source to account for the observed magnetospheric plasma population. However, due to spacecraft charging effects, it is difficult to measure the low-energy part of the plasma population. Recent advances in instrumentation and methodology, combined with more comprehensive measurements and auxiliary data, have provided far better opportunities to access the role of the cold ions. In this study, we have used measurements from the Cluster mission to quantify the amount of cold plasma supplied to the magnetosphere for various geomagnetic disturbance levels and solar wind conditions. The results show that the cold ion outflow flux does not respond substantially to changes in geomagnetic activity or changes in the solar wind or interplanetary magnetic field (IMF) but can vary by a factor of 3 as a result of enhanced solar irradiance and subsequent ionospheric ionization. Convection, and thus the transport of the cold plasma to the plasma sheet and inner magnetosphere, on the other hand, is largely controlled by the IMF. Our results indicate that a northward directed IMF and stagnant convection leads to a significant direct downtail loss of the cold plasma and, essentially, no supply of ionospheric ions to the magnetosphere. Correspondingly, a southward IMF results in enhanced convection, and the outflowing ions are almost completely recirculated to the plasma sheet. Under such conditions, there is a substantial loss of magnetospheric plasma through the ejection of plasmoids. This loss is estimated to be comparable or larger than the direct loss due to direct tailward escape along open lobe field lines during northward IMF. Overall, the average outflow of cold plasma is of the order of 10^{26} ions s^{-1}, and 90% of the outflow is returned to the plasma sheet.

[1]Max-Planck Institute for Solar Systems Research, Katlenburg-Lindau, Germany.

[2]Department of Physics and Technology, University of Bergen, Bergen, Norway.

Dynamics of the Earth's Radiation Belts and Inner Magnetosphere
Geophysical Monograph Series 199
© 2012. American Geophysical Union. All Rights Reserved.
10.1029/2012GM001317

[3]National Space Science Center, Chinese Academy of Sciences, Beijing, China.

[4]Swedish Institute of Space Physics, Uppsala, Sweden.

[5]Helmholtz Centre Potsdam, German Research Centre for Geosciences, Potsdam, Germany.

[6]Department of Physics, University of Oslo, Oslo, Norway.

[7]Swedish Institute of Space Physics, Kiruna, Sweden.

[8]Norwegian Defense Research Establishment, Kjeller, Norway.

1. INTRODUCTION

Ion outflow from the polar cap regions is believed to be a significant source of plasma for the terrestrial magnetosphere [*Chappell et al.*, 1987, 2000]. Early observations of ion outflow (see, e.g., the work of *Andre and Yau* [1997] for an overview) suggested that the outflow rate is of the order of 10^{26} ions s^{-1}. The outflowing ions are predominantly protons, but the outflow can have a significant oxygen component [e.g., *Yau and Andre*, 1997]. Other ions such as He$^+$ [e.g., *Abe et al.*, 1993], He^2, He^3 [e.g., *Axford*, 1968], NO$^+$ and O$_2^+$ [e.g., *Ogawa et al.*, 2011] have also been observed. In addition, there is a significant outflow of neutrals [*Engwall et al.*, 2009a].

Although a large number of outflowing ions are directly lost downtail along open lobe field lines into the solar wind, the majority is recirculated within the magnetosphere [*Haaland et al.*, 2012a] and may eventually contribute to the formation of the hot plasma sheet and ring current population [e.g., *Giang et al.*, 2009]. Ion outflow possibly also plays a role for substorm triggering [e.g., *Winglee and Harnett*, 2011], although the direct role of ion outflow and its composition for substorm processes is still not fully understood.

Obtaining reliable and accurate measurements of cold ion outflow is challenging: Measurements have to be taken above the polar caps or in the magnetically connected lobes, regions characterized by a very tenuous plasma. In such plasma environments, solar illumination will cause a significant spacecraft charging due to photoelectron emissions, and the low-energy part of the ion population will be shielded from the ion detectors. Correspondingly, electron detectors will be contaminated by secondary electrons caused by the photo emissions.

Some of these problems can be overcome by actively controlling the spacecraft potential by emitting heavy ions to counterbalance the photoemission of electrons. This technique has been applied in a number of missions, e.g., the Geotail [*Schmidt et al.*, 1995], Polar [*Moore et al.*, 1995], Interball spacecraft [*Riedler et al.*, 1998], and more recently by the Double-Star and Cluster missions [*Torkar et al.*, 2001]. Although it is not always possible to bring and keep the spacecraft to zero potential, this technique can provide a significant improvement of plasma measurements. One example of successful utilization of this technique to measure cold ion outflow was the study by *Su et al.* [1998]. They used ion moments from the Thermal Ion Dynamics Experiment on board the Polar spacecraft during a limited time period between April and May 1996 when the on board Plasma Source Instrument [see *Moore et al.*, 1995] was operating and stabilized the spacecraft potential to around 2 V.

Nevertheless, due to limitations in the instrumentation and measurement techniques, many of the earlier observations of ion outflow at high altitudes only contained measurements of ions with energies above a certain threshold. The contribution from the cold (energies up to a few tens of eV) part of the ion population was not included in the calculations. In the seminal paper by *Chappell et al.* [1987], claiming that the ionosphere was a fully adequate source of plasma for the plasma sheet, the cold ion population was actually termed "invisible" for this reason.

Until recently, much of our knowledge about the role of ion outflow for magnetospheric dynamics was therefore based on models and simulations. *Cully et al.* [2003a] used empirical models of the electric and magnetic field combined with a simplified model of the ionospheric source population (protons and oxygen ions with energies below 20 eV). By tracing the trajectories of the ions for various geomagnetic conditions and orientations of the interplanetary magnetic field (IMF), they were able to model the supply to the central plasma sheet. Their modeling showed that the transport of thermal plasma to the central plasma sheet was largely controlled by the convection electric field. Similar conclusions were also drawn by *Moore et al.* [1999] based on Polar observations, although their results were based on particle measurements with energies up to 450 eV. *Ebihara et al.* [2006] employed empirical models of the O$^+$ distribution combined with models of the electric and magnetic field to investigate the fate of suprathermal oxygen. Their results suggested greatly enhanced feed to the inner magnetosphere during active geomagnetic conditions.

Recent advances in instrumentation and techniques have allowed inclusion also of the *measured* contribution from cold plasma in ion outflow studies. In particular, the Cluster spacecraft quartet, with its comprehensive instrumentation has provided more accurate measurements of density [*Lybekk et al.*, 2012], velocity and flux of cold outflowing ions [*Engwall et al.*, 2006, 2009a], and convection [*Haaland et al.*, 2007; *Förster et al.*, 2007; *Haaland et al.*, 2008]. Results from these advances have shown that low-energy ions can dominate the density contribution in large regions of the Earth's magnetosphere. This has created a new awareness of the importance of cold plasma for the dynamics of the magnetosphere [*Engwall et al.*, 2009a; *Andre and Cully*, 2012].

The objective of this paper is to utilize these advantages and try to assess and quantify the importance of cold ions as a supplier of plasma to the plasma sheet. The results are based on measurements from the Cluster mission over a period of several years and, thus, cover a much larger volume of the magnetosphere than earlier studies. Better availability and more accurate measurements of the time-shifted solar wind,

IMF as well as solar EUV measurements have also enabled us to better assess the role of these parameters for ion outflow.

2. METHODOLOGY

In the present study, we apply the technique outlined in the work of *Haaland et al.* [2012a]. They used recent results from the Cluster mission to quantify the direct loss of cold plasma into the solar wind along open field lines. With knowledge of the total outflow of cold ions as well as the downtail loss, the amount of recirculated plasma can be determined. The present study is, thus, to some extent, the complementary of these results, sharing the same data basis and, to a great extent, the same methodology.

Basically, the technique combines the parallel and perpendicular bulk velocities of cold ions with a magnetic field model to calculate the trajectory of ions, to determine whether they end up in the Earth's nightside plasma sheet or escape directly downtail into the solar wind. By combining the velocities with corresponding measurements of cold ion densities, the number flux can be calculated. We can then quantify or at least give a lower limit on how much cold ion outflow can contribute to the formation of the near-Earth magnetosphere, including the ring current and radiation belt regions.

Except for an attempt to estimate the amount of plasma lost through substorm-associated plasmoid ejection, we will not try to address any other loss mechanisms, such as charge exchange, loss due to drift across the magnetopause, or precipitation, etc., nor will we discuss mechanisms responsible for the acceleration and heating of the plasma taking place in the plasma sheet; the present data are not suited to address these issues.

For convenience, we briefly explain the methodology used in the *Haaland et al.* [2012a] paper, but for a more comprehensive explanation, we refer to the original publication.

2.1. Determination of the Fate of Outflowing Cold Ions

Whether an ion escaping the Earth's polar cap ionosphere at a certain latitude is fed to the plasma sheet or directly lost downtail into the solar wind is essentially a competition between the outward field-aligned transport and the transverse transport due to convection.

Consider an ion escaping the ionosphere along a field line, which is open at time t_1 as schematically illustrated in Figure 1. The ion has an initial field-aligned velocity $V_\parallel$ and possibly an acceleration along the field line. As time goes, this field line will convect toward the central plasma sheet and eventually reconnect at the distant neutral line after some

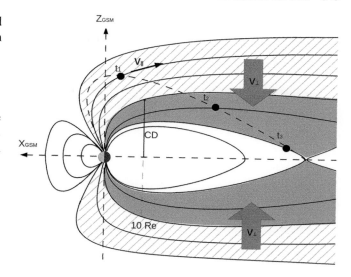

Figure 1. Definitions of open area (hatched), capture area (gray), trapped area (white), and convection distance (marked CD) in our model. Field lines on hatched background are open all the time it takes an ion to travel beyond the X line. Ions within flux tubes threading this area are thus lost downtail. Field lines on gray background are initially open, but during the time it takes for an ion to travel to the X line, the field line will convect a distance CD toward the plasma sheet. Ions at these latitudes or lower will thus be fed to the plasma sheet.

time, t_3. The gyrocenter of this ion will, therefore, follow the trajectory indicated as a dashed line in Figure 1.

If the ion is still earthward of the distant neutral line at time t_3, it will be fed to the plasma sheet where it can contribute to the formation of the hot plasma sheet. Other ions starting simultaneously, but at higher latitudes, or ions having larger parallel velocities and/or acceleration will end up tailward of the X line and escape from the magnetosphere directly into the solar wind.

For ions with a given outflow velocity, $V_\parallel$ and convection velocity $V_\perp$, we can then define a set of regions in the magnetosphere, which will determine the fate of these ions, as illustrated in Figure 1.

In our simplified model, the distant neutral line is assumed to be located at $X = -100\,R_E$ (see section 5.2 for a discussion about the validity of this assumption). We use the T01 magnetospheric field model [*Tsyganenko*, 2002a, 2002b] to find the foot point of the ions observed. From the data, we then calculate the average outflow and convection velocities and use the above method to determine the fate of the outflowing ions as a function of geomagnetic and solar wind conditions. Combined with measurements of cold ion densities, we can then get the total flux and total amount of cold ions supplied to the plasma sheet.

In some sense, this procedure resemble particle tracing done by, e.g., *Cladis* [1986], *Cully et al.* [2003a], and *Ebihara et al.* [2006] using the gyrocenter equation of motion [*Northrop*, 1963] with model fields. We do some simplifications though: First, gravity and mirror forces can be neglected in the altitudes discussed here. Also, since the objective is to determine whether the ions are directly lost (i.e., able to travel beyond distant neutral line) or end up somewhere in the plasma sheet, there is no need to do a detailed tracing of individual particles.

2.2. Size and Shape of the Source Regions

In their estimate of the total outflow, *Engwall et al.* [2009b] assumed that the source of the cold ions was at 1000 km altitude in the polar caps, defined as the area poleward of 70° magnetic latitude. Neither spatial inhomogeneities nor any expansions or contractions of the polar cap area in response to, e.g., IMF changes or geomagnetic activity were taken into account.

To get a more realistic estimate of the polar cap size, we make use of the results from *Sotirelis et al.* [1998]. They used characteristic particle signatures to identify the polar cap boundary. The rationale behind this approach is that the flux of energetic particles is much higher on closed field lines, and a sharp drop in intensity can be interpreted as a traversal from closed field lines (a trapped particle population) to open field lines (polar cap). The results from *Sotirelis et al.* [1998] demonstrate that both size and shape of the polar cap area can change dramatically, primarily as a response to the solar wind interaction with the dayside magnetopause.

Of relevance to us is Figure 8 of *Sotirelis et al.* [1998], which shows the amount of open flux, Φ, as a function of IMF B_z. For any given value of IMF B_z in our data sets and assuming a field magnitude of 37,000 nT in 1000 km altitude, we can consult this figure and derive the polar cap area as A [m^2] = 2 × 10e9 × Φ [Wb]/37,000 [nT] (the factor 2 indicates that we take both hemispheres into account).

In order to cross-check the validity of these calculations, *Li et al.* [2012] tried to trace the cold ion measurements from the *Engwall et al.* [2009b] data set back to the topside ionosphere using the full guiding center equation of motion [*Northrop*, 1963] and taking all relevant forces into account. Their results demonstrated that the open polar cap was the primary source region of the cold ions, though elevated fluxes in regions near the expected locations of the cusp and nightside auroral zone were also found. By sorting the data according to geomagnetic activity and IMF direction, *Li et al.* [2012] were able also to confirm the expansion and contraction of the polar cap, consistent with the *Sotirelis et al.* [1998] results. The source region during disturbed conditions was found to be up to a factor 3 larger than during quiet conditions.

3. DATA BASIS

The primary data basis for this study is ion outflow measurements and convection measurements from the Electric Field and Wave Experiment (EFW) [see *Gustafsson et al.*, 2001] and the electron drift instrument (EDI) [see *Paschmann et al.*, 2001], on board the Cluster spacecraft.

In addition to advances in techniques and instrumentation since the seminal papers by *Yau et al.* [1985] and *Chappell et al.*, 1987], the spatial coverage of our data and much better availability of accurate solar wind and IMF observations allow for a more comprehensive study of the fate of the cold outflowing ions.

The Cluster measurements used in this study span a period of 5 years for the cold ion outflow measurements and 6 years of convection measurements. Cluster's 57 h, 4 × 19 R_E polar orbit has provided extensive data coverage in the lobe and polar cap regions.

Once again, we provide a brief description of technique, data set characteristics, as well as an assessment of the accuracy of the data.

3.1. Cold Plasma Density and Outflow Velocity

Measurements of ion outflow velocity and density are the same as those used by *Engwall et al.* [2009b] and later in the work of *Haaland et al.* [2012a]. This data set consists of more than 1 million time-tagged records of electron density and 3-D plasma drift velocity, obtained over the full 4–19 R_E altitude range of Cluster. Each record also contains auxiliary parameters such as spacecraft position, solar wind, and IMF information and geomagnetic disturbance indices.

Figure 2 shows the distribution of density and velocity of the cold plasma outflow. The average (arithmetic mean) density is 0.18 ± 0.22 cm^{-3}, and the average outflow velocity is 26 ± 17.2 km s^{-1} (the spread in these numbers indicate standard deviations).

The electron density has been estimated from a known functional relation between spacecraft potential, V_s, and the ambient electron density, Ne:

$$Ne \sim Ae^{-V_s/B}, \tag{1}$$

where the coefficients A and B primarily depend on spacecraft properties such as surface area and surface conductivity as well as solar irradiance. Assuming quasineutrality and singly charged ions, the ion density is equal to the electron density given by the above equation. We refer to publications

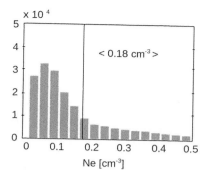
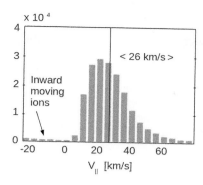
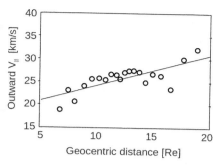

Figure 2. Data characteristics of the *Engwall et al.* [2009b] data set used for the present study. (left) Distribution of outflow densities. (center) Distribution of outflow velocity, $V_\|$. Horizontal axes show density and velocity, respectively, and vertical axes indicate number of samples. (right) Outflow velocity as a function of altitude.

by, e.g., *Pedersen et al.* [2001] or *Lybekk et al.* [2012] for a more comprehensive explanation of this method.

The outflow velocity and direction is calculated from simultaneous measurements of the convection electric field and an observed spacecraft wake. In the cold tenuous plasma of the polar cap and lobe regions, the spacecraft potential energy, eV_s, typically exceeds the bulk kinetic energy (E_k) of the ions, which, on the other hand, is larger than the thermal energy of the ions. The following inequality then exists:

$$kT_i < E_k < eV_s, \qquad (2)$$

where k is the Boltzmann constant, T_i is the ion temperature, and e is the elementary charge.

As a consequence of the above condition, a wake void of ions forms behind the spacecraft [see, e.g., *Eriksson et al.*, 2006]. The far more mobile electrons, on the other hand, will be able to fill the wake. Consequently, the EFW double probe instrument detects an artificial electric field along the wake direction. Combined with the magnetic field and the convection electric field, it is then possible to obtain the parallel velocity of the ions. For approximately 180,000 records of the full EFW data set, it was possible to calculate the outflow velocity by using this method.

Since the velocity determination relies on identification of this wake, the method is typically more sensitive to protons than to heavier ions. This tendency is usually weak, and any error in determined parallel flow velocity becomes significant only if the ion species have different parallel flow velocities, and the bulk kinetic energy of the heavy ions are close to or above the energy required to overcome the spacecraft potential.

Measurement errors due to the methodology is of the order ±40% or less for the velocity calculations and less than ±20% for the electron density calculations (details about error bounds of the data and methodology can be found in Sections 2.3 and 6.1 in the work of *Engwall et al.* [2009b])

Acceleration along the field is apparent from the right panel of Figure 2. We estimate an initial acceleration of the order of 0.6–0.7 km s$^{-1}R_E$. Acceleration due to gravity, magnetic mirror forces, or parallel electric fields do not play any role in this altitude range, so the observed increase in velocity is mainly attributed to centrifugal acceleration [e.g., *Cladis*, 1986; *Cladis et al.*, 2000; *Nilsson et al.*, 2010].

The calculation of outflow velocity rests on the simultaneous availability of EDI and EFW data of good quality. So far, only data from Cluster SC3 for the years 2001–2005 have been processed. Still, the number of records is large enough to allow us to extract subsets within certain geomagnetic activity levels, IMF directions, etc., to check correlations and dependencies.

3.2. Convection Toward the Plasma Sheet

Convection toward the plasma sheet using EDI data was studied in detail by *Haaland et al.* [2008, 2009], and we use the same data set for the present study. This data set consists of approximately 450,000 1 min averages (approximately 7600 h) mapped to an YZ_{GSM} plane at $X = -10 R_E$. At this distance and beyond, the convection is predominantly in the ±Z direction, i.e., toward the plasma sheet, as indicated in Figure 1.

EDI measures the gyrocenter drift velocity of an artificially emitted 1 keV electron beam. For a given magnetic field, this is equivalent to the convection electric field as long as there are no significant magnetic field gradients. EDI is not affected by the wake effects discussed in the previous section and works extremely well in regions of space with stable magnetic field and low ambient electron density such as the polar cap and lobe regions. Since both the magnetic field and the electron gyrotime can be determined with high accuracy, measurement errors with this technique are negligible. Uncertainties in statistical averages, therefore, come almost solely from the spread of the measurements.

For a southward-directed IMF, the variability in the convection data is typically low, with a well-defined convection direction and magnitude. For northward IMF conditions, the convection is more stagnant, and the spread in the data is comparable to the measurements itself. Since the convection velocities are vector quantities, there is spread in both magnitude and direction and, thus, no single parameter describing the variability. For a more detailed assessment of the spread in magnitude and direction of the EDI data set, we refer to Section 4.2 and Figure 4 in the work of *Haaland et al.* [2008].

4. RESULTS

The numbers given in the work of *Engwall et al.* [2009b, 2009a] and also the recent results from *Andre and Cully* [2012] suggest that ion outflow from the Earth's ionosphere is dominated by cold ions, predominantly protons. Outflow rates were found to be of the order of 10^{26} ions s^{-1}, consistent with earlier studies [e.g., *Yau and Andre*, 1997; *Cully et al.*, 2003b, 2003a; *Huddleston et al.*, 2005].

As demonstrated in the work of *Haaland et al.* [2012a], the direct downtail loss of cold ions can vary between essentially zero for periods with strong convection up to 2.5×10^{25} ions s^{-1} for quiet magnetospheric conditions and stagnant convection. Over a longer time span, approximately 10% of the outflowing cold ions are directly lost along open field lines into the solar wind, whereas 90% are recirculated.

In order to quantify the supply of cold ions to the plasma sheet in more detail, and investigate the role of solar wind parameters and geomagnetic activity, we created subsets of the full data set, containing only records within given ranges of selected driver parameters. Below, we discuss some of the results from this subsetting in detail.

4.1. Solar Wind and IMF Control

In order to assess the role of the IMF orientation for the outflow and loss balance, we created four subsets sorted according to the prevailing upstream IMF direction. Each subset contains data from a 90° clock angle sector centered around 0°, 90°, 180°, 270°, corresponding to northward, eastward, southward, and westward orientations of the IMF, respectively.

Table 1, which is essentially a reciprocal of the Table 2 given by *Haaland et al.* [2012a], summarizes the average key parameters for each of these IMF bins. Columns B to E show the average geomagnetic activity level and solar wind parameters for each subset. Columns F, G, and H show the corresponding average densities, velocities, and fluxes. Columns I to O show the average convection velocity and calculated values, explained in the footnotes.

Possible loss processes taking place after the ions have reached the plasma sheet, such as tailward retreat of a plasmoid, loss across the magnetopause, charge exchange, or precipitation are not taken into account in these calculations.

Table 1 reveals some interesting results. Whereas the convection (column L) varies from a few km s^{-1} for northward IMF to more than 12 km s^{-1} for southward IMF, neither outflow velocity nor density (columns F and G, respectively) changes dramatically with IMF direction.

Table 1. Averages From the Data Set for Four Different IMF Orientations

A	B	C	D	E	F	G	H	I	J	K	L	M	N	O
			Averages From *Engwall et al.* [2009b]					Calculated Values and Average Convection Velocity						
IMF[a]	Dst (nT)	Pdyn (nPa)	By (nT)	B_z (nT)	Ne (cm^{-2})	$V_\parallel$ (kms^{-1})	Flux (s^{-1} cm^{-2})	PCarea[b] (km^2)	Outflow[c] (s^{-1})	T_X[d] (min)	$V_\perp$[e] (km s^{-1})	CD[f] (R_E)	LossArea[g] (km^2)	PSFeed[h] (s^{-1})
B_z +	−21.9	2.43	0.7	3.7	0.196	24.6	1.35e8	0.8e7	1.1e25	212	1.7	3.4	7.29e6	0.38e25
B_z −	−39.2	2.41	−0.4	−4.6	0.233	26.9	1.48e8	4.3e7	6.4e25	204	12.2	23.1	0[i]	6.4-e25[j]
B_y+	−29.8	1.92	4.6	0.0	0.172	25.8	1.22e8	2.4e7	2.6e25	208	8.0	15.7	1.99e6	2.55e25
B_y −	−19.8	2.05	−5.4	−0.9	0.141	25.2	0.99e8	2.4e7	2.1e25	210	7.9	15.6	1.83e6	1.92e25

[a]The 90° sectors around 0°, 90°, 180°, and 270° clock angles.
[b]Polar cap area, based on Figure 8 in the work of *Sotirelis et al.* [1998] and an IGRF field of 37,000 nT at 1000 km altitude.
[c]Total outflow = polar cap area multiplied by flux.
[d]Travel time along field from $X_{GSE} = -10\ R_E$ to distant X line assumed to be located at 100 R_E.
[e]Average convection velocity based on the *Haaland et al.* [2008] data set.
[f]Convection distance: distance a field line convects during the travel time given in column K (see also Figure 1).
[g]Loss area: part of the polar cap where outflowing ions are lost downtail (see Figure 1).
[h]Feed to plasma sheet = total outflow − direct loss downtail.
[i]No direct downtail loss: all outflowing cold ions are fed to the plasma sheet.

A northward directed IMF is associated with almost stagnant convection (column L). During the 212 min (column K) it takes for an ion to travel along a field line from $X = -10 R_E$ to the distant X line 100 R_E downtail, the flux tube only convects around 3.4 R_E toward the plasma sheet (column M). As a result, nearly all cold ions emanating from the polar cap are able to travel downtail and escape into the solar wind, and the supply to the plasma sheet is minimal (column O). Even X line locations tailward of $X = -200 R_E$ do not prevent ions from escaping downtail into the solar wind for such slow convection velocities. During northward IMF, there is thus essentially no supply of cold plasma from the ionosphere to the plasma sheet.

The opposite can be observed during periods with southward directed IMF. The convection is now strong, presumably as a result of enhanced dayside reconnection. Hence, essentially, the entire population of cold ions from the ionosphere is convected to the plasma sheet. Also, a high convection velocity will prevent ions from traveling far downtail before reaching the plasma sheet. One can thus conclude that the supply of plasma of ionospheric origin is larger in the near-Earth and midtail regions than in the distant tail. Since both outflow density and velocity as well as the total polar cap area is larger during southward IMF, the total outflow and, thus, supply to the plasma sheet, is also much larger during southward IMF than for other IMF orientations. Some of the plasma supplied may be lost through tailward ejection of a plasmoid though. This will be further discussed in section 5.2.

A +/−By dominated IMF still has a fairly strong convection toward the plasma sheet, and the tailward loss is minimal. The polar cap area is about half of that for southward IMF, and the outflow density and outflow velocities are lower than for a purely southward IMF. The total outflow and, thus, supply to the plasma sheet is therefore smaller than for southward IMF.

By performing a similar subsetting, we also checked correlations between outflow and solar wind dynamical pressure (i.e., essentially a function of solar wind bulk velocity and solar wind ion density). These results are difficult to interpret, however. At first sight, the calculations suggest enhanced ion outflow during periods with high solar wind pressure, but this is probably a consequence of compression of the whole magnetosphere rather than additional supply from the ionosphere. The preliminary source estimates by *Li et al.* [2012] seem to confirm this, and the compression effect was also noted in the *Svenes et al.* [2008] and *Haaland et al.* [2012b] studies.

4.2. Geomagnetic Activity

To further facilitate comparison with earlier studies, we also tried to sort the data according to different geomagnetic activity levels. The *Engwall et al.* [2009b] and the *Haaland et al.* [2008] data sets both contain the *AE* and *Dst* indices to describe the prevailing geomagnetic conditions.

As a rule of thumb, the *Dst* index reflects geomagnetic storm processes with time scales of several hours to several days. *Dst* is primarily a measure of ring current enhancement, though there are some contributions from the solar wind pressure. The *AE* index, on the other hand, primarily reflects the strength of field-aligned currents connecting the magnetosphere with the auroral zone. These field-aligned currents are thought to arise from bursty bulk flow events and substorm activity in the near-Earth plasma sheet. The time scale of such effects is of the order of minutes. Note that there is often a high degree of mutual correlation between the *AE* and *Dst* indices, and also between these indices and the solar wind and IMF [e.g., *Förster et al.*, 2007]. The effects reflected by both *Dst* and *AE* are ultimately driven by the solar wind, in particular, the IMF.

Owing to the long transport times involved in cold ion outflow, and the focus on the polar cap rather than the auroral zone, the *Dst* index is therefore the best indicator of geomagnetic activity for our purpose.

The results are summarized in Table 2, which is similar to Table 1 but shows the calculated quantities for three different geomagnetic activity levels.

During disturbed conditions (here defined as *Dst* values less than −20 nT to get reliable statistics), both convection velocity and outflow flux are larger than during quiet or moderate conditions. In addition, the polar cap is more expanded, hence larger source area and a larger total outflow. Owing to the strong convection, almost all ions (4.03×10^{25} ions s^{-1} or ~96% of total outflow) are fed to the plasma sheet. Since substorm activity is more likely to take place during storm time conditions [e.g., *Kamide et al.*, 1998], downtail loss of plasma through escaping plasmoids is more likely to occur, though.

Moderate geomagnetic activity levels (*Dst* in the range 0 to −20 nT) are characterized by lower average convection velocities, thus enabling a larger fraction of the outflowing ions to escape downtail along open field lines. The polar cap size and thus source area is also smaller than for disturbed conditions. As a result, the total supply to the plasma sheet is only about 1/3 of that for storm time conditions.

During quiet times (positive *Dst* values), the observed cold ion density and thus flux seem to be somewhat higher than during moderate conditions. As with the solar wind ram pressure correlation mentioned above, we primarily interpret this as a compression of the magnetosphere rather than additional supply from the ionosphere. However, due to the more stagnant convection, a significant part of the outflowing ions is now directly lost downtail, and only about 1×10^{24} ions s^{-1} reach the plasma sheet.

Table 2. Similar to Table 1 But With Averages From the Data Set for Three Different Geomagnetic Activity Levels

A	B	C	D	E	F	G	H	I	J	K	L	M	N	O
	Averages From *Engwall et al.* [2009b]							Calculated Values and Average Convection Velocity						
Activity[a]	Dst (nT)	Pdyn (nPa)	By (nT)	B_z (nT)	Ne (cm^2)	$V_\parallel$ (km s^{-1})	Flux (s^{-1} cm^{-2})	PCarea[b] (km^2)	Outflow[c] (s^{-1})	T_X[d] (min)	$V_\perp$[e] (km s^{-1})	CD[f] (R_E)	LossArea[g] (km^2)	PSFeed[h] (s^{-1})
Quiet	7.5	3.02	−1.2	0.5	0.184	23.1	1.21e8	2.16e7	2.61e25	218	4.5	9.2	2.07e7	1.00e24
Moderate	−10.4	1.77	−0.5	−0.5	0.127	23.3	0.87e8	2.43e7	2.11e25	217	6.9	14.1	8.85e6	1.37e25
Disturbed	−43.0	2.24	1.4	−1.4	0.209	28.1	1.48e8	2.84e7	4.20e25	200	10.1	19.0	1.13e6	4.03e25

[a]Quiet, $Dst \geq 0$; moderate, Dst between −20 and 0; disturbed, Dst below −20 nT.

[b]Polar cap area: based on Figure 8 of *Sotirelis et al.* [1998] and an IGRF field of 37,000 nT at 1000 km altitude.

[c]Total outflow = polar cap area multiplied by flux.

[d]Travel time along field from $X_{GSE} = -10 \ R_E$ to distant X line.

[e]Average convection velocity based on the *Haaland et al.* [2008] data set.

[f]Convection distance: distance a field line convects during the travel time given in column K (see also Figure 1).

[g]Loss area: part of the polar cap where outflowing ions are lost downtail (see Figure 1).

[h]Feed to plasma sheet = total outflow − direct loss downtail.

4.3. Response to Changes in Solar Irradiance

Solar irradiance is believed to be the most prominent driver of ionization in the ionosphere [e.g., *Laakso et al.*, 2002; *Kitamura et al.*, 2011]. This is also evident from Cluster studies [*Svenes et al.*, 2008; *Lybekk et al.*, 2012], which shows a clear positive correlation between cold ion density and solar irradiance in the EUV range; high EUV values were associated with higher densities of cold plasma.

Engwall et al. [2009b] used the $F10.7$ index (a daily proxy for the total emission from the solar disc in the $F10.7$ cm wavelength) to investigate the relation between solar irradiance and cold ion outflux. Although the *Engwall et al.* [2009b] data set only covers a limited $F10.7$ range, their Figure 9 shows an increase by a factor of 3 between the lowest ($F10.7 = 100$) and highest ($F10.7 = 285$) values. More recent Cluster results by *Lybekk et al.* [2012], covering a longer time period of solar cycle 23 and larger ranges of the $F10.7$ index, suggests even larger variation in cold plasma density between low and high solar irradiance.

5. DISCUSSION

5.1. Feeding to Different Plasma Sheet Regions

From the above results, we can conclude that the fate of the outflowing ions is mainly controlled by the convection velocity, which can vary from essentially zero up to several tens of km s^{-1}, depending on geomagnetic disturbance level. Tables 1 and 2 show that the average outflow density and parallel velocity do not vary dramatically with solar wind conditions or geomagnetic activity level. An interesting consequence of this is that the convection will also strongly influence *where* (i.e., which radial distance) in the plasma sheet the recirculated cold ions will end up. We plan to follow up on this topic in a future study, but even our simplified model and the above results allow us to discuss this issue qualitatively.

Figure 3 schematically shows the typical transport paths for different magnetospheric conditions. As illustrated in Figure 3a, and also apparent from Table 1, stagnant convection will lead to substantial direct loss downtail. This also holds if the X line is located much further tailward than our assumptions.

The Cluster measurements of cold plasma outflow suggest that the outflow never completely subsides. As long as there is some convection toward the plasma sheet, this means that there is always some supply of cold plasma to the plasma sheet. Moderate convection velocities (a few km s^{-1}) will allow the ions to travel far downtail before reaching the plasma sheet (Figure 3b). When the ions eventually interact with the sharp reversal of the magnetic field in the tail midplane, they will be scattered and energized. Simulations have demonstrated that the closer this interaction takes place to the X line, the larger the energization [e.g., *Ashour-Abdalla et al.*, 1993, and references therein]. Ions fed at large downtail distances will spend more time in the plasma sheet where they can be energized.

During periods of strong convection, on the other hand, most of the outflowing cold plasma will be supplied closer to the Earth (illustrated in Figure 3c). With average outflow velocities around 25–30 km s^{-1} and convection velocities exceeding 15 km s^{-1}, essentially all ions are convected to the plasma sheet inside 50 R_E. These ions spend less time in the plasma sheet and will thus be less energized than ions supplied further downtail. Ions with very low parallel outflow

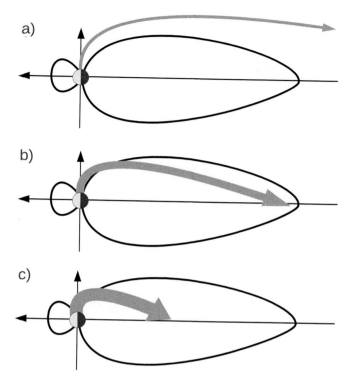

a)

b)

c)

Figure 3. Typical trajectories of cold ions for different disturbance levels. Thickness of arrows indicates flux. (a) Stagnant convection, typically northward IMF conditions and low geomagnetic activity. The outflow flux and velocity is low, and outflowing ions will be lost directly downtail to the solar wind. (b) Low to intermediate convection. Outflowing ions will be transported far downtail before reaching the plasma sheet. (c) Strong convection, typically associated with strong dayside reconnection and disturbed magnetospheric conditions. Outflowing ions will be returned to the plasma sheet closer to Earth.

velocities will even be convected almost directly to the plasmasphere without undergoing any significant energization in the tail at all.

In this sense, our results are largely in agreement with the Polar results reported by *Moore et al.* [1999], although their observations were based on particle measurements with energies up to 450 eV, with much higher average outflow velocities.

Whereas transport of plasma through the magnetotail lobes can reasonably be modeled as a steady convective motion, there are strong theoretical arguments against such a transport inside the plasma sheet [e.g., *Erickson and Wolf*, 1980; *Pritchett and Coroniti*, 1990]. Furthermore, measurements [e.g., *Angelopoulos et al.*, 1994] suggest that earthward transport of plasma in the plasma sheet primarily takes place through short-lived bursty bulk flows. With the present data

set, we are not well suited to investigate transport *inside* the plasma sheet or near the plasma sheet boundary layers, however.

5.2. Role of Magnetotail X Lines

The methodology used in this paper assumes the presence of a persistent X line, i.e., a reconnection region stretching across the magnetotail at a distance around 100 R_E downtail. This reconnection process is thought to be the counterpart of the dayside reconnection occurring as a result of interaction between the IMF and the geomagnetic field, in line with the large-scale plasma circulation envisaged by *Dungey* [1961]. Owing to the long transport times and the ability of the magnetotail to store magnetic flux [e.g., *Caan et al.*, 1975], the two reconnection processes do not necessarily need to be concurrent, though.

The position of the distant neutral line is based on studies by *Baker et al.* [1984] and *Birn et al.* [1992], suggesting an X line location around 100 R_E downtail.

It is assumed that magnetospheric plasma tailward of an X line is unable to be transported back to the near-Earth plasma sheet: it can be regarded as lost to the solar wind. The exact position of the distant X line does not play a big role for the total supply of cold ionospheric plasma to the plasma sheet. The largest supply occurs during strong convection. From our observations, and also illustrated in Figure 3c, the outflowing cold ions are brought to the plasma sheet close to the Earth, typically inside $X = -50 R_E$ in such cases. For slow or near stagnant convection, typically associated with quiet geomagnetic activity level, the outflowing ions are able to travel much longer downtail. However, in such cases, the total flux is also lower.

Reconnection and X line formation can also take place closer to the Earth, typically during geomagnetic active periods associated with bursty bulk flow activity and substorms [e.g., *Hones*, 1979; *Fuselier and Lewis*, 2011]. In this process, a fraction of the plasma sheet is thought to be pinched of and ejected downtail in the form of a plasmoid. Scale sizes, recurrence rate, and duration of near-Earth reconnection can vary greatly [e.g., *Newell and Gjerloev*, 2011], so any attempts to calculate the loss due to plasmoid ejections can only be crude estimates.

If we assume a substorm-associated 40 R_E wide X line at 25 R_E downtail [*Nagai et al.*, 1997] and a plasma sheet thickness of ~0.5 R_E [*Sergeev et al.*, 1990], we obtain a total plasmoid volume of approximately 4×10^{29} cm^3. Combined with an average plasma sheet density of 0.3 ions cm^{-3} [*Baumjohann et al.*, 1989], the loss in a single plasmoid is of the order of 10^{29} ions. Assuming up to five substorms per day [*Borovsky et al.*, 1993], we thus obtain a daily loss due to plasmoid ejection of the order $5–6 \times 10^{29}$ ions. Using the values from Tables 1 and 2 with supply rates of the order of a few times 10^{25} ions s^{-1}, we can then estimate that around 10%–50% of the ions supplied by the ionosphere eventually get lost due to plasmoid ejection.

In addition to the losses above, there are also various loss processes operating in the plasma sheet itself. *Matsui et al.* [1999] estimated that 3.6×10^{25} ions s^{-1} are lost through escape across the dayside magnetopause. There are also various other loss mechanism, such as charge exchange and precipitation.

5.3. Ion Outflow Composition

Our data set does not allow distinction between species, but we end with a few remarks about composition in this connection: Whereas outflow of H^+ is a fairly persistent feature of the polar cap ionosphere, heavier ions, e.g., O^+ have higher escape energies and require additional heating in order to escape. This heating is usually associated with elevated disturbance levels or solar activity. The O^+ to H^+ ratio will therefore vary with disturbance level and solar wind conditions [e.g., *Nosé et al.*, 2009, and references therein]. The composition of the plasma also plays an important role for fundamental plasma properties such as Alfvén speed, plasma pressure, plasma beta, and threshold levels for magnetospheric instabilities. Ion outflow can, therefore, play an important role for magnetospheric dynamics [e.g., *Glocer et al.*, 2009; *Brambles et al.*, 2011, and references therein].

6. SUMMARY

We have combined measurements of cold ion density and velocity from the high-altitude polar cap and lobe regions to investigate the role of cold ion outflow as a source of plasma for the Earth's plasma sheet. The data, obtained from Cluster spacecraft at $6–19$ R_E altitudes, are probably the most comprehensive in terms of spatial coverage and accuracy ever available for such a study. The availability of nearly continuous observations of solar wind parameters as well as measurements of geomagnetic activity have also allowed us to address the role of external influences for cold ion outflow and transport to the plasma sheet.

The key findings of this study are the following:

1. The supply of cold ions of ionospheric origin to the plasma sheet varies greatly. During quiet magnetic activity and low convection, our results indicate that the supply is of the order of 10^{24} ion s^{-1}. During disturbed conditions, typically associated with a southward oriented IMF and strong convection, the supply can be up to 100 times higher.

2. Over long time scales (i.e., much longer than disturbances such as geomagnetic storms, which affect the transport between the ionosphere and magnetosphere), approximately 90% of the outflowing cold ions are returned to the nightside plasma sheet, and thus 10% of the outflowing cold ions escape directly into the solar wind.

3. The outflow rate of cold ions from the source region in the ionosphere is predominantly modulated by solar irradiance and the size of this source region. The data show that the flux can change by a factor of 3 or more. Preliminary results from *Li et al.* [2012] also indicate that the source area in the polar cap vary by a factor 3.

4. Transport to the plasma sheet is almost exclusively controlled by the convection. During periods with strong convection, essentially all outflowing cold ions end up in the plasma sheet. The polar cap area is also typically larger during such conditions. Conversely, stagnant convection implies that a significant fraction of the total flux escapes downtail along open field lines and are directly lost to the solar wind.

5. The strength of the convection will also influence where (i.e., radial distances) the supply takes place. Strong convection leads to a significant supply close to the Earth, whereas weak convection will allow the outflowing cold ions to travel further downtail before reaching the plasma sheet.

6. Over a longer time period (much larger than typical substorm recurrence times), the loss due to tailward ejection of plasmoids is probably comparable or larger than the direct downtail loss of cold ions along open magnetic lobe field lines. Of the outflowing cold ions, 10%–50% escape to the solar wind after being heated in the plasma sheet.

Acknowledgments. SH acknowledges support from the World Universities Network and the Norwegian Research Council. The authors thank the International Space Science Institute, Bern, Switzerland, for providing computer resources and infrastructure for data exchange. The computer code used for the calculations in this paper has been made available as part of the QSAS science analysis system. QSAS is provided by the United Kingdom Cluster Science Centre (Imperial College London and Queen Mary, University of London) supported by the United Kingdom Science and Technology Facilities Council (STFC). Time-shifted solar wind data were obtained from the Coordinated Data Analysis Web (CDAWeb, see http://cdaweb.gsfc.nasa.gov/about.html).

REFERENCES

Abe, T., B. A. Whalen, A. W. Yau, R. E. Horita, S. Watanabe, and E. Sagawa (1993), EXOS D (Akebono) suprathermal mass spectrometer observations of the polar wind, *J. Geophys. Res.*, 98(A7), 11,191–11,203, doi:10.1029/92JA01971.

André, M., and C. M. Cully (2012), Low-energy ions: A previously hidden solar system particle population, *Geophys. Res. Lett.*, *39*, L03101, doi:10.1029/2011GL050242.

André, M., and A. Yau (1997), Theories and observations of ion energization and outflow in the high latitude magnetosphere, *Space Sci. Rev.*, *80*, 27–48, doi:10.1023/A:1004921619885.

Angelopoulos, V., C. F. Kennel, F. V. Coroniti, R. Pellat, M. G. Kivelson, R. J. Walker, C. T. Russell, W. Baumjohann, W. C. Feldman, and J. T. Gosling (1994), Statistical characteristics of bursty bulk flow events, *J. Geophys. Res.*, *99*(A11), 21,257–21,280, doi:10.1029/94JA01263.

Ashour-Abdalla, M., J. P. Berchem, J. Büchner, and L. M. Zelenyi (1993), Shaping of the magnetotail from the mantle: Global and local structuring, *J. Geophys. Res.*, *98*(A4), 5651–5676, doi:10.1029/92JA01662.

Axford, W. I. (1968), The polar wind and the terrestrial helium budget, *J. Geophys. Res.*, *73*(21), 6855–6859, doi:10.1029/JA073i021p06855.

Baker, D. N., et al. (1984), Direct observations of passages of the distant neutral line (80–140 R_E) following substorm pnsets: ISEE-3, *Geophys. Res. Lett.*, *11*(10), 1042–1045, doi:10.1029/GL011i010p01042.

Baumjohann, W., G. Paschmann, and C. A. Cattell (1989), Average plasma properties in the central plasma sheet, *J. Geophys. Res.*, *94*(A6), 6597–6606.

Birn, J., G. Yur, H. U. Rahman, and S. Minami (1992), On the termination of the closed field line region of the magnetotail, *J. Geophys. Res.*, *97*(A10), 14,833–14,840, doi:10.1029/92JA01145.

Borovsky, J. E., R. J. Nemzek, and R. D. Belian (1993), The occurrence rate of magnetospheric-substorm onsets: Random and periodic substorms, *J. Geophys. Res.*, *98*(A3), 3807–3813, doi:10.1029/92JA02556.

Brambles, O. J., W. Lotko, B. Zhang, M. Wiltberger, J. Lyon, and R. J. Strangeway (2011), Magnetosphere sawtooth oscillations induced by ionospheric outflow, *Science*, *332*, 1183–1186, doi:10.1126/science.1202869.

Caan, M. N., R. L. McPherron, and C. T. Russell (1975), Substorm and interplanetary magnetic field effects on the geomagnetic tail lobes, *J. Geophys. Res.*, *80*(1), 191–194, doi:10.1029/JA080i001p00191.

Chappell, C. R., T. E. Moore, and J. H. Waite Jr. (1987), The ionosphere as a fully adequate source of plasma for the Earth's magnetosphere, *J. Geophys. Res.*, *92*(A6), 5896–5910, doi:10.1029/JA092iA06p05896.

Chappell, C. R., B. L. Giles, T. E. Moore, D. C. Delcourt, P. D. Craven, and M. O. Chandler (2000), The adequacy of the ionospheric source in supplying magnetospheric plasma, *J. Atmos. Sol. Terr. Phys.*, *62*, 421–436, doi:10.1016/S1364-6826(00)00021-3.

Cladis, J. B. (1986), Parallel acceleration and transport of ions from polar ionosphere to plasma sheet, *Geophys. Res. Lett.*, *13*(9), 893–896, doi:10.1029/GL013i009p00893.

Cladis, J. B., H. L. Collin, O. W. Lennartsson, T. E. Moore, W. K. Peterson, and C. T. Russell (2000), Observations of centrifugal acceleration during compression of magnetosphere, *Geophys. Res. Lett.*, *27*(7), 915–918, doi:10.1029/1999GL010737.

Cully, C. M., E. F. Donovan, A. W. Yau, and H. J. Opgenoorth (2003a), Supply of thermal ionospheric ions to the central plasma sheet, *J. Geophys. Res.*, *108*(A2), 1092, doi:10.1029/2002JA009457.

Cully, C. M., E. F. Donovan, A. W. Yau, and G. G. Arkos (2003b), Akebono/Suprathermal Mass Spectrometer observations of low-energy ion outflow: Dependence on magnetic activity and solar wind conditions, *J. Geophys. Res.*, *108*(A2), 1093, doi:10.1029/2001JA009200.

Dungey, J. R. (1961), Interplanetary magnetic field and the auroral zones, *Phys. Rev. Lett.*, *6*, 47–48.

Ebihara, Y., M. Yamada, S. Watanabe, and M. Ejiri (2006), Fate of outflowing suprathermal oxygen ions that originate in the polar ionosphere, *J. Geophys. Res.*, *111*, A04219, doi:10.1029/2005JA011403.

Engwall, E., A. I. Eriksson, M. André, I. Dandouras, G. Paschmann, J. Quinn, and K. Torkar (2006), Low-energy (order 10 eV) ion flow in the magnetotail lobes inferred from spacecraft wake observations, *Geophys. Res. Lett.*, *33*, L06110, doi:10.1029/2005GL025179.

Engwall, E., A. I. Eriksson, C. M. Cully, M. André, R. Torbert, and H. Vaith (2009a), Earth's ionospheric outflow dominated by hidden cold plasma, *Nat. Geosci.*, *2*, 24–27, doi:10.1038/ngeo387.

Engwall, E., A. I. Eriksson, C. M. Cully, M. André, P. A. Puhl-Quinn, H. Vaith, and R. Torbert (2009b), Survey of cold ionospheric outflows in the magnetotail, *Ann. Geophys.*, *27*, 3185–3201.

Erickson, G. M., and R. A. Wolf (1980), Is steady convection possible in the Earth's magnetotail?, *Geophys. Res. Lett.*, *7*(11), 897–900, doi:10.1029/GL007i011p00897.

Eriksson, A. I., et al. (2006), Electric field measurements on Cluster: Comparing the double-probe and electron drift techniques, *Ann. Geophys.*, *24*, 275–289, doi:10.5194/angeo-24-275-2006.

Förster, M., G. Paschmann, S. E. Haaland, J. M. Quinn, R. B. Torbert, H. Vaith, and C. A. Kletzing (2007), High-latitude plasma convection form cluster EDI: Variances and solar wind correlation, *Ann. Geophys.*, *25*, 1691–1707.

Fuselier, S. A., and W. S. Lewis (2011), Properties of near-Earth magnetic reconnection from in situ observations, *Space Sci. Rev.*, *160*, 95–121, doi:10.1007/s11214-011-9820-x.

Giang, T. T., et al. (2009), Outflowing protons and heavy ions as a source for the sub-keV ring current, *Ann. Geophys.*, *27*, 839–849, doi:10.5194/angeo-27-839-2009.

Glocer, A., G. Tóth, T. Gombosi, and D. Welling (2009), Modeling ionospheric outflows and their impact on the magnetosphere, initial results, *J. Geophys. Res.*, *114*, A05216, doi:10.1029/2009JA014053.

Gustafsson, G., et al. (2001), First results of electric field and density measurements by Cluster EFW based on initial months of operation, *Ann. Geophys.*, *19*, 1219–1240.

Haaland, S., M. Förster, and G. Paschmann (2007), High-latitude plasma convection from cluster EDI measurements: Method and IMF-dependence, *Ann. Geophys.*, *25*, 239–253.

Haaland, S., G. Paschmann, M. Förster, J. Quinn, R. Torbert, H. Vaith, P. Puhl-Quinn, and C. Kletzing (2008), Plasma convection in the magnetotail lobes: Statistical results from Cluster EDI measurements, *Ann. Geophys.*, *26*, 2371–2382.

Haaland, S., B. Lybekk, K. Svenes, A. Pedersen, M. Förster, H. Vaith, and R. Torbert (2009), Plasma transport in the magnetotail lobes, *Ann. Geophys.*, *27*, 3577–3590.

Haaland, S., et al. (2012a), Estimating the capture and loss of cold plasma from ionospheric outflow, *J. Geophys. Res.*, *117*, A07311, doi:10.1029/2012JA017679.

Haaland, S., K. Svenes, B. Lybekk, and A. Pedersen (2012b), A survey of the polar cap density based on Cluster EFW probe measurements: Solar wind and solar irradiation dependence, *J. Geophys. Res.*, *117*, A01216, doi:10.1029/2011JA 017250.

Hones, E. W., Jr. (1979), Transient phenomena in the magnetotail and their relation to substorms, *Space Sci. Rev.*, *23*, 393–410, doi:10.1007/BF00172247.

Huddleston, M. M., C. R. Chappell, D. C. Delcourt, T. E. Moore, B. L. Giles, and M. O. Chandler (2005), An examination of the process and magnitude of ionospheric plasma supply to the magnetosphere, *J. Geophys. Res.*, *110*, A12202, doi:10.1029/ 2004JA010401.

Kamide, Y., et al. (1998), Current understanding of magnetic storms: Storm-substorm relationships, *J. Geophys. Res.*, *103* (A8), 17,705–17,728, doi:10.1029/98JA01426.

Kitamura, N., Y. Ogawa, Y. Nishimura, N. Terada, T. Ono, A. Shinbori, A. Kumamoto, V. Truhlik, and J. Smilauer (2011), Solar zenith angle dependence of plasma density and temperature in the polar cap ionosphere and low-altitude magnetosphere during geomagnetically quiet periods at solar maximum, *J. Geophys. Res.*, *116*, A08227, doi:10.1029/2011JA016631.

Laakso, H., R. Pfaff, and P. Janhunen (2002), Polar observations of electron density distribution in the Earth's magnetosphere. 1. Statistical results, *Ann. Geophys.*, *20*, 1711–1724, doi:10.5194/ angeo-20-1711-2002.

Li, K., et al. (2012), On the ionospheric source region of cold ion outflow, *Geophys. Res. Lett.*, *39*, L18102, doi:10.1029/2012GL 053297.

Lybekk, B., A. Pedersen, S. Haaland, K. Svenes, A. N. Fazakerley, A. Masson, M. G. G. T. Taylor, and J.-G. Trotignon (2012), Solar cycle variations of the Cluster spacecraft potential and its use for electron density estimations, *J. Geophys. Res.*, *117*, A01217, doi:10.1029/2011JA016969.

Matsui, H., T. Mukai, S. Ohtani, K. Hayashi, R. C. Elphic, M. F. Thomsen, and H. Matsumoto (1999), Cold dense plasma in the outer magnetosphere, *J. Geophys. Res.*, *104*(A11), 25,077–25,095, doi:10.1029/1999JA900046.

Moore, T. E., et al. (1995), The thermal ion dynamics experiment and plasma source instrument, *Space Sci. Rev.*, *71*, 409–458, doi:10.1007/BF00751337.

Moore, T. E., et al. (1999), Polar/TIDE results on polar ion out-flows, in *Sun-Earth Plasma Connections*, *Geophys. Monogr. Ser.*, vol. 109, edited by L. Burch, L. Carovillano and K. Antiochos,

pp. 87–101, AGU, Washington, D. C., doi:10.1029/GM 109p0087.

Nagai, T., R. Nakamura, T. Mukai, T. Yamamoto, A. Nishida, and S. Kokubun (1997), Substorms, tail flows and plasmoids, *Adv. Space Res.*, *20*, 961–971, doi:10.1016/S0273-1177(97)00504-8.

Newell, P. T., and J. W. Gjerloev (2011), Substorm and magnetosphere characteristic scales inferred from the SuperMAG auroral electrojet indices, *J. Geophys. Res.*, *116*, A12232, doi:10.1029/ 2011JA016936.

Nilsson, H., E. Engwall, A. Eriksson, P. A. Puhl-Quinn, and S. Arvelius (2010), Centrifugal acceleration in the magnetotail lobes, *Ann. Geophys.*, *28*, 569–576.

Northrop, T. (1963), *The Adiabatic Motion of Charged Particles*, Wiley, New York.

Nosé, M., S. Taguchi, S. P. Christon, M. R. Collier, T. E. Moore, C. W. Carlson, and J. P. McFadden (2009), Response of ions of ionospheric origin to storm time substorms: Coordinated observations over the ionosphere and in the plasma sheet, *J. Geophys. Res.*, *114*, A05207, doi:10.1029/2009JA014048.

Ogawa, Y., S. C. Buchert, I. Häggström, M. T. Rietveld, R. Fujii, S. Nozawa, and H. Miyaoka (2011), On the statistical relation between ion upflow and naturally enhanced ion-acoustic lines observed with the EISCAT Svalbard radar, *J. Geophys. Res.*, *116*, A03313, doi:10.1029/2010JA015827.

Paschmann, G., et al. (2001), The Electron Drift Instrument on Cluster: Overview of first results, *Ann. Geophys.*, *19*, 1273–1288.

Pedersen, A., P. Décréau, C.-P. Escoubet, G. Gustafsson, H. Laakso, P.-A. Lindqvist, B. Lybekk, A. Masson, F. Mozer, and A. Vaivads (2001), Four-point high time resolution information on electron densities by the electric field experiments (EFW) on Cluster, *Ann. Geophys.*, *19*, 1483–1489, doi:10.5194/angeo-19-1483-2001.

Pritchett, P. L., and F. V. Coroniti (1990), Plasma sheet convection and the stability of the magnetotail, *Geophys. Res. Lett.*, *17*(12), 2233–2236, doi:10.1029/GL017i012p02233.

Riedler, W., et al. (1998), Experiment RON for active control of spacecraft electric potential, *Cosmic Res.*, Engl. Transl., *36*, 49–58.

Schmidt, R., et al. (1995), Results from active spacecraft potential control on the Geotail spacecraft, *J. Geophys. Res.*, *100*(A9), 17,253–17,259, doi:10.1029/95JA01552.

Sergeev, V. A., P. Tanskanen, K. Mursula, A. Korth, and R. C. Elphic (1990), Current sheet thickness in the near-Earth plasma sheet during substorm growth phase, *J. Geophys. Res.*, *95*(A4), 3819–3828, doi:10.1029/JA095iA04p03819.

Sotirelis, T., P. T. Newell, and C.-I. Meng (1998), Shape of the open–closed boundary of the polar cap as determined from observations of precipitating particles by up to four DMSP satellites, *J. Geophys. Res.*, *103*(A1), 399–406, doi:10.1029/97JA 02437.

Su, Y.-J., J. L. Horwitz, T. E. Moore, B. L. Giles, M. O. Chandler, P. D. Craven, M. Hirahara, and C. J. Pollock (1998), Polar wind survey with the Thermal Ion Dynamics Experiment/Plasma Source Instrument suite aboard POLAR, *J. Geophys. Res.*, *103*(A12), 29,305–29,337, doi:10.1029/98JA02662.

Svenes, K., B. Lybekk, A. Pedersen, and S. Haaland (2008), Cluster observations of magnetospheric lobe plasma densities for different solar wind conditions – A statistical study, *Ann. Geophys.*, *26*, 2845–2852.

Torkar, K., et al. (2001), Active spacecraft potential control for Cluster: Implementation and first results, *Ann. Geophys.*, *19*, 1289–1302, doi:10.5194/angeo-19-1289-2001.

Tsyganenko, N. A. (2002a), A model of the near magnetosphere with a dawn-dusk asymmetry 1. Mathematical structure, *J. Geophys. Res.*, *107*(A8), 1179, doi:10.1029/2001JA 000219.

Tsyganenko, N. A. (2002b), A model of the near magnetosphere with a dawn-dusk asymmetry 2. Parameterization and fitting to observations, *J. Geophys. Res.*, *107*(A8), 1176, doi:10.1029/ 2001JA000220.

Winglee, R. M., and E. Harnett (2011), Influence of heavy ionospheric ions on substorm onset, *J. Geophys. Res.*, *116*, A11212, doi:10.1029/2011JA016447.

Yau, A. W., and M. André (1997), Sources of ion outflow in the high latitude ionosphere, *Space Sci. Rev.*, *80*, 1–25, doi:10.1023/ A:1004947203046.

Yau, A. W., E. G. Shelley, W. K. Peterson, and L. Lenchyshyn (1985), Energetic auroral and polar ion outflow at DE 1 altitudes: Magnitude, composition, magnetic activity dependence, and long-term variations, *J. Geophys. Res.*, *90*(A9), 8417–8432, doi:10.1029/JA090iA09p08417.

M. André, E. Engwall and A. Eriksson, Swedish Institute of Space Physics, Uppsala, Sweden.

M. Förster, Helmholtz Centre Potsdam, German Research Centre for Geosciences, Potsdam, Germany.

S. Haaland, Max-Planck Institute for Solar Systems Research, Katlenburg-Lindau, Germany. (stein.haaland@issi.unibe.ch)

C. Johnsen and N. Østgaard, Department of Physics and Technology, University of Bergen, Bergen, Norway.

K. Li, National Space Science Center, Chinese Academy of Sciences, Beijing, China.

B. Lybekk and A. Pedersen, Department of Physics, University of Oslo, Oslo, Norway.

H. Nilsson, Swedish Institute of Space Physics, Kiruna, Sweden.

K. Svenes, Norwegian Defense Research Establishment, Kjeller, Norway.

What Happens When the Geomagnetic Field Reverses?

J. F. Lemaire

Belgian Institute for Space Aeronomy, Brussels, Belgium

S. F. Singer

Department of Environmental Sciences, University of Virginia, Charlottesville, Virginia, USA
Science and Environmental Policy Project, Arlington, Virginia, USA

During geomagnetic field reversals, the radiation belt high-energy proton populations become depleted. Their energy spectra become softer, with the trapped particles of highest energies being lost first, and eventually recovering after a field reversal. The radiation belts rebuild in a dynamical way, with the energy spectra flattening on average during the course of many millennia, but without ever reaching complete steady state equilibrium between successive geomagnetic storm events determined by southward turnings of the interplanetary magnetic field (IMF) orientation. Considering that the entry of galactic cosmic rays (GCR) and the solar energetic particles (SEP) with energies above a given threshold are strongly controlled by the intensity of the northward component of the IMF, we speculate that at earlier epochs when the geomagnetic dipole was reversed, the entry of these energetic particles into the geomagnetic field was facilitated when the IMF was directed northward. Unlike in other complementary work where intensive numerical simulations have been used, our demonstration is based on a simple analytical extension of Störmer's theory. The access of GCR and SEP beyond geomagnetic cutoff latitudes is enhanced during epochs when the Earth's magnetic dipole is reduced, as already demonstrated earlier.

1. INTRODUCTION

Undoubtedly, geomagnetic field polarity reversals have had drastic effects on the inner radiation belts, as well on the access of galactic cosmic rays (GCR) and solar energetic particles (SEP) into the magnetosphere. Already, *Uffen* [1963, p. 143] speculated that "During those intervals the trapped corpuscular radiation may have been spilled on the Earth, ..." Solar proton events produce ozone depletions [e.g., *Jackman et al.*, 2005], and the access of GCR may possibly control the cloud coverage, and thus influence the Earth's climate [*Svensmark and Friis-Christensen*, 1997; *Kirkby et al.*, 2011]. The historical development and perspectives of magnetic polarity transitions on these geophysical phenomena, as well as on the biosphere have been updated by *Glassmeier and Vogt* [2010] (see also the book of *Glassmeier et al.* [2009]).

Directly related to the present study, *Stadelmann et al.* [2010] presented results based on thousands of trajectory calculations of GCR and SEP in a potential B field model of paleomagnetospheres where the dipole and quadrupole moments have been given prescribed values. Like *Smart et al.* [2000], they determined the geomagnetic cutoff latitudes and "impact areas," which quantify the percentage of the Earth's

Dynamics of the Earth's Radiation Belts and Inner Magnetosphere
Geophysical Monograph Series 199
10.1029/2012GM001307

surface that is accessible to charged particles of a given energy, from all incident directions and from all locations on a spheric shell in outer space. These detailed calculations as well as those of *Vogt and Glassmeier* [2000], *Vogt et al.* [2004, 2007], and *Zieger et al.* [2004, 2006] are interesting and suitable for nonsymmetrical magnetic field configurations. Emphasis in these earlier studies was mainly on the impact of GCR and SEP on the atmosphere. The main objectives of this study relate to the energy spectra of trapped radiation belt particles. The key role of the changes of orientation of the interplanetary magnetic field (IMF) during geomagnetic field reversals will be pointed out.

In the following, we use a less demanding analytical approach based on Störmer's theory, which was originally developed for a dipole magnetic distribution [*Störmer*, 1907, 1913, 1955]. This theory has been extended by *Lemaire* [2003] by adding a uniform IMF (**F**) to the magnetic dipole (**M**). As a consequence of its zonal symmetry φ, the azimuthal/longitudinal coordinate is a dummy variable in the expression of the Hamiltonian of charged particles moving in this B field configuration. Therefore, topologically allowed and forbidden zones can be defined for incoming and trapped particles of prescribed kinetic energy (KE) or magnetic rigidity. This property enables us to determine rather easily, analytically, without intensive numerical integration of particle trajectories, the geomagnetic cutoff surface, as well as the guiding center field line of trapped radiation belt particles (i.e., their "Thalweg") for different orientations of the IMF (**F**) and for different values of the Earth's dipole moment (**M**).

To set the stage, the Störmer theory will first be briefly reviewed. It will then be shown how the Thalweg of a trapped radiation belt particle of given energy breaks open by interconnecting to the IMF when the latter turns southward and becomes sufficiently negative. Conversely, interplanetary ions will then be able to enter in the inner part of the magnetosphere and possibly even impact the atmosphere.

Our analytical approach shows also that the energy for which trapped radiation belt particles can spill out of the trapping zones is gradually reduced when M becomes smaller during a geomagnetic field polarity transition.

2. BRIEF REVIEW OF STÖRMER'S THEORY

2.1. Coordinate Systems

Störmer [1907] found that trajectories of particles in a dipole B field are confined within regions corresponding to *inner and outer allowed zones* subsequently named after him. The allowed zones are separated by a *forbidden zone* whose shape is a function of the constants of motion of

charged particles: the energy, and the generalized or canonical angular momentum. These zones are well described by *Störmer* [1955], where he uses a coordinate system whose unit length is proportional to Ze, the electric charge of the particle, and inversely proportional to the modulus of generalized momentum, $p = mv$.

Another, but less well known, mathematical formulation of Störmer's theory was proposed many years later by *Dragt* [1965]. The latter is simpler and leads to more intuitive graphical representations of Störmer's allowed and forbidden zones. Dragt introduced the dimensionless time, t, and polar coordinates (ρ, z, ϕ), whose units are determined respectively by

$$t_u = \left| m(Ze\,M)^2/p_o^3 \right| \qquad (1)$$

$$r_u = \left| Ze\,M/p_o \right|, \qquad (2)$$

where m is the mass of the particle, and p_o is the (constant) azimuthal component of the generalized/canonical momentum (p_ϕ). The latter is a constant of motion due to the zonal/axial symmetry of the dipole magnetic field distribution. For trapped radiation belt particles, the value of r_u is larger than R_E, the radius of the Earth ($r_E = R_E/r_u < 1$). The relationship between r_u and t_u was illustrated by *Lemaire* [2003] in his Figure A4.

2.2. The Störmer Potential

Using Dragt's coordinate system, the distribution of the Störmer potential, $V(\rho, z)$, is illustrated in Figure 1 by isocontours ranging from $V_0 = 0$ to ∞ in a meridional plane. This 2-D representation is independent of ϕ, the longitude of the meridional plane. A corresponding 3-D landscape representation of the function $V(\rho, z)$ was illustrated in Figure 5a by *Lemaire* [2003].

In such 2- and 3-D plots for particles for which Ze and p_o have the same algebraic signs, there is a saddle point at $z = 0$ and $\rho = 2$, where two isocontours cross each other. At this saddle point, the equatorial potential distribution $V(\rho, 0)$ has a maximum value, $V(2, 0) = 1/32$. It can be seen from Figure 1 that beyond this saddle point, Störmer's potential tends to zero when $r = (\rho^2 + z^2)^{1/2} \rightarrow \infty$. In the opposite direction, $V(\rho, z)$ forms a deep "valley" whose steep earthward wall corresponds to the "geomagnetic cutoff surface."

The bottom of this "valley" indicated by the dotted line in Figure 1 is where $V = 0$. It corresponds to the minimum of Störmer's potential, which is called the Thalweg (path in the valley). Indeed, the function $V(\rho, z)$ can be compared to a geopotential surface whose "height" is proportional to $V(\rho, z)$. Any particle inside this valley is trapped within

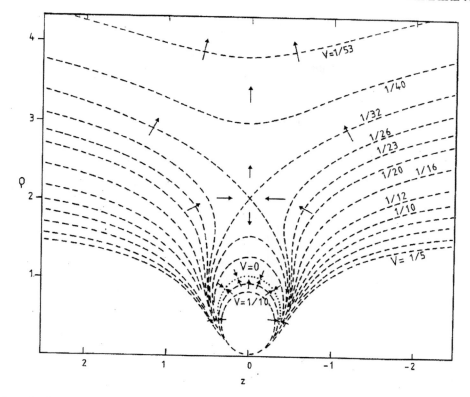

Figure 1. Störmer's dimensionless potential, $V(\rho, z)$. The dashed lines are the isocontours of $V(\rho, z)$. They determine the frontiers between the Störmer's allowed and forbidden zones for V ranging from $V = 0$ to $1/5$. The dotted line is the Thalweg, which coincides with the dipole magnetic field line around which trapped particle spiral and oscillate between conjugate mirror points. The arrows indicate the slopes of $V(\rho, z)$. The potential has maximum ($V = 1/32$) at $z = 0$ and $\rho_{max} = 2$: a saddle point where isocontours cross and where the inner and outer allowed zones interconnect when the kinetic energy of particles is larger than $1/32$, in Dragt's dimensionless length unit. When the energy of an interplanetary particle is larger than $1/32$, it can override the magnetic potential barrier and penetrate deep into the geomagnetic field between cusp-like dashed lines. These distributions are drawn for charged particle characterized by $\varepsilon = \text{sign}(-p_o/Z) = +1$. From *Dragt* [1965].

Störmer's inner allowed zone when its dimensionless KE is smaller than $1/32$. The projection of the trajectory of trapped particles in a meridional plane is then strictly confined between two of the dashed lines shown in Figure 1, which constitute the borders of the inner Störmer zone (e.g., $V = 1/1000$).

2.3. The Thalweg and the Guiding Center Field Line

The Thalweg at the very bottom of this "valley" coincides with the dipole magnetic field line along which trapped particles of small KE (KE < 1/32) spiral, bounce, and drift indefinitely. In Dragt's coordinate system, the Thalweg corresponds to the dipole magnetic field line traversing the equatorial plane at $\rho = r_T = 1$.

The Thalweg corresponds to the guiding center field line in Alfvén's adiabatic perturbation theory [*Alfvén*, 1940, 1950]. The equatorial crossing points of Thalwegs correspond to the central points of the bounce motion of trapped particles and

are part of what *Vogt and Glassmeier* [2000] label the "trapping center surface" in their paleomagnetosphere models.

2.4. Access of High-Energy Interplanetary Particles Into the Inner Geomagnetic Field

Interplanetary particles from infinity will move around the dipole either eastward or westward depending on the sign of their electric charge and of the angular momentum. When their dimensionless KE is smaller than $1/32$ their trajectories remain confined in the outer allowed zone, beyond a distant isocontours of Figure 1 (e.g., the dashed curve labeled $V = 1/53$).

Only interplanetary particles for which KE > 1/32 can transit over the saddle point region, i.e., over what happens to be a forbidden zone for particles of lower energies (KE < 1/32). Consequently, unless their KE exceeds the dimensionless threshold of $1/32$, GCR or SEP are unable to penetrate deep into the geomagnetic field and populate the inner allowed Störmer zone.

2.5. Sources of Radiation Belts Particles

Van Allen and Singer [1950a, 1950b] made good use of Störmer's theory to deduce the energy spectrum of primary cosmic rays by measuring their flux in rocket experiments at different geomagnetic latitudes. After the discovery of the radiation belts by *Van Allen et al.* [1958], *Singer* [1958a, 1958b, 1959a, 1959b] proposed cosmic ray albedo neutrons as a source for the corpuscular radiation populating Störmer's inner allowed zone. Note that this same injection mechanism was also proposed independently by *Vernov et al.* [1959] as a source of the trapped energetic particles detected both by the Sputnik and Explorer spacecraft. The source and loss processes for radiation belt particles were comprehensively reviewed by *Walt* [1996] and by *Singer and Lemaire* [2009].

2.6. The Renaissance and Fall of Störmer's Theory

Thus, Störmer's theory became, again, popular in 1958, after the discovery of the radiation belts. It looked to be a very promising tool until 1961, when McIlwain's invariant geomagnetic coordinate system proved to be very useful to map fluxes of energetic electrons and protons trapped in the geomagnetic field [*McIlwain*, 1961, 1966]. This gave new life to Alfvén's "first-order guiding center theory" also promoted by *Northrop* [1963]. Unfortunately, this contributed to the fall/decline of Störmer's theory from theoretical space physics. Nevertheless, at least for a few years, Störmer's theory had been successfully used in studies of trajectories of 0.1–10 MeV electrons and 0.5–300 MeV protons trapped within the radiation belts. See also *Akasofu* [2003] for a critical discussion.

3. AN EXTENSION OF STÖRMER'S THEORY

3.1. Magnetic Field Line Distribution

As has been recalled above, the cylindrical symmetry of this B field distribution allows to reduce from three to two the number of independent variables to describe the trajectory of a charged particle. Indeed, once $\rho(t)$ and $z(t)$ are calculated in a meridional plane as a function of time t, the azimuthal velocity, $v_\phi = \rho d\phi/dt$ is also determined. By integrating the azimuthal component of the equation of motion, $\phi(t)$ can then be directly calculated.

Less than a decade ago, it was found that the same procedure can apply in the cases of more general B field distributions or models: e.g., when a uniform IMF (**F**) is superimposed on the dipole magnetic field (**B**$_d$), as for instance in *Dungey's* [1961] magnetic field model. Indeed, when this IMF is either parallel or antiparallel to the dipole moment, **M**, the cylindrical/zonal symmetry of the B field and of the Hamiltonian is preserved. Under such circumstances, p_ϕ, the azimuthal component of the generalized momentum, is again a constant of motion $p_\phi = p_o$, and *Lemaire* [2003] pointed out that an extended Störmer potential, $V(\rho, z)$, can again be defined (see http://arxiv.org/abs/1207.5160).

Let b be a normalized value of F, the northward component of the IMF, namely,

$$b = F \ r_u^3/2M, \tag{3}$$

where M/r_u^3 is the magnetic field intensity in the equatorial plane at $r = r_u$. The relationship between b and F has been illustrated in Figure A5 of *Lemaire* [2003] for different values of r_u when the Earth reference dipole magnetic moment is $M_E = 8.06 \times 10^{15}$ T m^3.

For a fixed value of b, the dimensionless equation of magnetic field lines is given by

$$r = k\cos^2(\lambda)[1 - br^3]. \tag{4}$$

The whole family of magnetic field lines is generated by varying k from 0 to ∞.

For $b > 0$ (northward IMF) all magnetic field lines that are traversing the Earth surface are "closed." They cross, then, the equatorial plane at $r = \rho < k$. The equatorial distance of the Thalweg ρ_T is a solution of the algebraic equation: $\rho_T + b \ \rho_T^3 - 1 = 0$.

For southward IMF orientation $b < 0$, the total magnetic field, **B**$_d$ + **F**, vanishes along the equatorial circumference of radius $\rho_x = 1/(-2b)^{1/3}$. This circumference corresponds to the X line (or neutral line) of Dungey's magnetospheric model.

3.2. The Equatorial Cross Section of the Extended Störmer Potential

When the IMF (**F**) is parallel or antiparallel to the dipole magnetic moment (**M**), an axially symmetric magnetic potential, $V(\rho, z)$, can still be defined as in Störmer's theory [*Lemaire*, 2003].

Figure 2 shows the equatorial cross section of $V(\rho, 0)$ for $b = 0, -0.03, -0.05, +0.05,$ and $+0.03$.

The solid curve in Figure 2 corresponds to Störmer's equatorial potential distribution for a dipole, i.e., when $F = b = 0$. The four other curves correspond to nonzero values of the IMF; the dashed lines are for southward IMFs ($b < 0$), the dotted lines are for northward ones ($b > 0$).

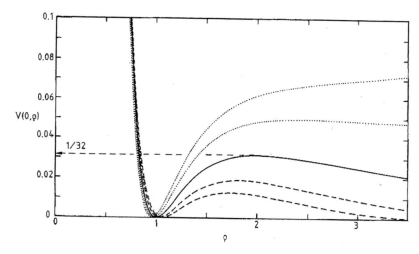

Figure 2. Equatorial cross-section of the extended Störmer potential, $V(\rho,0)$, as a function of the equatorial distance ρ. When the value of the interplanetary magnetic field (IMF) is equal to zero, the maximum value of Störmer's potential is equal to $V = 1/32 = 0.03125$. For $\mathbf{F} \neq 0$, this maximum value and its position vary as described in the text. The two lower curve lines correspond to southward IMF: $b = -0.03$ and -0.05. The two upper ones correspond to northward IMF: $b = +0.03$ and $+0.05$. From *Lemaire* [2003].

The maximum of $V(\rho, 0)$ is at an equatorial distance ρ_{max}, which is the solution of the equation: $\rho_{max} - b\rho_{max}^3 - 2 = 0$. When $b = 0$, Störmer's values are recovered: $\rho_{max} = 2$, $\rho_T = 1$, and $\rho_X = \infty$.

It can be seen that when b becomes more negative, the maximum "height" of the potential decreases below the value $V_P = 1/32$. Conversely, when the IMF is northward and assumes larger values, $V(\rho, 0)$ increases. This implies that larger kinetic energies are required for charged particles to spill out the inner trapping zone as well as to enter into it. Thus, Figure 2 illustrates how the IMF orientation and intensity controls the energy threshold of charged particles that can exit the radiation belts and enter the magnetosphere or a paleomagnetosphere whose magnetic moment would be reduced.

3.3. Critical Thalweg Field Lines

In Störmer's theory, the Thalwegs are always "closed." In Lemaire's extended theory, the Thalweg is not always "closed": it can become "open" when $b < b_T = -4/(3)^3 = -0.148148$. The Thalweg is then formed of two separate magnetic field line segments extending from the Earth's surface to infinity in both hemispheres.

The threshold value $b_T = -0.148148$ corresponds to a critical value for which $\rho_T = \rho_X$. The dashed line in Figure A5 of *Lemaire* [2003] indicates that this critical value of b_T corresponds to $F = -42.75$ nT, -9.2 nT, and -3.4 nT, respectively, for $r_u = 6\,R_E$, $10\,R_E$, and $14\,R_E$. The same critical

value b_T can as well be obtained by reducing M, the magnetic moment of a paleomagnetosphere by factors of 42.75, 9.2, and 3.4, respectively, with respect to $M_E = 8.06 \times 10^{15}$ T m^3.

For $b < b_T$, the isocontours of the extended Störmer potential are illustrated in Figure 3, using the same representation as in Figure 1. The two "open" branches of the Thalweg are then located in the middle of the dark-red cusp regions (valleys); these semi-infinite field line segments extend in both hemispheres from the Earth's surface to infinity ($z = \pm\infty$). They are not drawn in Figure 3.

Therefore, when $b < -0.148148$, the Thalweg magnetic field lines are interconnected to the interplanetary magnetic lines. Whether these field line segments are interconnected or reconnected is a matter of terminology or semantics beyond our preoccupation in the present study.

3.4. Geomagnetic Cutoff

The geomagnetic cutoff surface is the locus of points where the meridional component of the velocities becomes equal to zero: i.e., where $v_\rho^2 + v_z^2 = 0$ and $\mathbf{v} \cdot \mathbf{B} = 0$. When a GCR particle reaches the geomagnetic cutoff surface, its velocity is normal to the meridian plane, and $v_\phi = \rho\,d\phi/dt = v$. Note that in Alfvén's "*first-order guiding center theory or approximation*" the geomagnetic cutoff was not explicitly defined. It can, however, be assimilated to the "mirror points" where $v_\phi = v$ and $\mathbf{v} \cdot \mathbf{B} = 0$.

In Figures 2 and 3, the geomagnetic cutoff is located along the inner "slope" of the "valley." Its radial distance (r_G) is

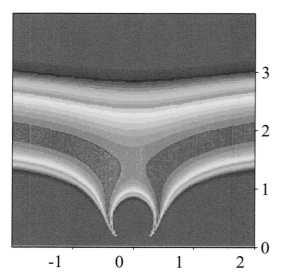

Figure 3. Extended Störmer potential, $V(\rho, z)$ for a large (negative) southward IMF: $b = -0.21$, (i.e., $\mathbf{F} = -13.4$ nT when $r_u = 10 \, R_E$). The neutral or X line is then at the equatorial distance $\rho_X = 1/(-2b)^{1/3} = 1.33$ in units r_u. In this case, the Thalweg is formed of two semi-infinite branches extending to infinity along two interconnected magnetic field lines. Along these open Thalweg segments (not drawn), interplanetary particles of all energies can enter and penetrate deep into the geomagnetic field. From *Lemaire* [2003].

always smaller than ρ_T. The latitude (λ_G) where it penetrates the Earth's atmosphere can be approximated by $\lambda_G \cong$ arc $\cos(r_E)^{1/2}$ for radiation belt particles for which $r_E^2/4\gamma_1^2 \ll 1$ and when $-br_E^3 \ll 1$: see equation (19) of *Lemaire* [2003; http://arxiv.org/abs/1207.5160].

It can be verified that λ_G and r_G are not very sensitive functions of the IMF component F or b. However, these quantities characterizing the geomagnetic cutoff positions depend significantly on the value of M, as well as on the magnetic rigidity, which is equal to p_o/Ze as verified by *Smart and Shea* [1967].

These results are well supported by GCR observations indicating that the geomagnetic cutoff latitude decreases as a function of the magnetic rigidity, but that their fluxes are almost not affected by the orientation of the IMF. These results from Lemaire's extended Störmer theory confirm also that λ_G and r_G decrease when M, the dipole moment of paleomagnetospheres is reduced, as also shown from trajectory calculations by *Vogt et al.* [2007].

4. WHAT HAPPENS WHEN THE EARTH'S MAGNETIC MOMENT REVERSES?

It has been considered above that the geomagnetic field distribution can be modeled (approximated) by a stationary superposition of (1) a dipole whose magnetic moment **M** is directed antiparallel to the oz-axis, and (2) a uniform IMF whose intensity **F** is either parallel or antiparallel to this axis. Of course, other cylindrically symmetric/zonal, magnetic field distributions can be considered for which F is an analytical function of ρ and z, but not of ϕ. The *Dst* field generated by a symmetrical ring current is just the next simplest example. Special paleomagnetospheres where the quadrupoles and multipole components have a zonal symmetry, as in studies of *Vogt and Glassmeier* [2000] and *Vogt et al.* [2004, 2007], are other B field distributions for which a generalized Störmer potential, $V(\rho, z)$ could be defined, in principle.

Such generalization of Störmer's potential would clearly improve the picture but would complicate the mathematics at the expense of analytical simplicity. Thus, for the sake of simplicity and illustration, we considered a uniform IMF as an ideal B field model to outline what might happen when **M** experiences a secular variation and possibly reverses.

Let us now slowly decrease the value of M, but without changing the value of F. According to equation (2), the unit length, $r_u = |ZeM/p_o|$ decreases with M for a particle of given charge (Ze) and a constant canonical angular momentum (p_o).

When M decreases, the modulus of $b = Fr_u^3/(2M)$ increases, and the extended potential displayed in Figure 2 will either grow or drop below the solid curve, depending on the initial orientation of the IMF. Remind that, according to equation (3), b is half the ratio between the IMF intensity and that of the Earth dipole at $\rho = 1$ in the equatorial plane.

Now, when the IMF changes from a northward to a southward orientation, $V(\rho, z)$ switches rapidly from one of the (upper) dotted curves for which $b > 0$, to a (lower) dashed curve for which $b < 0$, . . . and vice versa when the IMF turns back northward. Therefore, at each southward turning of the IMF, a certain fraction of trapped radiation, belt particles can escape from the geomagnetic field provided their KE exceeds the maximum of $V(\rho, z)$.

The smaller the value of M, the lower is the energy threshold of radiation belt particles, which can escape from the paleomagnetosphere. Extremely negative values of b are more frequently obtained when M is much smaller than its present-day value.

For sufficient small value of M, the value of b can become more frequently smaller than the critical value $b_T = -0.148148$, when the IMF fluctuates as usual. The Thalwegs of trapped radiation belt particles will then break open more often into two separate branches interconnecting the inner region of the geomagnetic field with the IMF. When such extreme situations occur, trapped radiation belt particles of lower energies and rigidities can escape along these open

Thalweg segments. They are lost to interplanetary space more frequently. Note that on such IMF polarity changes, solar wind particles might also more easily rain into the inner magnetosphere over the polar caps.

Not to say, of course, that our simplified scenario deliberately ignores additional effects that solar wind-induced electric fields and time-dependent magnetic fields will impose in the magnetopause and magnetosheath regions on escaping magnetospheric particles of a few tens of keV or less.

Anyway, all this implies that the trapping lifetime and the residual flux of energetic radiation belt particles must be dramatically perturbed and reduced at epochs of geomagnetic polarity transitions, . . . at least much more than nowadays. Similar conclusions had already been obtained by *Zieger et al.* [2004] and *Vogt et al.* [2007] from their detailed trajectory calculations in paleomagnetospheric models. See also the review by *Glassmeier and Vogt* [2010].

Note that after a geomagnetic field reversal, when M has eventually recovered the same magnitude, but of opposite sign, the unit length (r_u), the geomagnetic cutoff latitude, and Thalweg latitude (λ_G and λ_T) recover the same values as before the geomagnetic field polarity transition. Note that in the caption of Figure 2, the orientation of the IMF must be changed, however. Interplanetary particles will then have easier access to the inner magnetosphere for northward IMF orientation, instead of the southward one as happens to be the case today. Furthermore, storm time ring currents will then build up when the IMF is turning northward instead of southward as nowadays.

5. ENERGY SPECTRUM OF RADIATION BELT PARTICLES DURING GEOMAGNETIC REVERSAL

Based on this scenario, we can now infer what happens when the dipole field weakens and reverses as assumed above. Trapped protons are characterized by their KE and their rigidity, Br_L, which is equal to p_o/Ze. For GCR or radiation belt particle of given rigidity, the gyroradius, r_L, must increase when $M(t)$ and $B(r, t)$ decrease. This causes the proton to intersect the higher-density portions of the atmosphere at lower altitudes and lower latitudes. As a matter of consequence, this causes an enhanced loss of energy and momentum of the trapped particles and constitutes, therefore, an additional mechanism, which limits the lifetimes of radiation belt particles, as first pointed out and modeled by *Griem and Singer* [1955]. See also the review by *Singer and Lenchek* [1962].

On the outer edge of the proton belt, r_L increases until it becomes of the order of the Earth's radius, at which point adiabatic invariance breaks down [*Singer*, 1959a, 1959b]. According to *Lemaire* [1962, 1963], the equatorial plane is,

indeed, the most efficient place to violate the conservation of the first and second adiabatic invariants of particles of a given electric charge, mass, and KE. This implies that charged particles with equatorial pitch angles close to 90° are more prone to become untrapped than those mirroring at higher latitudes [*Lemaire*, 1962, 1963]. Note in Figures 1 and 3 that it is precisely close to the equatorial plane where the inner and outer allowed zones of Störmer are interconnected, that particles are able to spill out of the inner magnetosphere, nowhere else.

Of course, Alfvén's perturbation theory can also break down as a consequence of rapid time variations of the B field, for instance, due to the effect of hydromagnetic waves as originally discussed by *Dragt* [1961]. Resonant wave-particle interactions with magnetospheric ULF and whistler waves have such nonadiabatic effects on the trapped radiation belt particles as continuously proposed (see for instance the work of *Horne et al.* [2005], and references therein).

Whether these nonadiabatic processes are able to fully account for the rapid "nonadiabatic acceleration and losses" observed with Explorer 15 and reported by *McIlwain* [1963, 1996] remains to be proven and will demand additional independent investigations.

Anyway, from the discussion above, we conclude that during the geomagnetic field reversal, the highest energy protons will be lost first. The energy spectrum of trapped proton gradually steepens, as pointed in a review by *Glassmeier and Vogt* [2010].

As demonstrated by radiation belt measurements of *McIlwain* [1963, 1996] and others, the radiation belt particle population never reaches a stationary equilibrium spectrum. This is due to the endless variations of the IMF and the continuous pitch angle scattering of trapped particles by their nonresonant wave-particle interactions with magnetospheric ULF and whistler waves. The smaller the paleomagnetic dipole moment, (1) the larger will be the variations of particle fluxes remaining trapped, (2) the lower will be their maximum energy threshold, and (3) the softer will be their average energy spectrum.

When the geomagnetic field eventually recovers but with a reversed polarity, the reverse scenario takes place: average energy spectrum gradually flattens and becomes harder, maximum energy threshold for which radiation belt particles remain trapped is gradually increasing.

We hope that the present reconsideration of Störmer's theory will be useful from a historical perspective and will inspire younger researchers who may not have been exposed to Störmer's seminal theory. We expect that our extension of this theory, by adding a simple uniform IMF to Störmer's dipole B field model, has given some insight into the not yet fully exploited potentialities of this approach in investigations on the role of (1) southward/northward turning of the IMF, and (2) of geomagnetic field reversals on the energy spectrum of radiation belt particles, and on the geomagnetic cutoff of GCRs, or (3) on the formation of ring current, and (4) on the generation of *McIlwain's* [1974] substorm injection boundaries in today's magnetosphere as well as in paleomagnetospheres.

Acknowledgments. We wish to thank V. Pierrard, M. Echim, and D. Summers for editing various versions of this manuscript. We appreciate also the useful comments and suggestions of both referees.

REFERENCES

Akasofu, S. (2003), Chapman and Alfvén: A rigorous mathematical physicist versus an inspirational experimental physicist, *Eos Trans.*, *84*(29), 269.

Alfvén, H. (1940), On the motion of a charged particle in a magnetic field, *Ark. Math. Astron. Fys.*, *27A*(22), 1–20.

Alfvén, H. (1950), *Cosmical Electrodynamics*, 237 pp., Clarendon, Gloucestershire, U. K.

Dragt, A. J. (1961), Effect of hydromagnetic waves on the lifetime of Van Allen radiation protons, *J. Geophys. Res.*, *66*(6), 1641–1649.

Dragt, A. J. (1965), Trapped orbits in a magnetic dipole field, *Rev. Geophys.*, *3*(2), 255–298.

Dungey, J. W. (1961), Interplanetary magnetic field and the auroral zones, *Phys. Rev. Lett.*, *6*, 47–48.

Glassmeier, K. H., and J. Vogt (2010), Magnetic polarity transitions and biospheric effects, *Space Sci. Rev.*, *155*, 387–410.

Glassmeier, K. H., H. Soffel, and J. W. Negendank (Eds.) (2009), *Geomagnetic Variations, Space-Time Structure, Processes, and Effects on System Earth*, Springer, Heidelberg, Germany.

Griem, H., and S. F. Singer (1955), Geomagnetic albedo at rocket altitudes at the equator, *Phys. Rev.*, *99*, 608.

Horne, R. B., et al. (2005), Wave acceleration of electrons in the Van Allen radiation belts, *Nature*, *437*(7056), 227–230, doi:10.1038/nature03939.

Jackman, C. H., M. T. DeLand, G. J. Labow, E. L. Fleming, D. K. Weisenstein, M. K. W. Ko, M. Sinnhuber, J. Anderson, and J. M. Russell (2005), The influence of several very large solar proton events in years 2000–2003 on the neutral middle atmosphere, *Adv. Space Res.*, *35*, 445–450, doi:10.1016/j.asr.2004.09.006.

Kirkby, J., et al. (2011), Role of sulphuric acid, ammonia and galactic cosmic rays in atmospheric aerosol nucleation, *Nature*, *476*, 429–433, doi:10.1038/nature10343.

Lemaire, J. (1962), Note sur quelques conséquences des conditions d'Alfvén pour l'interprétation des zones de Van Allen et des déplacements ainsi que des courants azimutaux à grande distance, *Bull. Soc. R. Sci. Liége*, *31*, 556–574.

Lemaire, J. (1963), Note on some consequences of Alfvén's conditions for interpreting the Van Allen zones and their movements as

well as distant azimuthal currents, *AD/AF 61(052)-587, TN-4*, Cambridge Res. Lab., OAR, Eur. Off., Aerosp. Res., USAF, Bedford, Mass.

Lemaire, J. (2003), The effect of a southward interplanetary magnetic field on Störmer's allowed regions, *Adv. Space Res.*, *31*(5), 1131–1153, doi:10.1016/S0273-1177(03)00099-1.

McIlwain, C. E. (1961), Coordinates for mapping the distribution of magnetically trapped particles, *J. Geophys. Res.*, *66*(11), 3681–3691.

McIlwain, C. E. (1963), The radiation belts, natural and artificial, *Science*, *142*(3590), 355–361.

McIlwain, C. E. (1966), Magnetic coordinates, *Space Sci. Rev.*, *5*, 585–598.

McIlwain, C. E. (1974), Substorm injection boundaries, in *Magnetospheric Processes*, edited by B. M. McCormac, pp. 143–154, D. Reidel, Dordrecht, The Netherlands.

McIlwain, C. E. (1996), Processes acting upon outer zone electrons, in *Radiation Belts: Models and Standards, Geophys. Monogr. Ser.*, vol. 97, edited by J. F. Lemaire, D. Heynderickx and D. N. Baker, pp. 15–26, AGU, Washington, D. C., doi:10.1029/GM097p0015.

Northrop, T. G. (1963), *The Adiabatic Motion of Charged Particles*, Interscience, New York.

Singer, S. F. (1958a), Radiation belt and trapped cosmic ray albedo, *Phys. Rev. Lett.*, *1*, 171–173.

Singer, S. F. (1958b), Trapped albedo theory of the radiation belt, *Phys. Rev. Lett.*, *1*, 181–183.

Singer, S. F. (1959a), Artificial modifications of the Earth's radiation belts, *Adv. Astronaut. Sci.*, *4*, 335–354.

Singer, S. F. (1959b), Cause of the minimum in the Earth's radiation belts, *Phys. Rev. Lett.*, *3*, 188–190.

Singer, S. F., and J. F. Lemaire (2009), Geomagnetically trapped radiation: Half a century of research, in *Fifty Years of Space Research*, edited by O. Zakutnyaya and D. Odintsova, pp. 115–127, Space Res. Inst. of Russ. Acad. of Sci., Moscow, Russia.

Singer, S. F., and A. M. Lenchek (1962), Geomagnetically trapped radiation, in *Progress in Elementary Particle and Cosmic Ray Physics*, edited by J. G. Wilson and S. A. Wouthuysen, chap. 3, pp. 245–335, North-Holland, Amsterdam, The Netherlands.

Smart, D. F., and M. A. Shea (1967), A study of the effectiveness of the McIlwain coordinates in estimating cosmic-ray vertical cutoff rigidities, *J. Geophys. Res.*, *72*(13), 3447–3454.

Smart, D. F., M. A. Shea, and E. O. Flückiger (2000), Magnetospheric models and trajectory computations, *Space Sci. Rev.*, *93*, 305–333.

Stadelmann, A., J. Vogt, K. H. Glassmeier, M. B. Kallenrode, and G. H. Voigt (2010), Cosmic ray and solar energetic particle flux in paleomagnetospheres, *Earth Planets Space*, *62*, 333–345.

Störmer, C. (1907), Sur les trajectoires des corpuscules électrisés dans l'espace sous l'action du magnétisme terrestre avec application aux aurores boréales, *Arch. Sci. Phys. Nat.*, *24*, 317–364.

Störmer, C. (1913), *Résultats Des Calculs Numériques Des Trajectoires Des Corpuscules Électriques Dans Le Champ D'un Aimant Élémentaire, Skrifter (Norske videnskaps-akademi. I-Mat.-naturv. klasse)*, vol. 4, Kristiania, Oslo, Norway.

Störmer, C. (1955), *The Polar Aurora*, pp. 209–334, Clarendon Press, Oxford, U. K.

Svensmark, H., and E. Friis-Christensen (1997), Variation of cosmic ray flux and global cloud coverage—A missing link in solar-climate relationships, *J. Atmos. Sol. Terr. Phys.*, *59*(11), 1225–1232.

Uffen, R. J. (1963), Influence of the Earth's core on the origin and evolution of life, *Nature*, *198*(4876), 143–144, doi: 10.1038/198143b0.

Van Allen, J. A., and S. F. Singer (1950a), On the primary cosmic ray spectrum, *Phys. Rev.*, *78*, 819, doi:10.1103/PhysRev.78.819.

Van Allen, J. A., and S. F. Singer (1950b), Erratum: On the primary cosmic ray spectrum, *Phys. Rev.*, *80*, 116, doi:10.1103/PhysRev. 80.116.

Van Allen, J. A., G. H. Ludwig, E. C. Ray, and C. E. McIlwain (1958), Observation of high intensity radiation by satellites 1958 Alpha and Gamma, *Jet Propul.*, *28*, 588–592.

Vernov, S. N., N. L. Grigorov, I. P. Ivanenko, A. I. Lebedinski, V. S. Murzin, and A. E. Chudakov (1959), Possible mechanism of production of terrestrial corpuscular radiation under the action of cosmic rays, *Sov. Phys. Dokl.*, *4*, 154–157.

Vogt, J., and K. H. Glassmeier (2000), On the location of trapped particle populations in quadrupole magnetospheres, *J. Geophys. Res.*, *105*(A6), 13,063–13,071.

Vogt, J., B. Zieger, A. Stadelmann, K.-H. Glassmeier, T. I. Gombosi, K. C. Hansen, and A. J. Ridley (2004), MHD simulations of quadrupolar paleomagnetospheres, *J. Geophys. Res.*, *109*, A12221, doi:10.1029/2003JA010273.

Vogt, J., B. Zieger, K.-H. Glassmeier, A. Stadelmann, M.-B. Kallenrode, M. Sinnhuber, and H. Winkler (2007), Energetic particles in the paleomagnetosphere: Reduced dipole configurations and quadrupolar contributions, *J. Geophys. Res.*, *112*, A06216, doi:10.1029/2006JA012224.

Walt, M. (1996), Source and loss processes for radiation belt particles, in *Radiation Belts: Models and Standards*, *Geophys. Monogr. Ser.*, vol. 97, edited by J. F. Lemaire, D. Heynderickx and D. N. Baker, pp. 1–13, AGU, Washington, D. C., doi:10.1029/GM097p0001.

Zieger, B., J. Vogt, K.-H. Glassmeier, and T. I. Gombosi (2004), Magnetohydrodynamic simulation of an equatorial dipolar paleomagnetosphere, *J. Geophys. Res.*, *109*, A07205, doi:10.1029/2004JA010434.

Zieger, B., J. Vogt, A. J. Ridley, and K. H. Glassmeier (2006), A parametric study of magnetosphere ionosphere coupling in the paleomagnetosphere, *Adv. Space Res.*, *38*, 1707–1712.

J. F. Lemaire, Belgian Institute for Space Aeronomy, Brussels B-1180, Belgium. (joseph.lemaire@aeronomie.be)

S. F. Singer, Science and Environmental Policy Project, Arlington, VA 22202, USA. (singer@sepp.org)

What the Satellite Design Community Needs
From the Radiation Belt Science Community

T. P. O'Brien, J. E. Mazur, and T. B. Guild

The Aerospace Corporation, El Segundo, California, USA

The satellite design community depends on the radiation belt science community for knowledge of the radiation environment in which satellites will operate. We provide an up-to-date discussion of the current needs of the satellite design community and how those needs can be met by specific developments in radiation belt science. Generally, the satellite design community needs worst-case and mean radiation environment specifications. Providing those specifications touches on many areas of radiation belt science research. We recommend several changes to common practice in the science community that can increase the benefit that science provides to the satellite designers. The recommended changes fall in the areas of instrumentation, data presentation, global simulations, theoretical worst cases, and geophysical coordinates. In most cases, the changes we recommend are minor extensions to common practice.

1. INTRODUCTION

The links between the community of radiation belt researchers and the organizations that acquire and design satellite systems may not be obvious to either endeavor. There was a time when major satellite contractors had their own in-house radiation belt research staff, to which they looked for guidance in vehicle design and operations. This is no longer the case, and in many aspects, both communities have evolved away from each other. What we advocate is not a return to prior ways of doing business but, instead, a greater awareness in the scientific community of which aspects of the fast-paced developments in radiation belt research can be useful to satellite designers. With just a few, often minor, modifications, advanced scientific research results and approaches can yield huge benefits to both communities.

Satellite designers typically design and test a spacecraft and its subsystems to be "all weather." That is, they design

the satellite to perform adequately in all likely space environmental conditions. They do not typically design a system with known space environment susceptibilities, so that it depends operationally on space weather nowcasts and forecasts. Thus, at the highest level, the satellite design community needs, from the radiation belt science community, a specification of the mean and worst-case radiation environments, especially those that cause total dose, internal charging, and single-event effects, tailored to each individual mission orbit. Because a satellite design inherently involves trading various risks against each other and against budget, mass, and schedule constraints, a satellite designer needs indications of how likely the actual environment encountered on orbit is to exceed a selected design environment. Thus, radiation environment specifications must carry indications of uncertainty and probability of occurrence.

The above high level needs flow into a set of lower-level objectives that can sometimes be met by the scientific community with only slight changes to common practice. In other cases, where the change in practice imposes significant effort or cost, this document will provide some justification for the additional investment. The changes to common practice fall in the areas of instrumentation, data presentation, global simulations, theoretical worst cases, and geophysical

Dynamics of the Earth's Radiation Belts and Inner Magnetosphere
Geophysical Monograph Series 199
10.1029/2012GM001316

coordinates. We will review recommendations in each of these areas after a brief introduction to the space radiation environment hazards a satellite designer must consider and how that information feeds into the design and test of a vehicle or subsystem.

First, we begin with an enumeration of the hazards. Table 1 provides a list of space radiation environment hazards that must be considered in the design of a satellite or subsystem. Not all of the hazards apply to all orbits, and not all of the hazards are solely associated with the radiation belts. Still, radiation belt science has much to contribute to the satellite design specifications addressing these hazards. Figure 1 shows an example of how environmental specifications can inform the development of a satellite mission from concept to on-orbit success (for a thorough exposition, see the work of *Ladbury* [2007]). Initially, dose depth curves and worst-case linear energy transfer specifications can inform the concept development to enhance mission duration and reduce single-event effects risk through orbit selection, technology development, prototyping, and preliminary testing. Once the mission concept is firmly established, system environment specifications contribute to the overall design. The environment specification includes mission fluences and worst cases such as those in Table 1. These specifications are typically "flowed down" into derived subsystem requirements that influence subsystem and part selection. Such requirements might specify minimum box shielding to limit radiation effects, resistivity requirements to reduce the risk of charging, and baseline specifications for electrical, electronic, and electromechanical parts. When those subsystem specifications cannot be directly met, various mitigation strategies are employed, often using a more detailed approach to the environment specification. For example, spot shielding is often tailored using detailed ray tracing calculations exploiting the actual shielding mass provided by other components in the box and other structures on the vehicle. Once the design is fabricated and assembled, environmental specifications can even be used to determine what subsystem- and system-level testing is required (such as electrostatic discharge testing).

There are three main mechanisms for radiation belt science to feed into satellite designs: direct use of sensor data from prior missions in the same orbit as future missions (e.g., for geostationary orbit, see the work of *Sicard-Piet et al.* [2008], or for navigation orbits, see the work of *Bourdarie et al.* [2009b]), global or regional statistical climatology models derived from single or multiple missions [*Sawyer and Vette*, 1976; *Vette*, 1991; *Gussenhoven et al.*, 1993; *Brautigam and Bell*, 1995], and long-term data assimilative model runs, also known as reanalysis climatology [e.g., *Bourdarie et al.*, 2009a]. While it is a niche area to perform these knowledge transition activities, the capability to do that economically depends on cooperation from, and sometimes investment by, the larger radiation belt science community.

2. INSTRUMENTATION AND DATA PRESENTATION

Many radiation belt particle sensors suffer from penetrating backgrounds [e.g., *Gussenhoven et al.*, 1993]. This leads to the unfortunate situation that even if a vehicle carrying relevant particle sensors has already flown in a specific orbit of interest, sensor backgrounds make the data unusable for developing a whole-orbit radiation specification. Thus, whenever practical, particle sensors should be designed to be able to make robust measurements throughout the orbit of the host vehicle [see, e.g., *Vampola*, 1998]. This will enable straightforward use and interpretation of those data for future satellites designed for that orbit. Further, it will alleviate a problem that has long plagued efforts to develop new radiation environment specifications: systematic spatial holes in our knowledge of the particle environment. These holes have persisted for over 50 years of space exploration, in spite of continued collection of data in specific orbits.

Open data policies have led to wide release of many space radiation data sets. However, the best practices for supporting data (metadata) are still developing. It is well known that space particle data sets often disagree with each other regarding both the long-term statistical properties of the radiation environment and sometimes even regarding the flux intensity or energy

Table 1. Environmental Hazards Affecting Spacecraft[a]

Environmental Hazard	Particle Population	Specification Type
Total ionizing dose (shielded)	>100 keV H^+, e^-	mission average
Single-event effects	>10 MeV amu^{-1} H^+, ions	worst case
Displacement damage	>10 MeV H^+, Secondary neutrons	mission average
Nuclear activation	>50 MeV H^+, Secondary neutrons	both
Internal charging	100 keV to 10 MeV e^-	worst case
Surface charging	0.01–100 keV e^-	worst case
Surface dose (unshielded)	0.5–100 keV e^-, H^+, O^+	mission average

[a]After *O'Brien et al.*, [2007b]. Reprinted with permission of The Aerospace Corporation.

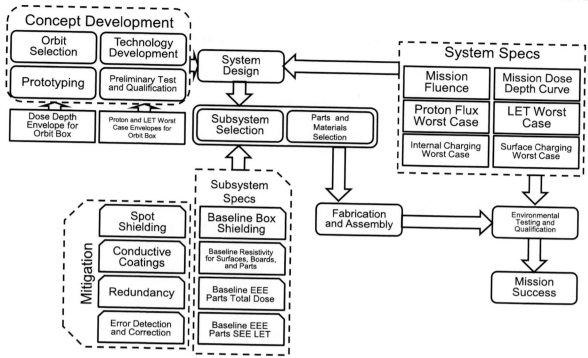

Figure 1. An example diagram of how environmental specifications inform the process from satellite mission concept to final success.

spectrum at specific times and locations. These disagreements must be managed as uncertainty or risk when the data flow directly or indirectly into satellite design specifications. Therefore, in order for a data set to be most useful for satellite design, whether being used directly in the same orbit as a future mission, being ingested into a larger climatological model, or being incorporated in a data assimilative climatology simulation, it must be accompanied by information about data errors. *Friedel et al.* [2005] described a systematic approach to estimating and, thereby, removing the systematic errors in data sets. Following that approach, version 1.1 of the Committee on Space Research (COSPAR) Panel on Radiation Belt Models (PRBEM) standard flux file format (http://irbem.svn.sourceforge.net/viewvc/irbem/docs/) included a data field to store daily cross-calibration factors for every particle flux variable stored in the file and another data field to store quality flags. An updated version of the standard (v1.2) also includes a field in which to store an indicator of the random or residual error resulting from cross calibration. Specifically, the PRBEM standard recommends the RMS residual error in the natural log of the fluxes. For small errors, this quantity is approximately the random component of the relative error in the fluxes (i.e., a value of 0.1 indicates a 10% random error).

Figure 2 shows an example of using solar proton events to intercalibrate a proton sensor in a highly elliptical orbit (HEO

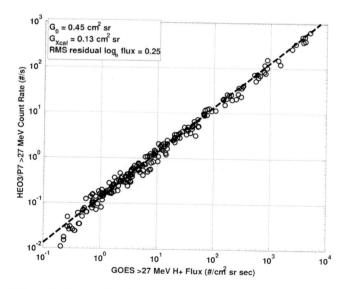

Figure 2. Cross-calibration between a HEO3/P7, a >27 MeV channel, and GOES protons (multiple vehicles) during a set of solar particle events reveals that the calculated geometric factor (G_0) is about a factor of 3 too large (compare to G_{Xcal}).

or 12 h Molniya) orbit with a reference proton sensor flying on NOAA's GOES vehicles in geostationary orbit (*Guild et al.* [2012] describe this process in more detail). During solar proton events, a HEO vehicle at apogee measures essentially the same energetic proton environment as a vehicle at geostationary orbit. Over a series of solar particle events, a scatter plot of the HEO sensor count rate versus the GOES sensor flux reveals the systematic and random errors. Using the GOES sensor as a reference, one can then compute a new geometric factor (G_{Xcal}) for HEO3/P7, which incorporates the systematic errors, as well as an RMS residual $\log_e$ flux (in this case 0.25 or a 25% relative error). The systematic error arises from simplifications and assumptions about the silicon target inside the sensor. The spread of the residuals in the log-log plot is not a strong function of particle flux, making a single relative error statistic an adequate description of the random error distribution. A more sophisticated treatment is possible, but it requires more information about the sensor.

Particle sensor data is typically published as particle fluxes as a function of energy and/or pitch angle. These fluxes are typically obtained by means of simple flux conversion factors (such as G above) or by means of inversion methods (for examples of both, see the work of *Selesnick and Blake* [2000]). For incorporation in data assimilative simulations and statistical climatology models, it is often necessary to use a more primitive version of the data, namely, the observed counts accompanied by a description of the sensor response as a function of energy, angle, and even species. Thus, provision of low-level sensor data and tabulated or parameterized response functions can enable better and broader use of space particle measurements. The COSPAR PRBEM has drafted standard formats for count rate files and sensor response files (http://irbem.svn. sourceforge.net/viewvc/irbem/docs/). These standards should simplify the process of adding new data sources into data assimilative global simulations.

3. GLOBAL SIMULATIONS

Global simulation models of the radiation belts offer two benefits to the satellite design community. The first is in the form of reanalysis climatology, by which we mean long-term data assimilative simulations that provide the best estimate of the time-dependent global or regional radiation belt state over several years, ideally, more than a solar cycle. These simulations provide satellite designers with a realistic long-term radiation environment, with real geomagnetic storms and other associated dynamic phenomena, with real intraannual and interannual variations that are difficult to capture in a purely statistical climatology.

For the outer zone, one such reanalysis [*Bourdarie et al.*, 2009a] has already been used to develop yearly averaged global states for the electron contribution to total dose. Climatologies at higher time resolution could similarly be used to develop worst-case specifications for internal charging or single-event effects. Further, reanalysis climatologies can provide information about spatial and spatiotemporal correlations of radiation belt fluxes that are needed to generate surrogate time series of the global environmental evolution [see, e.g., *O'Brien*, 2005].

Finally, global simulations, whether data assimilative or not, offer a means of specifying the typical shapes of the radiation belts, which can, in turn, be used to "fill in" holes in the flux maps obtained by binning in situ sensor data. These flux maps are at the heart of statistical models, and filling in the holes is a tremendous challenge that could benefit significantly from the physical insights provided by global simulations.

4. WORST CASES

Specification of extreme environments can sometimes be provided via statistical analysis for certain data-rich problems [*Xapsos et al.*, 1998, 1999; *Koons*, 2001; *O'Brien et al.*, 2007a]. However, other problems require the physical insight found in the radiation belt science community. A famous example of an extreme environment specification developed from physical insight is the Kennel-Petschek limit [*Kennel and Petschek*, 1966]. This limit specifies the most intense stably trapped fluxes one should expect to encounter in the radiation belts. Unfortunately, it does not appear to apply to the higher energies typically required for satellite design.

Another approach [e.g., *Shprits et al.*, 2011] is to run an established physics-based simulation under conditions of extreme driving. This latter approach has merit, but the results must be presented carefully: it is important to recognize that a possible extreme environment may not be forbidden by the laws of physics, but that it may yet be excluded from consideration during satellite design by the laws of probability. Only extreme environments that are sufficiently well known and sufficiently inevitable over the lifetime of a satellite mission will be accounted for by prudent satellite designers. An extreme environment that is theoretically possible, but with unknown likelihood, will often be rejected by prudent mission designers if it leads to any significant cost or performance impacts to the mission.

On the other hand, a physics- or statistics-based upper limit that only slightly exceeds a worst observed case can be used by a mission designer to make a relatively small design change to dramatically decrease mission risk from the specified hazard. For example, had the worst-case environments provided in the work of *O'Brien et al.* [2007a] been factors of 10 higher than the worst observed case, they would have been of little use to satellite designers; as it happened, the statistically derived limits were only slightly higher than the highest observations, leading some missions to adopt them as design specifications. When physical and statistical analysis provide no indications of usable finite limits or likely worst-case environments, satellite designers must fall back on a probabilistic approach to acceptable risk, such as designing to a 95th percentile worst-case environment for a given mission duration.

5. GEOPHYSICAL COORDINATES

In those cases, when radiation belt knowledge is to be utilized in an arbitrary Earth orbit, some set of geophysical coordinates is necessary to organize the observations and/or simulation results. Typically, simulations and data are organized in an adiabatic coordinate system that relies on calculated drift, bounce, and cyclotron invariants. Such systems reasonably form the basis for the "fly through" tools used to provide average and worst-case environment specifications to satellite designers by projecting the model environment from model coordinates onto a spacecraft orbit trajectory and performing any necessary integrals over energy, angle, or time. However, several issues must be addressed carefully in order to have a practical and accurate tool using these types of coordinates.

First, the computation speed must be fast. Determination of the drift invariant is very computer intensive and cannot be done practically in an interactive software tool. However, it is possible to reduce the apparent processing time by precomputing the results of drift shell tracing for the entire radiation belt domain and, then, representing those results in some kind of analytical or tabular form. The International Radiation Belt Models library (irbem.sourceforge.net) includes a polynomial expansion for a quiet field model, and *Koller et al.* [2009] and *Koller and Zaharia* [2011] have used a neural network for an active field model. These fast L^* (drift invariant) calculations are approximately 1 million times faster than drift shell tracing, making them suitable for use in an interactive software tool.

The next problem is how to define the boundaries of the model domain. The outer boundary of the radiation belts is often assumed to be the magnetopause. However, it may, in fact, be interior to the magnetopause in the region where bifurcated drift orbits lead to a complex quasitrapped behavior [see, e.g., *Ukhorskiy et al.*, 2011]. In this boundary region, adiabatic invariants are likely not appropriate coordinates, but we know of no appropriate coordinate system that has been developed. Conversely, at low altitude, near the loss cone, the interaction with the atmosphere dominates the spatial gradients. For particles mirroring below about 600 km anywhere on their drift orbit, an atmosphere-referenced quasi-invariant system is often more appropriate than a truly adiabatic invariant system [see, e.g., *Heynderickx and Lemaire*, 1996; *Cabrera and Lemaire*, 2007]. Further, for electrons, there is significant quasitrapped flux in the drift loss cone (i.e., on trajectories that will eventually intersect the atmosphere in the South Atlantic Anomaly). Thus, again the traditional three adiabatic invariants are inadequate: a drift phase coordinate is needed.

Finally, the satellite designer is looking 5–20 years into the future, but the geophysical models needed to produce geophysical coordinates are typically focused on the present and past. The widely used International Geophysical Reference Field (IGRF) model [*Finlay et al.*, 2010] specifies a secular variation of the Earth's internal magnetic field that can be extrapolated only 5 years into the future. There is no specification on how to extrapolate the IGRF coefficients farther into the future, and it is unknown how different extrapolation approaches will affect the relevance of the derived geophysical coordinates. For external magnetic field models, the problem is even more significant: external field models require geomagnetic indices (Kp, Dst) and solar wind parameters [e.g., *Tsyganenko and Sitnov*, 2005]. Similarly, atmospheric models typically require F10.7 and/or Ap index inputs [e.g., *Picone et al.*, 2002]. For external field models and atmosphere models, some kind of surrogate model is needed for each of the required inputs. For example, *Xapsos et al.* [2002] developed a confidence level approach to surrogate time series of F10.7 for driving a low-altitude trapped proton model. While these types of surrogate models may not, themselves, produce significant physical insight, they require physical insight as a starting point and are a mechanism for transferring that insight to the satellite designers who can translate it into improved satellite performance, cost, or longevity.

6. CONCLUSIONS

We have outlined a series of needs and recommendations that will facilitate transfer of hard-won advances in our understanding of the radiation belts to the satellite design community that depends on radiation belt science for improvements in radiation environment specifications. While the investments required to implement all of these recommendations may not be possible at present, most of the recommendations are simple changes to, or extensions of, common practice in the science community. By engaging the broader radiation belt science community to improve the quality and quantity of knowledge transfer to the satellite designers, we hope to ensure that society gets the greatest possible return for its investment in radiation belt research.

Acknowledgments. The authors acknowledge useful discussions over many years with G. Ginet, S. Huston, S. Bourdarie, R. Friedel, R. Johnston, and our colleagues in the Space Sciences Department.

This work was funded, in part, by The Aerospace Corporation's Independent Research and Development Program. This paper is also available as Aerospace Technical Report ATR-2012(8181)-3.

REFERENCES

Bourdarie, S., A. Sicard-Piet, R. Friedel, R. P. O'Brien, T. Cayton, B. Blake, D. Boscher, and D. Lazaro (2009a), Outer electron belt specification model, *IEEE Trans. Nucl. Sci.*, *56*(4), 2251–2257.

Bourdarie, S., D. Lazaro, A. Hands, K. Ryden, and P. Nieminen, (2009b) Electron environment specification models for navigation orbits, in *2009 European Conference on Radiation and Its Effects on Components and Systems (RADECS), 14–18 Sept.*, pp. 364–368, IEEE, Piscataway, N. J., doi:10.1109/RADECS.2009.5994677.

Brautigam, D. H., and J. T. Bell (1995), CRRESELE Documentation, *PL-TR-95-2128, Environ. Res. Pap., 1178*, Phillips Lab., Hanscom AFB, Mass.

Cabrera, J., and J. Lemaire (2007), Using invariant altitude (h_{inv}) for mapping of the radiation belt fluxes in the low-altitude environment, *Space Weather*, *5*, S04007, doi:10.1029/2006SW000263.

Finlay, C. C., et al. (2010) International Geomagnetic Reference Field: The eleventh generation, *Geophys. J. Int.*, *183*(3), 1216–1230, doi:10.1111/j.1365-246X.2010.04804.x.

Friedel, R. H. W., S. Bourdarie, and T. E. Cayton (2005), Inter-calibration of magnetospheric energetic electron data, *Space Weather*, *3*, S09B04, doi:10.1029/2005SW000153.

Guild, T. B., et al. (2012), On-orbit inter-calibration of proton observations during solar particle events, *Rep. TOR-2007 (3905)-22*, The Aerospace Corp., El Segundo, Calif., in press.

Gussenhoven, M. S., E. G. Mullen, M. D. Violet, C. Hein, J. Bass, and D. Madden (1993), CRRES high energy proton flux maps, *IEEE Trans. Nucl. Sci.*, *40*(6), 1450–1457.

Heynderickx, D., and J. Lemaire (1996), Coordinate systems for mapping low-altitude trapped particle fluxes, *AIP Conf. Proc.*, *383*, 187–192, doi:10.1063/1.51530.

Kennel, C. F., and H. E. Petschek (1966), Limit on stably trapped particle fluxes, *J. Geophys. Res.*, *71*(1), 1–28.

Koller, J., and S. Zaharia (2011), LANL* V2.0: Global modeling and validation, *Geosci. Model Dev.*, *4*(3), 669–675, doi:10.5194/gmd-4-669-2011.

Koller, J., G. D. Reeves, and R. H. W. Friedel (2009), LANL* V1.0: A radiation belt drift shell model suitable for real-time and reanalysis applications, *Geosci. Model Dev.*, *2*, 113–122.

Koons, H. C. (2001), Statistical analysis of extreme values in space science, *J. Geophys. Res.*, *106*(A6), 10,915–10,921.

Ladbury, R. (2007), Radiation hardening at the system level, in *2007 IEEE NSREC Short Course*, pp. 1–94, IEEE, Piscataway, N. J. [Available at http://radhome.gsfc.nasa.gov/radhome/papers/NSREC07_SC_Ladbury.pdf.]

O'Brien, T. P. (2005), A framework for next-generation radiation belt models, *Space Weather*, *3*, S07B02, doi:10.1029/2005SW000151.

O'Brien, T. P., J. F. Fennell, J. L. Roeder, and G. D. Reeves (2007a), Extreme electron fluxes in the outer zone, *Space Weather*, *5*, S01001, doi:10.1029/2006SW000240.

O'Brien, T. P., J. E. Mazur, J. B. Blake, M. D. Looper, J. T. Bell, G. P. Ginet, and A. B. Campbell (2007b), Mapping the inner Van Allen belt: Requirements and implementation strategy, *Aerosp. Tech. Rep. ATR-2006(8377)-1*, The Aerosp. Corp., El Segundo, Calif.

Picone, J. M., A. E. Hedin, D. P. Drob, and A. C. Aikin (2002), NRLMSISE-00 empirical model of the atmosphere: Statistical comparisons and scientific issues, *J. Geophys. Res.*, *107*(A12), 1468, doi:10.1029/2002JA009430.

Sawyer, D. M., and J. I. Vette (1976), AP-8 trapped proton environment for solar maximum and solar minimum, *NSSDC/WDC-A-R&S 76-06*, Natl. Space Sci. Data Cent., Natl. Aeronaut. and Space Admin., Goddard Space Flight Cent., Greenbelt, Md.

Selesnick, R. S., and J. B. Blake (2000), On the source location of radiation belt relativistic electrons, *J. Geophys. Res.*, *105*(A2), 2607–2624.

Shprits, Y., D. Subbotin, B. Ni, R. Horne, D. Baker, and P. Cruce (2011), Profound change of the near-Earth radiation environment caused by solar superstorms, *Space Weather*, *9*, S08007, doi:10.1029/2011SW000662.

Sicard-Piet, A., S. Bourdarie, D. Boscher, R. H. W. Friedel, M. Thomsen, T. Goka, H. Matsumoto, and H. Koshiishi (2008), A new international geostationary electron model: IGE-2006, from 1 keV to 5.2 MeV, *Space Weather*, *6*, S07003, doi:10.1029/2007SW000368.

Tsyganenko, N. A., and M. I. Sitnov (2005), Modeling the dynamics of the inner magnetosphere during strong geomagnetic storms, *J. Geophys. Res.*, *110*, A03208, doi:10.1029/2004JA010798.

Ukhorskiy, A. Y., M. I. Sitnov, R. M. Millan, and B. T. Kress (2011), The role of drift orbit bifurcations in energization and loss of electrons in the outer radiation belt, *J. Geophys. Res.*, *116*, A09208, doi:10.1029/2011JA016623.

Vampola, A. L. (1998), Measuring energetic electrons — What works and what doesn't, in *Measurement Techniques in Space Plasmas: Particles, Geophys. Monogr. Ser.*, vol. 102, edited by F. Pfaff, E. Borovsky and T. Young, pp. 339–355, AGU, Washington, D. C., doi:10.1029/GM102p0339.

Vette, J. I. (1991), The AE-8 trapped electron model environment, *NSSDC/WDC-A-R&S 91-24*, Natl. Space Sci. Data Cent., Natl. Aeronaut. and Space Admin., Goddard Space Flight Cent., Greenbelt, Md.

Xapsos, M. A., G. P. Summers, and E. A. Burke (1998), Probability model for peak fluxes of solar proton events, *IEEE Trans. Nucl. Sci.*, *45*(6), 2948–2953.

Xapsos, M. A., G. P. Summers, J. L. Barth, E. G. Stassinopoulos, and E. A. Burke (1999), Probability model for worst case solar proton event fluencies, *IEEE Trans. Nucl. Sci.*, *46*(6), 1481–1485.

Xapsos, M. A., S. L. Huston, J. L. Barth, and E. G. Stassinopoulos (2002), Probabilistic model for low-altitude trapped proton fluxes, *IEEE Trans. Nucl. Sci.*, *49*(6), 2776–2781.

T. B. Guild, J. E. Mazur, and T. P. O'Brien, The Aerospace Corporation, Mail Stop CH3/330, 15049 Conference Center Drive, Chantilly, VA 20151, USA. (paul.obrien@aero.org)

Storm Responses of Radiation Belts During Solar Cycle 23: HEO Satellite Observations

J. F. Fennell

The Aerospace Corporation, Los Angeles, California, USA

S. Kanekal

Goddard Space Flight Center, Greenbelt, Maryland, USA

J. L. Roeder

The Aerospace Corporation, Los Angeles, California, USA

We examined the energetic electron responses to storms with $Dst \leq -75$ nT that occurred in the 1998–2008 period deep in the inner magnetosphere $2 \leq L \leq 4$ using HEO3 data. These observations cover the peak of solar cycle 23. We show that the observed poststorm >1.5 MeV electron flux decay rates for $L \sim 3$ had multiple time scales. Early in the recovery phase, the distribution of e-folding decay time scales had two peaks near 6 and 14 days, while late recovery decay rates had a peak near 20 days. Examination of the times from Dst minimum to the peak of enhanced electron fluxes at $L \sim 3$ showed that seed population electrons (>230 keV) had a mean e-folding rise time of ~0.9 days, while for relativistic electrons (>1.5 MeV), it was ~3.3 days. Examination of the poststorm electron flux responses in the $2.5 < L < 4$ showed that (1) there were essentially no poststorm electron flux decreases observed in the slot region and that occurrence of flux increases rose with decreasing Dst, (2) poststorm electron fluxes at $L = 3$ showed decreases only for events with minimum $Dst > -175$, and (3) at $L = 4$ statistics of events that caused flux increases versus no change or flux decreases approached those of earlier results.

1. INTRODUCTION

Fennell et al. [2005] used HEO3 observations to examine the storm responses of electron fluxes taken at low and high altitudes over the same L range. They found that the ratio of fluxes at high B values (small pitch angles) to those taken at low B values (large pitch angles) by the same instruments remained relatively constant on time scale of a day or so over a wide range of L values. Later, *Fennell and Roeder* [2010] examined the electron response to a set of storms that occurred in July 2004. They found that the two smallest storms did not cause relativistic electron enhancements for $L < 6.5$ but did cause enhancements in the electron "source" populations at >130 and >230 keV down to $L \sim 3$, while the largest events caused electron enhancements at all energies down to $L \sim 2.5$. They observed that "seed population" electrons, which are expected to be the source population that wave-particle interactions accelerate to MeV energies, were injected deep into the inner magnetosphere during all the storms but that their presence did not always lead to subsequent MeV electron enhancements for

Dynamics of the Earth's Radiation Belts and Inner Magnetosphere
Geophysical Monograph Series 199
10.1029/2012GM001356

the smaller events (as measured by *Dst*). In addition, they did a preliminary examination of the decays of poststorm-enhanced electron fluxes at $L = 3$. Here we expand the earlier studies to include all storms in the 1998–2008 periods that corresponds to most of solar cycle 23. In addition, we examined whether the storms caused flux enhancements, no flux change, or net flux loss over the range $2.75 \leq L \leq 4$ for >230 keV, >500 keV, >1.5 MeV, and >3 MeV electrons. In addition, we examined the time to peak flux from *Dst* minimum for those events that caused flux enhancements at $L = 3$. For this paper, we used the *Olson and Pfitzer* [1974] quiet field model to calculate L values. Using this static model means that the data is essentially sampled in the same regions of geographic space on a daily basis, given the HEO3 orbit described below.

2. INSTRUMENTATION AND DATA

This study uses data taken during the 1998–2008 period by a high Earth orbiting, HEO3, satellite in a ~1.15 × 7.2 R_E orbit inclined at ~63° with a ~12 h period. HEO3 covered a wide range of L values at both low to high altitudes. For this study, we only used data taken at altitudes ≥1.75 R_E geocentric (see Figure 1). Since the HEO orbit has a 12 h period, the geographic spatial positions sampled repeat on a daily basis generating four traversals in *B-L* space as shown in Figure 1.

The electron data were obtained using integral energy electron sensors that measured electrons with energies >0.23, >0.45, >0.6, >1.5, and >3 MeV. The three highest energy sensors make omnidirectional flux measurements. The >0.23 MeV electron data are from a proton-electron telescope with a ~15° conical field of view. The >0.45 MeV data is taken by a short "telescope" with a 5 mil Al foil over

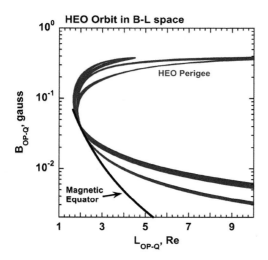

Figure 1. HEO orbit in *B-L* space. All data for study were taken where $B < 0.04$ gauss.

the detector to set its electron and proton thresholds [*Blake et al.*, 1997]. At times, we emphasize the results from the >1.5 MeV electrons since most results in the literature, concerning electron acceleration and transport in the inner magnetosphere, have emphasized the response of MeV electrons. However, we add in results from the lower energies, such as the >230 keV "seed" population electrons, often discussed in the literature, and the >0.45 and >0.6 MeV electrons, which show differences in their storm time responses from both the seed and MeV electrons.

3. OBSERVATIONS

3.1. Overview of Electron Responses During Solar Cycle 23

First, to set the stage, we show an overview of the 1998–2011 period using the HEO3 >1.5 MeV electron fluxes in Figure 2, displayed as an L versus time spectrogram. These data were taken from the higher altitude (lower B value) regions of the orbit shown in Figure 1. The top panel in Figure 2 shows the electron flux history over the $1.75 \leq L \leq 7$ range. The gray bars indicate where no data was available. The bottom panels in Figure 2 show 24 h running averages of *Dst* (middle) and the estimated plasma pause position, Lpp (bottom) [after *O'Brien and Moldwin*, 2003], respectively. There were multiple events where electron enhancements occurred in conjunction with storm time *Dst* depressions and Lpp reductions. Many of these electron enhancements penetrated deep into the slot region, and most were observed to occur at $L = 4$ and below. As the solar cycle progressed to solar minimum, the occurrence of electron enhancements at $L \leq 3$ was reduced, and the electron slot region widened and extended to larger L. Finally, in late 2008 through early 2010, the >1.5 MeV electron fluxes were reduced to extremely low levels throughout the whole outer zone.

The apparent >1.5 MeV fluxes, below the slot penetrations, extending down to $L = 1.75$ were contamination of the electron channel by energetic trapped protons at the outer edge of the inner zone and entrapped solar protons that penetrate into the slot region and diffuse inward. The contributions to low L trapped protons by Cosmic Ray Albedo Neutron Decay (CRAND) sourced protons and entrapped solar particles have been discussed in detail by *Selesnick et al.* [2007] where he showed the expected variability in the $L = 2$ trapped proton fluxes (his Figure 15) and relative contribution of solar and CRAND sourced protons (his Figure 19) as a function of L. These will not be discussed further here.

Figure 3 is a series of line plots (upper part of each panel) that more clearly show the >1.5 MeV electron enhancement magnitudes and decay profiles than can be inferred from Figure 2 because of the limitations of the color scale. Each

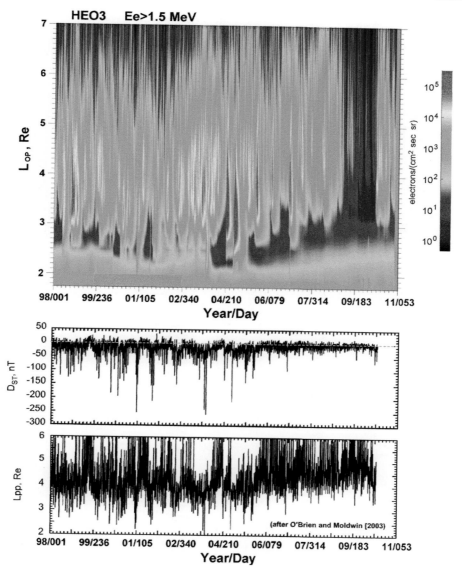

Figure 2. (top) HEO3 >1.5 MeV electron spectrogram covering the period 1998 through early 2011, (middle) plus 24 h running averages of *Dst*, and (bottom) the estimated plasmapause position Lpp.

panel is for a constant *L* value. The plots clearly show the order(s) of magnitude flux enhancements that occurred starting from the slot region (*L* = 2.75) out to *L* = 4. However, even this display does not allow for a detailed view of the flux decays because of the compressed time scale. (We select one storm interval for detailed examination in Figure 6 below.) Intercomparison of the panels in Figure 3 shows that there were differences in the electron enhancements among the different *L* values. One thing that is clear is that there was a reduction in the number of electron enhancements with decreasing *L* values. Also, we note that the lowest *L* values

often did not show dramatic flux drops (short duration downward "spikes" in the fluxes) prior to the poststorm enhancements. This will be discussed further below.

Figure 4 shows a spectrogram of the >230 keV electron fluxes in the top panel and the 24 h running average of *Dst* in the bottom panel. The electrons in Figure 4 are discussed in the literature as the seed population [e.g., *Baker et al.*, 1997; *Meredith et al.*, 2002]. These >230 keV electrons responded readily to magnetic disturbances and were often found to be enhanced at low *L* and in the electron slot regions in response to magnetic storms. Note that even during the solar minimum

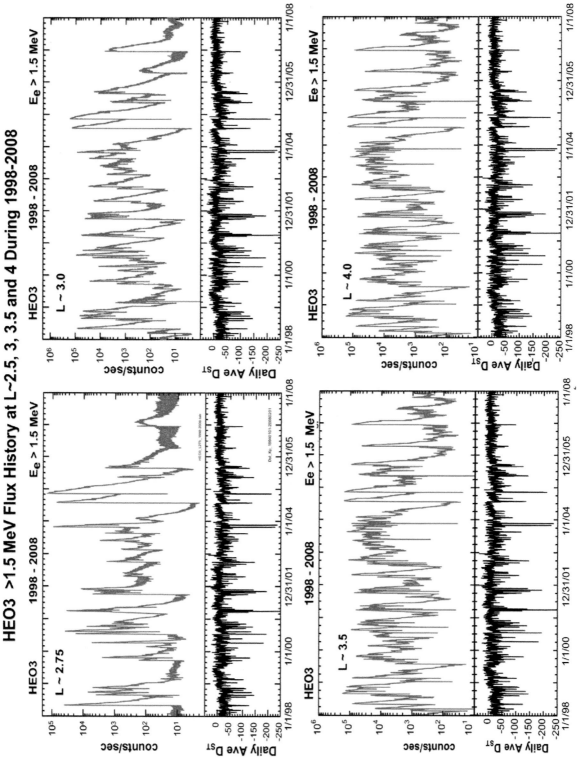

Figure 3. Plots of >1.5 MeV electron intensities at four constant *L* values for the interval 1 January 1998 through 1 January 2008. *Dst* is shown for reference in each plot. The very narrow downward "spikes" in the electron fluxes, like those seen in late 2003, at *L* = 3.5 and *L* = 4 are the signatures of storm main phase flux dropouts. Note these are not observed at *L* = 1.75 and *L* = 3.

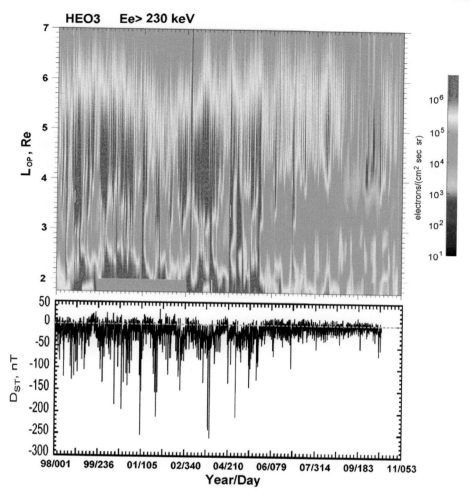

Figure 4. (top) HEO3 > 230 keV electron spectrogram and (bottom) *Dst* for the 2008 through early 2011 interval.

period, there were enhancements of these fluxes in the slot region in response to moderate magnetic activity. This indicates that significant fluxes of these seed electrons are generally present over a wide range of *L* values, including the slot region, whenever there is magnetic activity, and as we show below, the enhancements of these fluxes to their peak values generally lead the enhancements of the MeV fluxes to their peak values by many hours to days. However, the presence of the enhanced seed population does not guarantee that MeV fluxes will be enhanced also, especially in the slot region.

Below, we limit the discussion to electron observations at *L* ≤ 4 and to the time period 1998–2008 where most of the electron enhancements occurred. That is, we limit the presentation to the regions that are generally inside the plasmasphere during quiet periods. This provides a look at the regions where the storm time electron fluxes often peak

following moderate to large storms [*Obrien et al.*, 2003; *Tverskaya et al.*, 2003] and a view of the slot region responses. Figure 5 provides an overview of the activity levels experienced during the analysis period. It displays, in histogram form, the distribution of minimum *Dst* (top panel), the distribution of minimum estimated Lpp (middle), and the distribution of maximum *Kp* that occurred during the storms studied for reference. No storms with *Dst* ≥ −70 nT were used in this study. Most of the storms had minimum *Dst* of −75 to −150 nT. The majority of the storms had estimated Lpp ≥ 2.9. The magnetic activity, as indicated by the *Kp*, had maximum *Kp* ≥ 4+ with the majority having *Kp* > 6.

3.2. Electron Decays at L = 3

At the low *L* values, where storm main phase flux dropouts are relatively rare, the postenhancement losses are thought to

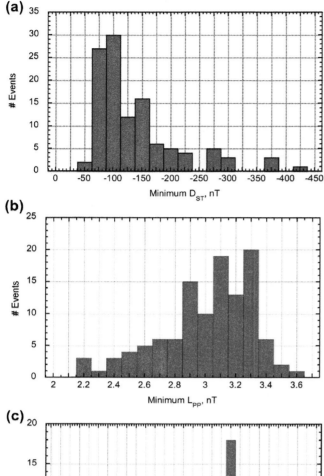

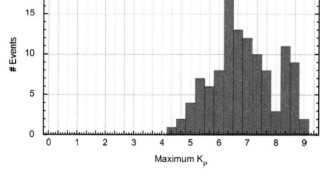

storms (as defined by *Dst*) and then only during the main and early recovery phases. Thus, we expect the loss rates at $L \leq 3$ to be mostly consistent with the wave particle interactions mentioned. With this in mind, we examined the electron decay rates at $L = 3$ to see how they may relate to the near constancy of the low-/high-altitude flux ratios noted by *Fennell et al.* [2005] and to expectations of plasmaspheric wave-particle–induced losses.

Figure 6 sets the stage for this part of the study. It shows the electron flux rise and subsequent decay following the May 1998 magnetic storm for three different energies. This storm had a minimum *Dst* of −205 nT. Figure 6 also shows where Solar Anomalous Magnetospheric Particle Explorer (SAMPEX) observed precipitation microbursts [*Lorentzen et al.*, 2001a, 2001b] and the *L* range over which they were observed during the ~80 day interval. The microbursts were intense and occurred primarily just before and during the poststorm flux increases at all energies [*Fennell et al.*, 2005].

An exponential fit to the flux decay was done for the energies shown in Figure 6. However, the $Ee > 0.6$ and >1.5 MeV data could not be fit with a single e-folding value

Figure 5. Distribution of (top) minimum *Dst* and (middle) Lpp plus (bottom) maximum *Kp* for the storms that occurred in the 1998–2008 interval.

be controlled by coulomb losses (d*E*/d*x* losses) and pitch angle scattering of the electrons by interactions with the plasmaspheric hiss and lightning-generated waves [*Lyons et al.*, 1972; *Lyons and Thorne*, 1973; *Meredith et al.*, 2007]. As Figure 2 (bottom panel) shows, the estimated plasmasphere boundary reached $L \leq 3$ in response to large

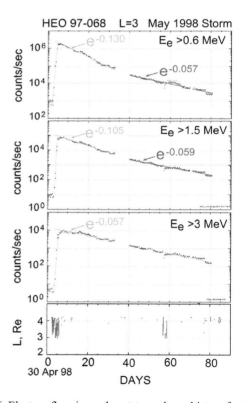

Figure 6. Electron flux rise and poststorm decay history for the May 1998 storm. (top) The electron response at 0.6, 1.5, and 3 MeV, respectively. (bottom) The *L* and time position of the electron microbursts observed by SAMPEX.

over the full period shown so the period beyond day 40 (9 June 1998) was fit separately. The exponential fit to the flux decay obtained early in the recovery phase of the storm is indicated by the blue line, and the late decay fit is shown by the red line. This change in the decay rate late in the decay period is a common feature in the HEO3 data. Figure 6 represents the way $L = 3$ electron decay data was gathered for the >1.5 MeV electron fluxes for events that occurred between 1 January 1998 and 1 January 2008. The intervals from a few days after the storm electron flux enhancements peaked to the next event were piecewise fit to represent the decay history like that shown in Figure 6.

Using the fits to the electron flux decays, described above and in the work of *Fennell et al.* [2005], we organized the decay times obtained into two groups, early time or initial decays, taken during the early recovery phase of the storms, and late decays, determined tens of days into the storm recovery. This was done for all storm time flux enhancement events observed by the HEO3 spacecraft in the 1998–2008 interval. The e-folding decay times at $L = 3$ derived from the analysis are plotted in histogram form in Figure 7. The set of e-folding times taken early in the postenhancement flux decay phase clustered around two distinct periods: ~5–6 and ~12 days, as shown in the top panel of Figure 7. The late decay intervals had e-folding times peaked near ~22 days, as shown in the bottom panel of Figure 7. The enhanced electron fluxes caused by the large storms in May, September, October, and November 1998 (refer to Figures 2 and 3) had early e-folding decay periods of 9.5–10.5 days, which is very similar to the results obtained with SAMPEX by *Baker et al.* [1994] and consistent with past theoretical modeling [*Lyons and Thorne*, 1973]. All the secondary or late decay intervals had decay periods longer than the early decays and were more variable. These contributed to the broad peak near ~22 days in the bottom panel of Figure 7.

3.3. Electron Enhancements at Different L Values

The data in Figure 3 are plotted as line plots instead of in L versus time spectrogram form so that differences from one L value to the next clearly stand out. Figure 3 shows that the electron fluxes were most variable at the larger L values shown with sharp sudden flux drops for short intervals of time. This variability was even greater for $L > 4$ (not shown). The sharp flux drops corresponded with periods of magnetic activity, in many cases, storm related.

At the lowest L values, one observed sudden electron flux enhancements followed by decay. At $L = 2.75$, the flux enhancements often rose out of the instrument background. At $L = 4$, there was a series of flux enhancements, most of

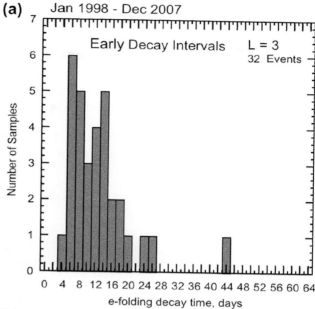

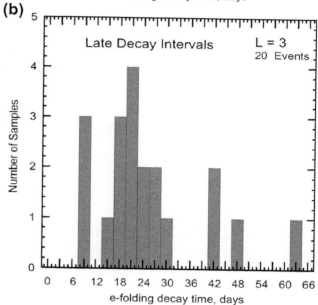

Figure 7. Histograms of e-folding decay times, in days, for the poststorm data from the 1998–2008 interval at $L = 3$: (a) early decay times and (b) late decay times. See text for definitions.

which were preceded by sharp flux drops as noted above. Generally, there was a strong association of the flux dropouts and subsequent rapid increases with the appearance of the microbursts [*Fennell et al.*, 2005].

Only the strongest storms caused flux enhancements in the electron slot region ($2 < L < 3$). Storms that caused slot region fillings were also associated with strong erosion of the plasmasphere, as shown at the bottom of Figure 2 and, for example, in the work of *Baker et al.* [2004] for the "Hallowe'en" 2003 storm. Each time *Dst* was <−150 nT, and the plasmapause was estimated to be below *L* = 3, the electron fluxes in the slot showed an increase. This indicates that erosion of the plasmapause to low *L* is one process that occurs simultaneous with electrons being transported to or accelerated at equally low *L*. However, it is not clear from the observations in Figures 2 through 4 whether the erosion of the plasmapause is a necessary condition for all low *L* electron flux enhancements. Shock acceleration of the type that occurred during the famous March 1991 event [*Li et al.*, 1993; *Blake et al.*, 1992] may occur without plasmapause erosion. However, the strong relationship between plasmapause erosion and low *L* electron enhancements is compelling and needs to be examined in detail [*Li et al.*, 2006]. The direct plasmapause position using the Polar satellite's frame potential [*Laakso et al.*, 2002] and IMAGE FUV [*Goldstein et al.*, 2003, 2005] measurements combined with HEO and SAMPEX data could be used to perform such a study in the future but are outside the scope of this report.

3.4. Electron Flux Rise Times at L = 3

One of the things of interest was the rate of poststorm flux rises following storms that caused a flux increase. As noted above, the flux rises generally occurred following a sudden flux dropout during the storm main phase, except in the slot region. Generally, the flux rises are said to take 2 to 3 days following the flux dropout. However, such rise times have not been carefully measured. One cannot determine the rise times by examining figures like Figures 2 through 4 because of the compressed display in time.

As noted above, the HEO observations provide four complete traversals through the *L* values of interest each day. We used the same data sets for the decay times in Figure 7 to estimate the rise times for the flux increases. We set the starting time for each determination to be the time of *Dst* minimum and the ending time to be the time at which the flux reached its peak value, prior to the onset of the flux decay. We performed an e-folding fit to the flux versus time at constant *L* for each flux increase in the data set at four different energies. The results for *L* = 3 are shown in Figure 8 where each panel shows the results for the energy indicated. A clear trend of the rise times was observed with the rise times being shortest for the low-energy electrons compared to the high-energy electrons. The lowest-energy electrons also provided a tighter distribution of rise times. The statistics for the individual energies are given in each panel of Figure 8.

The relatively rapid rise times for the low-energy population fluxes are necessary if this seed population is the source

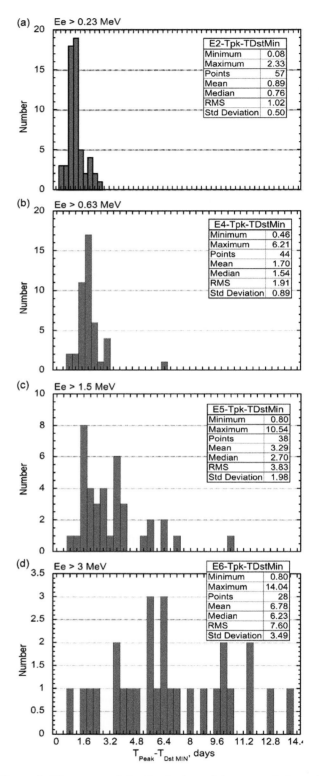

Figure 8. Histograms of the distribution of time to peak flux of electron enhancements at *L* = 3 relative to storm minimum *Dst* for (a) >0.23, (b) >0.63, (c) >1.5, and (d) >3 MeV.

for the higher-energy electrons that may be accelerated by wave-particle interactions. The unanswered question is how do these seed electrons reach $L = 3$ on time scales less than a day? Could the hundreds of keV electrons be transported by convection to the low L values? We should note that the rise time obtained for the >0.23 MeV electrons is at the limit of what can be measured using the HEO data.

The >0.63 MeV flux rise times were substantially longer than those of the >0.23 MeV fluxes. The >1.5 MeV rise times ranged from ~1.6 to >6 days with a median of ~2.7 days. The >3 MeV rise times were scattered over a wide range of times with a mean and median of >6 days but no clear preferred values. In many events, the >3 MeV flux was still rising when a second event occurred that caused a flux drop and initiated a new flux rise. Those events were deleted from the rise-time study.

3.5. Storm Time Electron Response

The storm recovery phase electron responses were examined for all the events. The poststorm fluxes were compared to the prestorm fluxes to determine whether the fluxes increased, decreased, or remained the same. For this comparison, the poststorm flux was measured at its peak value following the start of storm recovery as determined from Dst. This was then compared to the preonset flux. Only events that had a stable preonset flux level were used. The ratio of the recovery phase peak flux to the preonset flux was required to be ≥ 3 to be called an increase and $\leq 1/3$ to be called a decrease. Ratios with values of $1/3 <$ ratio < 3 were taken as a "no response" indication. Not all the energies showed the same responses at fixed L. There were indeed cases where some of the energies showed an increase, while others showed no response or a decrease. In general, the lowest energies were more likely to show an increase than the highest energies. The results of this comparison for the L values 2.75, 3, 3.5, and 4 are shown in Figures 9, 10, 11, and 12, respectively.

In Figures 9 through 12, we show the results of the storm time electron responses at four different energies (>0.23, >0.63, >1.5, and >3 MeV) at four different L values (2.75,

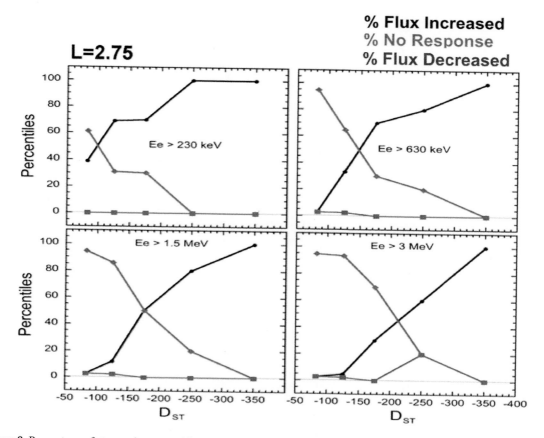

Figure 9. Percentage of storms that caused flux increases (black), no flux increase (red), or flux decrease (blue) at $L = 2.75$ for >0.23, >0.63, >1.5, and >3 MeV.

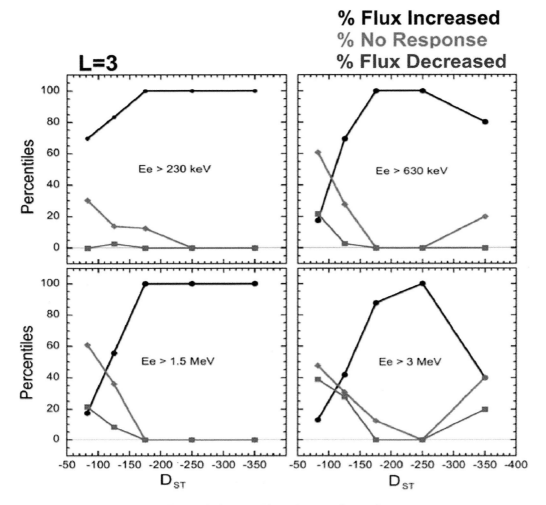

Figure 10. Same as Figure 9 except for $L = 3$.

3, 3.5, and 4). At $L = 2.75$ (Figure 9), the flux response trend was the same at the four electron energies examined. As the storm minimum Dst decreased, the occurrence of flux increases rose, while the occurrence of "no response" decreased. At this low L value, there was only one "flux decrease" observed. The responses at $L = 3$ (Figure 10) were similar to those at $L = 2.75$ except for the larger number of flux decreases for smaller $|Dst|$ and the reduced occurrence of flux increases for large $|Dst|$ at the highest energies. At $L = 3.5$ (Figure 11), the trend continued with greater occurrence of "no response" and "flux decrease" especially at the larger $|Dst|$. Finally, Figure 12 for $L = 4$ shows that the occurrence of "flux increase," "flux decrease," and "no response" was comparable to the results of *Reeves et al.* [2003] with roughly 40%–80% showing "flux increase," while a few to 30% show "no response" or a "flux decrease." The trend of greater relative occurrence of "flux increase" with decreasing energy

was also observed at $L = 4$. The occurrence of "flux decrease" was greatest for the larger $|Dst|$.

4. DISCUSSION

The flux decay times indicate that the combined transport and loss process along the field lines match expectations for wave-particle interaction models of the type discussed by *Lyons and Thorne* [1973], *Abel and Thorne* [1998], *Summers et al.* [2007], and *Meredith et al.* [2007]. The processes are primarily ones of pitch angle transport, coulomb drag (energy transport), and atmospheric losses. These are in competition with radial transport inside the plasmasphere and radial transport plus local acceleration via wave particle interactions outside the plasmasphere. Inside the plasmasphere, we expected the recovery phase decay rates to be relatively constant as long as there is no significant new activity. Thus, the late

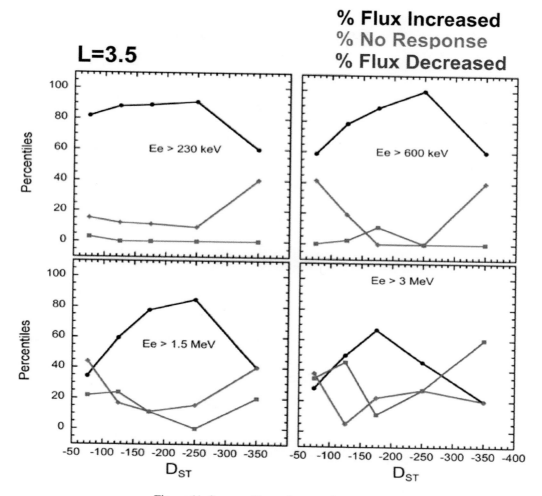

Figure 11. Same as Figure 9 except for $L = 3.5$.

decay rates shown in Figure 7 (bottom), which occur >10 days into the recovery, should be representative of plasmaspheric decays since these can only be observed if there are no new storms during the long decay period, and the plasmasphere should have recovered to $L > 3$ after 6–10 days.

The shortest early decays in Figure 7 (top) could have occurred prior to and during the plasmasphere refilling. Without detailed observed plasmapause position history, we cannot be sure. (As noted above, detailed comparisons of plasmasphere motion with electron flux decays may be possible for part of the study period, but was out of scope for this study.) The difference between the shorter initial decay rates and the longer ones in Figure 7 (top) could be a combination of pitch angle scattering by electromagnetic ion cyclotron and chorus when that L is outside the plasmasphere versus when $L = 3$ is inside the plasmasphere [O'Brien et al., 2008]. However, the plasmapause boundary recovers relatively rap-

idly to higher L values compared to the slow decay time scale indicated by the ~12 day peak in the top panel of Figure 7. The early and intermediate decay rates, represented in the top panel of Figure 7, may be caused by evolution of wave mode Eigen values and also different average magnetic activity and hiss production for those events.

The late decay period has not been examined previously and presents a new perspective on how electron decay proceeds. Previous studies have only considered the initial decay or used only a single e-folding time constant to represent the whole decay interval. The long decay times shown in the bottom panel of Figure 7 may also represent an evolution of wave mode Eigen values. While fits can be done to extract these late decay characteristic times, it is more likely that there is a smooth progression from the early faster decay times to the slower later ones. In fact, there is a hint of this in the top two panels of Figure 6. A next step would be to try

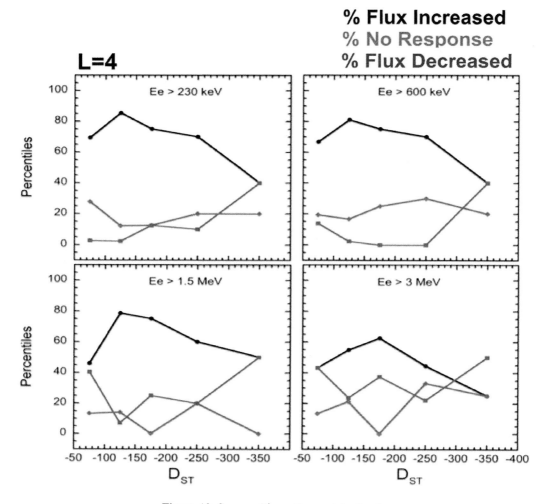

Figure 12. Same as Figure 9 except for *L* = 4.

to tease out this progression since it would be helpful for defining how to model the wave-particle interactions responsible for the resulting full decay time history.

The flux rise times at *L* = 3 show what we had generally expected. That is, the lowest-energy fluxes rose the fastest, while the highest energies rose significantly more slowly. We should also point out that the lower-energy fluxes (i.e., seed particles) would already be decaying, while the high-energy fluxes were rising. The >1.5 and >3 MeV flux rises take days and, in many cases, are still rising well into the late recovery phase of many storms. The spread in rise times at these energies may be related to whether the plasmasphere had recovered to *L* > 3 for the slower rising fluxes, inhibiting the ability of wave-particle interactions to raise the energy of the lower-energy seed populations that had peaked earlier. Future measurements of flux rise times for such energies over a

range of *L* values may provide significant constraint on storm time electron enhancement models.

In general, from the perspective of electron flux response to storms, it appears that at the lowest *L* values, there is either "no response" or a resultant "flux increase" from storms. The larger storms resulted in a higher probability of flux increases to low *L*. The same storms that provided flux increases at the lowest *L* values provided the greatest probability of flux decreases at the highest *L*, *L* = 4, studied. It is not clear what this means. Could the process that strongly dumps electrons penetrate down to *L* = 3.5–4.0? Is radial diffusion/transport of the electrons impeded in this region such that the scattering losses dominate? Or, do the losses from pitch angle scattering in this region often dominate over the chorus-generated acceleration? Either of these possibilities could explain why the relative occurrence of flux increases, for large $|Dst|$, is less in this region than at smaller L and why the occurrence of flux decreases is greater here.

There is much still to be learned from data in the lower L regions. A better combination of electron observations, like those reported here, and supportive information such as plasma density profiles (plasmasphere boundary position), plasma wave characterization, magnetopause boundary positions, and ring current penetration are needed to understand the processes in these regions. We expect that Radiation Belt Storm Probes (RBSP), which should provide such detailed observations, will provide a more complete picture.

Acknowledgments. This work was partially supported by grant 64361 from University of Colorado and NASA and partially under The Aerospace Corporation's Sustained Experimentation and Research for Program Applications (SERPA).

REFERENCES

Abel, B., and R. M. Thorne (1998), Electron scattering loss in Earth's inner magnetosphere: 1. Dominant physical processes, *J. Geophys. Res.*, *103*(A2), 2385–2396, doi:10.1029/97JA02919.

Baker, D. N., J. B. Blake, L. B. Callis, J. R. Cummings, D. Hovestadt, S. Kanekal, B. Klecker, R. A. Mewaldt, and R. D. Zwickl (1994), Relativistic electron acceleration and decay time scales in the inner and outer radiation belts: SAMPEX, *Geophys. Res. Lett.*, *21*(6), 409–412, doi:10.1029/93GL03532.

Baker, D. N., et al. (1997), Recurrent geomagnetic storms and relativistic electron enhancements in the outer magnetosphere: ISTP coordinated measurements, *J. Geophys. Res.*, *102*(A7), 14,141–14,148, doi:10.1029/97JA00565.

Baker, D. N., S. G. Kanekal, X. Li, S. P. Monk, J. Goldstein, and J. L. Burch (2004), An extreme distortion of the Van Allen belt arising from the 'Hallowe'en' solar storm in 2003, *Nature*, *432*, 878–881, doi:10.1038/nature03116.

Blake, J. B., W. A. Kolasinski, R. W. Fillius, and E. G. Mullen (1992), Injection of electrons and protons with energies of tens of MeV into $L < 3$ on 24 March 1991, *Geophys. Res. Lett.*, *19*(8), 821–824, doi:10.1029/92GL00624.

Blake, J. B., D. N. Baker, N. Turner, K. W. Ogilvie, and R. P. Lepping (1997), Correlation of changes in the outer-zone relativistic-electron population with upstream solar wind and magnetic field measurements, *Geophys. Res. Lett.*, *24*(8), 927–929, doi:10.1029/97GL00859.

Fennell, J. F., and J. L. Roeder (2010), Evolution and energization of energetic electrons in the inner magnetosphere, in *The Cluster Active Archive*, edited by H. Laakso, M. Taylor and C. P. Escoubet, chap. 34, pp. 467–489, Springer, Dordrecht, The Netherlands, doi:10.1007/978-90-481-3499-1_34.

Fennell, J., J. Blake, R. Friedel, and S. Kanekal (2005), The energetic electron response to magnetic storms: HEO satellite observations, in *The Inner Magnetosphere: Physics and Modeling*, *Geophys. Monogr. Ser.*, vol. 155, edited by T. I. Pulkkinen,

N. A. Tsyganenko, and R. H. Friedel, pp. 87–95, AGU, Washington, D. C., doi:10.1029/155GM10.

Goldstein, J., M. Spasojevic, P. H. Reiff, B. R. Sandel, W. T. Forrester, D. L. Gallagher, and B. W. Reinisch (2003), Identifying the plasmapause in IMAGE EUV data using IMAGE RPI in situ steep density gradients, *J. Geophys. Res.*, *108*(A4), 1147, doi:10.1029/2002JA009475.

Goldstein, J., S. G. Kanekal, D. N. Baker, and B. R. Sandel (2005), Dynamic relationship between the outer radiation belt and the plasmapause during March–May 2001, *Geophys. Res. Lett.*, *32*, L15104, doi:10.1029/2005GL023431.

Laakso, H., R. Pfaff, and P. Janhunen (2002), Polar observations of electron density distribution in the Earth's magnetosphere. 1. Statistical results, *Ann. Geophys.*, *20*, 1711–1724.

Li, X., I. Roth, M. Temerin, J. R. Wygant, M. K. Hudson, and J. B. Blake (1993), Simulation of the prompt energization and transport of radiation belt particles during the March 24, 1991 SSC, *Geophys. Res. Lett.*, *20*(22), 2423–2426, doi:10.1029/93GL02701.

Li, X., D. N. Baker, T. P. O'Brien, L. Xie, and Q. G. Zong (2006), Correlation between the inner edge of outer radiation belt electrons and the innermost plasmapause location, *Geophys. Res. Lett.*, *33*, L14107, doi:10.1029/2006GL026294.

Lorentzen, K. R., M. D. Looper, and J. B. Blake (2001a), Relativistic electron microbursts during the GEM storms, *Geophys. Res. Lett.*, *28*(13), 2573–2576, doi:10.1029/2001GL012926.

Lorentzen, K. R., J. B. Blake, U. S. Inan, and J. Bortnik (2001b), Observations of relativistic electron microbursts in association with VLF chorus, *J. Geophys. Res.*, *106*(A4), 6017–6027, doi:10.1029/2000JA003018.

Lyons, L. R., and R. M. Thorne (1973), Equilibrium structure of radiation belt electrons, *J. Geophys. Res.*, *78*(13), 2142–2149, doi:10.1029/JA078i013p02142.

Lyons, L. R., R. M. Thorne, and C. F. Kennel (1972), Pitch-angle diffusion of radiation belt electrons within the plasmasphere, *J. Geophys. Res.*, *77*(19), 3455–3474, doi:10.1029/JA077i019p03455.

Meredith, N. P., R. B. Horne, D. Summers, R. M. Thorne, R. H. A. Iles, D. Heynderickx, and R. R. Anderson (2002), Evidence for acceleration of outer zone electrons to relativistic energies by whistler mode chorus, *Ann. Geophys*, *20*, 967–979.

Meredith, N. P., R. B. Horne, S. A. Glauert, and R. R. Anderson (2007), Slot region electron loss timescales due to plasmaspheric hiss and lightning-generated whistlers, *J. Geophys. Res.*, *112*, A08214, doi:10.1029/2007JA012413.

O'Brien, T. P., and M. B. Moldwin (2003), Empirical plasmapause models from magnetic indices, *Geophys. Res. Lett.*, *30*(4), 1152, doi:10.1029/2002GL016007.

O'Brien, T. P., K. R. Lorentzen, I. R. Mann, N. P. Meredith, J. B. Blake, J. F. Fennell, M. D. Looper, D. K. Milling, and R. R. Anderson (2003), Energization of relativistic electrons in the presence of ULF power and MeV microbursts: Evidence for dual ULF and VLF acceleration, *J. Geophys. Res.*, *108*(A8), 1329, doi:10.1029/2002JA009784.

O'Brien, T. P., Y. Y. Shprits, and M. B. Moldwin (2008), Eigenmode analysis of pitch-angle diffusion of energetic electrons in the outer zone, *J. Atmos. Sol. Terr. Phys.*, *70*, 1738–1744, doi:10. 1016/j.jastp.2008.05.011.

Olson, W. P., and K. A. Pfitzer (1974), A quantitative model of the magnetospheric magnetic field, *J. Geophys. Res.*, *79*(25), 3739–3748, doi:10.1029/JA079i025p03739.

Reeves, G. D., K. L. McAdams, R. H. W. Friedel, and T. P. O'Brien (2003), Acceleration and loss of relativistic electrons during geomagnetic storms, *Geophys. Res. Lett.*, *30*(10), 1529, doi:10. 1029/2002GL016513.

Selesnick, R. S., M. D. Looper, and R. A. Mewaldt (2007), A theoretical model of the inner proton radiation belt, *Space Weather*, *5*, S04003, doi:10.1029/2006SW000275.

Summers, D., B. Ni, and N. P. Meredith (2007), Timescales for radiation belt electron acceleration and loss due to resonant wave-particle interactions: 2. Evaluation for VLF chorus, ELF hiss, and electromagnetic ion cyclotron waves, *J. Geophys. Res.*, *112*, A04207, doi:10.1029/2006JA011993.

Tverskaya, L. V., N. N. Pavlov, J. B. Blake, R. S. Selesnick, and J. F. Fennell (2003), Predicting the L-position of the storm-injected relativistic electron belt, *Adv. Space Res.*, *31*, 1039–1044, doi:10. 1016/S0273-1177(02)00785-8.

J. F. Fennell and J. L. Roeder, The Aerospace Corporation, MS M2-260, P. O. Box 92957, Los Angeles, CA 90009-2957, USA. (joseph.f.fennell@aero.org)

S. Kanekal, Goddard Space Flight Center, Code 672, Greenbelt, MD 20771-2400, USA.

Colorado Student Space Weather Experiment: Differential Flux Measurements of Energetic Particles in a Highly Inclined Low Earth Orbit

X. Li,[1,2] S. Palo,[2] R. Kohnert,[1] D. Gerhardt,[2] L. Blum,[1,2] Q. Schiller,[1,2]
D. Turner,[3] W. Tu,[4] N. Sheiko,[1] and C. Shearer Cooper[1,2]

The Colorado Student Space Weather Experiment (CSSWE) is a three-unit (10 cm × 10 cm × 30 cm) CubeSat mission funded by the National Science Foundation; it was launched into a low Earth, polar orbit on 13 September 2012 as a secondary payload under NASA's Educational Launch of Nanosatellites program. The science objectives of CSSWE are to investigate the relationship of the location, magnitude, and frequency of solar flares to the timing, duration, and energy spectrum of solar energetic particles reaching Earth and to determine the precipitation loss and the evolution of the energy spectrum of radiation belt electrons. CSSWE contains a single science payload, the Relativistic Electron and Proton Telescope integrated little experiment (REPTile), which is a miniaturization of the Relativistic Electron and Proton Telescope (REPT) built at the Laboratory for Atmospheric and Space Physics. The REPT instrument will fly onboard the NASA Radiation Belt Storm Probes mission, which consists of two identical spacecraft launched on 30 August 2012 that will go through the heart of the radiation belts in a low-inclination orbit. CSSWE's REPTile is designed to measure the directional differential flux of protons ranging from 10 to 40 MeV and electrons from 0.5 to >3 MeV. Such differential flux measurements have significant science value, and a number of engineering challenges were overcome to enable these clean measurements to be made under the mass and power limits of a CubeSat. The CSSWE is an ideal class project, providing training for the next generation of engineers and scientists over the full life cycle of a satellite project.

1. INTRODUCTION

A full understanding of energetic particle dynamics in the near-Earth space environment is of scientific significance as well as of practical importance. Particularly at higher-latitude regions, energetic particles from the interplanetary medium, such as solar energetic particles (SEPs), have direct access to the Earth. Additionally, existing energetic particles in the magnetosphere, such as relativistic electrons in the outer radiation belt, can reach low altitudes following the magnetic field lines whose foot points map to high latitudes (>40°). This high-latitude, low-altitude region is also populated with many satellites as well as the international space station, from which various extravehicle activities have been performed. Energetic particles with energies of MeV can have

[1]Laboratory for Atmospheric and Space Physics, University of Colorado, Boulder, Colorado, USA.
[2]Department of Aerospace Engineering Sciences, University of Colorado, Boulder, Colorado, USA.
[3]Department of Earth and Space Sciences, University of California, Los Angeles, California, USA.
[4]Los Alamos National Laboratory, Los Alamos, New Mexico, USA.

Dynamics of the Earth's Radiation Belts and Inner Magnetosphere
Geophysical Monograph Series 199
10.1029/2012GM001313

harmful radiation effects on the bodies of astronauts and various deleterious effects on satellite subsystems through either single event upset or deep dielectric discharging [*Baker*, 2001, 2002]. Better measurement and understanding of the energetic particles in a highly inclined low Earth orbit (LEO) are the science objectives of the Colorado Student Space Weather Experiment (CSSWE) mission. CSSWE has also provided a unique opportunity for students to acquire hands-on experience, under the tutelage of experienced scientists and engineers, throughout the entire engineering process, gaining experience in data analysis and modeling, and gaining scientific insight on solar activity and its effects on the near-Earth space environment.

We will first discuss the science background and motivation for this project, the science measurement requirement, and expected science results and impact, before we provide the system description of the mission, followed by planned data analysis and interpretation, and modeling efforts and then discussion and summary.

1.1. Science Background and Motivation

Humankind has long been fascinated with the Sun and its relationship to our planet. *Sabine* [1852] was first to note that geomagnetic activity tracks the 11 year solar activity cycle. The first solar flare ever observed, in white light, was followed about 18 h later by a very large geomagnetic storm [*Carrington*, 1860]. The existence of the Earth's radiation belts was established in 1958 by James Van Allen and coworkers using simple Geiger counters on board Explorer-1 and -3 spacecraft. Since then, more advanced space missions have provided insight into the phenomenology and range of processes active on the Sun and in the radiation belts.

We now understand that coronal mass ejections (CMEs), which are episodic ejections of material from the solar atmosphere into the solar wind, are the link between solar activity and large, nonrecurrent geomagnetic storms, during which the trapped radiation belt electrons have their largest variations. There is a strong correlation between CMEs and solar flares, but the correlation does not appear to be a causal one. Rather, solar flares and CMEs appear to be separate phenomena, both resulting from relatively rapid changes in the magnetic structure of the solar atmosphere [e.g., *Gosling*, 1993].

1.1.1. Solar flares and solar energetic particles. Solar flares are very violent processes in the solar atmosphere that are associated with large-energy releases ranging from 10^{22} J for subflares, to more than 10^{32} J for the largest flares [*Priest*, 1981]. The strongest supported explanation for the onset of the impulsive phase of a solar flare is that it is due to

magnetic reconnection of existing or recently emerged magnetic flux loops [*Aschwanden*, 2004, and references therein]. Magnetic reconnection accelerates particles, producing proton and electron beams that travel along flaring coronal loops. Some of the high-energy particles escape from the Sun and can reach the Earth's low-altitude, high-latitude regions.

Statistically, both the probability of observing energetic solar protons near the Earth as well as the maximum flux values observed are strongly dependent on the size of the flare and its position on the Sun. It is also now clear that the most intense and longest-lasting SEP events are produced by strong shocks in the solar wind driven by the fast CMEs [e.g., *Reames*, 1997]. It is currently believed that SEPs observed near Earth are of two basic populations. Events in one population, the so-called "impulsive" events, originate in flaring regions and typically last for several hours and have limited spatial extents (<30° in latitude and longitude) in the solar wind. In contrast, events in the other population, the so-called "gradual" events, tend to be more intense than the impulsive events, typically last for days, often spread over more than 180° in latitude and longitude and are strongly associated with CME-driven shock disturbances. In practice, since flares and CMEs often occur in conjunction with one another, many SEP events appear as hybrids of these two basic populations.

Crucial questions remain about exactly how and where both of the above populations are produced. In the case of flare events, a major uncertainty is how a given flare site connects magnetically to the interplanetary medium, i.e., the accessibility to, and extent of, open magnetic field lines in the vicinity of a flare site [*Cane and Lario*, 2006].

The time-intensity profiles of the SEP events observed in the ecliptic plane at 1 AU are organized in terms of the longitude of the observer with respect to the traveling CME-driven shock [*Cane et al.*, 1988; *Kanekal et al.*, 2008]. SEP events generated from the western longitudes have rapid rises followed by gradual decreasing intensities, while SEP events generated from eastern longitudes show slowly rising intensity enhancement structured around the arrival of the CME-driven shocks. It is clear that the location of the event is very important regarding how these SEPs affect the Earth's environment. This longitudinal dependence of the time-intensity profiles, together with the rate at which the particle intensities increase or decrease, have been used to predict the arrival of CME-driven shocks at 1 AU [*Smith et al.*, 2004; *Vandegriff et al.*, 2005].

1.1.2. Earth's radiation belts. Earth's radiation belts are usually divided into the inner belt, centered near 1.5 Earth radii (R_E) from the center of the Earth when measured in the equatorial plane, and the outer radiation belt that is most intense between 4 and 5 R_E. These belts form a torus around the Earth, and many important satellite orbits go through them, including GPS satellites, spacecraft at geosynchronous orbit (GEO), and those in highly inclined LEO.

The Earth's outer radiation belt consists of electrons in the energy range from keV to MeV. Compared to the inner radiation belt, which usually contains somewhat less energetic electrons but an extremely intense population of protons extending in energy up to several hundreds of MeV or even GeV, the outer belt consists of energetic electrons that show a great deal of variability that is well correlated with geomagnetic storms and high speed solar wind streams [*Williams*, 1966; *Paulikas and Blake*, 1979; *Baker et al.*, 1979]. Figure 1 shows measurements of radiation belt electrons and protons by the Solar, Anomalous, and Magnetospheric Particle Explorer (SAMPEX), ~550 km altitude and 82° inclination, from launch to the end of 2009 together with the sunspot number and the *Dst* index, which indicates the onset, duration, and magnitude of magnetic storms. The outer belt exhibits a strong seasonal and solar cycle variation. It was most intense, on average, during the descending phase of the sunspot cycle (1993–1995; 2003–2005), weakest during sunspot minimum (1996–1997; 2007–2009), and then became more intense

again during the ascending phase of the solar cycle (1997–1999). Seasonally, the outer belt is most intense around the equinoxes [*Baker et al.*, 1999] and also penetrates the deepest around the equinoxes [*Li et al.*, 2001]. In Figure 1, the vertical yellow bars along the horizontal axis mark equinoxes. Another remarkable feature of Figure 1 is the correlation of the inward extent of MeV electrons with the *Dst* index, which is also referred to as the magnetic storm index.

1.2. Science Measurement Requirements

CSSWE is a three-unit (10 cm × 10 cm × 30 cm) CubeSat mission. Resources are limited. Any design is subject to the constraints of mass, power, data rate, and budget. To reach the science objectives described earlier and take into consideration various limitations of a CubeSat and what existing measurements are already available, the following science measurement requirements are established: (1) electron differential flux measurements between 0.5 and 3 MeV, and integral flux measurements for >3 MeV, (2) proton differential flux measurements between 10 and 40 MeV, (3) time cadence: 6 s.

Even the above general requirements were not settled until various design work and trade studies were performed. The

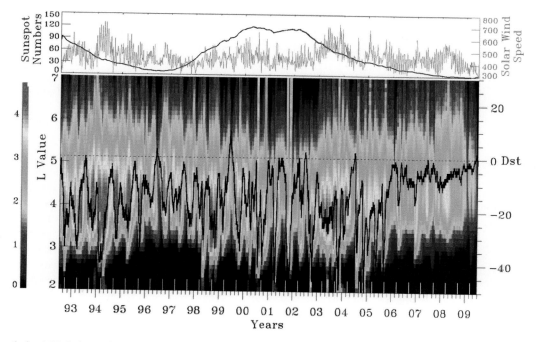

Figure 1. (top) Variations of yearly window-averaged sunspot numbers (black curve) and weekly window averaged solar wind speed (km s^{-1}, red curve). (bottom) Monthly window-averaged, color-coded in logarithm, and sorted in *L* (*L* bin: 0.1) electron fluxes of 2–6 MeV (# cm^{-2} s^{-1} sr^{-1}) by SAMPEX since its launch (3 July 1992) into a low-altitude (550 km × 600 km) and highly inclined (82°) orbit. The superimposed black curve represents monthly averaged *Dst* index. From *Li et al.* [2011].

time cadence is strictly limited by the downlink rate available based on one ground station built for this mission. The particle energy range is limited by the speed of electronic resolution and the available shielding mass, which is associated with the S/N ratio. Detailed spacecraft and instrument design will be described later.

1.3. Expected Science Results and Impact

1.3.1. Solar energetic particles. There are no existing differential flux measurements for protons in the tens of MeV range in LEO. NOAA/NPOES in LEO provide integral measurement of protons between 100 keV to low MeV. GOES at GEO have both integral and differential measurements of protons between 1 and 100 MeV. Relativistic Electron and Proton Telescope integrated little experiment (REPTile) on CSSWE will provide measurements of the differential flux of protons at LEO, which are critical for investigating the geomagnetic cutoff variations during SEP events and their implication for the radiation environment at the International Space Station [*Leske et al.*, 2001]. However, any significant science results have to be achieved with coordination with other available measurements and modeling efforts. For example, to study how the flare location, magnitude, and frequency relate to the timing, duration, and energy spectrum of SEPs reaching Earth, the information about the solar flare intensity and location, which are provided by other missions, namely, NASA Solar Dynamic Observatory (SDO) and/or NOAA GOES are required.

1.3.2. Outer radiation belt electrons. Instruments onboard SAMPEX, though a wonderful mission for its original objectives, were not designed to make accurate measurements of the outer radiation belt electrons. For example, they lack differential flux measurements for MeV electrons. With REPTile on CSSWE, we will have measurements necessary to better determine the electron energy spectrum. These measurements will help us to better understand the acceleration mechanisms and loss processes of the outer radiation belt electrons.

Measurements of outer belt electrons made by REPTile will also be useful for comparisons with those made by NASA's Radiation Belt Storm Probes (RBSP) mission. The RBSP satellites will travel through the heart of the outer belt, where they will make important measurements of outer belt fluxes and the various types of plasma waves that are important in electron acceleration and loss. However, for electrons and protons to precipitate into the atmosphere (a major loss mechanism), their equivalent equatorial pitch angle (PA) has to be very small, 2°–5° depending on their actual locations. The instruments onboard RBSP, sophisticated as they are,

cannot resolve the loss cone distribution because they stay close to the equator. Thus, it will be difficult to determine the precipitation loss from their measurements. Though REPTile measures a mixed population of precipitating as well as trapped radiation belt electrons and protons from its low-altitude high-inclination orbit, combining REPTile measurements at LEO with modeling efforts (to be discussed in detail later) will enable us to determine the precipitation loss. By comparing flux measurements made by RBSP and REPTile, better estimates can be made of the trapped electron population and the precipitating population.

1.3.2.1. Acceleration mechanisms. How the outer radiation belt is formed in the Earth's magnetosphere remains one of the most intriguing puzzles in space physics. For some time, it was thought to be well understood at least in its general outlines. However, recently, the paradigm for explaining the creation of the outer belt electrons has been shifting from one using almost exclusively the theory of radial diffusion to one emphasizing more the role of waves [*Li et al.*, 1999; *Horne et al.*, 2005; *Shprits et al.*, 2007; *Chen et al.*, 2007; *Bortnik and Thorne*, 2007; *Li et al.*, 2007; *Tu et al.*, 2009; *Turner et al.*, 2010], presumably chorus whistler waves, in local heating of radiation belt electrons. A key proof of this new paradigm is to see how the energy spectrum of the radiation belt electrons evolves. A hardening spectrum (higher-energy electrons increasing faster than lower-energy electrons) at a given location would support the theory of in situ heating of the electrons by waves. Because of its low-altitude orbit, CSSWE will measure outer belt electrons four times in one orbital period, ~1.5 h, or about 60 times in a day. With its differential flux measurements, REPTile will be able to provide the critical information of the evolution of the electrons' energy spectrum.

1.3.2.2. Loss mechanisms. Some waves, like electromagnetic ion cyclotron waves, can cause PA diffusion of electrons, sending some electrons into the loss cone. Other waves, like whistler mode chorus waves, can cause energy diffusion as well as PA diffusion. An important consequence of the PA variation is precipitation loss. REPTile measurements can help to determine how many of the outer radiation belt electrons are lost to the atmosphere. Also, when RBSP and CSSWE are at similar magnetic longitudes and L shells, comparative studies can be conducted in which waves and fluxes measured by RBSP near the equator and the heart of the belt are compared to the REPTile measurements in LEO to directly compare waves with electron loss.

In summary, the science impacts of CSSWE are to provide needed measurements of energetic protons and electrons at LEO, in combination with other available measurements, to better address the following science questions: (1) How do solar flare location, magnitude, and frequency relate to the timing, duration, and energy spectrum of SEPs reaching Earth? (2) How do the loss rate and energy spectrum of the Earth's radiation belt electrons evolve?

2. SYSTEM DESCRIPTION OF THE CSSWE MISSION

2.1. Overview

CSSWE, like most satellites, is a collection of subsystems. In order to organize the subsystems, the requirements' flow down was defined throughout the mission development, bolstered by mass, power, data, and link budgets, as well as a risk analysis. A 3 U CubeSat is defined as a small volume (10 cm × 10 cm × 30 cm), small mass (< 4 kg), and completely autonomous (i.e., power, communications) satellite. Despite these strict requirements, CSSWE was delivered with margin in all budgets. The CSSWE architecture reflects the "keep it simple" method of satellite development; the system design was always simplified to meet requirements rather than designed to "push the envelope." Only two microcontrollers (MCUs) are present in the system, and two subsystems (attitude control system and thermal) are almost entirely passive. The command and data handling (C&DH) board and communication (COMM) radio were commercial off-the-shelf (COTS) purchases in an effort

to minimize risk. Figure 2 shows the system block diagram of CSSWE. In the following sections, we will provide a general description of individual subsystems, with some more detailed description on the science payload, REPTile.

2.2. Structure and Thermal Design

The structural design of the CSSWE CubeSat began with the commercially available Pumpkin, Inc. 3 U aluminum chassis. The left image in Figure 3 shows a rendering of the aluminum shell of the CubeSat with custom-designed solar panels. The right side of Figure 3 shows the interior components of the CubeSat with the REPTile scientific instrument at the center and electronic boards (light blue) and battery (yellow) on the top of REPTile. The interior view shows the custom structural supports made to accommodate the relatively heavy instrument as well as all electrical components. The design of the interior structure was also driven by the CubeSat requirement that the satellite center of mass be within 2 cm of the geometric center, as well as the need for simple assembly and disassembly during integration and testing. All of the interior components assemble in a vertical stack, allowing the exterior shell to slide over the entire assembly during integration. Extensive finite element analysis was performed on the structural components to ensure that the CubeSat could survive a vibration environment three

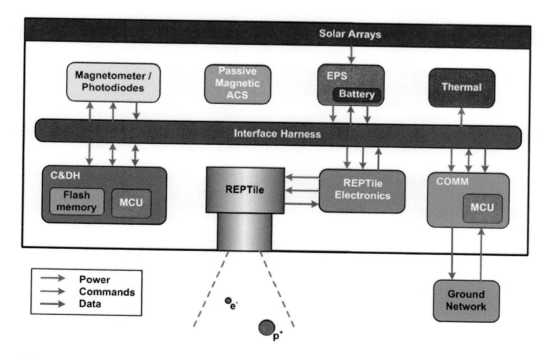

Figure 2. System block diagram of Colorado Student Space Weather Experiment (CSSWE), where e$^-$ and p$^+$ represent energetic electrons and protons to be measured.

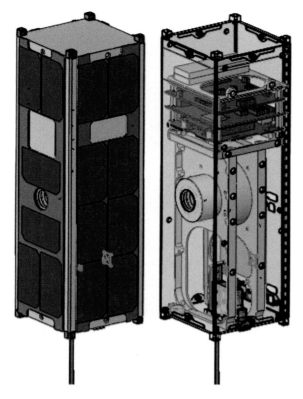

Figure 3. Three-dimensional rendering of the CSSWE CubeSat, (left) exterior view and (right) interior view.

sigma higher than expected during launch. Prelaunch qualification and acceptance random vibration testing confirmed that the structure met all strength requirements.

Figure 4 shows the completed flight model after final integration. The cutouts (e.g., the silver area between the solar cells in Figure 4) are radiative windows designed to keep the satellite electronics stack cool. The small slits visible in the top thermal window are two of four electronic ports used for preflight communication and battery charging. Detailed thermal analysis was performed on the entire system using Thermal Desktop, a software tool used to model thermal environments. Individual electronic boards were thermally modeled to provide a predicted on-orbit temperature profile for sensitive electronics. The radiative windows mentioned above were subsequently added to help radiate excess heat from the system, to maintain internal board temperatures within range of manufacturer specifications for all electronics components (between $-30°C$ to $+45°C$).

2.3. Electric and Power System and Solar Arrays

CSSWE employs a direct energy transfer system that was designed, fabricated, and tested by the University of

Colorado students. The CSSWE power system architecture was modeled after the design used for the University of Michigan's RAX mission [*Cutler et al.*, 2010].

Four independent solar array strings, one on each long-axis side of the CubeSat, are each fed into 8.8 V regulators with isolation diodes on the output. Solar arrays, designed and fabricated by the University of Colorado students, employ 28% efficient, uncovered, triple junction solar cells. The 6–8.4 V bus is driven by the voltage of the 8.4 V lithium polymer battery. The 3.3 and 5 V regulated buses are powered from the battery bus and supply power to most of the subsystem electronics. The exceptions to this are the antenna deployment module and the transmitter power amplifier, which are powered directly from the 6–8.4 V power bus. An external charge port allows the spacecraft to be powered by an external power supply for ground test and mission simulations in the laboratory. The battery can be disconnected from the bus by either of two series switches, the Remove Before Flight (RBF) switch or the deployment switch, as is required in the CalPoly CubeSat Design Specification. The RBF switch is used while testing in the laboratory and is closed upon final integration into the P-POD launcher. The deployment switch remains open, while the spacecraft is integrated in the P-POD launcher. The deployment switch closes upon spacecraft deployment, connecting the battery to the bus and turning the spacecraft on once on orbit. Figure 5 shows the schematics of the electrical and power system, with solar cell inputs (PVX and PVY) on the left; 3.3, 5, and 6–8.4 V outputs to various other subsystems are shown on the right.

Figure 4. Final flight model and P-POD launcher.

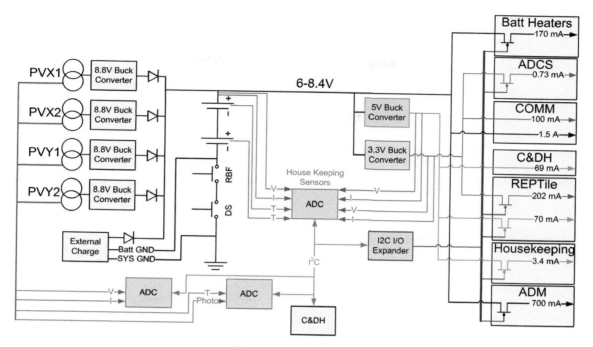

Figure 5. Electrical and power system electric diagram.

2.4. Command and Data Handling System

The C&DH system for the CSSWE CubeSat utilizes an off-the-shelf processor module from Pumpkin, Inc. that uses a 16 bit MCU, the MSP430 from Texas Instruments. The MCU runs at 8 MHz and has 8 kb of random-access memory; this allows CSSWE to meet mission requirements while providing ultralow power consumption to reduce load on the battery. The C&DH firmware was written in C, and run under the Salvo real-time operating system. It was developed using CrossStudio for MSP430 from Rowley Associates Ltd. The processor module involves additional hardware including an interface to a secure digital (SD) card that is used to store science and housekeeping data, log files, and configuration parameters.

C&DH communicates with nearly every other subsystem in the satellite; most interfaces use the Inter-Integrated Circuit protocol, an interface often used for data acquisition. The only exceptions are communication with the hardware supporting the SD card, which uses Serial Peripheral Interface, and communication with the radio, which uses asynchronous serial. The interface to the radio and SD card was dictated by the hardware manufacturers.

The C&DH firmware has five main tasks: (1) responding to commands from the ground, (2) acquiring and storing science data, (3) acquiring and storing housekeeping data, (4) deploying the antenna, and (5) controlling the battery

heaters. This simplicity is, in part, due to the overall design goal of minimal autonomy: to build the system in a way that reduces the number of decisions that C&DH makes on its own, while ensuring that enough information is available to operators on the ground to make informed decisions. C&DH is required to make a few autonomous choices; two of them have been mentioned already: deploying the antenna and controlling the battery heaters (to maintain the battery within operational temperature bounds even when out of range of the ground station). In addition, C&DH automatically drops into safe mode (and stops science operations) if it determines that the state of charge of the battery has fallen to a critically low level. Both the temperature settings of the battery heaters and the determination of whether the battery is "critically low" are controlled by parameters uploaded from the ground, allowing controllers to modify the spacecraft's operation in case of sensor malfunction on orbit. Additionally, CSSWE uses two independent watchdogs to ensure that neither the communications system (COMM) nor C&DH "locks up." If either subsystem does not respond to its associated watchdog, a soft reset is performed.

The storing of data onto the SD card takes place in two distinct streams. First of all, science data from the REPTile instrument and attitude data from the magnetometer are acquired and stored every 6 s. Second, housekeeping data (including voltages, currents, and temperatures from various locations on the spacecraft) are acquired every minute, and then every 10 min the minimum, maximum, and average values are stored to the SD card. Both streams of data are stored in time-offset files that allow C&DH to easily respond to ground requests for data from particular time ranges.

Commands exist to allow the ground to start and stop science mode, request transmission of specific subsets of the acquired data, update parameters controlling system operation, request transmission of housekeeping sensor values, and perform diagnostic functions. Communication packets are password protected to prevent unauthorized users from commanding the spacecraft.

2.5. Communications System

The CSSWE team chose a half-duplex communications architecture operating in the 70 cm band, primarily to reduce the complexity of the system. Given the size of the data product and the minimal amount of commanding required for CSSWE, sharing the uplink and downlink does not have a significant negative impact on our link budget.

The communications system onboard the spacecraft utilizes the AstroDev Lithium (Li-1) radio, which operates over a wide range of frequencies and temperatures, is 40% power efficient, and can output 34 dBm of RF power. Additionally, the Li-1 radio supports the AX.25 packet radio protocol at a rate of 9.6 kbit s^{-1} across the RF link and up to 115.2 kbit s^{-1} between the radio and C&DH over the serial link. Assuming

21.75 min of communications time per day, calculated using the Satellite Took Kit for our nominal orbit and a ground station in Boulder, we have the capacity to downlink 1.195 MB d^{-1}, providing almost 50% more link capacity than is required for the mission.

A monopole was selected for the satellite antenna configuration after testing numerous options. The total length of the deployed antenna is 48.3 cm. The matching and tuning of the antenna provided excellent performance over our operating frequency with a maximum antenna gain in excess of 2 dBi for a reasonably omnidirectional antenna. Figure 6 shows the measured gain pattern of the CSSWE COMM system before and after antenna deployment. The deployed gain drops below −5 dBi only in the regions along the axis of the antenna, as well as at the small nulls near ±125° from the exposed end of the antenna. The results of orbit simulations indicate that CSSWE is rarely in an attitude and at a range where the link cannot be closed through this null in the pattern. Given our testing and analysis, we are confident that the communications system will operate as designed and will meet all the mission requirements for commanding and data throughput.

2.6. Ground Network

To operate the CSSWE CubeSat, a ground station has been built on the rooftop of the Laboratory for Atmospheric and

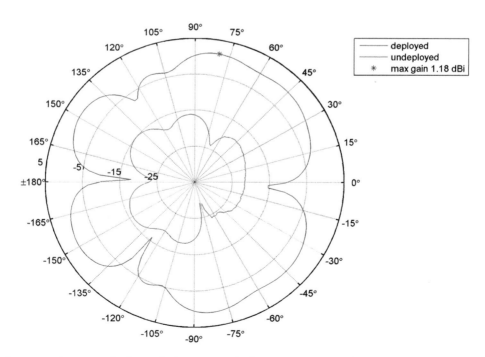

Figure 6. Antenna gain patterns before and after antenna deployment, as measured through anechoic chamber testing.

Space Physics (LASP). The CSSWE ground station operates in half-duplex mode, communicating at 437.345 MHz (UHF) for the uplink and downlink, the frequency designated to us by the International Amateur Radio Union. Commands are packetized and sent through Instrument and Spacecraft Interface Software (ISIS), commanding software that was inherited by LASP from the NOAA GOES-R program and has been customized for our uses. The Kantronics KAM XL terminal node controller modulates/demodulates the signal, and the Kenwood TS-2000 radio is used to communicate at the UHF band. Two M2 436CP42 cross Yagi antennas will be used, each with a gain of ~17 dBdc and a circular beamwidth of 21°. The antennas are pointed using a Yaesu G5500 azimuth-elevation rotator controlled by SatPC32, a software package developed for use with amateur satellites. This program also controls RF to account for Doppler shift during passes. The antennas and rotator are mounted on an 8 foot tower installed on the LASP roof and connected to the ground station control room with over ~200 feet of low-loss cabling, adding a total of −5.4 dBm loss to the RF signal. A block diagram of the ground station command and control chains is illustrated in Figure 7. The ground station has been fully tested and was used to command the satellite in a simulated on-orbit scenario before satellite delivery. We successfully sent commands and received data over the RF link with the CubeSat at an off-site location running off of its battery and solar panels. The ground station continues to be used and tested with an identical, spare version of the satellite built specifically for testing and calibration purposes.

2.7. Passive Magnetic Attitude Control System

CSSWE uses a passive magnetic attitude control system (MACS) that aligns the CubeSat to the Earth's local magnetic field line at all points in the orbit. The system is composed of two primary elements. The first is a bar magnet, which has a magnetic moment of $0.81\ \text{Am}^2$. Its dipole axis is parallel with the long axis of the spacecraft, and it provides a restoring torque toward the local magnetic field of the Earth. The second is an array of soft-magnetic hysteresis rods mounted perpendicular to the bar magnet, which are magnetized by the local earth field. As the satellite rotates, the relative orientation between the hysteresis rods and the local earth magnetic field changes, which changes the polarity of the rods. Energy is lost to heat as the magnetic domains within

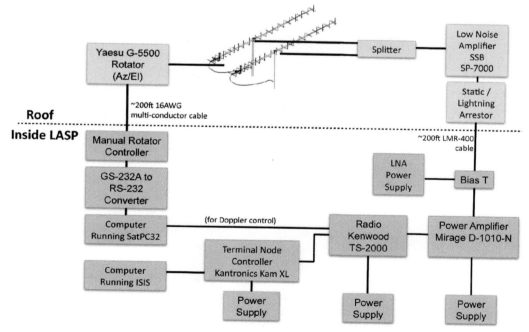

Figure 7. Ground station block diagram. Commands are packetized by the ground station software ISIS, passed through the terminal node controller and radio, where the signal is modulated about the assigned UHF frequency, 437.345 MHz, amplified ~10× by the Mirage D-1010-N power amplifier and transmitted through two Yagi antennas on the LASP roof. On the downlink, the signal is received and amplified ~24 dBm by the SSB SP-7000 low-noise amplifier and passed back through the RF chain to the ground station software. The pointing of the Yagi antennas is determined via a second computer running the tracking program SatPC32, which controls the azimuth-elevation rotator on the roof.

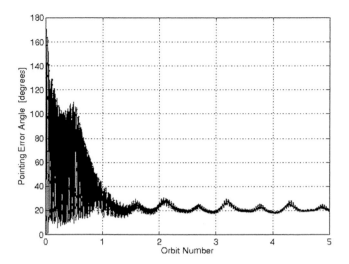

Figure 8. Expected CSSWE long-axis-pointing direction versus local magnetic field as a function of time (orbit number) after launch.

the hysteresis rods change direction. This energy loss serves to dampen the satellite rotation until the satellite bar magnet axis is roughly aligned with the local earth field direction.

The CSSWE team has developed a passive magnetic attitude control simulation to model the spacecraft attitude over time. The exact magnitude of the torque due to the hysteresis rods is of paramount importance to such a model. Thus, a Helmholtz cage test setup was built to measure the rod magnetic moment versus the axial earth field. This measurement method seeks to provide accurate inputs to the simulation. Figure 8 shows the simulated earth field to satellite bar magnet axis angle over the first five orbits, assuming an initial angular offset of 180° and an initial spin rate of 18° min^{-1} (the expected initial spin rate for our specific launch). As shown from simulations, the satellite is expected to settle to a constant offset from the local magnetic field within two orbits and to oscillations of ±10° from this offset. The expected settling time is short because the expected initial spin rate is low. A calibrated magnetometer is located on board to provide two-axis attitude knowledge during operations. Two-axis attitude knowledge (relative to the earth's magnetic field) is expected within ±3°. Photodiodes on each of the solar arrays have also been installed to provide three-axis attitude knowledge when the spacecraft is insolated.

2.8. Relativistic Electron and Proton Telescope Integrated Little Experiment

As mentioned previously, the instrument on board CSSWE will consist of a particle telescope to make differential mea-

surements of energetic protons and electrons. Solid state detectors are often used to measure energetic particles in space, although challenges for such instrument designs still remain. For example, relativistic electrons scatter erratically upon interacting with matter; therefore, the amount of energy they deposit into a specified volume of material, and thus the initial energy of the electron, must be determined statistically. Protons, on the other hand, deposit energy according to the Bethe-Bloch formula as they travel. At high-energy (several tens of MeV), they have the ability to penetrate through the instrument shielding and impact the detectors from all directions. These characteristics of both species of particles must be accounted for in order to design a reliable energetic particle instrument.

2.8.1. Telescope design. The REPTile detector stack consists of four solid state silicon detectors similar to those used for the RBSP/Relativistic Electron and Proton Telescope (REPT) instruments, which have been delivered for a launch that occurred on 30 August 2012 (D. N. Baker et al., The Relativistic Electron-Proton Telescope (REPT) instrument on board the Radiation Belt Storm Probes (RBSP) spacecraft: Characterization of Earth's radiation belt high-energy particle populations, submitted to *Space Science Review*, 2012, hereinafter referred to as Baker et al., submitted manuscript, 2012). To minimize contamination from particles outside the instrument field-of-view, these detectors are housed in a tungsten (atomic number $Z = 74$) chamber, which is encased in aluminum ($Z = 13$) shielding. The outer aluminum shield serves to absorb most electrons and lower-energy protons while significantly reducing the energy of incident higher-energy protons, which can produce showers of contaminating secondary particles in high-Z materials. The dense, high-Z tungsten shield significantly increases the energy threshold at which protons can fully penetrate into the detector stack, effectively reducing the background flux. This layered shielding effectively blocks electrons with energy less than ~15 MeV and protons with energy less than ~75 MeV. Additional tungsten shielding at the rear of the detector stack prevents protons with energy less than ~90 MeV from penetrating the end-cap shielding, where the geometric factor is large (see Figure 9).

A shielded, baffled collimator defines the instrument's 52° field-of-view and its 0.526 sr cm^2 geometric factor. Particles that enter the detector stack through the collimator are filtered by a thin beryllium ($Z = 4$) foil, which acts as a high-energy pass filter, absorbing electrons with energy less than ~400 keV and protons with energy less than ~8 MeV. These energies correspond approximately to the lowest detectable energy of the instrument. Knife-edged collimator baffles have been designed such that no particle can enter the

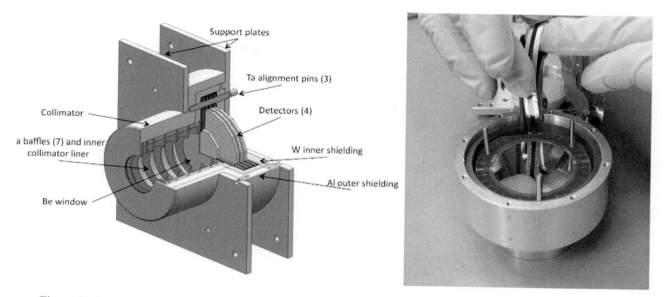

Figure 9. (left) Cross-sectional view of the instrument geometry. (right) Flight instrument during integration. The collimator is facing down in the image, and the back plate not yet attached, so the detector stack is visible.

detector stack without at least two reflections from an interior surface of the tantalum collimator. Tantalum ($Z = 73$) was chosen for the inner collimator lining and baffles as it provides a balance between stopping power and secondary particle characteristics.

2.8.2. Electronics design. The REPTile electronics perform three primary functions: (1) to recognize particles that hit the detectors, (2) to determine the particle species and incident energy, and (3) to convert the analog pulses to a digital signal to relay to C&DH. The system-level signal chain block diagram for one detector can be seen in Figure 10. The charge deposited into the detector by an incident particle (step 1) is swept from the silicon with a ~350 V bias voltage to the charge-sensitive amplifier (CSA, step 2).

Since the CSA must be capable of amplifying small signals from the detector, it is very sensitive to noise, and great care is taken to filter the signal and remove offsets from variations in temperature.

The second stage of amplification occurs at the pulse-shaping amplifier (step 3) and is used to further distinguish the voltage levels corresponding to electrons and protons. The analog pulse is converted to digital at a three-level discriminator chain (step 4), where the discriminator thresholds are set to the equivalent of 0.25, 1.5, and 4.5 MeV deposited in the detector. The discriminator chain is used to distinguish the species of particle, where particles depositing 0.25 MeV $< E \leq 1.5$ MeV are considered electrons, and particles depositing $E > 4.5$ MeV are binned as protons. The complex programmable logic device (step 5) simultaneously

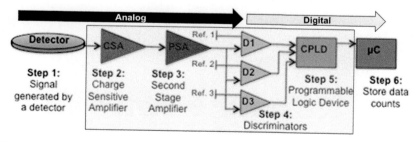

Figure 10. REPTile electronics block diagram for a single detector signal chain. The gray box corresponds to components on the REPTile electronics board.

Table 1. Coincidence Logic for Particle Binning[a]

Species	Energy (MeV)	Detector			
		1	2	3	4
Electron	0.5–1.5	100	000	000	000
Electron	1.5–2.2	X00	100	000	000
Electron	2.2–2.9	X00	100	100	000
Electron	>2.9	X00	100	100	100
Proton	8.5–18.5	111	000	000	000
Proton	18.5–25	1XX	111	000	000
Proton	25–30.5	1XX	111	111	000
Proton	30.5–40	1XX	111	111	111

[a]Each trio of bits represents the output of the three discriminators for that detector. A 1 signifies the threshold has been surpassed, and a 0 signifies the threshold has not been achieved. An X signifies that either a 1 or a 0 satisfies the logic. The bits correspond to, from left to right, the 0.25, 1.5, and 4.5 MeV references.

interprets the signals from all four detectors, and if the comparator outputs satisfy the binning logic outlined in Table 1, increments the appropriate counter. Every 6 s, the totals of each counter are sent to the C&DH (step 6).

2.8.3. Instrument characterization. The performance of the REPTile instrument is characterized using the Geant4 software package, which was designed by nuclear physicists at the European Organization for Nuclear Research. Geant4 was created to simulate particle beam tests and describe the passage of particles through matter, and it has been used to determine the performance of the Large Hadron Collider and the Tevatron collider at Fermilab.

Geant4 creates a virtual environment in which the instrument is assembled and bombarded with particles. The simulation quantifies the energy deposited into each detector for each incident particle. Figure 11 is a visualization of raw Geant4 data consisting of a wireframe instrument geometry and particle tracks through the environment. The electron tracks are red, protons are blue, and bremsstrahlung radiations are green. The left panel depicts a 2 MeV electron beam fired down the collimator from the left that, upon impacting the beryllium foil, begins to diverge into a scatter cone. The electrons then interact with the four silicon detectors, sometimes producing bremsstrahlung radiation. Some backscattering occurs: one electron reverses direction and embeds itself in the collimator wall. The right panel portrays a 20 MeV proton beam fired down the collimator. Proton scattering is minimal, and after passing through the first detector, the protons embed themselves in the second. The protons create low-energy electron showers when interacting with matter, so the proton tracks appear red when inside the silicon detectors.

Simulations are conducted for all incident angles and particle energies. The efficiency response of each detector is determined as a function of incident energy, as seen in Figure 12. For each energy increment, 10,000 particles are fired into the detector stack. The percent of particles that impact a detector are plotted in black, and the percent of particles that get logically binned in the corresponding energy channel (based on the logic outlined in Table 1) are plotted in red. The protons are relatively well behaved, and the channel thresholds are clearly defined. The electron channels, however, are more difficult to specify due to the random interactions inside the instrument. The energy channels are chosen

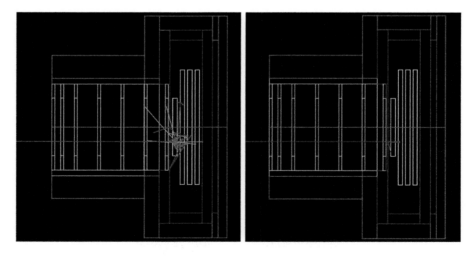

Figure 11. (left) A 2 MeV electron beam fired down the collimator of the REPTile instrument. Electrons (red) interact with the detectors and shielding, producing high-energy photons (green). (right) A 20 MeV proton beam fired down the collimator. Protons (blue) interact with the first detector and are embedded in the second detector.

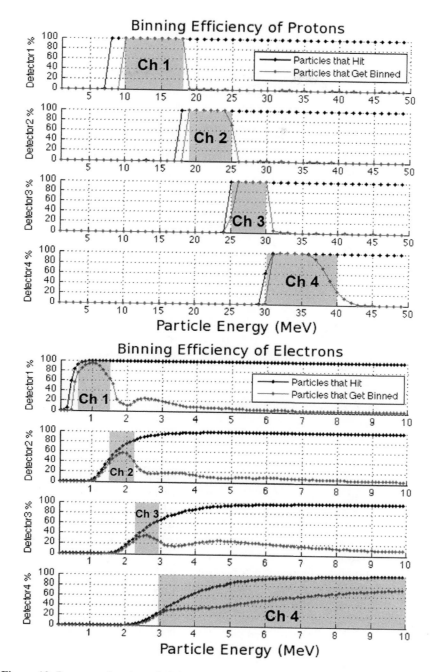

Figure 12. Response function of all four detectors for (left) protons and (right) electrons.

to maximize the response of each detector corresponding to the most efficient binning profile, which is shown in Figure 13. The energy deposited into each of the four detectors is plotted as a function of incident energy for electrons. The horizontal dashed lines correspond to discriminator thresholds of 0.25 and 1.5 MeV, between which the particle is classified as an electron. The vertical solid lines correspond to the electron energy range of the corresponding detector. The detectors discard 2.7%, 44.6%, 40.1%, and 30.4%, respectively, of the electrons in their appropriate energy range. The discriminator thresholds can be changed in-flight through uplink commands. Thus, if the ambient electron flux

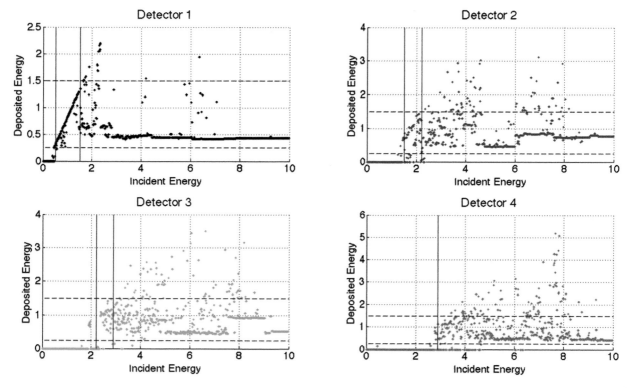

Figure 13. Energy deposited into all four detectors as a function of incident energy from a simulation of 20,000 electrons in Geant4. The horizontal dashed lines correspond to the 250 keV and 1.5 MeV discriminator thresholds, in between which the particle is binned as an electron. The vertical solid lines correspond to the energy range of the corresponding detector.

is very high, the upper-energy threshold can be lowered. With a lower threshold, fewer electrons are measured, but the measurements are cleaner, making the conversion from count rate to flux more reliable.

Additionally, the proficiency of the instrument's shields is determined by firing particle beams into the shielding. The particles that interact with the shielding or collimator before entering the detector stack are considered instrument noise. The field-of-view flux is contaminated largely by particles penetrating the rear shielding, motivating the additional tungsten shielding there. The analysis is performed assuming energy spectra during the most active times for each species: the electron spectrum from the AE8 solar max model (http://modelweb.gsfc.nasa.gov/models/trap.html) is modeled as $I(E) = 3.003 \times 10^5 E^{-2.3028}$, and the proton spectrum from data presented in the work of *Mewaldt et al.* [2005], from one of the largest SEP events in the last 50 years, is modeled as $I(0.1 \text{ MeV} \leq E \leq 26 \text{ MeV}) = 5.20 \times 10^4 E^{-1.1682}$ and $I(E > 26 \text{ MeV}) = 9.65 \times 10^8 E^{-4.2261}$. Despite the extreme spectrum assumptions, REPTile still meets the required S/N ratio of ≥ 2 for all channels. The detailed S/N ratio for each channel is outlined in Table 2.

The permanent magnet used for CSSWE's attitude control will alter incident particles' trajectories. Test-particle simulations were performed using the relativistic Lorentz force to simulate the possible effects on the REPTile measurements. In these simulations, a constant value for the Earth's magnetic field at LEO is used as a background field and a dipole magnetic field, centered directly behind the instrument (significantly closer than the actual magnet location), is included. The magnet is far enough from REPTile to assume a dipolar field, and the strength of the magnet's dipole is calculated using its magnetic moment. Test particles of both protons and electrons are fired down the bore sight of REPTile from a distance of 1 m. This initial position is small compared to the particles' gyroradii but large with respect to the ADCS magnetic field. The initial velocities for electrons corresponding to

Table 2. S/N Ratio for Electrons and Protons on All Four Detectors, Calculated for Extreme Energetic Particle Conditions

	Detector 1	Detector 2	Detector 3	Detector 4
Electron S/N	88.3	18.7	13.0	10.4
Proton S/N	13.6	7.0	5.3	2.0

10 eV to 5 MeV and for protons 1 keV to 50 MeV are used. Based on the results of this analysis, the Attitude Control Systems (ADCS) magnet alters the trajectories of only lower-energy electrons ($E < \sim 10$ keV) near REPTile. The effect of the ADCS magnet on relativistic electrons and energetic protons that enter through REPTile's field-of-view is negligible; thus, the instrument's performance is unaffected by its presence.

2.8.4. Instrument testing. Without access to particle accelerators due to budgetary constraints, the fully assembled spacecraft was tested with a strontium-90 radiation source fastened to the outside of the REPTile collimator. Strontium-90 has a half-life of 28 years and decays into yttrium-90, emitting an electron with maximum energy of 0.546 MeV. Yttrium-90 has a half-life of 2.7 days and decays into zirconium-90, emitting an electron with maximum energy of 2.28 MeV. Both isotopes emit electrons in a continuous kinetic energy spectrum from zero to the maximum. An independent empirical measurement of the strontium-90 spectrum was made and fit to a power law. Using the fit, the theoretical count rate for each of REPTile's differential energy channels was calculated. Data collected from the fully integrated strontium-90 test agreed with the theoretical count rate within expectation, confirming functionality of the instrument and despite the design challenges presented by an energetic particle telescope for a CubeSat platform.

3. MISSION OPERATION, DATA ANALYSIS, INTERPRETATION, AND MODELING

3.1. CSSWE Operational Scenario

The expected orbit is 480 km × 790 km, with an inclination of 65°. Once deployed from the launch vehicle, the spacecraft will power on, start charging batteries and begin to align itself with the Earth's magnetic field using a passive MACS, described earlier. Simulations have shown that, based on the orbit average power, the spacecraft will be power positive. The 8.4 W h batteries should charge to full capacity within 24 h on orbit. The system starts up in safe mode, transmitting a beacon every 18 s to aid in establishing contact with the ground station. The mission design allows 1 month for spacecraft contact and commissioning before the 3 month science mission, as illustrated in Figure 14. Student operators will establish contact and operate the spacecraft under the guidance of the experienced LASP mission operators using the ground station located at LASP.

The spacecraft passes over the Boulder ground station an average of 4.7 times each day with an average link time of 4.5 min available to download data on each pass. Accounting for an assumed 20% dropped packets, CSSWE can downlink 40% of the science data, and all housekeeping data generated in 1 day using only two (of the anticipated 4.7) 4.5 min passes. Because only high-latitude data is of interest for the mission, the science data may be selectively downlinked by accounting for satellite position when the data was recorded. However, because the entirety of the science mission can be stored on board the satellite SD card, data for any time can be requested during any pass. Thus, data from an interesting solar event that would be measured by REPTile even at low latitudes may be downlinked after the event has occurred.

3.2. Data Analysis and Interpretation

CSSWE will store science data on board until contact with the LASP ground station is made. Upon requests for specific time intervals, the satellite will return science data as time-stamped attitude information, spacecraft mode, detector

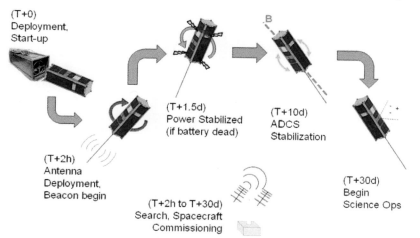

Figure 14. The early mission operations of the satellite are shown.

status, and binned electron and proton counts at 6 s cadence. The raw science packets will be received and parsed at the ground station at LASP using ISIS command and control software and saved as level 0 text files of raw count rates, raw magnetometer, and photodiode values (used later for attitude determination), spacecraft mode, and detector status. Corrections are then applied to these level 0 raw count rates to create level 1 data.

The conversion of level 0 count rate data into level 1 count rates will correct for background and dead time effects. Corrections will account for the charge collection time in the detector, which is 250~300 ns for 1.5 mm silicon. By assuming a Poisson distribution with 6 s cadence, a statistical correction can be made for the charge collection time. Additional corrections will be applied to account for electronic pileup, where the pileup time scale is ~8 μs. Pileup is dependent on the performance of specific electronic components.

The CSA, shown in Figure 10, has an inherent temperature-dependent exponential offset. The onboard electronics remove the offset to first order, but for periods where the first-order approximation breaks down, warning flags of various levels will be included with the data. Additionally, the accuracy of the CSA decreases during periods of high fluxes. Onboard corrections allow for some science to be recovered during these periods, but warning flags will also be included to indicate increasingly unreliable data. Additional warnings will be issued for other inconsistencies, such as improperly biased detectors or changes in discriminator threshold voltages. A copy of the flight hardware has been fabricated and will be used for additional calibration, as this flight spare behaves identically to the delivered CubeSat.

Level 2 data take the adjusted count rate (level 1) data and converts them into fluxes at designated energy ranges. We use bowtie analysis to derive an incident energy spectrum $f(E)$ for both electrons and protons based on a best fit to the data using the following equation:

$$C_i = \int \gamma f(E)\alpha_i(E)\mathrm{d}E, \qquad (1)$$

where C_i is the count rate for channel i, γ is the instrument geometric factor, f is the environmental particle flux, and α_i is the response function of detector i, as calculated from Geant4 simulations. The flux on each detector is then calculated based on the derived best fit energy spectrum. The differential fluxes will be provided as level 2 data, as well as the energy spectrum used, and a measure of the error in the spectral fit to the data.

Finally, level 3 data are the differential flux measurements converted into directional differential flux. This is done using the onboard magnetometer data (and photosensors mounted

Table 3. Data Level Description

Data Level	Description
Level 0	Raw 6 s electron and proton count rates
Level 1	Adjusted count rates, accuracy flags
Level 2	Differential flux per detector and species, estimated energy spectra, and error
Level 3	Directional differential flux, pitch angle, and L shell

on four sides of the solar array if the satellite is insolated) to determine the direction of the local background magnetic field relative to the alignment of the spacecraft. We then derive the look direction of the instrument to resolve the PA range of the measured particles. Using a magnetic field model, such as the *Tsyganenko* [2002] model, we map the magnetic field lines to the equator and determine the L value (equatorial radial distance in the Earth radii from the center of the Earth) and the equatorial PA of the particles (Table 3).

3.3. Modeling

Owing to the nondipolar nature of the Earth's magnetic field and CSSWE's orbit and orientation, the electrons measured by CSSWE can be categorized as a mixture of trapped, quasitrapped (in the drift loss cone (DLC)), and precipitating (in the bounce loss cone (BLC)) populations, depending on where (i.e., longitude and latitude in the Northern and Southern Hemispheres) the measurements are made [*Selesnick*, 2006; *Tu et al.*, 2010]. The BLC is defined as the range of equatorial PAs where an electron's mirror point reaches at or below an altitude of 100 km in either hemisphere (with electrons lost within one bounce period), and the DLC is defined as the range of equatorial PAs between the highest BLC angle at the South Atlantic Anomaly (SAA) region and the local BLC angle (electrons lost within one drift period). Figure 15a illustrates the identification of these three different populations based on a comparison of the equatorial PA for electrons mirroring at CSSWE's altitude (~600 km) at $L = 4$. The equatorial BLC angles across longitude (the upper boundary of the red area) are in the range of 5.2° to 6.8° in equatorial PA, with the highest near SAA (near 0° and 360°) and the lowest near 105° geomagnetic longitude. The equatorial PA of a data point is estimated by assuming it is locally mirroring at the satellite location, which is an approximation considering the wide field-of-view of the detector. Under this assumption, some of the so-called "trapped" electrons may actually be quasitrapped (if the actual local PA < 90°), or some "quasitrapped" electrons may actually be untrapped, but the "untrapped" electrons will be truly untrapped. Thus, using this approximation, we calculate the lower bound on

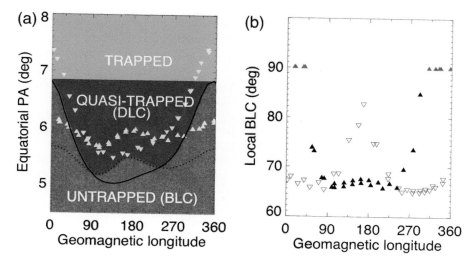

Figure 15. (a) Schematic illustration of three populations of energetic electrons that can be measured by CSSWE (600 km altitude): trapped, quasitrapped, and untrapped, depending on their equatorial pitch angle (PA) ranges (shown here at $L = 4$) and where the measurement was made (i.e., longitude and latitude in the Northern and Southern Hemisphere). When an electron reaches below 100 km altitude, it is assumed to be lost. The upward triangles represent measurements taken in the Northern Hemisphere and the downward ones in the Southern Hemisphere. The solid (dotted) curve represents the bounce loss cone (BLC) angle at $L = 4$ in the Southern (Northern) Hemisphere, so the final BLC at each longitude is the maximum of these two, with the range of equatorial PA inside the BLC filled by red color. (b) Local BLC angles at the measurement locations (upward solid and downward empty triangles) in Figure 15a, with the untrapped electron measurements in red (untrapped: these electrons, even mirroring at the measurement location, will be lost by reaching at 100 km at the other hemisphere).

precipitation loss, equal to or less than the actual flux of precipitating electrons.

This can be seen from Figure 15b, where the calculated local BLC angles, corresponding to the triangles in Figure 15a, are displayed, more of which are less than 90°. The corresponding untrapped electrons (red triangles) are measured at the location with local BLC at 90° (meaning they will be lost by reaching at or below 100 km in the other hemisphere even if they mirror at the measurement location).

Since the observed electron flux variation is a complicated balance between loss and energization for any quantitative study, physical modeling is needed. REPTile provides a 6 s integration measurement of the particle distributions, which contain a mixture of the three different populations, in varying proportions depending on longitude and hemisphere. To determine the precipitation loss from these data, modeling efforts are required. We have analyzed and modeled 6 s integration measurements of MeV electrons from SAMPEX, which was in a similar orbit and is expected to reenter the atmosphere soon [*Baker et al.*, 2012]. For example, Figure 16 shows energetic electron measurements by P1 and ELO channels on SAMPEX, represented by the filled triangles as a function of longitude when SAMPEX crossed $L = 4.5$ during a geomagnetic storm in 8–13 March 2008. The empty

triangles are simulation results, to be discussed later. Figure 16a shows a typical quiet time prior to a magnetic storm; Figure 16b immediately follows Figure 16a in time and includes the magnetic storm main phase, Figure 16c is during the storm early recovery phase, and Figure 16d commences the late recovery phase. The general pattern and variation of the data are the stably trapped population near 0° and 360° longitude in the south (green triangles pointing downward) has the highest count rates before the storm, decreases significantly during the storm main phase and stays low in the early recovery phase, and then returns to the prestorm level during the late recovery phase; the quasi-trapped population in the DLC from ~45° to 315° longitude (blue points) has intermediate data rates and increases eastward during the prestorm and late recovery phases because of the azimuthal drift, during the storm main phase it is relatively flatly distributed over the longitude; the untrapped population in the BLC near 0° and 360° in the north (red upward triangles) generally has the lowest count rates. Based on only the measurements, little physical information can be extracted. We have developed a drift-diffusion model at LASP that includes azimuthal drift and PA diffusion to simulate the low-altitude electron distribution observed at LEO to quantify the electron precipitation loss into the atmosphere [*Tu et al.*, 2010]. The model is governed by the equation:

$$\frac{\partial f}{\partial t} + \omega_d \frac{\partial f}{\partial \phi} = \frac{\omega_b}{x} \frac{\partial}{\partial x}\left(\frac{x}{\omega_b} D_{xx} \frac{\partial f}{\partial x}\right) + S, \qquad (2)$$

where $f(x, \phi, t)$ is the bounce-averaged electron distribution function at a given L shell and kinetic energy E, as a function of $x = \cos \alpha_0$ (where α_0 is the equatorial PA), drift phase ϕ, and time t; ω_d is the bounce-averaged drift frequency; ω_b is the bounce frequency; D_{xx} is the bounce-averaged PA diffusion coefficient, in the form

$$D_{xx} = D_{\text{dawn/dusk}} \tilde{E}^{-\alpha} \frac{1}{10^{-4} + x^{30}}, \qquad (3)$$

where $\tilde{E} = E/(1\ \text{MeV})$; and S is the source rate, defined as

$$S = S_0 \tilde{E}^{-\nu} \bar{g}_1(x)/p^2, \qquad (4)$$

where $\bar{g}_1$ is the lowest-order eigenfunction of the combined drift-diffusion operator (the terms in equation (1) involving $\partial/\partial\phi$ and $\partial/\partial x$), and p is the electron momentum for a given E. Free parameters include D_{dawn}, D_{dusk}, α, ν, and S_0.

By adjusting model-free parameters, we can fit the longitude dependence of the electron count rates in the model to the data. The best fit simulation results, shown as empty triangles in Figure 16, determine the PA diffusion coefficients of electrons at different energies for different intervals. Then, the electron lifetime at a specific energy can be estimated as: $\tau = 1/(100\bar{D})$, where $\bar{D}$ is the longitude-averaged model diffusion coefficient defined as $\bar{D} = (D_{\text{dawn}} + D_{\text{dusk}}) \tilde{E}^{-\mu}/2$.

4. SUMMARY

Here we have provided a detailed description of the upcoming Colorado Student Space Weather Experiment, an NSF-funded CubeSat mission launched on 13 September 2012 (our ground station was able to find, track, and receive beacon/housekeeping packets during the first pass around 04:00 LT next day, all appear nominal at this point). The CSSWE system architecture has been designed to maintain simplicity while meeting all of the well-defined and justified system and subsystem requirements. A "keep-it-simple" architecture mitigated risk and allowed the CSSWE team to design, manufacture, and test a fully functional satellite, which was successfully delivered on time to the launch provider with additional margin on the various system requirements defined for CubeSats. Housed within the 3 U CubeSat structure, the combined CSSWE subsystems provide the necessary platform to achieve CSSWE's primary

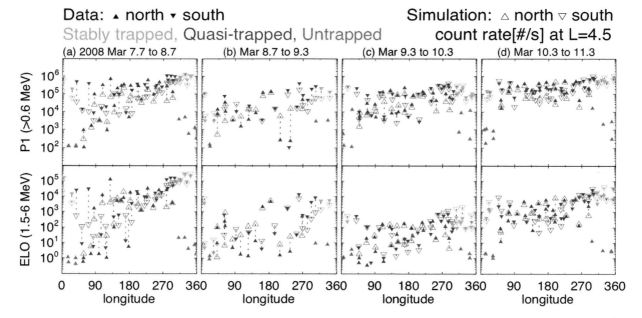

Figure 16. Electron count rate data (solid triangles) at $L = 4.5$ from two SAMPEX channels (P1 and ELO) versus geomagnetic longitude during (a) a quiet prestorm interval, (b) storm main phase, (c) early recovery phase, and (d) late recovery phase of the March 2008 storm. Data points are identified as trapped (green), quasitrapped (blue), and untrapped (red). Upward triangles are measured in the Northern Hemisphere and downward ones in the Southern Hemisphere. The simulation results are shown as empty triangles.

science goals to make differential flux measurement for energetic electrons and protons. Under the tutelage of Aerospace Engineering professors and LASP scientists and engineers, graduate students designed, built, and tested the internal structures, thermal, power, ground station, and attitude control subsystems, while COTS C&DH and communications subsystems and an external frame were integrated as well. Also, student designed and tested, CSSWE's primary science payload, REPTile, will observe solar energetic protons in the energy range 10–40 MeV and outer radiation belt electrons in the energy range 0.5 to >3 MeV. The necessary environmental tests and thorough end-to-end testing of each of the subsystems, and the fully integrated spacecraft with communication to the ground station, have been successful, providing confidence that CSSWE will perform as designed when on orbit.

Science data from REPTile will be processed and released using multiple levels of refinement, from raw, unprocessed count rates (level 0) to directional, differential energy fluxes with specified PA ranges and L shells (level 3). We have also discussed one application of how this data can be used to understand precipitation loss of outer radiation belt electrons based on the work of *Tu et al.* [2010]. This drift-diffusion model will use REPTile electron fluxes to quantify electron loss rates into the Earth's atmosphere. When SEPs occur, CSSWE will also be used to determine the energy spectra, intensity, and latitudinal extent of these ultraenergetic particles precipitating into the Earth's atmosphere. These are just two examples of how CSSWE science data will be used, but many more studies can be conducted, especially when the data are used in conjunction with data from other missions, such as SDO, RBSP, and/or Time History of Events and Macroscale Interactions during Substorms (THEMIS).

CubeSat missions are gaining popularity in the scientific community, and CSSWE is a prime example of their potential. Alongside the other NSF CubeSats (e.g., RAX [*Cutler et al.*, 2010] and CINEMA [*Lee et al.*, 2011]), CSSWE is proving how small, inexpensive, student-built and designed space missions are not only feasible, but fully practical for achieving valuable science objectives. CSSWE science observations will help to address unanswered questions concerning the nature and impact of solar energetic proton events at Earth. Additionally, by providing observations of pitch angle resolved relativistic electrons at LEO, CSSWE will complement NASA's RBSP mission to understand Earth's highly variable outer radiation belt. This demonstrates how, for an additional cost that is only a small fraction of the total mission cost, large, expensive science missions can benefit from one or more small spacecraft, like CubeSats, to provide additional points and types of measurements, particularly those that may be impossible for the larger mission to provide on its own.

Acknowledgments. The authors thank other CSSWE team members who are not listed here, as well as LASP scientists and engineers: D. N. Baker, T. Woods, J. Gosling, G. Tate, M. McGrath, W. Possell, V. Hoxie, S. Batise, C. Belting, K. Hubbell, J. Young, S. Worel, G. Allison, E. Wullschleger, P. Withnell, and V. George. We also thank Jamie Culter of U. of Michigan for consultation and support for the EPS and Comm system, R. Strieby, and R. Kile of Loveland Repeater Association for their help and support for the Ground Network and Comm system, and Jim White for his help and support on various aspects of the CSSWE mission. This work was supported by NSF (CubeSat program) grant AGSW 0940277.

REFERENCES

Aschwanden, M. J. (2004), *Physics of the Solar Corona: An Introduction*, Springer, Berlin.

Baker, D. N. (2001), Satellite anomalies due to space storms, in *Space Storms and Space Weather Hazards*, edited by I. A. Daglis, chap. 10, pp. 251–284, Springer, New York.

Baker, D. N. (2002), How to cope with space weather?, *Science*, 297, 1486–1487.

Baker, D. N., P. R. Higbie, R. D. Belian, and E. W. Hones Jr. (1979), Do Jovian electrons influence the terrestrial outer radiation zone?, *Geophys. Res. Lett.*, 6(6), 531–534.

Baker, D. N., S. G. Kanekal, T. I. Pulkkinen, and J. B. Blake (1999), Equinoctial and solstitial averages of magnetospheric relativistic electrons: A strong semiannual modulation, *Geophys. Res. Lett.*, 26(20), 3193–3196.

Baker, D. N., J. E. Mazur, and G. M. Mason (2012), SAMPEX to reenter atmosphere: Twenty-year mission will end, *Space Weather*, 10, S05006, doi:10.1029/2012SW000804.

Bortnik, J., and R. M. Thorne (2007), The dual role of ELF/VLF chorus waves in the acceleration and precipitation of radiation belt electrons, *J. Atmos. Sol. Terr. Phys.*, 69, 378–386.

Cane, H. V., and D. Lario (2006), An introduction to CMEs and energetic particles, *Space Sci. Rev.*, 123, 45–56, doi:10.1007/s11214-006-9011-3.

Cane, H. V., D. V. Reames, and T. T. von Rosenvinge (1988), The role of interplanetary shocks in the longitude distribution of solar energetic particles, *J. Geophys. Res.*, 93(A9), 9555–9567.

Carrington, R. C. (1860), Description of a singular appearance seen on the Sun on September 1, 1859, *Mon. Not. R. Astron. Soc.*, 20, 13–15.

Chen, Y., G. D. Reeves, and R. H. W. Friedel (2007), The energization of relativistic electrons in the outer Van Allen radiation belt, *Nat. Phys.*, 3, 614–617, doi:10.1038/nphys655.

Cutler, J., M. Bennett, A. Klesh, H. Bahcivan, and R. Doe (2010), The Radio Aurora Explorer – A bistatic radar mission to measure space weather phenomenon, paper presented at the 24th Annual Small Satellite Conference, Logan, Utah.

Gosling, J. T. (1993), The solar flare myth, *J. Geophys. Res.*, 98(A11), 18,937–18,949.

Horne, R. B., et al. (2005), Wave acceleration of electrons in the Van Allen radiation belts, *Nature*, 437, 227–230, doi:10.1038/nature03939.

Kanekal, S. G., M. Al-Dayeh, M. Desai, H. A. Elliott, and B. Klecker (2008) Relating solar energetic proton populations observed within the terrestrial magnetosphere to coronal mass ejections, magnetic flux ropes observations at 1 AU, *Eos Trans. AGU, 89*(53), Fall Meet. Suppl., Abstract SH23B-1643.

Lee, Y., et al. (2011), Development of CubeSat for space science mission: CINEMA, paper presented at the 62nd International Astronautical Congress, Capetown, South Africa.

Leske, R. A., R. A. Mewaldt, E. C. Stone, and T. T. von Rosenvinge (2001), Observations of geomagnetic cutoff variations during solar energetic particle events and implications for the radiation environment at the Space Station, *J. Geophys. Res., 106*(A12), 30,011–30,022.

Li, X., D. N. Baker, M. Teremin, T. E. Cayton, G. D. Reeves, R. S. Selesnick, J. B. Blake, G. Lu, S. G. Kanekal, and H. J. Singer (1999), Rapid enhancements of relativistic electrons deep in the magnetosphere during the May 15, 1997, magnetic storm, *J. Geophys. Res., 104*(A3), 4467–4476, doi:10.1029/1998JA900092.

Li, X., D. N. Baker, S. G. Kanekal, M. Looper, and M. Temerin (2001), Long term measurements of radiation belts by SAMPEX and their variations, *Geophys. Res. Lett., 28*(20), 3827–3830, doi:10.1029/2001GL013586.

Li, X., M. Temerin, D. N. Baker, and G. D. Reeves (2011), Behavior of MeV electrons at geosynchronous orbit during last two solar cycles, *J. Geophys. Res., 116*, A11207, doi:10.1029/2011JA016934.

Li, W., Y. Y. Shprits, and R. M. Thorne (2007), Dynamic evolution of energetic outer zone electrons due to wave-particle interactions during storms, *J. Geophys. Res., 112*, A10220, doi:10.1029/2007JA012368.

Mewaldt, R. A., C. M. S. Cohen, A. W. Labrador, R. A. Leske, G. M. Mason, M. I. Desai, M. D. Looper, J. E. Mazur, R. S. Selesnick, and D. K. Haggerty (2005), Proton, helium, and electron spectra during the large solar particle events of October–November 2003, *J. Geophys. Res., 110*, A09S18, doi:10.1029/2005JA011038.

Paulikas, G. A., and J. B. Blake (1979), Effects of the solar wind on magnetospheric dynamics: Energetic electrons at the synchronous orbit, in *Quantitative Modeling of Magnetospheric Processes, Geophys. Monogr. Ser.*, vol. 21, edited by W. P. Olson, pp. 180–202, AGU, Washington, D. C., doi:10.1029/GM021p0180.

Priest, E. R. (1981), *Solar Flare Magnetohydrodynamics*, Gordon and Breach, New York.

Reames, D. V. (1997), Energetic particles and the structure of coronal mass ejections, in *Coronal Mass Ejections, Geophys. Monogr. Ser.*, vol. 99, edited by N. Crooker, J. A. Joselyn and

J. Feynman, pp. 217–226, AGU, Washington, D. C., doi:10.1029/GM099p0217.

Sabine, E. (1852), On periodical laws discoverable in the mean effects of the larger magnetic disturbances, No. II, *Philos. Trans. R. Soc. London, 142*, 103–124.

Selesnick, R. S. (2006), Source and loss rates of radiation belt relativistic electrons during magnetic storms, *J. Geophys. Res., 111*(A4), A04210, doi:10.1029/2005JA011473.

Shprits, Y. Y., N. P. Meredith, and R. M. Thorne (2007), Parameterization of radiation belt electron loss timescales due to interactions with chorus waves, *Geophys. Res. Lett., 34*, L11110, doi:10.1029/2006GL029050.

Smith, Z., W. Murtagh, and C. Smithtro (2004), Relationship between solar wind low-energy energetic ion enhancements and large geomagnetic storms, *J. Geophys. Res., 109*, A01110, doi:10.1029/2003JA010044.

Tsyganenko, N. A. (2002), A model of the near magnetosphere with a dawn-dusk asymmetry 1. Mathematical structure, *J. Geophys. Res., 107*(A8), 1179, doi:10.1029/2001JA000219.

Tu, W., X. Li, Y. Chen, G. D. Reeves, and M. Temerin (2009), Storm-dependent radiation belt electron dynamics, *J. Geophys. Res., 114*, A02217, doi:10.1029/2008JA013480.

Tu, W., R. Selesnick, X. Li, and M. Looper (2010), Quantification of the precipitation loss of radiation belt electrons observed by SAMPEX, *J. Geophys. Res., 115*, A07210, doi:10.1029/2009JA014949.

Turner, D. L., X. Li, G. D. Reeves, and H. J. Singer (2010), On phase space density radial gradients of Earth's outer-belt electrons prior to sudden solar wind pressure enhancements: Results from distinctive events and a superposed epoch analysis, *J. Geophys. Res., 115*, A01205, doi:10.1029/2009JA014423.

Vandegriff, J., K. Wagstaff, G. Ho, and J. Plauger (2005), Forecasting space weather: Predicting interplanetary shocks using neural networks, *Adv. Space Res., 36*(12), 2323–2327.

Williams, D. J. (1966), A 27-day periodicity in outer zone trapped electron intensities, *J. Geophys. Res., 71*(7), 1815–1826.

L. Blum, C. Shearer Cooper, R. Kohnert, X. Li, Q. Schiller, and N. Sheiko, Laboratory for Atmospheric and Space Physics, University of Colorado, Boulder, CO 80303, USA. (lix@lasp.colorado.edu)

D. Gerhardt and S. Palo, Department of Aerospace Engineering Sciences, University of Colorado, Boulder, CO 80303, USA.

W. Tu, Los Alamos National Laboratory, Los Alamos, NM 8754, USA.

D. Turner, Department of Earth and Space Sciences, University of California, Los Angeles, CA 90095, USA.

Radiation Belts of the Solar System and Universe

B. H. Mauk

Applied Physics Laboratory, Johns Hopkins University, Laurel, Maryland, USA

Because of its likely tutorial value, I document here a personal journey regarding the hypothesized premise that studies of radiation belt physics at Earth and other solar system planets are informative for the study of extrasolar radiation regions. First, recent findings are summarized on comparisons between the radiation belts of all of the strongly magnetized planets of the solar system and what universal properties might be extracted. Then, because the diagnosis of extrasolar radiation regions occurs exclusively by means of synchrotron and other photon-emitting processes, Jupiter's powerful synchrotron emitting radiation belt is examined. Finally, tools developed at Jupiter are used to examine an example of a hyperenergetic radiation region outside of the solar system, that of the Crab Nebula. A significant challenge to understanding extrasolar radiation regions in the context of planetary radiation belt physics is that the processes that make such distant regions visible to us here at Earth, synchrotron and other photon emission processes, invariably play significant or dominant roles in the energetics of those regions. It is thus difficult to "see" extrasolar radiation systems like those that we study nearby.

1. INTRODUCTION

It is common in general audience presentations on the physics of planetary radiation belts to hear that we study, with our in situ sampling spacecraft, particle acceleration processes that are operating throughout the universe. Such statements are often accompanied by beautiful images of distant astrophysical objects such as the galactic center or the Crab Nebula. These claims are undoubtedly correct, but these assertions beg the question as to whether the processes that we study near home have direct application to understanding the structure, dynamics, and energetics of distant regions. Because of its likely tutorial value, I address that question here.

An important question for this task is: how do we compare radiation belts observed within very different planetary systems? Should a planet that is 10 times bigger or have a magnetic field that is 10 times stronger have a radiation belt that is 10 times more intense or 10 times more energetic? Recent findings that provide some answers to this question are presented.

Jupiter is the jumping-off point for any attempts to extrapolate planetary findings to extrasolar regions because Jupiter's radiation belt is the only planet that can be remotely imaged with radio wave synchrotron emissions. Thus, some space is devoted here to reviewing Jupiter's synchrotron emitting belt.

The lessons and tools developed for Jupiter are then directed to the radiation regions of the Crab Nebula (Figure 1) because it is one of the most canonical and familiar objects in the sky and because its characteristics are so well documented.

2. COMPARATIVE RADIATION BELTS

All the strongly magnetized planets of the solar system have robust radiation belts extending to relativistic energies [*Mauk and Fox*, 2010]. The question is: can we extrapolate what we learn between systems? For example, how can a system powered by internal planetary rotation (Jupiter) be compared to a system powered externally by the solar wind (Earth)?

Dynamics of the Earth's Radiation Belts and Inner Magnetosphere
Geophysical Monograph Series 199
10.1029/2012GM001305

Figure 1. Hubble Space Telescope images of the Crab Nebula in three different wavelengths overlaid. The red (OIII, 5007 A) and green (SII, 6717 A) components are thought to be from the expanding thermal components, and the blue (OI, 6300 A) is thought to represent the region of the electrons and positrons that emit synchrotron radiation. After *Hester* [2008] (NASA image published in *Annual Review of Astronomy and Astrophysics*).

2.1. The Kennel-Petschek Limit

One suggestion for a "standard" for comparing different planetary radiation belts [*Summers et al.*, 2009; *Mauk and Fox*, 2010] is the so-called Kennel-Petschek limit [*Kennel and Petschek*, 1966]. Considering only electrons, these authors noted that (1) magnetic flux tubes within the Earth's inner magnetosphere are robustly populated by some acceleration processes during magnetically active conditions, (2) accelerated particles trapped in magnetic bottle configurations are intrinsically unstable to the generation (net gain: G) of whistler waves near the magnetic equator, (3) a small fraction of the generated waves propagating along the magnetic field lines are reflected (R) from the ionosphere back toward the equatorial regions ($R \sim 5\%$ amplitude or 0.25% power flux is commonly assumed), and (4) if $GR > 1$, there occurs runaway growth of the waves, and electrons are lost at the very fast strong diffusion limit, whereby the loss cone of the magnetic configuration is filled by wave-particle scattering as fast as the loss cone particles are precipitated into the atmosphere. So, for robust acceleration processes, it is predicted that the particle distributions adjust themselves so that $GR \leq 1$. *Kennel and Petschek* [1966] derived an upper limit on the integral intensity of the particle distribution of the form $\int_{E*}^{\infty} I(E) \cdot dE \leq$ KPL, where $I(E)$ is particle intensity,

$E*$ is the energy above which the particles contribute to the positive growth of the waves, and "KPL" is short for "Kennel-Petschek limit," dependent on the radial position "L."

There have been refinements of the theory. *Schulz and Davidson* [1988] generated a differential version and predicted that particle distributions take on an $I(E) \sim E^{-1}$ spectral shape. This work and that of *Kennel and Petschek* [1966] assumed nonrelativistic particle energies. *Summers et al.* [2009] then developed a relativistic version of the integral KPL theory. *Mauk and Fox* [2010] then developed a relativistic differential Kennel-Petschek procedure and demonstrated that the $I(E) \sim E^{-1}$ spectral shape is roughly preserved with the relativistic theory. We summarize the *Mauk and Fox* [2010] findings here.

We characterize the net whistler mode wave gain (G) with the expression

$$G \approx exp\left[\frac{\gamma}{V_g}DR_p\right], \qquad (1)$$

where G is the ratio of final to initial wave amplitude as the waves propagates through the equatorial regions, γ is wave growth rate (s^{-1}), V_g is the wave group velocity (cm s^{-1}), D is the distance in planetary radii along the magnetic field direction, centered on the magnetic equator, where the wave growth rate remains positive and large, and R_p is the planetary radius (cm). Assuming that $R \sim 0.05$ such that $\ln[1/R] \sim 3$, one can manipulate $GR \leq 1$ using equation (1) to obtain

$$DR_p\gamma[\omega_r(E_r)] \leq 3V_g[\omega_r(E_r)]. \qquad (2)$$

Here both γ and V_g are functions of ω_r, the whistler mode wave frequency that is in gyroresonance with a specified set of electrons with minimum resonant electron energy E_r. The expressions that relate γ, V_g, ω_r, and E_r, are provided in an appendix of *Summers et al.* [2009].

Equation (2) is evaluated using analytic fits to measured radiation belt electrons:

$$I\left[\frac{1}{cm^2 \ s \ sr \ keV}\right]$$
$$= CE_{keV} \frac{[kT(\gamma1+1)+E_{keV}]^{(\gamma1+1)}}{\left(1+\left(\frac{E_{keV}}{E_2}\right)^{\gamma2}\right)\left(1+\left(\frac{E_{keV}}{E_3}\right)^{\gamma3}\right)} sin^{2S}(\alpha), \qquad (3)$$

which contains five to seven free fitting parameters: C, kT, $\gamma1$, $\gamma2$, and E_2, and sometimes $\gamma3$ and E_3, while the anisotropy parameter "S" is set to "0.3" based on high-intensity observations at Earth. We define two different values of C: C_m comes out of the spectral fitting process, and C_K is the value needed for the intensity to exactly match the KPL. Using the fact that $\gamma[\omega_r(E_r)]$ is linearly proportional to C, we can change equation (2) into an equality by replacing $\gamma[\omega_r(E_r)]$ with $(C_K/C_m)\,\gamma[\omega_r(E_r)]$ to yield

$$\frac{C_m}{C_K} = DR_p \frac{\gamma[\omega_r(E_r)]}{3V_g[\omega_r(E_r)]}. \qquad (4)$$

When this ratio is greater than 1, equal to 1, or less than 1 for a given resonant energy E_r, the electron intensity for that given resonant energy is greater than, equal to, or less than the differential KPL. Our prediction is that $C_m/C_K \leq 1$ for all E_r.

2.2. Comparing Planetary Radiation Belts

Figure 2 shows that that prediction is borne out for the most intense electron spectra at relativistic energies of all strongly magnetized planets of the solar system: Earth, Jupiter, Saturn, Uranus, and Neptune. The five spectra from five planets were chosen as being among the most intense for each planet at 1 MeV. The number next to each letter designating the planet indicates the radial L value where the measurement was made. This figure shows (a) at each resonant energy the spectra are all near or below the KPL (bottom), (b) the C_m/C_K profiles for Earth, Jupiter, and Uranus are flat and close to the value 1 from $E \sim 0.1$ to ~ 1 MeV (bottom), (c) the intensities in the top panel for Earth, Jupiter, and Uranus are correspondingly almost identical from 0.1 to 1 MeV and show the characteristic $I(E) \sim E^{-1}$ spectral shapes. The Jupiter spectrum extends to much higher energies while still remaining below the KPL.

Neptune and Saturn are different. The acceleration processes at Neptune appear not sufficient to populate the KP limiting profile at energies near 1 MeV. *Mauk and Fox* [2010] concluded that the absence at Neptune of dynamic injection phenomena may be key to the understanding of the absence of robust acceleration. At Saturn, it appears that the radiation belt intensities are dim because of absorption and degradation caused by dense clouds of gas and dust observed to be emanating from the Saturnian satellite Enceladus in the form of geyser-like plumes.

A caveat to the results discussed here is that, in the regions where intensities at relativistic energies (~ 1 MeV) are not intense, and at radial distance greater than that used for Figure 2, the particle intensities for $E_r \leq 0.2$ MeV can sometimes substantially exceed the Kennel-Petschek limit at Earth (see discussion in the work of *Mauk and Fox* [2010]). *Tang and Summers* [2012] discovered that this same condition prevails at Saturn for $L \geq 7\ R_S$. Dynamic injections appear sufficiently active there to overwhelm losses.

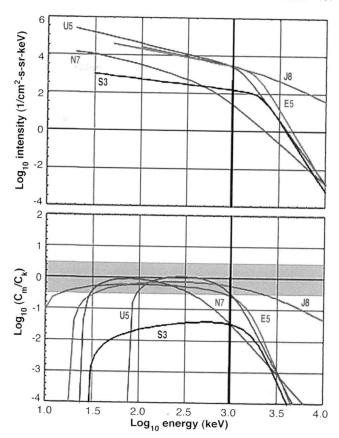

Figure 2. (top) Electron spectra from the five strongly magnetized planets (E, Earth; J, Jupiter; S, Saturn; U, Uranus; N, Neptune), representing the most intense spectra at 1 MeV. The numbers following the planet labels are the L value or radial distance where each spectrum was sampled. The spectra were sampled by the CRRES satellite for Earth, Voyager 2 for Uranus and Neptune, Cassini for Saturn, and a combination of Voyager 1 and the Pioneers for Jupiter. (bottom) Kennel-Petschek analysis of the five spectra in Figure 2, top. See the main text for an explanation of this panel. After *Mauk and Fox* [2010].

2.3. Implications for Extrasolar Radiation Regions

So, while not all of the planets populate their limiting spectra into the relativistic energies, the limiting spectrum, and $I(E) \sim E^{-1}$ spectral shapes, are good examples of what might be considered universal characteristics of robust relativistic planetary radiation belts. The question is: does this property or any other "universal" property for planetary magnetospheres, tell us anything about extrasolar radiation regions?

In order to be diagnosed, an extrasolar radiation region must be observed. Diagnosing such regions is mostly about the "synchrotron emission process" and related photon emission processes. We have a wonderful example of a synchrotron emitting region within planetary radiation belts: Jupiter.

3. JUPITER'S RADIATION BELT

The fact that Jupiter's radiation belt can be made "visible" by synchrotron radiation makes it very compelling (Figure 3). Synchrotron radiation is the photon emission process arising from relativistic electrons gyrating in a strong magnetic field. One sees from Figure 4, the synchrotron emission kernel for electrons with a single energy (see the appendix), that the intensity is a strong function of the magnetic field strength ($\sim B^2$), and the position of the peak frequency of emission is a strong function of both magnetic field ($\sim B$) and particle energy ($\sim \gamma^2$, where γ is the relativistic Lorentz factor with kinetic energy $E = (\gamma - 1) \cdot m_e c^2$; m_e is electron rest mass). The strong dependence on B is why the synchrotron radiation emanates from regions only close to Jupiter itself (Figure 3).

Synchrotron images of Jupiter are available over a variety of frequencies (related to electron energy), both from ground radio telescopes (Figure 3, top) and from space assets like Cassini (Figure 3, bottom). Also, there now exist ground techniques that allow observations over broad ranges of

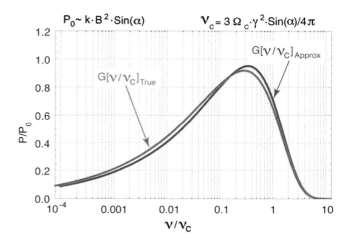

Figure 4. Shape of the synchrotron photon emission spectrum coming from relativistic electrons ($\gamma^2 \gg 1$) with energy $E \sim \gamma m_e c^2$ within a magnetic field of strength B and with a pitch angle α. The constant k can be derived in Appendix A. Two curves are shown, one representing the exact (true) theoretical expectation and another that is an analytic approximation. Expressions are provided in the appendix.

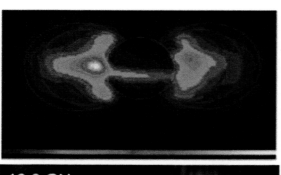

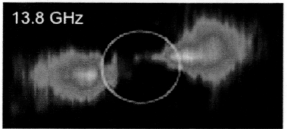

Figure 3. Synchrotron radio emission images of the inner radiation belt of Jupiter at radio frequencies (top) 1.4 GHz [from *Garrett et al.*, 2005], obtained from ground receivers, and (bottom) 13.8 GHz [from *Bolton et al.*, 2002] (NASA image published in *Nature*), obtained from the Cassini spacecraft during its flyby of Jupiter. These two frequencies correspond very roughly to emitting electron energies near 5 and 50 MeV, respectively.

frequency with associated energies extending to nearly 100 MeV (Figure 5, top) [*de Pater and Dunn*, 2003]. Detailed modeling of electron transport, adiabatic energization, and losses has resulted in a fairly good qualitative understanding of how the images obtain their characteristics (Figure 6) [*Santos-Costa and Bolton*, 2008]. Most interesting for later discussions is the role that the synchrotron emission process, itself, plays in the energetics of this inner radiation belt. Specifically, the synchrotron emission has been identified as a cause of the horn-like features in the synchrotron images (Figure 6; "synchrotron friction" causes loss of perpendicular energy, leaving only the parallel energy).

So, there is an important lesson that we have already begun to learn at Jupiter. Specifically, if the radiation region can be observed remotely, then the emission process, itself, plays a key role in the energetics and character of the radiation region.

The detailed modeling highlighted in Figure 6 goes far beyond what would be possible for an extrasolar radiation region. However, as an exercise, we can treat Jupiter as strictly a remote object and ignore all of the detailed information that we know about it. Such an exercise has been performed to generate Figure 5, using the synchrotron spectra in the top panel and the procedures described in the appendix using equation (3). The "forward-model" inversion of the synchrotron radio spectra yields electron spectra that compare favorably with spectra sampled in situ, in this case by Pioneer 11 (Figure 5, bottom). The continuous lines in the

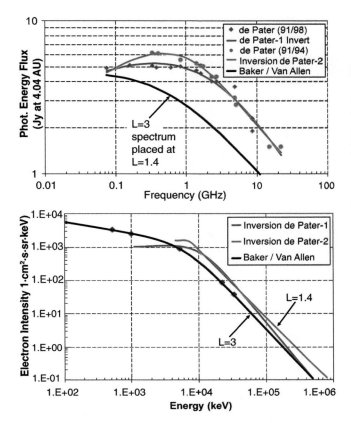

Figure 5. (top) Colored symbols and lines show the total image-integrated synchrotron emission spectra from Jupiter taken by ground receivers for different radio wave frequencies. Spectra from multiple time periods are shown. The discrete symbols are from a figure in the work of *de Pater and Dunn* [2003]. The colored lines are the result of my spectral inversion process to extract the electron spectra that are responsible for the synchrotron emissions. The black line is the synchrotron emissions that would come from an electron spectrum measured by Pioneer 11 at $L = 3$ R_J by *Baker and Van Allen* [1976] if that spectrum was positioned at $L = 1.4$. (bottom) The colored lines are the electron spectra extracted by inverting the synchrotron spectra, using the processes described in the appendix. The black line is a fit to the electron spectrum measured at $L = 3$ by Pioneer 11 [*Baker and Van Allen*, 1976].

top panel show how well our derived electron spectra reproduce the observed synchrotron radio spectra. To perform these inversions, just the Jovian magnetic field strength near the peak in the emission distributions has been used ($L \sim 1.4$, $B \sim 1.5G$), and the volume of the emission regions has been estimated from the visible images ($V \sim V_{\text{Jup}}$, where V_{Jup} is the volume Jupiter). One may do a reasonably good job of inferring the electron spectra by such inversions, provided an independent estimate of the magnetic field strength is available as, it turns out, we have for the Crab Nebula.

4. THE CRAB NEBULA

We turn now to a wonderful example of a distant astrophysical radiation region, the Crab Nebula (Figure 1). The Crab, generated by the famous supernova of 1054, has an active interior pulsar that generates an outward relativistic wind. The termination shock of that wind very close to the pulsar is thought to energize electrons and positrons to hyperrelativistic distributions that continue to expand outward (Figure 1; the blue emission region serves as a proxy) against a container of thermal gas and dust (Figure 1, red and green regions) that continues to expand. Crab characteristics are reviewed by *Hester* [2008]; e.g., the nominal magnetic field strength is 30–100 nT. The synchrotron spectrum has a dynamic range in emission frequency that is astounding (Figure 7). The crab is highly structured with filaments and "wisps," and is particularly structured in the innermost regions where the most energetic emissions likely arise. However, the degree of structuring of the synchrotron components in the majority of the visible volume (Figure 1, blue) is uncertain, as this component serves as the driver of expansion of the neutral components (red and green) further out.

4.1. Estimating the Crab's Electron Spectra

One may use the same tools used at Jupiter to invert the Crab's synchrotron radiation (Figure 8). An additional complexity comes from the fact that the emission volumes for those spectral components that have been imaged (radio, visible, X-ray) vary systematically with emission frequency [*Hester*, 2008]. Using the images in the three different spectral ranges, analogous to the analysis of *Amato et al.* [2000], yields an estimate of how the emission volumes vary with emission frequency, specifically

$$\text{Vol[cm}^3] = 10^{57} \left(\frac{\nu[\text{Hz}]}{10^7} \right)^{-0.22}. \tag{5}$$

Figure 8 shows two Crab spectra, one derived assuming that the emission volume is constant and the other assuming that the emission volume varies according to equation (5), which extrapolates what we know about the emission volumes from low to intermediate energies into the highest-energy regions. These procedures may well not generate a single spectrum that is accurate for any particular position within the Crab, but our purpose is to provide a quasiquantitative perspective on the electron distributions. The generation of quantitatively accurate representations most likely comes from a component-by-component analysis like that of *Meyer et al.* [2010].

Electrons-positrons participating with Crab synchrotron emissions cover energies from tens of MeV up to 10 PeV (10^{16} eV). At the lower energies (tens of MeV), one sees in

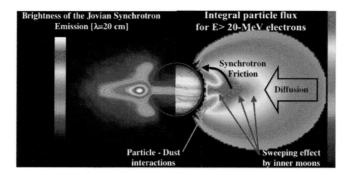

Figure 6. Schematic of electron transport, energization, and loss processes at Jupiter that contributes to the spatial and spectral characteristics of the synchrotron emissions from Jupiter's inner magnetosphere. Reprinted from *Santos-Costa and Bolton* [2008], with permission from Elsevier.

Figure 8 that the electron intensities at Jupiter are >3 orders of magnitude more intense than the Crab intensities, but the total energy density of electrons within the Crab is clearly far greater than that at Jupiter.

4.2. Kennel-Petschek Analysis of the Crab

It is of interest that the spectral index of the Crab spectrum at the lower energies is close to our ubiquitous E^{-1} spectrum anticipated from Kennel-Petschek theory. Even though the geometry of the Crab radiation region is likely nothing like that of a planetary magnetosphere (it is not clear that scattered electrons would have anywhere to go), it is irresistible to perform a Kennel-Petschek-like analysis of the Crab spectrum. Figure 9 shows that the KP theory provides no useful metric for understanding this electron radiation region. The Crab

spectrum is something like 4–5 orders of magnitude more intense than what would be anticipated from the KP metric. To align this analysis to a KP expectation, a wave growth length along the magnetic field of 10^{-5} to $10^{-6}D_{Crab}$ would be required rather than the 0.1 D_{Crab} value used for the figure. Since the gyroradii of a 10^8 keV electron in a 30 nT field is already 1 to 10 times bigger than that value perpendicular to the magnetic field, and since the major fraction of the synchrotron emitting populations serve as the driver of the outward expansion of the global thermal components (Figure 1), such structuring is not viable within the major fraction of the volume of the Crab. We conclude that the existence of the E^{-1} spectrum within the Crab is a coincidence and that the radiation within the Crab is more intense than one would extrapolate from planetary radiation belt studies.

4.3. Dynamics Within the Crab

A very intriguing finding was recently published showing that the Crab emission spectra are unexpectedly dynamic. *Abdo et al.* [2011] and *Tavani et al.* [2011] reported what they call "gamma ray flares" coming from the Crab (red and blue dots in Figure 10 to be compared to the nominal crab spectrum shown with a black line). The time variations over about 24 h (combined with velocity of light causality arguments) demand that these emissions come from a spatially confined region not more than 0.1% of the size of the Crab. (This size is still a factor of 100–1000 bigger than needed to make the KP analysis viable.) The published speculations about these flares center on diffusive shock acceleration and more esoteric suggestions.

I thought immediately of the CRRES observations of the creation of a brand new radiation belt [*Blake et al.*, 1992]. Could a simple model derived from planetary radiation belt studies provide the mechanism needed to explain the Crab flares?

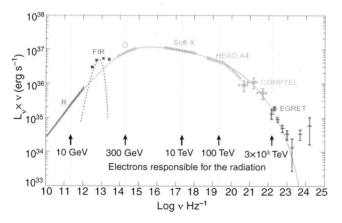

Figure 7. Synchrotron (mostly) emission spectrum for the Crab Nebula compiled from multiple sources by *Atoyan and Aharonion* [1996] and replotted and reviewed by *Hester* [2008]. The feature between 10^{12} and 10^{13} Hz is a contribution from thermal emission from dust. Reprinted with permission from Annual Reviews, Inc.

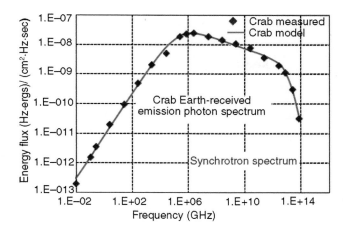

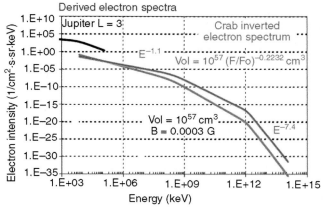

Figure 8. Result from the "forward modeling" inversion of the nominal synchrotron spectrum coming from the Crab Nebula. Inversion procedures are described in the appendix. (top) The solid points in the plot are sampled from the observed nominal spectrum (Figure 7 and similar figures). The solid line is our optimized model of the nominal emissions. (bottom) The optimized electron-positron spectra associated with the optimized synchrotron model shown in Figure 8, top. The blue-gray curve (the lowest) was obtained by assuming (clearly inaccurately) that all emitted frequency components come from the same emission volume (10^{57} cm^3). The red spectrum was obtained by assuming that the emission volume varies as a function of emission frequency according equation (5). The short black spectrum near the top is the electron spectrum measured within Jupiter's magnetosphere by the Pioneer 11 spacecraft near radial distance $L = 3$ R_J.

I naively modeled the gamma ray flares on the basis of adiabatic compression, using the spectrum derived for Figure 8 and the synchrotron inversion tools developed here. In Figure 10, the black curve is the nominal Crab emission spectrum. The purple curve is the emission from a very small spatial region, assumed to emit the same spectrum as does the Crab as a whole but with a much smaller emission volume. The parameter RVI labeling the purple curve is a free parameter that is the ratio of the volume-times-electron intensity within the small-scale region to be compressed compared to the same parameter for the nominal surrounding medium for the same electron energy. The parameter used is reasonable if the precompressed electron intensity within the small-scale region begins with an intensity that is a factor of 3 to 300 greater than the nominal broader Crab volume, depending on a range of reasonable assumptions about the nominal effective volume of emissions for the highest-energy portion of the nominal Crab spectrum. The green curve is what one gets when the small-scale emission region is adiabatically compressed by a canonical factor of 4, based on strong shock compression. Note that the effective compression ratio for the famous CRRES event was a much higher factor of ~10 because the shock-like structure propagated against a magnetic beach [*Li et al.*, 1993]. For adiabatic compression, we convert intensity to phase space density and assume conservation of the relativistic adiabatic invariant of gyration as B increases.

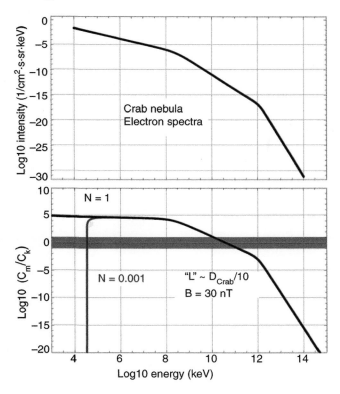

Figure 9. Kennel-Petschek analysis of the Crab Nebula electron spectrum shown in Figure 8. (top) A replotting of the "variable volume" spectrum in Figure 8. (bottom) Kennel-Petschek analysis of that spectrum assuming that the scale size (D) for whistler mode growth (commonly identified as "L" in planetary magnetospheres) is 0.1 times the diameter of the Crab Nebula.

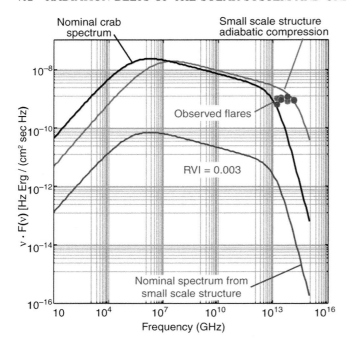

Figure 10. Observations and modeling related to the so-called gamma ray flares observed coming from the Crab Nebula by *Tavani et al.* [2011] and *Abdo et al.* [2011]. The colored dots are the observations from *Tavani et al.* [2011], which should be compared to the nominal crab spectrum shown with a black line. The rest of Figure 10 shows a naïve attempt to model the generation of the Crab gamma ray flares based on adiabatic energization. The violet curve is the presumed emission spectrum from a small-scale spatial structure that is subsequently compressed by shock wave dynamics. The green curve is the result of compressing the magnetic field within the small-scale structure by a factor of BR = 4, the canonical value for a strong shock. The parameter RVI is the ratio of the emission volume *V* of the small-scale structure times the initial electron-positron intensities within the small-scale structure, all normalized for the same parameters associated with the nominal Crab emissions for photon energies >100 MeV.

Figure 10, and similar plots using a range of possible compression ratios and RVI parameters, demonstrates that this radiation-belt-physics-motivated hypothesis could easily explain the Crab gamma ray flares if the fundamental assumptions used are valid. It is of substantial interest that the response of the synchrotron spectrum is so dramatic, a surprising ~3 orders of magnitude at the frequency of the gamma ray bursts. This strong response occurs because of the strong dependence of the emissions on magnetic field strength, and the strong dependence of the characteristic photon energy on both the magnetic field strength and electron energy. A modest disturbance generates a huge response in the emission properties, a feature that is valid irrespective of the validity of the particular assumptions used here for the Crab.

In this particular case, as emphasized by the referee of an ill-fated submitted paper, my radiation-belt-inspired intuition failed me. The referee demolished the assumption of adiabaticity. For synchrotron emission of gamma rays, the synchrotron cooling time of the emitting electrons is much faster than the cyclotron period. Also realized after the review was that there is a problem with the cyclotron period. The nonrelativistic cyclotron period within the 30 nT nominal magnetic field is ~1 ms. The cyclotron period of a 5 PeV electron (given the γ factor) is astoundingly over 100 days, much longer than the ~1 day acceleration period for gamma ray flares. For the synchrotron cooling, the textbook expression [e.g., *Tucker*, 1975] for frequency-integrated radiated power from a relativistic ($γ^2 \gg 1$) gyrating electron is

$$W[\text{ergs/s}] = \frac{2e^4 B^2 Sin^2(\alpha)}{3m_e^2 c^3} \gamma^2, \tag{6}$$

where α is particle pitch angle. One may compare the cyclotron period $T_{\text{cyc}} = 2\pi\gamma/\Omega_{\text{cyc}}$, where Ω_{cyc} is the nonrelativistic cyclotron frequency, with the timescale for synchrotron cooling: $T_{\text{synch}} = E/W$ (for $\alpha = \pi/2$), where E is the particle energy ($\sim \gamma \, m_e c^2$), and W is the radiated power from equation (6). One obtains

$$\frac{T_{\text{cyc}}}{T_{\text{synch}}} = \frac{4\pi}{3c} \gamma^2 \frac{e^2}{m_e c^2} \Omega_{cyc} = \frac{2\pi^2 e^2}{1125 \, c^3 \, hm_e} \varepsilon[\text{MeV}]$$

$$\sim 10\varepsilon[\text{MeV}], \tag{7}$$

where c is the speed of light, e is the electron charge, m_e is the electron mass, "h" is Planck's constant, and ε is the emitted photon energy in MeV. The expression is converted from using the energy of the emitting electron (left side) to using the energy of the emitted photon (right side), using the peak value of the single-electron energy response ($\sim 0.3v_c$ in Figure 4; v_c is defined in the appendix). For emitted photon energies greater than, say, 0.1 MeV, the electrons that emit them cannot be considered adiabatic because the timescale for synchrotron cooling is much faster than the gyroperiod. As emphasized by the referee cited in an earlier paragraph, when equation (7) is put in the form using emitted photon energy, the result does not depend on the magnetic field strength. Clearly, the electrons that emitted the gamma ray flares from the Crab (Figure 10; with photon energies >100 MeV) were not operating in the classical adiabatic regime. However, dynamics within radio wave and X-ray synchrotron emissions, with emission energies much lower than 0.1 MeV, could well be the consequence of adiabatic behavior.

5. DISCUSSION

Discussed here was the finding that relativistic radiation belts occur robustly at all of the strongly magnetized planets. An updated differential Kennel-Petschek limit provides one "standard" for comparing the radiation belts of the solar system. The radiation belts of Earth, Jupiter, and Uranus achieve this standard for MeV-class electrons; however, those of Neptune and Saturn do not. At least during the time of the Voyager 2 encounter with Neptune, the acceleration of electrons to relativistic energies was not as robust at Neptune as compared to Earth, Jupiter, and Uranus, possibly as the result of the absence of dynamic injection phenomena. The relativistic electrons within Saturn's radiation belt are likely depleted as a result of interactions with gas and dust from the plumes of the satellite Enceladus. Nonetheless, the Kennel-Petschek limit succeeds in representing one standard for comparing relativistic electron intensities between different planetary radiation belts.

Jupiter is used here as the jumping-off point for extrapolating radiation belt physics to the cosmos. Studies of Jupiter's synchrotron radiation provide a hint that the very processes that allow us to "see" distant radiation regions of the cosmos also play significant roles in the energetics of those distant regions. This strong coupling between our diagnostic capabilities and the physics of the regions being diagnosed limits our ability to extrapolate radiation belt physics to distant cosmic systems.

With our attempts to extrapolate radiation belt physics to the radiation regions of the canonical and beautiful Crab Nebula, it was found that the spectral intensities are much higher than anticipated with radiation belt processes, based on our KPL "standard." Also, some of the standard tools that we use for understanding dynamics within the radiation belts may not be as readily applicable within the distant cosmic radiation regions. Specifically, understanding dynamic variability within hyperenergetic electrons may not be as readily aided by standard invariant analysis that is so critical to understanding radiation belt physics.

Radiation belt physicists should be very careful in their claims about the applicability of radiation belt physics in solving problems faced by astrophysicists in understanding those cosmic radiation regions that are visible to us via photon emission processes.

APPENDIX A

Presented here are the expressions used to convert electron-positron intensity spectra into synchrotron emission spectra and vice versa. These expressions are derived using the works of *Tucker* [1975] and *Jackson* [1962]. The synchrotron power (ergs cm^{-3} sr^{-1}) emitted by a system is given by the expression

$$\frac{dR}{d\Omega}(\nu, \beta) = \iiint d\gamma d\Omega d\alpha \left(N(\gamma, \alpha) \frac{dp}{d\Omega}(\nu, \gamma, \alpha, \Theta) \right), \quad (A1)$$

where R is (power cm^{-3}), Ω is solid angle, ν is emission frequency, β is the viewing angle with respect to the magnetic field direction within the emitting volume, γ is the Lorentz factor for the emitting electrons, α is the emitting particle pitch angle (the angle between the particle momentum vector and the magnetic field), N is density (cm^3 $\gamma\alpha$)$^{-1}$, p is the emitting (power cm^{-3}) for each particle, and Θ is the angle between the plane of the gyrating particle and the viewing direction, specifically $\Theta = \alpha - \beta$. By assuming, because of the strong synchrotron beaming along the electron's momentum vector, that particles that significantly contribute have pitch angles (α) approximately equal to the viewing direction (β), one may derive the canonical expression

$$\frac{dR}{d\Omega}(\nu, \beta) \approx \int d\gamma N(\gamma, \beta) P(\nu, \gamma, \beta). \quad (A2)$$

P is the usual total power s^{-1} emitted by a single particle with Lorentz factor γ and pitch angle equal to the viewing direction β, given by the usual expression

$$P(\nu, \gamma, \beta) = \sqrt{\frac{3\nu_b B \sin(\beta)}{m_e c^2}} G\left(\frac{\nu}{\nu_c}\right), \quad (A3)$$

where

$$G\left(\frac{\nu}{\nu_c}\right) = \frac{\nu}{\nu_c} \int_{\nu/\nu_c}^{\infty} K_{5/3}(y) dy \quad (A4)$$

and where $K_{5/3}$ is the modified Bessel function of degree 5/3, B is the magnetic field strength (Gauss), ν_b is the nonrelativistic electron cyclotron frequency ($eB/(2\pi m_e c)$), and ν_c is the "critical frequency": $3\nu_b \gamma^2 \sin(\beta)/2$ (the definition given by *Tucker* [1975] but not *Jackson* [1962]).

Using the total density "n" as $n = \iint N(\gamma, \alpha) d\gamma d\alpha = \iiint f(\mathbf{p}) d^3 p$, where "f" is phase space density (PSD), p is momentum, and $f = I/p^2$, one can relate N to I, with

$$N(\gamma, \beta) = I[(\gamma, \alpha] \frac{m_e c}{(1 - 1/\gamma^2)^{1/2}}, \quad (A5)$$

where I is just our intensity expression converted to (erg^{-1}) and with the kinetic energy E_{keV} converted to ergs and then replaced with $E = (\gamma - 1) m_e c^2$). One finds, then, for $\gamma \gg 1$,

$$\frac{dR}{d\Omega}(\nu, \beta) = \int (d\gamma) m_e c I(\gamma, \beta) P(\nu, \gamma, \beta). \quad (A6)$$

The signal that is observed from the vicinity of Earth is

$$F(\nu, \beta) \left[\frac{\text{ergs}}{\text{cm}^2 \text{ s}}\right] = V_C \frac{1}{D^2} \frac{dR}{d\Omega}(\nu, \beta), \quad (A7)$$

where V_C is the volume of the emitting regions, D is the distance between Earth and the emitting object, and $1/D^2$ is the solid angle that a single point residing within the emitting volume sees when looking at a 1 cm^2 area in the vicinity of Earth. While F is the function that is plotted in Figure 5 (in units of Jansky [Jy]; 1 Jy = 10^{-23} ergs [s cm^2]$^{-1}$), Figures 8 and 10 use νF, which restricts the dynamic range of the vertical axis. Because we are looking for only a rough idea of the shape of the Crab electron spectrum, we have chosen here to simply set $\beta = \pi/2$ for the Crab rather than performing weighted averages over an assumed distribution of β values.

The intensity "I" in equation (A6) has five or seven free parameters, as defined in equation (3). We optimize these parameters using a forward modeling optimization procedures. For the Crab, we minimize:

$$Err(C, kT, \gamma 1, E_2, \gamma 2, E_3, \gamma 3)$$
$$= \sum_{i=1}^{m} \left[O(\nu_i) - \nu_i \frac{V_C}{D^2} \int (d\gamma) m_e c I\left(\gamma, \frac{\pi}{2}\right) P\left(\nu_i, \gamma, \frac{\pi}{2}\right) \right]^2,$$

$$(A8)$$

where $O(\nu_i)$ is the observed energy flux at each of the discrete frequencies shown in Figure 8 (top). Because of the large dynamic range for the observations over frequency, we actually minimize the square of the differences of the Logarithms of the observations and of the evaluated functions.

An additional trick that we use is to utilize an analytic approximation to the transcendental function $G(X)$ shown in equation (A4). Our analytic approximation, shown to be accurate to within ~12% or less at all values of "X," is:

$$G(X) \approx \sqrt{\frac{\pi}{2}} X^{1/2} e^{-X} \left(1 + \frac{4\pi \Gamma(1/3)}{\sqrt[3]{2}\sqrt{3}} \frac{1}{X} \right)^{1/6}, \qquad (A9)$$

where $\Gamma(1/3)$ is the gamma function evaluated at 1/3 (~2.6789). This analytic form was derived here by analytically joining the asymptotic limits for $G(X)$ for the conditions $X \gg 1$ and $X \ll 1$ from *Tucker* [1975]. A comparison between the approximated and the "true" functional form of $G(X)$ is shown in Figure 4.

Acknowledgments. We appreciate valuable discussions had with Aleksandr (Sasha) Ukhorskiy and Ralph L. McNutt Jr.

REFERENCES

Abdo, A. A., et al. (2011), Gamma-ray flares from the Crab Nebula, *Science*, *331*, 739–742, doi:10.1126/science.1199705.

Amato, E., M. Salvati, R. Bandiera, F. Pacini, and L. Woltjer (2000), Inhomogeneous models for plerions: The surface brightness profile of the Crab Nebula, *Astron. Astrophys.*, *359*, 1107–1110.

Atoyan, A. M., and F. A. Aharonian (1996), On the mechanisms of gamma radiation in the Crab Nebula, *Mon. Not. R. Astron. Soc.*, *278*, 525–541.

Baker, D. N., and J. A. Van Allen (1976), Energetic electrons in the Jovian magnetosphere, *J. Geophys. Res.*, *81*(4), 617–632.

Blake, J. B., W. A. Kolasinski, R. W. Fillius, and E. G. Mullen (1992), Injection of electrons and protons with energies of tens of MeV into L < 3 on 24 March 1991, *Geophys. Res. Lett.*, *19*(8), 821–824.

Bolton, S. J., et al. (2002), Ultra-relativistic electrons in Jupiter's radiation belts, *Nature*, *414*, 987–991.

de Pater, I., and D. E. Dunn (2003), VLA observations of Jupiter's synchrotron radiation at 15 and 22 GHz, *Icarus*, *163*, 449–455.

Garrett, H. B., S. M. Levin, S. J. Bolton, R. W. Evans, and B. Bhattacharya (2005), A revised model of Jupiter's inner electron belts: Updating the Divine radiation model, *Geophys. Res. Lett.*, *32*, L04104, doi:10.1029/2004GL021986.

Hester, J. J. (2008), The Crab Nebula: An astrophysical chimera, *Annu. Rev. Astron. Astrophys.*, *46*, 127–155.

Jackson, J. (1962), *Classical Electrodynamics*, Wiley, New York.

Kennel, C. F., and H. E. Petschek (1966), Limit on stably trapped particle fluxes, *J. Geophys. Res.*, *71*, 1–28.

Li, X., I. Roth, M. Temerin, J. R. Wygant, M. K. Hudson, and J. B. Blake (1993), Simulation of the prompt energization and transport of radiation belt particles during the March 24, 1991 SSC, *Geophys. Res. Lett.*, *20*(22), 2423–2426.

Mauk, B. H., and N. J. Fox (2010), Electron radiation belts of the solar system, *J. Geophys. Res.*, *115*, A12220, doi:10.1029/2010JA015660.

Meyer, M., D. Horns, and H.-S. Zechlin (2010), The Crab Nebula as a standard candle in very high-energy astrophysics, *Astron. Astrophys.*, *523*, A2, doi:10.1051/0004-6361/201014108.

Santos-Costa, D., and S. J. Bolton (2008), Discussing the processes constraining the Jovian synchrotron radio emission's features, *Planet. Space Sci.*, *56*, 326–345.

Schulz, M., and G. T. Davidson (1988), Limiting energy spectrum of a saturated radiation belt, *J. Geophys. Res.*, *93*(A1), 59–76.

Summers, D., R. Tang, and R. M. Thorne (2009), Limit on stably trapped particle fluxes in planetary magnetospheres, *J. Geophys. Res.*, *114*, A10210, doi:10.1029/2009JA014428.

Tang, R., and D. Summers (2012), Energetic electron fluxes at Saturn from Cassini observations, *J. Geophys. Res.*, *117*, A06221, doi:10.1029/2011JA017394.

Tavani, M., et al. (2011), Discovery of powerful gamma-ray flares from the Crab Nebula, *Science*, *331*, 736–739, doi:10.1126/science.1200083.

Tucker, W. H. (1975), *Radiation Processes in Astrophysics*, MIT Press, Cambridge, Mass.

B. H. Mauk, Applied Physics Laboratory, Johns Hopkins University, Laurel, MD 20723, USA. (Barry.Mauk@jhuapl.edu)

Plasma Wave Observations at Earth, Jupiter, and Saturn

G. B. Hospodarsky, K. Sigsbee, J. S. Leisner, J. D. Menietti, W. S. Kurth, D. A. Gurnett, and C. A. Kletzing

Department of Physics and Astronomy, University of Iowa, Iowa City, Iowa, USA

O. Santolík

Institute of Atmospheric Physics, Prague, Czech Republic

Department of Surface and Plasma Science, Charles University, Prague, Czech Republic

Plasma wave emissions have been detected at all of the planets that have been visited by spacecraft equipped with plasma wave instruments. (Mercury will be explored by the plasma wave instrument on BepiColombo Mercury Magnetospheric Orbiter in 2022.) Many of these emissions are believed to play a role in the acceleration of energetic particles, especially those observed in association with the radiation belts of Earth, Jupiter, and Saturn. Wave-particle interactions involving whistler mode chorus, hiss, equatorial noise, and electron cyclotron harmonics participate in both the acceleration and loss of these radiation belt particles and play a major role in radiation belt dynamics throughout the solar system. The effects of these wave modes, their occurrence probabilities, their amplitudes, and their relationships to solar wind properties and geomagnetic storm conditions have been investigated using a variety of spacecraft observations. This paper summarizes these studies and discusses the similarities and differences of the plasma waves detected at Earth, Jupiter, and Saturn.

1. INTRODUCTION

Plasma waves play an important role in radiation belt particle dynamics at Earth [*Horne and Thorne*, 1998; *Summers et al.*, 1998, 2007; *Thorne*, 2010], Jupiter [*Horne et al.*, 2008], and possibly Saturn [*Hospodarsky et al.*, 2008; *Mauk et al.*, 2009]. Wave-particle interactions are involved in both the acceleration and loss of radiation belt particles [*Kennel and Petschek*, 1966; *Omura and Summers*, 2006; *Mauk and Fox*, 2010; *Tang and Summers*, 2012]. In the next sections, we will briefly discuss the status of the current research for some of these emissions at Earth, Jupiter, and Saturn.

Dynamics of the Earth's Radiation Belts and Inner Magnetosphere
Geophysical Monograph Series 199
10.1029/2012GM001342

2. EARTH OBSERVATIONS

The first spacecraft observations of plasma waves in Earth's magnetosphere were obtained from Alouette I [*Barrington and Belrose*, 1963] and Injun III [*Gurnett and O'Brien*, 1964] in 1962. Since these early space-based observations, a number of spacecraft have provided ongoing observations of the relationships between the many plasma waves detected in Earth's inner magnetosphere and the Van Allen radiation belts. Figure 1 is a spectrogram showing plasma waves observed on the CRRES spacecraft, including whistler mode chorus, plasmaspheric hiss, electrostatic cyclotron harmonic (ECH) emissions, and equatorial noise emission, all of which are believed to play a role in radiation belt dynamics. The importance of these wave modes and their occurrence probabilities, intensities, and relationships to solar wind properties and geomagnetic storm conditions have been investigated using a variety of spacecraft observations

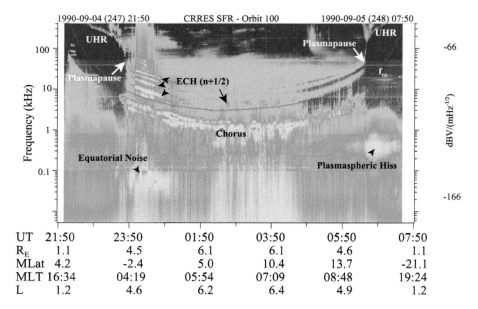

Figure 1. A spectrogram showing plasma waves that are believed to be important for radiation belt dynamics at the Earth observed on the CRRES spacecraft.

[*Meredith et al.*, 2001, 2003, 2004, 2006, 2007, 2008, 2009a; *Santolík et al.*, 2003, 2004, 2005, 2010; *Horne et al.*, 2007; *Breneman et al.*, 2009; *Sigsbee et al.*, 2010; *Li et al.*, 2011; *Bunch et al.*, 2011, 2012].

2.1. Whistler Mode Chorus Emissions

Chorus is an intense electromagnetic wave emission that propagates in the right-hand polarized whistler mode and is believed to be generated by nonlinear interactions of whistler mode waves with energetic electrons [*Storey*, 1953; *Allcock*, 1957; *Helliwell*, 1969; *Summers et al.*, 2007; *Katoh and Omura*, 2007; *Tao et al.*, 2012, and references therein]. Chorus observed in Earth's magnetosphere ranges in frequency from a few hundred Hz to a few kHz and typically occurs in two distinct frequency bands separated by a gap at one-half the electron cyclotron frequency, f_{ce} [*Tsurutani and Smith*, 1974]. The lower band starts at about 0.1 f_{ce}, and the upper band extends up to about 0.8 f_{ce} [*Meredith et al.*, 2001].

Chorus at Earth often contains a variety of structures, including rising and falling tones, and short impulsive bursts (individual whistler wave packets, commonly called chorus elements) with timescales of much less than a second [*Gurnett and O'Brien*, 1964; *Burtis and Helliwell*, 1969; *Sazhin and Hayakawa*, 1992; *LeDocq et al.*, 1998; *Lauben et al.*, 2002; *Santolík et al.*, 2003, and references therein]. The origin of these structures and their relationships to the source region of chorus is an active area of research [*Inan et al.*,

2004; *Chum et al.*, 2007; *Breneman et al.*, 2007; *Katoh and Omura*, 2011; *Omura et al.*, 2008, 2009; *Omura and Nunn*, 2011, and references therein]. Chorus is often observed during periods of disturbed magnetospheric conditions [*Tsurutani and Smith*, 1974, 1977; *Inan et al.*, 1992; *Lauben et al.*, 1998; *Sigsbee et al.*, 2008], and the occurrence of chorus is associated with energetic (10 to 100 keV), anisotropic ($T_\perp/T_\parallel > 1$) electron distributions [*Burton*, 1976; *Anderson and Maeda*, 1977; *Tsurutani et al.*, 1979; *Isenberg et al.*, 1982]. This association between energetic particles and chorus is consistent with a cyclotron resonance interaction [*Kennel and Petschek*, 1966; *Curtis*, 1978; *Sazhin and Hayakawa*, 1992].

A variety of studies have examined the propagation characteristics of chorus emissions at Earth [*Burtis and Helliwell*, 1969; *Burton and Holzer*, 1974; *Goldstein and Tsurutani*, 1984; *Hayakawa et al.*, 1984, 1990; *Nagano et al.*, 1996; *LeDocq et al.*, 1998; *Hospodarsky et al.*, 2001; *Lauben et al.*, 2002; *Santolík et al.*, 2005, 2009, 2010] and have suggested that chorus is generated very close to the magnetic equator. More recent work utilizing the Polar and the Cluster spacecraft [*Parrot et al.*, 2003; *Santolík and Gurnett*, 2003; *Santolík et al.*, 2005] have shown that the chorus source region for individual elements (measured along the magnetic field lines) is a few thousand km in size, and about 100 km transverse to the magnetic field.

The amplitudes of chorus emissions are important in understanding the possible wave-particle interactions. For example, at Earth, typical chorus amplitudes measured by

Cluster were <1 mV m^{-1}, but could reach ~30 mV m^{-1} during disturbed periods [*Santolík et al.*, 2003]. Magnetic wave amplitudes of chorus are typically a few pT to ~100 pT, with the largest fields detected by Cluster being a few nT (B_{wave}/B_o = ~0.01) (O. Santolík, personal communication, 2012). Using results from the STEREO spacecraft flybys of Earth, *Cattell et al.* [2008] reported even larger-amplitude electric field signals (240 mV m^{-1}) associated with high time resolution measurements of monochromatic whistler mode emissions in the Van Allen radiation belts. Similar large-amplitude emissions (~100 mV m^{-1}) have been reported in Time History of Events and Macroscale Interactions during Substorms (THEMIS) observations [*Cully et al.*, 2008; *Li et al.*, 2011] and WIND observations [*Kellogg et al.*, 2010; *Wilson et al.*, 2011]. These observations suggest that strong nonlinear wave-particle processes may be more important than previously thought in accelerating electrons to populate the radiation belts [*Bortnik et al.*, 2008a; *Yoon*, 2011].

2.2. Plasmaspheric Hiss

Plasmaspheric hiss is a broadband, structureless whistler mode emission found in Earth's plasmasphere with typical wave amplitudes of ~1 to 100 pT. Hiss is believed to be responsible for generating the slot region between the inner and outer radiation belts, by removing high-energy particles from the radiation belts after geomagnetic storms [*Lyons et al.*, 1972; *Lyons and Thorne*, 1973; *Spjeldvik and Thorne*, 1975; *Albert*, 1994; *Abel and Thorne*, 1998]. The characteristics of hiss detected by the CRRES spacecraft and the relationship of hiss to electron loss timescales within the plasmasphere have been examined in a series of studies [*Meredith et al.*, 2004, 2006, 2007, 2009b]. Intense hiss observed in plasmaspheric plumes has also been shown to be important in scattering outer zone electrons [*Summers et al.*, 2007, 2008].

The exact origin of the plasmaspheric hiss has been debated for many years. Three basic theories have been proposed: (1) growth from preexisting "weak" waves from free energy of unstable electron populations [*Thorne et al.*, 1973; *Church and Thorne*, 1983], (2) accumulation of whistler waves from lightning in Earth's atmosphere [*Draganov et al.*, 1992; *Green et al.*, 2005, 2006; *Thorne et al.*, 2006; *Meredith et al.*, 2006], and (3) inward propagation of chorus into the plasmasphere from a source near the magnetic equator and outside the plasmasphere [*Church and Thorne*, 1983; *Chum and Santolík*, 2005; *Santolík et al.*, 2006; *Bortnik et al.*, 2008b, 2009a, 2009b]. Recent ray-tracing studies [*Bortnik et al.*, 2008b, 2009a] and simultaneous observations of chorus and hiss from two THE-MIS spacecraft [*Bortnik et al.*, 2009b] have made a strong case for chorus producing the plasmaspheric hiss.

2.3. Equatorial Noise Emissions

Low-frequency electromagnetic emissions near Earth's geomagnetic equator were first reported by *Russell et al.* [1970], using the OGO 3 search coil magnetometer, and have been studied by more recent spacecraft [*Gurnett*, 1976; *Perraut et al.*, 1982; *Olsen et al.*, 1987; *Laakso et al.*, 1990; *Kasahara et al.*, 1992, 1994; *Boardsen et al.*, 1992; *Santolík et al.*, 2004]. The emissions propagate in the fast magnetosonic mode perpendicular to the ambient magnetic field, **B$_o$**, and are usually observed within a few degrees of the geomagnetic equator between the local proton gyrofrequency (f_{cH}) and the lower hybrid frequency (f_{LHR}).

Gurnett [1976] used IMP 6 and Hawkeye high-resolution wideband electric and magnetic field wave data to show that the equatorial noise emissions contain a great deal of small-scale structures consisting of a complex superposition of bands with frequency spacings from a few Hz to several tens of Hz. A similar structure was reported in the data from the GEOS spacecraft [*Perraut et al.*, 1982]. These spacings were shown to be similar to the proton and ion gyrofrequencies, suggesting that the waves play a role in controlling the distribution of energetic ions through ion gyrofrequency resonances with energetic ring current protons, alpha particles, and heavy ions.

Several recent studies have examined the properties of fast magnetosonic equatorial noise using modern instrumentation and analysis methods, as well as multispacecraft data. *Santolík et al.* [2004] performed a systematic analysis of equatorial noise from 781 perigees during the first 2 years of Cluster STAFF-SA data [*Cornilleau-Wehrlin et al.*, 1997]. Equatorial noise was found to have the most intense (typically ~1 to ~100 pT) wave magnetic fields of all natural emissions between f_{cH} and f_{LHR} and to occur about 60% of the time at radial distances between 3.9 and 5 R_E. *Němec et al.* [2005] examined the Cluster data in more detail and found that the most intense peaks in the power spectra occurred within 2° of the dipole magnetic equator and that the most likely emission frequency was between 4 and 5 times the local proton cyclotron frequency. Further analysis using a geomagnetic latitude based upon the location of the minimum-B equator [*Santolík et al.*, 2002] found that the most intense peaks occurred exactly at the magnetic equator [*Němec et al.*, 2006].

Meredith et al. [2008] presented a statistical survey of fast magnetosonic waves in the frequency range 0.5 $f_{LHR} < f < f_{LHR}$ and the occurrence of proton ring distributions using wave and particle data from the CRRES satellite. Fast magnetosonic waves were observed at most local times outside the plasmapause, but waves inside the plasmapause were restricted to the dusk sector. They found that the observed locations of the low-energy (E_R < 30 keV) proton ring distributions required for instability in the CRRES data closely matched the observed locations of magnetosonic waves on the duskside, both inside and outside the plasmapause. A statistical study of ion distributions thought to drive magnetosonic waves using 15 years of geosynchronous particle data found the peak occurrence of unstable ion distributions from midmorning to dusk for low geomagnetic activity periods. For high activity periods, the peak moved toward noon [*Thomsen et al.*, 2011].

The importance of these waves in radiation belt dynamics has been investigated by a number of authors [*Horne et al.*, 2000, 2007; *Chen et al.*, 2010; *Bortnik and Thorne*, 2010]. Using quasilinear theory, *Horne et al.* [2007] found that pitch angle and energy diffusion rates for equatorial noise were comparable to those for whistler mode chorus. Because of their importance, magnetosonic waves have been newly incorporated into ring current simulations using the Rice Convection Model and ring current-atmosphere interactions model [*Chen et al.*, 2010].

2.4. Electron Cyclotron Harmonics Emissions

ECH emissions are electrostatic emissions observed in bands between harmonics of the electron gyrofrequency, f_{ce}, and are sometimes known as $(n + 1/2)f_{ce}$ emissions [*Kennel et al.*, 1970; *Fredricks and Scarf*, 1973; *Shaw and Gurnett*, 1975; *Christiansen et al.*, 1978; *Horne and Thorne*, 2000; *Meredith et al.*, 2009a; *Thorne et al.*, 2010]. *Meredith et al.* [2009a] examined the ECH waves in CRRES data and showed that their intensity and occurrence rates were similar to chorus, and suggested that ECH emissions might be important in the production of the diffuse aurora. *Thorne et al.* [2010] examined these results in more detail and determined that the chorus emissions play a much larger role than the ECH waves in the auroral precipitation at lower L shells. However, at higher L shells, the ECH emissions may still dominate [*Liang et al.*, 2011; *Ni et al.*, 2012].

2.5. Upcoming Earth Missions

The Radiation Belt Storm Probes (RBSP) mission, to be launched late summer 2012, will consist of two identical spacecraft with a comprehensive suite of field and particle instruments. RBSP promises a rich new data set for studies of wave-particle interactions in Earth's radiation belts [*Staedter*, 2006; *Reeves*, 2007; *Kessel*, this volume]. In combination with other spacecraft assets and observational campaigns of the Balloon Array for RBSP Relativistic Electron Losses [*Millan et al.*, 2011], the RBSP mission will contribute to a significant increase in our understanding of Earth's Van Allen radiation belts.

3. JUPITER OBSERVATIONS

The Jovian magnetosphere is the largest and probably the most complex in the solar system. The Voyager flybys of Jupiter provided the first opportunities to directly measure the plasma waves in the Jovian magnetosphere [*Scarf et al.*, 1979a; *Gurnett et al.*, 1979; *Gurnett and Scarf*, 1983]. Figure 2 [*Kurth and Gurnett*, 1991] is a spectrogram obtained during the Voyager 1 flyby showing many of the plasma waves detected at Jupiter, including chorus, hiss, and ECH emissions. The orbiting Galileo mission provided additional opportunities to study plasma waves in the Jovian system [*Gurnett et al.*, 1996; J. D. Menietti et al., Chorus, ECH, and z-mode emissions observed at Jupiter and Saturn and a possible electron acceleration, submitted to the *Journal of Geophysical Research*, 2012, hereinafter referred to as Menietti et al., submitted manuscript, 2012]. Although many of the same plasma waves that are observed at Earth are also detected at Jupiter, the majority of studies have concentrated on the strong whistler mode chorus emission [*Scarf et al.*, 1979b; *Thorne and Tsurutani*, 1979; *Inan et al.*, 1983; *Thorne et al.*, 1997; *Bolton et al.*, 1997; *Xiao et al.*, 2003; *Katoh et al.*, 2011]. Chorus is detected primarily near the edge of the Io torus from L shells of about 6 to about 12 and has many of the characteristics of chorus observed at Earth, including two bands of chorus separated by a gap at $0.5 f_{ce}$ and small-scale structures (primarily rising tones) with timescales less than 1 s [*Coroniti et al.*, 1980, 1984]. Peak amplitudes of chorus detected by Voyager were found to be about 0.26 mV m^{-1} (10 pT assuming parallel propagation) and that the chorus scattered electrons with energy of a few keV [*Coroniti et al.*, 1980, 1984].

Statistical surveys of the occurrence and strength of the Jovian chorus detected by Galileo have been performed by *Horne et al.* [2008] and *Menietti et al.* [2008a]. These studies showed that chorus typically occurred between 400 and 8 kHz with peak amplitudes ~10^{-10} V^2 m^{-2} Hz^{-1} (~0.1 mV m^{-1} and ~3 pT). The maximum power and occurrence frequency was observed a few degrees above the magnetic equator with the amplitude and frequency decreasing rapidly as the spacecraft went to higher latitudes. *Horne et al.* [2008] demonstrated that interactions between electrons and whistler mode waves in the Jovian magnetosphere cause significant acceleration of MeV electrons. The volcanoes on the moon Io provide a source of particles that are ionized and form the Io plasma torus. Outward transport of cold dense plasma and inward transport of 1 to 100 keV plasma occur due to magnetic flux interchange instabilities. This transport develops temperature anisotropies that produce whistler mode chorus. Using wave power spectral density from the Galileo measurements, *Horne et al.* [2008] calculated

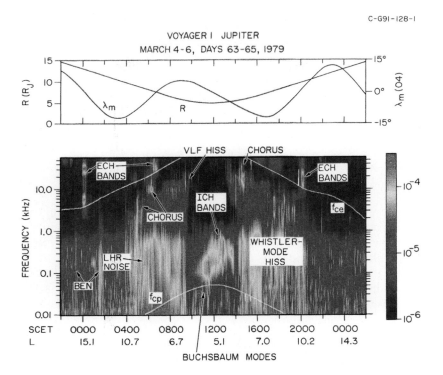

Figure 2. A spectrogram obtained during the Voyager 1 flyby showing many of the plasma waves detected at Jupiter. From *Kurth and Gurnett* [1991].

quasilinear pitch angle and energy diffusion coefficients and solved a 2-D diffusion equation over the region of 5 to 20 R_J. Fluxes of 1 to 6 MeV electrons were found to increase by more than an order of magnitude within 30 days. These electrons are then transported toward the planet via radial diffusion and accelerated to even higher (~50 MeV) energies by betatron and Fermi processes, resulting in the intense radiation belts detected close to the planet.

Tao et al. [2011] expanded on the work of *Horne et al.* [2008] using the intensity distribution of the chorus as determined by *Menietti et al.* [2008a] to examine the effect of different density models for the injection events and the effects of latitude-dependent wave normal angles for the chorus. They found that if the density inside the injection event is half that of the density outside, there is a negligible effect on the timescale of electron energization, but if the density inside the injection is one fourth of the density outside, the timescales of energization are significantly changed (1 and 3 MeV particle fluxes are decreased, 10 MeV electron fluxes is increased). The latitude-dependent wave normal angle distribution has no obvious effect on 10 MeV electron fluxes, but it reduces the fluxes of 1 and 3 MeV electrons by about a factor of 3. Thus, realistic density models of the injection events and the change of wave normal angle of chorus waves during propagation should be

taken into account in future studies to more accurately model electron flux evolution of Jovian electrons.

The next mission to visit Jupiter with a plasma wave instrument is the Juno mission, scheduled to arrive at Jupiter in July 2016. Although it is primarily an auroral physics mission (polar orbit), it will likely measure chorus and other plasma waves during its trajectory at lower magnetic latitudes, providing new insight on the plasma wave properties of emissions found at the inner edge of the Jovian radiation belts. Furthermore, the recently announced European Space Agency Jupiter Icy moon Explorer (JUICE) mission will carry sophisticated field and particle instruments that will provide additional plasma wave measurements in the region of interest at Jupiter.

4. SATURN OBSERVATIONS

A variety of plasma and radio waves detected by the Voyager spacecraft during their flybys of Saturn [*Gurnett et al.*, 1981; *Scarf et al.*, 1982, 1983] included whistler mode chorus and hiss [*Scarf et al.*, 1984], and ECH and upper hybrid resonance (UHR) emissions [*Kurth et al.*, 1983]. The Cassini mission to Saturn, with its many orbits and the opportunity to sample many different regions of the magnetosphere and a more capable Radio and Plasma Wave

Science (RPWS) instrument [*Gurnett et al.*, 2004], has allowed a much more detailed study of plasma waves to be accomplished. Many of these emissions have similar characteristics to emissions detected at Earth and Jupiter, but important differences have also been reported.

4.1. Whistler Mode Chorus

The Voyager spacecraft detected only a few short periods of chorus during their flybys of Saturn due primarily to the geometry of the flybys. Also, the lack of a search coil magnetometer and the Voyager plasma wave instrument response to dust impacts on the spacecraft in the equatorial region complicates the identification of chorus and other whistler mode waves in the low-resolution, low-rate data. Unfortunately, the amount of high-resolution wideband data obtained during the flybys was greatly limited due to telemetry constraints (see the work of *Scarf et al.* [1983] for a discussion of these issues). However, a wideband frame was obtained when chorus was present during the Voyager 1 flyby [*Scarf et al.*, 1983]. This high-resolution data showed a band of chorus present below 0.5 f_{ce} (from about 0.2 to 0.4 f_{ce}). The chorus was primarily hiss-like (diffuse) with some rising structures, but the temporal variations were "unusually slow" [*Scarf et al.*, 1983]. No chorus above 0.5 f_{ce} was detected by Voyager at Saturn. *Scarf et al.* [1983, 1984] examined the possibility of wave-particle interaction and pitch angle scattering of the electrons in Saturn's inner magnetosphere from the detected chorus, but the amplitudes were too small to play a significant role. However, during the period of the Voyager flybys, the Saturnian magnetosphere was not very active, and it is possible that the chorus emission could play a more significant role during more active periods [*Kurth and Gurnett*, 1991].

The Cassini spacecraft has detected a variety of wave emissions at Saturn [*Gurnett et al.*, 2005; *Hospodarsky et al.*, 2008; *Menietti et al.*, 2008b, 2008c; *Mauk et al.*, 2009; *Kurth et al.*, 2009; Menietti et al., submitted manuscript, 2012]. Figure 3 shows typical spectrograms of the electric and magnetic field intensities of the plasma wave emissions measured by RPWS during a pass through Saturn's inner magnetosphere. The intensities are plotted as decibels (dB) above background, and the white line shows f_{ce} derived from the magnetometer instrument [*Dougherty et al.*, 2004]. During this orbit, whistler mode emission is observed both during the inbound and the outbound trajectory. Following the definition of *Hospodarsky et al.* [2008], we will call this emission chorus, although at times, the emission may appear hiss-like. Electrostatic ECH and UHR waves are also present, and the Saturn kilometric radio (SKR) emission is observed above about 20 kHz throughout most of this period. The whistler

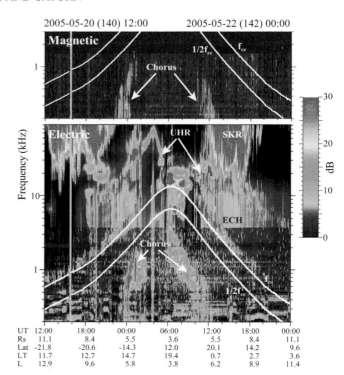

Figure 3. (top) Magnetic and (bottom) electric field frequency-time spectrograms from the Cassini Radio and Plasma Wave Science instrument. The white lines are f_{ce} and 0.5 f_{ce} determined from the magnetometer data.

mode chorus emissions shown in Figure 3 are similar to the chorus detected by the Voyager spacecraft with frequencies only below 0.5 f_{ce}. The majority of the chorus detected at Saturn has this characteristic and was called "magnetospheric" chorus by *Hospodarsky et al.* [2008]. "Magnetospheric" chorus is detected by the RPWS instrument during most orbits of Saturn when Cassini is within about 30° of the magnetic equator and between L shells of about 4 and 8.

Figure 4 shows two "magnetospheric" chorus spectrograms using the higher-resolution wideband receiver (WBR) data. A variety of small-scale structures, from hiss-like emission to rising tones, is detected but with larger timescales than observed at Earth or Jupiter (many seconds to many minutes). The emissions that have rising tones with periods on the order of 5 min (Figure 4, bottom) have been investigated in more detail (J. S. Leisner et al., manuscript in preparation, 2012). Although *Hospodarsky et al.* [2008] called the whistler mode emissions with ~5 min period rising tones "chorus," it is likely that they have a different origin than chorus at Earth. These whistler mode emissions with ~5 min periods, referred to as "rising whistler mode emission" and sometimes as "worms" (J. S. Leisner et al., manuscript in preparation) are detected about 5% of the time when Cassini is within 5.5 R_S,

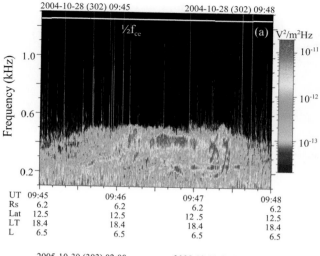

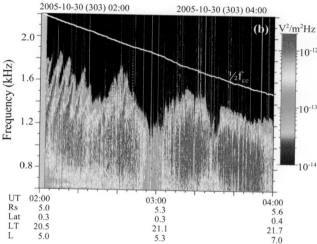

Figure 4. High-resolution wideband frequency-time spectrograms of "magnetospheric" chorus emissions measured by Cassini showing selected small-scale structure of the emissions. The white line is $0.5 f_{ce}$ determined from the magnetometer data. After *Hospodarsky et al.* [2008].

are observed near the magnetic equator and appear to be related to electrons with energies of a few keV, though what causes the 5 min periodicity is still not well understood. Although quasiperiodic emissions with similar spectral structure to the "worms" have been detected at low L shells and high latitudes at Earth [*Sato and Fukunishi*, 1981; *Pasmanik et al.*, 2004], it is unlikely they are generated by the same source mechanism due to the difference in where they are detected (high latitudes at Earth, near the equator at Saturn).

Chorus is also detected at Saturn in association with local plasma injections (defined as "injection event" chorus by *Hospodarsky et al.* [2008]). Injection events are planetward

injections of hot, tenuous plasma produced by the interchange instability as the colder, denser plasma from the inner magnetosphere flows outward in magnetospheres of rapidly rotating planets such as Saturn [*Mitchell et al.*, 2009b; *Mauk et al.*, 2009, and references therein]. Figure 5 shows two high-resolution examples of "injection event" chorus. For many plasma injection events, chorus emissions are detected both above and below $0.5 f_{ce}$ (white line), with a gap in the emission at $\sim 0.5 f_{ce}$. This chorus is usually detected for only the few minutes Cassini is in the injection event and is not observed outside of the injection event. These "injection event" chorus observations often contain structure (usually a series of rising tones) at a much smaller timescale (less than

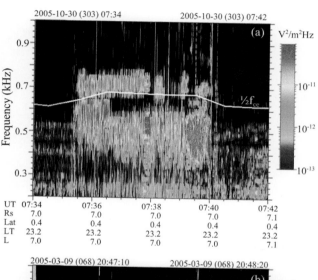

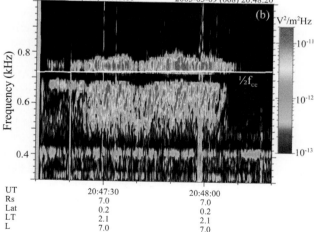

Figure 5. High-resolution wideband frequency-time spectrograms of "injection event" chorus emissions measured by Cassini showing selected small-scale structure of the emissions. The white line is $0.5 f_{ce}$ determined from the magnetometer data. After *Hospodarsky et al.* [2008].

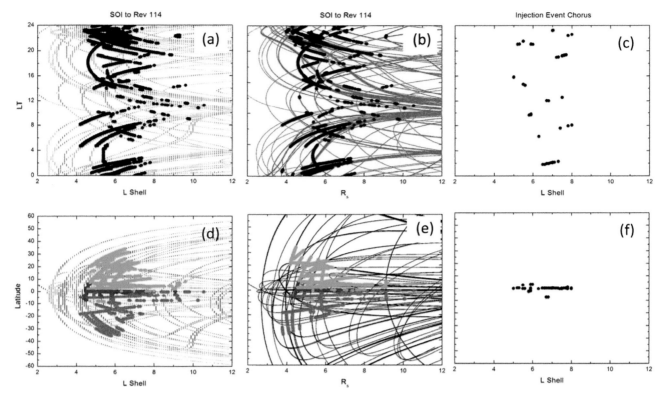

Figure 6. The position of Cassini during the periods that the (left and center) "magnetospheric" and (right) "injection event" chorus is detected. Figures 6a and 6c plot local time (LT) versus L shell and Figure 6b plots LT versus Saturn radius (R_s). Figures 6d and 6f plot magnetic latitude versus L shell, and Figure 6d plots magnetic latitude versus R_s. The propagation direction of the chorus emission detected at Saturn with respect to the magnetic field is shown by the color of the dots in Figures 6d and 6e, with green, propagating antiparallel to **B**, red, propagating parallel, and blue, a mixture of parallel and antiparallel.

a second to a few seconds) than the "magnetospheric" chorus and appear much more similar to chorus detected at Earth and Jupiter [*Hospodarsky et al.*, 2008].

Examination of the Cassini Plasma Spectrometer (CAPS) electron data [*Young et al.*, 2004] for the injection event shown in Figure 5a indicates that inside the injection, the low-energy (<100 eV) electron flux is reduced, while the high-energy component (>1 keV) is enhanced [*Rymer et al.*, 2008]. The higher-energy component forms pancake-like distributions while the lower-energy electrons are more field aligned [*Rymer et al.*, 2008, Figure 5]. *Menietti et al.* [2008c] examined this event and modeled the measured electron plasma distributions to conduct linear dispersion analysis of the plasma wave modes. They found that the higher-energy pancake-like distribution with temperature anisotropy can drive the observed whistler mode waves.

The relationships between the detection of whistler mode emission and the various orbital parameters of the Cassini spacecraft have also been examined for the first 115 orbits of

Cassini (1 July 2004 to 15 July 2009). This analysis is similar to the technique *Hospodarsky et al.* [2008] used for the first 45 orbits of Cassini. Whistler mode emission is identified only when the five-channel waveform receiver detects electromagnetic waves, and valid wave normal and Poynting vector calculations can be performed (see the work of *Hospodarsky et al.* [2008] for more details on the criteria used). As discussed above, we have not tried to distinguish between hiss and chorus for these studies, so we have defined all whistler mode emissions that we detect at low-latitude "chorus." Whistler mode auroral hiss is also often detected by Cassini at higher latitudes (typically >30°) [*Mitchell et al.*, 2009a; *Kurth et al.*, 2009; *Kopf et al.*, 2010; *Gurnett et al.*, 2011]. We distinguish between the whistler mode chorus and auroral hiss by the location at which it is detected (higher latitude for auroral hiss, low latitude for chorus), the different spectral characteristics (funnel shape for auroral hiss, band of emission between about 0.1 and 0.5 f_{ce} for chorus), and the direction of propagation (auroral hiss has only been seen by

Cassini coming up from the planet propagating toward the equator, chorus propagates away from the equatorial region). Figure 6 shows the occurrence of the Saturn "magnetospheric" chorus (solid dots) with respect to L shell versus LT (Figure 6a) and magnetic latitude (MLAT) (Figure 6d), and Saturn radius, R_s, versus LT (Figure 6b) and MLAT (Figure 6e) for Cassini's first 115 orbits (including Saturn orbit insertion). Figures 6c and 6f show the occurrence of "injection event" chorus with respect to L shell versus LT and MLAT, respectively. The dotted lines in the panels are the trajectory of Cassini during the orbits examined for chorus to show the orbital coverage for this period. As can be seen from Figure 6, both the "magnetospheric" and "injection event" chorus primarily occur between about 4 and 8 L shell (and about 4 to 8 R_s). There is no obvious correlation between the occurrence of the "magnetospheric" or "injection event" chorus and LT in this region. However, "magnetospheric" chorus that is detected at L shells above ~8 occurs primarily near local noon and near the magnetic equator.

Figures 6d, 6e, and 6f show that "magnetospheric" chorus can be detected at higher latitudes nearer the planet (inner cutoff is approximately at the Enceladus' L shell of ~4) and that "injection event" chorus is detected only near the magnetic equator. From the wave normal and Poynting vector analysis of these waves [Hospodarsky et al., 2001, 2008], the propagation characteristics of the chorus at Saturn with respect to the Saturn magnetic field are also shown in Figure 6. Solid green dots on Figure 6 refer to wave propagation antiparallel to the magnetic field (northward propagation away from the magnetic equator), red refers to parallel propagation (southward propagation), and blue shows mixed propagation directions. As can be seen, the chorus at Saturn propagates away from the magnetic equator.

An initial examination of the peak wave amplitudes by Hospodarsky et al. [2008] of the chorus emissions at Saturn found amplitudes (~1 mV m^{-1} and 0.04 nT) much smaller than the peak amplitudes that have been reported at Earth (~30 mV m^{-1} and 1 nT), but larger by at least an order of magnitude than the amplitudes detected at Saturn by Voyager. The detection of these larger amplitudes suggests that chorus may be responsible for some pitch angle and energy diffusion at Saturn, although this needs to be investigated in more detail.

4.2. ECH Emissions

Electrostatic emissions at Saturn include the ECH waves and the UHR emission (see Figure 3). The ECH emission typically occurs at multiple harmonics of f_{ce}, specifically $(n + 1/2)f_{ce}$, where n is an integer. The UHR emissions can be used to estimate the electron plasma density since the UHR

frequency, f_{UHF}, is given by $f_{pe}^2 + f_{ce}^2 = f_{uhr}^2$. By measuring the UHR frequency and obtaining f_{ce} from the magnetic field strength, the electron plasma density, n_e, can be calculated from $f_{pe} = 8980 \, n_e^{1/2}$. Persoon et al. [2005, 2006, 2009] have developed a plasma density model for the inner magnetosphere using UHR frequency observed by the Cassini spacecraft at Saturn.

ECH emissions show very different characteristics inside and outside of injection events at Saturn. Outside of injection events, the ECH emissions are primarily found in the first harmonic band centered near 1.5 f_{ce}, with weaker, more sporadic bands at the higher harmonics [Menietti et al., 2008b]. During injection events, the characteristics of the ECH emissions change drastically with an increase in intensity and harmonic structure [Menietti et al., 2008c; Tao et al., 2010]. This change in the ECH emission inside of an injection event can easily be seen from about 17:35:20 to about 17:40:15 UT in Figure 7 (from Menietti et al. [2008c]). This event is the same as shown in Figure 5a, and the change in the ECH waves occurs during the local plasma injection event observed in the CAPS data [Rymer et al., 2008].

4.3. Cassini Solstice Mission

The Cassini mission has been extended through September 2017, with the extended phase named for the Saturnian summer solstice, which occurs in May 2017. Since Cassini arrived at Saturn just after the northern winter solstice, the extension will allow for the first study of a complete seasonal

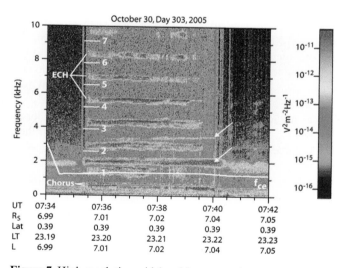

Figure 7. High-resolution wideband frequency-time spectrogram of the power spectra of the wave electric field during the injection event shown in Figure 5a. The enhanced electrostatic cyclotron harmonic wave spectra inside the injection event is easy to see starting at about 07:35:20 UT. From Menietti et al. [2008c].

Table 1. Summary of Wave Parameters

Emission	Earth	Jupiter	Saturn
Chorus	$0.1\,f_{ce}$ to $0.5\,f_{ce}$	$0.1\,f_{ce}$ to $0.5\,f_{ce}$	$0.1\,f_{ce}$ to $0.5\,f_{ce}$
	$0.5\,f_{ce}$ to $0.9\,f_{ce}$	$0.5\,f_{ce}$ to $0.9\,f_{ce}$	$0.5\,f_{ce}$ to $0.8\,f_{ce}$ (injections)
	$\sim$0.2 to $\sim$5 kHz	$\sim$0.1 to 8 kHz	$\sim$0.1 to 2 kHz
	$\sim$1 pT to $\sim$1 nT	$\sim$1 to 100 pT	$\sim$1 to 40 pT
Hiss	$\sim$100 to $\sim$5 kHz	not distinguished from chorus	not distinguished from chorus
	$\sim$1 to $\sim$100 pT		
Equatorial noise	$f_{cH} < f < f_{LHR}$	not observed	not observed
	$\sim$10 to $\sim$200 Hz		
	$\sim$1 to $\sim$100 pT		
Electrostatic cyclotron harmonic	$(n + 1/2){*}f_{ce}$	$(n + 1/2){*}f_{ce}$	$(n + 1/2){*}f_{ce}$
	$\sim$5 to 50 kHz	$\sim$3 to 200 kHz	$\sim$1 to 20 kHz
	$\sim$1 mV m^{-1}	$\sim$0.6 mV m^{-1}	$\sim$0.3 mV m^{-1}

period of Saturn by an orbiting spacecraft. Late in 2016/early 2017, Cassini will go into a series of high-inclination, "Juno-like" orbits that cross the equatorial region between the inner edge of the rings and the upper atmosphere (altitudes of few thousand km). These orbits should provide a rich data set of fields and particles in the region of Saturn's newly discovered inner radiation belt.

5. CONCLUSION

The last few years have seen major advances in our understanding of the role of plasma waves in the dynamics of planetary radiation belts, especially of Earth's Van Allen belts. Observations at Jupiter and Saturn have added to our understanding of the interaction of plasma waves and radiation belt particles in environments that are both very similar and very different from Earth. Table 1 summarizes some of the parameters of these waves at each planet. It should be noted that the majority of studies at Jupiter and Saturn do not distinguish between whistler mode hiss and chorus. While chorus, hiss, equatorial noise, and ECH emissions all appear to play a role in radiation belt dynamics at Earth, only chorus has been examined in detail at Jupiter and Saturn. Further work is needed to determine the importance of each type of plasma wave emissions at the outer planets. The RBSP mission at Earth, the Juno and JUICE missions to Jupiter, and the continued Cassini mission at Saturn, all with advanced plasma wave and particle instruments, will continue to increase our understanding of the role of plasma waves on the dynamics and evolution of the radiation belts.

Acknowledgments. The research at the University of Iowa was supported by NASA contracts 1415150 with the Jet Propulsion Laboratory and 921647 with the Applied Physics Laboratory at John Hopkins University. J.D.M. also acknowledges NASA grant NNX11AM36G.

REFERENCES

Abel, B., and R. M. Thorne (1998), Electron scattering loss in Earth's inner magnetosphere 1. Dominant physical processes, *J. Geophys. Res.*, *103*(A2), 2385–2396, doi:10.1029/97JA02919.

Albert, J. M. (1994), Quasi-linear pitch angle diffusion coefficients: Retaining high harmonics, *J. Geophys. Res.*, *99*(A12), 23,741–23,745, doi:10.1029/94JA02345.

Allcock, G. M. (1957), A study of the audio-frequency radio phenomenon known as "dawn chorus", *Aust. J. Phys.*, *10*, 286–298.

Anderson, R. R., and K. Maeda (1977), VLF emissions associated with enhanced magnetospheric electrons, *J. Geophys. Res.*, *82*(1), 135–146, doi:10.1029/JA082i001p00135.

Barrington, R. E., and J. S. Belrose (1963), Preliminary results from the very-low-frequency receiver aboard Canada's Alouette satellite, *Nature*, *198*(4881), 651–656.

Boardsen, S. A., D. L. Gallagher, D. A. Gurnett, W. K. Peterson, and J. L. Green (1992), Funnel-shaped, low-frequency equatorial waves, *J. Geophys. Res.*, *97*(A10), 14,967–14,976, doi:10.1029/92JA00827.

Bolton, S. J., R. M. Thorne, D. A. Gurnett, W. S. Kurth, and D. J. Williams (1997), Enhanced whistler-mode emissions: Signatures of interchange motion in the Io torus, *Geophys. Res. Lett.*, *24*(17), 2123–2126, doi:10.1029/97GL02020.

Bortnik, J., and R. M. Thorne (2010), Transit time scattering of energetic electrons due to equatorially confined magnetosonic waves, *J. Geophys. Res.*, *115*, A07213, doi:10.1029/2010JA015283.

Bortnik, J., R. M. Thorne, and U. S. Inan (2008a), Nonlinear interaction of energetic electrons with large amplitude chorus, *Geophys. Res. Lett.*, *35*, L21102, doi:10.1029/2008GL035500.

Bortnik, J., R. M. Thorne, and N. P. Meredith (2008b), The unexpected origin of plasmaspheric hiss from discrete chorus emissions, *Nature*, *452*, 62–66, doi:10.1038/nature06741.

Bortnik, J., R. M. Thorne, and N. P. Meredith (2009a), Plasmaspheric hiss overview and relation to chorus, *J. Atmos. Sol. Terr. Phys.*, *71*, 1636–1646, doi:10.1016/j.jastp.2009.03.023.

Bortnik, J., W. Li, R. M. Thorne, V. Angelopoulos, C. Cully, J. Bonnell, O. Le Contel, and A. Roux (2009b), First observation linking the origin of plasmaspheric hiss to discrete chorus emissions, *Science*, *324*, 775–778, doi:10.1126/science.1171273.

Breneman, A., C. A. Kletzing, J. Chum, O. Santolík, D. Gurnett, and J. Pickett (2007), Multispacecraft observations of chorus dispersion and source location, *J. Geophys. Res.*, *112*, A05221, doi:10.1029/2006JA012058.

Breneman, A. W., C. A. Kletzing, J. Pickett, J. Chum, and O. Santolík (2009), Statistics of multispacecraft observations of chorus dispersion and source location, *J. Geophys. Res.*, *114*, A06202, doi:10.1029/2008JA013549.

Bunch, N. L., M. Spasojevic, and Y. Y. Shprits (2011), On the latitudinal extent of chorus emissions as observed by the Polar Plasma Wave Instrument, *J. Geophys. Res.*, *116*, A04204, doi:10.1029/2010JA016181.

Bunch, N. L., M. Spasojevic, and Y. Y. Shprits (2012), Off-equatorial chorus occurrence and wave amplitude distributions as observed by the Polar Plasma Wave Instrument, *J. Geophys. Res.*, *117*, A04205, doi:10.1029/2011JA017228.

Burtis, W. J., and R. A. Helliwell (1969), Banded chorus—A new type of VLF radiation observed in the magnetosphere by OGO 1 and OGO 3, *J. Geophys. Res.*, *74*(11), 3002–3010, doi:10.1029/JA074i011p03002.

Burton, R. K. (1976), Critical electron pitch angle anisotropy necessary for chorus generation, *J. Geophys. Res.*, *81*(25), 4779–4781, doi:10.1029/JA081i025p04779.

Burton, R., and R. E. Holzer (1974), The origin and propagation of chorus in the outer magnetosphere, *J. Geophys. Res.*, *79*(7), 1014–1023, doi:10.1029/JA079i007p01014.

Cattell, C., et al. (2008), Discovery of very large amplitude whistler-mode waves in Earth's radiation belts, *Geophys. Res. Lett.*, *35*, L01105, doi:10.1029/2007GL032009.

Chen, L., R. M. Thorne, V. K. Jordanova, and R. B. Horne (2010), Global simulation of magnetosonic wave instability in the storm time magnetosphere, *J. Geophys. Res.*, *115*, A11222, doi:10.1029/2010JA015707.

Christiansen, P. J., M. P. Gough, G. Martelli, J. J. Bloch, N. Cornilleau, J. Etcheto, R. Gendrin, C. Beghin, P. Decreau, and D. Jones (1978), GEOS-1 observations of electrostatic waves, and their relationship with plasma parameters, *Space Sci. Rev.*, *22*, 383–400.

Chum, J., and O. Santolík (2005), Propagation of whistler-mode chorus to low altitudes: Divergent ray trajectories and ground accessibility, *Ann. Geophys.*, *23*, 3727–3738, doi:10.5194/angeo-23-3727-2005.

Chum, J., O. Santolík, A. W. Breneman, C. A. Kletzing, D. A. Gurnett, and J. S. Pickett (2007), Chorus source properties that produce time shifts and frequency range differences observed on different Cluster spacecraft, *J. Geophys. Res.*, *112*, A06206, doi:10.1029/2006JA012061.

Church, S. R., and R. M. Thorne (1983), On the origin of plasmaspheric hiss: Ray path integrated amplification, *J. Geophys. Res.*, *88*(A10), 7941–7957, doi:10.1029/JA088iA10p07941.

Cornilleau-Wehrlin, N., et al. (1997), The Cluster spatio-temporal analysis of field fluctuations (STAFF) experiment, *Space Sci. Rev.*, *79*, 107–136.

Coroniti, F. V., F. L. Scarf, C. F. Kennel, W. S. Kurth, and D. A. Gurnett (1980), Detection of Jovian whistler mode chorus; Implications for the Io torus aurora, *Geophys. Res. Lett.*, *7*(1), 45–48, doi:10.1029/GL007i001p00045.

Coroniti, F. V., F. L. Scarf, C. F. Kennel, and W. S. Kurth (1984), Analysis of chorus emissions at Jupiter, *J. Geophys. Res.*, *89*(A6), 3801–3820, doi:10.1029/JA089iA06p03801.

Cully, C. M., J. W. Bonnell, and R. E. Ergun (2008), THEMIS observations of long-lived regions of large-amplitude whistler waves in the inner magnetosphere, *Geophys. Res. Lett.*, *35*, L17S16, doi:10.1029/2008GL033643.

Curtis, S. A. (1978), A theory for chorus generation by energetic electrons during substorms, *J. Geophys. Res.*, *83*(A8), 3841–3848, doi:10.1029/JA083iA08p03841.

Dougherty, M. K., et al. (2004), The Cassini magnetic field investigation, *Space Sci. Rev.*, *114*, 331–383.

Draganov, A. B., U. S. Inan, V. S. Sonwalkar, and T. F. Bell (1992), Magnetospherically reflected whistlers as a source of plasmaspheric hiss, *Geophys. Res. Lett.*, *19*(3), 233–236, doi:10.1029/91GL03167.

Fredricks, R. W., and F. L. Scarf (1973), Recent studies of magnetospheric electric field emissions above the electron gyrofrequency, *J. Geophys. Res.*, *78*(1), 310–314, doi:10.1029/JA078i001p00310.

Goldstein, B. E., and B. T. Tsurutani (1984), Wave normal directions of chorus near the equatorial source region, *J. Geophys. Res.*, *89*(A5), 2789–2810, doi:10.1029/JA089iA05p02789.

Green, J. L., S. Boardsen, L. Garcia, W. W. L. Taylor, S. F. Fung, and B. W. Reinisch (2005), On the origin of whistler mode radiation in the plasmasphere, *J. Geophys. Res.*, *110*, A03201, doi:10.1029/2004JA010495.

Green, J. L., S. Boardsen, L. Garcia, S. F. Fung, and B. W. Reinisch (2006), Reply to "Comment on "On the origin of whistler mode radiation in the plasmasphere" by Green et al.", *J. Geophys. Res.*, *111*, A09211, doi:10.1029/2006JA011622.

Gurnett, D. A. (1976), Plasma wave interactions with energetic ions near the magnetic equator, *J. Geophys. Res.*, *81*(16), 2765–2770, doi:10.1029/JA081i016p02765.

Gurnett, D. A., and B. J. O'Brien (1964), High-latitude geophysical studies with satellite Injun 3 5. Very-low-frequency electromagnetic radiation, *J. Geophys. Res.*, *69*(1), 65–89, doi:10.1029/JZ069i001p00065.

Gurnett, D. A., and F. L. Scarf (1983), Plasma waves in the Jovian magnetosphere, in *Physics of the Jovian Magnetosphere*, edited by A. J. Dessler, pp. 285–316, Cambridge Univ. Press, Cambridge, U. K.

Gurnett, D. A., W. S. Kurth, and F. L. Scarf (1979), Plasma wave observations near Jupiter: Initial results from Voyager 2, *Science*, *206*, 987–991.

Gurnett, D. A., W. S. Kurth, and F. L. Scarf (1981), Plasma waves near Saturn: Initial results from Voyager 1, *Science*, *212*, 235–239, doi:10.1126/science.212.4491.235.

Gurnett, D. A., W. S. Kurth, A. Roux, S. J. Bolton, and C. F. Kennel (1996), Galileo plasma wave observations in the Io plasma torus and near Io, *Science*, *274*, 391–392.

Gurnett, D. A., et al. (2004), The Cassini radio and plasma wave investigation, *Space Sci. Rev.*, *114*, 395–463.

Gurnett, D. A., et al. (2005), Radio and plasma waves observed at Saturn from Cassini's approach and first orbit, *Science*, *307*, 1255–1259.

Gurnett, D. A., A. M. Persoon, J. B. Groene, W. S. Kurth, M. Morooka, J.-E. Wahlund, and J. D. Nichols (2011), The rotation of the plasmapause-like boundary at high latitudes in Saturn's magnetosphere and its relation to the eccentric rotation of the northern and southern auroral ovals, *Geophys. Res. Lett.*, *38*, L21203, doi:10.1029/2011GL049547.

Hayakawa, M., Y. Yamanaka, M. Parrot, and F. Lefeuvre (1984), The wave normals of magnetospheric chorus emissions observed on board GEOS 2, *J. Geophys. Res.*, *89*(A5), 2811–2821, doi:10.1029/JA089i05p02811.

Hayakawa, M., K. Hattori, S. Shimakura, M. Parrot, and F. Lefeuvre (1990), Direction finding of chorus emissions in the outer magnetosphere and their generation and propagation, *Planet. Space Sci.*, *38*, 135–143, doi:10.1016/0032-0633(90)90012-F.

Helliwell, R. A. (1969), Low-frequency waves in the magnetosphere, *Rev. Geophys.*, *7*, 281–303, doi:10.1029/RG007i001p00281.

Horne, R. B., and R. M. Thorne (1998), Potential waves for relativistic electron scattering and stochastic acceleration during magnetic storms, *Geophys. Res. Lett.*, *25*(15), 3011–3014, doi:10.1029/98GL01002.

Horne, R. B., and R. M. Thorne (2000), Electron pitch angle diffusion by electrostatic electron cyclotron harmonic waves: The origin of pancake distributions, *J. Geophys. Res.*, *105*(A3), 5391–5402, doi:10.1029/1999JA900447.

Horne, R. B., G. V. Wheeler, and H. S. C. K. Alleyne (2000), Proton and electron heating by radially propagating fast magnetosonic waves, *J. Geophys. Res.*, *105*(A12), 27,597–27,610, doi:10.1029/2000JA000018.

Horne, R. B., R. M. Thorne, S. A. Glauert, N. P. Meredith, D. Pokhotelov, and O. Santolík (2007), Electron acceleration in the Van Allen radiation belts by fast magnetosonic waves, *Geophys. Res. Lett.*, *34*, L17107, doi:10.1029/2007GL030267.

Horne, R. B., R. M. Thorne, S. A. Glauert, J. D. Menietti, Y. Y. Shprits, and D. A. Gurnett (2008), Gyro-resonant electron acceleration at Jupiter, *Nat. Phys.*, *4*, 301–304, doi:10.1038/nphys897.

Hospodarsky, G. B., T. F. Averkamp, W. S. Kurth, D. A. Gurnett, M. Dougherty, U. Inan, and T. Wood (2001), Wave normal and Poynting vector calculations using the Cassini radio and plasma wave instrument, *J. Geophys. Res.*, *106*(A12), 30,253–30,269, doi:10.1029/2001JA900114.

Hospodarsky, G. B., T. F. Averkamp, W. S. Kurth, D. A. Gurnett, J. D. Menietti, O. Santolík, and M. K. Dougherty (2008), Observations of chorus at Saturn using the Cassini Radio and Plasma Wave Science instrument, *J. Geophys. Res.*, *113*, A12206, doi:10.1029/2008JA013237.

Inan, U. S., R. A. Helliwell, and W. S. Kurth (1983), Terrestrial versus Jovian VLF chorus; A comparative study, *J. Geophys. Res.*, *88*(A8), 6171–6180, doi:10.1029/JA088iA08p06171.

Inan, U. S., Y. T. Chiu, and G. T. Davidson (1992), Whistler-mode chorus and morningside aurorae, *Geophys. Res. Lett.*, *19*(7), 653–656, doi:10.1029/92GL00402.

Inan, U. S., M. Platino, T. F. Bell, D. A. Gurnett, and J. S. Pickett (2004), Cluster measurements of rapidly moving sources of ELF/VLF chorus, *J. Geophys. Res.*, *109*, A05214, doi:10.1029/2003JA010289.

Isenberg, P. A., H. C. Koons, and J. F. Fennell (1982), Simultaneous observations of energetic electrons and dawnside chorus in geosynchronous orbit, *J. Geophys. Res.*, *87*(A3), 1495–1503, doi:10.1029/JA087iA03p01495.

Kasahara, Y., A. Sawada, M. Yamamoto, I. Kimura, S. Kokubun, and K. Hayashi (1992), Ion cyclotron emissions observed by the satellite Akebono in the vicinity of the magnetic equator, *Radio Sci.*, *27*(2), 347–362, doi:10.1029/91RS01872.

Kasahara, Y., H. Kenmochi, and I. Kimura (1994), Propagation characteristics of the ELF emissions observed by the satellite Akebono in the magnetic equatorial region, *Radio Sci.*, *29*(4), 751–767, doi:10.1029/94RS00445.

Katoh, Y., and Y. Omura (2007), Relativistic particle acceleration in the process of whistler-mode chorus wave generation, *Geophys. Res. Lett.*, *34*, L13102, doi:10.1029/2007GL029758.

Katoh, Y., and Y. Omura (2011), Amplitude dependence of frequency sweep rates of whistler mode chorus emissions, *J. Geophys. Res.*, *116*, A07201, doi:10.1029/2011JA016496.

Katoh, Y., F. Tsuchiya, Y. Miyoshi, A. Morioka, H. Misawa, R. Ujiie, W. S. Kurth, A. T. Tomás, and N. Krupp (2011), Whistler mode chorus enhancements in association with energetic electron signatures in the Jovian magnetosphere, *J. Geophys. Res.*, *116*, A02215, doi:10.1029/2010JA016183.

Kellogg, P. J., C. A. Cattell, K. Goetz, S. J. Monson, and L. B. Wilson III (2010), Electron trapping and charge transport by large amplitude whistlers, *Geophys. Res. Lett.*, *37*, L20106, doi:10.1029/2010GL044845.

Kennel, C. F., and H. E. Petschek (1966), Limit on stably trapped particle fluxes, *J. Geophys. Res.*, *71*(1), 1–28, doi:10.1029/JZ071i001p00001.

Kennel, C. F., F. L. Scarf, R. W. Fredricks, J. H. McGehee, and F. V. Coroniti (1970), VLF electric field observations in the magnetosphere, *J. Geophys. Res.*, *75*(31), 6136–6152, doi:10.1029/JA075i031p06136.

Kessel, R. L. (2012), NASA's Radiation Belt Storm Probes mission: From concept to reality, in *Dynamics of the Earth's Radiation Belts and Inner Magnetosphere*, Geophys. Monogr. Ser., doi:10.1029/2012GM001312, this volume.

Kopf, A. J., et al. (2010), Electron beams as the source of whistler-mode auroral hiss at Saturn, *Geophys. Res. Lett.*, *37*, L09102, doi:10.1029/2010GL042980.

Kurth, W. S., and D. A. Gurnett (1991), Plasma waves in planetary magnetospheres, *J. Geophys. Res.*, *96*, 18,977–18,991.

Kurth, W. S., F. L. Scarf, D. A. Gurnett, and D. D. Barbosa (1983), A survey of electrostatic waves in Saturn's magnetosphere, *J. Geophys. Res.*, *88*(A11), 8959–8970, doi:10.1029/JA088iA 11p08959.

Kurth, W. S., et al. (2009), Auroral processes, in *Saturn from Cassini-Huygens*, edited by M. K. Dougherty, L. W. Esposito and S. M. Krimigis, pp. 333–374, Springer, Dordrecht, The Netherlands, doi:10.1007/978-1-4020-9217-6_12.

Laakso, H., H. Junginger, A. Roux, R. Schmidt, and C. de Villedary (1990), Magnetosonic waves above $f_c(H^+)$ at geostationary orbit: GEOS 2 results, *J. Geophys. Res.*, *95*(A7), 10,609–10,621, doi:10.1029/JA095iA07p10609.

Lauben, D. S., U. S. Inan, T. F. Bell, D. L. Kirchner, G. B. Hospodarsky, and J. S. Pickett (1998), VLF chorus emissions observed by Polar during the January 10, 1997, magnetic cloud, *Geophys. Res. Lett.*, *25*(15), 2995–2998, doi:10.1029/98GL 01425.

Lauben, D. S., U. S. Inan, T. F. Bell, and D. A. Gurnett (2002), Source characteristics of ELF/VLF chorus, *J. Geophys. Res.*, *107* (A12), 1429, doi:10.1029/2000JA003019.

LeDocq, M. J., D. A. Gurnett, and G. B. Hospodarsky (1998), Chorus source location from VLF Poynting flux measurements with the Polar spacecraft, *Geophys. Res. Lett.*, *25*(21), 4063–4066, doi:10.1029/1998GL900071.

Li, W., J. Bortnik, R. M. Thorne, and V. Angelopoulos (2011), Global distribution of wave amplitudes and wave normal angles of chorus waves using THEMIS wave observations, *J. Geophys. Res.*, *116*, A12205, doi:10.1029/2011JA017035.

Liang, J., B. Ni, E. Spanswick, M. Kubyshkina, E. F. Donovan, V. M. Uritsky, R. M. Thorne, and V. Angelopoulos (2011), Fast earthward flows, electron cyclotron harmonic waves, and diffuse auroras: Conjunctive observations and a synthesized scenario, *J. Geophys. Res.*, *116*, A12220, doi:10.1029/2011JA017094.

Lyons, L. R., and R. M. Thorne (1973), Equilibrium structure of radiation belt electrons, *J. Geophys. Res.*, *78*(13), 2142–2149, doi:10.1029/JA078i013p02142.

Lyons, L. R., R. M. Thorne, and C. F. Kennel (1972), Pitch-angle diffusion of radiation belt electrons within the plasmasphere, *J. Geophys. Res.*, *77*(19), 3455–3474, doi:10.1029/JA 077i019p03455.

Mauk, B. H., and N. J. Fox (2010), Electron radiation belts of the solar system, *J. Geophys. Res.*, *115*, A12220, doi:10.1029/ 2010JA015660.

Mauk, B. H., et al. (2009), Fundamental plasma processes in Saturn's magnetosphere, in *Saturn from Cassini-Huygens*, edited by M. K. Dougherty, L. W. Esposito and S. M. Krimigis, pp. 281–331, Springer, Dordrecht, The Netherlands, doi:10.1007/978-1-4020-9217-6_11.

Menietti, J. D., R. B. Horne, D. A. Gurnett, G. B. Hospodarsky, C. W. Piker, and J. B. Groene (2008a), A survey of Galileo plasma wave instrument observations of Jovian whistler-mode chorus, *Ann. Geophys.*, *26*, 1819–1828, doi:10.5194/angeo-26-1819-2008.

Menietti, J. D., O. Santolík, A. M. Rymer, G. B. Hospodarsky, D. A. Gurnett, and A. J. Coates (2008b), Analysis of plasma waves observed in the inner Saturn magnetosphere, *Ann. Geophys.*, *26*, 2631–2644.

Menietti, J. D., O. Santolík, A. M. Rymer, G. B. Hospodarsky, A. M. Persoon, D. A. Gurnett, A. J. Coates, and D. T. Young (2008c), Analysis of plasma waves observed within local plasma injections seen in Saturn's magnetosphere, *J. Geophys. Res.*, *113*, A05213, doi:10.1029/2007JA012856.

Meredith, N. P., R. B. Horne, and R. R. Anderson (2001), Substorm dependence of chorus amplitudes: Implications for the acceleration of electrons to relativistic energies, *J. Geophys. Res.*, *106*(A7), 13,165–13,178, doi:10.1029/2000JA900156.

Meredith, N. P., M. Cain, R. B. Horne, R. M. Thorne, D. Summers, and R. R. Anderson (2003), Evidence for chorus-driven electron acceleration to relativistic energies from a survey of geomagnetically disturbed periods, *J. Geophys. Res.*, *108*(A6), 1248, doi:10. 1029/2002JA009764.

Meredith, N. P., R. B. Horne, R. M. Thorne, D. Summers, and R. R. Anderson (2004), Substorm dependence of plasmaspheric hiss, *J. Geophys. Res.*, *109*, A06209, doi:10.1029/ 2004JA010387.

Meredith, N. P., R. B. Horne, M. A. Clilverd, D. Horsfall, R. M. Thorne, and R. R. Anderson (2006), Origins of plasmaspheric hiss, *J. Geophys. Res.*, *111*, A09217, doi:10.1029/2006JA 011707.

Meredith, N. P., R. B. Horne, S. A. Glauert, and R. R. Anderson (2007), Slot region electron loss timescales due to plasmaspheric hiss and lightning-generated whistlers, *J. Geophys. Res.*, *112*, A08214, doi:10.1029/2007JA012413.

Meredith, N. P., R. B. Horne, and R. R. Anderson (2008), Survey of magnetosonic waves and proton ring distributions in the Earth's inner magnetosphere, *J. Geophys. Res.*, *113*, A06213, doi:10. 1029/2007JA012975.

Meredith, N. P., R. B. Horne, R. M. Thorne, and R. R. Anderson (2009a), Survey of upper band chorus and ECH waves: Implications for the diffuse aurora, *J. Geophys. Res.*, *114*, A07218, doi:10.1029/2009JA014230.

Meredith, N. P., R. B. Horne, S. A. Glauert, D. N. Baker, S. G. Kanekal, and J. M. Albert (2009b), Relativistic electron loss timescales in the slot region, *J. Geophys. Res.*, *114*, A03222, doi:10.1029/2008JA013889.

Millan, R. M., and the BARREL Team (2011), Understanding relativistic electron losses with BARREL, *J. Atmos. Sol. Terr. Phys.*, *73*(11–12), 1425–1434, doi:10.1016/j.jastp.2011. 01.006.

Mitchell, D. G., W. S. Kurth, G. B. Hospodarsky, N. Krupp, J. Saur, B. H. Mauk, J. F. Carbary, S. M. Krimigis, M. K. Dougherty, and D. C. Hamilton (2009a), Ion conics and electron beams associated with auroral processes on Saturn, *J. Geophys. Res.*, *114*, A02212, doi:10.1029/2008JA013621.

Mitchell, D. G., J. F. Carbary, S. W. H. Cowley, T. W. Hill, and P. Zarka (2009b), The dynamics of Saturn's magnetosphere, in *Saturn from Cassini-Huygens*, edited by M. K. Dougherty, L. W. Esposito and S. M. Krimigis, pp. 257–279, Springer, Dordrecht, The Netherlands, doi:10.1007/978-1-4020-9217-6_10.

Nagano, I., S. Yagitani, H. Kojima, and H. Matsumoto (1996), Analysis of wave normal and Poynting vectors of chorus emissions observed by Geotail, *J. Geomagn. Geoelectr.*, *48*, 299–307.

Němec, F., O. Santolík, K. Gereova, E. Macusova, Y. de Conchy, and N. Cornilleau-Wehrlin (2005), Initial results of a survey of equatorial noise emissions observed by the Cluster spacecraft, *Planet. Space Sci.*, *53*, 291–298.

Němec, F., O. Santolík, K. Gereova, E. Macusova, H. Laakso, Y. de Conchy, M. Maksimovic, and N. Cornilleau-Wehrlin (2006), Equatorial noise: Statistical study of its localization and the derived number density, *Adv. Space Res.*, *37*, 610–616.

Ni, B., J. Liang, R. M. Thorne, V. Angelopoulos, R. B. Horne, M. Kubyshkina, E. Spanswick, E. F. Donovan, and D. Lummerzheim (2012), Efficient diffuse auroral electron scattering by electrostatic electron cyclotron harmonic waves in the outer magnetosphere: A detailed case study, *J. Geophys. Res.*, *117*, A01218, doi:10.1029/2011JA017095.

Olsen, R. C., S. D. Shawhan, D. L. Gallagher, J. L. Green, C. R. Chappell, and R. R. Anderson (1987), Plasma observations at the Earth's magnetic equator, *J. Geophys. Res.*, *92*(A3), 2385–2407, doi:10.1029/JA092iA03p02385.

Omura, Y., and D. Nunn (2011), Triggering process of whistler mode chorus emissions in the magnetosphere, *J. Geophys. Res.*, *116*, A05205, doi:10.1029/2010JA016280.

Omura, Y., and D. Summers (2006), Dynamics of high-energy electrons interacting with whistler mode chorus emissions in the magnetosphere, *J. Geophys. Res.*, *111*, A09222, doi:10.1029/2006JA011600.

Omura, Y., Y. Katoh, and D. Summers (2008), Theory and simulation of the generation of whistler-mode chorus, *J. Geophys. Res.*, *113*, A04223, doi:10.1029/2007JA012622.

Omura, Y., M. Hikishima, Y. Katoh, D. Summers, and S. Yagitani (2009), Nonlinear mechanisms of lower-band and upper-band VLF chorus emissions in the magnetosphere, *J. Geophys. Res.*, *114*, A07217, doi:10.1029/2009JA014206.

Parrot, M., O. Santolík, N. Cornilleau-Wehrlin, M. Maksimovic, and C. C. Harvey (2003), Source location of chorus emissions observed by Cluster, *Ann. Geophys.*, *21*, 473–480.

Pasmanik, D. L., E. E. Titova, A. G. Demekhov, V. Y. Trakhtengerts, O. Santolík, F. Jiricek, K. Kudela, and M. Parrot (2004), Quasi-periodic ELF/VLF wave emissions in the Earth's magnetosphere: Comparison of satellite observations and modeling, *Ann. Geophys.*, *22*, 4351–4361, doi:10.5194/angeo-22-4351-2004.

Perraut, S., A. Roux, P. Robert, R. Gendrin, J. A. Savaud, J. M. Bosqued, G. Kremser, and A. Korth (1982), A systematic study of ULF waves above F_{H^+} from GEOS 1 and 2 measurements and their relationship with proton ring distributions, *J. Geophys. Res.*, *87*(A8), 6219–6236, doi:10.1029/JA087iA08p06219.

Persoon, A. M., D. A. Gurnett, W. S. Kurth, G. B. Hospodarsky, J. B. Groene, P. Canu, and M. K. Dougherty (2005), Equatorial electron density measurements in Saturn's inner magnetosphere, *Geophys. Res. Lett.*, *32*, L23105, doi:10.1029/2005GL024294.

Persoon, A. M., D. A. Gurnett, W. S. Kurth, and J. B. Groene (2006), A simple scale height model of the electron density in

Saturn's plasma disk, *Geophys. Res. Lett.*, *33*, L18106, doi:10.1029/2006GL027090.

Persoon, A. M., et al. (2009), A diffusive equilibrium model for the plasma density in Saturn's magnetosphere, *J. Geophys. Res.*, *114*, A04211, doi:10.1029/2008JA013912.

Reeves, G. D. (2007), Radiation belt storm probes: A new mission for space weather forecasting, *Space Weather*, *5*, S11002, doi:10.1029/2007SW000341.

Russell, C. T., R. E. Holzer, and E. J. Smith (1970), OGO 3 observations of ELF noise in the magnetosphere, 2. The nature of equatorial noise, *J. Geophys. Res.*, *75*(4), 755–768, doi:10.1029/JA075i004p00755.

Rymer, A. M., B. H. Mauk, T. W. Hill, C. Paranicas, D. G. Mitchell, A. J. Coates, and D. T. Young (2008), Electron circulation in Saturn's magnetosphere, *J. Geophys. Res.*, *113*, A01201, doi:10.1029/2007JA012589.

Santolík, O., and D. A. Gurnett (2003), Transverse dimensions of chorus in the source region, *Geophys. Res. Lett.*, *30*(2), 1031, doi:10.1029/2002GL016178.

Santolík, O., J. S. Pickett, D. A. Gurnett, M. Maksimovic, and N. Cornilleau-Wehrlin (2002), Spatiotemporal variability and propagation of equatorial noise observed by Cluster, *J. Geophys. Res.*, *107*(A12), 1495, doi:10.1029/2001JA009159.

Santolík, O., D. A. Gurnett, J. S. Pickett, M. Parrot, and N. Cornilleau-Wehrlin (2003), Spatio-temporal structure of storm-time chorus, *J. Geophys. Res.*, *108*(A7), 1278, doi:10.1029/2002JA009791.

Santolík, O., F. Němec, K. Gereova, E. Macusova, Y. de Conchy, and N. Cornilleau-Wehrlin (2004), Systematic analysis of equatorial noise below the lower hybrid frequency, *Ann. Geophys.*, *22*, 2587–2595.

Santolík, O., D. A. Gurnett, J. S. Pickett, M. Parrot, and N. Cornilleau-Wehrlin (2005), Central position of the source region of storm-time chorus, *Planet. Space Sci.*, *53*, 299–305, doi:10.1016/j.pss.2004.09.056.

Santolík, O., J. Chum, M. Parrot, D. A. Gurnett, J. S. Pickett, and N. Cornilleau-Wehrlin (2006), Propagation of whistler mode chorus to low altitudes: Spacecraft observations of structured ELF hiss, *J. Geophys. Res.*, *111*, A10208, doi:10.1029/2005JA011462.

Santolík, O., D. A. Gurnett, J. S. Pickett, J. Chum, and N. Cornilleau-Wehrlin (2009), Oblique propagation of whistler mode waves in the chorus source region, *J. Geophys. Res.*, *114*, A00F03, doi:10.1029/2009JA014586.

Santolík, O., J. S. Pickett, D. A. Gurnett, J. D. Menietti, B. T. Tsurutani, and O. Verkhoglyadova (2010), Survey of Poynting flux of whistler mode chorus in the outer zone, *J. Geophys. Res.*, *115*, A00F13, doi:10.1029/2009JA014925.

Sato, N., and H. Fukunishi (1981), Interaction between ELF-VLF emissions and magnetic pulsations: Classification of quasi-periodic ELF-VLF emissions based on frequency-time spectra, *J. Geophys. Res.*, *86*(A1), 19–29, doi:10.1029/JA086iA01p00019.

Sazhin, S. S., and M. Hayakawa (1992), Magnetospheric chorus emissions: A review, *Planet. Space Sci.*, *40*, 681–697, doi:10.1016/0032-0633(92)90009-D.

Scarf, F. L., D. A. Gurnett, and W. S. Kurth (1979a), Jupiter plasma wave observations: An initial Voyager 1 overview, *Science*, *204*, 991–995.

Scarf, F. L., F. V. Coroniti, D. A. Gurnett, and W. S. Kurth (1979b), Pitch-angle diffusion by whistler mode waves near the Io plasma torus, *Geophys. Res. Lett.*, *6*(8), 653–656, doi:10.1029/GL006i008p00653.

Scarf, F. L., D. A. Gurnett, W. S. Kurth, and R. L. Poynter (1982), Voyager 2 plasma wave observations at Saturn, *Science*, *215*, 587–594, doi:10.1126/science.215.4532.587.

Scarf, F. L., D. A. Gurnett, W. S. Kurth, and R. L. Poynter (1983), Voyager plasma wave measurements at Saturn, *J. Geophys. Res.*, *88*(A11), 8971–8984, doi:10.1029/JA088iA11p08971.

Scarf, F. L., L. A. Frank, D. A. Gurnett, L. J. Lanzerotti, A. Lazarus, and E. C. Sittler Jr. (1984), Measurements of plasma, plasma waves and suprathermal charged particles in Saturn's inner magnetosphere, in *Saturn*, edited by T. Gehrels and M. S. Matthews, pp. 318–353, Univ. of Ariz. Press, Tucson.

Shaw, R. R., and D. A. Gurnett (1975), Electrostatic noise bands associated with the electron gyrofrequency and plasma frequency in the outer magnetosphere, *J. Geophys. Res.*, *80*(31), 4259–4271, doi:10.1029/JA080i031p04259.

Sigsbee, K., J. D. Menietti, O. Santolík, and J. B. Blake (2008), Polar PWI and CEPPAD observations of chorus emissions and radiation belt electron acceleration: Four case studies, *J. Atmos. Sol. Terr. Phys.*, *70*(14), 1774–1788, doi:10.1016/j.jastp.2008.02.005.

Sigsbee, K., J. D. Menietti, O. Santolík, and J. S. Pickett (2010), Locations of chorus emissions observed by the Polar Plasma Wave Instrument, *J. Geophys. Res.*, *115*, A00F12, doi:10.1029/2009JA014579.

Spjeldvik, W. N., and R. M. Thorne (1975), The cause of storm after effects in the middle latitude D-region ionosphere, *J. Atmos. Terr. Phys.*, *37*(5), 777–795, doi:10.1016/0021-9169(75)90021-5.

Staedter, T. (2006), Teams chosen for radiation belt storm probes mission, *Space Weather*, *4*, S10009, doi:10.1029/2006SW000277.

Storey, L. R. O. (1953), An investigation of whistling atmospherics, *Philos. Trans. R. Soc. London, Ser. A*, *246*, 113–141, doi:10.1098/rsta.1953.0011.

Summers, D., R. M. Thorne, and F. Xiao (1998), Relativistic theory of wave-particle resonant diffusion with application to electron acceleration in the magnetosphere, *J. Geophys. Res.*, *103*(A9), 20,487–20,500, doi:10.1029/98JA01740.

Summers, D., B. Ni, and N. P. Meredith (2007), Timescales for radiation belt electron acceleration and loss due to resonant wave-particle interactions: 2. Evaluation for VLF chorus, ELF hiss, and electromagnetic ion cyclotron waves, *J. Geophys. Res.*, *112*, A04207, doi:10.1029/2006JA011993.

Summers, D., B. Ni, N. P. Meredith, R. B. Horne, R. M. Thorne, M. B. Moldwin, and R. R. Anderson (2008), Electron scattering by whistler-mode ELF hiss in plasmaspheric plumes, *J. Geophys. Res.*, *113*, A04219, doi:10.1029/2007JA012678.

Tang, R., and D. Summers (2012), Energetic electron fluxes at Saturn from Cassini observations, *J. Geophys. Res.*, *117*, A06221, doi:10.1029/2011JA017394.

Tao, X., R. M. Thorne, R. B. Horne, S. Grimald, C. S. Arridge, G. B. Hospodarsky, D. A. Gurnett, A. J. Coates, and F. J. Crary (2010), Excitation of electron cyclotron harmonic waves in the inner Saturn magnetosphere within local plasma injections, *J. Geophys. Res.*, *115*, A12204, doi:10.1029/2010JA015598.

Tao, X., R. M. Thorne, R. B. Horne, B. Ni, J. D. Menietti, Y. Y. Shprits, and D. A. Gurnett (2011), Importance of plasma injection events for energization of relativistic electrons in the Jovian magnetosphere, *J. Geophys. Res.*, *116*, A01206, doi:10.1029/2010JA016108.

Tao, X., J. Bortnik, R. M. Thorne, J. M. Albert, and W. Li (2012), Effects of amplitude modulation on nonlinear interactions between electrons and chorus waves, *Geophys. Res. Lett.*, *39*, L06102, doi:10.1029/2012GL051202.

Thomsen, M. F., M. H. Denton, V. K. Jordanova, L. Chen, and R. M. Thorne (2011), Free energy to drive equatorial magnetosonic wave instability at geosynchronous orbit, *J. Geophys. Res.*, *116*, A08220, doi:10.1029/2011JA016644.

Thorne, R. M. (2010), Radiation belt dynamics: The importance of wave-particle interactions, *Geophys. Res. Lett.*, *37*, L22107, doi:10.1029/2010GL044990.

Thorne, R. M., and B. T. Tsurutani (1979), Diffuse Jovian aurora influenced by plasma injection from Io, *Geophys. Res. Lett.*, *6*(8), 649–652, doi:10.1029/GL006i008p00649.

Thorne, R. M., E. J. Smith, R. K. Burton, and R. E. Holzer (1973), Plasmaspheric hiss, *J. Geophys. Res.*, *78*(10), 1581–1596, doi:10.1029/JA078i010p01581.

Thorne, R. M., T. P. Armstrong, S. Stone, D. J. Williams, R. W. McEntire, S. J. Bolton, D. A. Gurnett, and M. G. Kivelson (1997), Galileo evidence for rapid interchange transport in the Io torus, *Geophys. Res. Lett.*, *24*(17), 2131–2134, doi:10.1029/97GL01788.

Thorne, R. M., R. B. Horne, and N. P. Meredith (2006), Comment on "On the origin of whistler mode radiation in the plasmasphere" by Green et al., *J. Geophys. Res.*, *111*, A09210, doi:10.1029/2005JA011477.

Thorne, R. M., B. Ni, X. Tao, R. B. Horne, and N. P. Meredith (2010), Scattering by chorus waves as the dominant cause of diffuse auroral precipitation, *Nature*, *467*, 943–946, doi:10.1038/nature09467.

Tsurutani, B. T., and E. J. Smith (1974), Post midnight chorus: A substorm phenomenon, *J. Geophys. Res.*, *79*(1), 118–127, doi:10.1029/JA079i001p00118.

Tsurutani, B. T., and E. J. Smith (1977), Two types of magnetospheric ELF chorus and their substorm dependence, *J. Geophys. Res.*, *82*, 5112–5128, doi:10.1029/JA082i032p05112.

Tsurutani, B. T., E. J. Smith, H. I. West Jr., and R. M. Buck (1979), Chorus, energetic electrons and magnetospheric substorms, in *Wave Instabilities in Space Plasma*, edited by P. J. Palmadesso and K. Papadopoulos, pp. 55–62, D. Reidel, Dordrecht, The Netherlands.

Wilson, L. B., III, C. A. Cattell, P. J. Kellogg, J. R. Wygant, K. Goetz, A. Breneman, and K. Kersten (2011), The properties of large amplitude whistler mode waves in the magnetosphere:

Propagation and relationship with geomagnetic activity, *Geophys. Res. Lett.*, *38*, L17107, doi:10.1029/2011GL048671.

Xiao, F., R. M. Thorne, D. A. Gurnett, and D. J. Williams (2003), Whistler-mode excitation and electron scattering during an interchange event near Io, *Geophys. Res. Lett.*, *30*(14), 1749, doi:10.1029/2003GL017123.

Yoon, P. H. (2011), Large-amplitude whistler waves and electron acceleration, *Geophys. Res. Lett.*, *38*, L12105, doi:10.1029/2011GL047893.

Young, D. T., et al. (2004), Cassini plasma spectrometer investigation, *Space Sci. Rev.*, *114*, 1–112.

D. A. Gurnett, G. B. Hospodarsky, C. A. Kletzing, W. S. Kurth, J. S. Leisner, J. D. Menietti, and K. Sigsbee, Department of Physics and Astronomy, University of Iowa, Iowa City, IA 52242, USA. (george-hospodarsky@uiowa.edu)

O. Santolík, Institute of Atmospheric Physics AS CR, 14131 Prague 4, Czech Republic.

AGU Category Index